Abstract Algebra

Textbooks in Mathematics

Series editors:
Al Boggess, Kenneth H. Rosen

https://www.routledge.com/Textbooks-in-Mathematics/book-series/
CANDHTEXBOOMTH

Abstract Algebra
A First Course
Second Edition

Stephen Lovett
Wheaton College, USA

CRC Press
Taylor & Francis Group
Boca Raton London New York

CRC Press is an imprint of the
Taylor & Francis Group, an **informa** business

A CHAPMAN & HALL BOOK

Second edition published 2022
by CRC Press
6000 Broken Sound Parkway NW, Suite 300, Boca Raton, FL 33487-2742

and by CRC Press
4 Park Square, Milton Park, Abingdon, Oxon, OX14 4RN

© 2022 Taylor & Francis, LLC

First edition published by CRC Press 2016

CRC Press is an imprint of Taylor & Francis Group, LLC

Library of Congress Cataloging-in-Publication Data

Names: Lovett, Stephen (Stephen T.), author.
Title: Abstract algebra : a first course / authored by Stephen Lovett,
Wheaton College, USA.
Description: Second edition. | Boca Raton : Chapman & Hall, CRC Press,
2022. | Series: Textbooks in mathematics | Includes bibliographical
references and index.
Identifiers: LCCN 2021062210 (print) | LCCN 2021062211 (ebook) | ISBN
9781032289397 (hardback) | ISBN 9781032289410 (paperback) | ISBN
9781003299233 (ebook)
Subjects: LCSH: Algebra, Abstract--Textbooks.
Classification: LCC QA162 .L68 2022 (print) | LCC QA162 (ebook) | DDC
512/.02--dc23/eng20220415
LC record available at https://lccn.loc.gov/2021062210
LC ebook record available at https://lccn.loc.gov/2021062211

ISBN: 978-1-032-28939-7 (hbk)
ISBN: 978-1-032-28941-0 (pbk)
ISBN: 978-1-003-29923-3 (ebk)

DOI: 10.1201/9781003299233

Typeset in CMR10
by KnowledgeWorks Global Ltd.

Publisher's note: This book has been prepared from camera-ready copy provided by the authors.
Access the Support Material at: www.routledge.com/9781032289397.

Contents

Preface to Instructors

This textbook intends to serve as a first course in abstract algebra. Students are expected to possess a background in linear algebra and some basic exposure to logic, set theory, and proofs, usually in the form of a course in discrete mathematics or a transition course. The selection of topics serves both of the common trends in such a course: a balanced introduction to groups, rings, and fields; or a course that primarily emphasizes group theory. However, the book offers enough flexibility to craft many other pathways. By design, the writing style remains student-centered, conscientiously motivating definitions and offering many illustrative examples. Various sections, or sometimes just examples or exercises, introduce applications to geometry, number theory, cryptography, and many other areas.

Order and Selection of Topics

With 48 sections, each written to correspond to a one-hour contact period, this book offers various possible paths through a one-semester introduction to abstract algebra. Though Chapter 1 begins right away with the theory of groups, an instructor may wish to use some of the sections from the Appendix (complex numbers, set theory, elementary number theory) as preliminary content or review. To create a course that offers an approximately balanced introduction to groups, rings, and fields, the instructor could select to use Chapters 1 through 5 and perhaps opt to skip Sections 1.11, 1.12, and 1.13 in the chapter on group theory, which discusses groups in geometry, the Diffie-Hellman public key algorithm, and monoids, respectively. To design a course that emphasizes group theory, the author could complete all of Chapters 1 and 2, select an appropriate number of sections from the next three chapters on rings and fields, and then complete Chapter 6, which offers topics in group theory, including group actions and semi-direct products.

Besides the application sections, the instructor may wish to add color to the course by spending a little more time on some subsections marked as optional or by devoting a few days to student projects or various computer algebra systems designed for abstract algebra.

Software and Computer Algebra Systems

There exist a number of commercial computer algebra systems (CAS) (e.g., *Maple, Mathematica, MATLAB*) that provide packages that implement certain calculations that are useful in algebra. There also exist several free CAS that are specifically designed for computations in algebra (e.g., *SageMath, Magma*, and *Macaulay2*). It is impossible in this textbook to offer a tutorial on each one. Though the reader should visit the various CAS help pages, we occasionally end a section by discussing a few commands or libraries of commands that are relevant to that section.

This textbook will draw examples from *Maple*[1] and SAGEMATH or more briefly SAGE[2]. *Maple* comes with its own user interface that involves either a Worksheet Mode or Document Mode. The company MapleSoft offers excellent online help files and programming guides for every aspect of this CAS. (Though we do not explicitly describe the commands for *Mathematica*, they are similar in structure to *Maple*.)

To interact with SageMath, the user will employ either the SageMath shell or a Jupyter notebook. SAGE is built on Python[3] so it natively incorporates syntax from Python and itself can easily be called from a Python script. For example, Python's `Combinatorics` module offers commands for calculating the order of a permutation group, i.e., a group defined as a subgroup of the symmetric group S_n, but SAGE subsumes this.

Unless indicated by CAS, it is generally expected that the computations in the exercises be done by hand and not require the use of a CAS.

Projects

Another feature of this book is the project ideas, listed at the end of each chapter. The project ideas come in two flavors: investigative or expository.

The investigative projects briefly present a topic and pose open-ended questions that invite the student to explore the topic, asking them to try to answer their own questions. Investigative projects may involve computations, utilize some programming, or lead the student to try to prove something original (or original to them). For investigative projects, students should not consult outside sources, or only do so minimally for background. Indeed, it

---

[1] *Maple* is made by MapleSoft, whose website is https://www.maplesoft.com/

[2] Previously known as SAGE for "System for Algebra and Geometry Experimentation," the official website for SageMath is https://www.sagemath.org/

[3] The website https://docs.sympy.org/latest/modules/index.html lists all the modules available through SymPy, the Python library for symbolic mathematics.

is possible to find sources on many of the investigative projects, but finding such sources is not the point of an investigative project.

Expository projects invite the student to explore a topic with algebraic content or pertain to a particular mathematician's work through responsible research. The exploration should involve multiple sources and the paper should offer a report in the students' own words, and offering their own insights as they can. In contrast to investigative projects, expository projects rely on the use of outside (library) sources and cite them appropriately.

A possible rubric for grading projects involves the "4Cs of projects". The projects should be (1) Clear: Use proper prose, follow the structure of a paper, and provide proper references; (2) Correct: Proofs and calculations must be accurate; (3) Complete: Address all the questions or questions one should naturally address associated with the investigation; and (4) Creative: Evidence creative problem-solving or question-asking skills.

When they require letters of recommendation, graduate schools and sometimes other employers want to know a student's potential for research or other type of work (independent or team-oriented). If a student has not landed one of the few coveted REU spots during their undergraduate years, speaking to the student's participation in class does not answer that rubric well. A faculty person who assigns projects will have some speaking points for rubric on a letter of recommendation for a student.

Habits of Notation and Expression

This book regularly uses $\implies$ for logical implication and $\iff$ for logical equivalence. More precisely, if $P(x, y, \ldots)$ is a predicate with some variables and $Q(x, y, \ldots)$ is another predicate using the same variables, then

$$P(x, y, \ldots) \implies Q(x, y, \ldots) \quad \text{means} \quad \forall x \forall y \ldots (P(x, y, \ldots) \longrightarrow Q(x, y, \ldots))$$

and

$$P(x, y, \ldots) \iff Q(x, y, \ldots) \quad \text{means} \quad \forall x \forall y \ldots (P(x, y, \ldots) \longleftrightarrow Q(x, y, \ldots)).$$

As another habit of expression particular to this author, the textbook is careful to always and only use the expression "Assume [hypothesis]" as the beginning of a proof by contradiction. Like so, the reader can know ahead of time that whenever he or she sees this expression, the assumption will eventually lead to a contradiction.

Solutions Manual

A solutions manual for all exercises is available by request. Faculty may obtain one by contacting me directly at

`stephen.lovett@wheaton.edu`

or at

`www.routledge.com/9781032289397`

Furthermore, the author invites faculty using this textbook to email their suggestions for improvement and other project ideas for inclusion in future editions.

Acknowledgments

First, I must thank the reviewers and my publisher for many helpful suggestions for improvements. The book would not be what it is without their advice. Next, I must thank the mathematics majors at Wheaton College (IL) who served for many years as the test environment for many topics, exercises, and projects. I am indebted to Wheaton College (IL) for the funding provided through the Aldeen Grant that contributed to portions of this textbook. I especially thank the students who offered specific feedback on the draft versions of this book, in particular Kelly McBride, Roland Hesse, and David Garringer. Joel Stapleton, Caleb DeMoss, Daniel Bradley, and Jeffrey Burge deserve special gratitude for working on the solutions manual to the textbook. I also must thank Justin Brown for test running the book and offering valuable feedback. I also thank Africa Nazarene University for hosting my sabbatical, during which I wrote a major portion of this textbook.

Preface to Students

What is Abstract Algebra?

When a student of mathematics studies abstract algebra, he or she inevitably faces questions in the vein of, "What makes the algebra abstract?" or "What is it good for?" or, more teasingly, "I finished algebra in high school; why are you still studying it as a math major?" On the other hand, since undergraduate mathematics curriculum designers nearly always include an algebra requirement, then these questions illustrate an awareness gap by the general public about advanced mathematics. Consequently, we try to answer this question up front: "What is abstract algebra?"

Algebra, in its broadest sense, describes a way of thinking about classes of sets equipped with binary operations. In high school algebra, a student explores properties of operations ($+$, $-$, $\times$, and $\div$) on real numbers. In contrast, abstract algebra studies properties of operations without specifying what types of numbers or objects we work with. Hence, any theorem established in the abstract context holds not only for real numbers but for every possible algebraic structure that has operations with the stated properties.

Linear algebra follows a similar process of abstraction. After an introduction to systems of linear equations, it explores algebraic properties of vectors in $\mathbb{R}^n$ along with properties and operations of $m \times n$ matrices. Then, a linear algebra course generalizes these concepts to abstract (or general) vector spaces. Theorems in abstract linear algebra not only apply to n-tuples of real numbers but imply profound consequences for analysis, differential equations, statistics, and so on.

A typical first course in abstract algebra introduces the structures of groups, rings, and fields. There are many other interesting and fruitful algebraic structures but these three have many applications within other branches of mathematics – in number theory, topology, geometry, analysis, and statistics. Outside of pure mathematics, scientists have noted applications of abstract algebra to advanced physics, inorganic chemistry, methods of computation, information security, and various forms of art, including music theory. (See [18], [10], [9], or [12].)

Strategies for Studying

From a student's perspective, one of the biggest challenges to modern algebra is its abstraction. Calculus, linear algebra, and differential equations can be taught from many perspectives, but often most of the exercises simply require the student to carefully follow a certain prescribed algorithm. In contrast, in abstract algebra (and other advanced topics) students do not learn as many specific algorithms and the exercises challenge the student to prove new results using the theorems presented in the text. The student then becomes an active participant in the development of the field.

In this textbook, however, for many exercises (though not all) the student will find useful ideas either in a similar example or informative strategies in the proofs of the theorems in the section. Therefore, though the theorems are critical, it is very useful to spend time studying the examples and the proofs of theorems. Furthermore, to get a full experience of the material, we encourage the reader to peruse the exercises in order to see some of the interesting consequences of the theory.

Projects

This book offers a unique feature in the lists of projects at the end of each section. Graduate schools always, and potential employers sometimes, want to know about a mathematics student's potential for research. Even if an undergraduate student does not get one of the coveted spots in an official REU (Research Experience for Undergraduates) or write a senior thesis, the expository writing or the exploratory work required by a project offers a vehicle for a faculty person to assess the student's potential for research. So the author of this textbook does not view projects as just something extra or cute, but rather an opportunity for a student to work on and demonstrate his or her potential for open-ended investigation.

1

Groups

As a field in mathematics, group theory did not develop in the order that this book follows. Historians of mathematics generally credit Evariste Galois with first writing down the current definition of a group. Galois introduced groups in his study of symmetries among the roots of polynomials. Galois' methods turned out to be exceedingly fruitful and led to a whole area called Galois theory, a topic sometimes covered in a second course in abstract algebra.

As mathematicians separated the concept of a group from Galois' application, they realized two things. First, groups occur naturally in many areas of mathematics. Second, group theory presents many challenging problems and profound results in its own right. Like modern calculus texts that slowly lead to the derivative concept after a rigorous definition of the limit, we begin with the definition of a group and methodically prove results with a view toward as many applications as possible, rather than a single application.

Groups form another algebraic structure, like vector spaces. Unlike vector spaces, which require between 8 and 10 axioms (depending on how we organize the definition), the definition for a group only requires three axioms. Given the relative brevity of the definition for a group, the richness of the theory of groups and the vast number of applications may come as a surprise to readers.

In Section 1.1 we precede a general introduction to groups with an interesting example from geometry. Sections 1.2 through 1.4 introduce the axioms for groups, present many elementary examples, establish some elementary properties of group elements, and then discuss the concept of classification. Section 1.5 introduces the symmetric group, a family of groups that plays a central role in group theory.

Sections 1.6 and 1.7 study subgroups, how to describe them, or how to prove that a given subset is a subgroup, while Section 1.8 borrows from the Hasse diagrams of a partial order to provide a visual representation of subgroups within a group. Section 1.9 introduces the concept of a homomorphism between groups, functions that preserve the group structure. Section 1.10 introduces a particular method to describe the content and structure of a group and introduces the fundamental notion of a free group.

The last three sections are optional to a concise introduction to group theory. Sections 1.11 and 1.12 provide two applications of group theory, one to patterns in geometry and the other to information security. Finally, Section 1.13 introduces the concept of a monoid, offering an example of another common algebraic structure, similar to groups but with looser axioms.

DOI: 10.1201/9781003299233-1

1.1 Symmetries of a Regular Polygon

1.1.1 Dihedral Symmetries

Let $n \geq 3$ and consider a regular n-sided polygon, P_n. Call $V = \{v_1, v_2, \ldots, v_n\}$ the set of vertices of P_n as a subset of the Euclidean plane $\mathbb{R}^2$. For simplicity, we often imagine the center of P_n at the origin and that the vertex v_1 on the positive x-axis.

A *symmetry* of a regular n-gon is a bijection $\sigma : V \to V$ that is the restriction of a bijection $F : \mathbb{R}^2 \to \mathbb{R}^2$, that leaves the overall vertex-edge structure of P_n in place; i.e., if the unordered pair $\{v_i, v_j\}$ are the end points of an edge of the regular n-gon, then $\{\sigma(v_i), \sigma(v_j)\}$ is also an edge.

Consider, for example, a regular hexagon P_6 and the bijection $\sigma : V \to V$ such that $\sigma(v_1) = v_2$, $\sigma(v_2) = v_1$, and σ stays fixed on all the other vertices. Then σ is not a symmetry of P_6 because it fails to preserve the vertex-edge structure of the hexagon. As we see in Figure 1.1, though $\{v_2, v_3\}$ is an edge of the hexagon, while $\{\sigma(v_2), \sigma(v_3)\}$ are not the endpoints of an edge of the hexagon.

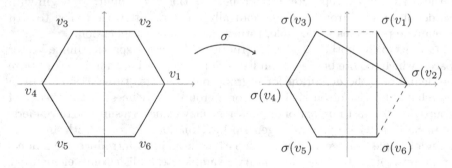

FIGURE 1.1: Not a hexagon symmetry.

In contrast, consider the bijection $\tau : V \to V$ defined by

$$\tau(v_1) = v_2, \ \tau(v_2) = v_1, \ \tau(v_3) = v_6, \ \tau(v_4) = v_5, \ \tau(v_5) = v_4, \ \tau(v_6) = v_3.$$

This bijection on the vertices is a symmetry of the hexagon because it preserves the edge structure of the hexagon. Figure 1.2 shows that τ can be realized as the reflection through the line L as drawn.

Definition 1.1.1

We denote by D_n the set of symmetries of the regular n-gon and call it the set of *dihedral* symmetries.

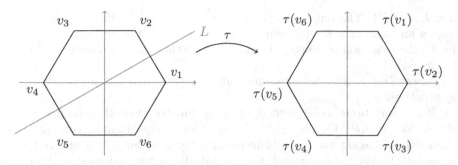

FIGURE 1.2: A reflection symmetry of the hexagon.

To count the number of bijections on the set $V = \{v_1, v_2, \ldots, v_n\}$, we note that a bijection $f : V \to V$ can map $f(v_1)$ to any element in V; then it can map $f(v_2)$ to any element in $V \setminus \{f(v_1)\}$; then it can map $f(v_3)$ to any element in $V \setminus \{f(v_1), f(v_2)\}$; and so on. Hence, there are

$$n \times (n-1) \times (n-2) \times \cdots \times 2 \times 1 = n!$$

distinct bijections on V.

However, a symmetry $\sigma \in D_n$ can map $\sigma(v_1)$ to any element in V (n options), but then σ must map v_2 to a vertex adjacent to $\sigma(v_1)$ (2 options). Once $\sigma(v_1)$ and $\sigma(v_2)$ are known, all remaining $\sigma(v_i)$ for $3 \le i \le n$ are determined. In particular, $\sigma(v_3)$ must be the vertex adjacent to $\sigma(v_2)$ that is not $\sigma(v_1)$; $\sigma(v_4)$ must be the vertex adjacent to $\sigma(v_3)$ that is not $\sigma(v_2)$; and so on. This reasoning leads to the following proposition.

Proposition 1.1.2

The cardinality of D_n is $|D_n| = 2n$.

It is not difficult to identify these symmetries by their geometric meaning. Half of the symmetries in D_n are rotations that shift vertices k spots counterclockwise, for k ranging from 0 to $n-1$. For the rest of this section, we denote by R_α the rotation around center of the polygon of angle α. The rotation $R_{2\pi k/n}$ of angle $2\pi k/n$ on the set of vertices, performs

$$R_{2\pi k/n}(v_i) = v_{((i-1+k) \bmod n)+1} \qquad \text{for all } 1 \le i \le n.$$

Note that R_0 is the identity function on V.

As remarked above, regular n-gons possess symmetries that correspond to reflections through lines that pass through the center of the polygon. Supposing the regular n-gon is centered at the origin and with a vertex on the x-axis, then there are n distinct reflection symmetries, each corresponding to a line through the origin and making an angle of $\pi k/n$ with the x-axis, for

$0 \leq k \leq n - 1$. The reflection symmetry in Figure 1.2 is through a line that makes an angle of $\pi/6$ with respect to the x-axis. In this section, we denote by F_β the reflection symmetry through the line that makes an angle of β with the x-axis.

Since $|D_n| = 2n$, the rotations and reflections account for all dihedral symmetries.

If two bijections on V preserve the polygon structure, then their composition does as well. Consequently, the function composition of two dihedral symmetries is again another dihedral symmetry and thus $\circ$ is a binary operation on D_n. However, having listed the dihedral symmetries as rotations or reflections, it is interesting to determine the result of the composition of two symmetries as another symmetry.

First, it is easy to see that rotations compose as follows:

$$R_\alpha \circ R_\beta = R_{\alpha+\beta},$$

where we subtract 2π from $\alpha + \beta$ if $\alpha + \beta \geq 2\pi$. However, the composition of a given rotation and a given reflection or the composition of two reflections is not as obvious. There are various ways to calculate the compositions. The first is to determine how the function composition acts on the vertices. For example, let $n = 6$ and consider the compositions of $R_{2\pi/3}$ and $F_{\pi/6}$.

v_i	v_1	v_2	v_3	v_4	v_5	v_6
$R_{2\pi/3}(v_i)$	v_3	v_4	v_5	v_6	v_1	v_2
$F_{\pi/6}(v_i)$	v_2	v_1	v_6	v_5	v_4	v_3
$(F_{\pi/6} \circ R_{2\pi/3})(v_i)$	v_6	v_5	v_4	v_3	v_2	v_1
$(R_{2\pi/3} \circ F_{\pi/6})(v_i)$	v_4	v_3	v_2	v_1	v_6	v_5

From this table and by inspection on how the compositions act on the vertices, we determine that

$$F_{\pi/6} \circ R_{2\pi/3} = F_{5\pi/6} \qquad \text{and} \qquad R_{2\pi/3} \circ F_{\pi/6} = F_{\pi/2}.$$

We notice with this example that the composition operation is not commutative.

Another approach to determining the composition of elements comes from linear algebra. Rotations about the origin by an angle α and reflections through a line through the origin making an angle β with the x-axis are linear transformations. With respect to the standard basis, these two types of linear transformations respectively correspond to the following 2×2 matrices

$$R_\alpha : \begin{pmatrix} \cos\alpha & -\sin\alpha \\ \sin\alpha & \cos\alpha \end{pmatrix} \qquad \text{and} \qquad F_\beta : \begin{pmatrix} \cos 2\beta & \sin 2\beta \\ \sin 2\beta & -\cos 2\beta \end{pmatrix}. \qquad (1.1)$$

For example, let $n = 6$ and consider the composition of $F_{\pi/6}$ and $F_{\pi/3}$.

The composition symmetry corresponds to the matrix product

$$F_{\pi/6} \circ F_{\pi/3} : \begin{pmatrix} \cos\frac{\pi}{3} & \sin\frac{\pi}{3} \\ \sin\frac{\pi}{3} & -\cos\frac{\pi}{3} \end{pmatrix} \begin{pmatrix} \cos\frac{2\pi}{3} & \sin\frac{2\pi}{3} \\ \sin\frac{2\pi}{3} & -\cos\frac{2\pi}{3} \end{pmatrix}$$

$$= \begin{pmatrix} \cos\frac{\pi}{3}\cos\frac{2\pi}{3} + \sin\frac{\pi}{3}\sin\frac{2\pi}{3} & \cos\frac{\pi}{3}\sin\frac{2\pi}{3} - \sin\frac{\pi}{3}\cos\frac{2\pi}{3} \\ \sin\frac{\pi}{3}\cos\frac{2\pi}{3} - \cos\frac{\pi}{3}\sin\frac{2\pi}{3} & \sin\frac{\pi}{3}\sin\frac{2\pi}{3} + \cos\frac{\pi}{3}\cos\frac{2\pi}{3} \end{pmatrix}$$

$$= \begin{pmatrix} \cos\frac{\pi}{3} & \sin\frac{\pi}{3} \\ -\sin\frac{\pi}{3} & \cos\frac{\pi}{3} \end{pmatrix}.$$

This matrix corresponds to a rotation and shows that

$$F_{\pi/6} \circ F_{\pi/3} = R_{-\pi/3} = R_{5\pi/3}.$$

Moving to symmetries of the square, whether we use one method or the other, the following table gives all their compositions.

$a \setminus b$	R_0	$R_{\pi/2}$	R_π	$R_{3\pi/2}$	F_0	$F_{\pi/4}$	$F_{\pi/2}$	$F_{3\pi/4}$	
R_0	R_0	$R_{\pi/2}$	R_π	$R_{3\pi/2}$	F_0	$F_{\pi/4}$	$F_{\pi/2}$	$F_{3\pi/4}$	
$R_{\pi/2}$	$R_{\pi/2}$	R_π	$R_{3\pi/2}$	R_0	$F_{\pi/4}$	$F_{\pi/2}$	$F_{3\pi/4}$	F_0	
R_π	R_π	$R_{3\pi/2}$	R_0	$R_{\pi/2}$	$F_{\pi/2}$	$F_{3\pi/4}$	F_0	$F_{\pi/4}$	
$R_{3\pi/2}$	$R_{3\pi/2}$	R_0	$R_{\pi/2}$	R_π	$F_{3\pi/4}$	F_0	$F_{\pi/4}$	$F_{\pi/2}$	(1.2)
F_0	F_0	$F_{3\pi/4}$	$F_{\pi/2}$	$F_{\pi/4}$	R_0	$R_{3\pi/2}$	R_π	$R_{\pi/2}$	
$F_{\pi/4}$	$F_{\pi/4}$	F_0	$F_{3\pi/4}$	$F_{\pi/2}$	$R_{\pi/2}$	R_0	$R_{3\pi/2}$	R_π	
$F_{\pi/2}$	$F_{\pi/2}$	$F_{\pi/4}$	F_0	$F_{3\pi/4}$	R_π	$R_{\pi/2}$	R_0	$R_{3\pi/2}$	
$F_{3\pi/4}$	$F_{3\pi/4}$	$F_{\pi/2}$	$F_{\pi/4}$	F_0	$R_{3\pi/2}$	R_π	$R_{\pi/2}$	R_0	

From this table, we can answer many questions about the composition operator on D_4. For example, if asked what $f \in D_4$ satisfies $R_{3\pi/2} \circ f = F_0$, we simply look in the row corresponding to $a = R_{3\pi/2}$ (the fourth row) for the b that gives the composition of F_0. A priori, without any further theory, there does not have to exist such an f, but in this case there does and $f = F_{\pi/4}$.

Some other properties of the composition operation on D_n are not as easy to identify directly from the table in (1.2). For example, by Proposition A.2.12, $\circ$ is associative on D_n. Verifying associativity from the table in (1.2) would require checking $8^3 = 512$ equalities. Also, $\circ$ has an identity on D_n, namely R_0. Indeed R_0 is the identity function on V. Finally, every element in D_n has an inverse: the inverse to $R_{2\pi k/n}$ is $R_{2\pi(n-k)/n}$ and the inverse to $F_{\pi k/n}$ is itself. We leave the proof of the following proposition as an exercise.

Proposition 1.1.3

Let n be a fixed integer with $n \geq 3$. Then the dihedral symmetries R_α and F_β satisfy the following relations:

$$R_\alpha \circ F_\beta = F_{\alpha/2+\beta}, \quad F_\alpha \circ R_\beta = F_{\alpha-\beta/2}, \quad \text{and} \quad F_\alpha \circ F_\beta = R_{2(\alpha-\beta)}.$$

Proof. (See Exercise 1.1.7.) □

1.1.2   Abstract Notation

We introduce a notation that is briefer and aligns with the abstract notation that we will regularly use in group theory.

Having fixed an integer $n \geq 3$, denote by r the rotation of angle $2\pi/n$, by s the reflection through the x-axis, and by ι the identity function. In other words,

$$r = R_{2\pi/n}, \quad s = F_0, \quad \text{and} \quad \iota = R_0.$$

In abstract notation, similar to our habit of notation for multiplication of real variables, we write ab to mean $a \circ b$ for two elements $a, b \in D_n$. Borrowing from a theorem in the next section (Proposition 1.2.13), since $\circ$ is associative, an expression such as $rrsr$ is well-defined, regardless of the order in which we pair terms to perform the composition. In this example, with $n = 4$,

$$rrsr = R_{\pi/2} \circ R_{\pi/2} \circ F_0 \circ R_{\pi/2} = R_\pi \circ F_0 \circ R_{\pi/2} = F_{\pi/2} \circ R_{\pi/2} = F_{\pi/4}.$$

To simplify notations, if $a \in D_n$ and $k \in \mathbb{N}^*$, then we write a^k to represent

$$a^k = \overbrace{aaa \cdots a}^{k \text{ times}}.$$

Hence, we write $r^2 sr$ for $rrsr$. Since composition $\circ$ is not commutative, $r^3 s$ is not necessarily equal to $r^2 sr$.

From Proposition 1.1.3, it is not hard to see that

$$r^k = R_{2\pi k/n} \quad \text{and} \quad r^k s = F_{\pi k/n},$$

where k satisfies $0 \leq k \leq n - 1$. Consequently, as a set

$$D_n = \{\iota, r, r^2, \ldots, r^{n-1}, s, rs, r^2 s, \ldots, r^{n-1} s\}.$$

The symbols r and s have a few interesting properties. First, $r^n = \iota$ and $s^2 = \iota$. These are obvious as long as we do not forget the geometric meaning of the functions r and s. Less obvious is the equality in the following proposition.

Proposition 1.1.4

Let n be an integer $n \geq 3$. Then in D_n equipped with the composition operation,

$$sr = r^{n-1} s.$$

Proof. We first prove that $rsr = s$. By Exercise 1.1.7, the composition of a rotation with a reflection is a reflection. Hence, $(rs)r$ is a reflection. For any n, r maps v_1 to v_2, then s maps v_2 to v_n, and then r maps v_n to $v - 1$. Hence, rsr is a reflection that keeps v_1 fixed. There is only one reflection in D_n that keeps v_1 fixed, namely s.

We now compose r^{n-1} on the left of the identity $rsr = s$. We get $r^n sr = r^{n-1} s$. Since $r^n = \iota$, we obtain $sr = r^{n-1} s$. $\square$

Corollary 1.1.5

Consider the dihedral symmetries D_n with $n \geq 3$. Then

$$sr^k = r^{n-k}s.$$

Proof. This follows by a repeated application of Proposition 1.1.4. □

1.1.3 Geometric Objects with Dihedral Symmetry

Regular polygons are not the only objects that possess dihedral symmetry. Any subset of the plane is said to have dihedral D_n symmetry if the set remains unchanged when the plane is transformed by all the rotations and reflections in D_n in reference to a given center C and an axis L.

For example, both shapes in Figure 1.3 possess D_6 dihedral symmetry.

FIGURE 1.3: Shapes with D_6 symmetry.

In Figure 1.4, the shape on the left displays D_7 symmetry while the shape on the right displays D_5 symmetry.

FIGURE 1.4: Shapes with D_7 and D_5 symmetry, respectively.

On the other hand, consider the curve in Figure 1.5. There does not exist an axis through which the shape is preserved under a reflection. Consequently, the shape does not possess D_3 symmetry. It does however, have rotational symmetry with the smallest rotation angle of $2\pi/3$.

If we know that a geometric pattern has a certain dihedral symmetry, we only need to draw a certain portion of the shape before it is possible to determine the rest of the object. Let $\mathcal{F}$ be a set of bijections of the plane and

FIGURE 1.5: Only rotational symmetry.

let S be a subset of the plane that is preserved by all the functions in $\mathcal{F}$, i.e., $f(S) = S$ for all $f \in \mathcal{F}$. We say that a subset S' *generates* S by $\mathcal{F}$ if

$$S = \bigcup_{f \in \mathcal{F}} f(S').$$

For example, consider the shapes shown in Figure 1.6. In both instances, S_0 is a different generating subset for the subset S that has dihedral symmetry.

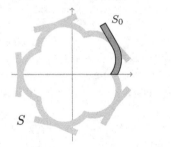
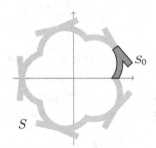

FIGURE 1.6: D_5 symmetry with generating subsets.

EXERCISES FOR SECTION 1.1

1. Use diagrams to describe all the dihedral symmetries of the equilateral triangle.

2. Write down the composition table for D_4.

3. Determine what $r^3 s r^4 s r$ corresponds to in dihedral symmetry of D_8.

4. Determine what $s r^6 s r^5 s r s$ corresponds to as a dihedral symmetry of D_9.

5. Let n be an even integer with $n \geq 4$. Prove that in D_n, the element $r^{n/2}$ satisfies $r^{n/2} w = w r^{n/2}$ for all $w \in D_n$.

6. Let n be an arbitrary integer $n \geq 3$. Show that an expression of the form

$$r^a s^b r^c s^d \cdots$$

is a rotation if and only if the sum of the powers on s is even.

7. Using (1.1) and linear algebra prove that

$$R_\alpha \circ F_\beta = F_{\alpha/2+\beta}, \quad F_\alpha \circ R_\beta = F_{\alpha-\beta/2}, \quad \text{and} \quad F_\alpha \circ F_\beta = R_{2(\alpha-\beta)}.$$

8. Describe the symmetries of an ellipse with unequal half-axes.

9. Determine the set of symmetries for each of the following shapes.

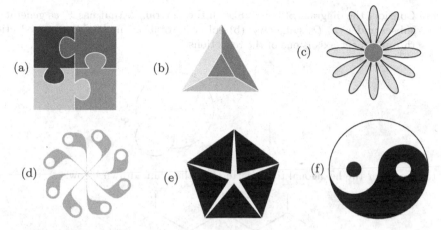

(a) (b) (c)

(d) (e) (f)

10. Sketch a pattern/shape (possibly a commonly known logo) that has D_8 symmetry but does not have D_n symmetry for $n > 8$.

11. Sketch a pattern/shape (possibly a commonly known logo) that has rotational symmetry of angle $\frac{\pi}{2}$ but does not have full D_4 symmetry.

12. Consider a regular tetrahedron. We call a *rigid motion* of the tetrahedron any rotation or composition of rotations in $\mathbb{R}^3$ that map a regular tetrahedron back into itself, though possibly changing specific vertices, edges, and faces. Rigid motions of solids do not include reflections through a plane.

   (a) Call $\mathcal{R}$ the set of rigid motions of a regular tetrahedron. Prove $|\mathcal{R}| = 12$.

   (b) Using a labeling of the tetrahedron, explicitly list all rigid motions of the tetrahedron.

   (c) Explain why function composition $\circ$ is a binary operation on $\mathcal{R}$.

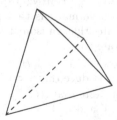

13. Consider the diagram S' below. Sketch the diagram S that has S' as a generating subset under (a) D_4 symmetry; (b) only rotational square symmetry. [Assume reflection through the x-axis is one of the reflections.]

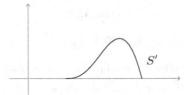

14. Consider the diagram S' below. Sketch the diagram S that has S' as generating subset under (a) D_3 symmetry; (b) only the rotations in D_3. [Assume reflection through the x-axis is one of the reflections.]

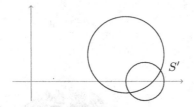

15. Consider the hexagonal tiling pattern on the plane drawn below.

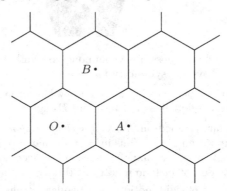

Define r as the rotation by $60°$ about O and t as the translation by the vector $\overrightarrow{OA}$, i.e., that maps the whole plane one hexagon to the right. We denote by $\circ$ the operation of composition on functions $\mathbb{R}^2 \to \mathbb{R}^2$ and we denote by r^{-1} and t^{-1} the inverse functions to r and t. [Recall that rotations and translations are examples of isometries (transformations that preserve distances) in the plane. A theorem in geometry states that an isometry is completely determined by its behavior on a nondegenerate triangle.]

(a) Determine the images O', A', and B' of O, A, and B respectively under the composition $r \circ t$. Deduce that $r \circ t$ is the symmetry of translation by the vector $\overrightarrow{OB}$. [Hint: Note that $t(O) = A$ and that $(r \circ t)(O) = r(A) = B$, so $O' = B$, but plot A' and B'.]

(b) Determine the images O'', A'', and B'' of O, A, and B respectively under the composition $t \circ r$ and deduce that $r \circ t$ and $t \circ r$ are not equal as symmetries of the plane.

(c) Show that $t \circ r \circ t^{-1}$ is the rotation of $60°$ about the point A. Find (and justify) a similar expression for the rotation of $60°$ about the point B.

(d) Describe in geometric terms the composition $r \circ t \circ r^2 \circ t^{-1} \circ r^{-1}$.

1.2 Introduction to Groups

1.2.1 Group Axioms

As we now jump into group theory with both feet, the reader might not immediately see the value in the definition of a group. The plethora of examples we provide subsequent to the definition will begin to showcase the breadth of applications.

Definition 1.2.1

A *group* is a pair $(G, *)$ where G is a set and $*$ is a binary operation on G that satisfies the following properties:

(1) associativity: $(a * b) * c = a * (b * c)$ for all $a, b, c \in G$;

(2) identity: there exists $e \in G$ such that $a * e = e * a = a$ for all $a \in G$;

(3) inverses: for all $a \in G$, there exists $b \in G$ such that $a * b = b * a = e$.

By Proposition A.2.16, if any binary operation has an identity, then that identity is unique. Similarly, any element in a group has exactly one inverse element.

Proposition 1.2.2

Let $(G, *)$ be a group. Then for all $a \in G$, there exists a unique inverse element to a.

Proof. Let $a \in G$ be arbitrary and suppose that b_1 and b_2 satisfy the properties of the inverse axiom for the element a. Then

$$
\begin{aligned}
b_1 &= b_1 * e && \text{by identity axiom} \\
&= b_1 * (a * b_2) && \text{by inverse axiom} \\
&= (b_1 * a) * b_2 && \text{by associativity} \\
&= e * b_2 && \text{by definition of } b_1 \\
&= b_2 && \text{by identity axiom.}
\end{aligned}
$$

Therefore, for all $a \in G$ there exists a unique inverse. $\qquad\square$

Since every group element has a unique inverse, our notation for inverses can reflect this. We denote the inverse element of a by a^{-1}.

The defining properties of a group are often called the *group axioms*. In common language, one often uses the term "axiom" to mean a truth that is self-evident. That is not the sense in which we use the term "axiom." In algebra, when we say that such and such are the axioms of a given algebraic structure, we mean the defining properties listed as (1)–(3) in Definition 1.2.1.

In the group axioms, there is no assumption that the binary operation $*$ is commutative. We say that two particular elements $a, b \in G$ commute (or commute with each other) if $a * b = b * a$.

Definition 1.2.3

A group $(G, *)$ is called *abelian* if for all $a, b \in G$, $a * b = b * a$.

The term *abelian* is named after Niels Abel (1802–1829), one of the founders of group theory.

Usually, the groups we encounter possess a binary operation with a natural description. Sometimes, however, it is useful or even necessary to list out all operation pairings. If $(G, *)$ is a finite group and if we label all the elements as $G = \{g_1, g_2, \ldots, g_n\}$, then the group's *Cayley table*[1] is the $n \times n$ array in which the (i, j)th entry is the result of the operation $g_i * g_j$. When listing the elements in a group it is customary that g_1 be the identity element of the group.

1.2.2   A Few Examples

It is important to develop a robust list of examples of groups that show the breadth and restriction of the group axioms.

Example 1.2.4. The pairs $(\mathbb{Z}, +)$, $(\mathbb{Q}, +)$, $(\mathbb{R}, +)$, and $(\mathbb{C}, +)$ are groups. In each case, addition is associative and has 0 as the identity element. For a given element a, the additive inverse is $-a$.   △

Example 1.2.5. The pairs $(\mathbb{Q}^*, \times)$, $(\mathbb{R}^*, \times)$, and $(\mathbb{C}^*, \times)$ are groups. Recall that A^* mean $A - \{0\}$ when A is a set that includes 0. In each group, 1 is the multiplicative identity, and, for a given element a, the (multiplicative) inverse is $\frac{1}{a}$. Note that $(\mathbb{Z}^*, \times)$ is not a group because it fails the inverse axiom. For example, there is no nonzero integer b such that $2b = 1$.

On the other hand $(\mathbb{Q}^{>0}, \times)$ and $(\mathbb{R}^{>0}, \times)$ are groups. Multiplication is a binary operation on $\mathbb{Q}^{>0}$ and on $\mathbb{R}^{>0}$, and it satisfies all the axioms.   △

Example 1.2.6. A vector space V is a group under vector addition with $\vec{0}$ as the identity. The (additive) inverse of a vector $\vec{v}$ is $-\vec{v}$. Note that the scalar multiplication of a vector spaces has no bearing on the group properties of vector addition.   △

---

[1] The Cayley table is named after the British mathematician, Arthur Cayley (1821–1895).

Example 1.2.7. In Section A.6, we introduced modular arithmetic. Recall that $\mathbb{Z}/n\mathbb{Z}$ represents the set of congruence classes modulo n and that $U(n)$ is the subset of $\mathbb{Z}/n\mathbb{Z}$ of elements with multiplicative inverses. Given any integer $n \geq 2$, both $(\mathbb{Z}/n\mathbb{Z}, +)$ and $(U(n), \times)$ are groups. The element $\overline{0}$ is the identity in $\mathbb{Z}/n\mathbb{Z}$ and the element $\overline{1}$ is the identity $U(n)$.

The tables for addition in (A.13) and (A.14) are the Cayley tables for $(\mathbb{Z}/5\mathbb{Z}, +)$ and $(\mathbb{Z}/6\mathbb{Z}, +)$. By ignoring the column and row for $\overline{0}$ in the multiplication table in Equation (A.13), we obtain the Cayley table for $(U(5), \times)$.

$\times$	$\overline{1}$	$\overline{2}$	$\overline{3}$	$\overline{4}$
$\overline{1}$	$\overline{1}$	$\overline{2}$	$\overline{3}$	$\overline{4}$
$\overline{2}$	$\overline{2}$	$\overline{4}$	$\overline{1}$	$\overline{3}$
$\overline{3}$	$\overline{3}$	$\overline{1}$	$\overline{4}$	$\overline{2}$
$\overline{4}$	$\overline{4}$	$\overline{3}$	$\overline{2}$	$\overline{1}$

$\triangle$

All the examples so far are of abelian groups. We began this chapter by introducing dihedral symmetries precisely because it offers an example of a nonabelian group.

Example 1.2.8 (Dihedral Groups). Let $n \geq 3$ be an integer. The pair $(D_n, \circ)$, where D_n is the set of dihedral symmetries of a regular n-gon and $\circ$ is function composition, is a group. We call $(D_n, \circ)$ the nth dihedral group. Since $rs = sr^{-1}$ and $r^{-1} \neq r$ for any $n \geq 3$, the group $(D_n, \circ)$ is not abelian. The table given in Equation (1.2) is the Cayley table for D_6. $\triangle$

Example 1.2.9. The pair $(\mathbb{R}^3, \times)$, where $\times$ is the vector cross product, is not a group. In fact, $\times$ fails all of the axioms for a group. First of all $\times$ is not associative. Indeed, if $\vec{\imath}$, $\vec{\jmath}$, and $\vec{k}$ are respectively the unit vectors in the x-, y-, and z- directions, then

$$\vec{\imath} \times (\vec{\imath} \times \vec{\jmath}) = \vec{\imath} \times \vec{k} = -\vec{\jmath} \neq (\vec{\imath} \times \vec{\imath}) \times \vec{\jmath} = \vec{0} \times \vec{\jmath} = \vec{0}.$$

Furthermore, $\times$ has no identity element. For any nonzero vector $\vec{a}$ and any other vector $\vec{v}$, the product $\vec{a} \times \vec{v}$ is perpendicular to $\vec{a}$ or is $\vec{0}$. Hence, for no vector $\vec{v}$ do we have $\vec{a} \times \vec{v} = \vec{a}$. Since $\times$ has no identity, the question about having inverses becomes moot. $\triangle$

Example 1.2.10. Let S be a set with at least 1 element. The pair $(\mathcal{P}(S), \cup)$ is not a group. The union operation $\cup$ on $\mathcal{P}(S)$ is both associative and has an identity $\emptyset$. However, if $A \neq \emptyset$, there does not exist a set $B \subseteq S$ such that $A \cup B = \emptyset$. Hence, $(\mathcal{P}(S), \cup)$ does not have inverses. $\triangle$

Example 1.2.11 (Matrix Groups). Let n be a positive integer. The set of $n \times n$ invertible matrices with real coefficients is a group with the multiplication operation. In this group, the identity is the identity matrix and the inverse of a matrix A is the matrix inverse A^{-1}. This group is called the *general linear group* of degree n over $\mathbb{R}$, and is denoted by $\mathrm{GL}_n(\mathbb{R})$.

In Section 3.3, we discuss properties of matrices in more generality. However, without yet providing the full algebraic theory, we point out that in order to perform matrix addition, matrix multiplication, matrix inverses, and even the Gauss-Jordan elimination, we need the entries of the matrix to come from a field. (See Definition 3.1.22.) Though we did not name them as such, the fields we have encountered so far are $\mathbb{Q}$, $\mathbb{R}$, $\mathbb{C}$, and $\mathbb{F}_p = \mathbb{Z}/p\mathbb{Z}$, where p is a prime number.

Suppose that F represents $\mathbb{Q}$, $\mathbb{R}$, $\mathbb{C}$, or $\mathbb{F}_p$. Of key importance is the fact that an $n \times n$ matrix A with coefficients in F is invertible if and only if the columns are linearly independent, which is also equivalent to $\det(A) \neq 0$. We denote by $\mathrm{GL}_n(F)$ the general linear group of degree n over F and we always denote the identity matrix by I.

As an explicit example, consider $\mathrm{GL}_2(\mathbb{F}_5)$. The number of 2×2 matrices over $\mathbb{F}_5$ is $5^4 = 625$. However, not all are invertible. To determine the cardinality $\mathrm{GL}_2(\mathbb{F}_5)$, we consider the columns of a matrix A, which must be linearly independent. The only condition on the first column is that it is not all 0. Hence, there are $5^2 - 1 = 24$ options for the first column. Given the first column, the only necessary condition on the second column is that it is not a $\mathbb{F}_5$-multiple of the first column. This accounts for $5^2 - 5 = 20$ (all columns minus the 5 multiples of the first column) options. Hence, $\mathrm{GL}_2(\mathbb{F}_5)$ has $24 \times 20 = 480$ elements.

An example of multiplication in $\mathrm{GL}_2(\mathbb{F}_5)$ is

$$\begin{pmatrix} \bar{1} & \bar{3} \\ \bar{2} & \bar{4} \end{pmatrix} \begin{pmatrix} \bar{1} & \bar{1} \\ \bar{3} & \bar{2} \end{pmatrix} = \begin{pmatrix} \overline{1+9} & \overline{1+6} \\ \overline{2+12} & \overline{2+8} \end{pmatrix} = \begin{pmatrix} \bar{0} & \bar{2} \\ \bar{4} & \bar{0} \end{pmatrix}$$

while the following illustrates calculating the inverse of a matrix

$$\begin{pmatrix} \bar{3} & \bar{3} \\ \bar{2} & \bar{1} \end{pmatrix}^{-1} = (\bar{3} - \bar{2} \cdot \bar{3})^{-1} \begin{pmatrix} \bar{1} & -\bar{3} \\ -\bar{2} & \bar{3} \end{pmatrix} = \bar{3} \begin{pmatrix} \bar{1} & \bar{2} \\ \bar{3} & \bar{3} \end{pmatrix} = \begin{pmatrix} \bar{3} & \bar{1} \\ \bar{4} & \bar{4} \end{pmatrix},$$

where the second equality holds because $\bar{2}^{-1} = \bar{3}$ in $\mathbb{F}_p$. The reader should verify that all the matrices involved in the above calculations have a nonzero determinant in $\mathbb{F}_5$. $\triangle$

1.2.3   Notation for Arbitrary Groups

In group theory, we will regularly discuss the properties of an arbitrary group. In this case, instead of writing the operation as $a * b$, where $*$ represents some unspecified binary operation, it is common to write the generic group operation as ab. With this convention of notation, it is also common to indicate the identity in an arbitrary group as 1 instead of e. In this chapter, however, we will continue to write e for the arbitrary group identity in order to avoid confusion. Finally, with arbitrary groups, we denote the inverse of an element a as a^{-1}.

This shorthand of notation should not surprise us too much. We already developed a similar habit with vector spaces. When discussing an arbitrary vector space, we regularly say, "Let V be a vector space." So though, in a strict sense, V is only the set of the vector space, we implicitly understand that part of the information of a vector space is the addition of vectors (some operation usually denoted $+$) and the scalar multiplication of vectors.

By a similar abuse of language, we often refer, for example, to "the dihedral group D_n," as opposed to "the dihedral group $(D_n, \circ)$." Similarly, when we talk about "the group $\mathbb{Z}/n\mathbb{Z}$," we mean $(\mathbb{Z}/n\mathbb{Z}, +)$ because $(\mathbb{Z}/n\mathbb{Z}, \times)$ is not a group. And when we refer to "the group $U(n)$," we mean the group $(U(n), \times)$. We will explicitly list the pair of set and binary operation if there could be confusion as to which binary operation the group refers. Furthermore, as we already saw with D_n, even if a group is equipped with a natural operation, we often just write ab to indicate that operation. Following the analogy with multiplication, in a group G, if $a \in G$ and k is a positive integer, by a^k we mean

$$a^k \stackrel{\text{def}}{=} \overbrace{aa \cdots a}^{k \text{ times}}.$$

We extend the power notation so that $a^0 = e$ and $a^{-k} = \left(a^{-1}\right)^k$, for any positive integer k.

Groups that involve addition give an exception to the above habit of notation. In that case, we always write $a + b$ for the operation, $-a$ for the inverse, and, if k is a positive integer,

$$k \cdot a \stackrel{\text{def}}{=} \overbrace{a + a + \cdots + a}^{k \text{ times}}. \tag{1.3}$$

We refer to $k \cdot a$ as a multiple of a instead of as a power. Again, we extend the notation to nonpositive "multiples" just as above with powers.

Proposition 1.2.12

Let G be a group and let $x \in G$. For all $n, m \in \mathbb{Z}$, the following identities hold

$$\text{(a) } x^m x^n = x^{m+n} \quad \text{(b) } (x^m)^n = x^{mn}$$

Proof. (Left as an exercise for the reader. See Exercise 1.3.11.) ☐

1.2.4 First Properties

The following proposition holds for any associative binary operation and does not require the other two axioms of group theory.

Proposition 1.2.13

Let S be a set and let $\star$ be a binary operation on S that is associative. In an operation expression with a finite number of terms,

$$a_1 \star a_2 \star \cdots \star a_n \qquad \text{with } n \geq 3, \qquad (1.4)$$

all possible orders in which we pair operations (i.e., parentheses orders) are equal.

Proof. Before starting the proof, we define a temporary but useful notation. Given a sequence $a_1, a_2, \ldots, a_k$ of elements in S, by analogy with the $\sum$ notation, we define

$$\bigstar_{i=1}^{k} a_i \stackrel{\text{def}}{=} (\cdots((a_1 \star a_2) \star a_3) \cdots a_{k-1}) \star a_k.$$

In this notation, we perform the operations in (1.4) from left to right. Note that if $k = 1$, the expression is equal to the element a_1.

We prove by (strong) induction on n, that every operation expression in (1.4) is equal to $\bigstar_{i=1}^{n} a_i$.

The basis step with $n \geq 3$ is precisely the assumption that $\star$ is associative.

We now assume that the proposition is true for all integers k with $3 \leq k \leq n$. Consider an operation expression (1.4) involving $n+1$ terms. Suppose without loss of generality that the last operation performed occurs between the jth and $(j+1)$th term, i.e.,

$$q = \overbrace{\left(\text{operation expression}_1\right)}^{j \text{ terms}} \star \overbrace{\left(\text{operation expression}_2\right)}^{n-j \text{ terms}}.$$

Since both operation expressions involve n terms or less, by the induction hypothesis

$$q = \left(\bigstar_{i=1}^{j} a_i\right) \star \left(\bigstar_{i=j+1}^{n} a_i\right).$$

Furthermore,

$$q = \left(\bigstar_{i=1}^{j} a_i\right) \star \left(a_{j+1} \star \left(\bigstar_{i=j+2}^{n} a_i\right)\right) \qquad \text{by the induction hypothesis}$$

$$= \left(\left(\bigstar_{i=1}^{j} a_i\right) \star a_{j+1}\right) \star \left(\bigstar_{i=j+2}^{n} a_i\right) \qquad \text{by associativity}$$

$$= \left(\bigstar_{i=1}^{j+1} a_i\right) \star \left(\bigstar_{i=j+2}^{n} a_i\right).$$

Repeating this $n - j - 2$ more times, we conclude that

$$q = \bigstar_{i=1}^{n+1} a_i.$$

The proposition follows. □

The following proposition lists properties of binary operations particular to the context of group theory.

Proposition 1.2.14

Let $(G, *)$ be a group.

(1) The identity in G is unique.

(2) For each $a \in G$, the inverse of a is unique.

(3) For all $a \in G$, $(a^{-1})^{-1} = a$.

(4) For all $a, b \in G$, $(a * b)^{-1} = b^{-1} * a^{-1}$.

Proof. We have already seen (1) and (2) in Proposition A.2.16 and Proposition 1.2.2, respectively.

For (3), by definition of the inverse of a we have $a * (a^{-1}) = (a^{-1}) * a = e$. However, this shows that a satisfies the inverse axiom for the element a^{-1}.

For (4), we have

$$
\begin{aligned}
(a * b)^{-1} * (a * b) &= e \\
\iff ((a * b)^{-1} * a) * b &= e & \text{associativity} \\
\iff (((a * b)^{-1} * a) * b) * b^{-1} &= e * b^{-1} & \text{operate } *b^{-1} \\
\iff ((a * b)^{-1} * a) * (b * b^{-1}) &= b^{-1} & \text{assoc and id} \\
\iff ((a * b)^{-1} * a) * e &= b^{-1} & \text{inverse axiom} \\
\iff (a * b)^{-1} * a &= b^{-1} & \text{identity axiom} \\
\iff (a * b)^{-1} &= b^{-1} * a^{-1}.
\end{aligned}
$$

$\square$

Proposition 1.2.15 (Cancellation Law)

A group G satisfies the left and right cancellation laws, namely

$$au = av \implies u = v \qquad \text{(Left cancellation)}$$
$$ub = vb \implies u = v. \qquad \text{(Right cancellation)}$$

Proof. If $au = av$, then operating on the left by a^{-1}, we obtain

$$a^{-1}au = a^{-1}av \implies eu = ev \implies u = v.$$

Similarly, if $ub = vb$, then by operating the equality on the right by b^{-1}, we obtain

$$ubb^{-1} = vbb^{-1} \implies ue = ve \implies u = v.$$

$\square$

It is important to note that Proposition 1.2.15 does not claim that $au = va$ implies $u = v$. In fact, $au = va$ implies $u = v$ if and only if a commutes with u or v.

The Cancellation Law leads to an interesting property about the Cayley table for a finite group. In combinatorics, a *Latin square* is an $n \times n$ array, filled with n different symbols in such a way that each symbol appears exactly once in each column and exactly once in each row. Since $au = av$ implies $u = v$, in the row corresponding to a, each distinct column has a different group element entry. Hence, each row of the Cayley table contains n different group elements. Similarly, right cancellation implies that in a given column, different rows have different group elements. Thus, the Cayley graph for every group is a Latin square.

1.2.5   Useful CAS Commands

In linear algebra, the theorem that every vector space has a basis gives us a standard method to describe elements in vector spaces, especially finite dimensional ones: (1) find a basis $\mathcal{B}$ of the vector space; (2) express a given vector by its coordinates with respect to $\mathcal{B}$. No corresponding theorem exists in group theory. Hence, one of the initial challenging questions of group theory is how to describe a group and its elements in a standard way. This is particularly important for implementing computational packages that study groups. There exist a few common methods and we will introduce them in parallel with the development of needed theory.

In *Maple* version 16 or below, the command `with(group);` accesses the appropriate package. In *Maple* version 17 or higher, the `group` package was deprecated in favor of `with(GroupTheory);`. The help files, whether provided by the program or those available online[2] provide a list of commands and capabilities. Doing a search on "GroupTheory" locates the help file for the `GroupTheory` package. The student might find useful the `LinearAlgebra` package or, to support Example 1.2.11, the linear algebra package for modular arithmetic.

Consider the following lines of *Maple* code, in which the left justified text is the code and the centered text is the printed result of the code.

――――――――――――――― Maple ――――――――――――――

$with(LinearAlgebra[Modular]) :$

$A := Mod(11, Matrix([[1, 2], [7, 9]]), integer[]);$

$$A := \begin{bmatrix} 1 & 2 \\ 7 & 9 \end{bmatrix}$$

$B := Mod(11, Matrix([[2, 3], [1, 10]]), integer[]);$

$$B := \begin{bmatrix} 2 & 3 \\ 1 & 10 \end{bmatrix}$$

―――――――――――――――――――――――――――――――――――

[2]See https://www.maplesoft.com/support/help/category.aspx?cid=162

$Determinant(11, A);$

$$6$$

$Multiply(11, A, B);$

$$\begin{bmatrix} 4 & 1 \\ 1 & 1 \end{bmatrix}$$

$MatrixPower(11, A, 5);$

$$\begin{bmatrix} 8 & 5 \\ 1 & 6 \end{bmatrix}$$

The first line makes active the linear algebra package for modular arithmetic. The next two code lines define matrices A and B respectively, both defined in $\mathbb{Z}/11\mathbb{Z}$. The next three lines calculate respective the determinant of A, the produce of AB, and the power A^5, always assuming we work in $\mathbb{Z}/11\mathbb{Z}$.

For SAGE, a browser search for "SageMath groups" will bring up references manuals and tutorials for group theory. Perhaps the gentlest introductory tutorial is entitled "Group theory and Sage."[3] We show here below the commands and approximate look for the same calculations in the console for SageMath for those we did above in *Maple*.

———————————— Sage ————————————

```
sage: M=MatrixSpace(GF(11),2,2)
sage: A=M([1,2, 7,9])
sage: B=M([2,3, 1,10])
sage: A.determinant()
6
sage: A*B
[4 1]
[1 1]
sage: A^5
[8 5]
[1 6]
```

In each of the lines, sage: is the command line prompt for Sage, where the user inputs their commands. The lines without that prompt indicate the results of calculations. The first line defines the set of 2×2 matrices defined over the field $\mathbb{F}_{11}$. (Note that GF(11) reads as general field mod 11.) The M in the lines of code defining A and B identify them as elements of the matrix space M.

Because of the abstract nature of group theory, the available commands in various CAS implement computations ranging from elementary to specialized for group theorists. In subsequent sections, we will occasionally discuss useful commands in one or another CAS that assist with group theory investigations.

—————————————

[3] https://doc.sagemath.org/html/en/thematic_tutorials/group_theory.html

EXERCISES FOR SECTION 1.2

In Exercises 1.2.1 through 1.2.14, decide whether the given set and the operation pair forms a group. If it is, prove it. If it is not, decide which axioms fail. You should always check that the symbol is in fact a binary operation on the given set.

1. The pair $(\mathbb{N}, +)$.

2. The pair $(\mathbb{Q} - \{-1\}, *)$, where $*$ is defined by $a * b = a + b + ab$.

3. The pair $(\mathbb{Q} - \{0\}, \div)$, with $a \div b = \frac{a}{b}$.

4. The pair $(A, +)$, where $A = \{x \in \mathbb{Q} \mid |x| < 1\}$.

5. The pair $(\mathbb{Z} \times \mathbb{Z}, *)$, where $(a, b) * (c, d) = (ad + bc, bd)$.

6. The pair $([0, 1), \boxplus)$, where $x \boxplus y = x + y - \lfloor x + y \rfloor$.

7. The pair $(A, +)$, where A is the set of rational numbers that when reduced have a denominator of 1 or 3.

8. The pair $(A, +)$, where $A = \{a + b\sqrt{5} \mid a, b \in \mathbb{Q}\}$.

9. The pair $(A, \times)$, where $A = \{a + b\sqrt{5} \mid a, b \in \mathbb{Q}\}$.

10. The pair $(A, \times)$, where $A = \{a + b\sqrt{5} \mid a, b \in \mathbb{Q} \text{ and } (a, b) \neq (0, 0)\}$. [Hint: First show that $\sqrt{5}$ is irrational.]

11. The pair $(U(20), +)$.

12. The pair $(\mathcal{P}(S), \triangle)$, where S is any set and $\triangle$ is the symmetric difference of two sets.

13. The pair $(G, \times)$, where $G = \{z \in \mathbb{C} \mid |z| = 1\}$.

14. The pair $(\mathcal{D}, *)$, where $\mathcal{D}$ is the set of open disks in $\mathbb{R}^2$, including the empty set $\emptyset$, and where $D_1 * D_2$ is the unique open disk of least radius that encloses both D_1 and D_2.

15. Construct the Cayley table for $U(15)$.

16. Prove that a group is abelian if and only if its Cayley table is symmetric across the main diagonal.

17. Prove that $S = \{2^a 5^b \mid a, b \in \mathbb{Z}\}$, as a subset of rational numbers, is a group under multiplication.

18. Prove that the set $\{\overline{1}, \overline{13}, \overline{29}, \overline{41}\}$ is a group under multiplication modulo 42.

19. Prove that if a group G satisfies $x^2 = e$ for all $x \in G$, then G is abelian.

20. Prove that if a group G satisfies $(xy)^{-1} = x^{-1}y^{-1}$ for all $x, y \in G$, then G is abelian.

21. Let $g_1, g_2, g_3 \in G$. What is $(g_1 g_2 g_3)^{-1}$? Generalize your result.

22. Prove that every cyclic group is abelian.

23. Prove that $\mathrm{GL}_n(\mathbb{F}_p)$ contains

$$(p^n - 1)(p^n - p)(p^n - p^2) \cdots (p^n - p^{n-1})$$

elements. [Hint: Use the fact the $\mathrm{GL}_n(\mathbb{F}_p)$ consists of $n \times n$ matrices with coefficients in $\mathbb{F}_p$ that have columns that are linearly independent.]

24. Write out the Cayley table for $GL_2(\mathbb{F}_2)$.

25. In the given general linear group, for the given matrices A and B, calculate the products A^2, AB, and B^{-1}.

(a) $GL_2(\mathbb{F}_3)$ with $A = \begin{pmatrix} 0 & 2 \\ 2 & 1 \end{pmatrix}$ and $B = \begin{pmatrix} 1 & 1 \\ 1 & 2 \end{pmatrix}$.

(b) $GL_2(\mathbb{F}_5)$ with $A = \begin{pmatrix} 1 & 3 \\ 4 & 1 \end{pmatrix}$ and $B = \begin{pmatrix} 0 & 1 \\ 2 & 3 \end{pmatrix}$.

(c) $GL_2(\mathbb{F}_7)$ with $A = \begin{pmatrix} 4 & 6 \\ 3 & 2 \end{pmatrix}$ and $B = \begin{pmatrix} 5 & 4 \\ 3 & 2 \end{pmatrix}$.

26. Let F be $\mathbb{Q}$, $\mathbb{R}$, $\mathbb{C}$, or $\mathbb{F}_p$. The *Heisenberg group* with coefficients in F is

$$H(F) = \left\{ \begin{pmatrix} 1 & a & b \\ 0 & 1 & c \\ 0 & 0 & 1 \end{pmatrix} \in GL_3(F) \,\middle|\, a, b, c \in F \right\}.$$

(a) Show that $H(F)$ is a group under matrix multiplication.

(b) Explicitly show the inverse of an element in $H(F)$.

1.3 Properties of Group Elements

As we progress through group theory, we will encounter more and more internal structure to groups that is not readily apparent from the three axioms for groups. This section introduces a few elementary properties of group operations.

1.3.1 Order of Elements

Definition 1.3.1

Let G be a group.

(1) If G is finite, we call the cardinality of $|G|$ the *order* of the group.

(2) Let $x \in G$. If $x^k = e$ for some positive integer k, then we call the *order* of x, denoted $|x|$, the smallest positive value of n such that $x^n = e$. If there exists no positive n such that $x^n = e$, then we say that the order of x is infinite.

Recall that if G is finite, $|G|$ is simply the number of elements in G. If G is infinite, the set theoretic notion of cardinality is more general.

Note that the order of a group element g is $|g| = 1$ if and only if g is the group's identity element.

As a reminder, we list the orders of a few groups encountered so far:

$|D_n| = 2n,$ (Proposition 1.1.2)

$|\mathbb{Z}/n\mathbb{Z}| = n,$

$|U(n)| = \phi(n),$ (Corollary A.6.9)

$|\operatorname{GL}_n(\mathbb{F}_p)| = (p^n - 1)(p^n - p)(p^n - p^2)\cdots(p^n - p^{n-1}).$ (Exercise 1.2.23)

Example 1.3.2. Consider the group $G = (\mathbb{Z}/20\mathbb{Z}, +)$. We calculate the orders of $\bar{5}$ and $\bar{3}$. Since the group has $+$ as its operation, "powers" means multiples. Instead of writing $\bar{a}^k$ with $\bar{a} \in \mathbb{Z}/20\mathbb{Z}$ and $k \in \mathbb{N}$ (which evokes multiplicative powers), we write $k \cdot \bar{a}$.

For $\bar{5}$, we calculate directly that

$$2 \cdot \bar{5} = \overline{10}, \quad 3 \cdot \bar{5} = \overline{15}, \quad 4 \cdot \bar{5} = \overline{20} = \bar{0}.$$

Hence, $|\bar{5}| = 4$. However, for $\bar{3}$ we notice the pattern

$$k \cdot \bar{3} = \overline{3k} \qquad \text{for } 0 \leq k \leq 6,$$
$$k \cdot \bar{3} = \overline{3(k-7)+1} \qquad \text{for } 7 \leq k \leq 13,$$
$$k \cdot \bar{3} = \overline{3(k-14)+2} \qquad \text{for } 14 \leq k \leq 20.$$

This shows that the first positive integer k such that $k \cdot \bar{3} = \bar{0}$ is 20. Hence, $|\bar{3}| = 20$. $\triangle$

Example 1.3.3. Consider the group $\operatorname{GL}_2(\mathbb{F}_3)$ and we calculate the order of

$$g = \begin{pmatrix} \bar{2} & \bar{1} \\ \bar{1} & \bar{0} \end{pmatrix}$$

by evaluating various powers of g:

$$g = \begin{pmatrix} \bar{2} & \bar{1} \\ \bar{1} & \bar{0} \end{pmatrix}, \qquad g^2 = g\begin{pmatrix} \bar{2} & \bar{1} \\ \bar{1} & \bar{0} \end{pmatrix} = \begin{pmatrix} \bar{2} & \bar{2} \\ \bar{2} & \bar{1} \end{pmatrix},$$

$$g^3 = g^2\begin{pmatrix} \bar{2} & \bar{1} \\ \bar{1} & \bar{0} \end{pmatrix} = \begin{pmatrix} \bar{0} & \bar{2} \\ \bar{2} & \bar{2} \end{pmatrix}, \qquad g^4 = g^3\begin{pmatrix} \bar{2} & \bar{1} \\ \bar{1} & \bar{0} \end{pmatrix} = \begin{pmatrix} \bar{2} & \bar{0} \\ \bar{0} & \bar{2} \end{pmatrix},$$

$$g^5 = g^4\begin{pmatrix} \bar{2} & \bar{1} \\ \bar{1} & \bar{0} \end{pmatrix} = \begin{pmatrix} \bar{1} & \bar{2} \\ \bar{2} & \bar{0} \end{pmatrix}, \qquad g^6 = g^5\begin{pmatrix} \bar{2} & \bar{1} \\ \bar{1} & \bar{0} \end{pmatrix} = \begin{pmatrix} \bar{1} & \bar{1} \\ \bar{1} & \bar{2} \end{pmatrix},$$

$$g^7 = g^6\begin{pmatrix} \bar{2} & \bar{1} \\ \bar{1} & \bar{0} \end{pmatrix} = \begin{pmatrix} \bar{0} & \bar{1} \\ \bar{1} & \bar{1} \end{pmatrix}, \qquad g^8 = g^7\begin{pmatrix} \bar{2} & \bar{1} \\ \bar{1} & \bar{0} \end{pmatrix} = \begin{pmatrix} \bar{1} & \bar{0} \\ \bar{0} & \bar{1} \end{pmatrix}.$$

From these calculations, we determine that the order of g is $|g| = 8$. $\triangle$

When studying a specific group, determining the orders of elements in groups is particularly useful. We present a number of propositions concerning the powers and orders of elements.

Proposition 1.3.4

> If G is a finite group and $g \in G$, then $|g|$ is finite.

Proof. Let $g \in G$ be any element. Consider the sequence of powers $\{g^i \mid i \in \mathbb{N}\}$. This is a subset of G. By the pigeon-hole principle, two elements in the set of powers must be equal. Hence, there exist nonnegative integers $i < j$ with $g^j = g^i$. Operating on both sides by $(g^{-1})^i$, we get $g^{j-i} = e$. Hence, g has finite order. $\square$

Proposition 1.3.5

> Let G be any group and let $x \in G$. Then $\left|x^{-1}\right| = |x|$.

Proof. (Left as an exercise for the reader. See Exercise 1.3.19.) $\square$

We point out that if $x^n = e$, then $x^{-1} = x^{n-1}$.

Proposition 1.3.6

> Let $x \in G$ with $x^n = e$ and $x^m = e$, then $x^d = e$ where $d = \gcd(m, n)$.

Proof. From Proposition A.5.10, the greatest common divisor $d = \gcd(m, n)$ can be written as a linear combination $sm + tn = d$, for some $s, t \in \mathbb{Z}$. Then

$$x^d = x^{sm+tn} = (x^m)^s (x^n)^t = e^s e^t = e. \qquad \square$$

Corollary 1.3.7

> Suppose that x is an element of a group G with $x^m = e$. Then the order $|x|$ divides m.

Proof. If $|x| = n$, then $x^n = x^m = e$. By Proposition 1.3.6, $x^{\gcd(m,n)} = e$. However, n is the least positive integer k such that $x^k = e$. Furthermore, since $1 \leq \gcd(m, n) \leq n$, this minimality condition $\gcd(m, n) = n$. This implies that n divides m. $\square$

Proposition 1.3.8

> Let G be a group and let $a \in \mathbb{N}^*$. Then we have the following results about orders:
>
> (1) If $|x| = \infty$, then $|x^a| = \infty$.
>
> (2) If $|x| = n < \infty$, then $|x^a| = \dfrac{n}{\gcd(n, a)}$.

Proof. For (1), assume that x^a has finite order k. Then $(x^a)^k = x^{ak} = e$. This contradicts $|x| = \infty$. Hence, x^a has infinite order.

For (2), let $y = x^a$ and $d = \gcd(n, a)$. Writing $n = db$ and $a = dc$, by Exercise A.5.11, we know that $\gcd(b, c) = 1$. Then

$$y^b = x^{ab} = x^{bcd} = x^{nc} = e^c = e.$$

By Corollary 1.3.7, the order $|y|$ divides b. Conversely, suppose that $|y| = k$. Then k divides b. Since $y^k = x^{ak} = e$, then $n|ak$. Thus $db|dck$, which implies that $b|ck$. However, $\gcd(b, c) = 1$, so we conclude that $b|k$. However, since $b|k$ and $k|b$ and $b, k \in \mathbb{N}^*$, we can conclude that $b = k$.

Hence, $|y| = |x^a| = b = n/d = n/\gcd(a, n)$. $\square$

Proposition 1.3.8 presents two noteworthy cases. If $\gcd(a, n) = 1$, then $|x^a| = n$. Second, if a divides n, then $|x^a| = n/a$.

It is important to point out that in a general group, there is very little that can be said about the relationship between $|xy|$, $|x|$, and $|y|$ for two elements $x, y \in G$. For example, consider the dihedral group D_n. Both s and rs refer to reflections through lines and hence have order 2. However, $s(sr) = r$, and r has order n. Thus, given any integer n, there exists a group where $|g_1| = 2$ and $|g_2| = 2$ and $|g_1 g_2| = n$.

Example 1.3.9. In D_{10}, we know that $|r| = 10$. Proposition 1.3.8 allows us to immediately determine the orders of all the elements r^k. For $k = 1, 3, 7, 9$, which are relatively prime to 10, we have $|r^k| = 10$. For $k = 2, 4, 6, 8$, we have $\gcd(10, k) = 2$ so $|r^k| = 10/2 = 5$. And finally, $|r^5| = 10/\gcd(10, 5) = 2$. $\triangle$

Example 1.3.10. In $\mathbb{Z}/20\mathbb{Z}$, we know that $|\overline{1}| = 20$. Then $\overline{12} = 12 \cdot \overline{1}$, so $a = 12$ in Proposition 1.3.8(2). Hence the order of $\overline{12}$ in $\mathbb{Z}/20\mathbb{Z}$ is $|\overline{12}| = 20/\gcd(20, 12) = 20/4 = 5$. $\triangle$

1.3.2 Cyclic Groups

The process of simply considering the successive powers of an element gives rise to an important class of groups.

Definition 1.3.11

A group G is called *cyclic* if there exists an element $x \in G$ such that every element of $g \in G$ we have $g = x^k$ for some $k \in \mathbb{Z}$. The element x is called a *generator* of G.

For example, we notice that for all integers $n \geq 2$, the group $\mathbb{Z}/n\mathbb{Z}$ (with addition as the operation) is a cyclic group because all elements of $\mathbb{Z}/n\mathbb{Z}$ are multiples of $\overline{1}$. As we saw in Section A.6, one of the main differences with usual arithmetic is that $n \cdot \overline{1} = \overline{0}$. The intuitive sense that the powers (or multiples) of an element "cycle back" motivate the terminology of cyclic group.

Remark 1.3.12. We point out that a finite group G is cyclic if and only if there exists an element $g \in G$ such that $|g| = |G|$. $\triangle$

Cyclic groups do not have to be finite though. The group $(\mathbb{Z}, +)$ is also cyclic because every element in $\mathbb{Z}$ is obtained by $n \cdot 1$ with $n \in \mathbb{Z}$.

Example 1.3.13. Consider the group $U(14)$. The elements are

$$U(14) = \{\overline{1}, \overline{3}, \overline{5}, \overline{9}, \overline{11}, \overline{13}\}.$$

This group is cyclic because, for example, the powers of $\overline{3}$ gives all the elements of $U(14)$:

i	1	2	3	4	5	6
$\overline{3}^i$	$\overline{3}$	$\overline{9}$	$\overline{13}$	$\overline{11}$	$\overline{5}$	$\overline{1}$

We note that the powers of $\overline{3}$ will then cycle around, because $\overline{3}^7 = \overline{3}^6 \cdot \overline{3} = \overline{3}$, then $\overline{3}^8 = \overline{9}$, and so on. $\triangle$

Example 1.3.14. At first glance, someone might think that to prove that a group is not cyclic we would need to calculate the order of every element. If no element has the same order as the cardinality of the group, only then we could say that the group is not cyclic. However, by an application of the theorems in this section, we may be able to conclude the group is not cyclic with much less work.

As an example, suppose we wish to determine if $U(200)$ is cyclic. Note that $|U(200)| = \phi(200) = 80$. The most obvious thing to try is to start calculating the powers of $\overline{3}$. Without showing all the powers here, we can check that $|\overline{3}| = 20$. So we conclude immediately that $\overline{3}$ is not a generator of $U(200)$. In this list, we would find that $\overline{3}^{10} = \overline{49}$. This implies (and also using Proposition 1.3.8) that $|\overline{49}| = 2$. It is easy to see that $\overline{199}^2 = (\overline{-1})^2 = \overline{1}$, so $|\overline{199}| = 2$. Now if $U(200)$ were cyclic with generator $\overline{a}$, then $|\overline{a}| = 80$. Also by Proposition 1.3.8, an element $\overline{a}^k$ has order 2 if and only if $\gcd(k, 80) = 40$. However, the only integer k with $1 \leq k \leq 80$ such that $\gcd(k, 80) = 40$ is 40 itself. Hence, a cyclic of order 80 has at most one element of order 2. Since $U(200)$ has more than one element of order 2, it is not a cyclic group. $\triangle$

Definition 1.3.15 (Finite Cyclic Groups)

Let n be a positive integer. We denote by Z_n the group with elements $\{e, x, x^2, \ldots, x^{n-1}\}$, where x has the property that $x^n = e$. This group is sometimes generically called *the cyclic group* of order n.

It is important to emphasize in this notation that we do not define the group as existing in any previously known arithmetic context. The element x does not represent some complex number or matrix or any other object. We have simply defined how it operates symbolically.

1.3.3 Useful CAS Commands

The `GroupTheory` package in *Maple* has the commands `GroupOrder` and `ElementOrder`, which implement a calculation of the order of a group and the order of an element in a group. We will wait a few sections to illustrate `ElementOrder` because we need a little more group theory background to describe how CASs implement groups and elements of groups. However, we can already give the following example of `GroupOrder`.

─────────────────────── Maple ───────────────────────

$with(GroupTheory)$:

$G := GL(3, 5)$;

$$G := GL(3, 5)$$

$GroupOrder(G)$;

$$1488000$$

$IsAbelian(G)$;

$$false$$

───

This defines G as the general linear group of invertible 3×3 matrices with coefficients in $\mathbb{Z}/5\mathbb{Z}$. Here the object class of G is the whole group. The next command finds the order (cardinality) of the group. Finally, the last line illustrates a command that asks whether the group is abelian or not.

The following code in SAGE implements the same thing.

─────────────────────── Sage ───────────────────────

```
sage: G=GL(3,GF(5))
sage: G.order()
1488000
sage: G.is_abelian()
False
```

───

Many CASs offer similar commands that test true/false properties about groups as a whole (e.g., finite, simple, perfect, solvable, nilpotent). These commands always look like *IsAbelian* in *Maple* or `is_abelian()` in SAGE.

Without a `GroupElement()` command, we can find the orders of elements in a group simply by running a for loop to calculate all the powers of an element up to a certain point and verifying when we reach the identity. Suppose we revisit the example of *Maple* code in Section 1.2.5 the command

$seq(A^\wedge i, i = 1..20)$;

prints out a list of all the powers of the matrix A, from A to A^{20} in the group $GL_2(\mathbb{F}_{11})$. We can accomplish the same thing in SAGE with the following code

```
sage: G=GL(2,GF(11))
sage: A=G([1,2,7,9])
sage: [A^n for n in range(1,20)]
      (not listed)
sage: A.order()
40
```

(For the sake of space, we did not list output of the third line.) In particular, SAGE offers a command to calculate the order of any group element.

EXERCISES FOR SECTION 1.3

1. Find the orders of $\bar{5}$ and $\bar{6}$ in $(\mathbb{Z}/21\mathbb{Z}, +)$.

2. Calculate the order $\overline{20}$ in $\mathbb{Z}/52\mathbb{Z}$.

3. Calculate the order of $\overline{285}$ in the group $\mathbb{Z}/360\mathbb{Z}$.

4. Calculate the order of r^{16} in D_{24}.

5. Find the orders of all the elements in $(\mathbb{Z}/18\mathbb{Z}, +)$.

6. Find the orders of all the elements in $U(21)$. Deduce that $U(21)$ is not cyclic.

7. Find the orders of all the elements in Z_{20}.

8. Show that for all integers $n \geq 1$, the element

$$R = \begin{pmatrix} \cos(2\pi/n) & -\sin(2\pi/n) \\ \sin(2\pi/n) & \cos(2\pi/n) \end{pmatrix}$$

   in $GL_2(\mathbb{R})$ has order n.

9. In $GL_2(\mathbb{F}_3)$, find the orders of the following elements:

$$\text{(a) } A = \begin{pmatrix} \bar{1} & \bar{2} \\ \bar{0} & \bar{1} \end{pmatrix}, \quad \text{(b) } B = \begin{pmatrix} \bar{2} & \bar{1} \\ \bar{1} & \bar{0} \end{pmatrix}, \quad \text{(c) } C = \begin{pmatrix} \bar{2} & \bar{2} \\ \bar{0} & \bar{2} \end{pmatrix}.$$

10. Prove that if $A \in GL_n(\mathbb{R})$ has order $|A| = k$, then all eigenvalues λ of A satisfy $\lambda^k = 1$.

11. Prove Proposition 1.2.12. [Hint: Pay careful attention to when powers are negative, zero, or positive.]

12. Determine if $U(11)$ a cyclic group.

13. Determine if $U(10)$ a cyclic group.

14. Determine if $U(36)$ a cyclic group.

15. Find all the generators of the cyclic group Z_{40}.

16. Find all the generators of the cyclic group $\mathbb{Z}/36\mathbb{Z}$.

17. Let p be an odd prime.
    (a) Use the Binomial Theorem to prove that $(1+p)^{p^n} \equiv 1 \pmod{p^{n+1}}$ for all positive integers n.
    (b) Prove also that $(1+p)^{p^{n-1}} \not\equiv 1 \pmod{p^{n+1}}$ for all positive integers n.
    (c) Conclude that $\overline{1+p}$ has order p^n in $U(p^{n+1})$.

18. Prove that $(\mathbb{Q}, +)$ is not a cyclic group.

19. Prove Proposition 1.3.5.

20. Let G be a group such that for all $a, b, c, d, x \in G$, the identity $axb = cxd$ implies $ab = cd$. Prove that G is abelian.

21. Let a and b be elements in a group G. Prove that if a and b commute, then the order of ab divides $\mathrm{lcm}(|a|, |b|)$.

22. Find a nonabelian group G and two elements $a, b \in G$ such that $|ab|$ does not divide $\mathrm{lcm}(|a|, |b|)$.

23. Let $x \in G$ be an element of finite order n. Prove that $e, x, x^2, \ldots, x^{n-1}$ are all distinct. Deduce that $|x| \leq |G|$.

24. Prove that for elements x and y in any group G we have $|x| = |yxy^{-1}|$.

25. Use the preceding exercise to show that for any elements g_1 and g_2 in a group $|g_1 g_2| = |g_2 g_1|$, even if g_1 and g_2 do not commute.

26. Prove that the group of rigid motions of the cube has order 24.

27. Prove that the group of rigid motions of the octahedron has order 24.

28. Prove that the group of rigid motions of a classic soccer ball has order 60.

29. (CAS) Using a computer algebra system find all the orders of all the elements in $\mathrm{GL}_2(\mathbb{F}_3)$.

30. (CAS) Using a CAS and some programming, find the number of elements in $\mathrm{GL}_2(\mathbb{F}_5)$ of any given order. [Hint: Use a dictionary in SAGE or a hash table in *Maple*.]

1.4 Concept of a Classification Theorem

One of the recurring themes in group theory is whether we can find all groups with certain stated properties. A theorem that answers a question like this is called a classification theorem. In fact, such classification questions play key roles throughout abstract algebra and regularly drive directions of investigation. Often, these theorems are quite profound. Though we are just at the beginning of group theory and such problems require many more concepts than we have developed yet, it is valuable to keep classification questions in the back of our minds for motivation.

1.4.1 Direct Sum of Groups

A useful perspective behind considering classification question involves thinking about how to create new (larger) groups, from smaller ones. The direct sum of two groups is one such method.

Definition 1.4.1 (Direct Sum)

Let $(G, *)$ and $(H, \star)$ be two groups. The *direct sum* of the groups is a new group $(G \times H, \cdot)$ where the set operation is defined by

$$(g_1, h_1) \cdot (g_2, h_2) = (g_1 * g_2, h_1 \star h_2).$$

The direct sum is denoted by $G \oplus H$.

We observe that the identity of $G \oplus H$ is (e_G, e_H), the identity of G paired with the identity of H.

The direct sum generalizes to any finite number of groups. For example, the group $(\mathbb{R}^3, +)$ is the triple direct sum of group $(\mathbb{R}, +)$ with itself. The reader should feel free to imagine all sorts of possibilities now, e.g., $\mathbb{Z}/9\mathbb{Z} \oplus D_{10}$, $\mathbb{Z} \oplus Z_2 \oplus Z_2$, $\mathrm{GL}_2(\mathbb{F}_7) \oplus U(21)$, and so on.

Example 1.4.2. Consider the group $Z_4 \oplus Z_2$. We can list the elements of this group as

$$Z_4 \oplus Z_2 = \{(x^m, y^n) \,|\, x^4 = e \text{ and } y^2 = e\}.$$

(To avoid confusion, we should not use the same letter for the generator of Z_4 and of Z_2. After all, they are different elements and from different groups.) We point out that $Z_4 \oplus Z_2$ is not another cyclic group. The size is $|Z_4 \oplus Z_2| = 8$. However, in a cyclic group of order 8 generated by some element z, only the element z^4 has order 2. However, in $Z_4 \oplus Z_2$, both (x^2, e) and (e, y) have order 2. Thus $Z_4 \oplus Z_2$ is not cyclic.

It is easy to prove that $Z_4 \oplus Z_2$ is abelian. Let $(x^i, y^j), (x^a, y^b) \in Z_4 \oplus Z_2$. Then

$$(x^i, y^j)(x^a, y^b) = (x^i x^a, y^j y^b) = (x^{i+a}, y^{j+b}) = (x^a x^i, y^b y^j) = (x^a, y^b)(x^i, y^j).$$

Since $Z_4 \oplus Z_2$ is abelian, it does not behave like D_4, which is not abelian. Hence, we have found three groups of order 8 with different properties: Z_8, $Z_4 \oplus Z_2$ and D_4. △

In the previous section, we pointed out that knowing the orders of two elements g_1, g_2 in a group G, we do not readily know the order of $g_1 g_2$. However, with direct sums, a theorem gives us a relationship among orders.

Theorem 1.4.3

Let $G_1, \ldots, G_n$ be groups and let $(g_1, \ldots, g_n) \in G_1 \oplus \cdots \oplus G_n$ be an element in the direct sum. Then the order of $(g_1, \ldots, g_n)$ is

$$|(g_1, \ldots, g_n)| = \text{lcm}(|g_1|, |g_2|, \ldots, |g_n|).$$

Proof. Call $M = \text{lcm}(|g_1|, |g_2|, \ldots, |g_n|)$ and let e_i be the identity in G_i for each $i = 1, 2, \ldots, n$. By definition, $(g_1, g_2, \ldots, g_n)^m = (e_1, e_2, \ldots, e_n)$ if and only if $g_i^m = e_i$ for all i. By Corollary 1.3.7, $|g_i|$ divides m for all i and thus $M \mid m$.

Suppose now that k is the order of $(g_1, g_2, \ldots, g_n)$, namely the least positive integer m such that $(g_1, g_2, \ldots, g_n)^m = (e_1, e_2, \ldots, e_n)$. Then by the preceding paragraph, $M \mid k$. However, M is a multiple of $|g_i|$ for each i, and hence,

$$(g_1, g_2, \ldots, g_n)^M = (g_1^M, g_2^M, \ldots, g_n^M) = (e_1, e_2, \ldots, e_n).$$

Hence, by Corollary 1.3.7, k divides M. Since M and k are positive numbers that divide each other, $M = k$. The theorem follows. $\square$

1.4.2   Classification Theorems

Classification theorems usually require theorems not yet at our disposal, e.g., Lagrange's Theorem, Cauchy's Theorem, Sylow's Theorem, and so on. Nonetheless, this section explores what we can discover for possibilities of small groups using the Cayley table or orders of elements.

Example 1.4.4 (Groups of Order 4). We propose to find all groups of order 4 by filling out all possible Cayley tables. Suppose that $G = \{e, a, b, c\}$ with a, b, and c distinct nonidentity group elements. A priori, all we know about the Cayley graph is the first column and first row.

	e	a	b	c
e	e	a	b	c
a	a			
b	b			
c	c			

Note that if a group G contains an element g of order n, then $\{e, g, g^2, \ldots, g^{n-1}\}$ is a subset of n distinct elements. (See Exercise 1.3.23.) Hence, a group of order 4 cannot contain an element of order 5 or higher.

Suppose that G contains an element of order 4, say, the element a. Then $G = \{e, a, a^2, a^3\}$. Without loss of generality, we can call $b = a^2$ and $c = a^3$

and the Cayley table becomes the following.

	e	a	b	c
e	e	a	b	c
a	a	b	c	e
b	b	c	e	a
c	c	e	a	b

We recognize this table as corresponding to the cyclic group Z_4.

Assume that G does not contain an element of order 4 but contains one of order 3. Without loss of generality, suppose that $|a| = 3$. Then $G = \{e, a, a^2, c\}$, with all elements distinct. The element ac cannot be equal to a^k for any k for then $c = a^{k-1}$, a contradiction. Furthermore, we cannot have $ac = c$ for then $a = e$, again a contradiction. Hence, a group of order 4 cannot contain an element of order 3.

Suppose now that all nonidentity elements in G have order 2. Filling out the Cayley table, this means that we have e in the entries on the main diagonal. We claim that $ab = c$. This is because ab cannot be e because a is its own inverse and $a \neq b$; ab cannot be a because this would imply that $b = e$; and ab cannot be b because this would imply that $a = e$. Once, we place c in the table corresponding to ab, by virtue of the Cayley table being a Latin square, we can fill out every other entry. We get:

$$
\begin{array}{c|cccc}
 & e & a & b & c \\
\hline
e & e & a & b & c \\
a & a & e & c & b \\
b & b & c & e & a \\
c & c & b & a & e \\
\end{array}
\qquad (1.5)
$$

This group is often denoted by V_4 and called the Klein-4 group. Our approach covered all possible cases for orders of elements in a group of order 4 so we conclude that Z_4 and V_4 are the only two groups of order 4. $\triangle$

The conclusion of the previous example might seem striking at first. In particular, we already know two groups of order 4 that have different properties: Z_4 and $Z_2 \oplus Z_2$. Consequently, V_4 and $Z_2 \oplus Z_2$ must in some sense be the same group. However, we do not yet have the background to fully develop an intuitive concept of sameness for groups. We return to that issue in Section 1.9. Nevertheless, if we write $Z_2 = \{e, x\}$ with $x^2 = e$, then it is an easy check to see (1.5) is the Cayley table for $Z_2 \oplus Z_2$ with $a = (x, e)$ and $b = (e, x)$.

Example 1.4.5. Though a more complete problem would be to find all groups of order 8, we propose to tackle a slightly simpler problem: to find all groups G of order 8 that contain an element x of order 4. Let y be another element in G that is distinct from any power of x. With these criteria, we know so far that G contains the distinct elements e, x, x^2, x^3, y. The element xy cannot be

e because that would imply $y = x^3$;

x because that would imply $y = e$;

x^2 because that would imply $y = x$;

x^3 because that would imply $y = x^2$;

y because that would imply $x = e$.

So xy is a new element of G. By similar reasonings that we leave to the reader, the elements x^2y and x^3y are distinct from all the others. Hence, G must contain the 8 distinct elements

$$\{e, x, x^2, x^3, y, xy, x^2y, x^3y\}. \tag{1.6}$$

We now consider various possibilities for the order of y.

Suppose first that $|y| = 2$. Consider the element yx. With a reason by cases similar to that above, yx cannot be e, x, x^2, x^3, or y. Thus, there are three cases: (1) $yx = x^3y$; (2) $yx = xy$; and (3) $yx = x^2y$.

Case 1. If $yx = x^3y$, then the group is in fact D_4, the dihedral group of the square, where x serves the role of r and y serves the role of s.

Case 2. If $yx = xy$, then x and y commute, so $x^sy^t = y^tx^s$ for all s, t and so G is abelian. We leave it up to the reader to show that this group is $Z_4 \oplus Z_2$.

Case 3. If $yx = x^2y$, then $yxy = x^2$. Hence,

$$x = y^2xy^2 = y(yxy)y = yx^2y = yxy^2xy = (yxy)(yxy) = x^4 = e.$$

We conclude that $x = e$, which contradicts the assumption that x has order 4. Hence, there exists no group of order 8 with an element x of order 4 and an element y of order 2 with $yx = x^2y$.

Suppose that $|y| = 3$. Consider the element y^2 in G. A quick proof by cases shows that y^2 cannot be any of the eight distinct elements listed in (1.6). Hence, there exists no group of order 8 containing an element of order 4 and one of order 3.

Suppose that $|y| = 4$. Again, we consider the element y^2. If there exists a group with all the conditions we have so far, then y^2 must be equal to an element in (1.6). Now $|y^2| = 4/2 = 2$ so y^2 cannot be e, x, x^3, or y, which have orders $1, 4, 4, 4$, respectively. Furthermore, y^2 cannot be equal to xy, (respectively x^2y or x^3y) because that would imply $x = y$ (respectively $x^2 = y$ or $x^3 = y$), which is against the assumptions on x and y. The only remaining possibility is $y^2 = x^2$.

We focus on this latter possibility, namely a group G containing x and y with $|x| = 4$, $|y| = 4$, $y \notin \{e, x, x^2, x^3\}$ and $x^2 = y^2$. If we now consider the element yx, we can quickly eliminate all possibilities except xy and x^3y. If $G = Z_4 \oplus Z_2 = \{(z, w) \mid z^4 = e \text{ and } w^2 = e\}$, then setting $x = (z, e)$ and $y = (z, w)$, it is easy to check that $Z_4 \oplus Z_2$ satisfies $x^2 = y^2$ and $yx = xy$.

So that is not new. On the other hand, if $yx = xy^3$, then G is a nonabelian group in which $x, x^3, y, y^3 = x^2y$ are elements of order 4. However, D_4 is the only nonabelian group of order 8 that we have encountered so far and in D_4 only r and r^3 have order 4. Hence, G must be a new group. $\triangle$

We now introduce the new group identified in this example but using the symbols traditionally associated with it.

Example 1.4.6 (The Quaternion Group). The quaternion group, denoted by Q_8, contains the following eight elements:

$$1, -1, i, -i, j, -j, k, -k.$$

The operations on the elements are in part inspired by how the imaginary number operates on itself. In particular, 1 is the identity element and multiplication by (-1) changes the sign of any element. We also have

$$\begin{aligned}
i^2 &= -1, & i^3 &= -i, & i^4 &= 1, \\
j^2 &= -1, & j^3 &= -j, & j^4 &= 1, \\
k^2 &= -1, & k^3 &= -k, & k^4 &= 1, \\
ij &= k, & jk &= i, & ki &= j, \\
ji &= -k, & kj &= -i, & ik &= -j.
\end{aligned}$$

Furthermore, if a and b are i, j, or k, then

$$(-a)b = (-1)(ab) \quad \text{and} \quad (-a)(-b) = ab.$$

Matching symbols to Example 1.4.5, note that $i^4 = j^4 = 1$, $i^2 = -1 = j^2$, and $ji = -k = (-i)j = i^3j$. This shows that Q_8 is indeed the new group of order 8 discovered at the end of the previous example. $\triangle$

A bigger classification question would involve finding all the groups of order 8. We could solve this problem at this point, but we will soon encounter theorems that establish more internal structure on groups that would make such questions easier. Consequently, we delay most classification questions until later.

Exercises for Section 1.4

1. Find the orders of all the elements in $Z_4 \oplus Z_2$.

2. What is the largest order of an element in $Z_{75} \oplus Z_{100}$? Illustrate with a specific element.

3. Show that $Z_5 \oplus Z_2$ is cyclic.

4. Show that $Z_4 \oplus Z_2$ is not cyclic.

5. Construct the Cayley table for $Z_3 \oplus Z_3$.

6. Let A and B be groups. Prove that the direct sum $A \oplus B$ is abelian if and only if A and B are both abelian.

7. Let G and H be two finite groups. Prove that $G \oplus H$ is cyclic if and only if G and H are both cyclic with $\gcd(|G|, |H|) = 1$.

8. Let $G_1, G_2, \ldots, G_n$ be n finite groups. Prove that $G_1 \oplus G_2 \oplus \cdots \oplus G_n$ is cyclic if and only if G_i is cyclic for all $i = 1, 2, \ldots, n$ and $\gcd(|G_i|, |G_j|) = 1$ for all $i \neq j$. [Hint: Use Exercise 1.4.7 and induction.]

9. Find all groups of order 5.

10. We consider groups of order 6. We know that Z_6 is a group of order 6. We now look for all the others. Let G be any group of order 6 that is not Z_6, i.e., does not contain an element of order 6.

    (a) Show that G cannot have an element of order 7 or greater.

    (b) Show that G cannot have an element of order 5.

    (c) Show that G cannot have an element of order 4.

    (d) Show that the nonidentity elements of G have order 2 or 3.

    (e) Conclude that there exist only two subgroups of order 6. In particular, there exists one abelian group of order 6 (the cyclic group Z_6) and one nonabelian group of order 6 (D_3 is such a group).

    [Comment: We will encounter a number of nonabelian groups of order 6 but the result of this exercise establishes that they are all in some sense the same. We will make precise this notion of sameness in Section 1.9.3.]

11. Let $G = \{e, v, w, x, y, z\}$ be a group of order 6. For the following partial table, decide if it can be completed to the Cayley table of some G and if so fill it in. [Hint: You may need to use associativity.]

	e	v	w	x	y	z
e	–	–	–	–	–	–
v	–	–	–	–	w	–
w	–	–	–	–	–	e
x	–	–	–	–	–	–
y	–	z	–	–	–	–
z	–	–	–	v	–	–

12. Let $G = \{e, t, u, v, w, x, y, z\}$ be a group of order 8. For the following partial table, decide if it can be completed to the Cayley table of some G and if so fill it in. [Hint: You may need to use associativity.]

	e	t	u	v	w	x	y	z
e	–	–	–	–	–	–	–	–
t	–	–	–	–	–	–	–	e
u	–	–	e	–	–	y	x	t
v	–	–	–	u	–	t	–	–
w	–	x	v	–	–	–	–	y
x	–	–	–	z	–	–	–	–
y	–	–	–	t	z	–	–	–
z	–	–	–	–	x	–	–	u

1.5 Symmetric Groups

Symmetric groups play a key role in group theory and in countless applications. This section introduces the terminology and elementary properties of symmetric groups.

1.5.1 Permutations

Definition 1.5.1

Let A be a nonempty set. Define S_A as the set of all bijections from A to itself.

Proposition 1.5.2

The pair $(S_A, \circ)$ is a group, where the operation $\circ$ is function composition.

Proof. The composition of two bijections is a bijection so composition is a binary operation on S_A.

Proposition A.2.12 establishes that $\circ$ is associative in S_A.

The function $id_A : A \to A$ such that $id_A(x) = x$ for all $x \in A$ is the group identity.

Since $f \in S_A$ is a bijection, there exists an inverse function, denoted $f^{-1} : A \to A$. By definition of the inverse function,

$$f \circ f^{-1} = f^{-1} \circ f = id_A.$$

Hence, $(S_A, \circ)$ has inverses. S_A satisfies all the axioms of a group. $\square$

We call S_A the *symmetric group* on A. In the case that $A = \{1, 2, \ldots, n\}$, then we write S_n instead of the cumbersome $S_{\{1,2,\ldots,n\}}$. We call the elements of S_A *permutations* of A.

Proposition 1.5.3

$|S_n| = n!$.

Proof. The order of S_n is the number of distinct bijections on $\{1, 2, \ldots, n\}$. To enumerate the bijections, we first count the injections f from $\{1, 2, \ldots, n\}$ to itself. Note that there are n options for $f(1)$. Since $f(2) \neq f(1)$, for each choice of $f(1)$, there are $n-1$ choices for $f(2)$. Given values for $f(1)$ and $f(2)$, there are $n-2$ possible choices for $f(3)$ and so on. Since an enumeration of injections requires an n-part decision, we use the product rule, giving us

$$n(n-1)(n-2) \cdots 3 \times 2 \times 1 = n!.$$

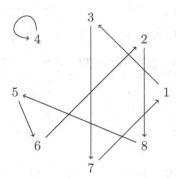

FIGURE 1.7: A permutation as a directed graph.

However, for any such injection f has n distinct images, so is also a bijection. Thus, $|S_n| = n!$. $\square$

Symmetric groups arise in a variety of natural contexts. In a 100-meter Olympic race, eight runners are given lane numbers. The function from the runner's lane number to the rank they place in the race is a permutation of S_8. A cryptogram is a word game in which someone writes a message but replacing each letter of the alphabet with another letter and a second person attempts to recover the original message. The first person's choice of how to scramble the letters of the alphabet is a permutation in S_a, where a is the number of letters in the alphabet used. When someone shuffles a deck of 52 cards, the resulting reordering of the cards represents a permutation in S_{52}.

We need a few convenient ways to visualize and represent a permutation on $\{1, 2, \ldots, n\}$.

Directed Graph. A visual method of representing a permutation $\sigma \in S_n$ involves using a directed graph. Each element of $\{1, 2, \ldots, n\}$ is written as a point on the plane and we draw an arrow from a to b if $f(a) = b$. In this way, a permutation will create a directed graph in which one arrow leaves each point and arrives at each point. See Figure 1.7 for an example.

Chart Notation. Another way of writing a permutation is to record in a chart or matrix the outputs like

$$\sigma = \begin{pmatrix} 1 & 2 & \cdots & n \\ \sigma(1) & \sigma(2) & \cdots & \sigma(n) \end{pmatrix}.$$

Using the chart notation, the permutation in Figure 1.7 is written as

$$\sigma = \begin{pmatrix} 1 & 2 & 3 & 4 & 5 & 6 & 7 & 8 \\ 3 & 8 & 7 & 4 & 6 & 2 & 1 & 5 \end{pmatrix}.$$

***n*-tuple.** If the value n is clear from context, then the top row of the chart notation is redundant. Hence, we can represent the permutation σ by the n-tuple $(\sigma(1), \sigma(2), \ldots, \sigma(n))$. Using the n-tuple notation, the permutation in Figure 1.7 is written as $\sigma = (3, 8, 7, 4, 6, 2, 1, 5)$.

Cycle Notation. A different notation turns out to be more useful for the purposes of group theory. In cycle notation, the expression

$$\sigma = (a_1 \, a_2 \cdots a_{m_1})(a_{m_1+1} \, a_{m_1+2} \cdots a_{m_2}) \cdots (a_{m_{k-1}+1} \, a_{m_{k-1}+2} \cdots a_{m_k}),$$

where a_ℓ are distinct elements in $\{1, 2, \ldots, n\}$, means that for any index i,

$$\sigma(a_i) = \begin{cases} a_{i+1} & \text{if } i \neq m_j \text{ for all } j \\ a_{m_{j-1}+1} & \text{if } i = m_j \text{ for some } j, \end{cases}$$

where $m_0 = 0$. Any of the expressions $(a_{m_{j-1}+1} \, a_{m_{j-1}+2} \cdots a_{m_j})$ is called a *cycle* because σ "cycles" through these elements in order as σ iterates. Using the cycle notation for the permutation in Figure 1.7 is

$$\sigma = (1\,3\,7)(2\,8\,5\,6)(4).$$

There are many different ways of expressing a permutation using the cycle notation. For example, as cycles, $(1\,3\,7) = (3\,7\,1) = (7\,1\,3)$. Standard cycle notation imposes four additional habits. (1) If σ is the identity function, we just write $\sigma = \text{id}$. (Advanced texts commonly refer to the identity permutation as 1 but, for the moment, we will use id in order to avoid confusion.) (2) We write each cycle of σ starting with the lowest integer in the cycle. (3) The order in which we list the cycles of σ is such that initial elements of each cycle are in increasing order. (4) Finally, we omit any cycle of length 1. We say that a permutation in S_n is written in *standard* cycle notation if it satisfies these requirements. The standard cycle notation for the permutation in Figure 1.7 is $\sigma = (1\,3\,7)(2\,8\,5\,6)$.

Definition 1.5.4

- An *m-cycle* is a permutation that in standard cycle notation consists of only one cycle of length m.

- Two cycles are called *disjoint* if they involve no common integers.

- A 2-cycle is also called a *transposition*.

We use the special term transposition for a 2-cycle because it simply interchanges (transposes) two elements and leaves the rest fixed. By the construction, the cycles that appear in the standard cycle notation for a permutation are all disjoint.

Example 1.5.5. To illustrate the cycle notation, we list all the permutations in S_4 in standard cycle notation:

$(1\,2\,3\,4)$, $(1\,2\,4\,3)$, $(1\,3\,2\,4)$, $(1\,3\,4\,2)$, $(1\,4\,2\,3)$, $(1\,4\,3\,2)$,
$(1\,2\,3)$, $(1\,3\,2)$, $(1\,2\,4)$, $(1\,4\,2)$, $(1\,3\,4)$, $(1\,4\,3)$, $(2\,3\,4)$, $(2\,4\,3)$,
$(1\,2)$, $(1\,3)$, $(1\,4)$, $(2\,3)$, $(2\,4)$, $(3\,4)$,
$(1\,2)(3\,4)$, $(1\,3)(2\,4)$, $(1\,4)(2\,3)$,
id .

As a simple counting exercise, we verify that we have all the 3-cycles by calculating how many there should be. Each cycle consists of 3 integers. The number of ways of choosing 3 from 4 integers is $\binom{4}{3}$. For each selection of 3 integers, after writing the least integer first, there are $2! = 2$ options for how to order the remaining two integers in the 3-cycle. Hence, there are $2 \cdot \binom{4}{3} = 8$ three-cycles in S_4.   △

The *cycle type* of a permutation describes how many disjoint cycles of a given length make up the standard cycle notation of that permutation. Hence, we say that $(1\,3)(2\,4)$ is of cycle type $(a\,b)(c\,d)$, or of type $2, 2$.

Example 1.5.6. As another example, consider the symmetric group S_6. There are $6! = 720$ elements in S_6. We count how many permutations there are in S_6 of a given cycle type. In order to count the 6-cycles, note that every integer from 1 to 6 appears in the cycle notation of a 6-cycle. In standard cycle notation, we write 1 first and then all $5! = 120$ orderings of $\{2, 3, 4, 5, 6\}$ give distinct 6-cycles. Hence, there are 120 6-cycles in S_6.

We now count the permutations in S_6 of the form $\sigma = (a_1\,a_2\,a_3)(a_4\,a_5\,a_6)$. The standard cycle notation of a permutation has $a_1 = 1$. To choose the values in a_2 and a_3, there are 5 choices for a_2 and then 4 remaining choices for a_3. With a_1, a_2, and a_3 chosen, we know that $\{a_4, a_5, a_6\} = \{1, 2, 3, 4, 5, 6\} - \{a_1, a_2, a_3\}$. The value of a_4 must be the minimum value of $\{a_4, a_5, a_6\}$. Then there are two ways to order the two remaining elements in the second 3-cycle. Hence, there are $5 \cdot 4 \cdot 2 = 40$ permutations that consist of the product of two disjoint 3-cycles.

Cycle type	Number of elements
id	1
$(a\,b)$	$\binom{6}{2} = 15$
$(a\,b\,c)$	$2 \cdot \binom{6}{3} = 40$
$(a\,b\,c\,d)$	$3!\binom{6}{4} = 90$
$(a\,b\,c\,d\,e)$	$4!\binom{6}{5} = 144$
$(a\,b\,c\,d\,e\,f)$	$5! = 120$
$(a\,b\,c\,d)(e\,f)$	$3!\binom{6}{4} = 90$
$(a\,b\,c)(d\,e\,f)$	40
$(a\,b\,c)(d\,e)$	$2\binom{6}{3} = 120$
$(a\,b)(c\,d)$	$\frac{1}{2}\binom{6}{2}\binom{4}{2} = 45$
$(a\,b)(c\,d)(e\,f)$	$\binom{6}{2}\binom{4}{2}/3! = 15$
Total:	720

The above table counts all the different permutations in S_6, organized by lengths of cycles in standard cycle notation. △

1.5.2 Operations in Cycle Notation

When calculating operations in the symmetric group, we must remember that permutations are bijective functions and that function composition is read from right to left. Because of this, if $\sigma, \tau \in S_A$, then the composition $\sigma\tau$ (short for $\sigma \circ \tau$) means the bijection where we apply τ first and then σ.

Consider the following example in which we determine the cycle notation for a product. Suppose that we are in S_6 and $\sigma = (1\,4\,2\,6)(3\,5)$ and $\tau = (2\,6\,3)$. We write

$$\sigma\tau = (1\,4\,2\,6)(3\,5)(2\,6\,3)$$

and read from right to left how $\sigma\tau$ maps the integers as a composition of cycles, not necessarily disjoint now. (In the diagrams below, the arrows beneath the permutation indicate the direction of reading and the arrows above indicate the action of a cycle on an element along the way.)

$$(1\ 4\ 2\ 6)\ (3\ 5)\ (2\ 6\ 3) \qquad \text{so } \sigma\tau = (1\,4\,\ldots$$
$$4 \leftarrow 4 \qquad\qquad\qquad 1$$

$$(1\ 4\ 2\ 6)\ (3\ 5)\ (2\ 6\ 3) \qquad \text{so } \sigma\tau = (1\,4\,2\,\ldots$$
$$2 \leftarrow 2 \qquad\qquad 4$$

$$(1\ 4\ 2\ 6)\ (3\ 5)\ (2\ 6\ 3) \qquad \text{so } \sigma\tau = (1\,4\,2)(3\,\ldots$$
$$1 \leftarrow \qquad 6 \qquad 2 \quad 2$$

Note that since $\sigma\tau(2) = 1$, we closed the first cycle and start a new cycle with the smallest integer not already appearing in any previous cycle of $\sigma\tau$.

$$(1\ 4\ 2\ 6)\ (3\ 5)\ (2\ 6\ 3) \qquad \text{so } \sigma\tau = (1\,4\,2)(3\,6\,\ldots$$
$$6 \leftarrow 6 \qquad 2 \qquad\qquad 3$$

$$(1\ 4\ 2\ 6)\ (3\ 5)\ (2\ 6\ 3) \qquad \text{so } \sigma\tau = (1\,4\,2)(3\,6\,5).$$
$$5 \leftarrow \qquad 5 \qquad\quad 3 \qquad 6$$

And we are done. We closed the cycle at the end of 5 because all integers 1 through 6 already appear in the standard cycle notation of $\sigma\tau$ so the cycle must be closed. However, it is a good practice to verify by the same method that $\sigma\tau(5) = 3$.

It should be obvious that the cycle notation expresses a permutation as the composition of disjoint cycles.

Proposition 1.5.7

Disjoint cycles in S_n commute.

Proof. Let $\sigma, \tau \in S_n$ be disjoint cycles. We have three cases, depending on i.

If i is one of the integers in the cycle of τ, then $\tau(i)$ is another integer in the cycle of τ. Since the cycles are disjoint, $\sigma(\tau(i)) = \tau(i)$. On the other hand, since i is not in the cycle σ, $\sigma(i) = i$ so $\tau(\sigma(i)) = \tau(i)$.

If i is one of the integers in the cycle of σ, then $\sigma(i)$ is another integer in the cycle of σ. Since the cycles are disjoint, $\tau(\sigma(i)) = \sigma(i)$. On the other hand, since i is not in the cycle τ, as above, $\tau(i) = i$, so $\sigma(\tau(i)) = \sigma(i)$.

If i is neither one of the integers in the cycle of σ nor one of the integers in the cycle of τ, then $\sigma(\tau(i)) = \sigma(i) = i$ and $\tau(\sigma(i)) = \tau(i) = i$.

Hence, $\sigma(\tau(i)) = \tau(\sigma(i))$ for all $i \in \{1, 2, \ldots, n\}$ so $\sigma\tau = \tau\sigma$. $\qquad\square$

The fact that disjoint cycles commute implies that to understand powers and inverses of permutations, understanding of how powers and inverses work on cycles is sufficient. Indeed, if $\tau_1, \tau_2, \ldots, \tau_k$ are disjoint cycles and $\sigma = \tau_1\tau_2 \cdots \tau_k$, then

$$\sigma^m = \tau_1^m \tau_2^m \cdots \tau_k^m, \text{ for all } m \in \mathbb{Z}, \quad \text{including} \quad \sigma^{-1} = \tau_1^{-1}\tau_2^{-1} \cdots \tau_k^{-1}.$$

Consequently, some properties about a permutation σ and its powers depend only on the cycle type of σ. We leave a number of these results for the exercises but we state one proposition here because of its importance.

Proposition 1.5.8

For all $\sigma \in S_n$, the order $|\sigma|$ is the least common multiple of the lengths of the disjoint cycles in the standard cycle notation of σ.

Proof. (Left as an exercise for the reader. See Exercise 1.5.19.) $\qquad\square$

The cycle notation also makes it easy to find the inverse of a permutation. The inverse function to a permutation σ simply involves reading the cycle notation backwards.

Example 1.5.9. Let $\sigma = (1\,3\,7)(2\,5\,4)(6\,10)$ in S_{10}. We propose to calculate σ^{-1} and then to determine the order of σ by calculating all the powers of σ.

To calculate σ^{-1}, we read the cycles backwards so

$$\sigma^{-1} = (7\,3\,1)(4\,5\,2)(10\,6) = (1\,7\,3)(2\,4\,5)(6\,10).$$

The second equals follows by rewriting the cycle with the lowest integer first. This is equivalent to starting at the lowest integer in the cycle and reading the cycle backwards.

For the powers of σ we have

$$\sigma = (1\,3\,7)(2\,5\,4)(6\,10),$$
$$\sigma^2 = (1\,3\,7)(2\,5\,4)(6\,10)(1\,3\,7)(2\,5\,4)(6\,10) = (1\,7\,3)(2\,4\,5),$$
$$\sigma^3 = \sigma^2\sigma = (1\,7\,3)(2\,4\,5)(1\,3\,7)(2\,5\,4)(6\,10) = (6\,10),$$
$$\sigma^4 = \sigma^3\sigma = (6\,10)(1\,3\,7)(2\,5\,4)(6\,10) = (1\,3\,7)(2\,5\,4),$$
$$\sigma^5 = \sigma^4\sigma = (1\,3\,7)(2\,5\,4)(1\,3\,7)(2\,5\,4)(6\,10) = (1\,7\,3)(2\,4\,5)(6\,10),$$
$$\sigma^6 = \sigma^5\sigma = (1\,7\,3)(2\,4\,5)(6\,10)(1\,3\,7)(2\,5\,4)(6\,10) = \text{id}.$$

This shows that $|\sigma| = 6$ and thereby illustrates Proposition 1.5.8. $\triangle$

We briefly consider the product of cycles that are not disjoint. Some of the simplest products involve two transpositions,

$$(1\,2)(1\,3) = (1\,3\,2) \quad \text{and} \quad (1\,3)(1\,2) = (1\,2\,3).$$

The fact that these products are different establishes the following proposition.

Proposition 1.5.10

The group S_n is nonabelian for all $n \geq 3$.

1.5.3 Inversions of a Permutation

Permutations play a central role in combinatorics, a field that studies techniques for counting the possible arrangements in any kind of discrete structure. We end this section with a brief discussion on the inversions of a permutation. This concept will come in handy when we discuss even and odd permutations in the next sections.

As a motivating example, suppose that we consider 5 events in history and attempt to remember the order in which they occurred. There are $5! = 120$ possible orderings of this time line. Suppose that we number the events in historical order as E_1, E_2, E_3, E_4, E_5 and suppose that someone guesses the historical order as G_1, G_2, G_3, G_4, G_5. Any guess about their historical order corresponds to a permutation $\sigma \in S_5$ via

$$G_{\sigma(i)} = E_i \quad \text{for all } i \in \{1, 2, 3, 4, 5\}.$$

This means that the person guessed the actual ith historical event to be the $\sigma(i)$th event in chronological order.

Suppose that someone guesses the chronological order of the births of five mathematicians and puts them in the following order.

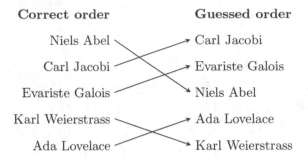

	Correct order		Guessed order

The corresponding permutation is $\sigma = (1\,3\,2)(4\,5)$.

What is a natural way to evaluate how incorrect the guess is? If a guess was correct except for interchanging the first two, i.e., $\sigma = (1\,2)$, that should not be considered egregious. The worst guess would completely reverse the chronological order, i.e., $\sigma = (1\,5)(2\,4)$. A measure of incorrectness for the guessed ordering is to count the number of *inversions*.

Definition 1.5.11

Let n be an integer with $n \geq 2$. Define T_n as the set of ordered pairs

$$T_n = \left\{ (i,j) \in \{1, 2, \ldots, n\}^2 \,|\, i < j \right\}.$$

The number of *inversions* of $\sigma \in S_n$ is number

$$\mathrm{inv}(\sigma) = \left| \{ (i,j) \in T_n \,|\, \sigma(i) > \sigma(j) \} \right|.$$

In other words, T_n consists of all possible pairs (i,j) of indices, where the first index is less than the second. Then $\mathrm{inv}(\sigma)$ is the number of times σ reverses the order of the pair. The set T_n has cardinality

$$|T_n| = \sum_{i=1}^{n-1} (n-i) = n(n-1) - \sum_{i=1}^{n-1} i = n(n-1) - \frac{(n-1)n}{2} = \frac{(n-1)n}{2}.$$

Hence, $0 \leq \mathrm{inv}(\sigma) \leq \frac{1}{2} n(n-1)$.

In the above example about birth order of five mathematicians, σ acts on the pairs of T_5. The table lists (i,j) on top and $(\sigma(i), \sigma(j))$ on the bottom row, showing the inversions in bold.

$(1,2)$	$(1,3)$	$(1,4)$	$(1,5)$	$(2,3)$	$(2,4)$	$(2,5)$	$(3,4)$	$(3,5)$	$(4,5)$
$(\mathbf{3,1})$	$(\mathbf{3,2})$	$(3,5)$	$(3,4)$	$(1,2)$	$(1,5)$	$(1,4)$	$(2,5)$	$(2,4)$	$(\mathbf{5,4})$

Hence, we count that $\mathrm{inv}(\sigma) = 3$.

Definition 1.5.12

A permutation $\sigma \in S_n$ is called *even* (resp. *odd*) if $\mathrm{inv}(\sigma)$ is an even (resp. odd) integer. The designation even or odd is called the *parity* of the permutation σ.

1.5.4   Useful CAS Commands

Both *Maple* and SAGE offer commands to determine the order, the parity, the cycle type and many other properties of permutations. We encourage the reader to explore these.

In *Maple*, the command `Perm` to define a permutation is immediately available but some of the commands for computing with permutations are in the `GroupTheory` package. The *Maple* help files provide a tutorial entitled *Working with Permutations*. The following code illustrates a few commands that are relevant to the content of this section.

─────────────────── Maple ───────────────────

$s := Perm([[1, 4, 3], [2, 6]])$:
$t := Perm([6, 1, 3, 4, 2, 5])$:
$s[2]$;

$$6$$

$with(GroupTheory)$:
$s^{(-1)}$;

$$(1,3,4)(2,6)$$

$s.t$;

$$(1,4,3,6)(2,5)$$

───

The first and second lines define the permutations s and t, the first in standard cycle notation, the second using the n-tuple notation. The third line shows how to apply the permutation s as a function to the input of 2. The next line brings in the `GroupTheory` package that contains commands and methods to operate on permutations. The last two lines calculate the inverse s^{-1} and composition st.

Illustrating *Maple*'s programming language, the next block of code defines a procedure that counts the number of inversions of a permutation.

─────────────────── Maple ───────────────────

```
CountInversions := proc(p)
   local n, i, j, sum := 0;
   n := PermDegree(p);
   for i from 1 to n − 1 do
    for j from i + 1 to n do
     if p[i] > p[j] then sum := sum + 1; fi:
    od:
   od:
   return sum;
  end:

  CountInversions(t);
```

$$7$$

───

There are a number of ways to define permutations in SAGE and we encourage the reader to consult the documentation files online entitled "Permutations" or "Permutation group elements." The first of the webpages describes methods associated with permutations that are more relevant for combinatorics with the latter focus more on applications to group theory. The following code illustrates the same commands as the *Maple* code, but then shows a few commands related to inversions.

_____ Sage _____

```
sage: G=SymmetricGroup(6)
sage: s=G("(1,4,3)(2,6)")
sage: t=G([6,1,3,4,2,5])
sage: s(2)
6
sage: s.inverse()
(1,3,4)(2,6)
sage: s*t
(1,4,3,6)(2,5)
sage: tp=Permutation(t)
sage: tp
[6, 1, 3, 4, 2, 5]
sage: tp.number_of_inversions()
7
sage: tp.inversions()
[(1, 2), (1, 3), (1, 4), (1, 5), (1, 6), (3, 5), (4, 5)]
```

The first sage: line defines the group G as S_6. The next five lines implement the same commands as the first block of *Maple* code above. However, note that in SAGE, the permutations are understood as elements of the group G. The command tp=Permutation(t) defines tp as the same permutation as t but expressed in a permutation object class. In this class, SAGE always writes permutations as n-tuples, illustrated by the line sage: tp. The last two lines illustrate useful methods from this object class. The following code gives an example of SAGE using Python code to calculate

$$\sum_{\sigma \in S_6} \text{inv}(\sigma).$$

_____ Sage _____

```
sage: G=SymmetricGroup(6)
sage: sum=0
sage: for p in G:
....:        pp=Permutation(p)
```

```
....:          sum=sum+pp.number_of_inversions()
....:
sage: sum
5400
```

In the above code, "....:" represents the SAGE console prompt for an indentation in Python code. In this case, the indentation represents the `for` loop block.

EXERCISES FOR SECTION 1.5

1. Write the standard cycle notation for the following permutations expressed in chart notation.

   (a) $\sigma = \begin{pmatrix} 1 & 2 & 3 & 4 & 5 & 6 & 7 \\ 6 & 3 & 2 & 4 & 7 & 1 & 5 \end{pmatrix}$; (b) $\tau = \begin{pmatrix} 1 & 2 & 3 & 4 & 5 & 6 & 7 \\ 3 & 5 & 7 & 2 & 1 & 6 & 4 \end{pmatrix}$.

2. Suppose that $\sigma \in S_8$ given in n-tuple notation is $(3, 2, 6, 4, 5, 8, 7, 1)$. Depict σ with a directed graph and express it in standard cycle notation.

3. Suppose that $\sigma \in S_9$ given in n-tuple notation is $(4, 6, 5, 2, 3, 1, 8, 7, 9)$. Depict σ with a directed graph and express it in standard cycle notation.

4. Show that the function $f : \mathbb{Z}/10\mathbb{Z} \to \mathbb{Z}/10\mathbb{Z}$ defined by $f(\bar{a}) = \bar{a}^3$ is a permutation on $\mathbb{Z}/10\mathbb{Z}$ and write f in cycle notation as an element of $S_{\{0,1,\dots,9\}}$.

5. Repeat the previous question with $\mathbb{Z}/11\mathbb{Z}$ instead of $\mathbb{Z}/10\mathbb{Z}$.

6. In S_6, with $\sigma = (1\,3\,5)(2\,6)$ and $\tau = (1\,3\,4\,5\,6)$, calculate: (a) $\sigma\tau$; (b) $\tau\sigma$; (c) σ^2; (d) τ^{-1}; (e) $\sigma\tau\sigma^{-1}$.

7. In S_7, with $\sigma = (1\,4)(2\,6)(3\,5\,7)$ and $\tau = (1\,6\,7)$, calculate: (a) $\sigma\tau$; (b) $\tau\sigma$; (c) $\tau^{-1}\sigma^2$; (d) $\sigma\tau\sigma^{-1}$.

8. List all the cycle types in S_7.

9. Let $\sigma = (a_0\,a_1\,a_2\,\cdots\,a_{m-1})$ be an m cycle in S_n. Prove that $\sigma^k(a_i) = a_{(i+k)\bmod m}$. Conclude that the order of σ is m.

10. Let $\sigma \in S_n$ be an m-cycle. Prove that σ^k is an m-cycle if and only if k and m are relatively prime.

11. Suppose that a_i are distinct positive integers of $i = 1, 2, \dots, m$. Prove that

$$(a_1\,a_2\,a_3\,\cdots\,a_m) = (a_1\,a_m)\cdots(a_1\,a_4)(a_1\,a_3)(a_1\,a_2).$$

Use this to show that an m-cycle is even if and only if m is odd.

12. Suppose that a_i are distinct positive integers of $i = 1, 2, \dots, m$. Prove that

$$(a_1\,a_2\,a_3\,\cdots\,a_m) = (a_1\,a_2)\cdots(a_{m-2}\,a_{m-1})(a_{m-1}\,a_m).$$

Use this to show that an m-cycle is even if and only if m is odd.

13. Use Exercise 1.5.11 (or Exercise 1.5.12) to show that every permutation can be written as a product of transpositions.

14. Let σ be an m-cycle and suppose that $d < m$ divides m. Prove that the standard cycle notation of σ^d is the product of d disjoint cycles of length m/d.

15. What is the highest possible order for permutations in S_{11}? Illustrate your answer with a specific element having that order.

16. What is the highest possible order of an element in each of the following groups. Illustrate with a specific element. (a) $S_5 \oplus D_{11}$ (b) $S_5 \oplus S_5$

17. What is the highest possible order of an element in $S_7 \oplus S_7 \oplus S_7$? Illustrate your answer with a specific element having that order.

18. Prove that a permutation $\sigma \in S_n$ satisfies $\sigma^{-1} = \sigma$ if and only if σ is the identity or, in standard cycle notation, consists of a product of disjoint 2-cycles.

19. Prove Proposition 1.5.8. [Hint: Use Exercise 1.5.9.]

20. In some S_n, find two elements σ and τ such that $|\sigma| = |\tau| = 2$ and $|\sigma\tau| = 3$.

21. In some S_n, find two elements σ and τ such that $|\sigma| = |\tau| = 2$ and $|\sigma\tau| = 4$.

22. How many permutations of order 4 does S_7 have?

23. How many even permutations of order 5 does S_8 have? Odd permutations?

24. Suppose that $m \leq n$. Prove that the number of m-cycles in S_n is
$$\frac{n(n-1)(n-2)\cdots(n-m+1)}{m}.$$

25. Suppose that $n \geq 4$. Prove that the number of permutations of cycle type $(ab)(cd)$ in S_n is
$$\frac{n(n-1)(n-2)(n-3)}{8}.$$

26. Calculate the set $\{|\sigma| \mid \sigma \in S_7\}$, i.e., the set of orders of elements in S_7.

27. We work in the group S_n for $n \geq 3$. Prove or disprove that if σ_1 and σ_2 have the same cycle type and τ_1 and τ_2 have the same cycle type, then $\sigma_1\tau_1$ has the same cycle type as $\sigma_2\tau_2$. Does your answer depend on n?

28. In a six-contestant steeple race, the horses arrived in the order C, B, F, A, D, E. Suppose someone predicted they would arrive in the order, F, E, C, B, D, A. How many inversions are in the guessed ordering?

29. Count the number of inversions in the permutation of Exercise 1.5.4.

30. In S_5, count the number of inversions of the following permutations: (a) $\sigma = (1\,4\,2\,5)$; (b) $\tau = (1\,4\,3)(2\,5)$; (c) $\rho = (1\,5)(2\,3)$.

31. In S_6, count the number of inversions of the following permutations: (a) $\sigma = (1\,3\,5\,6\,2)$; (b) $\tau = (1\,6)(2\,3\,4)$; (c) $\rho = (1\,3\,5)(2\,4\,6)$.

32. Consider a 2-cycle in S_n of the form $\tau = (a\,b)$. Prove that $\text{inv}(\tau) = 2(b-a) - 1$. [Hint: Do a proof by cases involving the pairs $(i,j) \in T_n$, depending on whether i or j or both is equal or not to a or b.]

33. (CAS) Use a CAS to calculate $\sum_{\sigma \in S_n} \text{inv}(\sigma)$ for $n = 1, 2, 3, 4, 5$. Then use the Online Encyclopedia of Integer Sequences (OEIS) to identify this sequence.

1.6 Subgroups

In any algebraic structure, it is common to consider a subset that carries the same algebraic structure. In linear algebra, for example, we encounter subspaces of a vector space. In Section A.4, we discuss subposets. This section presents subgroups.

1.6.1 Subgroup: Definition and Examples

Definition 1.6.1

Let G be a group. A nonempty subset $H \subseteq G$ is called a *subgroup* if

(1) $\forall x, y \in H$, $xy \in H$ (closed under operation);

(2) $\forall x \in H$, $x^{-1} \in H$ (closed under taking inverses).

If H is a subgroup of G, we write $H \leq G$.

Since a subgroup H of a group G must be nonempty, there exists some $x \in H$. Since H is closed under taking inverses, then $x^{-1} \in H$. Since H is closed under the group operation, then $e = xx^{-1} \in H$. Hence, a subgroup contains the identity element. The property of associativity is inherited from associativity in G and so since $e \in H$ and H is closed under taking inverses, H equipped with the binary operation on G is a group in its own right. (As a point of terminology, it is important to understand that we do not say that "a group is closed under an operation." Such a statement is circular since a binary operation on G by definition maps any pair of elements in G back into G. The terminology of "closed under an operation" is a matter of concern only for strict subsets of G.)

Example 1.6.2. With the usual addition operation, $\mathbb{Z} \leq \mathbb{Q} \leq \mathbb{R} \leq \mathbb{C}$. With the multiplication operation we have $\mathbb{Q}^* \leq \mathbb{R}^* \leq \mathbb{C}^*$. However, $\mathbb{Z}^*$ is not a subgroup of $\mathbb{Q}^*$, written $\mathbb{Z}^* \not\leq \mathbb{Q}^*$, because even though $\mathbb{Z}^*$ is closed under multiplication, it is not closed under taking multiplicative inverses. For example $2^{-1} = \frac{1}{2} \notin \mathbb{Z}^*$. $\triangle$

Example 1.6.3. Any group G always has at least two subgroups, the trivial subgroup $\{e\}$ and all of G. $\triangle$

Example 1.6.4. If $G = D_n$, then $R = \{\iota, r, r^2, \ldots, r^{n-1}\}$ is a subgroup. This is the subgroup of rotations. Also for all integers i between 0 and $n - 1$, the subsets $H_i = \{\iota, sr^i\}$ are subgroups. These subgroups of two elements correspond to reflection about various lines of symmetry. $\triangle$

Example 1.6.5. Let $G = S_n$ and consider the subset of permutations that leave the elements $\{m + 1, m + 2, \ldots n\}$ fixed. This is a subgroup of S_n that consists of all elements in S_m. $\triangle$

Example 1.6.6 (A Nonexample). As a nonexample, note that $U(n)$ is not a subgroup of $\mathbb{Z}/n\mathbb{Z}$. Even though $U(n)$ is a subset of $\mathbb{Z}/n\mathbb{Z}$, the former involves the multiplication operation in modular arithmetic while the latter involves the addition. If we considered the pair $(U(n), +)$, we have $\overline{1}$ and $\overline{n-1}$ in $U(n)$ but $\overline{1} + \overline{n-1} = \overline{0} \notin U(n)$. Hence, $U(n)$ is not closed under addition.   △

The definition of a subgroup has two criteria. It turns out that these two can be combined into one. This result shortens a number of subsequent proofs.

Proposition 1.6.7 (One-Step Subgroup Criterion)

A nonempty subset H of a group G is a subgroup if and only if $\forall x, y \in H$, $xy^{-1} \in H$.

Proof. ($\Longrightarrow$) If H is a subgroup, then for all $x, y \in H$, the element $y^{-1} \in H$ and hence $xy^{-1} \in H$.

($\Longleftarrow$) Suppose that H is a nonempty subset with the condition described in the statement of the proposition. First, since H is nonempty, $\exists x \in H$. Using the one-step criterion, $xx^{-1} = e \in H$. Second, since e and $x \in H$, using the one-step criterion, $ex^{-1} = x^{-1} \in H$. We have now proven that H is closed under taking inverses. Finally, for all $x, y \in H$, we know that $y^{-1} \in H$ so, using the one-step criterion again, $x(y^{-1})^{-1} = xy \in H$. Hence, H is closed under the group operation.   □

The following example introduces an important group but uses the One-Step Subgroup Criterion to prove it is a group.

Example 1.6.8 (Special Linear Group). Let F denote $\mathbb{Q}$, $\mathbb{R}$, $\mathbb{C}$, or $\mathbb{F}_p$. Define the subset of $\mathrm{GL}_n(F)$ by

$$\mathrm{SL}_n(F) = \{A \in \mathrm{GL}_n(F) \mid \det A = 1\}.$$

Obviously, $\mathrm{SL}_n(F)$ is not empty because the identity matrix has determinant 1 so it is in $\mathrm{SL}_n(F)$. Furthermore, according to properties of the determinant,

$$\det(AB^{-1}) = \det(A)\det(B^{-1}) = \det(A)\det(B)^{-1} = 1$$

for all $A, B \in \mathrm{SL}_n(F)$. By the One-Step Subgroup Criterion, $\mathrm{SL}_n(F) \leq \mathrm{GL}_n(F)$. We call $\mathrm{SL}_n(F)$ the *special linear group*.   △

If the subset H is finite, there is another simplification.

Proposition 1.6.9 (Finite Subgroup Test)

Let G be a group and let H be a nonempty finite subset of G. If H is closed under the operation, then H is a subgroup of G.

Proof. Given the hypotheses of the proposition, in order to establish that H is a subgroup of G, we only need to show that it is closed under taking inverses.

Let $x \in H$. Since H is closed under the operation, $x^n \in H$ for all positive integers n. Thus, the set $S = \{x^n \mid n \in \mathbb{N}^*\}$ is a subset of H and hence finite. By the pigeonhole principle, there exists $m, n \in \mathbb{N}^*$ with $n \neq m$ such that $x^m = x^n$. Without loss of generality, suppose that $n > m$. Then $n - m \in \mathbb{N}^*$ and $x^{n-m} = e$. Thus, x has finite order, say $|x| = k$. If $x = e$, then it is its own inverse. If $x \neq e$, then $k \geq 2$. Since $k - 1$ is a positive integer, $x^{k-1} = x^{-1} \in S \subseteq H$. This shows that H is closed under taking inverses. $\square$

It is also useful to be aware of the interaction between subgroups and subset operations.

Proposition 1.6.10

Let G be a group and let H and K be two subgroups of G. Then $H \cap K \leq G$.

Proof. Note that $1 \in H \cap K$ so $H \cap K$ is not empty. Let $x, y \in H \cap K$. Then since $x, y \in H$, by the One-Step Subgroup Criterion, $xy^{-1} \in H$. Similarly for $xy^{-1} \in K$. Consequently, $xy^{-1} \in H \cap K$ and by the One-Step Subgroup Criterion again, we conclude that $H \cap K$ is a subgroup of G. $\square$

On the other hand, the union of two subgroups is not necessarily another subgroup and the set difference of two subgroups is never another subgroup. Consequently, in relation to the operations of union and intersection on subsets, subgroups behave in a similar way as subspaces of a vector space do.

By an induction reasoning, knowing that the intersection of two subgroups is again a subgroup implies that any intersection of a finite number of subgroups is again a subgroup. However, it is also true that a general intersection (not necessarily finite) of a collection of subgroups is again a subgroup. (See Exercise 1.6.28.)

1.6.2 The Alternating Subgroup

We illustrate our discussion of subgroups to this point by introducing an important subgroup in S_n. Instead of building up toward a main theorem, we present the key result and use the rest of the section to prove it.

Theorem 1.6.11

The subset of even permutations in S_n is a subgroup of order $n!/2$.

This subgroup carries a special name, stemming from applications of this group in other areas of mathematics.

Definition 1.6.12

The subgroup of even permutations in S_n is called the *alternating group*, and is denoted by A_n.

Recall (Definition 1.5.12) that a permutation $\sigma \in S_n$ is called even (resp. odd) if $\mathrm{inv}(\sigma)$ is even (resp. odd). Exercise 1.5.32 in the previous section established this first proposition.

Proposition 1.6.13

The number of inversions of a transposition $(a\,b)$ is odd. More precisely, $\mathrm{inv}((a\,b)) = 2(b - a) - 1$.

The following lemmas form the stepping stones to the main theorem of this section.

Lemma 1.6.14

Let $\sigma, \tau \in S_n$. Then

$$\mathrm{inv}(\sigma\tau) \equiv \mathrm{inv}(\sigma) + \mathrm{inv}(\tau) \pmod 2.$$

Proof. Let $\sigma, \tau \in S_n$. We partition T_n in four subsets:

$U_1 = \{(i,j) \in T_n \mid \tau \text{ inverts } (i,j) \text{ and } \sigma \text{ inverts } (\tau(i), \tau(j))\}$,
$U_2 = \{(i,j) \in T_n \mid \tau \text{ inverts } (i,j) \text{ and } \sigma \text{ does not invert } (\tau(i), \tau(j))\}$,
$U_3 = \{(i,j) \in T_n \mid \tau \text{ does not invert } (i,j) \text{ and } \sigma \text{ inverts } (\tau(i), \tau(j))\}$,
$U_4 = \{(i,j) \in T_n \mid \tau \text{ does not invert } (i,j) \text{ and } \sigma \text{ does not invert } (\tau(i), \tau(j))\}$.

Notice that $\mathrm{inv}(\sigma\tau) = |U_2| + |U_3|$, $\mathrm{inv}(\sigma) = |U_1| + |U_3|$, and $\mathrm{inv}(\tau) = |U_1| + |U_2|$. Hence,

$$\mathrm{inv}(\sigma) + \mathrm{inv}(\tau) = 2|U_1| + |U_2| + |U_3| \equiv |U_2| + |U_3| \equiv \mathrm{inv}(\sigma\tau) \pmod 2. \quad \square$$

Lemma 1.6.15

For a given $n \geq 2$, the number of even permutations in S_n is equal to the number of odd ones.

Proof. For this proof, call O_n the subset of odd permutations and consider the function $\psi : A_n \to O_n$ defined by $\psi(\sigma) = \sigma(1\,2)$. This function is injective because

$$\psi(\sigma) = \psi(\tau) \iff \sigma(1\,2) = \tau(1\,2) \iff \sigma = \tau.$$

It is also surjective. Let $\tau \in O_n$. By Proposition 1.6.13, $\mathrm{inv}((1\,2)) = 1$ and by Lemma 1.6.14, $\mathrm{inv}(\tau(1\,2)) \equiv \mathrm{inv}(t) + 1 \pmod 2$. Thus $\tau(1\,2)$ is even and $\psi(\tau(1\,2)) = \tau$. Since ψ is bijective, $|A_n| = |O_n|$. $\quad \square$

We are now in a position to prove the main theorem of this subsection.

Proof (of Theorem 1.6.11). First, we note that $A_n \neq \emptyset$ since the identity has zero inversions, and hence is an even permutation.

Next, to prove that A_n is closed under the operation, let $\sigma, \tau \in A_n$. Then $\mathrm{inv}(\sigma) \equiv \mathrm{inv}(\tau) \equiv 0 \pmod 2$. By Lemma 1.6.14, $\mathrm{inv}(\sigma\tau) \equiv 0+0 \equiv 0 \pmod 2$, so $\sigma\tau \in A_n$.

To show that A_n is closed under taking inverses, Lemma 1.6.14 implies that

$$0 = \mathrm{inv}(\mathrm{id}) = \mathrm{inv}(\sigma\sigma^{-1}) \equiv \mathrm{inv}(\sigma) + \mathrm{inv}(\sigma^{-1}) \pmod 2.$$

for all $\sigma \in S_n$. Thus σ and σ^{-1} have the same parity. In particular, for all $\sigma \in A_n$, we also have $\sigma^{-1} \in A_n$.

Finally, note that $A_n \cup O_n = S_n$ and $A_n \cap O_n = \emptyset$. So by Lemma 1.6.15, $|S_n| = 2|A_n|$ and so $|A_n| = n!/2$. □

The propositions that we have developed lead to yet another way to characterize even and odd permutations.

Theorem 1.6.16

> A permutation is even (resp. odd) if and only if it can be written as a product of an even (resp. odd) number of transpositions.

Proof. By Exercise 1.5.13, every permutation $\sigma \in S_n$ can be expressed as a product of transpositions $\sigma = \tau_1 \tau_2 \cdots \tau_m$. By Lemma 1.6.14,

$$\mathrm{inv}(\sigma) \equiv \mathrm{inv}(\tau_1) + \mathrm{inv}(\tau_2) + \cdots + \mathrm{inv}(\tau_m) \pmod 2.$$

By Proposition 1.6.13, $\mathrm{inv}(\tau_i)$ is odd for all i and hence $\mathrm{inv}(\sigma)$ is even if and only if m is even. The theorem follows. □

Example 1.6.17. As a specific example, the elements of A_4 are

$$A_4 = \{\mathrm{id}, (123), (124), (132), (134), (142), (143),$$
$$(234), (243), (12)(34), (13)(24), (14)(23)\}.$$

As already noted, id is an even permutation. Then by Theorem 1.6.16 any permutation of the form $(a\,b)(c\,d)$ in standard cycle notation is even. Finally, since a 3-cycle $(a\,b\,c)$ can be written as $(a\,b)(b\,c)$, by the same theorem, we see that all 3-cycles are even. △

Example 1.6.18 (Vandermonde Polynomials). The *Vandermonde polynomial* of n variables $x_1, x_2, \ldots, x_n$ is the multivariable polynomial

$$\prod_{1 \leq i < j \leq n} (x_j - x_i).$$

This product has $\binom{n}{2}$ terms (and hence has degree $\binom{n}{2}$). Note that each term in the product corresponds uniquely to one pair in T_n.

For a given $\sigma \in S_n$ and a given pair $(i, j) \in T_n$, the term $(x_{\sigma(j)} - x_{\sigma(i)})$ is equal to $\pm(x_{j'} - x_{i'})$ for some other pair $(i', j') \in T_n$ and with the sign being negative if and only if σ inverts the pair (i, j). Hence,

$$\prod_{1 \le i < j \le n} (x_{\sigma(j)} - x_{\sigma(i)}) = (-1)^{\text{inv}(\sigma)} \prod_{1 \le i < j \le n} (x_j - x_i).$$

Some authors use this property on how permutations act on the Vandermonde polynomial as determining the parity of the permutation. By this approach, a permutation is called even if and only if it does not change the sign of the Vandermonde polynomial. $\triangle$

This application of the number of inversions of a permutation leads to one more concept.

Definition 1.6.19

The *sign* of $\sigma \in S_n$ is $\text{sign}(\sigma) = (-1)^{\text{inv}(\sigma)}$. In other words,

$$\text{sign}(\sigma) = \begin{cases} 1 & \sigma \text{ is even} \\ -1 & \sigma \text{ is odd.} \end{cases}$$

EXERCISES FOR SECTION 1.6

In Exercises 1.6.1 through 1.6.16, prove or disprove that the given subset A of the given group G is a subgroup.

1. $G = \mathbb{Z}$ with addition and A are the multiples of 5.

2. $G = (\mathbb{Q}, +)$ and A is the set of rational numbers with odd denominators (when written in reduced form).

3. $G = (\mathbb{Q}^*, \times)$ and A is the set of rational numbers of the form $\frac{p^2}{q^2}$.

4. $G = (\mathbb{C}^*, \times)$ and $A = \{a + ai \mid a \in \mathbb{R}\}$.

5. $G = \mathbb{Z}/12\mathbb{Z}$ and $A = \{\overline{0}, \overline{4}, \overline{8}\}$.

6. $G = U(11)$ and $A = \{\overline{1}, \overline{2}, \overline{9}, \overline{10}\}$.

7. $G = S_5$ and A is the set of transpositions.

8. $G = D_6$ and $A = \{\iota, s, r^2, sr^2\}$.

9. $G = D_6$ and $A = \{\iota, s, r^3, sr^3\}$.

10. $G = (\mathbb{R}^*, \times)$ and $A = \{\sqrt{\frac{p}{q}} \mid \frac{p}{q} \in \mathbb{Q}^{>0}\}$.

11. $G = (\mathbb{R}, +)$ and $A = \{\sqrt{\frac{p}{q}} \mid \frac{p}{q} \in \mathbb{Q}^{>0}\}$.

12. $G = U(30)$ and $A = \{\overline{1}, \overline{7}, \overline{13}, \overline{19}\}$.

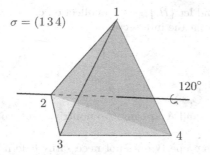

$\sigma = (1\,3\,4)$

$120°$

FIGURE 1.8: A rigid motion of a tetrahedron.

13. $G = \mathrm{GL}_2(\mathbb{R})$ and A is the subset of matrices with integer coefficients.

14. $G = S_{\mathbb{R}^{>0}}$, i.e., the set of bijections on $\mathbb{R}^{>0}$ with the operation of composition, and $A = \{f \in G \mid f(x) = x^{p/q}$, where p and q are odd integers$\}$.

15. $G = S_{\mathbb{R}}$ and let $A = \{f \in G \mid f(\mathbb{Z}) \subseteq \mathbb{Z}\}$.

16. $G = S_{\mathbb{R}}$ and let $A = \{f \in G \mid f(\mathbb{Z}) = \mathbb{Z}\}$.

17. Find an example to illustrate that the union of two subgroups is not necessarily another subgroup.

18. Prove that a finite group G with order greater than 2 cannot have a subgroup H with $|H| = |G| - 1$.

19. Let G_1 and G_2 be two groups. Prove that $\{(x, e_2) \mid x \in G_1\}$ and that $\{(e_1, y) \mid y \in G_2\}$ are subgroups of $G_1 \oplus G_2$.

20. Determine all the finite subgroups of $(\mathbb{R}^*, \times)$.

21. Let F be $\mathbb{Q}$, $\mathbb{R}$, $\mathbb{C}$, or $\mathbb{F}_p$ (or more generally any field). Prove that the subset in $\mathrm{GL}_n(F)$ of upper triangular matrices is a subgroup. [Recall that a matrix is upper triangular if all the entries below the main diagonal are 0.]

22. Show by example that a permutation can be written in more than one way as a product of transpositions. Prove that if $\sigma = \tau_1 \tau_2 \cdots \tau_m$ and $\sigma = \varepsilon_1 \varepsilon_2 \cdots \varepsilon_n$ are two different expressions of σ as a product of transpositions, then m and n have the same parity.

23. Show that for all $\sigma, \tau \in S_n$, the element $\sigma \tau \sigma^{-1}$ has the same parity as τ.

24. Label the vertices of a tetrahedron with integers $\{1, 2, 3, 4\}$. Prove that the group of rigid motions of a tetrahedron is A_4. (See Figure 1.8.)

25. Let A be an $n \times n$ matrix and let $\sigma \in S_n$. Suppose that A' (respectively A'') is the matrix obtained from A by permuting the columns (respectively rows) of A according to the permutation σ. Prove that $\det(A') = \det(A'') = \mathrm{sign}(\sigma)\det(A)$.

26. Let H and K be two subgroups of G. Prove that $H \cup K$ is a subgroup of G if and only if $H \subseteq K$ or $K \subseteq H$.

27. Let H be a subgroup of a group G and let $g \in G$. Prove that if n is the smallest positive integer such that $g^n \in H$, then n divides $|g|$.

28. Let G be a group and let $\{H_i\}_{i\in\mathcal{C}}$ be a collection of subgroups of G, not necessarily finite. Prove that the intersection

$$\bigcap_{i\in\mathcal{C}} H_i$$

   is a subgroup of G.

29. Consider the group $(\mathbb{Q}, +)$.
   (a) Prove that if H and K are any two nontrivial subgroups of $\mathbb{Q}$, then $H \cap K$ is nontrivial.
   (b) Prove that this property does not necessarily hold if $\mathbb{Q}$ is replaced with $\mathbb{R}$.

1.7 Abstract Subgroups

The previous section introduced the notion of a subgroup of a group and offered examples from specific groups. In this section we introduce general methods to define subgroups, applicable to any group.

1.7.1 Subgroups Generated by a Subset

Let S be a subset of a group G. Since subgroups of G are closed under the operation and under taking inverses, any subgroup of G that contains S must contain all elements obtained by repeated operations or inverses from elements in S.

Definition 1.7.1

Let S be a nonempty subset of a group G. We define $\langle S \rangle$ as the subset of *words* made from elements in S, that is to say

$$\langle S \rangle = \{ s_1^{\alpha_1} s_2^{\alpha_2} \cdots s_n^{\alpha_n} \mid n \in \mathbb{N},\ s_i \in S,\ \alpha_i \in \mathbb{Z} \}.$$

Note that the s_i are not necessarily distinct.

Proposition 1.7.2

For any nonempty subset S of a group G, $\langle S \rangle \leq G$.

Proof. First of all $\langle S \rangle$ is nonempty since it contains S. For any two elements $x = s_1^{\alpha_1} s_2^{\alpha_2} \cdots s_n^{\alpha_n}$ and $y = t_1^{\beta_1} t_2^{\beta_2} \cdots t_m^{\beta_m}$ in $\langle S \rangle$, we have

$$xy^{-1} = s_1^{\alpha_1} s_2^{\alpha_2} \cdots s_n^{\alpha_n} t_m^{-\beta_m} \cdots t_2^{-\beta_2} t_1^{-\beta_1}.$$

This product is again an element of $\langle S \rangle$ so by the One-Step Subgroup Criterion $\langle S \rangle \leq G$. $\qquad\square$

By virtue of Proposition 1.7.2, $\langle S \rangle$ is called the *subgroup generated* by S. (In the analogy with vector spaces, a subgroup generated by a subset is like the span of a set of elements in a vector space, which is a subspace.)

Example 1.7.3. Let $G = D_6$ be the dihedral group on the hexagon. The subgroup $\langle r \rangle$ consists of all powers of r, so is $\langle r \rangle = \{\iota, r, r^2, r^3, r^4, r^5\}$. Notice that $\langle r \rangle$ is the subgroup of rotations.

The subgroup $\langle s \rangle = \{1, s\}$ consists of only two elements, reflection across the reference axis of symmetry and the identity transformation.

The subgroup $\langle s, r^2 \rangle$ contains the elements ι, s, r^2, and r^4, by virtue of taking powers of elements in $\{s, r^2\}$. However, $\langle s, r^2 \rangle$ also contains sr^2 and sr^4. The defining relation on s and r give $r^a s = sr^{6-a}$. Hence, as we apply this relation, the parity on the power of r does not change. Hence,

$$\langle s, r^2 \rangle = \{\iota, r^2, r^4, s, sr^2, sr^4\}.$$

Finally, consider the subgroup $\langle s, sr \rangle$. Obviously this subgroup contains s but it also contains $r = s(sr)$. Hence, $\langle s, sr \rangle = D_6$ because it contains all rotations and all reflections. $\triangle$

For any element a in a group G, the subgroup $\langle a \rangle$ is a cyclic subgroup of G whose order is precisely the order $|a|$. It is important to note that distinct sets of generators may give that same subgroup. In the previous example, we noted that $\langle r, sr \rangle = \langle r, s \rangle$. This occurs even with cyclic subgroups. For example, in D_6, the rotation subgroup is $\langle r \rangle = \langle r^5 \rangle$. In D_6, we also have $\langle r^2 \rangle = \langle r^4 \rangle$.

Example 1.7.4. Consider the group S_4. Let $H = \langle (1\,3), (1\,2\,3\,4) \rangle$. We list out all the elements of H. By taking powers of the generators, we know that

$$\text{id}, \quad (1\,3), \quad (1\,2\,3\,4), \quad (1\,2\,3\,4)^2 = (1\,3)(2\,4), \quad (1\,2\,3\,4)^3 = (1\,4\,3\,2)$$

are all in H. By operating among these, H also contains

$$(1\,3)(1\,2\,3\,4) = (1\,2)(3\,4), \quad (1\,3)(1\,2\,3\,4)^2 = (2\,4), \quad (1\,3)(1\,2\,3\,4)^3 = (1\,4)(2\,3).$$

It is also easy to calculate that $(1\,2\,3\,4)(1\,3) = (1\,4)(2\,3)$, $(1\,2\,3\,4)^2(1\,3) = (1\,3)(2\,4)$, and $(1\,2\,3\,4)^3(1\,3) = (1\,2)(3\,4)$. At this point, we may suspect that

$$H = \{\text{id}, (1\,3), (1\,2\,3\,4), (1\,3)(2\,4), (1\,4\,3\,2), (2\,4), (1\,2)(3\,4), (1\,4)(2\,3)\}$$

but we have not yet proven that H does not have any other elements. However, the identity $(1\,3)(1\,2\,3\,4) = (1\,4\,3\,2)(1\,3)$ shows that though $(1\,3)$ and $(1\,2\,3\,4)$ do not commute, it is possible to pass $(1\,2\,3\,4)$ to the left of $(1\,3)$ by changing the power on the 4-cycle. Hence, every element in H can be written as $(1\,2\,3\,4)^a (1\,3)^b$ where $a = 0, 1, 2, 3$ and $b = 0, 1$. Thus, we have indeed found all the elements in H. $\triangle$

Definition 1.7.5

A group (or a subgroup) is called *finitely generated* if it is generated by a finite subset.

A finite group is always finitely generated. Indeed, a finite group G is generated (not minimally so) by G itself. On the other hand, in the group $(\mathbb{Z}, +)$ we have $\langle 1 \rangle = \mathbb{Z}$ which gives an example of an infinite group that is finitely generated. It is not hard to find a group that is not finitely generated.

Example 1.7.6. Let $(\mathbb{Q}^{>0}, \times)$ be the multiplicative group of positive rational numbers and let P be the set of prime numbers. Every positive rational number r can be written in the form

$$r = p_1^{\alpha_1} p_2^{\alpha_2} \cdots p_m^{\alpha_m}$$

where $\alpha_i \in \mathbb{Z}$. In our usual way of writing fractions, any prime p_i with $\alpha_i < 0$ would be in the prime factorization of the denominator and with $\alpha_i > 0$ would be in the prime factorization of the numerator. Consequently, P is a generating set of $\mathbb{Q}^{>0}$.

This does not yet imply that $(\mathbb{Q}^{>0}, \times)$ is not finitely generated. We now show this by contradiction. Assume that it is generated by a finite set $\{r_1, r_2, \ldots, r_k\}$ of rational numbers. The prime factorizations of the numerators and denominators of all the r_i (written in reduced form) involve a finite number of primes, say $\{p_1, p_2, \ldots, p_n\}$. Let p_0 be a prime not in $\{p_1, p_2, \ldots, p_n\}$. Then $p_0 \in \mathbb{Q}^{>0}$ but $p_0 \notin \langle r_1, r_2, \ldots, r_k \rangle$. Hence, $(\mathbb{Q}^{>0}, \times)$ is not finitely generated. $\triangle$

1.7.2 Center, Centralizer, Normalizer

Proposition 1.7.2 gave us a way to construct subgroups of a group. However, a number of subsets defined in terms of equations also turn out to always be subgroups. Many play central roles in understanding the internal structure of a group so we present a few such subgroups here.

Definition 1.7.7

The *center* $Z(G)$ is the subset of G consisting of all elements that commute with every other element in G. In other words,

$$Z(G) = \{x \in G \mid xg = gx \text{ for all } g \in G\}.$$

Proposition 1.7.8

Let G be any group. The center $Z(G)$ is a subgroup of G.

Proof. Note that $1 \in Z(G)$, so $Z(G)$ is nonempty. Let $x, y \in Z(G)$. Then

$$(xy)g = x(yg) = x(gy) = (xg)y = (gx)y = g(xy)$$

so $Z(G)$ is closed under the operation. Let $x \in Z(G)$. By definition $xg = gx$ so $g = x^{-1}gx$ and $gx^{-1} = x^{-1}g$. Thus, $x^{-1} \in Z(G)$ and we conclude that $Z(G)$ is closed under taking inverses. □

Note that $Z(G) = G$ if and only if G is abelian. On the other hand, $Z(G) = \{1\}$ means that the identity is the only element that commutes with every other element. Intuitively speaking, $Z(G)$ gives a measure of how far G is from being abelian. The center itself is an abelian subgroup. However, $Z(G)$ is not necessarily the largest abelian subgroup of G.

Example 1.7.9. Let F be $\mathbb{Q}$, $\mathbb{R}$, $\mathbb{C}$, or $\mathbb{F}_p$ (where p is prime). In this example, we prove that

$$Z(GL_n(F)) = \{aI \mid a \neq 0\}, \tag{1.7}$$

where I is the identity matrix in $GL_n(F)$.

By properties of matrix multiplication, for all matrices $B \in GL_n(F)$ we have $B(aI) = a(BI) = aB = (aI)B$. Hence, $\{aI \mid a \neq 0\} \subseteq Z(GL_n(F))$. The difficulty lies is proving the reverse inclusion.

Suppose $1 \leq i, j \leq n$ with $i \neq j$. Let E_{ij} be the $n \times n$ matrix consisting of zeros in all entries except for a 1 in the (i, j)th entry. The matrix E_{ij} is not in $GL_n(F)$ but $I + E_{ij}$ is, since $\det(I + E_{ij}) = 1$. Since $BI = IB$ for all $B \in GL_n(F)$, then $B(I + E_{ij}) = (I + E_{ij})B$ if and only if $BE_{ij} = E_{ij}B$. Thus, all $B \in Z(GL_n(F))$ satisfy the matrix product $BE_{ij} = E_{ij}B$.

The matrix product BE_{1j} is the matrix of zeros everywhere except for its jth column being the first column of B. Similarly, $E_{1j}B$ is the matrix of zeros everywhere except for its first row being the jth row of B. (See Exercise 1.7.16.) Thus, for a particular $j \geq 2$, the identity $BE_{1j} = E_{1j}B$ implies that

$$b_{jk} = \begin{cases} 0 & \text{if } k \neq j \\ b_{11} & \text{if } k = j. \end{cases}$$

If $B \in Z(GL_n(F))$, then $BE_{1j} = E_{1j}B$ for all pairs $2 \leq j \leq n$. Therefore, all off-diagonal elements of B are zero and $b_{jj} = b_{11}$ for all j, i.e., all diagonal elements of B are equal. This establishes $Z(GL_n(F)) \subseteq \{aI \mid a \neq 0\}$ and we deduce (1.7). △

Definition 1.7.10

Let G be a group. The *centralizer* of a nonempty subset A in G is

$$C_G(A) = \{g \in G \mid gag^{-1} = a \text{ for all } a \in A\}.$$

In other words, $g \in C_G(A)$ if and only if g commutes with every element in A.

The operation gag^{-1} occurs so often in group theory that it bears a name: the *conjugation* of a by g. The condition that $gag^{-1} = a$ is tantamount to $ga = ag$. Consequently, the centralizer consists of all elements in G that commute with every element of the subset A.

The center $Z(G)$ of a group is a particular example of a centralizer, namely $Z(G) = C_G(G)$. If $A = \{a\}$, i.e., is a singleton set, then we write $C_G(a)$ instead of $C_G(\{a\})$.

Proposition 1.7.11

For any subset $A \neq \emptyset$ of a group G, we have $C_G(A) \leq G$.

Proof. Since $1 \in C_G(A)$, we know $C_G(A) \neq \emptyset$.

Let $x, y \in C_G(A)$ be arbitrary. By definition, $xa = ax$ and $ya = ay$ for all $a \in A$. Hence,

$$
\begin{aligned}
(xy)a = x(ya) = x(ay) \quad &\text{since } y \in C_G(A) \\
= (xa)y = (ax)y \quad &\text{since } x \in C_G(A) \\
= a(xy).
\end{aligned}
$$

Thus, $xy \in C_G(A)$.

Let $x \in C_G(A)$. Then for all $a \in A$ since $xax^{-1} = a$, then $xa = ax$ and $a = x^{-1}ax$. Thus, $x^{-1} \in C_G(A)$.

We conclude that $C_G(A)$ is a subgroup of G. $\square$

In order to present the next construction that always gives a subgroup, we introduce some notation. Let $A \subseteq G$ and let $g \in G$. Then we define the subsets gA, Ag, and gAg^{-1} as

$$
gA = \{ga \mid a \in A\}, \quad Ag = \{ag \mid a \in A\}, \quad gAg^{-1} = \{gag^{-1} \mid a \in A\}.
$$

For the set gA (and similarly for the other two sets), the function $f : A \to gA$ defined by $f(a) = ga$ is a bijection with inverse function $f^{-1}(x) = g^{-1}x$. Consequently, gA, Ag, and gAg^{-1} have the same cardinality as A.

Definition 1.7.12

Let G be a group. The *normalizer* of a nonempty subset A in G is

$$
N_G(A) = \{g \in G \mid gAg^{-1} = A\}.
$$

Proposition 1.7.13

For any subset $A \neq \emptyset$ of a group G, we have $N_G(A) \leq G$.

Proof. (Left as an exercise for the reader. See Exercise 1.7.27.) $\square$

If A consists of a single element, then $C_G(A) = N_G(A)$. Otherwise, the condition for normalizer is looser than the condition for centralizer: $g \in N_G(A)$ means that $a \mapsto gag^{-1}$ permutes the elements of A, whereas $g \in C_G(A)$ means that $a \mapsto gag^{-1}$ is the identity on elements of A. So, for any subset A of G, we have

$$Z(G) \le C_G(A) \le N_G(A) \le G.$$

If G is abelian, then equality holds at each $\le$, regardless of the subset A.

1.7.3   Useful CAS Commands

Given a group G, both *Maple* and SAGE offer methods to define the subgroup $\langle S \rangle$ generated by the subset $S \subseteq G$. Because of their central importance, subgroups of symmetric groups are called *permutation groups*. In most computer algebra systems, permutation groups have their own constructors and methods.

--------------------------------- Maple ---------------------------------

$with(GroupTheory):$

$G1 := Group([[1,6]], [[1,4,2], [3,5,6]]):$

$GroupOrder(G1);$

$$24$$

$IsAbelian(G1);$

$$false$$

$Center(G1);$

$$Z(\langle (1,6), (1,4,2)(3,5,6) \rangle)$$

$Elements(Center(G1));$

$$\{(1,6)(2,5)(3,4), ()\}$$

(We remind the reader that in Maple, ending a line with ':' tells *Maple* to perform the calculation without displaying the result, whereas ending with ';' does display the result.) This code defines the subgroup $G1$ of S_6 as

$$G1 = \langle (1\,6)\,(1\,4\,2)(3\,5\,6) \rangle.$$

Then we find the order of $G1$, determine if it is abelian, define the center of $G1$, and then list the elements in the center of $G1$. The difference between the last two lines is that calculating the center of a group defines another group object, but displaying the list of elements is another command. (The reader should notice in this example that though $G1$ is generated by an element of order 2 and an element of order 3, it has order 24.)

The following commands in SAGE implement the same thing.

———————————————————— Sage ————————————————————

```
sage: G=PermutationGroup(['(1,6)','(1,4,2)(3,5,6)'])
sage: G.order()
24
sage: G.is_abelian()
False
sage: G.center()
Subgroup generated by ... (not listed)
sage: G.center().list()
[(), (1,6)(2,5)(3,4)]
```

(We did not show line 4.) In SAGE, we can also define subgroups of any group when generated by a set of elements.

———————————————————— Sage ————————————————————

```
sage: G=GL(2,GF(5))
sage: m1=G([[2,1],[3,3]])
sage: m2=G([[1,3],[1,1]])
sage: H=G.subgroup([m1,m2])
sage: H.order()
48
```

This code constructs the subgroup

$$\left\langle \begin{pmatrix} 2 & 1 \\ 3 & 3 \end{pmatrix}, \begin{pmatrix} 1 & 3 \\ 1 & 1 \end{pmatrix} \right\rangle$$

of $GL_2(\mathbb{F}_5)$. Interestingly enough, replacing GF(5) with QQ, SAGE's label for $\mathbb{Q}$, defines 2×2 matrices with coefficients in $\mathbb{Q}$, making G a subgroup of $GL_2(\mathbb{Q})$. Then H.order() would return +Infinity.

Both *Maple* and SAGE offer commands related to the normalizer and centralizer of a set, conjugates of elements, and many other topics.

EXERCISES FOR SECTION 1.7

1. In $(\mathbb{Z}, +)$, list all the elements in the subgroup $\langle 12, 20 \rangle$.

2. In D_8, list all the elements in the subgroup $\langle sr^2, sr^6 \rangle$.

3. In $U(40)$, list all the elements in $\langle \overline{3}, \overline{31} \rangle$.

4. Find all the subgroups of Z_{20}. [Hint: They call all be expressed as $\langle z^a \rangle$, where z is the generator of Z_{20}.]

5. Find all the subgroups of D_5, expressed as subgroups generated by small subsets.

6. In $GL_2(\mathbb{F}_3)$, list all the elements in $\left\langle \begin{pmatrix} 1 & 1 \\ 0 & 1 \end{pmatrix}, \begin{pmatrix} 1 & 0 \\ 1 & 1 \end{pmatrix} \right\rangle$.

7. In $GL_2(\mathbb{F}_5)$, list all the elements in $\left\langle \begin{pmatrix} 1 & 1 \\ 0 & 1 \end{pmatrix}, \begin{pmatrix} 2 & 0 \\ 0 & 1 \end{pmatrix} \right\rangle$.

8. Prove that in S_4, the subgroup $\langle (1\,2\,3), (1\,2)(3\,4) \rangle$ is A_4.

9. This exercise finds generating subsets of S_n.
    (a) Prove that S_n is generated by $\{(1\,2), (2\,3), (3\,4), \ldots, (n-1\,n)\}$.
    (b) Prove that S_n is generated by $\{(1\,2), (1\,3), \ldots, (1\,n)\}$.
    (c) Prove that S_n is generated by $\{(1\,2), (1\,2\,3\,\cdots\,n)\}$.
    (d) Show that S_4 is not generated by $\{(1\,2), (1\,3\,2\,4)\}$.

10. Prove that for any prime number p, the symmetric group S_p is generated by any transposition and any p-cycle.

11. Show that if p is prime, then $A_p = \langle (1\,2\,3), (1\,2\,3\cdots p) \rangle$.

12. Prove that if $A \subseteq B$ are subsets in a group G, then $\langle A \rangle \le \langle B \rangle$.

13. Let G be a group and let $S \subseteq G$ be a subset. Let $\mathcal{C}$ be the collection of subgroups of G that contain S. Prove that

$$\langle S \rangle = \bigcap_{H \in \mathcal{C}} H.$$

[Hint: Use Exercise 1.6.28. This exercise allows us to conclude that $\langle S \rangle$ is the smallest subgroup of G that contains the subset S.]

14. Let G be a group that is finitely generated. Prove that any subgroup is also finitely generated.

15. Let F be $\mathbb{Q}$, $\mathbb{R}$, $\mathbb{C}$, or $\mathbb{F}_p$, with p a prime number. Prove that the subset of $GL_n(F)$ of diagonal matrices is an abelian subgroup. Explain why this abelian subgroup is strictly larger than $Z(GL_n(F))$.

16. Prove the assertion in Example 1.7.9 that BE_{ij} is the matrix of zeros everywhere except that it has the ith column of B as its jth column, and that $E_{ij}B$ is the matrix of zeros everywhere except that it has the jth row of B in the ith row.

17. Let G be an abelian group. Prove that the following two subsets are subgroups
    (a) $\{g^n \mid g \in G\}$, where n is a fixed integer.
    (b) $\{g \in G \mid g^n = e\}$, where n is a fixed integer.

18. Let G be an abelian group that is not necessarily finite. Define the subset $\text{Tor}(G)$ to be the subset of elements that have finite order. Prove that $\text{Tor}(G)$ is a subgroup. [$\text{Tor}(G)$ is called the *torsion subgroup* of G.]

19. Describe the elements in $\text{Tor}(\mathbb{C}^*)$, the torsion subgroup of $\mathbb{C}^*$. (See Exercise 1.7.18.) Show that $\text{Tor}(\mathbb{C}^*)$ is not finitely generated.

20. Prove that $Z(G) = \bigcap_{a \in G} C_G(a)$.

21. Prove that the center $Z(D_n)$ of the dihedral group is $\{\iota\}$ if n is odd and $\{\iota, r^{n/2}\}$ if n is even.

22. Prove that for all $n \ge 3$, the center of the symmetric group is $Z(S_n) = \{\text{id}\}$.

23. In the group D_n, calculate the centralizer and the normalizer for each of the subsets (a) $\{s\}$, (b) $\{r\}$, and (c) $\langle r \rangle$.

24. For the given group G, find the $C_G(A)$ of the respective sets A.
    (a) $G = S_3$ and $A = \{(1\,2\,3)\}$.
    (b) $G = S_5$ and $A = \{(1\,2\,3)\}$.
    (c) $G = S_4$ and $A = \{(1\,2)\}$.

25. For the given group G, find the $C_G(A)$ and $N_G(A)$ of the respective sets A.
    (a) $G = D_6$ and $A = \{s, r^2\}$.
    (b) $G = Q_8$ and $A = \{i, j\}$.
    (c) $G = S_4$ and $A = \{(1\,2), (3\,4)\}$.

26. Let $V_1 \subseteq V_2$ be subsets of a group G.
    (a) Prove that $C_G(V_2) \leq C_G(V_1)$.
    (b) Prove that $N_G(V_2) \leq N_G(V_1)$.

27. Prove Proposition 1.7.13.

28. Let H be a subgroup of a group G.
    (a) Show that $H \leq N_G(H)$.
    (b) Show that if A is a subset of G, then A is not necessarily a subset of $N_G(A)$.

1.8   Lattice of Subgroups

In order to develop an understanding of the internal structure of a group, listing all the subgroups of a group has some value. However, showing how these subgroups are related to each other carries more information. The lattice of subgroups offers a visual representation of containment among subgroups.

Denote by $\mathrm{Sub}(G)$ the set of all subgroups of the group G. Obviously, $\mathrm{Sub}(G) \subseteq \mathcal{P}(G)$. For any two subgroups $H, K \in \mathrm{Sub}(G)$, $H \subseteq K$ if and only if $H \leq K$. Consequently, $(\mathrm{Sub}(G), \leq)$ is a poset, namely the subposet of $(\mathcal{P}(G), \subseteq)$ on the subset $\mathrm{Sub}(G)$.

Proposition 1.8.1

> For all groups G, the poset $(\mathrm{Sub}(G), \leq)$ is a lattice.

Proof. We know that $(\mathcal{P}(G), \subseteq)$ is a lattice: the least upper bound of any two subsets A and B is $A \cup B$ and the greatest lower bound is $A \cap B$. By Proposition 1.6.10, for any two $H, K \in \mathrm{Sub}(G)$, we also have $H \cap K \in \mathrm{Sub}(G)$, so $H \cap K$ their greatest lower bound in the poset $(\mathrm{Sub}(G), \leq)$.

In contrast, for $H, K \leq G$, the union $H \cup K$ is generally not a subgroup. By Exercise 1.7.13, $\langle H \cup K \rangle$ is the smallest (by inclusion) subgroup of G that contains both H and K, and thus the least upper bound of H and K in $(\mathrm{Sub}(G), \leq)$. Since every pair of subgroups of G has a least upper bound and a greatest lower bound in $(\mathrm{Sub}(G), \leq)$, the poset is a lattice. $\square$

The construction given in the above proof for a least upper bound of H and K, namely $\langle H \cup K \rangle$ is called the *join* of H and K.

Since $(\mathrm{Sub}(G), \le)$ is a poset, we can create the Hasse diagram for it. By a common abuse of language, we often say "draw the lattice of G" for "draw the Hasse diagram of the poset $(\mathrm{Sub}(G), \le)$." The lattice of a group shows all subgroups and their containment relationships.

Example 1.8.2 (Prime Cyclic Groups). Let p be a prime number and consider the group Z_p, generated by z. Every non-identity element $g \in Z_p$ has $g = z^a$ for some a with $p \nmid a$. The order of g is $p/\gcd(p,a) = p$. Hence, every nontrivial subgroup of Z_p is all of Z_p. Thus, the lattice of subgroups of Z_p is:

$$Z_p$$
$$|$$
$$\{e\}$$

This is the simplest lattice of any nontrivial group. △

Example 1.8.3. Consider the cyclic group Z_8 with generator z. It has a total of 4 subgroups, namely $\{e\}$, $\langle z^4 \rangle$, $\langle z^2 \rangle$, and Z_8. The lattice of Z_8 is:

$$Z_8$$
$$|$$
$$\langle z^2 \rangle$$
$$|$$
$$\langle z^4 \rangle$$
$$|$$
$$\{e\}$$

Note that $\langle z^3 \rangle = \{z^3, z^6, z, z^4, z^7, z^2, z^5, e\} = Z_8$ and that $\langle z^6 \rangle = \{z^6, z^4, z^2, e\} = \langle z^2 \rangle$. Also, $\langle z^4, z^6 \rangle = \langle z^2 \rangle$. So, though it is not immediately obvious, all subgroups of Z_8 do appear in the above diagram. △

Example 1.8.4. Figure 1.9 gives us the lattice of Z_{24}. △

The above examples hint at a trend connecting the divisors of n and the subgroups of a cyclic group. The following proposition formalizes this.

Proposition 1.8.5

Every subgroup of a cyclic group G is cyclic. Furthermore, if G is finite with $|G| = n$, then all the subgroups of G consist of $\langle z^d \rangle$ for all divisors d of n and where z is a generator of G.

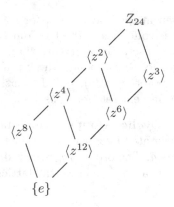

FIGURE 1.9: The lattice of Z_{24}.

Proof. Let G be a cyclic group (not necessarily finite) generated by an element z. Let H be a nontrivial subgroup of G and let $S = \{a \in \mathbb{N}^* \mid z^a \in H\}$, the set of positive powers of z in H. By the well-ordering principle, S has a least element, say c. We prove by contradiction that $H = \langle z^c \rangle$.

Suppose that H contains z^k where c does not divide k. Then by integer division, there exist $q, r \in \mathbb{Z}$ with $0 < r < c$ such that $k = cq + r$. Then the element $z^r = z^{k-qc} = z^k(z^c)^{-q}$ is in H, which contradicts the minimality of c. The $c \mid k$ and so $H = \langle z^c \rangle$.

Now consider the case with G finite and $|G| = n$. Since $e = z^n \in H$, using the argument in the previous paragraph, we see that c must divide n. □

We give a few more examples of lattices of noncyclic groups.

Example 1.8.6 (Quaternion Group). The following diagram gives the lattice of Q_8:

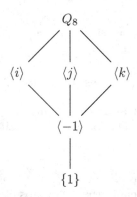

$\triangle$

Finite subgroups of equal cardinality will be incomparable. Hence, though it is not necessary to do so, when drawing lattices of groups it is common to place the subgroups of equal cardinality on the same horizontal level.

Example 1.8.7. The lattice of A_4, the alternating group on four elements (which has order 12):

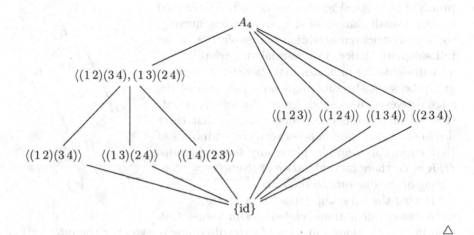

Example 1.8.8. The lattice of D_6, the hexagonal dihedral group:

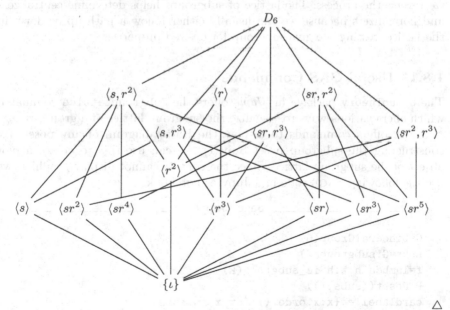

When we begin to determine the subgroup lattice of a group, we quickly discover the challenge of determining whether or not we have found all the subgroups. As we go further along, Lagrange's Theorem (Theorem 2.1.10) will help us considerably. However, in the meantime, we must content ourselves by checking many possibilities.

As the size or internal complexity of groups increases, the subgroup lattice diagrams become unwieldy. Furthermore, in specific examples and in proofs of theoretical results, we are often interested in only a small number of subgroups. Consequently, we often restrict our attention to a sublattice of the full subgroup lattice. In a normal subgroup lattice, an unbroken edge indicates that there exists no subgroup between the subgroups on either end of the edge. However, in a subdiagram of a subgroup lattice, an unbroken edge no longer means that there does not exist a subgroup between the endpoints of that edge. For example, given any two subgroups $H, K \leq G$, there is a sublattice of $(\mathrm{Sub}(G), \leq)$ consisting of the diagram to the right.

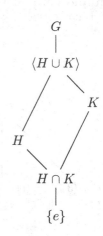

Having the subgroup lattice of a group G facilitates many calculations related to subgroups. Calculating intersections and joins of two subgroups is easy. For the intersection of H and K, we find the highest subgroup L of G in which there is a path up from L to H and a path up from L to K. To find the join of two subgroups, we reverse the process. The lattice of subgroups helps determine centralizers and normalizers because we can usually either follow a path up or down in the lattice, testing the give subgroup for desired properties.

1.8.1   Useful CAS Commands

The `GroupTheory` package in *Maple* offers the `SubgroupLattice` command, which offers various ways to visualize the subgroup lattice of a group.

SAGE offers commands to display the Hasse diagram of any poset. By constructing the subgroup poset $\mathrm{Sub}(G)$, we can use this to create a plot object of the subgroup lattice. Since the console cannot display graphics, we use the following SAGE code in a Jupyter notebook.

———————————— Sage (Jupyter) ————————————

```
G=DihedralGroup(6)
subs=G.subgroups()
f=lambda h,k:h.is_subgroup(k)
P=Poset((subs,f))
cardlabel = {x:x.order() for x in subs}
P.plot(element_labels=cardlabel, vertex_size = 800)
```

In this code, we define G as D_6 and then define the set *subs*, which is a set of all the subgroups of G. The third line for f defines the partial order of containment, while the fourth line constructs P which is now the poset $(\mathrm{Sub}(G), \subseteq)$. The user defined variable `cardlabel` is a dictionary type that to each subgroup in *subs* associates its order. Finally, the last line plots the corresponding Hasse diagram. To run the code, we click the ▶Run button on the Jupyter notebook. This will create a PNG picture of the group lattice. Admittedly, this method does not place the subgroups at levels corresponding to their cardinality. By changing the `cardlabel` dictionary variable, we can put different labels on the vertices of the diagram. For example, using `str(x.gens())[1:-1]` instead of `x.order()` will label each subgroup vertex by the generators of the subgroup.

EXERCISES FOR SECTION 1.8

1. Determine the subgroup lattice of Z_{20}.

2. Determine the subgroup lattice of $\mathbb{Z}/105\mathbb{Z}$.

3. Determine the subgroup lattice of Z_{100}.

4. Determine the subgroup lattice of $U(20)$. Also answer whether $U(20)$ is cyclic.

5. Determine the subgroup lattice of $U(35)$. Underline in the lattice the subgroups of $U(35)$ that are cyclic.

6. Determine the subgroup lattice for $Z_{p^2 q}$ where p and q are prime numbers.

7. Determine the subgroup lattice for Z_{p^n} where p is a prime number and n is a positive integer.

8. Determine the subgroup lattice of $Z_3 \oplus Z_9$.

9. Determine the subgroup lattice of D_4.

10. Determine the subgroup lattice of D_p where p is a prime number.

11. Determine the subgroup lattice of D_8.

12. Determine the subgroup lattice of S_3.

13. Determine the subgroup lattice of $Z_2 \oplus Z_2 \oplus Z_2$.

14. Determine the subgroup lattice of $Z_p \oplus Z_p$ where p is a prime number. Also show that all nontrivial strict subgroups are cyclic.

15. Show that there is no group whose subgroup lattice has the shape of a 3-cube, i.e., has the same lattice as the poset $(\mathcal{P}(\{1, 2, 3\}), \subseteq)$.

16. Prove that the lattice of A_4 is that which is given in Example 1.8.7.

17. Let $m, n \in \mathbb{Z}$ where we consider $\mathbb{Z}$ as a group equipped with addition.
    (a) Find a generator of $\langle m \rangle \cap \langle n \rangle$.
    (b) Find a generator of the join of $\langle m, n \rangle$.

18. Determine (without creating the subgroup lattice) the number and order of all cyclic subgroups of $Z_5 \oplus Z_{15}$.

19. Determine (without creating the subgroup lattice) the number and order of all cyclic subgroups of $Z_{15} \oplus Z_{15}$.

20. Let $G = D_6$. Use the lattice in Example 1.8.8 to determine:
    (a) $C_{D_6}(s)$
    (b) $N_{D_6}(sr)$
    (c) $N_{D_6}(s, r^3)$

21. Let $g, h \in G$ such that $|g| = 12$ and $|h| = 5$. Prove that $\langle g \rangle \cap \langle h \rangle = \{e\}$.

22. Let x and y be elements in a group G such that $|x|$ and $|y|$ are relatively prime. Prove that $\langle x \rangle \cap \langle y \rangle = \{e\}$.

23. Prove the converse of Example 1.8.2, namely prove that the only groups that only have two subgroups are Z_p, where p is prime.

24. Let G be a group that has exactly one nontrivial proper subgroup. Prove that G is a cyclic group of order p^2 where p is a prime number.

25. Let G be an abelian group that has exactly two nontrivial proper subgroups H and K, neither of which is contained in the other.
    (a) Prove that both H and K must be cyclic an generated by an element of prime order. Call these generators x and y respectively.
    (b) Show that if $|x| = |y|$, then $\langle xy \rangle$ is a subgroup also of order $|x|$ but that is not equal to H or K.
    (c) Show that $\langle x, y \rangle$ is strictly larger than both H and K, so $G = \langle x, y \rangle$.
    (d) Deduce that G is a cyclic group of order pq where p and q are distinct primes.

26. Prove that if G is a finite group whose entire lattice of subgroups consists of one chain, then G is cyclic and of order p^n where p is prime and n is a positive integer.

1.9   Group Homomorphisms

The concept of a function is ubiquitous in mathematics. However, in different branches we often impose conditions on the functions we consider. For example, in linear algebra we do not study arbitrary functions from one vector space to another but limit our attention to linear transformations. As exhibited in Section A.4, when studying posets it is common to restrict attention to monotonic functions.

Given two objects A and B with a particular algebraic structure, if we consider an arbitrary function $f : A \to B$, a priori, the only type of information that f carries is set theoretic. In other words, information about algebraic properties of A would be lost under f. However, if we impose certain properties on the function, it can, intuitively speaking, preserve the structure.

1.9.1 Homomorphisms

Definition 1.9.1

Let $(G, *)$ and $(H, \bullet)$ be two groups. A function $\varphi : G \to H$ is called a *homomorphism* from G to H if for all $g_1, g_2 \in G$,

$$\varphi(g_1 * g_2) = \varphi(g_1) \bullet \varphi(g_2). \tag{1.8}$$

The operation inside the function on the left-hand side is an operation in G while the operation on the right-hand side occurs in the group H. With abstract group notation, we write (1.8) as

$$\varphi(g_1 g_2) = \varphi(g_1)\varphi(g_2)$$

but must remember that the group operations occur in different groups.

Example 1.9.2. Fix a positive real number b and consider the function $f(x) = b^x$. Power rules state that for all $x, y \in \mathbb{R}$, $b^{x+y} = b^x b^y$. In the language of group theory, this identity can be restated by saying that the exponential function $f(x) = b^x$ is a homomorphism from $(\mathbb{R}, +)$ to $(\mathbb{R}^*, \times)$. $\triangle$

Example 1.9.3. The function of inclusion $f : (\mathbb{Z}, +) \to (\mathbb{R}, +)$ given by $f(x) = x$ is a homomorphism. $\triangle$

Example 1.9.4. The function $f : Z_n \to Z_n$ given by $f(x) = x^2$ is a homomorphism. Let z be a generator of Z_n. Then for all $z^a, z^b \in Z_n$,

$$f(z^a z^b) = \left(z^a z^b\right)^2 = \left(z^{a+b}\right)^2 = z^{2(a+b)} = z^{2a+2b} = z^{2a} z^{2b} = f(z^a) f(z^b). \quad \triangle$$

Example 1.9.5. Consider the direct sum $Z_2 \oplus Z_2$, where each Z_2 has generator z. Consider the function $\varphi : Q_8 \to Z_2 \oplus Z_2$ defined by

$$\varphi(\pm 1) = (e, e) \quad \varphi(\pm i) = (z, e) \quad \varphi(\pm j) = (e, z) \quad \varphi(\pm k) = (z, z).$$

This is a homomorphism but in order to verify it, we must check that φ satisfies (1.8) for all 64 products of terms in Q_8. However, we can cut down the work. First notice that for all terms $a, b \in \{1, i, j, k\}$, the products $(\pm a)(\pm b) = \pm(ab)$ with the sign as appropriately defined. The following table shows $\varphi(ab)$ with a in the columns and b in the rows.

	± 1	$\pm i$	$\pm j$	$\pm k$
± 1	(e, e)	(z, e)	(e, z)	(z, z)
$\pm i$	(z, e)	(e, e)	(z, z)	(e, z)
$\pm j$	(e, z)	(z, z)	(e, e)	(z, e)
$\pm k$	(z, z)	(e, z)	(z, e)	(e, e)

All the entries of the table are precisely $\varphi(a)\varphi(b)$, which confirms that φ is a homomorphism. $\triangle$

Example 1.9.6. Let n be an integer greater than 1. The function $\varphi(a) = \bar{a}$ that maps an integer to its congruence class in $\mathbb{Z}/n\mathbb{Z}$ is a homomorphism. This holds because of Proposition A.6.4 and the definition

$$\bar{a} + \bar{b} = \overline{a + b}.$$

$\triangle$

Example 1.9.7 (Determinant). Let F be $\mathbb{Q}$, $\mathbb{R}$, or $\mathbb{C}$. The determinant function $\det : \mathrm{GL}_n(F) \to (F^*, \times)$ is a homomorphism. This is precisely the content of the theorem in linear algebra that

$$\det(AB) = \det(A)\det(B).$$

This result also applies with modular arithmetic base p, when p is a prime number. In this case, we already denoted $(\mathbb{F}_p^*, \times)$ by $U(p)$.

$\triangle$

Example 1.9.8 (Sign Function). In Definition 1.6.19 we defined the sign function on S_n as

$$\mathrm{sign}(\sigma) = (-1)^{\mathrm{inv}(\sigma)}.$$

As a function $\mathrm{sign} : S_n \to \{1, -1\}$. The set $\{1, -1\}$ becomes a group when equipped with the usual multiplication of integers. By Lemma 1.6.14, for all $\sigma, \tau \in S_n$,

$$\mathrm{sign}(\sigma\tau) = (-1)^{\mathrm{inv}(\sigma\tau)} = (-1)^{\mathrm{inv}(\sigma) + \mathrm{inv}(\tau)}$$
$$= (-1)^{\mathrm{inv}(\sigma)}(-1)^{\mathrm{inv}(\tau)} = \mathrm{sign}(\sigma)\,\mathrm{sign}(\tau).$$

Thus, the sign function is a homomorphism $S_n \to (\{1, -1\}, \times)$.

$\triangle$

As a first property preserved by homomorphisms, the following proposition shows that homomorphisms map powers of elements to powers of corresponding elements.

Proposition 1.9.9

Let $\varphi : G \to H$ be a homomorphism of groups.

(1) $\varphi(e_G) = e_H$.

(2) For all $x \in G$, $\varphi(x^{-1}) = \varphi(x)^{-1}$.

(3) For all $x \in G$ and all $n \in \mathbb{Z}$, $\varphi(x^n) = \varphi(x)^n$.

(4) For all $x \in G$, if $|x|$ is finite, then $|\varphi(x)|$ divides $|x|$.

Proof. For (1), denote $\varphi(e_G) = u$. Since $e_G e_G = e_G$, using the homomorphism property, we have

$$u = \varphi(e_G) = \varphi(e_G e_G) = \varphi(e_G)\varphi(e_G) = u^2.$$

Applying u^{-1} to both sides of this equation $u^2 = u$, we deduce that $u = e_H$.

For (2), let $x \in G$. Applying φ to both sides of $e_G = xx^{-1}$ gives

$$e_H = \varphi(e_G) = \varphi(xx^{-1}) = \varphi(x)\varphi(x^{-1}).$$

Similarly, $\varphi(x^{-1})\varphi(x) = e_H$. From the definition of inverse, $\varphi(x^{-1}) = \varphi(x)^{-1}$.

For (3), note that parts (1) and (2) establish the result for $n = 0, -1$. For all other integers n, a simple induction proof gives us the result.

For (4), suppose that $|x| = k$. Then $x^k = e_G$. Using (3) and (1), we deduce that $\varphi(x)^k = e_H$. By Corollary 1.3.7, $|\varphi(x)|$ divides k. □

1.9.2 Kernel and Image

In linear algebra, the concepts of kernel and image of a linear transformation play a central role in many applications, in particular finding and describing solutions to systems of linear equations. The concepts of kernel and image play an equally important role in group theory.

Definition 1.9.10

Let $\varphi : G \to H$ be a homomorphism between groups.

(1) The *kernel* of φ is $\mathrm{Ker}\, \varphi = \{g \in G \,|\, \varphi(g) = e_H\}$, where e_H is the identity in H.

(2) The *image* of φ is $\mathrm{Im}\, \varphi = \{h \in H \,|\, \exists g \in G, \varphi(g) = h\}$. The image is also called the *range* of φ.

Proposition 1.9.11

Let $\varphi : G \to H$ be a homomorphism of groups. The kernel $\mathrm{Ker}\, \varphi$ is a subgroup of G.

Proof. $\mathrm{Ker}\, \varphi$ is nonempty since $e_G \in \mathrm{Ker}\, \varphi$. Now let $x, y \in \mathrm{Ker}\, \varphi$. Then

$$\begin{aligned}
\varphi(xy^{-1}) &= \varphi(x)\varphi(y^{-1}) && \text{since } \varphi \text{ is a homomorphism} \\
&= \varphi(x)\varphi(y)^{-1} && \text{by Proposition 1.9.9(2)} \\
&= e_H e_H^{-1} = e_H && \text{since } x, y \in \mathrm{Ker}\, \varphi.
\end{aligned}$$

Hence, $xy^{-1} \in \mathrm{Ker}\, \varphi$. Thus, $\mathrm{Ker}\, \varphi \le G$ by the One-Step Subgroup Criterion. □

Proposition 1.9.12

Let $\varphi : G \to H$ be a homomorphism of groups. The image $\mathrm{Im}\, \varphi$ is a subgroup of H.

Proof. (Left as an exercise for the reader. See Exercise 1.9.1.) □

Example 1.9.13. Consider the homomorphism det : $GL_n(\mathbb{R}) \to \mathbb{R}^*$. (See Example 1.9.7.) The kernel of the determinant homomorphism is the set of matrices whose determinant is 1, namely $SL_n(\mathbb{R})$, the special linear group. $\triangle$

Example 1.9.14. Consider the sign function sign : $S_n \to (\{1, -1\}, \times)$ as defined in Example 1.9.8. The kernel Ker(sign) is precisely the alternating group A_n as a subgroup of S_n. The function φ is surjective so the image is all of $\{1, -1\}$. $\triangle$

The kernel and the image of a homomorphism are closely connected to whether the homomorphism is injective or surjective.

Proposition 1.9.15

Let $\varphi : G \to H$ be a homomorphism of groups. Then
(1) φ is injective if and only if $\operatorname{Ker} \varphi = \{e_G\}$.
(2) φ is surjective if and only if $H = \operatorname{Im} \varphi$.

Proof. (Left as an exercise for the reader. See Exercise 1.9.10.) $\square$

1.9.3 Isomorphisms

We have seen some examples where groups, though presented differently, may actually look strikingly the same. For example $(Z_n, \cdot)$ and $(\mathbb{Z}/n\mathbb{Z}, +)$ behave identically and likewise for $(\mathbb{Z}, +)$ and $(2\mathbb{Z}, +)$, where $2\mathbb{Z}$ means all even numbers. This raises the questions (1) when should we call two groups the same and (2) what would doing so mean.

Definition 1.9.16

Let G and H be two groups. A function $\varphi : G \to H$ is called an *isomorphism* if (1) φ is a homomorphism and (2) φ is a bijection. If there exists an isomorphism between two groups G and H, then we say that G and H are *isomorphic* and we write $G \cong H$.

When two groups are isomorphic, they are for all intents and purposes of group theory the same. We could have defined an isomorphism as a bijection φ such that both φ and φ^{-1} are both homomorphisms. However, this turns out to be heavier than necessary as the following proposition shows.

Proposition 1.9.17

If φ is an isomorphism (as defined in Definition 1.9.16), then $\varphi^{-1} : H \to G$ is a homomorphism.

Proof. (Left as an exercise for the reader. See Exercise 1.9.25.) $\square$

Example 1.9.18. Let $G = Z_2 \oplus Z_3$ and $H = Z_6$. Suppose that Z_2 is generated by x, Z_3 is generated by y and Z_6 is generated by z. Consider the function $f : H \to G$ defined by $f(z^k) = (x^k, y^k)$ for all $0 \leq k \leq 5$. Now if $a, a' \in \mathbb{Z}$, then $z^a = z^{a'}$ if and only if $6 \mid (a' - a)$. Consequently, $2 \mid (a' - a)$ and $3 \mid (a' - a)$. Thus $z^a = z^{a'}$ implies that $x^a = x^{a'}$ and $y^b = y^{b'}$. Consequently, given the definition of f, it is also true that $f(z^k) = (x^k, y^k)$ for all $k \in \mathbb{Z}$. With this established, we have

$$f(z^a z^b) = f(z^{a+b}) = (x^{a+b}, y^{a+b}) = (x^a, y^a)(x^b, y^b) = f(z^a)f(z^b),$$

so f is a homomorphism. By Theorem 1.4.3, the element (x, y) has order $\mathrm{lcm}(2, 3) = 6$. Hence, $\mathrm{Im}\, f = G$ and so f is a surjective. Since f is a surjection between finite sets of the same cardinality, f is a bijection. We conclude that f is an isomorphism. We can now write that $Z_6 \cong Z_2 \oplus Z_3$. △

Example 1.9.19. Let b be a positive real number. We know that $f(x) = b^x$ is a bijection between $\mathbb{R}$ and $\mathbb{R}^{>0}$ with inverse function $f^{-1}(x) = \log_b x = (\ln x)/(\ln b)$. Example 1.9.2 showed that f is a homomorphism and thus it is an isomorphism between $(\mathbb{R}, +)$ and $(\mathbb{R}^{>0}, \times)$. Proposition 1.9.17 implies that $f^{-1}(x) = \log_b x$ is a homomorphism from $(\mathbb{R}^{>0}, \times)$ to $(\mathbb{R}, +)$. This is tantamount to the logarithm rule that $\log_b(xy) = \log_b(x) + \log_b(y)$. △

Example 1.9.20. In this example, we provide an isomorphism between $\mathrm{GL}_2(\mathbb{F}_2)$ and S_3. Consider the following function:

A	$\begin{pmatrix} 1 & 0 \\ 0 & 1 \end{pmatrix}$	$\begin{pmatrix} 1 & 1 \\ 0 & 1 \end{pmatrix}$	$\begin{pmatrix} 1 & 0 \\ 1 & 1 \end{pmatrix}$	$\begin{pmatrix} 0 & 1 \\ 1 & 0 \end{pmatrix}$	$\begin{pmatrix} 1 & 1 \\ 1 & 0 \end{pmatrix}$	$\begin{pmatrix} 0 & 1 \\ 1 & 1 \end{pmatrix}$
$\varphi(A)$	id	$(1\,2)$	$(1\,3)$	$(2\,3)$	$(1\,2\,3)$	$(1\,3\,2)$

If we compare the group table on $\mathrm{GL}_2(\mathbb{F}_2)$ and the group table on S_3 (see Exercise 1.9.16), we find that this particular function φ preserves how group elements operate, establishing that φ is a isomorphism. △

Proposition 1.9.21

Let $\varphi : G \to H$ be an isomorphism of groups. Then

(1) $|G| = |H|$.

(2) G is abelian if and only if H is abelian.

(3) φ preserves orders of elements, i.e., $|x| = |\varphi(x)|$ for all $x \in G$.

Proof. Part (1) follows immediately from the requirement that φ is a bijection.

For (2), suppose that G is abelian. Let $h_1, h_2 \in H$. Since φ is surjective, there exist $g_1, g_2 \in G$ such that $\varphi(g_1) = h_1$ and $\varphi(g_2) = h_2$. Then

$$h_1 h_2 = \varphi(g_1)\varphi(g_2) = \varphi(g_1 g_2) = \varphi(g_2 g_1) = \varphi(g_2)\varphi(g_1) = h_2 h_1.$$

Hence, we have shown that if G is abelian, then H is abelian. Repeating the argument with φ^{-1} establishes the converse, namely that if H is abelian, then G is abelian.

For (3), we consider two cases, whether the order $|x|$ is finite or infinite. First suppose that $|x| = n$ is finite. By Proposition 1.9.9(3), $\varphi(x^n) = \varphi(x)^n$. Then $1_H = \varphi(1_G) = \varphi(x^n) = \varphi(x)^n$. So by Corollary 1.3.7, $|\varphi(x)|$ is finite and divides $|x|$. Applying the same argument to φ^{-1} and the element $\varphi(x)$, we deduce that $|x|$ divides $|\varphi(x)|$. Since $|x|$ and $|\varphi(x)|$ are both positive and divide each other, $|x| = |\varphi(x)|$.

Suppose now that $|x|$ is infinite. Assume that $\varphi(x)^m = 1_H$ for some $m > 0$. Then $\varphi(x^m) = 1_H$ and since φ is injective, we again deduce that $x^m = 1_G$. Hence, the order of x is finite, which gives a contradiction. Thus, the order $|\varphi(x)|$ must be infinite too. $\qquad\square$

Proposition 1.9.21 is particularly useful to prove that two groups are not isomorphic. If either condition (1) or (2) fails, then the groups cannot be isomorphic. Also, if two groups have a different number of elements of a given order, then Proposition 1.9.21(3) cannot hold for any isomorphism and thus the two groups are not isomorphic.

However, we underscore that the three conditions in Proposition 1.9.21 are necessary conditions but not sufficient: just because all three conditions hold, we cannot deduce that φ is an isomorphism. Example 1.9.24 illustrates this.

Remark 1.9.22. If two groups are isomorphic, then they have isomorphic subgroup lattices. However, the converse is not true. There are many pairs of nonisomorphic groups with subgroup lattices that are isomorphic as posets. See Figure 1.10 for an example. $\qquad\triangle$

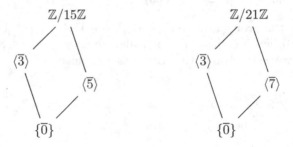

FIGURE 1.10: Nonisomorphic groups with identical lattices.

Example 1.9.23. We prove that D_4 and Q_8 are not isomorphic. They are both of order 8 and they are both nonabelian. However, in D_4 only the elements r and r^3 have order 4 while in Q_8, the elements $i, -i, j - j, k, -k$ are all of order 4. Consequently, there cannot exist a bijection between D_4 and Q_8 that satisfies Proposition 1.9.21(3). Hence, $D_4 \not\cong Q_8$. $\qquad\triangle$

Example 1.9.24. There exist various clever strategies to show that $\mathbb{Q}^{>0}$ is countable. From the required bijections with $\mathbb{N}^*$, it is easy to prove that there is a bijection between $\mathbb{Q}$ and $\mathbb{Z}$. In this example, we show that there does not exist an isomorphism between $(\mathbb{Z}, +)$ and $(\mathbb{Q}, +)$. Interestingly enough, this result does not follow from Proposition 1.9.21 because $|\mathbb{Z}| = |\mathbb{Q}|$, $\mathbb{Z}$ and $\mathbb{Q}$ are both abelian, and all nonzero elements of both groups have infinite order.

Suppose there does exist an isomorphism $f : \mathbb{Q} \to \mathbb{Z}$. If r is a rational number and $n \in \mathbb{Z}$, then by Proposition 1.9.9(3) with addition, $f(n \cdot r) = n \cdot f(r)$. Suppose that we define $f(1) = a$, for some nonzero a. Then

$$a = f(1) = f\left(\frac{2a}{2a}\right) = 2a \cdot f\left(\frac{1}{2a}\right).$$

This implies that $1 = 2f(1/2a)$. But this is a contradiction since $f(1/2a) \in \mathbb{Z}$. Thus, as groups $(\mathbb{Z}, +) \not\cong (\mathbb{Q}, +)$. $\triangle$

The motivating example that $Z_n \cong \mathbb{Z}/n\mathbb{Z}$ offered at the beginning of this subsection generalizes to a broader result for any cyclic groups.

Proposition 1.9.25

Two cyclic groups of the same cardinality are isomorphic.

Proof. Recall from Exercise 1.2.22 that cyclic groups are abelian.

First suppose that G and H are finite cyclic groups, both of order n. Suppose that G is generated by x and H is generated by an element y. Define the function $\varphi : G \to H$ by $\varphi(x^a) = y^a$. Since H is abelian, φ is a homomorphism; we need to prove that it is a bijection.

The image of φ is $\{\varphi(x^k) \,|\, 0 \le k \le n-1\} = \{y^k \,|\, 0 \le k \le n-1\} = H$, so φ is a surjection. A surjection between finite sets is a bijection. Hence, φ is an isomorphism.

The proof is similar if G and H are infinite cyclic groups. (We leave the proof to the reader. See Exercise 1.9.26.) $\square$

Because of Proposition 1.9.25, we talk about *the* cyclic group of order n.

In Section 1.4.2, we introduced the notion of classification theorems of groups. A classification theorem in group theory is a theorem that, given a property of groups, lists all nonisomorphic groups with that property. Example 1.4.4 established that every group of order 4 is isomorphic to either Z_4 or $Z_2 \oplus Z_2$. In this statement, it is implied that Z_4 and $Z_2 \oplus Z_2$ are nonisomorphic. (We can easily see this from the fact that Z_4 contains elements of order 4, while $Z_2 \oplus Z_2$ does not.)

Exercise 1.4.10 established that every group of order 6 is isomorphic to either Z_6 or D_3. Revisiting Example 1.9.20 in light of this exercise, we deduce that since S_3 and $\mathrm{GL}_2(\mathbb{F}_2)$ are both nonabelian of order 6, then they are both isomorphic to D_3 and hence to each other.

The question "What is the isomorphism type of G?" means to find a well-known group that is isomorphic to G. The expression *isomorphism type* is imprecise because there does not exist a standard list of nonisomorphic groups against which to compare every group. Appendix A.7 provides a list of the first few groups listed by order.

If a homomorphism $\varphi : G \to H$ is injective, then when restricting the codomain, the function $\varphi : G \to \operatorname{Im}\varphi$ is an isomorphism. Since φ sets up an isomorphism between G and a subgroup of H, then an injective homomorphism from G to H is often called an *embedding* of G in H.

Definition 1.9.26

An isomorphism $\psi : G \to G$ is called an *automorphism* of G. The set of all automorphisms on a group G is denoted by $\operatorname{Aut}(G)$.

The set $\operatorname{Aut}(G)$ of a group G is interesting because it forms a group under function composition and is a subgroup of S_G, the group of all permutations on the set G. (See Exercise 1.9.37)

1.9.4   Cayley's Theorem and its Applications

In earlier sections, we mentioned the difficulty of describing all groups and their elements in a consistent manner. Cayley's Theorem offers one approach.

Theorem 1.9.27 (Cayley's Theorem)

Every finite group G is isomorphic to a subgroup of S_n for some n.

Before giving the proof of this theorem and illustrating it with examples, we mention the importance for computational methods. Because of Cayley's Theorem implementing operations and methods for permutation groups (subgroups of some S_n) ultimately encompasses all finite groups. For example, in SAGE, the command G=DihedralGroup(5) defines the group G as the subgroup of S_5 expressed as

$$\langle (1\,2\,3\,4\,5), (2\,5)(3\,4) \rangle.$$

In particular, every finite group is isomorphic to one constructed as described in Subsection 1.7.3.

Both *Maple* (GROUPTHEORY package) and SAGE offer many commands that construct specific groups. For example, both have commands for the nth dicyclic group, for the Baby Monster Group, for a projective symplectic group over a finite field, and many more.

Proof (of Theorem 1.9.27). Write the elements of G as $G = \{g_1, g_2, \ldots, g_n\}$. Define the function $\psi : G \to S_n$ by $\psi(g) = \sigma$ where

$$g g_i = g_{\sigma(i)} \qquad \text{for all } i \in \{1, 2, \ldots, n\}.$$

So $\psi(g)$ is the permutation on G of how left "multiplication" by g permutes the elements of G. We need to show that ψ is a homomorphism. Let $g, h \in G$ with $\psi(g) = \sigma$ and $\psi(h) = \tau$. Then for all $1 \le i \le n$, we have $gg_i = g_{\sigma(i)}$ and $hg_i = g_{\tau(i)}$. Then

$$(hg)g_i = h(gg_i) = hg_{\sigma(i)} = g_{\tau(\sigma(i))} = g_{\tau\sigma(i)} \qquad \text{for all } i \in \{1, 2, \ldots, n\}.$$

Hence, $\psi(hg) = \tau\sigma = \psi(h)\psi(g)$ and thus ψ is a homomorphism.

We next show that φ is an injection. Suppose that $g, h \in G$ with $\psi(g) = \psi(h)$. Then in particular $gg_1 = hg_1$. By operating on the right by g_1^{-1}, we deduce that $g = h$.

If we restrict the codomain of ψ to the subgroup $\operatorname{Im}\psi$ in S_n, we see that ψ becomes an isomorphism between G and $\operatorname{Im}\psi$. $\qquad\square$

The proof of Cayley's Theorem shows that any group G is isomorphic to a subgroup of S_n where $|G| = n$. However, as we illustrate below, $|G|$ is usually not the least integer m such that G is isomorphic to a subgroup of S_m.

Example 1.9.28. Let $G = D_4$, the dihedral group on the square. The proof of Cayley's Theorem establishes an isomorphism between D_4 and a subgroup of S_8. To find one such isomorphism φ, label the elements of D_4 with

$$g_1 = \iota, \ g_2 = r, \ g_3 = r^2, \ g_4 = r^3, \ g_5 = s, \ g_6 = sr, \ g_7 = sr^2, \ g_8 = sr^3.$$

We easily calculate that

$$rg_1 = r = g_2, \quad rg_2 = r^2 = g_3, \quad rg_3 = r^3 = g_4, \quad rg_4 = \iota = g_1,$$

$$rg_5 = rs = g_8, \quad rg_6 = rsr = g_5, \quad rg_7 = rsr^2 = g_6, \quad rg_8 = rsr^3 = g_7.$$

Hence, $\varphi(r) = (1\,2\,3\,4)(5\,8\,7\,6)$. Furthermore,

$$sg_1 = s = g_5, \quad sg_2 = sr = g_6, \quad sg_3 = sr^2 = g_7, \quad sg_4 = sr^3 = g_8,$$

$$sg_5 = \iota = g_1, \quad sg_6 = r = g_2, \quad sg_7 = r^2 = g_3, \quad sg_8 = r^3 = g_4.$$

Hence, $\varphi(s) = (1\,5)(2\,6)(3\,7)(4\,8)$. Consequently, by the isomorphism in the proof of Cayley's Theorem, we deduce that

$$D_4 \cong \langle (1\,2\,3\,4)(5\,8\,7\,6), (1\,5)(2\,6)(3\,7)(4\,8) \rangle.$$

This is not the only natural isomorphism between D_4 and a subgroup of a symmetric group. Consider how we introduced D_n in Section 1.1.2 as depicted below.

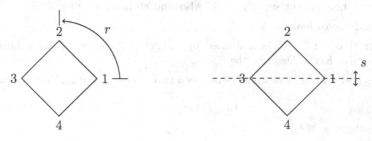

By considering how r and s operate on the vertices, we can view r and s as elements in S_4. The appropriate way to understand this perspective is to define a homomorphism $\varphi : D_4 \to S_4$ by $\varphi(r) = (1\,2\,3\,4)$ and $\varphi(s) = (2\,4)$. For all elements $r^a s^b \in D_4$, the function φ is

$$\varphi(r^a s^b) = (1\,2\,3\,4)^a (2\,4)^b.$$

By exhaustively checking all operations, we can verify that this function is a homomorphism. However, we know that r and s were originally defined as the functions $(1\,2\,3\,4)$ and $(2\,4)$ so this homomorphism is natural. The homomorphism φ is injective and hence establishes an isomorphism between D_4 and the subgroup $\langle (1\,2\,3\,4), (2\,4) \rangle$ in S_4. $\triangle$

Computer algebra systems that describe groups through an isomorphism with a subgroup of a symmetric group S_n often do not use the embedding described by the proof of Cayley's Theorem. Usually systems implement the embedding with the smallest possible n or with a value of n that is natural for the specific group.

EXERCISES FOR SECTION 1.9

1. Let $\varphi : G \to H$ be a homomorphism between groups. Prove that $\operatorname{Im} \varphi \leq H$.

2. Find all homomorphisms from Z_{15} to Z_{20}.

3. Find all homomorphisms from Z_4 to $Z_2 \oplus Z_2$.

4. Prove that the function $\varphi : G \to G$ defined by $\varphi(g) = g^{-1}$ is a homomorphism if and only if G is abelian.

5. Let Z_{33} be generated by an element y and Z_{12} by an element z. Suppose that $\varphi : Z_{33} \to Z_{12}$ is a homomorphism satisfying $\varphi(y^7) = z^8$. Find $\varphi(y^a)$; determine $\operatorname{Ker} \varphi$; and determine $\operatorname{Im} \varphi$.

6. Consider the function $f : U(12) \oplus (\mathbb{Z}/16\mathbb{Z}) \to \mathbb{Z}/4\mathbb{Z}$ defined by $f(a,b) = ab$ with elements considered now modulo 4. Prove that f is a well-defined function. Determine whether or not f is a homomorphism. If it is, determine $\operatorname{Ker} f$; if it is not, give a counterexample.

7. What common fallacy in elementary algebra is addressed by the statement that "the function $f : (\mathbb{R}, +) \to (\mathbb{R}, +)$ with $f(x) = x^2$ is not a homomorphism."

8. Let G be a group. Determine whether or not the function $\varphi : G \oplus G \to G$ defined by $\varphi(x,y) = x^{-1}y$ is a homomorphism.

9. Find all homomorphisms $Z_5 \to S_5$. Also find all homomorphisms $S_5 \to Z_5$.

10. Prove Proposition 1.9.15.

11. Prove that $\varphi : \mathbb{Z} \oplus \mathbb{Z} \to \mathbb{Z}$ defined by $\varphi(x,y) = 2x + 3y$ is a homomorphism. Determine $\operatorname{Ker} \varphi$. Describe the fiber $\varphi^{-1}(6)$.

12. Let φ and ψ be homomorphisms between two groups G and H. Prove or disprove that

$$\{g \in G \mid \varphi(g) = \psi(g)\}$$

is a subgroup of G.

13. Consider the function $f : U(33) \to U(33)$ defined by $f(x) = x^2$. Show that f is a homomorphism and find the kernel and the image of f.

14. Prove that the function $f : \mathrm{GL}_2(\mathbb{R}) \to \mathrm{GL}_3(\mathbb{R})$ with

$$f \begin{pmatrix} a & b \\ c & d \end{pmatrix} = \begin{pmatrix} a^2 & 2ab & b^2 \\ ac & ad + bc & bd \\ c^2 & 2cd & d^2 \end{pmatrix}.$$

   (a) To show that this is in fact a function, prove that for all $A \in \mathrm{GL}_2(\mathbb{R})$, we have $\det(f(A)) = (\det A)^3$. Hence, since $\det(A) \neq 0$, then $f(A) \in \mathrm{GL}_3(\mathbb{R})$.
   (b) Prove that f is a homomorphism.
   (c) Determine the kernel of f.

15. Consider the multiplicative group $\mathbb{R}^\times$ and consider the function $f : \mathbb{R}^\times \to \mathbb{R}^{>0} \times \{1, -1\}$ given by $f(x) = (|x|, \mathrm{sign}(x))$. We assume that all groups have multiplication as their operation. Show that f is an isomorphism.

16. Construct both the Cayley tables for both S_3 and $\mathrm{GL}_2(\mathbb{F}_2)$ to show that the function given in Example 1.9.20 establishes an isomorphism between S_3 and $\mathrm{GL}_2(\mathbb{F}_2)$.

17. Find all homomorphisms $\mathbb{Z} \to \mathbb{Z}$. Determine which ones are isomorphisms.

18. Prove that $Z_m \oplus Z_n \cong Z_{mn}$ if m and n are relatively prime.

19. Prove that $Z_9 \oplus Z_3 \not\cong Z_{27}$.

20. Prove that $Z_m \oplus Z_n \cong Z_{\mathrm{lcm}(m,n)} \oplus Z_{\gcd(m,n)}$. [Hint: Use Exercise 1.9.18.]

21. Let G_1 and H be two arbitrary groups. Prove that $G \oplus H \cong H \oplus G$.

22. Let G_1 and G_2 be two arbitrary groups. Prove that the "projection" maps $p_1 : G_1 \oplus G_2 \to G_1$ and $p_2 : G_1 \oplus G_2 \to G_2$ defined by $p_1(x, y) = x$ and $p_2(x, y) = y$ are homomorphisms. Find the kernels and images of p_1 and p_2.

23. Prove that if a homomorphism $\varphi : G \to H$ is injective, then $G \cong \varphi(G)$.

24. Prove that for every homomorphism $f : (\mathbb{R}, +) \to (\mathbb{R}^{>0}, \times)$ there exists a positive real number a such that $f(q) = a^q$ for all $q \in \mathbb{Q}$.

25. Prove Proposition 1.9.17.

26. Prove that if two groups G and H are infinite and cyclic, then $G \cong H$. [Hint: See the proof of Proposition 1.9.25. Since G and H are infinite, you must prove separately that f is surjective and injective.]

27. Prove that $Z_2 \oplus Z_2 \oplus Z_2$ and Q_8 are not isomorphic.

28. Prove that D_6 and A_4 are not isomorphic.

29. Prove that D_{12} and S_4 are not isomorphic.

30. Prove that $(\mathbb{R}^*, \times)$ and $(\mathbb{C}^*, \times)$ are not isomorphic.

31. Prove or disprove that $U(7)$ and $U(9)$ are isomorphic.

32. Prove or disprove that $U(44)$ and $U(25)$ are isomorphic.

33. Let d be a divisor of a positive integer n. First prove that the function $\varphi : U(n) \to U(d)$ defined by $\varphi(\bar{a}) = \overline{a \bmod d}$ is a homomorphism and then determine the kernel and the image of φ.

34. Prove that $U(21) \cong Z_2 \oplus Z_6$.

35. Let G be the group of rigid motions of a cube. In Exercise 1.3.26 we saw that G has order 24. Prove that each rigid motion of the cube corresponds uniquely to a permutation of the 4 main diagonals through the cube. Conclude that $G \cong S_4$.

36. Let G be a group, let H be a subgroup, and let $g \in G$.
    (a) Prove that $gHg^{-1} = \{ghg^{-1} \mid h \in H\}$ is a subgroup of G.
    (b) Prove that the function $\varphi_g : H \to gHg^{-1}$ by $\varphi_g(h) = ghg^{-1}$ is an isomorphism between H and gHg^{-1}.
    [The subgroup gHg^{-1} is said to be a *conjugate* to H.]

37. Prove that $\text{Aut}(G)$ is a group with the operation of composition. Conclude that $\text{Aut}(G) \leq S_G$.

38. Let G be a group and let $g \in G$ be any group element. Define the function $\psi_g : G \to G$ by $\psi_g(x) = gxg^{-1}$.
    (a) Prove that ψ_g is an automorphism of G with inverse $(\psi_g)^{-1} = \psi_{g^{-1}}$.
    (b) Prove that the function $\Psi : G \to \text{Aut}(G)$ defined by $\Psi(g) = \psi_g$ is a homomorphism.

    [An automorphism of the form ψ_g is called an *inner automorphism*. The image of Ψ in $\text{Aut}(G)$ is called the *group of inner automorphisms* and is denoted $\text{Inn}(G)$.]

39. Prove that $\text{Aut}((\mathbb{Q}, +)) \cong (\mathbb{Q}^*, \times)$.

40. This exercise determines the automorphism group $\text{Aut}(Z_n)$. Suppose that Z_n is generated by the element z.
    (a) Prove that every homomorphism $\psi : Z_n \to Z_n$ is completely determined by where it sends the generator z.
    (b) Prove that every homomorphism $\psi : Z_n \to Z_n$ is of the form $\psi(g) = g^a$ for some integer a with $0 \leq a \leq n - 1$. For the scope of this exercise, denote by ψ_a the homomorphism such that $\psi_a(g) = g^a$.
    (c) Prove that $\psi_a \in \text{Aut}(Z_n)$ if and only if $\gcd(a, n) = 1$.
    (d) Show that the function $\Psi : U(n) \to \text{Aut}(Z_n)$ with $\Psi(\bar{a}) = \psi_a$ is an isomorphism to conclude that $U(n) \cong \text{Aut}(Z_n)$.

1.10   Group Presentations

Previous sections alluded to the desirable goal of having a consistent way to describe groups and their elements. Cayley's Theorem established that every finite group is isomorphic to a subgroup of some symmetric group S_n. The proof of Cayley's Theorem embeds G into $S_{|G|}$, which has size $|G|!$. Since $n!$ grows so much faster than n, this approach has some limitations. In this section, we introduce another approach to describing groups that gives a consistent notation.

Up to this point in the textbook, we used the notation e to refer to the identity element of an arbitrary group. We did this in order to avoid confusion. However, it is common in the literature to use the symbol 1 for the identity of an arbitrary group so we adopt this practice from now on. Of course, if we know that the group operation is some form of addition, then we refer to that identity as 0.

1.10.1   Generators and Relations

In Section 1.1.2, where we introduced the abstract notation for the dihedral group D_n, we saw that all dihedral symmetries can be obtained as various compositions of r (the rotation by angle $2\pi/n$) and s (the reflection through the x-axis). We also know that $r^n = 1$ because r has order n, and that $s^2 = 1$ because s has order 2. However, we also proved that r and s satisfy $sr = r^{-1}s$.

Since every element in D_n can be created from operations on r and s, we say that r and s *generate* D_n and that $r^n = 1$, $s^2 = 1$, and $sr = rs^{-1}$ are relations on r and s. In group theory, we write

$$D_n = \langle r, s \mid r^n = s^2 = 1, \ sr = r^{-1}s \rangle \tag{1.9}$$

to express that every element in D_n can be obtained by a finite number of repeated operations between r and s and that every algebraic relation between r and s can be deduced from the relations $r^n = 1$, $s^2 = 1$, and $sr = r^{-1}s$. The expression (1.9) is the standard presentation of D_n.

The relation $sr = r^{-1}s$ shows that in any term $s^k r^l$ it is possible to "move" the s to the left of the term by appropriately changing the power on r. Hence, in any word involving the generators r and s, it is possible, by appropriate changes on powers, to move all the powers of s to the right and all the powers of r to the left. Hence, every expression in the r and s can be rewritten as $r^l s^k$ with $0 \le k \le 1$ and $0 \le l \le n - 1$. Though we already knew $|D_n| = 2n$ from geometry, this reasoning shows that there are at most $2n$ terms in the group given by this presentation.

Definition 1.10.1

Let G be a group. A *presentation* of G is an expression

$$G = \langle g_1, g_2, \ldots, g_k \mid R_1 \ R_2 \ \cdots \ R_s \rangle$$

where each R_i is an equation in the elements $g_1, g_2, \ldots, g_k$. This means that every element in G can be obtained as a combination of operations on the *generators* $g_1, g_2, \ldots, g_k$ and that any relation between these elements can be deduced from the *relations* $R_1, R_2, \ldots, R_s$.

Example 1.10.2 (Cyclic Groups). The nth order cyclic group has the presentation

$$Z_n = \langle z \mid z^n = 1 \rangle.$$

The distinct elements are $\{1, z, z^2, \ldots, z^{n-1}\}$ and the operation is $z^a z^b = z^{a+b}$, where $z^n = 1$. △

Example 1.10.3. Consider an alternate presentation for D_5, namely $\langle a, b \mid a^2 = b^2 = 1, (ab)^5 = 1 \rangle$, where $a = rs$ and $b = s$. The relation $(ab)^5 = 1$ corresponds to $r^5 = 1$ and the relation $a^2 = 1$ is equivalent to $rsrs = 1$, which can be rewritten as $sr = r^{-1}s$. This shows that $D_5 = \langle a, b \mid a^2 = b^2 = 1, (ab)^5 = 1 \rangle$. △

This last example illustrates two important properties. First, Example 1.10.3 and the standard presentation for D_5 give two distinct presentations for the same group. Second, in the standard presentation of the dihedral group D_5, the relations $r^5 = 1$ and $s^2 = 1$ coupled with the fact that $|D_5| = 10$, may lead a reader to speculate that the order of a group is equal to the product of the order of the generators. This is generally not the case as Example 1.10.3 shows.

It is important to note that if we expect/want generators to commute, the presentation of the group must explicitly state this. So for example,

$$\langle a, b \mid a^2 = b^2 = 1, ab = ba \rangle$$

contains exactly, four elements $\{1, a, b, ab\}$ and is isomorphic to $Z_2 \oplus Z_2$. On the other hand,

$$\langle a, b \mid a^2 = b^2 = 1 \rangle \tag{1.10}$$

contains an infinite number of elements since there is no relation that allows us to simplify a word like $ababab$. The group in (1.10) is called the *infinite dihedral group* and is denoted D_∞.

In the examples given so far, we began with a well-defined group and gave a presentation of it. More importantly, we can define a group by a presentation. However, before providing examples, it is useful to introduce free groups. These are groups with certain generators but with no relations between them. Free groups possess interesting properties but force us to be precise in how we use symbols in abstract groups.

1.10.2 Free Groups

In the previous paragraphs, we transitioned subtly from a group having a presentation, to a group being defined by a presentation. This is a subtle difference but we must put the latter concept on a rigorous footing. In order to do so, we introduce free groups. The following discussion about strings and words of symbols feels rather technical, but it is the foundation for how students are taught to deal with symbols early in their algebra experience.

Let S be a set. We usually think of S as a set of symbols. A *word* in S consists either of the expression 1, called the *empty word*, or an expression of the form

$$w = s_1^{\alpha_1} s_2^{\alpha_2} \cdots s_m^{\alpha_m}, \tag{1.11}$$

where $m \in \mathbb{N}^*$, $s_i \in S$ not necessarily distinct and $\alpha_i \in \mathbb{Z} - \{0\}$ for $1 \leq i \leq m$. (The order of the s_i matters.) If the word is not the empty word 1, we call m the length of the word. For example, if $S = \{x, y, z\}$, then

$$x^2 y^{-4} z x^{13} x^{-2} y y y z^2, \quad y z z z y z x^{-2}, \quad y z^2 y^{-3} z$$

are examples of words in S. We call a word *reduced* if $s_{i+1} \neq s_i$ for all $1 \leq i \leq m - 1$. In the above examples of words from $\{x, y, z\}$, only the last word is reduced.

We define $F(S)$ as the set of all reduced words of S. We define the operation $\cdot$ of concatenation of reduced words by concatenating the expressions and then eliminating adjacent symbols with powers that collect or cancel. More precisely, for all $w \in F(S)$, $w \cdot 1 = 1 \cdot w = w$ and for two nonempty reduced words $a = s_1^{\alpha_1} s_2^{\alpha_2} \cdots s_m^{\alpha_m}$ and $b = t_1^{\beta_1} t_2^{\beta_2} \cdots t_n^{\beta_n}$, then the concatenation

$$(s_1^{\alpha_1} s_2^{\alpha_2} \cdots s_m^{\alpha_m}) \cdot (t_1^{\beta_1} t_2^{\beta_2} \cdots t_n^{\beta_n})$$

is as follows:

Case 1. The empty string 1 if $m = n$, $s_{m+1-i} = t_i$ and $\beta_i = -\alpha_{m+1-i}$ for all i with $1 \leq i \leq m$;

Case 2. The reduced word $s_1^{\alpha_1} s_2^{\alpha_2} \cdots s_{m-k}^{\alpha_{m-k}} t_{k+1}^{\beta_{k+1}} t_{k+2}^{\beta_{k+2}} \cdots t_n^{\beta_n}$ if $s_{m-k} \neq t_{k+1}$ and if $s_{m+1-i} = t_i$ and $\beta_i = -\alpha_{m+1-i}$ for all i with $1 \leq i \leq k$;

Case 3. The reduced word $s_1^{\alpha_1} s_2^{\alpha_2} \cdots s_{m-k}^{\alpha_{m-k}} t_k^{\alpha_{m+1-k}+\beta_k} t_{k+1}^{\beta_{k+1}} \cdots t_n^{\beta_n}$ if $s_{m-k} \neq t_{k+1}$ and if $s_{m+1-i} = t_i$ for all i with $1 \leq i \leq k$ and if $\beta_k \neq -\alpha_{m+1-k}$ but $\beta_i = -\alpha_{m+1-i}$ for all i with $1 \leq i \leq k - 1$.

We call k the *overlap* of the pair (a, b). A few examples of this concatenation-reduction are

$$(x^3 y^{-2}) \cdot (x^{10} y^2) = x^3 y^{-2} x^{10} y^2 \quad \text{case 2 with } k = 0,$$
$$(x y^2 z^{-3}) \cdot (z^3 y x^{-1}) = x y^3 x^{-1} \quad \text{case 3 with } k = 1,$$
$$(x y^3 x^{-2} z) \cdot (z^{-1} x^2 y^{-3} z y^2) = x z y^2 \quad \text{case 2 with } k = 3,$$
$$(z y x^{-2}) \cdot (x^2 y^{-1} z^{-1}) = 1 \quad \text{case 1.}$$

Theorem 1.10.4

The operation $\cdot$ of concatenation-reduction on $F(S)$ is a binary operation and the pair $(F(S), \cdot)$ is a group with identity 1, and for $w \neq 1$ expressed as in (1.11),

$$w^{-1} = s_m^{-\alpha_m} \cdots s_2^{-\alpha_2} s_1^{-\alpha_1}.$$

Proof. In all three cases for the definition of concatenation of reduced words, the resulting word is such that no successive symbol is equal. Hence, the word is reduced and $\cdot$ is a binary operation on $F(S)$.

That the empty word is the identity is built into the definition of concatenation. Furthermore, Case 1 establishes that $(s_1^{\alpha_1} s_2^{\alpha_2} \cdots s_m^{\alpha_m})^{-1} = s_m^{-\alpha_m} \cdots s_2^{-\alpha_2} s_1^{-\alpha_1}$.

The difficulty of the proof resides in proving that concatenation is associative. For the rest of this proof, we denote the length of a word $w \in F(S)$ as $L(w)$. Let a, b, and c be three arbitrary reduced words in $F(S)$. Let h be the overlap of (a, b) and let k be the overlap of (b, c).

First, suppose that $h + k < L(b)$. Then we can write $a = a'a''$, $b = b'b''b'''$, and $c = c'c''$, where $a''b'$ are the symbols reduced out or reduced to a word of length 1 in the concatenation $a \cdot b$, and suppose that $b'''c'$ are the symbols reduced out or reduced to a word of length 1 in the concatenation $b \cdot c$. We write $a \cdot b = a'[a''b']b''b'''$, where $[a''b']$ stands for removed or reduced to a word of length 1 depending on Case 2 or Case 3 of the concatenation-reduction. Then, as reduced words

$$(a \cdot b) \cdot c = (a'[a''b']b''b''') \cdot c = a'[a''b']b''[b'''c']c'' = a \cdot (b'b''[b'''c']c'') = a \cdot (b \cdot c).$$

Suppose next that $h + k = L(b)$. We write $a = a'a''$, $b = b'b''$ and $c = c'c''$ so that $a \cdot b = a'[a''b']b''$ and $b \cdot c = b'[b''c']c''$. Then

$$(a \cdot b) \cdot c = (a'[a''b']b'') \cdot c = (a'[a''b']) \cdot ([b''c']c'') = a \cdot (b'[b''c']c'') = a \cdot (b \cdot c).$$

Finally, suppose that $h + k > L(b)$. Now we subdivide each reduced word into three parts as $a = a'a''a'''$, $b = b'b''b'''$, and $c = c'c''c'''$ where

$$a \cdot b = a'[a''a'''b'b'']b''' \qquad \text{and} \qquad b \cdot c = b'[b''b'''c'c'']c'''.$$

Now any of these subwords can be the empty word. However, in order for the reductions to occur as these subwords are defined, a few relations must hold. For the lengths of the subwords, we must have

$$L(b'') = L(a'') = L(c'') = h + k - L(b),$$
$$L(a''') = L(b') = h - L(b''),$$
$$L(c') = L(b''') = k - L(b'').$$

Furthermore, we must have $b' = (a''')^{-1}$ and $c' = (b''')^{-1}$ and

$$a'' = s_1^{\alpha_1} s_2^{\alpha_2} \cdots s_{m-1}^{\alpha_{m-1}} s_m^{\alpha_m}, \qquad b'' = s_m^{-\alpha_m} s_{m-1}^{-\alpha_{m-1}} \cdots s_2^{-\alpha_2} s_1^{\beta},$$
$$c'' = s_1^{-\beta} s_2^{\alpha_2} \cdots s_{m-1}^{\alpha_{m-1}} s_m^{\gamma}.$$

Then we can calculate that

$$\begin{aligned}
(a \cdot b) \cdot c &= (a's^{\alpha_1 + \beta} b''') \cdot c = (a's^{\alpha_1 + \beta} b''') \cdot (c's_1^{-\beta} s_2^{\alpha_2} \cdots s_{m-1}^{\alpha_{m-1}} s_m^{\gamma} c''') \\
&= a's_1^{\alpha_1} s_2^{\alpha_2} \cdots s_{m-1}^{\alpha_{m-1}} s_m^{\gamma} c''' \\
&= (a's_1^{\alpha_1} s_2^{\alpha_2} \cdots s_{m-1}^{\alpha_{m-1}} s_m^{\alpha_m} a''') \cdot (b's_m^{-\alpha_m + \gamma} c''') = a \cdot (b's_m^{-\alpha_m + \gamma} c''') \\
&= a \cdot (b \cdot c).
\end{aligned}$$

In all three possible cases of combinations of overlap, associativity holds.

We conclude that $(F(S), \cdot)$ is a group. $\qquad\qquad\qquad\qquad\qquad\square$

Definition 1.10.5

Let S be a set of symbols. The set $F(S)$ equipped with the operation $\cdot$ of concatenation, is called the *free group* on S. The cardinality of S is called the *rank* of the free group $F(S)$.

The term "free" refers to the fact that its generators do not have any relations among them besides the relations imposed by power rules,

$$x^\alpha x^\beta = x^{\alpha+\beta}, \qquad \text{for all } x \in G, \text{ and all } \alpha, \beta \in \mathbb{Z}. \qquad (1.12)$$

In fact, we defined the concatenation operation on reduced words as we did in order to satisfy (1.12).

All free groups are infinite because each symbol in a reduced word can carry powers of any nonzero integer.

1.10.3 Defining Groups from a Presentation

A group defined by a presentation is similar to a free group in that it is first understood through its symbols, rather than the symbols representing some function, matrix, or number. In a group defined by a presentation, the elements are simply reduced words in the generators but the relations impose additional simplifications beyond just the power rules that hold in any group.

Example 1.10.6. To illustrate similarities and differences in various sets of relations, consider the following three groups.

$$G_1 = \langle x, y \,|\, x^3 = y^7 = 1, \ xy = yx \rangle,$$
$$G_2 = \langle a, b \,|\, a^3 = b^7 = 1, \ ab = b^2a \rangle,$$
$$G_3 = \langle u, v \,|\, u^3 = v^7 = 1, \ uv = v^2u^2 \rangle.$$

In G_1, since $xy = yx$, in any word in the generators x and y, all the x symbols can be moved to the left. Thus, all elements in G_1 can be written as $x^k y^\ell$. Furthermore, since $x^3 = y^7 = 1$, then $x^i y^j$ with $0 \le i \le 2$ and $0 \le j \le 6$ give all the elements of G_1. We claim that all 21 of these elements are distinct. To prove this, we must show that $x^i y^j$ are distinct for $0 \le i \le 2$ and $0 \le j \le 6$. If $x^k y^\ell = x^m y^n$, we have

$$x^k y^\ell = x^m y^n \iff x^{k-m} = y^{n-\ell}.$$

By Corollary 1.3.7, since $x^3 = 1$, the order of x^{k-m} divides 3 and, since $y^7 = 1$, the order of $y^{n-\ell}$ divides 7. Since $x^{k-m} = y^{n-\ell}$, then the order of this element must divide $\gcd(3,7) = 1$. Hence, $x^{k-m} = y^{n-\ell} = 1$. Thus, 3 divides $k - m$

and 7 divides $n - \ell$, but if we assume that $0 \le k, m \le 2$ and $0 \le \ell, n \le 6$, then we conclude that $k = m$ and $n = \ell$. This proves the claim. Hence, G_1 is a group of order 21 in which the elements operate as $(x^k y^l)(x^m y^n) = x^{k+m} y^{l+n}$. It is easy to see that $G_1 \cong Z_3 \oplus Z_7$ and by Exercise 1.9.18, we deduce that $G_1 \cong Z_{21}$.

In G_2, from the relation $ab = b^2 a$, we see that all the a symbols may be moved to the right of any b symbols, though possibly changing the power on b. In particular,

$$a^n b = a^{n-1} b^2 a = a^{n-2} (ab) ba = a^{n-2} b^2 aba = a^{n-2} b^4 a^2 = \cdots = b^{2^n} a^n,$$

and also

$$a^n b^k = b^{2^n} a^n b^{k-1} = b^{2^n} b^{2^n} a^n b^{k-2} = \cdots = b^{k2^n} a^n.$$

Thus, every element in G can be written as $b^m a^n$. Also, since $a^3 = b^7 = 1$, then $b^i a^j$ with $0 \le i \le 6$ and $0 \le j \le 2$ give all the elements of the group. The same reasoning used for G_1 shows that all 21 of these elements are distinct. Hence, G_2 is a group of order 21 but in which the group elements operate according to

$$(b^k a^\ell)(b^m a^n) = b^{k + m 2^\ell} a^{\ell + n}.$$

In G_3, the relation $uv = v^2 u^2$ does not readily show that in every expression of u and v, the v's can be moved to the left of the u's. Consider the following equalities coming from the relation

$$u^2 v = u(uv) = uv^2 u^2 = (uv)vu^2 = v^2 u^2 vu^2$$
$$= v^2 uv^2 u^4 = v^4 u^2 vu^4 = v^4 uv^2 u^6 = v^6 u^2 vu^6.$$

Now $u^6 = 1$ so we have $u^2 v = v^6 u^2 v$ which leads to $v^6 = 1$. By Proposition 1.3.6, we conclude that $v = 1$. Then the relation $uv = v^2 u^2$ becomes $u = u^2$, which implies that $u = 1$. Hence, we conclude that G_3 is the trivial group $G_3 = \{1\}$.

We should not surmise that a group generated by elements x and y that satisfy $x^3 = 1$ and $y^7 = 1$ must have an order less than 21. Exercise 1.7.11 with $p = 7$ gives a group of order 2520 obtained from two generators of order 3 and 7, respectively. $\triangle$

The presentation of G_3 in Example 1.10.6 shows that the combination of relations in a presentation can lead to what we sometimes loosely call *implicit relations*. By this, we mean that the group could be described using much simpler relations.

The following example offers a more extreme example of the size of a group in relation to the orders of the generators.

Example 1.10.7. In (1.10) we encountered the infinite dihedral group D_∞. Using the presentation given there, the element ab has infinite order. If we set $x = a$ and $z = ab$, then clearly $xz = b$, so D_∞ can also be generated by the set

$\{x, z\}$. However, $xz = b = b^{-1} = (xz)^{-1} = z^{-1}x$. Thus another presentation for D_∞ is

$$D_\infty = \langle x, z \mid x^2 = 1,\ xz = z^{-1}x \rangle.$$

This presentation resembles the standard presentation of a dihedral group, except that z (which is similar to r) has infinite order. △

Example 1.10.8 (Direct Sums of Cyclic Groups). It is easy to find a presentation for direct sums of cyclic groups. As a simple example,

$$Z_{10} \oplus Z_5 \cong \langle x, y \mid x^{10} = y^5 = 1,\ xy = yx \rangle.$$

In any presentation, if the only relations on the generators are relations associated with the orders of each generator, relations of commutation among all generators, then the group is a direct sum of cyclic groups. △

It is always possible to find a presentation of any finite group. As an extreme approach, we can take the set of generators to be the set of all elements in the group and the set of relations as all the calculations in the Cayley table. More often than not, however, we are interested in describing the group with a small list of generators and relations. Depending on the group, it may be a challenging problem to find a minimal generating subset.

A profound result that illustrates the complexity of working with generators and relations, the Novikov-Boone Theorem [26, 7] states that in the context of a given presentation, given two words w_1 and w_2, there exists no algorithm to decide if $w_1 = w_2$. With certain specific relations, it may be possible to decide if two words are equal. For example, with the dihedral group, there is an algorithm to reduce any word to one in a complete list of distinct words.

Remark 1.10.9. Some authors write the relations in a presentation as words w_i and understand the relation to be $w_i = 1$. For example, in this habit of notation, we express the commuting relation $xy = yx$ as $xyx^{-1}y^{-1} = 1$. The presentation for $Z_{10} \oplus Z_5$ as given in Example 1.10.8 would be

$$Z_{10} \oplus Z_5 \cong \langle x, y \mid x^{10},\ y^5,\ xyx^{-1}y^{-1} \rangle.$$ △

1.10.4 Presentations and Homomorphisms

Suppose that G has a presentation $\langle g_1, g_2, \ldots, g_k \mid R_1\ R_2\ \cdots\ R_s \rangle$. Every element $w \in G$ is a word in the generators, $w = u_1^{\alpha_1} u_2^{\alpha_2} \cdots u_\ell^{\alpha_\ell}$, with $u_i \in \{g_1, g_2, \ldots, g_k\}$, so for a homomorphism $\varphi : G \to H$ we have

$$\varphi(w) = \varphi(u_1)^{\alpha_1} \varphi(u_2)^{\alpha_2} \cdots \varphi(u_\ell)^{\alpha_\ell}.$$

Hence, φ is entirely determined by the values of $\varphi(g_1), \varphi(g_2), \ldots, \varphi(g_k)$.

When trying to construct a homomorphism from G to a group H, it is not possible to associate arbitrary elements in H to the generators of G and always obtain a homomorphism. The following theorem makes this precise.

Theorem 1.10.10 (Extension Theorem on Generators)

Let $G = \langle g_1, g_2, \ldots, g_k \mid R_1 \quad R_2 \quad \cdots \quad R_s \rangle$ and let $h_1, h_2, \ldots, h_k \in H$ be elements that satisfy the relations R_i as the generators of G when replacing g_i with h_i for $i = 1, 2, \ldots, k$. Then the function $\{g_1, g_2, \ldots, g_k\} \to H$ that maps $g_i \mapsto h_i$ for $i = 1, 2, \ldots, k$ can be extended to a unique homomorphism $\varphi : G \to H$ that has $\varphi(g_i) = h_i$.

Proof. We define the function $\varphi : G \to H$ by $\varphi(g_i) = h_i$ for $i = 1, 2, \ldots, k$ and for each element $g \in G$, if $g = u_1^{\alpha_1} u_2^{\alpha_2} \cdots u_\ell^{\alpha_\ell}$ with $u_j \in \{g_1, g_2, \ldots, g_k\}$, then

$$\varphi(g) \overset{\text{def}}{=} \varphi(u_1)^{\alpha_1} \varphi(u_2)^{\alpha_2} \cdots \varphi(u_\ell)^{\alpha_\ell}.$$

By construction, φ satisfies the homomorphism property $\varphi(xy) = \varphi(x)\varphi(y)$ for all $x, y \in G$. However, since different words can be equal, we have not yet determined if φ is a well-defined function.

Two words v and w in the generators $g_1, g_2, \ldots, g_k$ are equal if and only if there is a finite sequence of words $w_1, w_2, \ldots, w_n$ such that $v = w_1$, $w = w_n$, and w_i to w_{i+1} are related to each other by either one application of a power rule (as given in Proposition 1.2.12) or one application of a relation R_j. Since the elements $h_1, h_2, \ldots, h_k \in H$ satisfy the same relations $R_1, R_2, \ldots, R_s$ as $g_1, g_2, \ldots, g_k$, then the same equalities apply between the words $\varphi(w_i)$ and $\varphi(w_{i+1})$ as between w_i and w_{i+1}. This establishes the chain of equalities

$$\varphi(v) = \varphi(w_1) = \varphi(w_2) = \cdots = \varphi(w_n) = \varphi(w).$$

Hence, if $v = w$ are words in G, then $\varphi(v) = \varphi(w)$. Thus, φ is a well-defined function and hence is a homomorphism. $\qquad\square$

Example 1.10.11. We use Theorem 1.10.10 to prove that $\langle (1\,2\,3)(4\,5), (1\,2) \rangle$ in S_5 is isomorphic to D_6. We set up a function from $\{r, s\}$, the standard generators of D_6 to S_5 by

$$r \mapsto (1\,2\,3)(4\,5) \qquad \text{and} \qquad s \mapsto (1\,2).$$

Obviously $r^6 = 1$ and $((1\,2\,3)(4\,5))^6 = 1$ while $s^2 = 1$ and $(1\,2)^2 = 1$. In D_6, we also have the relation $rs = sr^{-1}$ while in S_5,

$$(1\,2\,3)(4\,5)(1\,2) = (1\,3)(4\,5) \qquad \text{and}$$

$$(1\,2)\,((1\,2\,3)(4\,5))^{-1} = (1\,2)(1\,3\,2)(4\,5) = (1\,3)(4\,5).$$

Thus, $(1\,2\,3)(4\,5)$ and $(1\,2)$ satisfy the same relations as r and s. Hence, by Theorem 1.10.10, this mapping on generators extends to a homomorphism $\varphi : D_6 \to S_5$. Obviously, $\varphi(D_6) = \langle (1\,2\,3)(4\,5), (1\,2) \rangle$. However, it is not hard to verify that $\langle (1\,2\,3)(4\,5), (1\,2) \rangle$ consists of exactly 12 elements. Hence, φ is injective and by Exercise 1.9.23, $D_6 \cong \varphi(D_6) = \langle (1\,2\,3)(4\,5), (1\,2) \rangle$. $\qquad\triangle$

1.10.5 Useful CAS Commands

We saw earlier that the constructor *Group* in *Maple*'s *GroupTheory* package allows us to define permutation groups. However, this same constructor also allows to define a group using generators and relations. The constructor uses the alternate method to write relations as described in Remark 1.10.9.

—————————————————— Maple ——————————————————

$with(GroupTheory)$:

$G2 := Group \left((a, b, \left[[a, a, a], [b, b, b, b, b, b, b], \left[a, b, a, \dfrac{1}{b}, \dfrac{1}{b} \right] \right] \right)$:

$GroupOrder(G2)$;

$$21$$

$IsAbelian(G2)$;

$$false$$

$H := Subgroup([[a, b]], G2)$:

$GroupOrder(H)$;

$$3$$

———————————————————————————————————————

The above code shows how to use the *Group* constructor to define the group G_2 from Example 1.10.6. The relation $ab = b^2a$ is equivalent to $aba^{-1}b^{-2} = 1$, which gives the third relation. This code also illustrates the *Subgroup* constructor. Here we constructed $H = \langle ab \rangle$ and subsequently had *Maple* calculate the order of this subgroup. If G is a group defined by a presentation, the command `Simplify(G)` attempts to simplify the presentation.

The following code does the same thing in SAGE.

—————————————————— Sage ——————————————————

```
sage: F.<a,b>=FreeGroup()
sage: G=F/[a^3,b^7,a*b*a^(-1)*b^(-2)]
sage: G.order()
21
sage: H=G.subgroup([a*b])
sage: H.order()
3
```

———————————————————————————————————————

Though the notation seems strange at first, we will be able to explain it better once we have encountered the concept of quotient groups.

1.10.6   Cayley Graph of a Presentation

The Cayley graph of a presentation of a group is another visual tool to understand the internal structure of a group, particularly a group of small order. The vertices of the Cayley graph are the elements of the group G and the edges are pairs $\{x, y\}$ if there is a generator g such that $y = gx$. One variant of the Cayley graph colors the edges accordingly to distinguish which generator corresponds to which edge. Yet another variant is a directed graph that places an arrow from x to y if there is a generator g such that $y = gx$.

It is important to note that Cayley graph depends on the set of generators in the presentation. So if $G = \langle S \rangle = \langle S' \rangle$, where S and S' are different subsets of G, the set of vertices will be the same, corresponding to elements of G, but the edges of the graph will be different.

As an example, it is not hard to show that $S_4 = \langle (1\,2\,3), (1\,2\,3\,4) \rangle$. Figure 1.11 shows the Cayley graph for S_4 using these generators. The double edges correspond to left multiplication by $(1\,2\,3)$ and the single edges to left multiplication by $(1\,2\,3\,4)$. This Cayley graph has the adjacency structure of the Archimedean solid named a rhombicuboctahedron.

EXERCISES FOR SECTION 1.10

1. Find a set of generators and relations for S_3. Prove that $D_3 \cong S_3$.

2. Find a set of generators and relations for the group $Z_2 \oplus Z_2 \oplus Z_2$.

3. Prove that in any group, given any two elements $x, y \in G$, the subgroups $\langle x, y \rangle$ and $\langle x, xy \rangle$ are isomorphic.

4. Consider the group presentation $G = \langle x, y \,|\, x^4 = y^3 = 1, \ xy = y^2 x^2 \rangle$. Prove that G is the trivial group $G = \{1\}$.

5. Prove that $\langle i, j \,|\, i^4 = j^4 = 1, \ i^2 = j^2, \ ij = j^3 i \rangle$ is a presentation of the quaternion group Q_8. In particular, show how to write -1, $-i$, $-j$, k, and $-k$ as operations on the generators i and j.

6. Show that the group $G = \langle x, y \,|\, x^5 = 1, \ y^2 = 1, \ x^2 yxy = 1 \rangle$ is isomorphic to Z_2.

7. Prove that D_7 and $\langle x, y \,|\, x^7 = y^3 = 1, \ yx = x^2 y \rangle$ have the same lattice structure.

8. Let $G = \langle a, b \,|\, a^6 = b^7 = 1, \ ab = b^3 a \rangle$.

   (a) Prove that every element in G can be written uniquely as $b^m a^n$ for $0 \le m \le 6$ and $0 \le n \le 5$.

   (b) Write the element $a^4 b^2 a^{-2} b^5$ in the form $b^m a^n$.

   (c) Find the order of ab and list all the powers $(ab)^i$.

9. The *quasidihedral group* of order 16 is defined by

$$QD_{16} = \langle a, b \,|\, a^8 = b^2 = 1, \ ab = ba^3 \rangle.$$

   Show that QD_{16} has 16 elements and then draw the subgroup lattice of QD_{16}.

10. Prove that if all nonidentity elements in a finite group G have order 2, then there exists a nonnegative integer n such that

$$G = \langle a_1, a_2, \ldots, a_n \,|\, a_i^2 = 1 \ a_i a_j = a_j a_i \text{ for all } i, j = 1, 2, \ldots, n \rangle.$$

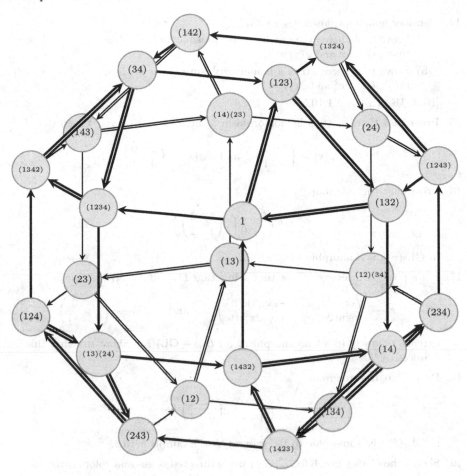

FIGURE 1.11: Cayley graph for $S_4 = \langle (1\,2\,3), (1\,2\,3\,4) \rangle$.

$$\overbrace{\qquad}^{n \text{ times}}$$

Deduce that $G \cong Z_2^n = \overbrace{Z_2 \oplus Z_2 \oplus \cdots \oplus Z_2}^{n \text{ times}}$. [Hint: Use induction on the minimum number of generators of G.]

11. Consider the infinite dihedral group as presented in (1.10).

    (a) Show that $b(ab)^n = (ab)^{-n-1}a$ for all integers n.

    (b) Conclude that every element in D_∞ can be written as $(ab)^n$ or $(ab)^n x$ for some $n \in \mathbb{Z}$.

12. Prove that the subgroup $\langle (1\,2\,3\,4), (2\,4) \rangle$ in S_4 is isomorphic to D_4. [Hint: See Example 1.9.28.]

13. We work in the group S_5.

    (a) Find a subgroup of S_5 isomorphic to D_5.

    (b) Prove that S_5 does not have any subgroups isomorphic to D_n for $n \geq 7$.

14. Consider homomorphisms $Q_8 \to Z_4$.

   (a) Prove that there exists no homomorphism $\varphi : Q_8 \to Z_4 = \langle z \mid z^4 = 1 \rangle$ such that $\varphi(i) = z$ and $\varphi(j) = 1$.

   (b) Prove that there exists a homomorphism $\psi : Q_8 \to Z_4 = \langle z \mid z^4 = 1 \rangle$ such that $\varphi(i) = z^2$ and $\varphi(j) = 1$.

   [Hint: Use Exercise 1.10.5.]

15. Prove that there exists a homomorphism from $\varphi : D_3 \to GL_2(\mathbb{F}_3)$ such that

$$\varphi(r) = \begin{pmatrix} 1 & 1 \\ 0 & 1 \end{pmatrix} \quad \text{and} \quad \varphi(s) = \begin{pmatrix} 2 & 0 \\ 0 & 1 \end{pmatrix}.$$

16. Prove that the subgroup

$$\left\langle \begin{pmatrix} 0 & 2 \\ 1 & 0 \end{pmatrix}, \begin{pmatrix} 1 & 1 \\ 1 & 2 \end{pmatrix} \right\rangle$$

in $GL_2(\mathbb{F}_3)$ is isomorphic to Q_8.

17. Fix a positive integer n. Show that a function $D_n \to GL_2(\mathbb{R})$ that maps

$$r \longmapsto \begin{pmatrix} \cos(2\pi/n) & -\sin(2\pi/n) \\ \sin(2\pi/n) & \cos(2\pi/n) \end{pmatrix} \quad \text{and} \quad s \longmapsto \begin{pmatrix} 1 & 0 \\ 0 & -1 \end{pmatrix}$$

extends uniquely to a homomorphism $\varphi : D_n \to GL_2(\mathbb{R})$. Show also that this φ is injective.

18. Prove that the subgroup

$$\left\langle \begin{pmatrix} 2 & 0 \\ 0 & 1 \end{pmatrix}, \begin{pmatrix} 1 & 1 \\ 0 & 1 \end{pmatrix} \right\rangle$$

in $GL_2(\mathbb{F}_7)$ is isomorphic to the group G_2 in Example 1.10.6.

19. Sketch the Cayley graph for $Z_6 \oplus Z_2$ using directed edges and colors corresponding to generators.

20. Sketch the Cayley graph of D_4 using the generating set $\{s, sr\}$.

21. Show that A_4 is generated by $\{(1\,2\,3), (1\,2)(3\,4)\}$ and then sketch the Cayley graph of A_4 using these generators.

22. Decide whether some group has a tetrahedron as its Cayley graph for some set of generators.

23. Let S be a set and let $F(S)$ be the free group on S. Prove that two reduced words $w_1 = s_1^{\alpha_1} s_2^{\alpha_2} \cdot s_m^{\alpha_m}$ and $w_2 = t_1^{\beta_1} t_2^{\beta_2} \cdots t_n^{\beta_n}$ are equal in $F(S)$ if and only if $m = n$, $s_i = t_i$ for all $1 \le i \le n$ and $\alpha_i = \beta_i$ for all $1 \le i \le n$. (This was not obvious from the definition of reduced words.)

For Exercises 1.10.24 and 1.10.25, refer to the exercise block beginning with Exercise 1.9.37 for a definition of the automorphism group of a group.

24. Prove that $\text{Aut}(Z_2 \oplus Z_2) \cong S_3$. [Hint: Use the presentation $Z_2 \oplus Z_2 = \langle x, y \mid x^2 = y^2 = 1, xy = yx \rangle$.]

25. This exercise guides the reader to prove that $\text{Aut}(Q_8) \cong S_4$.

FIGURE 1.12: The *Quintrino* sculpture (Courtesy of Bathsheba Grossman).

(a) Let φ be an arbitrary automorphism of $\mathrm{Aut}(Q_8)$. Prove that (i) $\varphi(i) \in \{i, -i, j, -j, k, -k\}$; (ii) $\varphi(-1) = -1$; (iii) $\varphi(j) \in \{i, -i, j, -j, k, -k\} - \{\varphi(i), \varphi(-i)\}$; (iv) all values of $\varphi(x)$ are determined by $\varphi(i)$ and $\varphi(j)$.

(b) Prove that all functions $\varphi : \{i, j\} \to Q_8$ that are allowed by the above conditions extend to valid automorphisms. Conclude that $|\mathrm{Aut}(Q_8)| = 24$. [Hint: Use the presentation given in Exercise 1.10.5.]

(c) By labeling the faces of a cube with $\{i, -i, j, -j, k, -k\}$ where x and $-x$ are opposite faces, show that $\mathrm{Aut}(Q_8)$ is isomorphic to the group of rigid motions of a cube.

(d) Use Exercise 1.9.35 to conclude that $\mathrm{Aut}(Q_8) \cong S_4$.

26. Consider the sculpture entitled *Quintrino* depicted in Figure 1.12 and let G be its group of symmetry.

 (a) Show that $|G| = 60$.

 (b) Show that G can be generated by two elements σ and τ.

 (c) Show that G can be viewed as a subgroup of S_{12} by writing σ and τ explicitly as elements in S_{12}.

 (d) (*) Show that G is isomorphic to A_5.

[Other sculptures by the same artist can be found at http://bathsheba.com.]

1.11   Groups in Geometry

Groups arise naturally in many areas of mathematics. Geometry in particular offers many examples of groups. The dihedral group, which we introduced in Section 1.1 as a motivation for groups, comes from geometry. However, there are countless connections between group theory and geometry, ranging from generalizations of dihedral groups (reflection groups, e.g., [20]) to applications in advanced differential geometry and topology. In fact, group theory became so foundational to geometry that in 1872, Felix Klein proposed the Erlangen program: to classify all geometries using projective geometry and groups of allowed transformations.

In this section, we introduce a few instances in which groups arise in geometry in an elementary way. This section only offers a glimpse into these topics.

1.11.1   The Groups of Isometries

In the real plane $\mathbb{R}^2$, the Euclidean distance between two points $P = (x_P, y_P)$ and $Q = (x_Q, y_Q)$ is

$$d(P, Q) = \sqrt{(x_Q - x_P)^2 + (y_Q - y_P)^2}.$$

The plane equipped with this distance function is called the *Euclidean plane*. More generally, the Euclidean n-space is the set $\mathbb{R}^n$ equipped with the Euclidean distance function

$$d(\vec{x}, \vec{y}) = \sqrt{\sum_{i=1}^{n} (y_i - x_i)^2}.$$

Definition 1.11.1

An *isometry* of Euclidean space is a function $f : \mathbb{R}^n \to \mathbb{R}^n$ that preserves the distance, namely f satisfies

$$d(f(\vec{x}), f(\vec{y})) = d(\vec{x}, \vec{y}) \qquad \text{for all } \vec{x}, \vec{y} \in \mathbb{R}^n. \tag{1.13}$$

The Greek etymology of isometry is "same measure." It is possible to broaden the concept of Euclidean distance to more general notions of distance. This is formalized by metric spaces.

Definition 1.11.2

A *metric space* is a pair (X, d) where X is a set and d is a function $d : X \times X \to \mathbb{R}^{\geq 0}$ satisfying

(1) (identity of equal elements) $d(x, y) = 0$ if and only if $x = y$;

(2) (symmetry) $d(x, y) = d(y, x)$ for all $x, y \in X$;

(3) (triangle inequality) $d(x, y) + d(y, z) \geq d(x, z)$ for all $x, y, z \in X$.

More generally than Definition 1.11.1, an isometry between metric spaces (X, d) and (Y, d') is a bijection $f : X \to Y$ that satisfies (1.13) for all $x, y \in X$. However, for the rest of this section, we only discuss isometries of Euclidean spaces.

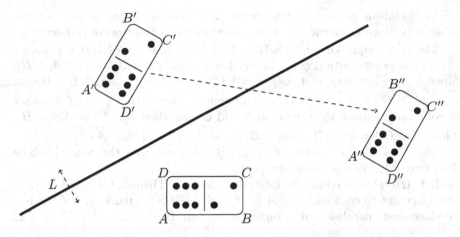

FIGURE 1.13: An reflection through a line L followed by a translation.

Some examples of isometries in the plane include a rotation about some fixed point A by an angle α and reflection about a line L. Many others exist. Figure 1.13 shows an isometry obtained by reflecting about a line L followed by a translation. However, without knowing all isometries, we can nonetheless establish the following proposition.

Proposition 1.11.3

The set of isometries on Euclidean space $\mathbb{R}^n$ is a group under composition.

The following properties of isometries are helpful for the proof of Proposition 1.11.3.

Proposition 1.11.4

Let f be an isometry of Euclidean space. For arbitrary points $A, B, C \in \mathbb{R}^n$ we denote by $A' = f(A)$, $B' = f(B)$, and $C' = f(C)$.

(1) f preserves betweenness;

(2) f preserves collinearity;

(3) if $\overrightarrow{AC} = \lambda \overrightarrow{AB}$, then $\overrightarrow{A'C'} = \lambda \overrightarrow{A'B'}$;

(4) $\triangle A'B'C'$ is congruent to $\triangle ABC$;

(5) f maps parallelograms to nondegenerate parallelograms.

Proof. A point B is said to be *between* A and C if $d(A,B) + d(B,C) = d(A,C)$. Since an isometry preserves the distance, then $d(A',B') + d(B',C') = d(A',C')$. Hence, B' is between A' and C'.

In Euclidean geometry, three points are collinear if one is between the other two. Since isometries preserve betweenness, they preserve collinearity.

The vector equation $\overrightarrow{AC} = \lambda\overrightarrow{AB}$ with $\lambda \geq 0$ holds when $\{A, B, C\}$ is a set of collinear points, with A not between B and C and when $d(A, C) = \lambda d(A, B)$. Since f is an isometry, $d(A', C') = d(A, C) = \lambda d(A, B) = \lambda d(A', B')$. Hence, $\overrightarrow{A'C'} = \lambda\overrightarrow{A'B'}$. When $\lambda \leq 0$, this vector equality means that $\{A, B, C\}$ is a set of collinear points, with A between B and C and when $d(A, C) = |\lambda|d(A, B)$. Then $d(A', C') = d(A, C) = \lambda d(A, B) = \lambda d(A', B')$ and thus $\overrightarrow{A'C'} = \lambda\overrightarrow{A'B'}$.

That $\triangle A'B'C'$ is congruent to $\triangle ABC$ follows from the Side-Side-Side Theorem of Euclidean Geometry.

Let $ABCD$ be a nondegenerate parallelogram. Then $\triangle ABD$ and $\triangle BDC$ are congruent to each and to $\triangle A'B'D'$ and $\triangle B'D'C'$. Hence, $A'B'C'D'$ is a nondegenerate parallelogram, congruent to $ABCD$. $\qquad\square$

Proof (of Proposition 1.11.3). We provide the proof for the Euclidean plane ($n = 2$) but the proof generalizes to arbitrary n.

We first must check that function composition is a binary operation on the set of isometries of $\mathbb{R}^n$. Let f, g be two isometries of $\mathbb{R}^n$. Then for any $P, Q \in \mathbb{R}^n$,

$$
\begin{aligned}
d\left((f \circ g)(P), (f \circ g)(Q)\right) &= d\left(f(g(P)), f(g(Q))\right) \\
&= d(g(P), g(Q)) \qquad \text{since } f \text{ is an isometry} \\
&= d(P, Q) \qquad \text{since } g \text{ is an isometry.}
\end{aligned}
$$

Thus, $f \circ g$ is an isometry.

Function composition is always associative (Proposition A.2.12).

The identity function id : $\mathbb{R}^n \to \mathbb{R}^n$ is an isometry and satisfies the group axioms for an identity element.

In order to show that the set of isometries is closed under taking inverses, we need to show that an arbitrary isometry f is a bijection and that the inverse function is again an isometry. Suppose that $f(P) = f(Q)$ for two points $P, Q \in \mathbb{R}^2$. Then $d(f(P), f(Q)) = 0$. Since f is an isometry, then $d(P, Q) = 0$ and hence $P = Q$. This shows that every isometry is injective.

Establishing that an isometry is a surjection requires the most work. Let O be a point in the domain and let $A_1, A_2, \ldots, A_n$ be points so that $\overrightarrow{OA_1}, \overrightarrow{OA_2}, \ldots, \overrightarrow{OA_n}$ form a basis of $\mathbb{R}^n$. Let $O' = f(O)$ and $A'_i = f(A_i)$. By Proposition 1.11.4(5) and an induction argument, we deduce that $\overrightarrow{O'A'_1}, \overrightarrow{O'A'_2}, \ldots, \overrightarrow{O'A'_n}$ is a basis of the codomain. Let Q be an arbitrary point in $\mathbb{R}^n$. There exist real numbers $\lambda_1, \lambda_2, \ldots, \lambda_n$ such that

$$
\overrightarrow{O'Q} = \lambda_1\overrightarrow{O'A'_1} + \lambda_2\overrightarrow{O'A'_2} + \cdots + \lambda_n\overrightarrow{O'A'_n}.
$$

We define a sequence of points $P_1, P_2, \ldots, P_n$ by $\overrightarrow{OP_1} = \lambda_1\overrightarrow{OA_1}$ and

$$
\overrightarrow{OP_i} = \overrightarrow{OP_{i-1}} + \lambda_i\overrightarrow{OA_i} \qquad \text{for } 1 < i \leq n.
$$

This is equivalent to $\overrightarrow{P_{i-1}P_i} = \lambda_i \overrightarrow{OA_i}$. Denote $P = P_n$. By Proposition 1.11.4(3), $\overrightarrow{O'P'}_1 = \lambda_1 \overrightarrow{O'A'}_1$ and $\overrightarrow{P'_{i-1}P'_i} = \lambda_i \overrightarrow{O'A'}_i$ for $1 < i \le n$. Proposition 1.11.4(5) implies that

$$\overrightarrow{O'P'}_i = \overrightarrow{O'P'}_{i-1} + \overrightarrow{P_{i-1}P_i} = \overrightarrow{O'P'}_{i-1} + \lambda_i \overrightarrow{O'A'}_i$$

for $1 < i \le n$. Then

$$\overrightarrow{O'P'} = \overrightarrow{O'P'}_n = \lambda_1 \overrightarrow{O'A'}_1 + \lambda_2 \overrightarrow{O'A'}_2 + \cdots + \lambda_n \overrightarrow{O'A'}_n$$

and hence $P' = f(P) = Q$. This shows that isometries are surjective

Finally, since an isometry is a bijection, the inverse function f^{-1} exists and, for arbitrary points $P' = f(P)$ and $Q' = f(Q)$ in the codomain, we have

$$d(P', Q') = d(f(P), f(Q)) = d(P, Q) = d(f^{-1}(P), f^{-1}(Q)).$$

Thus, the inverse function is also an isometry. This proves that the set of isometries is closed under taking inverses. □

Proposition 1.11.4 leads to a characterization of all isometries in $\mathbb{R}^n$.

Theorem 1.11.5

A function $f : \mathbb{R}^n \to \mathbb{R}^n$ is an isometry of $\mathbb{R}^n$ if and only if there exists an $n \times n$ matrix A satisfying $A^\top A = I$ and some constant vector $\vec{b}$ such that

$$f(\vec{x}) = A\vec{x} + \vec{b}.$$

In particular, isometries of the Euclidean plane are of the form

$$\vec{x} \longmapsto \begin{pmatrix} a_1 & -\varepsilon a_2 \\ a_2 & \varepsilon a_1 \end{pmatrix} \vec{x} + \begin{pmatrix} b_1 \\ b_2 \end{pmatrix},$$

where $\varepsilon \in \{1, -1\}$, $a_i \in \mathbb{R}$ with $a_1^2 + a_2^2 = 1$, and $b_i \in \mathbb{R}$.

Proof. Let $f : \mathbb{R}^n \to \mathbb{R}^n$ be an isometry. Let $\vec{b} = f(\vec{0})$ and define the function $T : \mathbb{R}^n \to \mathbb{R}^n$ by $T(\vec{x}) = f(\vec{x}) - \vec{b}$. This is again an isometry since it consists of an isometry composed with a translation by the vector $\vec{b}$. The isometry T now has the property that $T(O) = O$, where O is the origin. Let $P, Q \in \mathbb{R}^n$ be two points and call $P' = T(P)$ and $Q' = T(Q)$. Then by Proposition 1.11.4(5), the diagonal of nondegenerate parallelograms is preserved, so

$$T(\overrightarrow{OP} + \overrightarrow{OQ}) = \overrightarrow{OP'} + \overrightarrow{OQ'} = T(\overrightarrow{OP}) + T(\overrightarrow{OQ}).$$

By Proposition 1.11.4(3), we also have $T(\lambda \overrightarrow{OP}) = \lambda \overrightarrow{OP'} = \lambda T(\overrightarrow{OP})$. We conclude that T is a linear transformation. Hence, on coordinate vectors, an isometry has the form

$$f(\vec{x}) = A\vec{x} + \vec{b},$$

for some $n \times n$ matrix A and some fixed vector $\vec{b}$.

Let P and Q be arbitrary points with position vectors $\vec{x}$ and $\vec{y}$ respectively. Then $d(P, Q) = \|\vec{y} - \vec{x}\|$. Since f is an isometry,

$$\|\vec{y} - \vec{x}\| = d(P, Q) = d(f(P), f(Q)) = \|(A\vec{y} + \vec{b}) - (A\vec{x} + \vec{b})\| = \|A(\vec{y} - \vec{x})\|. \tag{1.14}$$

Using properties of the dot product, it is easy to show that

$$\vec{x} \cdot \vec{y} = \frac{1}{2} \left(\|\vec{x}\|^2 + \|\vec{y}\|^2 - \|\vec{y} - \vec{x}\|^2 \right).$$

By (1.14), since $\|A\vec{x}\| = \|\vec{x}\|$ for all vectors $\vec{v} \in \mathbb{R}^n$, we also have

$$\vec{x} \cdot \vec{y} = (A\vec{x}) \cdot (A\vec{y}) = (A\vec{x})^\top (A\vec{y}) = \vec{x}^\top A^\top A \vec{y}.$$

Since this holds for arbitrary vectors, setting $\vec{x} = \vec{e}_i$ and $\vec{y} = \vec{e}_j$, the standard basis vectors, we deduce that the ij'th entry of $A^\top A$ is $\vec{e}_i \cdot \vec{e}_j$. Thus $A^\top A = I$.

If $n = 2$, we consider the identity $A^\top A = I$ with a generic matrix

$$\begin{pmatrix} 1 & 0 \\ 0 & 1 \end{pmatrix} = \begin{pmatrix} a & b \\ c & d \end{pmatrix} \begin{pmatrix} a & c \\ b & d \end{pmatrix} = \begin{pmatrix} a^2 + b^2 & ac + bd \\ ac + bd & c^2 + d^2 \end{pmatrix}.$$

Hence,

$$\begin{cases} a^2 + b^2 & = 1 \\ ac + bd & = 0 \\ c^2 + d^2 & = 1. \end{cases}$$

Note first that if $a = 0$, then $b = \pm 1$, $d = 0$ and $c = \pm 1$, with the $\pm$'s on b and c independent. If $a \neq 0$, then $c = -bd/a$ and the third equation gives, $(-bd/a)^2 + d^2 = 1$, which gives $b^2 d^2 + a^2 d^2 = a^2$ and then $a^2 = d^2$, using the first equation. But then $0 = (a^2 + b^2) - (c^2 + d^2) = b^2 - c^2$, so $b^2 = c^2$ also. Finally using the second equation, we deduce that $d = \varepsilon a$, while $c = -\varepsilon b$, with $\varepsilon \in \{-1, 1\}$ with the condition $a^2 + b^2 = 1$, still holding. The theorem follows. $\square$

A matrix $A \in \mathrm{GL}_n(\mathbb{R})$ satisfying $A^\top A = I$ is called an *orthogonal* matrix. The set of all real orthogonal $n \times n$ matrices is a subgroup of $\mathrm{GL}_n(\mathbb{R})$, called the *orthogonal group*. We denote it by $\mathrm{O}(n)$.

If we set a point as the origin, then by Theorem 1.11.5 we see that $\mathrm{O}(n)$ is the subgroup of isometries that leaves the origin fixed. However, let $\vec{p} \in \mathbb{R}^n$ be any other point. The set of isometries that leave $\vec{p}$ fixed is the set

$$t_{\vec{p}} \mathrm{O}(n) t_{\vec{p}}^{-1} = \{ t_{\vec{p}} \circ f \circ t_{\vec{p}}^{-1} \mid f \in \mathrm{O}(n) \},$$

where $t_{\vec{p}}$ is a translation by the vector $\vec{p}$. By Exercise 1.9.36, we see that the subgroup of isometries that leave a given point fixed is conjugate and hence isomorphic to the subgroup of isometries that leave the origin fixed.

Because $\det(A^\top) = \det(A)$ for all $n \times n$ real matrices, then every orthogonal matrix satisfies $\det(A)^2 = 1$. This gives two possibilities for the determinant of an orthogonal matrix, namely 1 or -1.

Definition 1.11.6

An isometry $f : \mathbb{R}^n \to \mathbb{R}^n$ as described in Definition 1.11.5, is called *direct* (resp. *indirect*) if $\det(A) = +1$ (resp. $\det(A) = -1$).

Example 1.11.7. In linear algebra, we find that equations of transformation for rotation of an angle of α about the origin are

$$\begin{pmatrix} x \\ y \end{pmatrix} \longmapsto \begin{pmatrix} \cos \alpha & -\sin \alpha \\ \sin \alpha & \cos \alpha \end{pmatrix} \begin{pmatrix} x \\ y \end{pmatrix}.$$

It is easy to check that rotations are direct isometries. The equations of transformation for reflection through a line through the origin making an angle of β with the x-axis are

$$\begin{pmatrix} x \\ y \end{pmatrix} \longmapsto \begin{pmatrix} \cos 2\beta & \sin 2\beta \\ \sin 2\beta & -\cos 2\beta \end{pmatrix} \begin{pmatrix} x \\ y \end{pmatrix}.$$

In contrast to rotations, reflections through lines are indirect isometries. △

Example 1.11.8. To find the equations of transformation for the rotation of angle α about a point $\vec{p} = (p_1, p_2)$ besides the origin, we obtain it as a composition of a translation by $-\vec{p}$, followed by rotation about the origin of angle α, then followed with a translation by $\vec{p}$. The equations then are

$$\begin{pmatrix} x \\ y \end{pmatrix} \longmapsto \begin{pmatrix} p_1 \\ p_2 \end{pmatrix} + \begin{pmatrix} \cos \alpha & -\sin \alpha \\ \sin \alpha & \cos \alpha \end{pmatrix} \begin{pmatrix} x - p_1 \\ y - p_2 \end{pmatrix}.$$

△

From Theorem 1.11.5 applied to the case of the Euclidean plane, we can deduce (see Exercise 1.11.6) that an isometry is uniquely determined by how it maps a triangle $\triangle ABC$ into its image $\triangle A'B'C'$. In other words, knowing how an isometry maps three noncollinear points is sufficient to determine the isometry uniquely.

The orthogonal group is an important subgroup of the group of isometries. In the rest of the section, we consider two other types of subgroups of the group of isometries of the Euclidean plane.

1.11.2 Frieze Groups

For hundreds of years, people have adorned the walls of rooms with repetitive patterns. Borders as a crown to a wall, as a chair rail, as molding to a door, or as a frame to a picture are particularly common artistic and architectural details. Frieze patterns are the patterns of symmetries used in borders.

Definition 1.11.9

A (discrete) *frieze group* is a subgroup of the group of Euclidean plane isometries whose subgroup of translations is isomorphic $\mathbb{Z}$.

In the usual group of isometries in the plane, the translations form a subgroup isomorphic to $\mathbb{R}^2$. We sometimes use the description of "discrete" for frieze groups in contrast to "continuous" because there is a translation of least positive displacement.

Example 1.11.10. Consider for example the following pattern and let G be the group of isometries of the plane that preserve the structure of the pattern.

The subgroup of translations of G consists of all translations that are an integer multiple of $2\overrightarrow{PQ}$. Some other transformations in G include

- reflections through a vertical line L_1 through P or any line parallel to L_1 displaced by an integer multiple of $\overrightarrow{PQ}$;

- reflection through the horizontal line $L_3 = \overleftrightarrow{PQ}$;

- rotations by an angle of π about P, Q, or any point translated from P by an integer multiple of $\overrightarrow{PQ}$.

It is possible to describe G with a presentation. Let s_i be the reflection through L_i, for $i = 1, 2, 3$. We claim that

$$G = \langle s_1, s_2, s_3 \mid s_1^2 = s_2^2 = s_3^2 = 1, \ (s_1 s_3)^2 = 1, \ (s_2 s_3)^2 = 1 \rangle. \tag{1.15}$$

In order to prove the claim, we first should check that s_1, s_2, s_3 do indeed generate all of G. By Exercise 1.11.8, $s_1 s_3$ corresponds to rotation by π about P and $s_2 s_3$ corresponds to rotation by π about Q. By Exercise 1.11.7 $s_2 s_1$ corresponds to a translation by $2\overrightarrow{PQ}$. In order to obtain a reflection through another vertical line besides L_1 or L_2, or a rotation about another point besides P or Q, we can translate the strip to center it on P and Q, apply the desired transformation (s_1, s_2, $s_1 s_3$, or $s_2 s_3$), and then translate back. For example, the rotation by an angle of π about Q_3, can be described by $(s_2 s_1)^2 s_2 s_3 (s_2 s_1)^{-2}$. This shows that our choice of generators is sufficient to generate G.

Next, we need to check that we have found all the relations. The relations $s_i^2 = 1$ are obvious. The last two relations are from the fact that $s_1 s_3$ and $s_2 s_3$ are rotations by π so have order 2. Note that $\langle s_1 s_2 \rangle \cong \mathbb{Z}$ is the infinite subgroup of translation and $\langle s_1, s_2 \rangle \cong D_\infty$ (see Example 1.10.7). From the relation, $(s_1 s_3)^2 = 1$, we see that

$$s_1 s_3 = (s_1 s_3)^{-1} = s_3^{-1} s_1^{-1} = s_3 s_1$$

because $s_1^2 = 1$ and $s_3^2 = 1$. Hence, s_3 commutes with s_1. Similarly s_3 commutes with s_2. Hence, all elements in G can be written as an alternating string of s_1 and s_2 or an alternating string of s_1 and s_2 followed by s_3. For example, rotation by π about Q_3 is

$$(s_2 s_1)^2 s_2 s_3 (s_2 s_1)^{-2} = s_2 s_1 s_2 s_1 s_2 s_3 s_1 s_2 s_1 s_2 = s_2 s_1 s_2 s_1 s_2 s_1 s_2 s_1 s_2 s_3$$
$$= (s_2 s_1)^4 s_2 s_3.$$

If we set $t = s_2 s_1$, then every element can be written as

- t^k for $k \in \mathbb{Z}$ (translation by $2k\overrightarrow{PQ}$);

- $t^k s_1$ for $k \in \mathbb{Z}^{\leq 0}$ (reflection through the line $k\overrightarrow{PQ}$ from L_1);

- $t^k s_2$ for $k \in \mathbb{Z}^{\geq 0}$ (reflection through the line $k\overrightarrow{PQ}$ from L_2);

- $t^k s_3$ for $k \in \mathbb{Z}$ (reflection through L_3 composed with a translation by $2k\overrightarrow{PQ}$);

- $t^k s_1 s_3$ for $k \in \mathbb{Z}^{\leq 0}$ (rotation through the point $k\overrightarrow{PQ}$ from P); or

- $t^k s_2 s_3$ for $k \in \mathbb{Z}^{\geq 0}$ (rotation through the point $k\overrightarrow{PQ}$ from Q).

This gives us a full description of all elements in G and it also shows that there exist no relations in G not implied by those in the presentation in (1.15). $\triangle$

Following the terminology in Section 1.1.3, if a pattern has a Frieze group G of symmetry, we call a subset of the pattern a fundamental region (or fundamental pattern) if the entire pattern is obtained from the fundamental region by applying elements from G and that this region is minimal among all subsets that generate the whole pattern. For example, a fundamental region for the pattern in Example 1.11.10 is the following.

Frieze patterns are ubiquitous in artwork and architecture throughout the world. Figures 1.14 through 1.17 show a few such patterns.

FIGURE 1.14: A traditional Celtic border.

FIGURE 1.15: A 17th century European chair rail.

FIGURE 1.16: A modern wood molding.

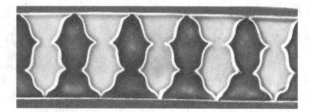

FIGURE 1.17: The border on a Persian rug.

1.11.3   Wallpaper Groups

Though wallpaper designers could use a frieze pattern to cover an entire wall, it is more common to use a fundamental pattern that is bounded (or in practice small with respect to the size of the wall). Then to cover this entire plane, we would require two independent directions of translation.

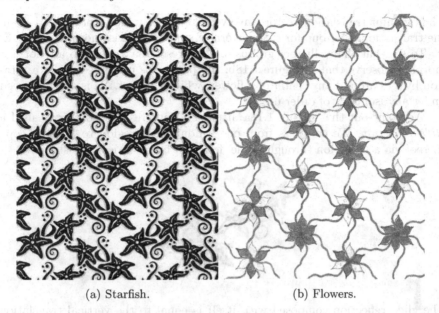

(a) Starfish.   (b) Flowers.

FIGURE 1.18: Some wallpaper patterns.

Definition 1.11.11

A *wallpaper group* is a subgroup of the group of Euclidean plane isometries whose subgroup of translations is isomorphic $\mathbb{Z} \oplus \mathbb{Z}$.

Consider for example, Figures 1.18a and 1.18b. These two figures illustrate patterns covering the Euclidean plane, each preserved by a different wallpaper group.

To first see that each is a wallpaper group, notice that in Figure 1.18a, the fundamental region consisting of two starfish

can be translated to an identical pattern along any translation $\vec{t} = a\vec{i} + b\vec{j}$, where $a, b \in \mathbb{Z}$ and $\vec{i}$ corresponds to one horizontal unit translation and $\vec{j}$ to one vertical unit. Clearly, the translation subgroup of isometries preserving the pattern is $\mathbb{Z} \oplus \mathbb{Z}$. In Figure 1.18b, a flower can be translated into any other flower (ignoring shading) by

$$a\left(\frac{\sqrt{3}}{2}\vec{i} + \frac{1}{2}\vec{j}\right) + b\vec{j} \qquad \text{for } a, b \in \mathbb{Z}.$$

The two unit translation vectors involved are linearly independent and again the translation subgroup for this pattern of symmetry is isomorphic to $\mathbb{Z} \oplus \mathbb{Z}$.

To see that the wallpaper groups for Figures 1.18a and 1.18b are not isomorphic, observe that in Figure 1.18b, the pattern is preserved under a rotation by $\pi/3$ around any center of a flower, while Figure 1.18a is not preserved under any isometry of order 6.

We point out that Figure 1.18a displays an interesting isometry, called a *glide reflection*. We can pass from one fundamental region to another copy thereof via a reflection through a line, followed by a vertical translation.

The glide reflection composed with itself is equal to the vertical translation with distance equal to the least vertical gap between identical regions. Note that in a glide reflection, the translation (glide) is always assumed to be parallel to the line through which the reflection occurs. One must still specify the length of the translation. For example, a glide reflection f through the x-axis with a translation distance of $+2$ has for equations

$$f \begin{pmatrix} x \\ y \end{pmatrix} = \begin{pmatrix} 1 & 0 \\ 0 & -1 \end{pmatrix} \begin{pmatrix} x \\ y \end{pmatrix} + \begin{pmatrix} 2 \\ 0 \end{pmatrix} = \begin{pmatrix} x + 2 \\ -y \end{pmatrix}.$$

The Dutch artist M. C. Escher (1898–1972) is particularly well-known for his exploration of interesting artwork involving wallpaper symmetry groups.[4] Part of the genius of his artwork resided in that he devised interesting and recognizable patterns that were tessellations, patterns in which, unlike Figure 1.18a, there is no blank space. In a tessellation, the fundamental region tiles and completely covers $\mathbb{R}^2$.

It is possible to classify all wallpaper groups. Since this section only offers a glimpse of applications of group theory to geometry, we do not offer a proof here or give the classification.

Theorem 1.11.12 (Wallpaper Group Theorem)

There exist exactly 17 nonisomorphic wallpaper groups.

Over the centuries, the study of symmetry patterns of the plane has drawn considerable interest both by artists and mathematicians alike. Many books

---

[4]M. C. Escher's official website: *http://www.mcescher.com/*.

study this topic from a variety of directions. From a geometer's perspective, [17] offers a careful and encyclopedic analysis of planar symmetry patterns.

Frieze groups and wallpaper groups are examples of *crystallographic groups*. A crystallographic group of $\mathbb{R}^n$ is a subgroup G of the group of isometries on $\mathbb{R}^n$ such that the subgroup of translations in G is isomorphic to $\mathbb{Z}^n$. The adjective "crystallographic" is motivated by the fact that regular crystals fill Euclidean space in a regular pattern whose translation subgroup is isomorphic to $\mathbb{Z}^3$.

EXERCISES FOR SECTION 1.11

1. Calculate the equations of transformation for the rotation by $\pi/3$ about the point $(3, 2)$.

2. Determine the equations of transformation for the reflection through the line that makes an angle of $\pi/6$ with the x-axis and goes through the point $(3, 4)$.

3. Determine the equations of transformation for the reflection through the line $2x + 3y = 6$.

4. Determine the equations of the isometry of reflection-translation through the line $x = y$ with a translation distance of 1 (in the positive x-direction).

5. We work in $\mathbb{R}^3$. Show by direct matrix calculations that rotation by θ about the x-axis, followed by rotation by α about the y-axis is not generally equal to the rotation by α about the y-axis followed by rotation by θ about the x-axis.

6. Use Theorem 1.11.5 to prove the claim that it suffices to know how an isometry f on the Euclidean plane maps three non-collinear points to know f uniquely.

7. Let L_1 and L_2 be two parallel lines in the Euclidean plane. Prove that reflection through L_1 followed by reflection through L_2 is a translation of vector $\vec{v}$ that is twice the perpendicular displacement from L_1 to L_2.

8. Let L_1 and L_2 be two lines in the plane that are not parallel. Prove that reflection through L_1 followed by reflection through L_2 is a rotation about their point of intersection of an angle that is double the angle from L_1 to L_2 (in a counterclockwise direction).

9. Let $A = (0, 0)$, $B = (1, 0)$, and $C = (0, 1)$. Determine the isometry obtained by composing a rotation about A of angle $2\pi/3$, followed by a rotation about B of angle $2\pi/3$, followed by a rotation about C of angle $2\pi/3$. Find the equations of transformation and describe it in simpler terms.

10. Let f be the isometry of rotation by α about the point $A = (a_1, a_2)$ and let g be the isometry of rotation by β about the point $B = (b_1, b_2)$. Show that $f \circ g$ may be described by a rotation by $\alpha + \beta$ about B followed by a translation and give this translation vector.

11. Let f be the plane Euclidean isometry of rotation by α about a point A and let t be a translation by a vector $\vec{v}$. Prove that the conjugate $f \circ t \circ f^{-1}$ is a translation and determine this translation explicitly.

12. Let g be the plane Euclidean isometry of reflection through a line L and let t be a translation by a vector $\vec{v}$. Prove that the conjugate $g \circ t \circ g^{-1}$ is a translation and determine this translation explicitly.

13. Prove that orthogonal matrices have determinant 1 or -1. Prove also that

$$SO(n) = \{A \in O(n) \mid \det(A) = 1\}$$

is a subgroup of $O(n)$ and hence of $GL_n(\mathbb{R})$. [The subgroup $SO(n)$ is called the *special orthogonal group*.]

14. Prove that $SO(2)$ (see Exercise 1.11.13) consists of matrices

$$\left\{ \begin{pmatrix} \cos\theta & -\sin\theta \\ \sin\theta & \cos\theta \end{pmatrix} \,\Big|\, \theta \in [0, 2\pi) \right\}.$$

15. Prove that the function $d_t : \mathbb{R}^2 \times \mathbb{R}^2 \to \mathbb{R}$ given by

$$d_t((x_1, y_1), (x_2, y_2)) = |x_2 - x_1| + |y_2 - y_1|$$

satisfies the conditions of a metric on $\mathbb{R}^2$ as defined in Definition 1.11.2. Prove also that the set of surjective isometries for d_t is a group and show that it is not equal to the group of Euclidean isometries. [Hint: This metric on $\mathbb{R}^2$ is called the *taxi metric* because it calculates that distance a taxi would travel between two points if it could only drive along north-south and east-west streets.]

16. List all the symmetries and describe the compositions between them for the infinitely long pattern shown below:

17. List all the symmetries and describe the compositions between them for the infinitely long sine curve shown below:

18. List all the symmetries and describe the compositions between them for the infinitely long pattern shown below:

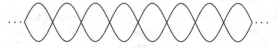

19. Find a presentation for the frieze group associated with Figure 1.14.

20. Find a presentation for the frieze group associated with Figure 1.15.

21. Find a presentation for the frieze group associated with Figure 1.17. Show that it is the same as the frieze group associated with the pattern in Figure 1.16.

22. Find a presentation for the frieze group associated with the following pattern and then sketch the fundamental pattern.

23. (*) Prove that there are only 7 nonisomorphic frieze groups.

24. Prove that the wallpaper group for the following pattern is not isomorphic to the wallpaper groups for either Figure 1.18a or 1.18b.

25. Prove that the wallpaper group for the following pattern is not isomorphic to the wallpaper groups for either Figure 1.18a or 1.18b.

For Exercises 1.11.26 through 1.11.29, sketch a reasonable portion of the pattern generated by the following fundamental pattern and the group indicated in each exercise. Assume the minimum distance for any translation is 2.

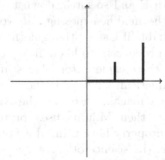

26. The wallpaper group corresponding to Figure 1.18a.

27. The wallpaper group corresponding to Figure 1.18b.

28. The wallpaper group corresponding to the pattern in Exercise 1.11.32.

29. The wallpaper group corresponding to the pattern in Exercise 1.11.24.

30. Find a presentation for the wallpaper group corresponding to Figure 1.18a.

31. Find a presentation for the wallpaper group corresponding to the pattern in Exercise 1.11.25.

32. Find a presentation for the wallpaper group corresponding to the wallpaper pattern here below.

1.12   Diffie-Hellman Public Key

1.12.1   A Brief Background on Cryptography

In this section, we will study an application of group theory to *cryptography*, the science of keeping information secret.

Cryptography has a long history, with one of the first documented uses of cryptography attributed to Caesar. When writing messages he wished to keep in confidence, the Roman emperor would shift each letter by 3 to the right, assuming the alphabet wraps around. In other words, he would substitute a letter of A with D, B with E and so forth, down to replacing Z with C. To anyone who intercepted the modified message, it would look like nonsense. This was particularly valuable if Caesar thought there existed a chance that an enemy could intercept orders sent to his military commanders.

After Caesar's cipher, there came letter wheels in the early Renaissance, letter codes during the American Civil War, the Navajo windtalkers during World War II, the Enigma machine used by the Nazis, and then a whole plethora of techniques since then. Military uses, protection of financial data, and safety of intellectual property have utilized cryptographic techniques for centuries. For a long time, the science of cryptography remained the knowledge of a few experts because both governments and companies held that keeping their cryptographic techniques secret would make it even harder for "an enemy" to learn one's information security tactics.

Today, electronic data storage, telecommunication, and the Internet require increasingly complex cryptographic algorithms. Activities that are commonplace like conversing on a cellphone, opening a car remotely, purchasing something online, all use cryptography so that a conversation cannot be intercepted, someone else cannot easily unlock your car, or an eavesdropper cannot intercept your credit card information.

Because of the proliferation of applications of cryptography in modern society, no one should assume that the cryptographic algorithm used in any given instance remains secret. In fact, modern cryptographers do not consider an information security algorithm at all secure if part of its effectiveness relies on the algorithm remaining secret. But not everything about a cryptographic algorithm can be known to possible eavesdroppers if parties using the algorithm hope to keep some message secure. Consequently, most, if not all, cryptographic techniques involve an algorithm but also a "key," which can be a letter, a number, a string of numbers, a string of bits, a matrix or some other mathematical object. The security of the algorithm does not depend on the algorithm staying secret but rather on the key remaining secret. Users can change keys from time to time without changing the algorithm and have confidence that their messages remain secure.

A basic cryptographic system involves the following objects.

(1) A message space $\mathcal{M}$. This can often be an n-tuple of elements from some alphabet $\mathcal{A}$ (so $\mathcal{M} = \mathcal{A}^n$) or any sequence from some alphabet $\mathcal{A}$ (so $\mathcal{M} = \mathcal{A}^{\mathbb{N}}$). The original message is called *plaintext*.

(2) A *ciphertext* space $\mathcal{C}$. This is the set of all possible hidden messages. It is not uncommon for $\mathcal{C}$ to be equal to the plaintext space $\mathcal{M}$.

(3) A *keyspace* $\mathcal{K}$ that provides the set of all possible keys to be used in a cryptographic algorithm.

(4) An encryption procedure $E \in \text{Fun}(\mathcal{K}, \text{Fun}(\mathcal{M}, \mathcal{C}))$ such that for each key $k \in \mathcal{K}$, there is an injective function $E_k : \mathcal{M} \to \mathcal{C}$.

(5) A decryption procedure $D \in \text{Fun}(\mathcal{K}, \text{Fun}(\mathcal{C}, \mathcal{M}))$ such that for each key $k \in \mathcal{K}$, there exists a key $k' \in \mathcal{K}$, with a function $D_{k'} : \mathcal{C} \to \mathcal{M}$ satisfying

$$D_{k'}(E_k(m)) = m \qquad \text{for all } m \in \mathcal{M}.$$

In many algorithms, $k' = k$ but that is not necessarily the case. (The requirement that E_k be injective makes the existence of $D_{k'}$ possible.)

In an effective cryptographic algorithm, it should be very difficult to recover the keys k or k' given just ciphertext $c = E_k(m)$ (called a "ciphertext only attack") or even given ciphertext $c = E_k(m)$ and the corresponding plaintext m (called a "ciphertext and known plaintext attack").

From the mathematician's viewpoint, it is interesting that all modern cryptographic techniques rely on number theory and advanced abstract algebra that is beyond the understanding of the vast majority of people. Companies involved in designing information security products or protocols must utilize advanced mathematics.

Now imagine that you begin a communication with a friend at a distance (electronically or otherwise) and that other people can listen in on everything that you communicate to each other. Would it be possible for that communication to remain secret? More specifically, would it be possible to together choose a key k (for use in a subsequent cryptographic algorithm) so that people who are eavesdropping on the whole communication do not know what that key is. It seems very counterintuitive that this should be possible but such algorithms do exist and are called *public key* cryptography techniques.

In this section, we present the Diffie-Hellman protocol. Devised in 1970, it was one of the first public key algorithms. An essential component of the effectiveness of the Diffie-Hellman protocol is the Fast Exponentiation Algorithm.

1.12.2   Fast Exponentiation

Let G be a group, let g be an element in G, and let n be a positive integer. To calculate the power g^n, one normally must calculate

$$g^n = \overbrace{g \cdot g \cdots g}^{n \text{ times}},$$

which involves $n - 1$ operations. (If fact, when we implement this into a computer algorithm, since we must take into account the operation of incrementing a counter, the above direct calculation takes a minimum of $2n - 1$ computer operations.) If the order $|g|$ and the power n are large, one may not notice any patterns in the powers of g that would give us any shortcuts to determining g^n with fewer than $n - 1$ group operations.

The Fast Exponentiation Algorithm allows one to calculate g^n with many fewer group operations than n, thus significantly reducing the calculation time.

Algorithm 1.12.1: FASTEXPONENTIATION(g, n)

$(b_k b_{k-1} \cdots b_1 b_0)_2 \leftarrow$ CONVERTTOBINARY(n)
$x \leftarrow g$
for $i \leftarrow (k - 1)$ **downto** 0
$\quad$ **do** $\begin{cases} \text{if } b_i = 0 \\ \quad \text{then } x \leftarrow x^2 \\ \quad \text{else } x \leftarrow x^2 g \end{cases}$
return (x)

The reason that x has the value of g^n at the end of the **for** loop is because when the algorithm terminates,

$$x = g^{b_k 2^k + b_{k-1} 2^{k-1} + \cdots + b_1 2 + b_0},$$

which is precisely g^n. Note that in the binary expansion $n = (b_k b_{k-1} \cdots b_1 b_0)_2$, there is an assumption that $b_k = 1$.

Each time through the `for` loop, we do either one or two group operations. Hence, we do at most $2k = 2\lfloor \log_2 n \rfloor$ groups operations. In practice, when implementing this algorithm, getting the binary expansion for n takes $k + 1 = \lfloor \log_2 n \rfloor + 1$ operations (integer divisions) and the operation of decrementing the counter i takes a total of k operations. This gives a total of at most $4\lfloor \log_2 n \rfloor + 1$ computer operations.

Example 1.12.1. Let $G = U(311)$. Note that 311 is a prime number. We propose to calculate $\overline{7}^{39}$.

The binary expansion of 39 is $39 = (100111)_2 = 2^5 + 2^2 + 2^1 + 2^0$. Following the steps of the algorithm, we

- assign $x := \overline{7}$;
- for $i = 4$, since $b_4 = 0$, assign $x := x^2 = \overline{7}^2 = \overline{49}$;
- for $i = 3$, since $b_3 = 0$, assign $x := x^2 = \overline{49}^2 = \overline{224}$;
- for $i = 2$, since $b_2 = 1$, assign $x := \overline{7}x^2 = \overline{7} \times \overline{224}^2 = \overline{113}$;
- for $i = 1$, since $b_1 = 1$, assign $x := \overline{7}x^2 = \overline{7} \times \overline{113}^2 = \overline{126}$;
- for $i = 0$, since $b_0 = 1$, assign $x := \overline{7}x^2 = \overline{7} \times \overline{126}^2 = \overline{105}$.

We conclude that in $U(311)$, we have $\overline{7}^{39} = \overline{105}$. $\triangle$

In the above example, we performed 8 group calculations as opposed to the necessary 38 had we simply multiplied $\overline{7}$ to itself 38 times. This certainly sped up the process for calculating $\overline{7}^{39}$ by hand. However, running a `for` loop with 39 iterations obviously does not come close to straining a computer's capabilities.

Example 1.12.2. Let $G = U(435465768798023)$. Again, 435465768798023 is a prime number (though that is not necessary for the algorithm). We propose to calculate

$$\overline{379}^{1234567890123}$$

in $U(435465768798023)$. Using the standard way of finding powers, it would take 1234567890122 operations in G, a number that begins to require a significant computing time. However, using the Fast Exponentiation Algorithm, we only need to do at most $2(\lfloor \log_2 1234567890123 \rfloor + 1) = 82$ group operations. For completeness, we give the result of the calculation

$$\overline{379}^{1234567890123} = \overline{370162048004176}.$$ $\triangle$

Both of the above examples involve groups of the form $U(p)$ but Fast Exponentiation applies in any group.

1.12.3 Diffie-Hellman Algorithm

Diffie-Hellman is a protocol (an algorithm involving two or more agents) for two people who communicate publicly and never get together in secret to

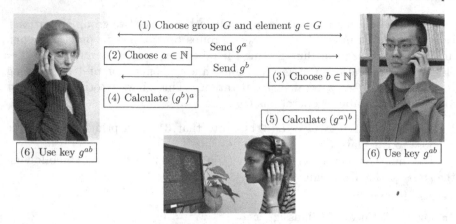

FIGURE 1.19: Diffie-Hellman.

select a key in such a way that an eavesdropper cannot easily determine what that key is.

In order to describe how Diffie-Hellman works, we will introduce three players: Alice, Bob, and Eve. Alice and Bob want to talk secretly while Eve wants to eavesdrop on their conversation. We assume that Eve can hear everything that Alice and Bob say to each other. In the following diagram, that which is boxed stays secret to the individual and that which is not boxed is heard by everyone, including Eve. The Diffie-Hellman public key protocol works as follows (Figure 1.19).

(1) Alice (on the left) and Bob (on the right) settle on a group G and a group element g, called the *base*. Ideally, the order of g should be very large. (If you are doing calculations by hand, the order of g should be in the hundreds. If we are using computers, the order of g is ideally larger than what a typical `for` loop runs through, say 10^{15} or much more.)

(2) Alice chooses a relatively large integer a and sends to Bob the group element g^a, calculated with Fast Exponentiation.

(3) Bob chooses a relatively large integer b and sends to Alice the group element g^b, calculated with Fast Exponentiation.

(4) Alice calculates $(g^b)^a = g^{ab}$ using Fast Exponentiation.

(5) Bob calculates $(g^a)^b = g^{ab}$ using Fast Exponentiation.

(6) Alice and Bob will now use the group element g^{ab} as the key.

The reason why Eve cannot easily figure out g^{ab} is simply a matter of how long it would take her to do so. We should assume that Eve intercepts the group G, the base g, the element g^a, and the element g^b. However, Eve does not know the integers a or b. In fact, Alice does not know b and Bob does not know a. The reason why this is safe is that Fast Exponentiation makes it possible to calculate powers of group elements quickly while, on the other hand, the problem of determining n from knowing g and g^n could take as long as n. If the power n is very large, it simply takes far too long. Not knowing a or b, Eve cannot quickly determine the key g^{ab} if she only knows g, g^a, and g^b.

The fact that the security of the Diffie-Hellman public key exchange relies on the speed of one calculation versus the relative slowness of a reverse algorithm may seem unsatisfactory at first. However, suppose that it takes only microseconds for Alice and Bob to establish the key g^{ab} but centuries (at the present computing power) for Eve to recover a from g and g^a. The secret message between Alice and Bob would have lost its relevance long before Eve could recover the key.

The problem of finding n given g and g^n in some group is called the *Discrete Log Problem*. Many researchers in cryptography study the Discrete Log Problem, especially in groups of the form $U(n)$, for $n \in \mathbb{N}$. For reasons stemming from modular arithmetic, there are algorithms that calculate n from g and g^n using fewer than n operations. However, Diffie-Hellman using some other groups, do not have such weaknesses. A popular technique called *elliptic curve cryptography* is one such example of the Diffie-Hellman protocol that does not possess some of the weaknesses of $U(n)$.

In any Diffie-Hellman public key exchange, some choices of group G and base g are wiser than others. It is preferable to choose a situation in which $|g|$ is high. Otherwise, an exhaustive calculation on the powers of g would make a brute-force solution to the Discrete Log Problem a possibility for Eve. Furthermore, given a specific g, some choices of a and b are not wise either. For example, if g^a (or g^b) has a very low order, then g^{ab} would also have a very low order. Then, even if Eve does not know the key for certain, she would only be a small number of possible keys in the form $(g^a)^k$.

1.12.4 ElGamal Encryption

ElGamal encryption is a cryptographic protocol based on Diffie-Hellman. We describe now with the assumption that Alice wants to send an encrypted message to Bob.

(1) Alice and Bob settle on a group G and on the base, a group element g. The plaintext space $\mathcal{M}$ can be any set, the ciphertext space $\mathcal{C}$ will be sequences of elements in G, and the keyspace is $\mathcal{K} = G$.

(2) Alice and Bob also choose a method of encoding the message into a sequence of elements in G; i.e., they choose an injective function $h : \mathcal{M} \to \text{Fun}(\mathbb{N}^*, G)$.

(3) They run Diffie-Hellman to obtain their public key $k = g^{ab}$.

(4) Alice encodes her message m with a sequence of group elements $h(m) = (m_1, m_2, \ldots, m_n)$.

(5) Alice sends to Bob the group elements $E_k(m) = (km_1, km_2, \ldots, km_n) = (c_1, c_2, \ldots, c_n)$. Note that for each i, we have $c_i = g^{ab}m_i$.

(6) To decipher the ciphertext, we have $k' = k = g^{ab}$. Bob calculates the m_i from $m_i = c_i k^{-1} = c_i(g^{ab})^{-1}$.

[We point out that with a group element of large order, it is not always obvious how to determine the inverse of a group element. We use Corollary 2.1.12, which says that $g^{|G|} = 1$. Hence, to calculate g^{-ab}, without knowing a but only knowing g^a, using Fast Exponentiation, Bob calculates

$$(g^a)^{|G|-b} = g^{a|G|-ab} = g^{-ab}.]$$

(7) Since h is injective, Bob can find Alice's plaintext message $(m_1, m_2, \ldots, m_n)$.

Example 1.12.3. We use the group $G = U(3001)$ and choose for a base the group element $g = \bar{2}$. (It turns out that $|g| = 1500$. In general, one does not have to know the order of g, but merely hope that it is high.) The message space is the set of sequences $\mathcal{M} = \mathrm{Fun}(\mathbb{N}^*, \mathcal{A})$ where $\mathcal{A}$ is the set consisting of the 26 English letters and the space character.

Alice and Bob decide to encode their messages (define $h : \mathcal{M} \to \mathrm{Fun}(\mathbb{N}^*, G)$) as follows. Ignore all punctuation, encode a space with the integer 0 and each letter of the alphabet with its corresponding ordinal, so A is 1, B is 2, and so on, where we allow for two digits for each letter. Hence, a space is actually 00 and A is 01. Then group pairs of letters simply by adjoining them to make (up to) a four-digit number. Thus, "GOOD-BYE CRUEL WORLD" becomes the finite sequence

$$715, \ 1504, \ 225, \ 500, \ 318, \ 2105, \ 1200, \ 2315, \ 1812, \ 400,$$

where we completed the last pair with a space. We now view these numbers as elements in $U(3001)$.

Alice chooses $a = 723$ while Bob chooses $b = 1238$. In binary, $a = (1011010011)_2$ and $b = (10011010110)_2$. Using Fast Exponentiation, Alice calculates that $g^a = \bar{2}^{723} = \overline{1091}$ and Bob calculates that $g^b = \bar{2}^{1238} = \overline{1056}$.

Alice just wants to say, "HI BOB." She first calculates $g^{ab} = \overline{1056}^{723} = \overline{2442}$. Her corresponding message in group elements is: $\overline{809}, \bar{2}, \overline{1502}$. The ciphertext $m_i g^{ab}$ for $i = 1, 2, 3$ is the string of group elements: $\overline{920}, \overline{1883}, \overline{662}$.

On his side, Bob now first calculates the element $(g^a)^{|G|-b} = \overline{1091}^{1762} = \overline{102}$. The deciphered code is

$$\overline{920} \times \overline{102} = \overline{809}, \quad \overline{1883} \times \overline{102} = \bar{2}, \quad \overline{662} \times \overline{102} = \overline{1502}.$$

Bob then easily recovers "HI BOB" as the original message.   △

In the following exercises, the reader should understand that the situations presented all are small enough to make it simple for a computer to perform a brute force attack to find a from g^a or b from g^b. In real applications, the group G, the base g, the elements a and b are chosen large enough by parties so make the Discrete Log Problem not tractable using a brute force attack.

EXERCISES FOR SECTION 1.12

1. Show all your steps as you perform Fast Exponentiation to calculate by hand $\bar{5}^{73}$ in the group $U(153)$.

2. Show all your steps as you perform Fast Exponentiation to calculate by hand $\bar{2}^{111}$ in the group $U(501)$.

3. Show all your steps as you perform Fast Exponentiation to calculate by hand $\bar{3}^{795}$ in the group $U(7703)$.

4. Consider the group $G = \mathrm{GL}_2(\mathbb{F}_{199})$. Show all your steps as you perform Fast Exponentiation to calculate by hand

$$\begin{pmatrix} \bar{0} & \overline{23} \\ \overline{10} & \bar{3} \end{pmatrix}^{42}.$$

5. Show all your steps as you perform Fast Exponentiation to calculate by hand

$$\begin{pmatrix} 3 & 2 \\ 1 & 1 \end{pmatrix}^{9}$$

in $\mathrm{GL}_2(\mathbb{R})$.

6. Suppose that Alice and Bob decide to use the group $U(3001)$ as their group G and the element $\bar{5}$ as the base. If Alice chooses $a = 73$ and Bob chooses $b = 129$, then what will be their common key using the Diffie-Hellman public key exchange algorithm?

7. Suppose that Alice and Bob decide to use the group $U(4237)$ as their group G and the element $\overline{11}$ as the base. If Alice chooses $a = 100$ and Bob chooses $b = 515$, then what will be their common key using the Diffie-Hellman public key exchange algorithm?

In Exercises 1.12.8 to 1.12.11, use the method in Example 1.12.3 to take strings of letters to strings of numbers. Each time, use the same method to change a letter to a numbers and only collect two letters to make an integer that is at most 4 digits.

8. Play the role of Alice. Use the group $G = U(3001)$ and the base $g = \bar{2}$. Change a and b to $a = 437$ and $b = 1000$. Send to Bob the ciphertext for the following message: "MEET ME AT DAWN."

9. Play the role of Alice. Use the group $G = U(3001)$ but use the base $g = \bar{7}$. Bob sends you $g^b = \overline{2442}$ and you decide to use $a = 2319$. Send to Bob the ciphertext for the following message: "SELL ENRON STOCKS NOW."

10. Play the role of Bob. Use the group $G = U(3517)$ and use the base $g = \overline{11}$. Alice sends you $g^a = \overline{1651}$ and you tell Alice you will use $b = 789$. You receive the following ciphertext from Alice:

$$\overline{369}, \quad \overline{665}, \quad \overline{1860}, \quad \overline{855}, \quad \overline{3408}, \quad \overline{1765}, \quad \overline{1820}, \quad \overline{1496}.$$

Show the steps using fast exponentiation to recover the decryption key g^{-ab} and use this to recover the plaintext for the message that Alice sent you.

11. Play the role of Bob. Use the group $G = U(7522)$ and use the base $g = \overline{3}$. Alice sends you $g^a = \overline{2027}$ and you tell Alice you will use $b = 2013$. You receive the following ciphertext from Alice:

$$\overline{4433}, \quad \overline{5198}, \quad \overline{6996}, \quad \overline{3275}, \quad \overline{7067}, \quad \overline{2568}, \quad \overline{1894}, \quad \overline{6037}, \quad \overline{7208}.$$

Show the steps using fast exponentiation to recover the decryption key g^{-ab} and use this to recover the plaintext for the message that Alice sent you.

12. We design the following Diffie-Hellman/ElGamal setup to encipher strings of bits (0 or 1). We will choose a prime number p with $2^{10} < p < 2^{11}$ and use the group $G = U(p)$ and choose a base $g \in G$. We break up the string of bits into blocks of 10 bits and map a block of ten bits into a number between 0 and 1023 by using a binary expression, hence,

$$(b_0, b_1, \dots, b_9) \longrightarrow b_0 + b_1 \cdot 2 + \dots + b_k \cdot 2^k + \dots + b_9 \cdot 2^9 = m_i.$$

This is a plaintext unit m_i and we view it as an element in $U(p)$. With this setup to convert strings of bits to elements in $U(p)$, we then apply the usual Diffie-Hellman key exchange and the ElGamal encryption. As one extra layer, when Alice sends the cipher text to Bob, she writes it as a bit string, but with the difference that since numbers $c = mg^{ab}$ can be expressed uniquely as an integer less than 2^{11}, blocks of 10 bits become blocks of 11 bits.

Here is the exercise. You play the role of Alice. You and Bob decide to use $p = 1579$ and the base of $g = \overline{7}$. Bob sends you $g^b = 993$ and you decide to use $a = 78$. Show all the steps to create the Diffie-Hellman key g^{ab}. Use this to create the ciphertext as described in the previous paragraph for the following string of bits:

$$1011010110 \ 1011111001 \ 1111100010.$$

Show all the work.

13. Use the setup as described in Exercise 1.12.12 to encipher bit strings. This time, however, you play the role of Bob. You and Alice use the group $U(1777)$ and the base $g = \overline{10}$, Alice sends you $g^a = \overline{235}$, and you choose to use $b = 1573$. Alice has sent you the following string of bits:

$$00100110110 \ 01110101001 \ 10011011100.$$

Show all the steps to create the Diffie-Hellman decryption key g^{-ab}. Turn the ciphertext into strings of elements in $U(1777)$. By multiplying by g^{-ab}, recover the list of elements in $U(1777)$ that correspond to plaintext. (These should be integers m_i with $0 \le m_i \le 1023$.) Convert them to binary to recover the plaintext message in bit strings.

14. We design the following application of Diffie-Hellman. We choose to encrypt 29 characters: 26 letters of the alphabet, space, the period ".", and the comma ",". We associate the number 0 to a space character, 1 through 26 for each of the letters, 27 to the period, and 28 to the comma. We use the group $G = GL_2(\mathbb{F}_{29})$, the general linear group on modular arithmetic base 29. Given a message in English, we write the numerical values of the characters in a $2 \times n$ matrix, reading the characters of the alphabet by successive columns. Hence, "SAY FRIEND AND ENTER" would become the matrix

$$M = \begin{pmatrix} 19 & 25 & 6 & 9 & 14 & 0 & 14 & 0 & 14 & 5 \\ 1 & 0 & 18 & 5 & 4 & 1 & 4 & 5 & 20 & 18 \end{pmatrix} \in M_{2 \times 10}(\mathbb{F}_{29}),$$

where we have refrained from putting the congruence bars over the top of the elements only for brevity. Then given a key $K \in GL_2(\mathbb{F}_{29})$, we encrypt the message into ciphertext by calculating the matrix $C = KM$. Hence, with $K = \begin{pmatrix} 3 & 4 \\ 5 & 9 \end{pmatrix}$ the ciphertext matrix becomes

$$C = \begin{pmatrix} 3 & 17 & 3 & 18 & 0 & 4 & 0 & 20 & 6 & 0 \\ 17 & 9 & 18 & 3 & 19 & 9 & 19 & 16 & 18 & 13 \end{pmatrix}$$

and "CQQICRRI SDI STPFR M" is the ciphertext message in characters. [Note that this enciphering scheme is not an ElGamal enciphering scheme as described in Section 1.12.4. Here, $\mathcal{C} = \mathcal{M}$ and we do not use a function h so that the enciphering function E_k does not involve products in the group G.]

Here is the exercise. You play the role of Alice. You and Bob decide on the above enciphering scheme. You will choose the key K in the usual Diffie-Hellman manner. You use the group $G = GL_2(\mathbb{F}_{29})$ and the base $g = \begin{pmatrix} 1 & 2 \\ 3 & 5 \end{pmatrix}$. Bob sends you $g^b = \begin{pmatrix} 27 & 24 \\ 7 & 17 \end{pmatrix}$ and you choose to use $a = 17$. Calculate the Diffie-Hellman key and use this to determine the ciphertext corresponding to "COME HERE, NOW." (Since there is an odd number of characters in the message, append a space on the end to make a message of even length.)

15. We use the communication protocol as described in Exercise 1.12.14. You use the same group and the same base but this time you play the role of Bob. Alice sends you $g^a = \begin{pmatrix} 5 & 27 \\ 26 & 1 \end{pmatrix}$ and you choose to use $b = 12$. After you send your g^b to Alice, she then creates the public key and sends you the message "LPSKQIMBW.ECRBHL" in ciphertext. Show all the steps with fast exponentiation to calculate the deciphering key g^{-ab} and recover the plaintext message. (Note that since we know how to take inverses of matrices in $GL_2(\mathbb{F}_{28})$, it suffices to calculate g^{ab} and then find the inverse as opposed to calculating $(g^a)^{|G|-b}$.)

16. We design the following Diffie-Hellman/ElGamal setup. We choose to encrypt 30 characters: 26 letters of the alphabet, space, the period ".", the comma "," and the exclamation point "!". We associate the number 0 to a space character, 1 through 26 for each of the letters, 27 to the period, 28 to the comma and 29 to "!". We choose to compress triples of characters as follows: (b_1, b_2, b_3), where each $b_i \in \{0, 1, \ldots, 29\}$, corresponds to the number

$$b_1 \times 30^2 + b_2 \times 30 + b_3.$$

The resulting possible numbers are between 0 and $30^3 - 1 = 26,999$. Now, the smallest prime bigger than 30^3 is $p = 27011$. We will work in the group $U(27011)$ and we will view messages as sequences of elements in G encoded as described above. For example: "HI FRANK!" uses the compression of

$$8 \times 30^2 + 9 \times 30 + 0 = 7,470 \qquad 6 \times 30^2 + 18 \times 30 + 1 = 5,941$$
$$14 \times 30^2 + 11 \times 30 + 29 = 12,959$$

and hence corresponds to the sequence in G of $\overline{7470}, \overline{5941}, \overline{12959}$.

Here's the exercise. You are Bob. Alice and you use the system described above and we use the base of $g = \bar{2}$. Alice wants to send a message to you. She first sends her $g^a = \overline{5,738} \in G$. You (Bob) choose the integer $b = 10,372$ and send back $g^b = \overline{255}$. You receive the following ciphertext from Alice:

$$\overline{11336} \quad \overline{8377} \quad \overline{17034} \quad \overline{688} \quad \overline{1031} \quad \overline{13929}.$$

Using Fast Exponentiation, determine the inverse of the public key, g^{-ab}. Decipher the sequence of numbers corresponding to Alice's plaintext message. From the message coding scheme, determine Alice's original message (in English).

17. Use the text to strings of groups elements as described in Exercise 1.12.16. Use the same group $G = U(27011)$ but use the base $g = \bar{5}$. Play the role of Alice. Select $a = 10,000$ while Bob sends you $g^b = \overline{15128}$. Show all the steps to create the string of elements in G that are the ciphertext for the message "I WILL SURVIVE."

1.13   Semigroups and Monoids

In this section, we present two more algebraic structures closely related to groups. They possess value in their own right but we present them here to illustrate examples of structures that are close to groups but have few axioms.

1.13.1   Semigroups

Definition 1.13.1

A *semigroup* is a pair $(S, \circ)$, where S is a set and $\circ$ is an associative binary operation on S.

Having introduced groups already, we can see that a semigroup resembles a group but with only the associativity axiom. Note that Proposition 1.2.13 holds in any semigroup. Obviously, every group is a semigroup.

In every semigroup $(S, \circ)$, because of associativity, the order in which we group the operations in an expression of the form $a \circ a \circ \cdots \circ a$ does not change the result. Hence, we denote by a^k the (unique) element $\overbrace{a \circ a \circ \cdots \circ a}^{k \text{ times}}$.

Example 1.13.2. The set of positive integers equipped with addition $(\mathbb{N}^{>0}, +)$ is a semigroup. $\triangle$

Example 1.13.3. All integers equipped with multiplication $(\mathbb{Z}, \times)$ is also a semigroup. That not all elements have inverses prevented $(\mathbb{Z}, \times)$ from being a group but that does not matter for a semigroup. $\triangle$

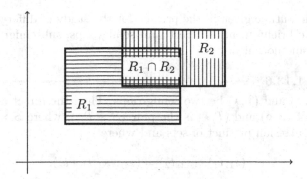

FIGURE 1.20: Operation in Example 1.13.5.

Example 1.13.4. Let $S = \mathbb{Z}$ and suppose that $a \circ b = \max\{a, b\}$. It is easy to see that for all integers a, b, c,

$$a \circ (b \circ c) = \max\{a, \max\{b, c\}\} = \max\{a, b, c\} = \max\{\max\{a, b\}, c\} = (a \circ b) \circ c.$$

Hence, $(\mathbb{Z}, \circ)$ is a semigroup. There is no integer e such that $\forall a \in \mathbb{Z}, \max\{e, a\} = a$, because r would need to be less than every integer. Hence, $(\mathbb{Z}, \circ)$ does not have an identity and consequently it cannot have inverses to elements. $\triangle$

Example 1.13.5. Consider the following set of rectangles in the plane

$$S = \{[a, b] \times [c, d] \subseteq \mathbb{R}^2 \mid a, b, c, d \in \mathbb{R}\}.$$

Note that this includes finite vertical (if $a = b$) or horizontal (if $c = d$) lines as well as the empty set $\emptyset$ (if $b < a$ of $d < c$). The intersection of two elements in S is

$$([a_1, b_1] \times [c_1, d_1]) \cap ([a_2, b_2] \times [c_2, d_2])$$
$$= [\max(a_1, a_2), \min(b_1, b_2)] \times [\max(c_1, c_2), \min(d_1, d_2)]$$

so $\cap$ is a binary operation on S. (See Figure 1.20.) We know that $\cap$ is associative so the pair $(S, \cap)$ is a semigroup. As in Example 1.13.4, there is no identity element. If an identity element U existed, then $U \cap R = R$ for all $R \in S$. Hence, $R \subseteq U$ for all $R \in S$ and thus,

$$\left(\bigcup_{R \in S} R \right) = \mathbb{R}^2 \subseteq U.$$

Hence, $U = \mathbb{R}^2$ but $\mathbb{R}^2$ is not an element of S. $\triangle$

Out of the outline given in the preface for the study of different algebraic structures, we briefly mention direct sum semigroups, subsemigroups, generators, and homomorphisms.

Definition 1.13.6

Let $(S, \circ)$ and $(T, \star)$ be two semigroups. Then the direct sum semigroup of $(S, \circ)$ and $(T, \star)$ is the pair $(S \times T, \cdot)$, where $S \times T$ is the usual Cartesian product of sets and where

$$(s_1, t_1) \cdot (s_2, t_2) = (s_1 \circ s_2, t_1 \star t_2).$$

The direct sum of S and T is denoted by $S \oplus T$.

Definition 1.13.7

A subset A a semigroup $(S, \circ)$ is called a *subsemigroup* if $a \circ b \in A$ for all $a, b \in A$.

A subsemigroup is a semigroup in its own right, using the operation inherited from the containing semigroup. In the theory of groups, in order for a subset H of a group G to be a group in its own right, H needed to be closed under the operation and taking inverses. For semigroups, the condition on taking an inverse does not apply.

Definition 1.13.8

Let $(S, \circ)$ and $(T, \star)$ be two semigroups. A function $f : S \to T$ is called a *semigroup homomorphism* if

$$f(a \circ b) = f(a) \star f(b) \qquad \text{for all } a, b \in S.$$

A semigroup homomorphism that is also a bijection is called a *semigroup isomorphism*.

Example 1.13.9. Let p be a prime number. We define the prime order function $\mathrm{ord}_p : \mathbb{N}^* \to \mathbb{N}$ as $\mathrm{ord}_p(n) = \alpha$, where α is the power on p in the prime factorization of n. This is well-defined by the Fundamental Theorem of Arithmetic. Note that if $p \nmid n$, then $\mathrm{ord}_p(n) = 0$. It is easy to show that,

$$\mathrm{ord}_p(ab) = \mathrm{ord}_p(a) + \mathrm{ord}_p(b) \quad \text{for all } a, b \in \mathbb{N}^*.$$

This means that ord_p is a semigroup homomorphism from $(\mathbb{N}^*, \times)$ to $(\mathbb{N}, +)$. $\triangle$

Example 1.13.10. Consider the following four functions $f_i : \{1, 2, 3\} \to \{1, 2, 3\}$ expressed in n-tuple notation (see Section 1.5.1) as

$$f_1 = (1, 2, 2), \quad f_2 = (3, 2, 3), \quad f_3 = (2, 2, 2), \quad f_4 = (3, 2, 2).$$

We claim that the set $S = \{f_1, f_2, f_3, f_4\}$ together with function composition is a semigroup. We know that function composition is associative. To prove the claim, we simply need to show that $\circ$ is a binary operation on S. The table of composition is:

	f_1	f_2	f_3	f_4
f_1	f_1	f_3	f_3	f_3
f_2	f_4	f_2	f_3	f_4
f_3	f_3	f_3	f_3	f_3
f_4	f_4	f_3	f_3	f_3

This shows that $\circ$ is a binary operation on S, making $(S, \circ)$ into a semigroup.$\triangle$

This last example illustrates the use of a Cayley table to easily see the operations in a semigroup. Though we introduced Cayley tables in Section 1.2.1 in reference to groups, it makes sense to discuss a Cayley table in the context of any finite set S equipped with a binary operation. However, for semigroups, we see that there need not be a row and column indicating the identity operation. Furthermore, since a semigroup does not necessarily contain inverses to elements, the Cayley table of a semigroup is not necessarily a Latin square.

1.13.2 Monoids

Definition 1.13.11

A *monoid* is a pair $(M, *)$, where M is a set and $*$ is a binary operation on M, such that

(1) associativity: $a * (b * c) = (a * b) * c$ for all $a, b, c \in M$;

(2) identity: there exists $e \in M$ such that $a * e = e * a = a$ for all $a \in M$.

Definition 1.13.12

A monoid $(M, *)$ is called *commutative* if $a * b = b * a$ for all $a, b \in M$.

If a monoid is commutative, we typically write the operation with an addition symbol and denote the identity by 0. Consequently, it is not uncommon to say "let $(M, +)$ be a commutative monoid."

Definition 1.13.13

A monoid $(M, *)$ is said to possess the *cancellation property* if for all $a, b, c \in M$,
$$a * c = b * c \Longleftrightarrow a = b \text{ and}$$
$$c * a = c * b \Longleftrightarrow a = b.$$

It is important to note that a monoid may have the cancellation property without possessing inverses, as we will see in some examples below.

Example 1.13.14. The nonnegative integers with addition $(\mathbb{N}, +)$ is a monoid. The presence of 0 in $\mathbb{N}$ gives the nonnegative integers the needed identity. This monoid is commutative. This monoid also possess the cancellation property. $\triangle$

Example 1.13.15. Positive integers with multiplication $(\mathbb{N}^*, \times)$ is a monoid. This possesses the necessary identity but has no inverses. Note that $(\mathbb{N}, \times)$ is also a monoid: Associativity still holds and, $1 \times 0 = 0$, so 1 is still an identity for all $\mathbb{N}$. This monoid is commutative.

Observe that $(\mathbb{N}^*, \times)$ has the cancellation property but $(\mathbb{N}, \times)$ does not. A counterexample for the latter is $0 \times 1 = 0 \times 2$ but $1 \neq 2$. $\triangle$

Example 1.13.16. Let A be a set and let $F = \mathrm{Fun}(A, A)$ be the set of all functions $f : A \to A$. Since function composition is associative, $(F, \circ)$ is a (noncommutative) monoid under function composition $\circ$ with the identity function id_A as the identity element. Functions that are bijections have inverses, but in a monoid not every element must have an inverse. $\triangle$

Example 1.13.17. Let $(M, *)$ be a monoid. If S, T are subsets of M, then we define

$$S * T \stackrel{\text{def}}{=} \{s * t \mid s \in S \text{ and } t \in T\}.$$

Then $(\mathcal{P}(M), *)$ is a monoid. This result is not obvious and we leave it as an exercise for the reader (Exercise 1.13.9). This is called the *power set monoid*.$\triangle$

Example 1.13.18 (Free Monoid, Monoid of Strings). Let Σ be a set of characters. Consider the set of finite strings of elements from Σ, including the empty string, denoted by 1. Authors who work with this structure in the area of theoretical computer science usually denote this set of strings by Σ^*. We define the operation of concatenation $\cdot$ on Σ^* by

$$a_1 a_2 \cdots a_m \cdot b_1 b_2 \cdots b_n \stackrel{\text{def}}{=} a_1 a_2 \cdots a_m b_1 b_2 \cdots b_n.$$

The pair $(\Sigma^*, \cdot)$ is a monoid where the empty string is the identity. This is called the *free monoid* on the set of characters Σ. $\triangle$

The concepts of direct sums, subobjects, and homomorphisms are nearly identical as with groups or semigroups. We give an explicit description here for completeness.

Definition 1.13.19

Let $(M, *)$ and $(N, \circ)$ be two monoids. Then the direct sum monoid of $(M, *)$ and $(N, \circ)$ is the pair $(M \times N, \cdot)$, where $M \times N$ is the usual Cartesian product of sets and where

$$(m_1, n_1) \cdot (m_2, n_2) = (m_1 * m_2, n_1 \circ n_2).$$

The direct sum of M and N is denoted by $M \oplus N$. The identity in $M \oplus N$ is $(1_M, 1_N)$.

We say that the operation occurs componentwise. As with groups and other algebraic structures, Definition 1.13.19 generalizes to the direct sum of any finite number of monoids. It is also possible to make sense of the direct sum of any collection of monoids but this requires some technical care.

Definition 1.13.20

Let $(M, *)$ be a monoid. A nonempty subset $A \subseteq M$ is called a *submonoid* of M if $a * b \in A$ for all $a, b \in A$ (closed under $*$) and if $1_M \in A$.

A submonoid is a monoid in its own right, using the same operation. If we compare this definition to Definition 1.6.1, we observe that the latter required a subgroup to be closed under the operation and closed under taking inverses. Then a simple proof, using inverses, showed that the identity must be in every subgroup. Since elements do not necessarily have inverses in a monoid, the definition of a submonoid must explicitly require the identity to be in the submonoid.

Definition 1.13.21

Let $(M, *)$ and $(N, \circ)$ be two monoids. A function $f : M \to N$ is called a *monoid homomorphism* if

(1) $f(a * b) = f(a) \circ f(b)$ for all $a, b \in M$;

(2) $f(1_M) = 1_N$.

A bijective monoid homomorphism is called a *monoid isomorphism*.

Definition 1.13.22

Let $(M, *)$ be a monoid. We define the *opposite monoid* as the pair $(M^{\mathrm{op}}, *^{\mathrm{op}})$ where, as sets, $M^{\mathrm{op}} = M$ and the operation is

$$a *^{\mathrm{op}} b = b * a.$$

Obviously, if $(M, *)$ is a commutative monoid, then $(M^{\mathrm{op}}, *^{\mathrm{op}}) = (M, *)$. More precisely, the identity function $\mathrm{id} : M \to M$ is a monoid isomorphism if and only if the monoid is commutative.

Example 1.13.23 (State Machine). A *state machine* or *semiautomaton* is a triple (Q, Σ, T) where

- Q is a nonempty set whose elements are called the *states*;
- Σ is a nonempty set whose elements are called the *input symbols*;
- T is a function $T : Q \times \Sigma \to Q$ called a *transition function*.

This set theoretic construct models a machine that can possess various states and depending on an input from Σ changes from one state to another.

Now for every word $w \in \Sigma^*$ we define a function $T_w : Q \to Q$ as follows:

(1) $T_1 : Q \to Q$ is the identity function;

(2) if $w = s \in \Sigma$, then $T_w(q) = T(q, s)$ for all $q \in Q$;

(3) if $w = s_1 s_2 \cdots s_n$ for $s_i \in \Sigma$, then $T_w = T_{s_n} \circ \cdots \circ T_{s_2} \circ T_{s_1}$.

The set of functions gives all possible finite compositions of transition functions arising from transitions produced from one input symbol at a time. This set of functions is denoted by $M(Q, \Sigma, T)$. It is a monoid under function composition and it is called the *transition monoid* or *input monoid*.

The association $t : \Sigma^* \to M(Q, \Sigma, T)^{\mathrm{op}}$ defined by $t(w) = T_w$ is a monoid isomorphism. The function t is a bijection by definition of $M(Q, \Sigma, T)$. Furthermore, if $w_1 = s_1 s_2 \cdots s_m$ and $w_2 = \sigma_1 \sigma_2 \cdots \sigma_n$, then

$$
\begin{aligned}
t(w_1 \cdot w_2) &= T_{s_1 s_2 \cdots s_m \sigma_1 \sigma_2 \cdots \sigma_n} \\
&= T_{\sigma_n} \circ \cdots \circ T_{\sigma_2} \circ T_{\sigma_1} \circ T_{s_m} \circ \cdots \circ T_{s_2} \circ T_{s_1} \\
&= T_{w_2} \circ T_{w_1} \\
&= T_{w_1} \circ^{\mathrm{op}} T_{w_2}.
\end{aligned}
$$

This shows that t is a monoid homomorphism. Hence, we have shown t is a monoid isomorphism. Note that the use of the opposite monoid was necessary primarily because we are reading a word $w = s_1 s_2 \cdots s_m$ from left to right, while we apply the functions T_{s_i} from right to left. $\triangle$

1.13.3   The Grothendieck Group

We end this section with a description of the *Grothendieck group*, which associates a unique group G to any commutative monoid M such that M is isomorphic to a submonoid of G. This construction generalizes the algebraic process of constructing $(\mathbb{Z}, +)$ from $(\mathbb{N}, +)$.

Let M be a commutative monoid. Consider the direct sum monoid $M \oplus M$ and define an equivalence relation $\sim$ on $M \oplus M$ by

$$(m_1, m_2) \sim (n_1, n_2) \iff m_1 + n_2 + k = m_2 + n_1 + k \quad \text{for some } k \in M. \quad (1.16)$$

We leave the proof that this is an equivalence relation as an exercise. (See Exercise 1.13.17.) The presence of the "$+k$ for some $k \in M$" may seem strange but we explain it later. We write $[(m_1, m_2)]$ as the equivalence class for the pair (m_1, m_2) and we denote by $\widetilde{M} = (M \oplus M)/\sim$ the set of all equivalence classes of $\sim$ on $M \times M$.

Now suppose that $(a_1, a_2) \sim (b_1, b_2)$ and $(c_1, c_2) \sim (d_1, d_2)$. Then

$$
\begin{aligned}
a_1 + b_2 + k &= a_2 + b_1 + k \\
c_1 + d_2 + \ell &= c_2 + d_1 + \ell
\end{aligned}
$$

for some $k, \ell \in M$. Thus,

$$(a_1 + c_1) + (b_2 + d_2) + (k + \ell) = (a_2 + c_2) + (b_1 + d_2) + (k + \ell)$$

and hence $(a_1, a_2) + (c_1, c_2) \sim (b_1, b_2) + (d_1, d_2)$. Thus,

$$[(a_1, a_2)] + [(c_1, c_2)] \stackrel{\text{def}}{=} [(a_1, a_2) + (c_1, c_2)] = [(a_1 + c_1, a_2 + c_2)] \quad (1.17)$$

is a well-defined binary operation on $\widetilde{M}$.

Proposition 1.13.24

The set $\widetilde{M}$ equipped with the operation $+$ defined in (1.17) is a group. The identity in $\widetilde{M}$ is $[(0,0)]$ and the inverse of $[(m,0)]$ is $[(0,m)]$. Furthermore, if M possesses the cancellation property, then M is isomorphic to the submonoid $\{[(m,0)] \in \widetilde{M} \mid m \in M\}$.

Proof. That $+$ is associative on $\widetilde{M}$ follows from $+$ being associative on $M \oplus M$. According to (1.17), the element $[(0,0)]$ is an identity on $\widetilde{M}$. Furthermore, we notice that $(0,0) \sim (m,m)$ for all $m \in M$. Hence,

$$[(m,n)] + [(n,m)] = [(m+n, m+n)] = [(0,0)].$$

Thus, $[(n,m)]$ is the inverse of $[(m,n)]$.

Suppose now that M possesses the cancellation property. Because of (1.17), the function $f : M \to \widetilde{M}$ with $f(m) = [(m,0)]$ is a monoid homomorphism. The image of f is $\{[(m,0)] \in \widetilde{M} \mid m \in M\}$. However, $f(m) = f(n)$ is equivalent to $(m,0) \sim (n,0)$ which is equivalent to

$$m + k = n + k$$

for some $k \in M$. Thus, $m = n$ by the cancellation property. Then f is an injective function and is thus a monoid isomorphism between M and $\{[(m,0)] \in \widetilde{M} \mid m \in M\}$. $\square$

We now see why we needed the "$+k$ for some $k \in M$" in (1.16). It is possible in M with two unequal elements m and n for there to exist a k such that $m + k = n + k$. In an abelian group $m + k = n + k$ implies $m = n$. Hence, in $\widetilde{M}$, we would need $f(m)$ and $f(n)$ to be the same element.

Definition 1.13.25

Let $(M, +)$ be a commutative monoid. The group $(\widetilde{M}, +)$ described in Proposition 1.13.24 is called the *Grothendieck group* associated with $(M, +)$.

In the Grothendieck group associated with $(\mathbb{N}, +)$, elements in $\widetilde{\mathbb{N}}$ are equivalence classes of pairs (a, b). Since $\mathbb{N}$ has the cancellation property, $(a, b) \sim (c, d)$ if and only if $a + d = b + c$. We can think of these pairs as a displacement of magnitude the difference of a and b to the right (positive) if $a \geq b$ and to the left (negative) if $b \geq a$. Identifying $\mathbb{Z} = \widetilde{\mathbb{N}}$, we view positive and negative integers as displacement integers with a direction.

EXERCISES FOR SECTION 1.13

1. Prove that $(\mathbb{N}^*, \gcd)$ is a semigroup but not a monoid.

2. Prove that $(\mathbb{N}^*, \mathrm{lcm})$ is a semigroup and is a monoid.

3. Let p be a prime number. Recall the ord_p function defined in Example 1.13.9. In Exercise 1.13.1 you showed that $(\mathbb{N}^*, \gcd)$ is a semigroup. Prove that ord_p : $(\mathbb{N}^*, \gcd) \to (\mathbb{N}, \min)$ is a semigroup homomorphism.

4. Let $(L, \preccurlyeq)$ be a lattice.
   (a) Prove that $(L, \circ)$, where $a \circ b = \mathrm{lub}(a, b)$ is a semigroup.
   (b) Prove that $(L, \star)$, where $a \star b = \mathrm{glb}(a, b)$ is a semigroup.

5. Let $(L, \preccurlyeq)$ be a lattice and let $(L^{\mathrm{op}}, \preccurlyeq^{\mathrm{op}})$ be the lattice defined by

$$a \preccurlyeq^{\mathrm{op}} b \Longleftrightarrow b \preccurlyeq a.$$

Prove that the function $f : L \to L^{\mathrm{op}}$ defined by $f(a) = a$ is a semigroup isomorphism between $(L, \circ)$, where $a \circ b$ gives the least upper bound between a and b in $(L, \preccurlyeq)$ and $(L^{\mathrm{op}}, \star)$, where $a \star b$ gives the greatest lower bound between a and b in $(L^{\mathrm{op}}, \preccurlyeq^{\mathrm{op}})$.

6. Let U be a set and $\mathcal{P}(U)$ the power set of U. (a) Prove that $(\mathcal{P}(U), \cap)$ is a monoid but not a group. (b) Prove that $(\mathcal{P}(U), \cup)$ is a monoid but not a group.

7. Let U be a set and consider the function $f : \mathcal{P}(U) \to \mathcal{P}(U)$ defined by $f(A) = \overline{A}$, set complement. Prove that f is a monoid isomorphism from $(\mathcal{P}(U), \cap)$ to $(\mathcal{P}(U), \cup)$.

8. Let F be the set of functions from $\{1, 2, 3, 4\}$ to itself and consider the monoid $(F, \circ)$ with the operation of function composition. Let S be the smallest sub-semigroup of F that contains the functions expressed in 4-tuple notation

$$f_1 = (1, 1, 2, 3), \quad f_2 = (4, 2, 2, 3), \quad f_3 = (4, 1, 3, 3), \quad f_4 = (4, 1, 2, 4).$$

Prove that S is not a monoid. Prove that S contains no bijections.

9. Prove that the power set of a monoid $(M, *)$ is indeed a monoid as claimed in Example 1.13.17.

10. Let U be a set and let $\mathrm{Fun}(U, U)$ be the monoid of functions from U to U, with the operation of composition. Let A be a subset of U and let $N_A = \{f \in \mathrm{Fun}(U, U) \mid f(A) \subseteq A\}$.
    (a) Show that N_A is a submonoid of $\mathrm{Fun}(U, U)$.
    (b) Give an example that shows that not every submonoid of $\mathrm{Fun}(U, U)$ is N_A for some $A \subseteq U$.

11. Call M the set of nonzero polynomials with real coefficients. The pair $(M, \times)$ is a monoid with the polynomial 1 as the identity. Decide which of the following subsets of M are submonoids and justify your answer.
    (a) Nonconstant polynomials.
    (b) Polynomials whose constant coefficient it 1.
    (c) Palindromic polynomials. [A polynomial $a_n x^n + \cdots + a_1 x + a_0$ is called *palindromic* if $a_{n-i} = a_i$ for all $0 \leq i \leq n$.]

(d) Polynomials with odd coefficients.

(e) Polynomials with $a_{n-1} = 0$.

(f) Polynomials with no real roots.

12. Let $\varphi : M \to N$ be a monoid homomorphism. We define the *kernel* of the monoid homomorphism as

$$\operatorname{Ker}\varphi = \{m \in M \mid \varphi(m) = 1_N\}.$$

Show that $\operatorname{Ker}\varphi$ is a submonoid of M.

13. Let $\varphi : M \to N$ be a monoid homomorphism. We define the *image* of the monoid homomorphism as

$$\operatorname{Im}\varphi = \{n \in N \mid n = \varphi(m) \text{ for some } m \in M\}.$$

Show that $\operatorname{Im}\varphi$ is a submonoid of N.

14. Prove that $C^0(\mathbb{R}, \mathbb{R})$, the set of continuous real-valued function from $\mathbb{R}$, is a submonoid of $\operatorname{Fun}(\mathbb{R}, \mathbb{R})$ (equipped with composition).

15. Consider the monoid $M = \operatorname{Fun}(\mathbb{R}, \mathbb{R})$ and consider the function

$$f(x) = \begin{cases} -x & \text{if } x < 0 \\ 0 & \text{if } 0 \leq x < 1 \\ 1 - x & \text{if } 1 \leq x < 2 \\ x & \text{if } 2 \leq x. \end{cases}$$

(a) Prove that f^1, f^2, f^3, f^4 are all distinct.

(b) Prove that $f^n = f^4$ for $n \geq 4$.

(c) This is an example of a sequence of the form $(f^k)_{k \geq 1}$ that is not constant but eventually constant. Explain why eventually constant sequences do not occur in groups.

16. Let $(S, \cdot)$ be a semigroup. Let e be some object not in S and consider the set $S' = S \cup \{e\}$. Define the operation $*$ on S' by $a * b = a \cdot b$ for all $a, b \in S$ and $a * e = e * a = a$ for all $a \in S'$. Prove that $(S', *)$ is a monoid.

17. Prove that the relation $\sim$ defined (1.16) is indeed an equivalence relation.

18. Show that the Grothendieck group associated with the monoid $(\mathbb{N}^*, \times)$ is the group $(\mathbb{Q}^{>0}, \times)$.

19. Consider the monoid $(\mathbb{N}, +)$. Let $M = \langle 4, 7 \rangle$ denote the smallest submonoid that contains the subset $\{4, 7\}$. [We say that M is the submonoid *generated* by $\{4, 7\}$.] List the first 15 elements in M. Prove that the Grothendieck group associated with $(M, +)$ is group-isomorphic to $(\mathbb{Z}, +)$.

20. Let A be an abelian group. Prove that the Grothendieck group associated with A as a monoid is a group that is isomorphic to A.

1.14 Projects

Investigative Projects

PROJECT I. **Rubik's Cube.** Let K be the group of operations on the Rubik's Cube. Can you determine a set of generators and explore some of the relations among them? Is K naturally a subgroup of some S_n? If so, how and for what n? What are the orders of some elements that are not your generators? Can you determine the size of K either from the previous question or from other reasoning? Explore any other properties of the Rubik's Cube group.

PROJECT II. **Matrix Groups.** Consider the family of groups $GL_n(\mathbb{F}_p)$, where n is some positive integer with $n \geq 2$ and p is a prime number. Can you provide generators for some of them? What are some subgroups? Can you give presentations for some of these? Can you show some of them as isomorphic to subgroups of some S_m. [Hint: Use a CAS.]

PROJECT III. **Shuffling Cards and S_{52}.** A shuffle of a card deck is an operation that changes the order of a deck and hence can be modeled by an element of S_{52}. In this project, study patterns of shuffling cards. Two popular kinds of shuffles are the random riffle shuffle and the overhand shuffle. A perfect riffle involves cutting the deck in half and interlacing one card from one half with exactly one card from the other half. How would you model shuffling styles by certain permutations in S_{52}? Might patterns occur in shuffling?

PROJECT IV. **Sudoku and Group Theory.** Let S be the set of all possible solutions to a Sudoku puzzle, i.e., all possible ways of filling out the grid according to the rules. There exists a group of transformations $\mathcal{G}$ on specific numbers, rows, and columns such that given any solution s and any $g \in \mathcal{G}$, we can know for certain that $g(s)$ is another solution. Determine and describe concisely this group $\mathcal{G}$. Can you describe it as a subgroup of some large permutation group? Can you find the size of $\mathcal{G}$?

We will call two fillings s_1 and s_2 of the Sudoku grid equivalent if $s_2 = g(s_1)$ for some $g \in \mathcal{G}$. Explore if there exists nonequivalent Sudoku fillings.

PROJECT V. **The 15 Puzzle.** Find a source that describes the so-called 15-puzzle. Here is an interesting question about this puzzle. Suppose that you start with tiles 1 through 15 going left to right, top row down to bottom row, with the empty square in the lower right corner. Is it possible to obtain every theoretical configuration of tiles on the board? (For reference, label the empty slot as number 16.) If it is not, try to find the subgroup of S_{16} or S_{15} of transformations that you can perform on this puzzle. Make sure what you work with is a group. Can you generalize?

PROJECT VI. **Groups of Functions.** Consider the set of functions of the form $f(x) = \dfrac{ax+b}{cx+d}$, where $a, b, c, d \in \mathbb{Z}$. (Do not worry about domains of definition or, if you insist, just think of the domains of these functions as on $\mathbb{R} - \mathbb{Q}$.) Show that (with perhaps a restriction on what a, b, c, d are) this set of functions is a group. Try to find generators and relations for this group or any other way to describe the group.

PROJECT VII. **Groups of Rigid Motions of Polyhedra.** Let Π be a polyhedron. Call $G(\Pi)$ the group of rigid motions in $\mathbb{R}^3$ that map Π into Π. For example, If Π is the cone over a pentagon, then $G(\Pi) = Z_5$. Does $G(\Pi)$ consist of only transformations that are rotations about an axis? Find $G(\Pi)$ for the regular polyhedra. Find $G(\Pi)$ for some irregular polyhedra that do not have $G(\Pi) = \{1\}$. For a given group G, does there exist a polyhedron Π such that $G(\Pi) = G$?

PROJECT VIII. **Probabilities on S_n.** Explore probabilities and expected values (or variances) of various random variables associated with permutations. For example, for various n, determine the probability that a permutation has no fixed points (a fixed point of $\sigma \in S_n$ is an integer a such that $\sigma(a) = a$); calculate the expected value of the number of the number of fixed points, of the order of the permutations, or of the number of inversions of permutations. Feel free to concoct your own random variables on S_n. Attempt to prove formulas that are true for all n and then calculate the limit as $n \to \infty$.

PROJECT IX. **Cayley Graphs.** Construct the Cayley graphs for various presentations of groups that we have encountered so far. Discuss some of the properties of the Cayley graphs. Can non-isomorphic groups have the same Cayley graph? Can you find an Archimedean solid that is not the Cayley graph of any group?

PROJECT X. **Diffie-Hellman and ElGamal.** Program and document a computer program that implements the ElGamal encryption scheme on a file using a key that is created via the Diffie-Hellman procedure. Use a group G and a base $g \in G$ such that the order of g is larger than current computers will run a `for` loop in a reasonable amount of time. Feel free to choose $\mathcal{M}$ as you think is effective and $h : \mathcal{M} \to \text{Fun}(\mathbb{N}^*, G)$ as you wish.

PROJECT XI. **Sós Permutations.** Let $\alpha, \beta \in [0,1)$ and let $f_{\alpha,\beta} : \mathbb{N} \to [0,1)$ be the function that returns the fractional part of $\alpha m + \beta$. Fix a positive integer n and additionally assume that $q\alpha \notin \mathbb{N}$ for all $1 \le q \le n$. The $(n+1)$-tuple of real numbers $(f(0), f(1), \ldots, f(n))$ are all distinct (show this), so we define the permutation $\pi = \pi_{\alpha,\beta}^{(n)}$ on $\{0, 1, \ldots, n\}$ such that

$$f(\pi(0)) < f(\pi(1)) < \cdots < f(\pi(n)).$$

We denote the set of permutations arising in this way as

$$\text{Sos}_n = \{\pi_{\alpha,\beta}^{(n)} \mid \alpha, \beta \in [0,1) \text{ and } \nexists q \in \{1,\ldots,n\}, q\alpha \in \mathbb{N}\}.$$

Study this subset of permutations in $S_{\{0,1,\ldots,n\}} \cong S_{n+1}$. for small n, find all the permutations in Sos_n. What thought do you have about $|\text{Sos}_n|$ for general n. Is Sos_n a subgroup? Is it all of S_{n+1}? Discuss inversions or cycles of such permutations based on α and β.

Expository Projects

PROJECT XII. **Niels Abel, 1802–1829.** The term *abelian* honors Danish born mathematician Niels Abel. Using reliable sources, explore Abel's personal background, his mathematical work, and the reception of his work in the mathematical community. Investigate the aspect of his work that encouraged later mathematicians to dub commutative groups as "abelian groups."

PROJECT XIII. **Evariste Galois, 1811–1832.** Many sources on the history of mathematics state that Evariste Galois' study on symmetries among roots of polynomials laid the foundation for group theory. Using reliable sources, explore Galois' personal background, his mathematical work, and the reception of his work in the mathematical community. Identify the sources that support the claim made by so many historians of mathematics.

PROJECT XIV. **Arthur Cayley, 1821–1895.** In this chapter, we saw the name Cayley attached to various objects that helped us study groups, e.g., the Cayley table of a group (or monoid), the Cayley graph (of a presentation). Using reliable sources, explore Cayley's personal background, his mathematical work, and the reception of his work in the mathematical community. Examine the role of group theory in his work.

PROJECT XV. **Music and Group Theory.** There are a number of interesting connections between group theory and tonal music theory. Explore this connection.

PROJECT XVI. **Wallpaper Groups.** Find, discuss, and illustrate a proof the Wallpaper Groups Theorem. Examine methods that artists use to create a wallpaper pattern that is invariant under one of the 17 wallpaper symmetry groups.

PROJECT XVII. **Group Theory in Chemistry.** Explain various applications of group theory to chemistry. Argue whether the use of group theory in these contexts contribute significantly.

2

Quotient Groups

One of the most fascinating aspects of group theory is how much internal structure follows by virtue of the three group axioms. Groups possess much more internal structure than we have yet seen in Chapter 1. The internal properties will often permit us to create a "quotient group," which is a smaller group that retains some of the group information and conflates other information.

The process of creating a quotient group arises in the creation of modular arithmetic from arithmetic on the integers. See Section A.6. Consequently, we briefly review that process.

Fix a positive integer n greater than 1. Let $G = (\mathbb{Z}, +)$ and consider the subgroup $H = n\mathbb{Z}$. We defined the congruence relation as

$$a \equiv b \iff n \mid (b - a) \iff b - a \in H.$$

We proved that $\equiv$ was an equivalence relation and we defined the congruence class $\bar{a}$ as the set of all elements that are congruent to a. Note that

$$\bar{a} = \{\dots, a - 2n, a - n, a, a + n, a + 2n, \dots\} = a + n\mathbb{Z} = a + H.$$

We defined $\mathbb{Z}/n\mathbb{Z}$ as the set of equivalence classes modulo n. Explicitly, $\mathbb{Z}/n\mathbb{Z} = \{\bar{0}, \bar{1}, \bar{2}, \dots, \overline{n-1}\}$. Furthermore, we showed that addition behaves well with respect to congruence, by which we mean that $a \equiv c$ and $b \equiv d$ imply that $a + b \equiv c + d$. Then, defining addition on $\mathbb{Z}/n\mathbb{Z}$ as

$$\bar{a} + \bar{b} \stackrel{\text{def}}{=} \overline{a + b}$$

is in fact well-defined. From this, we were able to create the group $(\mathbb{Z}/n\mathbb{Z}, +)$.

We will soon say that $\mathbb{Z}/n\mathbb{Z}$ is the quotient group of $\mathbb{Z}$ by its subgroup $n\mathbb{Z}$. We will also notice that the group $(\mathbb{Z}, +)$ has many properties (e.g., abelian, cyclic, infinite) that make the construction simpler than in the arbitrary case.

Section 2.1 introduces the concept of cosets, which immediately leads to Lagrange's Theorem, a deep theorem about the internal structure of groups. Section 2.2 defines normal subgroups, which are necessary for the construction of quotient groups. Section 2.3 gives the construction for quotient groups and provides many examples. Section 2.4 develops a number of theorems that illustrate how to understand the internal structure of a group from knowing the structure of a quotient group. Finally, in Section 2.5, using the quotient process, we prove a classification theorem for all abelian groups that are finitely generated.

DOI: 10.1201/9781003299233-2

2.1 Cosets and Lagrange's Theorem

2.1.1 Cosets

Following the guiding example of modular arithmetic provided at the beginning of the chapter, we first consider what should play the role of $\bar{a} = a + n\mathbb{Z}$ in groups in general.

Definition 2.1.1

Let G be a group and let H be a subgroup. The set of elements defined by gH (respectively Hg) is called the left (respectively right) *coset* of H by g.

Example 2.1.2. Let $G = D_4$ and consider the subgroups $R = \langle r \rangle$ and $H = \langle s \rangle$. The left cosets of R are

$$
\begin{aligned}
1R &= \{1, r, r^2, r^3\}, & sR &= \{s, sr, sr^2, sr^3\}, \\
rR &= \{r, r^2, r^3, 1\}, & srR &= \{sr, sr^2, sr^3, s\}, \\
r^2R &= \{r^2, r^3, 1, r\}, & sr^2R &= \{sr^2, sr^3, s, sr\}, \\
r^3R &= \{r^3, 1, r, r^2\}, & sr^3R &= \{sr^3, s, sr, sr^2\}.
\end{aligned}
$$

Note that $1R = rR = r^2R = r^3R$ and $sR = srR = sr^2R = sr^3R$. Hence, there are only two distinct left cosets of R, namely $1R = R$, which consists of all the rotations in D_4, and sR, which consists of all the reflections. The right cosets of R are

$$
\begin{aligned}
R &= \{1, r, r^2, r^3\}, & Rs &= \{s, rs, r^2s, r^3s\} = \{s, sr^3, sr^2, sr\}, \\
Rr &= \{r, r^2, r^3, 1\}, & Rsr &= \{sr, rsr, r^2sr, r^3sr\} = \{sr, s, sr^3, sr^2\}, \\
Rr^2 &= \{r^2, r^3, 1, r\}, & Rsr^2 &= \{sr^2, rsr^2, r^2sr^2, r^3sr^2\} = \{sr^2, sr, s, sr^3\}, \\
Rr^3 &= \{r^3, 1, r, r^2\}, & Rsr^3 &= \{sr^3, rsr^3, r^2sr^3, r^3sr^3\} = \{sr^3, sr^2, sr, s\}.
\end{aligned}
$$

Note that $R = Rr = Rr^2 = Rr^3$ and $Rs = Rsr = Rsr^2 = Rsr^3$. Again, there are only two distinct right cosets of R, namely R and Rs. We also note that for all $g \in D_4$, we have $gR = Rg$. This resembles commutativity, but we must recall that D_4 is not commutative. Before we are tempted to think this happens for all subgroups (and we should be inclined to doubt that it would), let us do the same calculations with the subgroup H.

The left cosets of H are

$$
\begin{aligned}
1H &= \{1, s\}, & sH &= \{s, 1\} = \{1, s\}, \\
rH &= \{r, rs\} = \{r, sr^3\}, & srH &= \{sr, srs\} = \{sr, r^3\}, \\
r^2H &= \{r^2, r^2s\} = \{r^2, sr^2\}, & sr^2H &= \{sr^2, sr^2s\} = \{sr^2, r^2\}, \\
r^3H &= \{r^3, r^3s\} = \{r^3, sr\}, & sr^3H &= \{sr^3, sr^3s\} = \{sr^3, r\}.
\end{aligned}
$$

There are four distinct left cosets: $H = sH$, $rH = sr^3H$, $r^2H = sr^2H$, and $r^3H = srH$. The right cosets are

$$
\begin{aligned}
H &= \{1, s\}, & Hs &= \{s, 1\} = \{1, s\}, \\
Hr &= \{r, sr\}, & Hsr &= \{sr, r\}, \\
Hr^2 &= \{r^2, sr^2\}, & Hsr^2 &= \{sr^2, r^2\}, \\
Hr^3 &= \{r^3, sr^3\}, & Hsr^3 &= \{sr^3, r^3\}.
\end{aligned}
$$

Notice again that there are four distinct right cosets: $H = Hs$, $Hr = Hsr$, $Hr^2 = Hsr^2$, and $Hr^3 = Hsr^3$. However, with the subgroup H, it is not true that $gH = Hg$ for all $g \in G$. For example, $Hr = \{r, sr\}$ while $rH = \{r, sr^3\}$. △

This first example underscores that left cosets are not necessarily equal to right cosets, but that depending on the subgroup, they might be equal. Of course, if G is an abelian group then $gH = Hg$ for all $H \le G$ and for all $g \in G$.

Proposition 2.1.3

> Let H be a subgroup of a group G and let $g \in G$ be arbitrary. Then there exists a bijection between H and gH and between H and Hg. Furthermore, if H is a finite subgroup of G, then $|H| = |gH| = |Hg|$.

Proof. Consider the function $f : H \to gH$ defined by $f(x) = gx$. This function is injective because $f(x_1) = f(x_2)$ implies that $gx_1 = gx_2$ so that $x_1 = x_2$. Furthermore, the function is surjective by definition of gH. Thus, f is a bijection between H and gH. If H is finite, then $|H| = |gH|$.

Similarly, the function $\varphi : H \to Hg$ defined by $\varphi(x) = xg$ is a bijection between H and Hg. □

Recall that in the motivating example of modular arithmetic, the cosets $a + n\mathbb{Z}$ corresponded to the congruence classes modulo n. In general groups, cosets correspond to certain equivalence relations. However, because groups are not commutative in general, we must consider two equivalence relations.

Proposition 2.1.4

> Let G be a group and let H be a subgroup. The relations $\sim_1$ and $\sim_2$, defined respectively as
>
> $$ a \sim_1 b \iff a^{-1}b \in H \quad \text{and} \quad a \sim_2 b \iff ba^{-1} \in H $$
>
> are equivalence relations. Furthermore, the equivalence classes for $\sim_1$ (resp. $\sim_2$) are the left (resp. right) cosets of H.

Proof. We prove the proposition for $\sim_1$. The proof for $\sim_2$ is identical in form.

Let G be a group and let H be any subgroup. For all $a \in G$, $a^{-1}a = 1 \in H$, so $\sim_1$ is reflexive. Suppose that $a \sim_1 b$. Then $a^{-1}b \in H$. Since H is closed under taking inverses, $(a^{-1}b)^{-1} = b^{-1}a \in H$, and so $b \sim_1 a$. Thus, $\sim_1$ is symmetric. Now suppose that $a \sim_1 b$ and $b \sim_1 c$. By definition, $a^{-1}b \in H$ and $b^{-1}c \in H$. Since H is a subgroup, $(a^{-1}b)(b^{-1}c) = a^{-1}c \in H$, which means that $a \sim_1 c$. Hence, $\sim_1$ is transitive. We conclude that $\sim_1$ is an equivalence relation.

The $\sim_1$ equivalence class of $a \in G$ consists of all elements g such that $a \sim_1 g$, i.e., all elements g such that there exists $h \in H$ with $a^{-1}g = h$. Thus, $g = ah$ so $g \in aH$. Conversely, if $g \in aH$, then $a^{-1}g \in H$ and thus $a \sim_1 g$. This shows that the $\sim_1$ equivalence class of a is aH. $\qquad\square$

The equivalence relations $\sim_1$ and $\sim_2$ are not necessarily distinct. The relations are identical if and only if all the left cosets of H match up with right cosets of H.

If the group is abelian, left and right cosets are equal for all subgroups H and hence the relations $\sim_1$ and $\sim_2$ are equal. This is why we did not need to define two concepts of congruence relations on $\mathbb{Z}$. Indeed, for all $a, b \in \mathbb{Z}$,

$$-a + b = b - a.$$

Proposition 2.1.4 leads immediately to the following corollary.

Proposition 2.1.5

Let H be a subgroup of a group G and let $g_1, g_2 \in G$. Then

$$g_1 H = g_2 H \iff g_1^{-1} g_2 \in H \iff g_2^{-1} g_1 \in H \qquad \text{and}$$
$$Hg_1 = Hg_2 \iff g_2 g_1^{-1} \in H \iff g_1 g_2^{-1} \in H.$$

It is also possible to deduce Proposition 2.1.5 from the following reasoning. The equality holds $g_1 H = g_2 H$ if and only if $H = g_1^{-1} g_2 H$ if and only if for all $h \in H$, there exists $h' \in H$ such that $g_1^{-1} g_2 h = h'$. This implies that $g_1^{-1} g_2 = h' h^{-1} \in H$. Conversely, if $g_1^{-1} g_2 \in H$, then writing $g_1^{-1} g_2 = h''$, for all h in H, we have $g_1^{-1} g_2 h = h'' h \in H$, so $g_1^{-1} g_2 H = H$ and thus $g_2 H = g_1 H$. A similar reasoning holds with right cosets.

Because equivalence classes on a set partition that set, Proposition 2.1.4 leads immediately to the following corollary.

Corollary 2.1.6

Let H be a subgroup of a group G. The set of left (respectively right) cosets form a partition of G.

Figure 2.1 illustrates how left and right cosets of a subgroup H partition a group G. It is important to note that the subgroup H is both a left and a right coset, and the remaining left and right cosets partition $G - H$. In the figure, the left cosets are shown to overlap the right cosets. In general, it is possible for each left coset to be a right coset or for only some left cosets to intersect with some right cosets.

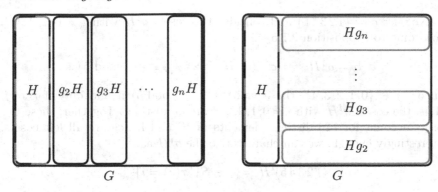

FIGURE 2.1: Illustration of left and right cosets of H in G.

In the example of modular arithmetic, though, the original group $G = \mathbb{Z}$ was infinite, for any given modulus n, there were exactly n (a finite number of) left cosets, $\overline{0}, \overline{1}, \ldots, \overline{n-1}$. This may happen in a general group setting.

Definition 2.1.7

Let H be a subgroup of a group G that is not necessarily finite. If the number of distinct left cosets is finite, then this number is denoted by $|G : H|$ and is called the *index* of H in G.

For any subgroup H of a group G, the set of inverse elements H^{-1} is equal to H. Consequently, the inverse function $f(x) = x^{-1}$ on the group G, maps the left coset gH to the right coset $(gH)^{-1} = H^{-1}g^{-1} = Hg^{-1}$. Thus, the inverse function gives a bijection between the set of left cosets and the set of right cosets of H. In particular, if $|G : H|$ is finite, then it also counts the number of right cosets.

Example 2.1.8. Let $G = S_5$ and let $H = \{\sigma \in G \,|\, \sigma(5) = 5\}$. Since H is the symmetric group on $\{1, 2, 3, 4\}$, we have $H \cong S_4$. Hence, $|G| = 120$ and $|H| = 24$. Note that the index of H in S_5 is $|S_5 : H| = 5$, so whether we consider left or right cosets, we will find 5 of them. Obviously, H is both a left coset and a right coset of H. We investigate a few other cosets of H.

Consider the left coset $(1\,2)H$. Since $(1\,2) \in H$, then $(1\,2)H = H$. Furthermore, any transposition $(a\,b)$ in which $a, b < 5$ satisfies $(a\,b)H = H$. Another left coset $(1\,5)H \neq H$ because $(1\,5) \notin H$. Now consider a third coset $(2\,5)H$. Since $(2\,5) \notin H$, we know that $(2\,5)H \neq H$. However, by Proposition 2.1.5, since $(2\,5)^{-1}(1\,5) = (1\,2\,5) \notin H$ so $(2\,5)H \neq (1\,5)H$. In fact, for any a and b satisfying $1 \le a < b \le 5$, we have $(b\,5)^{-1}(a\,5) = (a\,b\,5) \notin H$ so $(a\,5)H \neq (b\,5)H$ for $a \neq b$. Each coset consists of 24 elements and we have found 5 distinct cosets. Since $5 \times 24 = 120 = |S_5|$, we have found all the left cosets, which implies that $(a\,5)H$ for $1 \le a \le 5$ is a complete list of left cosets of H.

Now let $\sigma = (1\,2\,3\,4\,5)$ and consider the cosets $\sigma^i H$ with $i = 0, 1, 2, 3, 4$. According to Proposition 2.1.5,

$$\sigma^i H = \sigma^j H \Longleftrightarrow \sigma^{i-j} \in H \Longleftrightarrow i - j \equiv 0 \pmod{5}.$$

Since $i, j \in \{0, 1, 2, 3, 4\}$, the equality $i \equiv j \pmod 5$ if and only if $i = j$. Thus, the cosets $\sigma^i H$ with $i = 0, 1, 2, 3, 4$ are all distinct. Together, these five cosets account for 120 distinct elements of S_5 and hence are all left cosets. Interestingly enough, we can characterize the $\sigma^i H$ as

$$(1\,2\,3\,4\,5)^i H = \{\tau \in S_5 \,|\, \tau(5) = i\}. \qquad \triangle$$

Previously, we emphasized that though both $\mathbb{Z}$ and its subgroup $n\mathbb{Z}$ are infinite, the index $|\mathbb{Z} : n\mathbb{Z}| = n$ is finite. The following example illustrates some other possibilities of subgroup indices in the context of an infinite group.

Example 2.1.9. Consider the multiplicative group of nonzero reals, $(\mathbb{R}^*, \times)$. The subset $H = \mathbb{R}^{>0}$ of positive real numbers is a subgroup. The subgroup $\mathbb{R}^{>0}$ has only two distinct cosets, namely $\mathbb{R}^{>0}$ and $(-1)\mathbb{R}^{>0} = \mathbb{R}^{<0}$. Hence, the index of $\mathbb{R}^{>0}$ in $\mathbb{R}^*$ is $|\mathbb{R}^* : \mathbb{R}^{>0}| = 2$. As a point of terminology, it is not proper to say that "$\mathbb{R}^{>0}$ consists of half of the real numbers" because $\mathbb{R}^{>0}$ is the same cardinality as $\mathbb{R}$. However, the concept of index makes precise our intuitive sense that "$\mathbb{R}^{>0}$ consists of half of the real numbers."

As another example, consider the subgroup $\mathbb{Q}^*$. We show by contradiction that the index of $\mathbb{Q}^*$ in $\mathbb{R}^*$ is not only infinite, but uncountable. Assume that $\mathbb{Q}^*$ has a countable number of cosets in $\mathbb{R}^*$. Let S be a complete set of distinct representatives of the partition formed by the cosets. The Cartesian product of two countable sets is countable so $S \times \mathbb{Q}^*$ is countable. By construction of cosets, the function $f : S \times \mathbb{Q}^* \to \mathbb{R}^*$ defined by $f(s, q) = sq$ is a surjection. The existence of a surjective function from a countable set onto $\mathbb{R}^*$ implies that $\mathbb{R}^*$ is countable, which is a contradiction. Hence, the index $|\mathbb{R}^* : \mathbb{Q}^*|$ is uncountable. $\qquad \triangle$

2.1.2   Lagrange's Theorem

Proposition 2.1.3 leads to a profound theorem in group theory. A perceptive student might have conjectured the following result from some examples, but none of the theorems provided in Chapter 1 lead to it. With the concept of cosets, this theorem is hiding in plain sight. We uncover it now.

Theorem 2.1.10 (Lagrange's Theorem)

Let G be a finite group. If $H \leq G$, then $|H|$ divides $|G|$. Furthermore,

$$\frac{|G|}{|H|} = |G : H|.$$

Proof. By Proposition 2.1.3, each left coset of H has the same cardinality of $|H|$. Since the set of left cosets partitions G, then the sum of cardinalities of the distinct cosets is equal to $|G|$. Thus,

$$|G| = |H| \cdot |G : H|,$$

and the theorem follows. $\square$

In the language of posets, Lagrange's Theorem can be rephrased by saying that if G is a finite group, then the cardinality function from $(\text{Sub}(G), \leq)$ to $(\mathbb{N}^*, |)$ is monotonic.

A number of corollaries follow from Lagrange's Theorem.

Corollary 2.1.11

For every element g in a finite group G, the order $|g|$ divides $|G|$.

Proof. The order $|g|$ is the order of the subgroup $\langle g \rangle$. Hence, $|g|$ divides $|G|$ by Lagrange's Theorem. $\square$

Corollary 2.1.12

For every element g in a finite group G, we have $g^{|G|} = 1$.

Proof. By Corollary 2.1.11, if $|g| = k$, then k divides $|G|$. So $|G| = km$ for some $m \in \mathbb{Z}$. Since $g^k = 1$ by definition of order, then $g^{|G|} = g^{km} = (g^k)^m = 1^m = 1$. $\square$

Lagrange's Theorem and its corollaries put considerable restrictions on possible subgroups of a group G. For example, in Exercise 1.4.10 concerning the classification of groups of order 6, that G does not contain elements of order 4 or of order 5 follows immediately from Lagrange's Theorem.

As a simple application, knowing that the size of a subgroup can only be a divisor of the size of the group may tell us whether or not we have found all the elements in a subgroup. In particular, as the following example illustrates, if we know that a subgroup H has $|H|$ strictly greater than the largest strict divisor of $|G|$, then we can deduce that $|H| = |G|$ and hence that $H = G$.

Example 2.1.13. Consider the group A_4 and the subgroup $H = \langle (1\,2\,3), (1\,2\,4) \rangle$. Since $|A_4| = 12$, by Lagrange's Theorem, the subgroups of A_4 can only be of order 1, 2, 3, 4, 6, or 12. By taking powers of the generators of H, we know that

$$1, \quad (1\,2\,3), \quad (1\,3\,2), \quad (1\,2\,4), \quad (1\,4\,2)$$

are in H. Furthermore, $(1\,2\,3)(1\,4\,2) = (1\,4\,3)$ is in H as must be its square $(1\,3\,4)$. This shows that H contains at least 7 elements. Since $|H|$ is greater than 6, by Lagrange's Theorem, $|H| = 12$ and hence $H = A_4$. $\triangle$

Lagrange's Theorem also leads immediately to the following important classification theorem.

Proposition 2.1.14

Let G be a group such that $|G| = p$, a prime number. Then $G \cong Z_p$.

Proof. Let $g \in G$ be a nonidentity element. Then $\langle g \rangle$ is a subgroup of G that has at least 2 elements. By Lagrange's Theorem, $|g| = |\langle g \rangle|$ divides p. Hence, $|g| = p$ and $\langle g \rangle = G$. Therefore, G is cyclic. The proposition follows from Proposition 1.9.25. □

2.1.3   Partial Converses to Lagrange's Theorem

Given the profound nature of Lagrange's Theorem, it is natural to wonder whether the theorem or any of its corollaries have converses. Consider the following questions. (1) Given a finite group G, if d is a divisor of $|G|$, does there necessarily exist a subgroup of order d? (2) Given a finite group G, if d is a divisor of $|G|$, does there necessarily exist an element of order d? The answer to both of these questions is "no."

The answer to question (2) is obvious in that if every group contained an element of order $|G|$, then every group would be cyclic, which we know to be false. For question (1), the smallest counter-example occurs with A_4. In Example 1.8.7, we drew the lattice of A_4 and found that it possesses subgroups of order 1, 2, 3, 4, and 12, but none of order 6.

Some partial converses to Lagrange's Theorem do exist. What we mean by this is that, for some divisors of $|G|$, we may be able to know whether there exists a subgroup of that order or an element of that order. We mention two such theorems as examples, the proofs of which we give in Chapter 6, using techniques from group actions.

Theorem 2.1.15 (Cauchy's Theorem)

Let G be a group and p a prime divisors of $|G|$. Then G contains an element of order p.

Theorem 2.1.16 (Sylow's Theorem, First Part)

Let G be a group and let p^n be the highest power of a prime p dividing $|G|$. Then G contains a subgroup of order p^n.

Even without Cauchy's Theorem, it is sometimes possible to determine whether a group G contains elements of certain orders by virtue of Corollary 2.1.11. The following example illustrates the reasoning.

Example 2.1.17. Let G be a group of order 35. We prove that G must have an element of order 5 and an element of order 7. If G contains an element z

of order 35 (which would imply that G is cyclic), then z^5 has order 7 and z^7 has order 5.

Assume that G has no elements of order 7. By Corollary 2.1.11, the only allowed orders of elements would be 1 and 5. Obviously, the identity is the only element of order 1. But if two elements a, b are of order 5 and not powers of each other, then $\langle a \rangle \cap \langle b \rangle = \{1\}$. Hence, each nonidentity element would be in a cyclic subgroup of order 5 but containing 4 nonidentity elements. If there are k such subgroups, then we would have $4k + 1 = 35$. This is a contradiction. Hence, G must have an element of order 7.

Similarly, assume that G has no elements of order 5. Again, by Corollary 2.1.11, the only allowed orders of elements would be 1 and 7. Any element of order 7 generates a cyclic subgroup, containing the identity element and 6 elements of order 7. If there are h such subgroups, then we would have $6h + 1 = 35$. Again, this is a contradiction. Hence, G must contain an element of order 5. $\triangle$

2.1.4 Products of Subgroups

As another application of cosets, we consider subsets of a group G of the form HK, where H and K are two subgroups of G. We define the subset HK as

$$HK = \{hk \mid h \in H \text{ and } k \in K\}.$$

This subset is, in general, not a subgroup of G. It is, however, a union of certain cosets, in particular a union of right cosets of H and also a union of left cosets of K via

$$HK = \bigcup_{k \in K} Hk = \bigcup_{h \in H} hK. \tag{2.1}$$

In either of the above expressions, it is possible that many terms in the union are redundant as some of the cosets may be equal. By an analysis of cosets of H and of $H \cap K$, we can prove the following proposition.

Proposition 2.1.18

If H and K are finite subgroups of a group G, then

$$|HK| = \frac{|H| \, |K|}{|H \cap K|}.$$

Proof. Consider HK as the union of left cosets of K as given in (2.1). Each left coset of K has $|K|$ elements, so $|HK|$ is a multiple of $|K|$. We simply need to count the number of distinct left cosets of K in HK. By Proposition 2.1.5, $h_1 K = h_2 K$ if and only if $h_2^{-1} h_1 \in K$. Since $h_2^{-1} h_1 \in H$, then $h_2^{-1} h_1 \in H \cap K$, which again by Proposition 2.1.5 is equivalent to $h_1 (H \cap K) = h_2 (H \cap K)$.

However, $H \cap K \leq H$. By the above reasoning, the number of distinct left cosets of K in HK is the number of distinct left cosets of $H \cap K$ in H. By Lagrange's Theorem, this number is $|H|/|H \cap K|$. Thus,

$$|HK| = \frac{|H|}{|H \cap K|}|K| = \frac{|H||K|}{|H \cap K|}.$$

□

Recall that the join $\langle H \cup K \rangle$ of H and K is the smallest (by inclusion) subgroup of G that contains both H and K. Obviously, $\langle H \cup K \rangle$ must contain all products of the form hk with $h \in H$ and $k \in K$ but perhaps much more. Hence, HK is a subset of the join of H and K. If HK happens to be a subgroup of G, then $HK = \langle H \cup K \rangle$. By Lagrange's Theorem, we can deduce that

$$\frac{|H||K|}{|H \cap K|} \leq |\langle H \cup K \rangle| \quad \text{and} \quad |\langle H \cup K \rangle| \,\big|\, |G|. \tag{2.2}$$

Note that when HK is not a subgroup of G, we cannot use Lagrange's Theorem to deduce that $|HK|$ divides $|\langle H \cup K \rangle|$.

Example 2.1.19. As an application of the result in (2.2), let $G = S_5$ and let H be the subgroup of G that fixes 4 and 5, while K is the subgroup of G that fixes 2 and 3. Note that $H \cong S_3$ and $K \cong S_3$ so $|H| = |K| = 6$. Furthermore, if $\sigma \in H \cap K$, then σ fixes 2, 3, 4, and 5 and hence must also fix 1. Thus, $H \cap K = \{1\}$. By Proposition 2.1.18, $|HK| = 36$. Since $36 \nmid 120 = |S_5|$, then by Lagrange's Theorem, we know that HK is not a subgroup of S_5. By (2.2), we can also deduce that the join of H and K is greater than 36, but a divisor of 120. Given only this information, we know that $|\langle H \cup K \rangle|$ is 40, 60, or 120. $\triangle$

2.1.5   Useful CAS Commands

Both *Maple* and SAGE offer commands to construct the list of cosets, either left or right, for a subgroup of a group. The following examples illustrate working with the cosets of $H = \langle (1,2)(3,4), (1,3)(2,4) \rangle$ in A_4.

—————————————— Maple ——————————————

$with(GroupTheory):$

$G := AlternatingGroup(4):$

$H := Subgroup([[1,2],[3,4]],[[1,3],[2,4]], G):$

$LeftCosets(H, G);$

$\{((2,3,4) \cdot \langle (1,2)(3,4), (1,3)(2,4) \rangle, ((2,4,3) \cdot \langle (1,2)(3,4), (1,3)(2,4) \rangle,$
$(() \cdot \langle (1,2)(3,4), (1,3)(2,4) \rangle\}$

$LeftCoset([[1,3,4]], H);$

$((1,3,4)) \cdot \langle (1,2)(3,4), (1,3)(2,4) \rangle$

Elements(%);

$$\{(1,4,2),(2,4,3),(1,3,4),(1,2,3)\}$$

The command to generate the set of right cosets of a subgroup group is RightCosets(H,G). The commands LeftCoset and RightCoset return objects c that respond to the following methods: Representative(c), numelems(c), x in c (returns a boolean), or Elements(c). On the last line of code, we remind the reader that in *Maple* the symbol % references the result of the previously executed command.

In SAGE, the command cosets() creates a list of left (or right) cosets, with each coset described as a list of elements in the group. We can select an individual coset as we would select any element from a list.

--- Sage ---

```
sage: G=AlternatingGroup(4)
sage: r1=G("(1,2)(3,4)")
sage: r2=G("(1,3)(2,4)")
sage: H=G.subgroup([r1,r2])
sage: A=G.cosets(H,side='left')
sage: A
[[(), (1,2)(3,4), (1,3)(2,4), (1,4)(2,3)],
 [(2,3,4), (1,3,2), (1,4,3), (1,2,4)],
 [(2,4,3), (1,4,2), (1,2,3), (1,3,4)]]
sage: A[0]
[(), (1,2)(3,4), (1,3)(2,4), (1,4)(2,3)]
```

By changing the option side to 'right' we change whether we want left or right cosets of the subgroup.

EXERCISES FOR SECTION 2.1

1. Find all distinct left cosets of $H = \langle \bar{3} \rangle$ in $\mathbb{Z}/21\mathbb{Z}$ and list the elements in each.

2. Suppose that Z_{20} is generated by the element z. List all the distinct left cosets of $\langle z^5 \rangle$ and list all the elements in each coset.

3. In the group Q_8, list the distinct left cosets of the following subgroups and in each case list all the elements in each coset: (a) $\langle i \rangle$; (b) $\langle -1 \rangle$.

4. Find all distinct left cosets of $H = \langle \bar{4} \rangle$ in $U(35)$ and list elements in each.

5. Find all distinct left cosets of $H = \langle \bar{7}, \bar{47} \rangle$ in $U(48)$ and list elements in each.

6. Find all distinct left cosets of $H = \langle (1\,2\,3) \rangle$ in A_4 and list elements in each. Following the interpretation given in Exercise 1.6.24 of A_4 as the group of rigid motions of a tetrahedron, describe each left coset by what all the elements of the coset do to the tetrahedron.

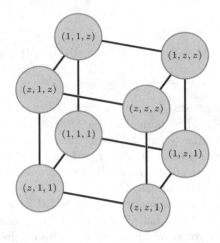

FIGURE 2.2: Cayley graph of $Z_2 \oplus Z_2 \oplus Z_2$.

7. Consider the group $G = Z_2 \oplus Z_2 \oplus Z_2$ with $Z_2 = \langle z \mid z^2 = 1 \rangle$. There is a bijection between the elements of G and the (vertices of the) unit cube C via $f(z^a, z^b, z^c) = (a, b, c)$, where a, b, and c are 0 or 1. (See Figure 2.2.) For each of the following subgroups, list all the distinct left cosets, describe them as subsets of the unit cube (via the mapping f), and label the cosets on a sketch of the unit cube.

   (a) $H = \langle (1, z, 1) \rangle$
   (b) $K = \langle (z, z, z) \rangle$
   (c) $L = \langle (z, 1, 1), (1, z, 1) \rangle$

8. Consider the group $\mathbb{C}^\times$ with multiplication and define the subgroup $H = \{z \in \mathbb{C}^\times \mid |z| = 1\}$. Describe geometrically (as a subset of the plane) each coset of H.

9. Consider the group $\mathbb{C}^\times$ with multiplication and consider the subgroup $H = \mathbb{R}^\times$. Describe geometrically (as a subset of the plane) each coset of H.

10. Prove the last claim in Example 2.1.8. With the group $G = S_5$ and the same subgroup H, prove also that the right cosets $H(1\,2\,3\,4\,5)^i$ correspond to

$$H(1\,2\,3\,4\,5)^i = \{\tau \in S_5 \mid \tau(5 - i) = 5\}.$$

11. Consider the group $(\mathbb{Q}, +)$.

   (a) Show that $|\mathbb{Q} : \mathbb{Z}|$ is infinite.
   (b) Show that there is no proper subgroup of $\mathbb{Q}$ of finite index.

12. Consider the cosets of $H = \langle s \rangle$ in D_4 as described in Example 2.1.2. We write $AB = \{ab \mid a \in A \text{ and } b \in B\}$ for any subsets $A, B \subseteq G$. Prove that $(rH)(r^2H)$ is a left coset of H, but that $(rH)(r^3H)$ is not.

13. Let H be a subgroup of G. Suppose one attempted to define the function F from the set of left cosets of H to the set of right cosets of H via $F(gH) = Hg$. Explain why F is not a function. [Hint: We also say that F is not well-defined.]

14. Show that a complete set of distinct representative of $\sim_1$ is not necessarily a complete set of distinct representatives of $\sim_2$.

15. Let G be a group and do not assume it is finite. Prove that if $H \leq K \leq G$, then
$$|G : H| = |G : K| \cdot |K : H|.$$

16. Let G be a group and let H and K be subgroups with $|G : H| = m$ and $|G : K| = n$ finite. Prove that
    (a) $|G : H \cap K| \leq mn$;
    (b) $\mathrm{lcm}(m, n) \leq |G : H \cap K|$.
    [Hint: Use Exercise 2.1.15.]

17. Let $\varphi : G \to H$ be a group homomorphism.
    (a) Prove that the left cosets of $\mathrm{Ker}\,\varphi$ are the fibers of φ. [Recall that a fiber $f : A \to B$ is a subset of A of the form $f^{-1}(b)$ for some $b \in B$.]
    (b) Deduce that for all $g \in G$, the left coset $g(\mathrm{Ker}\,\varphi)$ is equal to the right coset $(\mathrm{Ker}\,\varphi)g$.

18. Consider the Cayley graph for S_4 given in Figure 1.11.
    (a) Prove that the triangles (whose edges are double edges) correspond to right cosets of $\langle (1\,2\,3) \rangle$.
    (b) Prove that the squares with all single edges correspond to the right cosets of $\langle (1\,2\,3\,4) \rangle$.
    (c) Prove that the squares with mixed edge styles are not the left or right cosets of any subgroup.

19. If a group G has order $|G| = 105$, list all possible orders of subgroups of G.

20. If a group G has order $|G| = 48$, list all possible orders of subgroups of G.

21. Prove that $\overline{n - 1} \in U(n)$ for all integers $n \geq 3$. Apply Lagrange's Theorem to $\langle n - 1 \rangle$ to deduce that Euler's totient function $\phi(n)$ is even for all $n \geq 3$.

22. Prove or disprove that $\mathbb{Z}_6 \oplus \mathbb{Z}_{10}$ has (a) a subgroup of order 4; (b) a subgroup isomorphic to $\mathbb{Z}_4$.

23. Suppose that G is a group with $|G| = pq$, where p and q are primes, not necessarily distinct. Prove that every proper subgroup of G is cyclic.

24. Let $G = \mathrm{GL}_2(\mathbb{F}_5)$.
    (a) Use Lagrange's Theorem to determine all the possible sizes of subgroups of G.
    (b) Show that the orders of the following elements are respectively 3, 4, and 5,
$$A = \begin{pmatrix} 0 & 2 \\ 2 & 4 \end{pmatrix}, \quad \text{and} \quad B = \begin{pmatrix} 3 & 0 \\ 0 & 3 \end{pmatrix}, \quad \text{and} \quad C = \begin{pmatrix} 1 & 1 \\ 0 & 1 \end{pmatrix}.$$
    (c) Determine the order of AB and BC without performing any matrix calculations.

25. Let $G = \mathrm{GL}_2(\mathbb{F}_5)$ and let H be the subgroup of upper triangular matrices. Prove that $|G : H| = 6$ and find 6 different matrices $g_1, g_2, \ldots, g_6$ such that the cosets $g_i H$ for $i = 1, 2, \ldots, 6$ are all the left cosets of H.

26. Let p be a prime number. Prove that the subgroups of $Z_p \oplus Z_p$ consist of $\{1\}$, $Z_p \oplus Z_p$ and $p+1$ subgroups that are cyclic and of order p.

27. Let G be a group of order 21. (a) Prove that G must have an element of order 3. (b) By the strategy of Example 2.1.17, can we determine whether G must have an element of order 7?

28. Let G be a group of order $3p$, where p is a prime number. Prove that G has an element of order 3.

29. Let G be a group of order pq, where p and q are distinct odd primes such that $(p-1) \nmid (q-1)$ and $(q-1) \nmid (p-1)$. Prove that G contains an element of order p and an element of order q.

30. Let $H, K \leq G$, a group. Prove that if $\gcd(|H|, |K|) = 1$, then $H \cap K = \{1\}$.

31. Let G be a group and let H be a subgroup with $|G : H| = p$, a prime number. Prove that if K is a subgroup of G that strictly contains H, then $K = G$.

32. Let G be a group of order pqr, where p, q, r are distinct primes. Let A be a subgroup of order pq and let B be a subgroup of order qr. Prove that $AB = G$ and that $|A \cap B| = q$.

33. Use Lagrange's Theorem applied to $U(n)$ to prove Euler's Theorem (a generalization of Fermat's Little Theorem), which states that if $\gcd(a, n) = 1$, then

$$a^{\varphi(n)} \equiv 1 \pmod{n}.$$

34. Show that there exists a subgroup of order d for each divisor d of $|S_4| = 24$ and give an example of a subgroup for each divisor.

35. *Classification of Groups of Order* $2p$. Let p be a prime number. This exercise guides the proof that a group of order $2p$ is isomorphic to Z_{2p} or D_p. Let G be an arbitrary group with $|G| = 2p$.

    (a) Without using Cauchy's Theorem, prove that G contains an element a of order 2 and an element b of order p.

    (b) Prove that if $ab = ba$, then $G \cong Z_{2p}$.

    (c) Prove that if a and b do not commute, then $aba = b^{-1}$. Deduce in this case that $G \cong D_p$.

36. Let G be a group and let $H, K \leq G$. Prove that $HK \leq G$ if and only if $HK = KH$ as sets.

2.2   Conjugacy and Normal Subgroups

In the previous section, we discussed left cosets and right cosets in a group G. By considering simple examples, we found that generally a left coset of a subgroup H is not necessarily equal to a right coset. However, in Example 2.1.2 we saw that every left coset of $\langle r \rangle$ in D_4 is a right coset. This property that subgroups may have, called *normal*, plays a vital role for the construction of quotient groups.

This section studies normal subgroups. Some of the constructions or criteria developed in this section may at first pass seem unnecessarily complicated if we are simply trying to generalize the construction employed to create modular arithmetic. The important difference between general groups and $(\mathbb{Z}, +)$ is that the latter is abelian, while groups in general are not.

2.2.1 Normal Subgroups

Let G be a group and let $H \leq G$ be a subgroup. If a left coset gH is equal to a right coset Hg', then since $1 \in H$, we know that $g \in Hg'$. If $g \in Hg'$, then $Hg' = Hg$ as right cosets. Hence, the criterion that a subgroup is such that every left coset is equal to a right coset can be summarized in the following definition.

Definition 2.2.1

> Let G be a group. A subgroup $N \leq G$ is called *normal* if $gN = Ng$ for all $g \in G$. If N is a normal subgroup of G, we write $N \trianglelefteq G$.

In Example 2.1.2, we saw that while $\langle r \rangle \trianglelefteq D_4$, in contrast $\langle s \rangle$ is not a normal subgroup of D_4. In notation, we write $\langle s \rangle \ntrianglelefteq D_4$.

The criterion for a normal subgroup is equivalent to a variety of other conditions on the subgroup. Before we list these conditions in Theorem 2.2.4, we mention a few results that are immediate from the definition. The first observation is that every group G has at least two normal subgroups: the trivial group $\{1\}$ and itself G. The next proposition gives another sufficient condition for a subgroup to be normal.

Proposition 2.2.2

> Let G be a group (not necessarily finite). If H is a subgroup such that $|G : H| = 2$, then $H \trianglelefteq G$.

Proof. By definition, if $|G : H| = 2$, then H has two left cosets, just as it has two right cosets. Now, $H = 1H$ is a left coset. Since the collection of left cosets form a partition of G, then $G - H$ is the other left coset. Similarly, $H = H1$ is a right coset and, by the same reason as before, $G - H$ is the other right coset. Hence, every left coset is equal to a right coset, and thus H is a normal subgroup of G. $\square$

Example 2.2.3. From Proposition 2.2.2, we immediately see that $\langle r \rangle$, $\langle r^2, s \rangle$, and $\langle r^2, rs \rangle$ are normal subgroups of D_4, simply because each of those subgroups has order 4 and $|D_4| = 8$. $\triangle$

Theorem 2.2.4

Let N be a subgroup of G. The following are equivalent:

(1) $N \trianglelefteq G$.

(2) $gNg^{-1} = N$ for all $g \in G$.

(3) $N_G(N) = G$.

(4) For all $g \in G$ and all $n \in N$, $gng^{-1} \in N$.

(5) $\sim_1$ and $\sim_2$ as defined in Proposition 2.1.4 are equal relations.

Proof. (1) $\Longleftrightarrow$ (2): If $gN = Ng$ for all $g \in G$, then by multiplication on the right by g^{-1}, we obtain $gNg^{-1} = N$. Conversely, given the condition $gNg^{-1} = N$ for all $g \in G$, by multiplying on the right by g, we recover $gN = Ng$ for all $g \in G$.

(2) $\Longleftrightarrow$ (3): This is automatic from the definition of normalizers.

(3) $\Longleftrightarrow$ (4): If $gNg^{-1} = N$ for all $g \in G$, then for all $n \in N$, we know that $gng^{-1} \in N$. Conversely, assuming (4), we can immediately conclude that $gNg^{-1} \subseteq N$ for all $g \in G$. Note that given $g \in G$, since g^{-1} also must satisfy condition (4), we also know that $g^{-1}Ng \subseteq N$. By multiplying on the left by g and on the right by g^{-1}, we deduce that $N \subseteq gNg^{-1}$. Since we already had $gNg^{-1} \subseteq N$, we conclude that $gNg^{-1} = N$.

(1) $\Longleftrightarrow$ (5): The condition that N is normal in G means that every left coset is a right coset. By Proposition 2.1.4, this is equivalent to saying that every $\sim_1$-equivalence class is equal to a $\sim_2$-equivalence class. By Proposition A.3.15, the equivalence relations $\sim_1$ and $\sim_2$ are equal. $\qquad\square$

We underscore that the condition $gNg^{-1} = N$ does not imply $gng^{-1} = n$ for all $g \in G$ and all $n \in N$. It merely implies that the process of operating on the left by g and right by g^{-1} produces a bijection on N. We explore this issue more in Section 2.2.2.

In practice, as we explore properties of normal subgroups, the various criteria of Theorem 2.2.4 may be more useful in some contexts than in others. Proposition 2.2.2 illustrated a situation where the original definition is sufficiently convenient to establish the result. The following two propositions, which are important in themselves, illustrate situations where a different criteria is more immediately useful in the proof.

Proposition 2.2.5

If G is abelian, then every subgroup $H \leq G$ is normal.

Proof. For all $g \in G$ and $h \in H$, we have $ghg^{-1} = gg^{-1}h = h \in H$. By Theorem 2.2.4(4), $H \trianglelefteq G$. $\qquad\square$

This proposition hints at why generalizing the construction of modular arithmetic to all groups poses some subtleties that were not apparent in modular arithmetic: $(\mathbb{Z}, +)$ is abelian. By a similar reason, we can also conclude the more general proposition.

Proposition 2.2.6

Let G be a group. Any subgroup H in the center $Z(G)$ is a normal subgroup $H \trianglelefteq G$.

Proof. (Left as an exercise for the reader. See Exercise 2.2.5.) □

Proposition 2.2.7

Let $\varphi : G \to H$ be a homomorphism between groups. Then $\operatorname{Ker} \varphi \trianglelefteq G$.

Proof. Let $n \in \operatorname{Ker} \varphi$ and let $g \in G$. Then

$$
\begin{aligned}
\varphi(gng^{-1}) &= \varphi(g)\varphi(n)\varphi(g)^{-1} && \text{by homomorphism properties} \\
&= \varphi(g)1\varphi(g)^{-1} && \text{since } n \in \operatorname{Ker} \varphi \\
&= 1_H.
\end{aligned}
$$

By Theorem 2.2.4(4), we conclude that $\operatorname{Ker} \varphi \trianglelefteq G$. □

This proposition, though easy to prove, leads to some surprising and profound consequences. For example, if $n \geq 3$, there exists no homomorphism $\varphi : D_n \to G$ such that $\operatorname{Ker} \varphi = \langle s \rangle$. Also, a fiber of any homomorphism $\varphi : G \to H$, i.e., a set $\varphi^{-1}(\{h\})$, is either the empty set or a coset of $\operatorname{Ker} \varphi$. If $\varphi(g) = h$, then $\varphi^{-1}(\{h\}) = g(\operatorname{Ker} \varphi)$. Note that if G is finite, then all nonempty fibers have the same cardinality as $|\operatorname{Ker} \varphi|$.

When attempting to determine whether a subgroup of a group G is normal, using Definition 2.2.1 or Theorem 2.2.4(4) requires a large number of calculations. For example, checking the latter criterion requires $|G| \cdot |N|$ number of calculations gng^{-1} to determine if $N \trianglelefteq G$. However, if a finite group and its subgroup are both presented by generators, the following theorem provides a quick shortcut.

Theorem 2.2.8

Let G be a group generated by a subset T. Let $N = \langle S \rangle$ be the subgroup generated by the subset S. Then $N \trianglelefteq G$ if and only if for all $t \in T$ and all $s \in S$, $tst^{-1} \in N$.

Proof. ($\Longrightarrow$) This direction is obvious as a consequence of Theorem 2.2.4(4).
($\Longleftarrow$) Suppose that $tst^{-1} \in N$ for all generators $s \in S$ and all $t \in T$.

Let $n \in N$. Then by definition of generating subset, $n = s_1^{\varepsilon_1} s_2^{\varepsilon_2} \cdots s_k^{\varepsilon_k}$ for some nonnegative integer k, and where for each $i = 1, \ldots, k$, we have $s_i \in S$ and $\varepsilon_i = \pm 1$. For any $g \in G$,

$$gng^{-1} = g(s_1^{\varepsilon_1} s_2^{\varepsilon_2} \cdots s_k^{\varepsilon_k})g^{-1} = (gs_1^{\varepsilon_1}g^{-1})(gs_2^{\varepsilon_2}g^{-1}) \cdots (gs_k^{\varepsilon_k}g^{-1})$$
$$= (gs_1g^{-1})^{\varepsilon_1}(gs_2g^{-1})^{\varepsilon_2} \cdots (gs_kg^{-1})^{\varepsilon_k}.$$

Consequently, if $gSg^{-1} \subseteq N$, then $gng^{-1} \in N$ for all $n \in N$. Therefore, $gSg^{-1} \subseteq N$ implies that $g \in N_G(N)$.

Since, by hypothesis, $tSt^{-1} \subseteq N$ for all $t \in T$, then $T \subseteq N_G(N)$. Since $N_G(N) \leq G$ the we also have $\langle T \rangle \subseteq N_G(N)$. However, $\langle T \rangle = G$ so we deduce that $N_G(N) = G$, and hence that N is a normal subgroup of G. $\qquad \square$

Example 2.2.9. Consider the group D_8 and consider the subgroup $H = \langle r^4, s \rangle$. We test to see if H is a normal subgroup in D_8. Notice first that $H = \{1, r^4, s, sr^4\}$. We only need to perform four calculations:

$$r(r^4)r^{-1} = r^4 \qquad sr^4s^{-1} = r^4 \qquad rsr^{-1} = r^2s \qquad sss^{-1} = s.$$

By Theorem 2.2.8, the third calculation $rsr^{-1} = r^2s \notin H$ shows that H is not a normal subgroup. $\qquad \triangle$

It is important to remark that, in contrast to the subgroup relation $\leq$ on the set of subgroups of G, the relation of normal subgroup $\trianglelefteq$ is not transitive, and hence is not a partial order on $\mathrm{Sub}(G)$. The easiest illustration comes from the dihedral group D_4. By Proposition 2.2.2, we see that $\langle r^2, s \rangle \trianglelefteq D_4$ and $\langle s \rangle \trianglelefteq \langle r^2, s \rangle$. However, $\langle s \rangle$ is not a normal subgroup of D_4. Therefore,

$$K \trianglelefteq H \trianglelefteq G \nRightarrow K \trianglelefteq G.$$

One intuitive reason for this property of the relation of normal subgroup is that even if $hKh^{-1} \subseteq K$ for all $h \in H$, the condition $gKg^{-1} \subseteq K$ for all $g \in G$ is a stronger condition and might not still hold.

We can now put into context the terminology of "normalizer" of a subgroup. Recall that

$$N_G(H) = \{g \in G \mid gHg^{-1} = H\}.$$

Consequently, $N_G(H)$ is the largest subgroup K of G such that $H \trianglelefteq K$. The normalizer gives some way of measuring how far a subgroup H is from being normal. For all $H \leq G$, we have

$$H \leq N_G(H) \leq G,$$

with $N_G(H) = G$ if and only if $H \trianglelefteq G$. Intuitively speaking, we can say that H is farthest from being normal when $N_G(H) = H$.

Exercise 2.1.36 establishes that if $H, K \leq G$ are two subgroups, then HK is a subgroup of G if and only if $HK = KH$ as sets.

Corollary 2.2.10

If $H \leq N_G(K)$, then HK is a subgroup of G. In particular, if $K \trianglelefteq G$, then HK is a subgroup of G for all $H \leq G$.

Proof. (Left as an exercise for the reader. See Exercise 2.2.12.) □

2.2.2 Conjugacy

The expression in criteria (4) of Theorem 2.2.4 is not new. We encountered it before when discussing centralizers and normalizers. (See Definitions 1.7.10 and 1.7.12.) We remind the reader of a definition given in Section 1.6.

Definition 2.2.11

Let G be a group and let $g \in G$.

- If $x \in G$, the element gxg^{-1} is called the *conjugate* of x by g.
- If $S \subseteq G$, the subset gSg^{-1} is called the *conjugate* of S by g.

We have seen the conjugate of a group element in other contexts previously. In Example 1.11.10, we found a presentation for the Frieze group of a certain pattern. In that example, we noted that rotation by π about Q_3 is equal to trt^{-1}, where t corresponds to translation along $\overrightarrow{QQ_3}$ and r is rotation by π about Q. In linear algebra, one encounters the change of basis formula. If A is the $n \times n$ matrix associated with a linear transformation $T : \mathbb{R}^n \to \mathbb{R}^n$ with respect to a basis $\mathcal{B}$, and if B is the matrix associated with the same linear transformation, but with respect to the basis $\mathcal{B}'$, then

$$B = MAM^{-1},$$

where M is the coordinate transition matrix from $\mathcal{B}$ to $\mathcal{B}'$ coordinates. In this latter example, A and B need not be invertible, but, if T is an isomorphism, then A and B are invertible and the conjugation $B = MAM^{-1}$ occurs entirely in the group $\mathrm{GL}_n(\mathbb{R})$.

The above examples give us the intuitive sense that conjugation corresponds to a change of origin, a change of basis, or some change of perspective more generally. Consider the conjugation rsr^{-1} in the dihedral group D_n. Explicitly, $rsr^{-1} = sr^{-2}$, which is the reflection through the line L' that is related to the s-reflection line L by a rotation by r. (See Figure 2.3.)

Proposition 2.2.12

Let G be a group and define the relation $\sim_c$ on G by $x \sim_c y$ if $y = gxg^{-1}$ for some $g \in G$. Then $\sim_c$ is an equivalence relation. The relation $\sim_c$ is called the *conjugacy* relation.

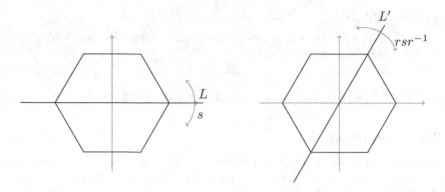

FIGURE 2.3: A conjugacy operation in D_n.

Proof. (Left as an exercise for the reader. See Exercise 2.2.11.)   □

The *conjugacy class* of an element $x \in G$ is the set $[x] = \{gxg^{-1} \mid g \in G\}$. By Proposition 2.2.12 and properties of equivalence relations, the conjugacy classes in G partition G.

Example 2.2.13 (Conjugacy Classes in S_n). In order to determine all the conjugacy classes in S_n, we first prove the following claim. For all m-cycles $(a_1\ a_2\ \cdots\ a_m)$ and for all permutations $\sigma \in S_n$,

$$\sigma(a_1\ a_2\ \cdots\ a_m)\sigma^{-1} = (\sigma(a_1)\ \sigma(a_2)\ \cdots\ \sigma(a_m)). \tag{2.3}$$

Write $\tau = \sigma(a_1\ a_2\ \cdots\ a_m)\sigma^{-1}$ and let $i = 1, 2, \ldots, m-1$. The permutation τ applied to $\sigma(a_i)$ gives

$$\sigma(a_1\ a_2\ \cdots\ a_m)\sigma^{-1}\sigma(a_i) = \sigma(a_1\ a_2\ \cdots\ a_m) \cdot a_i = \sigma(a_{i+1}).$$

If $i = m$, then similarly we easily see that τ applied to $\tau(\sigma(a_m)) = \sigma(a_1)$. We have seen that τ permutes the set $\{\sigma(a_1), \sigma(a_2), \ldots, \sigma(a_m)\}$. If $b \in \{1, 2, \ldots, n\}$, but $b \notin \{\sigma(a_1), \sigma(a_2), \ldots, \sigma(a_m)\}$, then there exists $c \in \{1, 2, \ldots, n\} \setminus \{a_1, a_2, \ldots, a_m\}$ such that $b = \sigma(c)$. Then

$$\tau(b) = \sigma(a_1\ a_2\ \cdots\ a_m)\sigma^{-1}\sigma(c) = \sigma(a_1\ a_2\ \cdots\ a_m) \cdot c = \sigma(c) = b.$$

We have calculated where τ sends all the elements and found that τ is given by (2.3).

Now every element $\omega \in S_n$ is a product of disjoint cycles, $\omega = \tau_1\tau_2\cdots\tau_m$. Furthermore, we can write

$$\sigma\omega\sigma^{-1} = (\sigma\tau_1\sigma^{-1})(\sigma\tau_2\sigma^{-1})\cdots(\sigma\tau_m\sigma^{-1}), \tag{2.4}$$

where each $\sigma\tau_i\sigma^{-1}$ is calculated from (2.3).

As a numerical example, only using (2.3) and (2.4) we determine that

$$(1\,3\,5\,4)(1\,4\,2)(3\,5)(1\,3\,5\,4)^{-1} = (3\,1\,2)(5\,4) = (1\,2\,3)(4\,5).$$

Consequently, we can see that if a permutation ω has a given cycle type, then for all $\sigma \in S_n$, $\sigma\omega\sigma^{-1}$ will have the same cycle type. Conversely, if two permutations $\omega_1, \omega_2 \in S_n$ have the same cycle type, then by using (2.3) and (2.4), we can find a $\sigma \in S_n$ such that $\omega_2 = \sigma\omega_1\sigma^{-1}$. Thus, the conjugacy classes in S_n are precisely the sets of permutations that have the same cycle type. The table in Example 1.5.6 lists the conjugacy classes in S_6. $\triangle$

Conjugacy classes and normal subgroups are closely related in the following way. If N is a normal subgroup of a group G and $x \in N$, then by Theorem 2.2.4, $gxg^{-1} \in N$ for all $g \in G$. Consequently, the conjugacy class of x is in N. This leads to the following proposition.

Proposition 2.2.14

A subgroup $H \leq G$ is normal if and only if it is the union of some conjugacy classes.

In Exercise 1.9.36, we proved that if H is a subgroup of a group G, and $g \in G$, then gHg^{-1} is also a subgroup of G that is isomorphic to H. Furthermore, define the relation $\sim_c$ on the set of subgroups $\mathrm{Sub}(G)$ defined by $H \sim_c K$ if and only if $K = gHg^{-1}$ for some $g \in G$. It is easy to show that $\sim_c$ is an equivalence relation on $\mathrm{Sub}(G)$. If H is a normal subgroup, then $gHg^{-1} = H$ for all $g \in G$. In the language of equivalence classes, $H \trianglelefteq G$ if and only if the $\sim_c$-equivalence class of H is the single element set $\{H\}$.

By Theorem 2.2.4, the conjugation operation on a normal subgroup by an element g is a function $N \to N$. However, more can be said.

Proposition 2.2.15

Let H be any subgroup of a group G. Then for all $g \in N_G(H)$, the function $\psi_g : H \to H$ defined by $\psi_g(h) = ghg^{-1}$ is an automorphism of H. Furthermore, the association $\Psi : N_G(H) \to \mathrm{Aut}(H)$ defined by $\Psi(g) = \psi_g$ is a homomorphism.

Proof. Let $h_1, h_2 \in H$. Then

$$\psi_g(h_1 h_2) = gh_1 h_2 g^{-1} = gh_1 g^{-1} gh_2 g^{-1} = \psi_g(h_1)\psi_g(h_2).$$

This proves that ψ_g is a homomorphism. It is easy to check that $\psi_g^{-1} = \psi_{g^{-1}}$, so the function ψ_g is a bijection and, hence, an automorphism of H.

Now let $a, b \in N_G(H)$ be arbitrary. Then for all $h \in H$,

$$\psi_{ab}(h) = (ab)h(ab)^{-1} = abhb^{-1}a^{-1} = a(bhb^{-1})a^{-1} = \psi_a(\psi_b(h)).$$

Hence, $\Psi(ab) = \Psi(a) \circ \Psi(b)$, which establishes that $\Psi : N_G(H) \to \mathrm{Aut}(H)$ is a homomorphism. $\qquad\qquad\qquad\qquad\qquad\qquad\qquad\qquad\qquad\qquad\qquad\qquad\square$

In particular, if $N \trianglelefteq G$, then $N_G(N) = G$ so $\Psi : G \to \mathrm{Aut}(N)$ defined as above is a homomorphism.

2.2.3   Simple Groups

Every group always has at least two normal subgroups, namely the trivial group $\{1\}$ and itself G. It is possible for the group not to contain any other normal subgroups. For example, if p is prime, then Z_p only has the two subgroups $\{1\}$ and Z_p so these are the only two normal subgroups.

> **Definition 2.2.16**
>
> A group G is called *simple* if it contains no normal subgroups besides $\{1\}$ and itself.

Simple groups play an important role in more advanced topics in group theory. By what we said above, Z_p is a simple group whenever p is a prime number. Determining if a group is simple is not always a "simple" task. We have encountered one other family of groups that is simple, namely A_n with $n \geq 5$. Exercise 2.2.28 guides the reader to prove that A_5 is simple. The proof that A_n is simple for $n \geq 6$ is more challenging and is beyond the scope of this book.

2.2.4   Useful CAS Commands

Both *Maple* and SAGE offer commands to construct the construct conjugacy classes and to determine whether a subgroup of a group is normal. They also both have commands to determine if a given group G is simple. Some of these commands only work for permutation groups, rather than groups described by a presentation.

Maple Function	SAGE
`IsNormal(H,G);`	`H.is_normal(G)`
Returns `true` or `false` whether the subgroup H of G is normal.	
`ConjugacyClasses(G);`	`G.conjugacy.classes()`
Returns the list of conjugacy classes of the group G.	
`ConjugacyClass(g,G);`	`G.conjugacy_class(g)`
Returns the conjugacy class of the element g in the group G.	
`IsSimple(G);`	`G.is_simple()`
Returns a boolean for whether the group G is simple.	

In *Maple*, as with cosets, the conjugacy class object cc has four methods associated with it: Representative(c), numelems(c), x in c (returns a boolean), or Elements(c). SAGE also offers the command conjugacy_classes_representative(), which returns a complete set of distinct representatives of the conjugacy classes. For examples, the following code in *Maple* and in SAGE produces a list whose values are the sizes of the conjugacy classes in S_7.

──────────────── Maple ────────────────

$G := SymmetricGroup(7):$

$ccs := ConjugacyClasses(G):$

$cclist := []:$

for i **from** 1 **to** $nops(ccs)$ **do**

$\quad cclist := [op(cclist), numelems(ccs[i])]:$

od:

$cclist;$

$$[1, 630, 420, 70, 21, 105, 210, 105, 504, 720, 210, 504, 840, 420, 280]$$

Sage does not have a command to directly obtain the size of a conjugacy class. However, by the Orbit-Stabilizer Theorem discussed in Section 6.4.2, the size of the conjugacy class of g in G is $|G|/|C_G(g)|$. So the following code in SAGE produces the same effect as the above code in *Maple*.

──────────────── Sage ────────────────

```
sage: G=SymmetricGroup(7)
sage: g_order=G.order()
sage: reps=G.conjugacy_classes_representatives()
sage: cclist=[]
sage: for g in reps:
....:     cclist.append(g_order/G.centralizer(g).order())
....:
sage: cclist
[1, 21, 105, 105, 70, 420, 210, 280, 210, 630, 420, 504,
 504, 840, 720]
```

EXERCISES FOR SECTION 2.2

1. Prove that A_n is a normal subgroup of S_n.

2. Determine whether $\langle r^2 \rangle$ is normal in D_8.

3. Determine whether $\langle (1\,2)(3\,4), (1\,3)(2\,4) \rangle$ is normal in A_4.

4. Find all normal subgroups of D_6. [Hint: See Example 1.8.8.]

5. Prove that if $H \leq Z(G)$, then $H \trianglelefteq G$.

6. Prove that every subgroup of Q_8 is a normal subgroup. [This shows that the converse to Proposition 2.2.5 is false.]

7. Let $F = \mathbb{Q}, \mathbb{R}, \mathbb{C}$, or $\mathbb{F}_p$ where p is prime. Prove that for all positive integers n,

$$\mathrm{SL}_n(F) \trianglelefteq GL_n(F).$$

8. Let $F = \mathbb{Q}, \mathbb{R}, \mathbb{C}$, or $\mathbb{F}_p$ where p is prime. Denote by $T_n(F) \leq \mathrm{GL}_n(F)$ the upper triangular matrices in $GL_n(F)$.
   (a) Prove that the function $\varphi : T_n(F) \to (F^\times)^n$ defined by

   $$\varphi(A) = (a_{11}, a_{22}, \ldots, a_{nn})$$

   is a homomorphism, where by $(F^\times)^n$ we mean the nth direct sum of the multiplicative group $F^\times = F - \{0\}$.
   (b) Conclude that the subgroup of $T_n(F)$ consisting of matrices with 1s down the diagonal is a normal subgroup.

9. Let n be a positive integer and let $G = \mathrm{GL}_n(\mathbb{R})$.
   (a) Prove that $H = \mathrm{GL}_n(\mathbb{Q})$ is a subgroup that is not a normal subgroup.
   (b) Define the subset $K = \{A \in \mathrm{GL}_n(\mathbb{R}) \,|\, \det(A) \in \mathbb{Q}\}$. Prove that K is a normal subgroup that contains H.

10. Consider the group $S_\mathbb{N}$ of permutations on $\mathbb{N}$. Let $K \subseteq S_\mathbb{N}$ be the set of permutations that fix all bu a finite number of elements; i.e., $\sigma \in K$ if and only if $\{n \in \mathbb{N} \,|\, \sigma(n) \neq n\}$ is a finite set. Prove that K is a subgroup of $S_\mathbb{N}$ and that it is a normal subgroup.

11. Prove Proposition 2.2.12.

12. Prove Corollary 2.2.10.

13. Let G be a group. Prove that if $H \leq G$ is the unique subgroup of a given order n, then $H \trianglelefteq G$.

14. Let $\varphi : G \to H$ be a group homomorphism and let $N \trianglelefteq H$. Prove that $\varphi^{-1}(N) \trianglelefteq G$. [Note that this generalizes the fact that kernels of homomorphisms are normal subgroups of the domain group.]

15. Let G be a group, $H \leq G$ and $N \trianglelefteq G$. Prove that $H \cap N \trianglelefteq H$.

16. Prove that the intersection of two normal subgroups N_1, N_2 of a group G is again a normal subgroup $N_1 \cap N_2 \trianglelefteq G$.

17. Let N_1 and N_2 be normal subgroups is G. Prove that $N_1 N_2$ is the join of N_1 and N_2 and that it is a normal subgroup in G.

18. Let $\{N_i\}_{i \in I}$ be a collection of normal subsets of G. Prove that the intersection

$$\bigcap_{i \in I} N_i$$

is a normal subgroup. Do not assume that I is finite.

19. Suppose that a subgroup $H \leq G$ is such that if $h \in H$ with $|h| = n$, then H contains all the elements in G of order n. Prove that H is a normal subgroup.

20. Let A be any subset of a group G. Prove that $C_G(A) \trianglelefteq N_G(A)$.

21. We say that a subgroup H of a group G is a *characteristic subgroup* if $\psi(H) = H$ for all automorphisms $\psi \in \text{Aut}(G)$. Prove that every characteristic subgroup is a normal subgroup.

22. Prove that if $g \in Z(G)$, then the conjugacy class of g is the singleton set $\{g\}$.

23. Prove that in an abelian group, all the conjugacy classes consist of a single element.

24. Prove that the conjugacy classes of D_n are:

    (a) if n is even: $\{1\}$, $\{r^{n/2}\}$, $\{r^a, r^{-a}\}$ for $1 \leq a \leq \frac{n}{2} - 1$, $\{s, sr^2, \ldots, sr^{n-2}\}$, and $\{sr, sr^3, \ldots, sr^{n-1}\}$;

    (b) if n is odd: $\{1\}$, $\{r^a, r^{-a}\}$ for $1 \leq a \leq \frac{n-1}{2}$, and $\{s, sr, sr^2, \ldots, sr^{n-1}\}$.

25. List all the conjugacy classes in A_4.

26. List all the conjugacy classes in the group G_2 of Example 1.10.6.

27. Let G be a group. Consider the group of automorphisms $\text{Aut}(G)$. Prove that the group $\text{Inn}(G)$ of inner automorphisms (see Exercise 1.9.38) is a normal subgroup of $\text{Aut}(G)$.

28. A_5 *is Simple.* In this exercise, we guide the reader to prove that A_5 is a simple group.

    (a) Prove that, as a partition of A_5, the set of conjugacy classes in A_5 is a strict refinement of the partition of A_5 into cycle types.

    (b) Prove that A_5 has 5 conjugacy classes: 2 of equal size whose union is the subset of 5-cycles; all of the 3-cycles; the permutations with cycle type $(2, 2)$, and the identity.

    (c) After determining the sizes of all of the conjugacy classes of A_5, use Proposition 2.2.14 and Lagrange's Theorem to conclude that A_5 has no normal subgroups besides $\{1\}$ and the whole group.

2.3 Quotient Groups

We began Chapter 2 by proposing to generalize to all groups the construction that led from $(\mathbb{Z}, +)$ to addition in modular arithmetic, $(\mathbb{Z}/n\mathbb{Z}, +)$. Our discussion sent us far afield, but we never constructed something analogous to modular arithmetic. As promised in Section 2.2, we are in a position to generalize the modular arithmetic construction to general groups.

2.3.1   Quotient Groups

We jump right in with the main theorem of this section. We recall that if $N \trianglelefteq G$, then left cosets and right cosets are the same. Hence, we simply refer to "cosets" without specifying right or left.

Theorem 2.3.1

Let G be a group and N a normal subgroup. The set of cosets of N with the operation

$$(xN) \cdot (yN) \stackrel{\text{def}}{=} (xy)N \tag{2.5}$$

has the structure of a group with identity N and inverses given by $(xN)^{-1} = x^{-1}N$.

Proof. The biggest concern of this proof is the show that the group operation proposed in (2.5) is well-defined.

Let $xn_1 \in xN$ and $yn_2 \in yN$, be arbitrary elements in the cosets xN and yN respectively. Since $N \trianglelefteq G$, then $y^{-1}n_1y \in N$, which we call n_3. Then

$$xn_1yn_2 = xy(y^{-1}n_1y)n_2 = xyn_3n_2 \in (xy)N.$$

In other words, any element in xN operated with any element in yN is an element in xyN. Hence, (2.5) is a well-defined binary operation on the set of cosets of N.

If xN, yN, and zN are left cosets, then

$$\begin{aligned}
(xN) \cdot ((yN) \cdot (zN)) &= (xN) \cdot ((yz)N) = (x(yz))N \\
&= ((xy)z)N = ((xy)N) \cdot (zN) = ((xN) \cdot (yN)) \cdot (zN).
\end{aligned}$$

Thus, the associativity of this operation on left cosets follows from associativity in the original group operation.

Clearly $1N$ serves the identity. Finally $(xN) \cdot (x^{-1}N) = (xx^{-1})N = 1N$ and the reverse order is also true, so $x^{-1}N = (xN)^{-1}$. We conclude that the set of cosets of N is a group. $\square$

Definition 2.3.2

Let G be a group and let N be a normal subgroup. The group defined as the set of cosets with the operation defined in Theorem 2.3.1, is called the *quotient group* of G by N, and is denoted by G/N.

The curious reader might wonder whether there might be some other way to define an equivalence relation $\sim$ on G (i.e., partition G) so that we could also define an operation on the set of equivalence classes by $[x] \cdot [y] = [xy]$ for all $x, y \in G$. Exercise 2.3.30 leads the reader to show that the above quotient group construction is the *only* way we could do this.

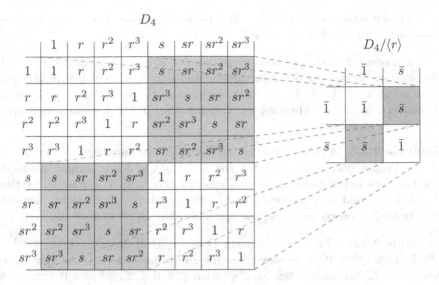

FIGURE 2.4: Cayley graphs of the quotient $D_4/\langle r \rangle$.

2.3.2 Examples of Quotient Groups

Example 2.3.3. As a first example, observe that $(\mathbb{Z}/n\mathbb{Z}, +)$ is the quotient group of $\mathbb{Z}$ by $n\mathbb{Z}$. Since $\mathbb{Z}$ is abelian, every subgroup is normal. In fact, the notation for modular arithmetic inspired the notation for quotient groups in general. △

Example 2.3.4. Consider the dihedral group on a square, D_4, and consider $R = \langle r \rangle$, the subset of rotations. Since R has index 2, $R \trianglelefteq D_4$. The quotient group consists of two elements $D_4/R = \{R, sR\}$. Following the habit of notation with modular arithmetic, it is not uncommon to write $\bar{1}$ for R and $\bar{s}$ for sR.

Consider the Cayley tables of D_4 and the quotient group D_4/R. When the Cayley table of a group G is organized so that elements in the same coset of N are contiguous, then blocks of size $|N| \times |N|$ in the original Cayley table will correspond to a single entry in the Cayley table of the quotient group G/N. (See Figure 2.4.)

In this particular situation, the Cayley table of D_4/R carries an interesting geometric interpretation. Since the coset R corresponds to all the rotations in D_4 and since sR corresponds to all the reflections through lines of symmetry, we can interpret the above Cayley table for $D_4/\langle r \rangle$ as:

·	rotation	reflection
rotation	rotation	reflection
reflection	reflection	rotation

So every rotation composed with a rotation is a rotation; every rotation composed with a reflection is a reflection, and so forth.   △

As illustrated in the previous two examples, it is not uncommon to mimic the notation used in modular arithmetic and denote a coset gN in the quotient group G/N by $\bar{g}$. In modular arithmetic, the modulus is understood by context. Similarly, when we use this notation $\bar{g}$, the normal subgroup is understood by context.

Example 2.3.5. As another example, consider the subgroup $N = \langle -1 \rangle$ in Q_8. By Proposition 2.2.6, since N is the center of Q_8, it is a normal subgroup. The elements in the quotient group Q_8/N are $\{\bar{1}, \bar{i}, \bar{j}, \bar{k}\}$. It is easy to see that $\bar{i}^2 = \overline{-1} = \bar{1}$ and similarly for $\bar{j}$ and $\bar{k}$. Hence, all the nonidentity elements have order 2. Consequently, we can conclude that $Q_8/N \cong Z_2 \oplus Z_2$.   △

Example 2.3.6. Consider the group $G = \langle x, y \,|\, x^3 = y^7 = 1, \quad xyx^{-1} = y^2 \rangle$. In Example 1.10.6, we saw that this group has order 21 and that every element in G can be written as $x^a y^b$ with $a = 0, 1, 2$ and $b = 0, 1, 2, \ldots, 6$. By Theorem 2.2.8, since $xyx^{-1} = y^2 \in \langle y \rangle$ and $yyy^{-1} = y \in \langle y \rangle$, then $\langle y \rangle$ is normal in G.

The cosets of $\langle y \rangle$ are $\bar{1} = \langle y \rangle$, $\bar{x} = x\langle y \rangle$, and $\overline{x^2} = x^2 \langle y \rangle$. The corresponding quotient group is

$$G/\langle y \rangle = \{\bar{1}, \bar{x}, \overline{x^2}\} \cong Z_3.$$ △

Definition 2.3.7

Let $N \trianglelefteq G$. The function $\pi : G \to G/N$ defined by $\pi(g) = gN$ is called the *canonical projection* of G onto G/N.

By the definition of the operation on cosets in (2.5), the canonical projection is a homomorphism. In Proposition 2.2.7, we saw that kernels of homomorphisms are normal subgroups; the converse is in fact true, which leads to the following proposition.

Proposition 2.3.8

A subgroup N of G is normal if and only if it is the kernel of some homomorphism.

Proof. This follows from Proposition 2.2.7 and the fact that N is the kernel of the canonical projection $\pi : G \to G/N$.   □

By Proposition 1.9.9(4), this proposition implies that for all $g \in G$, the order of gN in G/N divides the order of g in G. We can prove this same result in another fashion and obtain more precise information. By Exercise 2.3.14,

$$|gN| = \frac{|\langle g \rangle N|}{|N|}.$$

By Proposition 2.1.18, in G we have

$$|\langle g \rangle N| = \frac{|g|\,|N|}{|\langle g \rangle \cap N|},$$

which implies that

$$|g| = \frac{|\langle g \rangle N|}{|N|}|\langle g \rangle \cap N| = |\langle g \rangle \cap N| \cdot |gN|.$$

The subgroup order $|\langle g \rangle \cap N|$ is the divisibility factor between $|gN|$ and $|g|$.

As a final set of examples of quotient groups, by Proposition 2.2.6, $Z(G) \trianglelefteq G$ for all groups G. In Exercise 2.3.21, the reader is asked to prove the important result that if $G/Z(G)$ is cyclic, then G is abelian. In some intuitive sense, the quotient group $G/Z(G)$ "removes" any elements that commute with everything else. This intuitive manner of thinking may be misleading because the center of the quotient group $G/Z(G)$ is not necessarily trivial (Exercise 2.3.20). Nonetheless, given a group G, the quotient group $G/Z(G)$ often tells us something important about the group G. As specific examples, we mention in the so-called *projective linear groups*.

Example 2.3.9 (Projective Linear Groups). Suppose that F is $\mathbb{C}$, $\mathbb{R}$, $\mathbb{Q}$, or $\mathbb{F}_p$, where p is prime. In Example 1.7.9, we proved that the center of $\mathrm{GL}_n(F)$ consists of matrices of the form aI, where $a \in F^\times = F - \{0\}$. A similar result still holds for the center of $\mathrm{SL}_n(F)$, except not all diagonal matrices of the form aI are in $\mathrm{SL}_n(F)$.

The *projective general linear* group of order n is

$$\mathrm{PGL}_n(F) = \mathrm{GL}_n(F)/Z(\mathrm{GL}_n(F)),$$

while the *projective special linear* group of order n is

$$\mathrm{PSL}_n(F) = \mathrm{SL}_n(F)/Z(\mathrm{SL}_n(F)). \qquad \triangle$$

2.3.3 Direct Sum Decomposition

Early on, we introduced the concept of a direct sum of a finite number of groups. Recall that if $G_1, G_2, \ldots, G_n$ are groups, then the direct sum $G = G_1 \oplus G_2 \oplus \cdots \oplus G_n$ is the group in which the set is the Cartesian product of the groups and in which the operation on n-tuples is performed componentwise.

By a generalization of Exercise 2.3.25, for all i, the subset

$$\tilde{G}_i = \{(1, 1, \ldots, 1, g_i, 1, \ldots, 1) \mid g_i \in G_i\} \subseteq G_1 \oplus G_2 \oplus \cdots \oplus G_n$$

is isomorphic to G_i and is a normal subgroup of G. By the result of Exercise 2.2.17, the product set of normal subgroups is the join of the subgroups. In this situation, as a join of subgroups,

$$G = \tilde{G}_1 \tilde{G}_2 \cdots \tilde{G}_n$$

and for all $k = 1, 2, \ldots, n - 1$,

$$\tilde{G}_1 \tilde{G}_2 \cdots \tilde{G}_k \cap \tilde{G}_{k+1} = \{(1, 1, \ldots, 1)\}.$$

Furthermore,

$$G_1 \oplus G_2 \oplus \cdots \oplus G_n / \tilde{G}_i \cong G_1 \oplus \cdots \oplus G_{i-1} \oplus \widehat{G_i} \oplus G_{i+1} \oplus \cdots \oplus G_n,$$

where the $\widehat{G_i}$ notation indicates that the corresponding term is omitted.

In the above discussion, we assumed that we started with a collection of groups, constructed the direct sum group, and studied some of its properties. In contrast, suppose that G is an arbitrary group; then it is possible that G is isomorphic to a direct sum of groups $G_1 \oplus G_2 \oplus \cdots \oplus G_n$. If it is, then these groups G_i would be isomorphic to subgroups of G and possess the properties mentioned above. The following theorem states when and how a group G may be isomorphic to a direct sum of its own subgroups.

Theorem 2.3.10 (Direct Sum Decomposition)

Let $N_1, N_2, \ldots, N_k$ be a finite collection of normal subgroups of a group G satisfying:

(1) $G = N_1 N_2 \cdots N_k$;

(2) $N_1 N_2 \cdots N_i \cap N_{i+1} = \{1\}$ for all $i = 1, 2, \ldots, k - 1$.

Then $G \cong N_1 \oplus N_2 \oplus \cdots \oplus N_k$.

Proof. We prove the theorem by induction on k, starting with $k = 2$.

So suppose that G is any group with normal subgroups N_1 and N_2 such that $N_1 N_2 = G$ and $N_1 \cap N_2 = \{1\}$. Let $n_1 \in N_1$ and $n_2 \in N_2$ be arbitrary and consider the element $n_1^{-1} n_2^{-1} n_1 n_2$. This element is called the *commutator* of n_1 and n_2 and is denoted by $[n_1, n_2]$. (See Exercise 2.3.24.) Since $n_2^{-1} \in N_2 \trianglelefteq G$, then $n_1^{-1} n_2^{-1} n_1 \in N_2$ and $[n_1, n_2] \in N_2$. Similarly, $N_1 \trianglelefteq G$, so $n_2^{-1} n_1 n_2 \in N_1$ so again $[n_1, n_2] \in N_1$. Thus, $[n_1, n_2] \in N_1 \cap N_2$. However, $N_1 \cap N_2 = \{1\}$, so

$$n_1^{-1} n_2^{-1} n_1 n_2 = 1 \implies n_2^{-1} n_1 n_2 = n_1 \implies n_1 n_2 = n_2 n_1.$$

Since $G = N_1 N_2$, then for all $g \in G$ we have $g = n_1 n_2$ for some $n_1 \in N_1$ and $n_2 \in N_2$. Suppose that there exist $n_1' \in N_1$ and $n_2' \in N_2$ such that $n_1 n_2 = n_1' n_2'$. Then $(n_1')^{-1} n_1 = n_2' n_2^{-1}$. Since $(n_1')^{-1} n_1 \in N_1$ and $n_2' n_2^{-1} \in N_2$, both are in $N_1 \cap N_2$, which implies that $1 = (n_1')^{-1} n_1 = n_2' n_2^{-1}$. Thus $n_1 = n_1'$ and $n_2 = n_2'$. Hence, there is a unique way of writing g as $n_1 n_2$ with $n_1 \in N_1$ and $n_2 \in N_2$.

We now define the function $\varphi : G \to N_1 \oplus N_2$ by $\varphi(g) = (n_1, n_2)$ whenever $g = n_1 n_2$. We have just shown that this function is well-defined. Furthermore, is $g = n_1 n_2$ and $h = n_1' n_2'$, then by the commutativity we showed in the first paragraph, we have

$$\varphi(gh) = \varphi(n_1 n_2 n_1' n_2') = \varphi(n_1 n_1' n_2 n_2')$$
$$= (n_1 n_1', n_2 n_2') = (n_1, n_2)(n_1', n_2') = \varphi(g)\varphi(h),$$

showing that φ is a homomorphism. This function is also bijective with inverse $\varphi^{-1}(n_1, n_2) = n_1 n_2$. Thus φ is an isomorphism and hence $G \cong N_1 \oplus N_2$. This proves the theorem for $k = 2$.

Now suppose that the theorem is true for some integer k, namely that for any group G, if there is a list of normal subgroups $N_1, N_2, \ldots, N_k$ in G satisfying (1) and (2), then $G = N_1 \oplus N_2 \oplus \cdots \oplus N_k$. Consider a group G with normal subgroups $N_1, N_2, \ldots, N_{k+1}$ satisfying (1) and (2). By the induction hypothesis,

$$N_1 N_2 \cdots N_k \cong N_1 \oplus N_2 \oplus \cdots \oplus N_k.$$

Then with (1) and (2) and what we proved for the basis step,

$$G = (N_1 N_2 \cdots N_k) N_{k+1} = (N_1 N_2 \cdots N_k) \oplus N_{k+1} \cong N_1 \oplus N_2 \oplus \cdots \oplus N_k \oplus N_{k+1}.$$

The theorem follows by induction. $\qquad\square$

Example 2.3.11. As an example of the Direct Sum Decomposition Theorem, consider the group $U(35)$. This is an abelian group with $\phi(35) = 24$ elements. Again, recall that in any abelian group, all subgroups are normal. Consider the subgroups $N_1 = \langle \overline{6} \rangle$ and $N_2 = \langle \overline{2} \rangle$. For elements, we have

$$\langle \overline{6} \rangle = \{\overline{1}, \overline{6}\},$$
$$\langle \overline{2} \rangle = \{\overline{1}, \overline{2}, \overline{4}, \overline{8}, \overline{16}, \overline{32}, \overline{29}, \overline{23}, \overline{11}, \overline{22}, \overline{9}, \overline{18}\}.$$

We note that $N_1 \cap N_2 = \{\overline{1}\}$. Furthermore, by Proposition 2.1.18, $|N_1 N_2| = 24$, so $N_1 N_2 = G$. The subgroups N_1 and N_2 satisfy the conditions of the Direct Sum Decomposition Theorem. Hence,

$$U(35) \cong \langle \overline{6} \rangle \oplus \langle \overline{2} \rangle \cong Z_2 \oplus Z_{12}.$$

The group $U(35)$ can be decomposed even further. Consider the three subgroups $\langle \overline{6} \rangle$, $\langle \overline{8} \rangle$, and $\langle \overline{16} \rangle$. We leave it to the reader (Exercise 2.3.26) to verify that these subgroups satisfy the conditions of the Direct Sum Decomposition Theorem. Then, we conclude that

$$U(35) \cong \langle \overline{6} \rangle \oplus \langle \overline{16} \rangle \oplus \langle \overline{8} \rangle \cong Z_2 \oplus Z_3 \oplus Z_4.$$

Note that since $\gcd(3, 4) = 1$, then $Z_{12} \cong Z_4 \oplus Z_3$, so the above two decompositions are equivalent. In this latter decomposition, depicted visually in Figure 2.5, we could write

$$U(35) = \{\overline{6}^a \overline{16}^b \overline{8}^c \mid a = 0, 1; \ b = 0, 1, 2; \ c = 0, 1, 2, 3\}. \qquad \triangle$$

2.3.4   Useful CAS Commands

The quotient group construction poses a number of decision challenges for computer algebra systems.

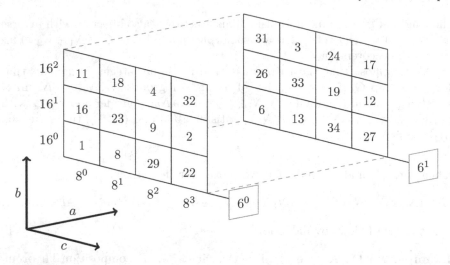

FIGURE 2.5: A visualization of the decomposition of $U(35)$.

Maple offers the command `Factor(g,H)`, which factors the group element g into a pair `[h,r]`, where $h \in H$ and r is a coset representative of the right coset Hg, where r is the representative returned by the `RightCosets` command applied to H. So where $H \trianglelefteq G$, we can use these representatives and the `Factor` command to perform operations in the quotient group G/H.

Some CASs with group theory implement a quotient group construction when the group is defined as a permutation group. In order to perform quotient group calculations with a finite group G on these systems, we first must have an isomorphism of G into some S_n. For the same of an example we consider the group

$$G = \langle a, b \mid a^3 = 1, \ b^7 = 1, \ ab = b^2 a \rangle.$$

We had seen in Example 1.10.6 that this group has order 21. So Cayley's Theorem says we can embed G into S_{21}. However, it turns out that we can embed G into S_7. Note that the relation $ab = b^2 a$ can be rewritten as $aba^{-1} = b^2$, which is a conjugacy relationship. Suppose that we defined $\varphi(b) = (1\,2\,3\,4\,5\,6\,7)$. Then

$$\varphi(b)^2 = (1\,3\,5\,7\,2\,4\,6).$$

Using (2.4), we note that

$$(2\,3\,5)(4\,7\,6) \cdot (1\,2\,3\,4\,5\,6\,7) \cdot ((2\,3\,5)(4\,7\,6))^{-1} = (1\,3\,5\,7\,2\,4\,6).$$

So if we set $\varphi(a) = (2\,3\,5)(4\,7\,6)$, it is clear that

$$\varphi(a)^3 = 1, \quad \varphi(b)^7 = 1, \quad \varphi(a)\varphi(b)\varphi(a)^{-1} = \varphi(b)^2.$$

By the Extension Theorem on Generators, $\varphi : \{a, b\} \to S_7$ extends uniquely to a homomorphism $\varphi : G \to S_7$. We are not yet sure that φ is injective. However, it is easy to see that $\langle \varphi(a) \rangle$ is a cyclic subgroup of S_7 of size 3 and $\langle \varphi(b) \rangle$ is a cyclic subgroup of S_7 of size 7. By Lagrange's Theorem, $\operatorname{lcm}(3, 7) = 21$ divides the order of $\varphi(G)$, but $|\varphi(G)| \leq 21$. We deduce that $|\varphi(G)| = 21$ so φ is injective. From this, we deduce that

$$G \cong \langle (1\,2\,3\,4\,5\,6\,7), (2\,3\,5)(4\,7\,6) \rangle.$$

From now on, we can work with this embedding of G to study it with a computer algebra system.

SAGE offers a command to create a quotient group of G/N, but it returns a group that is isomorphic to the quotient group. Consider the following example that works with the group G from the above paragraph and takes the quotient of G by $\langle b \rangle$.

―――――――――――――― Sage ――――――――――――――

```
sage: G=PermutationGroup(["(1,2,3,4,5,6,7)","(2,3,5)(4,7,6)"])
sage: b=G("(1,2,3,4,5,6,7)")
sage: H=G.subgroup([b])
sage: H.is_normal(G)
True
sage: Q=G.quotient(H)
sage: Q.list()
[(),(1,2,3), (1,3,2)]
```

EXERCISES FOR SECTION 2.3

1. Prove that $S_n/A_n \cong Z_2$.

2. Prove that $\langle r^2 \rangle \trianglelefteq D_4$, list the elements in $D_4/\langle r^2 \rangle$, and show that $D_4/\langle r^2 \rangle \cong Z_2 \oplus Z_2$.

3. Consider the group $(\mathbb{R}^\times, \times)$ and the subgroup $\{1, -1\}$. Prove that $\mathbb{R}^\times/\{1, -1\} \cong \mathbb{R}^{>0}$, where the latter set is given the group structure of multiplication.

4. Consider the group $G = U(33)$ and consider the subgroup $H = \langle \overline{4} \rangle$. Since G is abelian, $H \trianglelefteq G$. List the elements of G/H and determine its isomorphism type (i.e., find to which group in the table of Section A.7 it is isomorphic.)

5. Explicitly list the elements in the quotient group $U(33)/\langle \overline{10} \rangle$. Show that this quotient group is cyclic and find an explicit generator.

6. Consider the quotient group construction described in Example 2.3.5. Fill out the Cayley table of Q_8 and shade boxes of this Cayley table with 4 different colors to mimic the visual illustration of the quotient group process as shown in Example 2.3.4.

7. Consider the function $\pi : \mathbb{R}^2 \to \mathbb{R}$ given by $\pi(x, y) = 2x + 3y$.

   (a) Show that π is a homomorphism from $(\mathbb{R}^2, +)$ to $(\mathbb{R}, +)$ and describe the fibers geometrically.

(b) Determine $\operatorname{Ker} \pi$.

(c) Interpret geometrically the addition operation in $\mathbb{R}^2/\operatorname{Ker} \pi$.

8. Let N be a normal subgroup of G. Prove that for all $g \in G$ and all $k \in \mathbb{Z}$, $(gN)^k = (g^k)N$ in G/N.

9. Prove that the quotient group of a cyclic group by any subgroup is again a cyclic group. Deduce that if $d|n$, then in $\mathbb{Z}/n\mathbb{Z}/\langle \bar{d} \rangle \cong \mathbb{Z}/d\mathbb{Z}$.

10. Consider the group $(\mathbb{R}, +)$ and the quotient group $\mathbb{R}/\mathbb{Z}$. Recall that the torsion subgroup of an abelian group is the subgroup of elements of finite order. (See Exercise 1.7.18.)

(a) Prove that $\operatorname{Tor}(\mathbb{R}/\mathbb{Z}) = \mathbb{Q}/\mathbb{Z}$.

(b) Prove that $\mathbb{R}/\mathbb{Z}$ is isomorphic to the circle group, defined as

$$\left\{ \begin{pmatrix} \cos\theta & -\sin\theta \\ \sin\theta & \cos\theta \end{pmatrix} \in \operatorname{GL}_2(\mathbb{R}) \,\middle|\, \theta \in \mathbb{R} \right\}.$$

11. Prove that if G is generated by a subset $\{g_1, g_2, \ldots, g_n\}$, then the quotient group G/N is generated by $\{\bar{g}_1, \bar{g}_2, \ldots, \bar{g}_n\}$.

12. Consider the dihedral group D_n and let d be a divisor of n.

(a) Prove that $\langle r^d \rangle \trianglelefteq D_n$.

(b) Show that $D_n/\langle r^d \rangle \cong D_d$.

(c) Give a geometric interpretation of this last result. (What information is conflated when taking the quotient group?)

13. Let G be a group and let $\sim$ be an equivalence relation on G. Give necessary and sufficient conditions on $\sim$ so that $(G/\sim, \cdot)$ is a group, where $[a] \cdot [b] = [ab]$.

14. Let N be a normal subgroup of a group G and write $\bar{g}$ for the coset gN in the quotient group G/N.

(a) Show that for all $g \in G$, if the order of $\bar{g}$ is finite, then $|\bar{g}|$ is the least positive integer k such that $g^k \in N$.

(b) Deduce that the element order $|\bar{g}|$ is equal to $|\langle g \rangle N|/|N|$.

15. Consider the group called G, which is given in generators and relations as

$$G = \langle x, y \mid x^4 = y^3 = 1, \ x^{-1}yx = y^{-1} \rangle.$$

(a) Prove that G is a nonabelian group of order 12.

(b) Prove that $\langle y \rangle$ is a normal subgroup and that $G/\langle y \rangle \cong \mathbb{Z}_4$.

(c) Prove that G is not isomorphic to A_4 or D_6. [This group is an example of a *semidirect product*, a topic that is covered in Section 6.6. The common notation for this group is $\mathbb{Z}_3 \rtimes \mathbb{Z}_4$.]

16. Consider the group G given in generators and relations as

$$G = \langle x, y \mid x^4 = y^5 = 1, \ x^{-1}yx = y^2 \rangle.$$

(a) Prove that G is a nonabelian group of order 20.

(b) Prove that $\langle y \rangle$ is a normal subgroup and that $G/\langle y \rangle \cong \mathbb{Z}_4$.

(c) Prove that G is not isomorphic to D_{10}.

[Note: This group is called the *Frobenius group* of order 20. It is denoted by F_{20}.]

17. Let G be the group defined by

$$G = \langle x, y \,|\, x^2 = y^8 = 1, \, yx = xy^5 \rangle.$$

   (a) Show that $\langle y^4 \rangle \trianglelefteq G$.
   (b) Show that $G/\langle y^4 \rangle$ is a group of order 8.
   (c) Find the isomorphism type of $G/\langle y^4 \rangle$ (i.e., find to which group in the table of Section A.7 it is isomorphic.) Explain.

   [Note: This group is called the *modular group* of order 16.]

18. Consider the group $SL_2(\mathbb{F}_3)$, i.e., the special linear group of 2×2 matrices with determinant 1, with entries in $\mathbb{F}_3$, modular arithmetic base 3.

   (a) Show that this group is nonabelian with 24 elements.
   (b) Prove that the center of $SL_2(\mathbb{F}_3)$ is the subgroup of two elements

$$Z(SL_2(\mathbb{F}_3)) = \left\{ \begin{pmatrix} 1 & 0 \\ 0 & 1 \end{pmatrix}, \begin{pmatrix} 2 & 0 \\ 0 & 2 \end{pmatrix} \right\}.$$

   (c) The projective special linear group

$$PSL_2(\mathbb{F}_3) = SL_2(\mathbb{F}_3)/Z(SL_2(\mathbb{F}_3))$$

   has order 12. Determine, with proof, to which group in the table of Section A.7 it is isomorphic.

19. Consider the group $GL_2(\mathbb{F}_3)$, i.e., the general linear group of 2 by 2 invertible matrices with elements in modular arithmetic modulo 3. (By the result of Exercise 1.2.23, this group has 48 elements.) We consider the group $G = PGL_2(\mathbb{F}_3)$, the projective general linear group of order 2 on $\mathbb{F}_3$.

   (a) Prove that $|G| = 24$ and show that G and S_4 have the same number of elements of any given order.
   (b) Show explicitly that $G \cong S_4$. [Showing that they have the same number of elements of a given order is evidence but not sufficient for a proof of the isomorphism.]

20. Find an example of a group G in which the center of $G/Z(G)$ is not trivial.

21. Prove that if $G/Z(G)$ is cyclic, then G is abelian. Give an example to show that G is not necessarily abelian if we only assume that $G/Z(G)$ is abelian.

22. Prove that if $|G| = pq$, where p and q are two primes, not necessarily distinct, then G is either abelian or $Z(G) = \{1\}$. [Hint: Use Exercise 2.3.21.]

23. Let N be a normal subgroup of a finite group G and let $g \in G$. Prove that if $\gcd(|g|, |G/N|) = 1$, then $g \in N$.

24. Let G be a group. The *commutator subgroup*, denoted G', is defined as the subgroup generated by all products $x^{-1}y^{-1}xy$, for any $x, y \in G$. In other words,

$$G' = \langle x^{-1}y^{-1}xy \,|\, x, y \in G \rangle.$$

   (a) Prove that $G' \trianglelefteq G$.
   (b) Without using part (a), prove that G' is a characteristic subgroup of G. (See Exercise 2.2.21.)

(c) Prove that G/G' is abelian.

25. Let A and B be two groups and let $G = A \oplus B$. The subgroup $A \times \{1\} \leq G$ is isomorphic to A. Prove that $A \times \{1\} \trianglelefteq G$ and that $G/(A \times \{1\}) \cong B$.

26. In Example 2.3.11, prove that the subgroups $\langle \overline{6} \rangle$, $\langle \overline{8} \rangle$, and $\langle \overline{16} \rangle$ satisfy the conditions of the Direct Sum Decomposition Theorem.

27. Use the direct sum decomposition to show that $U(100) \cong Z_{20} \oplus Z_2 \cong Z_5 \oplus Z_4 \oplus Z_2$.

28. Let p be a prime number and let k be a positive integer. Prove that Z_{p^k} is not isomorphic to the direct product of any other groups.

29. Prove that Q_8 is not isomorphic to the direct sum of any other groups.

30. This exercise constructs an alternate approach to defining a quotient group from a group G.

   (a) Let $\sim$ be an equivalence relation on G. Show that the operation $\cdot$ on $(G/\sim)$ given by $[x] \cdot [y] = [xy]$ is well-defined if and only if $\sim$ satisfies the following property

$$\forall g_1, g_2, x_1, x_2 \in G, \ g_1 \sim x_1 \text{ and } g_2 \sim x_2 \implies g_1 g_2 \sim x_1 x_2. \qquad \text{(P)}$$

   (b) Prove that if $\sim$ satisfies Property (P), then $(G/\sim, \cdot)$ is a group.

   (c) Call N the equivalence class $[1_G]$. Prove that $N \trianglelefteq G$.

   (d) Prove that all the equivalence classes of $\sim$ have the form gN for some $g \in G$.

   (e) Deduce that for any $\sim$ satisfying Property (P) there exists a normal subgroup $N \trianglelefteq G$ such that $\sim$ is the equivalence relation $\sim_1$ for N defined in Proposition 2.1.4 (which is also equal to $\sim_2$ because N is normal).

2.4 Isomorphism Theorems

As we saw in the previous section, properties of quotient groups of a group G may imply relationships between certain subgroups of G. Much more can be said, however. In this section, we discuss some theorems, the four isomorphism theorems, that describe further structure within a group.

2.4.1 First Isomorphism Theorem

When we introduced the concept of a homomorphism, we gave the intuitive explanation that a homomorphism preserves the group structure. The first of the isomorphism theorems shows more precisely in what sense a homomorphism carries the structure of one group into another.

Theorem 2.4.1 (First Isomorphism Theorem)

Let $\varphi : G \to H$ be a homomorphism between groups. Then $\operatorname{Ker} \varphi \trianglelefteq G$ and $G / \operatorname{Ker} \varphi \cong \operatorname{Im} \varphi$.

Proof. Proposition 2.2.7 established that $\operatorname{Ker} \varphi \trianglelefteq G$.

The elements in the quotient group $G / \operatorname{Ker} \varphi$ are left cosets of the form $g(\operatorname{Ker} \varphi)$. We define the function

$$\overline{\varphi} : G / \operatorname{Ker} \varphi \longrightarrow \operatorname{Im} \varphi = \varphi(G)$$
$$g(\operatorname{Ker} \varphi) \longmapsto \varphi(g).$$

For any $g \in G$, the element g is only one of many possible representatives of the coset $g \operatorname{Ker} \varphi$. Thus, in order to verify that this is even a function, we first need to check that $\overline{\varphi}(g(\operatorname{Ker} \varphi))$ gives the same output for every representative of $g(\operatorname{Ker} \varphi)$. Suppose gh is any element in the coset $g \operatorname{Ker} \varphi$. Then $\varphi(gh) = \varphi(g)\varphi(h) = \varphi(g)$. Thus, the choice of representative has no effect on the stated output $\varphi(g)$. This simply means that $\overline{\varphi}$ is well-defined as a function.

However, it is easy to check that $\overline{\varphi}$ is a homomorphism. Furthermore, $\overline{\varphi}$ is surjective since every element $\varphi(g)$ in $\varphi(G)$ is obtained as the output $\overline{\varphi}(g \operatorname{Ker} \varphi)$. To prove injectivity, let $g_1, g_2 \in G$. Then

$$\overline{\varphi}(g_1 \operatorname{Ker} \varphi) = \overline{\varphi}(g_2 \operatorname{Ker} \varphi) \Longleftrightarrow \varphi(g_1) = \varphi(g_2)$$
$$\Longleftrightarrow \varphi(g_2)^{-1}\varphi(g_1) = 1 \Longleftrightarrow \varphi(g_2^{-1}g_1) = 1$$
$$\Longleftrightarrow g_2^{-1}g_1 \in \operatorname{Ker} \varphi \Longleftrightarrow g_1 \operatorname{Ker} \varphi = g_2 \operatorname{Ker} \varphi.$$

This proves injectivity of $\overline{\varphi}$. Consequently, $\overline{\varphi}$ is bijective and thus an isomorphism and the theorem follows. $\square$

The First Isomorphism Theorem shows that the image of an isomorphism, as a subgroup of the codomain, must already exist within the structure of the domain group, not as a subgroup but as a quotient group. This theorem also shows how any homomorphism $\varphi : G \to H$ can be factored into the surjective (canonical projection) map $\pi : G \to G / \operatorname{Ker} \varphi$ and an injective homomorphism $\overline{\varphi} : G / \operatorname{Ker} \varphi \to H$ so that $\varphi = \overline{\varphi} \circ \pi$. We often depict this relationship by the following diagram.

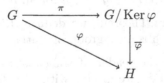

The First Isomorphism Theorem leads to many consequences about groups, some elementary and some more profound. One implication is that

if $\varphi : G \to H$ is an injective homomorphism, then $\operatorname{Ker}\varphi = \{1\}$ and so $G/\operatorname{Ker}\varphi = G \cong \varphi(G)$. In this situation, we sometimes say that G is *embedded* in H or that φ is an embedding of G into H because φ maps G into an exact copy of itself as a subgroup of H.

As another example, suppose that G is a simple group. By definition, it contains no normal subgroups besides $\{1\}$ and itself. Hence, by the First Isomorphism Theorem, any homomorphism φ from G is either injective ($\operatorname{Ker}\varphi = \{1\}$) or trivial ($\varphi(G) = \{1\}$). Thus, under any homomorphism, a simple group is either an embedding into any other group or its image is trivial.

Combining the First Isomorphism Theorem with Lagrange's Theorem, we are able to deduce the following not obvious corollary.

Corollary 2.4.2

Let G and H be finite groups with $\gcd(|G|, |H|) = 1$. Then the only homomorphism $\varphi : G \to H$ is the trivial homomorphism, $\varphi(g) = 1_H$.

Proof. Since $\varphi(G) \le H$, then by Lagrange's Theorem $|\varphi(G)|$ divides $|H|$. By the First Isomorphism Theorem, $|\varphi(G)| = |G|/|\operatorname{Ker}\varphi|$. Consequently $|\varphi(G)|$ divides $|G|$. Hence, $|\varphi(G)|$ divides $\gcd(|G|, |H|) = 1$, so $|\varphi(G)| = 1$. The only subgroup of H that has only 1 element is $\{1_H\}$. $\square$

The First Isomorphism Theorem leads to many other more subtle results in group theory. The following Normalizer-Centralizer Theorem is an immediate consequence but is important in its own right. In future sections, we will see how this theorem implies more subtle constraints on the internal structure of a group, leading to consequences for the classification of groups.

Theorem 2.4.3 (Normalizer-Centralizer Theorem)

Let H be a subgroup of a group G. Then $N_G(H)/C_G(H)$ is isomorphic to a subgroup of $\operatorname{Aut}(H)$.

Proof. By Proposition 2.2.15, the function $\Psi : N_G(H) \to \operatorname{Aut}(H)$, defined by $\Psi(g) = \psi_g$, where $\psi_g : N \to N$ with $\psi_g(n) = gng^{-1}$ is a homomorphism. The image subgroup $\Psi(N_G(H)) \le \operatorname{Aut}(H)$ could be strictly contained in $\operatorname{Aut}(H)$.

Now the kernel of Ψ is precisely

$$\operatorname{Ker}\Psi = \{g \in N_G(H) \mid ghg^{-1} = h \text{ for all } h \in H\} = C_G(H).$$

By the First Isomorphism Theorem, we deduce that $N_G(H)/C_G(H)$ is isomorphic $\Psi(N_G(H))$, which is a subgroup of $\operatorname{Aut}(H)$. $\square$

As a general hint to the reader, if someone ever asks for a proof that $G/N \cong H$, where G and H are groups with $N \trianglelefteq G$, then there must exist a surjective homomorphism $\varphi : G \to H$ such that $\operatorname{Ker}\varphi = N$. Hence, when attempting to prove such results, the First Isomorphism Theorem offers the strategy of looking for an appropriate homomorphism.

2.4.2   Second Isomorphism Theorem

Theorem 2.4.4 (The Second Isomorphism Theorem)

Let G be a group and let A and B be subgroups such that $A \leq N_G(B)$. Then AB is a subgroup of G, $B \trianglelefteq AB$, $A \cap B \trianglelefteq A$ and

$$AB/B \cong A/A \cap B.$$

Proof. Since $A \leq N_G(B)$, then A normalizes B and AB, which is also a subgroup of G, normalizes B. Thus, $B \trianglelefteq AB$.

Define the function $\phi : A \to AB/B$ by $\phi(a) = aB$. This is a homomorphism precisely because the group operation in AB/B is well-defined:

$$\phi(a_1 a_2) = (a_1 a_2)B = (a_1 B)(a_2 B) = \phi(a_1)\phi(a_2).$$

Clearly ϕ is surjective so $\phi(A) = AB/B$, but we would like to determine the kernel. Now $\phi(a) = 1B$ if and only if $aB = 1B$ if and only if $a \in B$. This means that $\operatorname{Ker} \phi = A \cap B$. Thus, $A \cap B \trianglelefteq A$ and by the First Isomorphism Theorem

$$A/A \cap B = AB/B. \qquad \square$$

The Second Isomorphism Theorem is also called the Diamond Isomorphism Theorem because it concerns the relative sides of particular "diamonds" inside the lattice structure of a group.

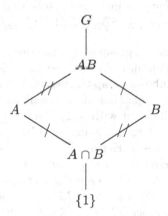

In the above diagram, assuming $A \leq N_G(B)$, then the opposite /-sides not only have the same index, $|AB : B| = |A : A \cap B|$, but correspond to normal subgroups and satisfy $AB/B \cong A/A \cap B$. On the other hand, if $B \leq N_G(A)$, then the opposite //-sides satisfy the same property. Finally, in the special case that A and B are both normal subgroups of G, then the Second Isomorphism Theorem applies to both pairs of opposite sides of the diamond.

2.4.3 Third Isomorphism Theorem

Theorem 2.4.5 (The Third Isomorphism Theorem)

Let G be a group and let H and K be normal subgroups of G with $H \leq K$. Then $K/H \trianglelefteq G/H$ and $(G/H)/(K/H) \cong G/K$.

Proof. Consider the mapping $\varphi : G/H \to G/K$ with $\varphi(gH) = gK$. We first need to show that it is a well-defined function. Suppose that $g_1 H = g_2 H$. Then $g_2^{-1} g_1 \in H$. But since $H \leq K$, then $g_2^{-1} g_1 \in H$ implies that $g_2^{-1} g_1 \in K$. Thus, $g_1 K = g_2 K$. Hence, φ is well-defined. For all $gK \in G/K$ we have $\varphi(gH) = gK$ so φ is surjective. The kernel of φ is

$$\operatorname{Ker} \varphi = \{gH \in G/H \mid g \in K\} = K/H.$$

By the First Isomorphism Theorem, $\operatorname{Ker} \varphi = K/H \trianglelefteq G/H$ and

$$(G/H)/(K/H) = (G/H)/\operatorname{Ker} \varphi \cong \operatorname{Im} \varphi = G/K. \qquad \square$$

Example 2.4.6. A simple example of the Third Isomorphism Theorem concerns subgroups of $G = \mathbb{Z}$. Let $H = 48\mathbb{Z}$ and $K = 8\mathbb{Z}$. We have $H \leq K$ and, since G is abelian, both H and K are normal subgroups. We have

$$G/H = \mathbb{Z}/48\mathbb{Z}, \quad K/H = 8\mathbb{Z}/48\mathbb{Z} = \langle \overline{8} \rangle, \quad \text{and} \quad G/K = \mathbb{Z}/8\mathbb{Z},$$

where in K/H, the element $\overline{8}$ is in $\mathbb{Z}/48\mathbb{Z}$. Note that $K/H \cong Z_6$. The Third Isomorphism Theorem gives $(\mathbb{Z}/48\mathbb{Z})/(8\mathbb{Z}/48\mathbb{Z}) \cong \mathbb{Z}/8\mathbb{Z}$. $\triangle$

Example 2.4.7. Let G be the group of symmetries on the vertices of the cube that preserve the cube structure. We can see that $|G| = 48$: in the following way. A symmetry of the cube can map the vertex 1 to any of the eight vertices. Then σ can map the three edges that are incident with vertex 1 in any way to the three edges that are incident with $\sigma(1)$. There are $3! = 6$ possibilities for this mapping of incident edges. Knowing $\sigma(1)$ and how σ maps its incident edges, completely determines σ. Hence, $|G| = 8 \times 6 = 48$.

Exercise 1.9.35 discussed the group R of rigid motions of the cube. We observe that $R \leq G$ but $R \neq G$ because reflections through planes are not rigid motions. The exercise should that $R \cong S_4$ as the permutation group on the 4 maximal diagonals. Since $|G : R| = 48/24 = 2$, then $R \trianglelefteq G$. Figure 2.6 illustrates three symmetries of a cube, two reflections through a plane, and one rotation about a maximal diagonals. The reflection through a maximal diagonal is a rigid motion but there are many other rigid motions.

Consider also the subgroup H of G generated by the rotations of $120°$ about the maximal diagonals. It is not hard to see that all symmetries of the cube are generated by reflections through planes. If f is a reflection through a plane and r is a rotation by $2\pi/3$ through a maximal diagonal L, then frf^{-1}

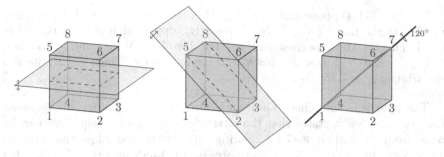

FIGURE 2.6: Symmetries of a cube.

is the rotation through the maximal diagonal $L' = f(L)$. From Theorem 2.2.8, we conclude that $H \trianglelefteq G$. If we view R as S_4 via how it permutes the maximal diagonals of the cube, H is generated by 3-cycles, which are even, and hence H corresponds to the subgroup A_4 in R.

In this example, $H \leq R \leq G$ and with $H, R \trianglelefteq G$. The quotient group $G/R = \{R, fR\} \cong Z_2$ carries the information of whether a symmetry is a rigid motion (R) or a reflected rigid motion (fR). The quotient group $R/H \cong Z_2$ carries information about whether a rigid motion is odd or even in the identification of R with S_4. Finally, $G/H \cong Z_2 \oplus Z_2$ contains both information. The Third Isomorphism Theorem, which states that $(G/H)/(R/H) \cong G/R$, says that information about orientation, G/R, is contained in G/H as R/H. $\triangle$

2.4.4 Fourth Isomorphism Theorem

Theorem 2.4.8 (The Fourth Isomorphism Theorem)

Let G be a group and $N \trianglelefteq G$. The function

$$f : \{A \in \mathrm{Sub}(G) \mid N \subseteq A\} \to \mathrm{Sub}(G/N)$$

defined by $f(A) = A/N$ is a bijection. Furthermore, for all subgroups A, B with $N \leq A, B \leq G$ we have:

(1) $A \leq B$ if and only if $A/N \leq B/N$.

(2) If $A \leq B$, then $|B/N : A/N| = |B : A|$.

(3) $\langle A/N, B/N \rangle = \langle A, B \rangle / N$.

(4) $A/N \cap B/N = (A \cap B)/N$.

(5) $A \trianglelefteq G$ if and only if $A/N \trianglelefteq G/N$.

Proof. (Parts (2) through (5) are left as exercises for the reader. See Exercise 2.4.13.)

For part (1), suppose first that $A \leq B$. Then for all $gN \in A/N$, we have $g \in A \subseteq B$ and hence $gN \in B/N$. Conversely, suppose that $A/N \leq B/N$. Let $a \in A$. Then by the hypothesis, $aN \in B/N$. If $aN = bN$ for some $b \in B$, then $b^{-1}a \in N$. But $N \leq B$ so $b^{-1}a = b'$ and hence $a = bb'$. Thus, $a \in B$. Since a was arbitrary, $A \leq B$. $\qquad\square$

The Fourth Isomorphism Theorem is also called the *Lattice Isomorphism Theorem* because it states that the lattice of a quotient group G/N can be found from the lattice of G by ignoring all vertices and edges that are not above N in the lattice of G. Furthermore, part (2) indicates that if we labeled each edge in the lattice of subgroups with the index between groups, then even these indices are preserved when passing to the quotient group.

Example 2.4.9 (Quaternion Group). Consider the quaternion group Q_8. Note that $Z(Q_8) = \langle -1 \rangle$ and hence this is a normal subgroup. The following lattice diagram of Q_8 depicts with double edges all parts of the diagram above $\langle -1 \rangle$. Hence, according to the Fourth Isomorphism Theorem, the lattice of $Q_8/\langle -1 \rangle$ is the sublattice involving double edges.

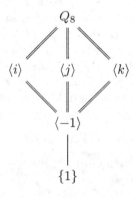

The Fourth Isomorphism Theorem parallels how lattices interact with subgroups. The lattice of a subgroup H of a group G can be found from the lattice of G by ignoring all vertices and edges that are not below H in the lattice of G. If this similarity tempts someone to guess that a group might be completely determined from knowing G/N and N, the reader should remain aware that this is not the case. We do not need to look further than $G = D_3$ with $N = \langle r \rangle$ as compared to Z_6 with $N = \langle z^2 \rangle$ for an example. In both cases $G/N \cong Z_2$ and $N \cong Z_3$ but $D_3 \not\cong Z_6$.

EXERCISES FOR SECTION 2.4

1. Show that the only homomorphism from D_7 to Z_3 is trivial.

2. Show that the only homomorphism from S_4 to Z_3 is trivial.

3. Let $\varphi : G \to H$. Prove that $|\varphi(G)|$ divides $|G|$.

4. Let F be $\mathbb{Q}$, $\mathbb{R}$, $\mathbb{C}$, or $\mathbb{F}_p$. Prove that $\mathrm{GL}_n(F)/\mathrm{SL}_n(F) \cong F^{\times}$.

5. Let p be an odd prime and let H be any group of odd order. Prove that the only homomorphism from D_p to H is trivial.

6. Let G be a group and consider homomorphisms $\varphi : G \to Z_2$.

   (a) Prove that if φ is surjective, then exactly half of the elements of G map to the identity in Z_2.

   (b) Show that there does not exist a surjective homomorphism from A_5 to Z_2.

7. Let G_1 and G_2 be groups and let N_1 and N_2 be normal subgroups respectively in G_1 and G_2. Prove that $N_1 \oplus N_2 \trianglelefteq G_1 \oplus G_2$ and that $(G_1 \oplus G_2)/(N_1 \oplus N_2) \cong (G_1/N_1) \oplus (G_2/N_2)$.

8. Let N_1 and N_2 be normal subgroups of G. Prove that $N_1N_2/(N_1 \cap N_2) \cong (N_1N_2/N_1) \oplus (N_1N_2/N_2)$.

9. Let $F = \mathbb{Q}, \mathbb{R}, \mathbb{C}$, or $\mathbb{F}_p$ where p is prime. Let $T_2(F)$ be the set of 2×2 upper triangular matrices with a nonzero determinant. We consider $T_2(F)$ as a group with the operation of multiplication.

   (a) Prove that $U_2(F) = \left\{ \begin{pmatrix} 1 & b \\ 0 & 1 \end{pmatrix} \mid b \in F \right\}$ is a normal subgroup that is isomorphic to $(F, +)$.

   (b) Prove that the corresponding quotient group is $T_2(F)/U_2(F) \cong U(F) \oplus U(F)$, where $U(F)$ is the group of elements with multiplicative inverses in F equipped with multiplication.

10. Let G be a group and let $N \trianglelefteq G$ such that $|G|$ and $|\operatorname{Aut}(N)|$ are relatively prime. Then $N \leq Z(G)$. [Hint: Use Proposition 2.2.15.]

11. Suppose that H and K are distinct subgroups of G, each of index 2. Prove that $H \cap K$ is a normal subgroup of G and that $G/(H \cap K) \cong Z_2 \oplus Z_2$.

12. Let p be a prime number and suppose that $\operatorname{ord}_p(|G|) = a$. Assume $P \leq G$ has order p^a and let $N \trianglelefteq G$ with $\operatorname{ord}_p(|N|) = b$. Prove that $|P \cap N| = p^b$ and $|PN/N| = p^{a-b}$.

13. Prove all parts of the Fourth Isomorphism Theorem.

14. Let $\mathcal{G}$ be the group of isometries of $\mathbb{R}^n$. Prove that the subgroup $\mathcal{D}$ of direct isometries is a normal subgroup and that $\mathcal{G}/\mathcal{D} \cong Z_2$. [See Definitions 1.11.5 and 1.11.6.]

2.5   Fundamental Theorem of Finitely Generated Abelian Groups

Early on in the study of groups, we discussed classifications theorems, which are theorems that list all possible groups with some specific property. The First Isomorphism Theorem leads to the Fundamental Theorem of Finitely Generated Abelian Groups (abbreviated by FTFGAG), which, among other things, provides a complete classification of all finite abelian groups. The proof begins with a study of free abelian groups.

2.5.1   Free Abelian Groups

The concept of a free abelian group is modeled after aspects of linear algebra. In this section, we will write abelian groups additively so that nx corresponds to the usual n-fold addition as defined by (1.3). For any subset X of an abelian group $(G, +)$, a *linear combination* of X is an expression of the form

$$c_1 x_1 + c_2 x_2 + \cdots + c_r x_r,$$

where $r \in \mathbb{N}^*$, $c_i \in \mathbb{Z}$, and $x_i \in S$.

Definition 2.5.1

A subset $X \subseteq G$ of an abelian group is called *linearly independent* if for every finite subset $\{x_1, x_2, \ldots, x_r\} \subseteq X$,

$$c_1 x_1 + c_2 x_2 + \cdots + c_r x_r = 0 \quad \Longrightarrow \quad c_1 = c_2 = \cdots = c_r = 0. \quad (2.6)$$

A *basis* of an abelian group G is a linearly independent subset X that generates G.

The cyclic group $\mathbb{Z}$ has $\{1\}$ as a basis. The direct sum $\mathbb{Z} \oplus \mathbb{Z}$ has $\{(1,0), (0,1)\}$ as a basis since every element (m, n) can be written as $m(1,0) + n(0,1)$. However, $\{(3,1), (1,0)\}$ is a basis as well because $(3,1) - 3(1,0) = (0,1)$, so again $\langle (3,1), (1,0) \rangle = \mathbb{Z} \oplus \mathbb{Z}$ and the set is also linearly independent. In contrast, $\mathbb{Z}/10\mathbb{Z}$ does not have a basis since for all $x \in \mathbb{Z}/10\mathbb{Z}$, we have $10x = 0$ and $10 \neq 0$ in $\mathbb{Z}$.

Definition 2.5.2

An abelian group $(G, +)$ is called a *free abelian* group if it has a basis.

In particular, $\mathbb{Z}$, $\mathbb{Z} \oplus \mathbb{Z}$, and more generally $\mathbb{Z}^r$ for a positive integer r are free abelian groups. A free abelian group could have an infinite basis. For example, $\mathbb{Z}[x]$ is a free abelian group with basis $S = \{1, x, x^2, \ldots\}$ because every polynomial in $\mathbb{Z}[x]$ is a (finite) linear combination of elements in S. If a free abelian group has an infinite basis, every element must still be a finite linear combination of basis elements.

Proposition 2.5.3

Let $(G, +)$ be a free abelian group with basis X. Every element $g \in G$ can be expressed uniquely as a linear combination of elements in X.

Proof. By definition, every element $g \in G$ is a linear combination of elements in X. Suppose that

$$g = c_1 x_1 + c_2 x_2 + \cdots + c_r x_r \qquad \text{and} \qquad g = c_1' x_1' + c_2' x_2' + \cdots + c_s' x_s'$$

are two linear combination of the element g. By allowing some c_i and some c'_j to be 0, and by taking the union $\{x_1, x_2, \ldots, x_r, x'_1, x'_2, \ldots, x'_s\}$, we can assume that $r = s$ and that $x_i = x'_i$ for $i = 1, 2, \ldots, r$. Then subtracting these two expressions, gives

$$0 = (c_1 - c'_1)x_1 + (c_2 - c'_2)x_2 + \cdots + (c_r - c'_r)x_r.$$

Since the set of elements $\{x_1, x_2, \ldots, x_r\}$ is linearly independent, $c_i - c'_i = 0$ so $c_i = c'_i$ for all i. Hence, there is a unique expression for g as a linear combination of elements in X. $\qquad\square$

Example 2.5.4. The abelian group $(\mathbb{Q}, +)$ is not free. Assume that $\mathbb{Q}$ has a basis X. As a first case, suppose that $X = \{a/b\}$, where the fraction is expressed in reduce form. Then for any integer $n \in \mathbb{Z}$, the fraction na/b has a denominator that is a divisor of b, so multiples of a/b cannot give all of $\mathbb{Q}$, which means that X cannot generate $\mathbb{Q}$. As a second case, suppose that X contains at least two nonzero elements $\frac{a}{b}$ and $\frac{c}{d}$. Then

$$(-bc)\frac{a}{b} + (da)\frac{c}{d} = -ac + ac = 0.$$

This contradicts linear independence. We conclude by contradiction that $\mathbb{Q}$ does not have a basis. $\qquad\triangle$

Theorem 2.5.5

Let G be a nonzero free abelian group with a basis of r elements. Then G is isomorphic to $\mathbb{Z} \oplus \mathbb{Z} \oplus \cdots \oplus \mathbb{Z} = \mathbb{Z}^r$.

Proof. Let $X = \{x_1, x_2, \ldots, x_r\}$ be a finite basis of G. Consider the function $\varphi : \mathbb{Z}^r \to G$ given by

$$\varphi(c_1, c_2, \ldots, c_r) = c_1 x_1 + c_2 x_2 + \cdots + c_r x_r.$$

This function satisfies

$$
\begin{aligned}
\varphi((c_1, c_2, &\ldots, c_r) + (d_1, d_2, \ldots, d_r)) \\
&= \varphi(c_1 + d_1, c_2 + d_2, \ldots, c_r + d_r) \\
&= (c_1 + d_1)x_1 + (c_2 + d_2)x_2 + \cdots + (c_r + d_r)x_r \\
&= c_1 x_1 + c_2 x_2 + \cdots + c_r x_r + d_1 x_1 + d_2 x_2 + \cdots + d_r x_r \\
&= \varphi(c_1, c_2, \ldots, c_r) + \varphi(d_1, d_2, \ldots, d_r)
\end{aligned}
$$

so it is a homomorphism. Since the basis generates G, then φ is surjective. Furthermore, since

$$
\begin{aligned}
\operatorname{Ker} \varphi &= \{(c_1, c_2, \ldots, c_r) \in \mathbb{Z}^r \mid c_1 x_1 + c_2 x_2 + \cdots + c_r x_r = 0\} \\
&= \{(0, 0, \ldots, 0)\},
\end{aligned}
$$

the homomorphism is also injective. Thus, φ is an isomorphism. $\qquad\square$

Theorem 2.5.5 leads immediately to this important corollary.

Proposition 2.5.6

Let G be a finitely generated free abelian group. Then every basis of G has the same number of elements.

Proof. Suppose that G has a basis X with r elements. Then G is isomorphic to $\mathbb{Z}^r$. The subgroup $2G = \{g + g \,|\, g \in G\}$ is isomorphic to $(2\mathbb{Z})^r$ so by Exercise 2.4.7,

$$G/2G = (\mathbb{Z} \oplus \mathbb{Z} \oplus \cdots \oplus \mathbb{Z})/(2\mathbb{Z} \oplus 2\mathbb{Z} \oplus \cdots \oplus 2\mathbb{Z}) \cong Z_2^r.$$

Thus, $|G/2G| = 2^r$.

First, assume that G also has a finite basis with $s \neq r$ elements. Then $|G/2G| = 2^s \neq 2^r$, a contradiction.

Second, assume that G has an infinite basis Y. Let $y_1, y_2 \in Y$. Assume also that $\overline{y}_1 = \overline{y}_2$ in $G/2G$, then $y_1 - y_2 \in 2G$, so $y_1 - y_2$ is a finite linear combination of elements in Y (with even coefficients). In particular, Y is a linearly dependent set, which contradicts Y being a basis. Thus, in the quotient group $G/2G$, the elements $\{\overline{y} \,|\, y \in Y\}$ are all distinct. Since Y is an infinite set, so is $G/2G$. This contradicts $|G/2G| = 2^r$.

Consequently, if G has a basis of r elements, then every other basis is finite and has r elements. $\square$

Definition 2.5.7

If G is a finitely generated free abelian group, then the common number r of elements in a basis is called the *rank*. The rank is also called the *Betti number* of G and is denoted by $\beta(G)$.

In our efforts to classify all finitely generated abelian groups, we introduced free groups as a stepping stone to the general classification theorem (Theorem 2.5.11). We are still faced with a few questions:

- Is an abelian group with elements of finite order free?

- What is the structure of a subgroup of a free abelian group?

Answers to these questions become key ingredients in what follows.

Lemma 2.5.8

Let $X = \{x_1, x_2, \ldots, x_r\}$ be a basis of a free abelian group G. Let i be an index $1 \leq i \leq r$ with $i \neq j$ and let $t \in \mathbb{Z}$. Then

$$X' = \{x_1, \ldots, x_{j-1}, x_j + tx_i, x_{j+1}, \ldots, x_r\}$$

is also a basis of G.

Proof. Since $x_j = (-t)x_i + x_j + tx_i$, then $x_j \in \langle X' \rangle$ and so $X \subseteq \langle X' \rangle$. Since $\langle X \rangle = G \leq \langle X' \rangle$, then $G = \langle X' \rangle$ so X' generates G.

Furthermore, suppose that

$$c_1 x_1 + \cdots + c_{j-1} x_{j-1} + c_j(x_j + tx_i) + c_{j+1} x_{j+1} + \cdots + c_r x_r = 0$$

for some $c_1, c_2, \ldots, c_r \in \mathbb{Z}$. Since X is linearly independent

$$\begin{cases} c_k = 0 & \text{if } k \neq i \\ c_k = -tn_j & \text{if } k = i. \end{cases}$$

However, since $i \neq j$, we have $c_j = 0$, so $c_i = -t\,0 = 0$. Thus, X' is a linearly independent set. Hence, X' is another basis of G. $\qquad\square$

The reader may notice the similarity between the basis described in Lemma 2.5.8 and the replacement row operation used in the Gauss-Jordan elimination algorithm from linear algebra.

Theorem 2.5.9

Let G be a nonzero free abelian group of finite rank s and let $H \leq G$ be a nontrivial subgroup. Then H is a free abelian group of rank $t \leq s$. There exists a basis $\{x_1, x_2, \ldots, x_s\}$ for G and positive integers $n_1, n_2, \ldots, n_t$, where n_i divides n_{i+1} for all $1 \leq i \leq t-1$ such that $\{n_1 x_1, n_2 x_2, \ldots, n_t x_t\}$ is a basis of H.

Proof. We prove the theorem by starting from a basis of G and repeatedly adjusting it using Lemma 2.5.8.

By the well-ordering of the integers, there exists a minimum value n_1 in the set

$$\{c_1 \in \mathbb{N}^* \mid c_1 y_1 + c_2 y_2 + \cdots + c_s y_s \in H \text{ for } \textit{any} \text{ basis } \{y_1, y_2, \ldots, y_s\} \text{ of } G\}.$$

Let $z_1 \in H$ be an element that instantiates this minimum value, and write $z_1 = n_1 y_1 + c_2 y_2 + \cdots + c_s y_s$. By integer division, for all $i \geq 2$, we can write $c_i = n_1 q_i + r_i$ with $0 \leq r_i < n_1$. Set $x_1 = y_1 + q_2 y_2 + \cdots + q_s y_s$. By Lemma 2.5.8, $\{x_1, y_2, \ldots, y_s\}$ is a basis of G. Furthermore,

$$z_1 = n_1 x_1 + r_2 y_2 + \cdots + r_s y_s.$$

However, since n_1 is the least positive coefficient that occurs in any linear combination over any basis of G and, since $0 \leq r_i < n_1$, we have $r_2 = \cdots = r_s = 0$. So, in fact $z_1 = n_1 x_1$.

If $\{n_1 x_1\}$ generates H, then we are done and $\{n_1 x_1\}$ is a basis of H. If not, then there is a least positive value of c_2 for linear combinations

$$z_2 = a_1 x_1 + c_2 y_2 + \cdots + c_s y_s \in H,$$

where $y_2, \ldots, y_s \in G$ such that $\{x_1, y_2, \ldots, y_s\}$ is a basis of G. Call this least positive integer n_2. Note that we must have $n_1 | a_1$, because otherwise, since $n_1 x_1 \in H$ by subtracting a suitable multiple of $n_1 x_1$ from $a_1 x_1 + c_2 y_2 + \cdots + c_s y_s$ we would obtain a linear combination of $\{x_1, y_2, \ldots, y_s\}$ that is in H and has a lesser positive coefficient for x_1, which would contradict the minimality of n_1. Again, if we take the integer division of c_i by n_2, $c_i = n_2 q_i + r_i$ with $0 \le r_i < n_2$ for all $i \ge 3$, then

$$z_2 = a_1 x_1 + n_2(y_2 + q_3 y_3 + \cdots + q_s y_s) + r_3 y_3 + \cdots + r_s y_s.$$

We denote $x_2 = y_2 + q_3 y_3 + \cdots + q_s y_s$. Also, all the $r_i = 0$ because any $r_i > 0$ would contradict the minimality of n_2. Thus, $z_2 = a_1 x_1 + n_2 x_2 \in H$ and also $n_2 x_2 = z_2 - (a_1/n_1)n_1 x_1 \in H$. By Lemma 2.5.8, $\{x_1, x_2, y_3, \ldots, y_s\}$ is a basis of G. Furthermore, if we consider the integer division of n_2 by n_1, written as $n_2 = n_1 q + r$ with $0 \le r < n_1$, then

$$n_1 x_1 + n_2 x_2 = n_1(x_1 + q x_2) + r x_2 \in H.$$

But then, since $r < n_1$, by the minimal positive condition on n_1, we must have $r = 0$. Hence, $n_1 \mid n_2$.

If $\{n_1 x_2, n_2 x_2\}$ generates H, then we are done because the set is linearly independent since by construction $\{x_1, x_2, y_3, \ldots, y_s\}$ is a basis of G so $\{x_1, x_2\}$ is linearly independent and $\{n_1 x_1, n_2 x_2\}$ is then a basis of H. The pattern continues and only terminates when it results in a basis

$$\{x_1, \ldots, x_t, y_{t+1}, \ldots, y_s\}$$

of G such that $\{n_1 x_1, n_2 x_2, \ldots, n_t x_t\}$ is a basis of H for some positive integers n_i such that $n_i \mid n_{i+1}$ for $1 \le i \le t - 1$. $\square$

This proof is not constructive since it does not provide a procedure to find the n_i, which is necessary to construct the x_1, x_2, and so on. We merely know the n_i exist by the well-ordering of integers. In some instances it is easy to find a basis of the subgroup as in the following example.

Example 2.5.10. Consider the free abelian group $G = \mathbb{Z}^3$ and the subgroup $H = \{(x, y, z) \in \mathbb{Z}^3 \mid x + 2y + 3z = 0\}$. If we considered the equation $x + 2y + 3z = 0$ in $\mathbb{Q}^3$, then the Gauss-Jordan elimination algorithm gives H as the span of $\text{Span}(\{(-2, 1, 0), (-3, 0, 1)\})$. Taking only integer multiples of these two vectors does give all points (x, y, z) with y and z taking on every pair of integers. Hence, $\{(-2, 1, 0), (-3, 0, 1)\}$ is a basis of H and we see clearly that H is a free abelian group of rank 2. $\triangle$

2.5.2   Invariant Factors Decomposition

The difficult work in the proof of Theorem 2.5.9 leads to the Fundamental Theorem for Finitely Generated Abelian Groups.

Theorem 2.5.11 (FTFGAG)

Let G be a finitely generated abelian group. Then G can be written uniquely as

$$G \cong \mathbb{Z}^r \oplus Z_{d_1} \oplus Z_{d_2} \oplus \cdots \oplus Z_{d_k} \qquad (2.7)$$

for some nonnegative integers $r, d_1, d_2, \ldots, d_k$ satisfying $d_i \geq 2$ for all i and $d_{i+1} \mid d_i$ for $1 \leq i \leq k-1$.

Proof. Since G is finitely generated, there is a finite subset $\{g_1, g_2, \ldots, g_s\}$ that generates G. Define the function $h : \mathbb{Z}^s \to G$ by

$$h(n_1, n_2, \ldots, n_s) = n_1 g_1 + n_2 g_2 + \cdots + n_s g_s.$$

By the same reasoning as the proof of Theorem 2.5.5, h is a surjective homomorphism. Then $\operatorname{Ker} h$ is a subgroup of $\mathbb{Z}^s$ and by the First Isomorphism Theorem, since h is surjective, $\mathbb{Z}^s / (\operatorname{Ker} h) \cong G$. By Theorem 2.5.9, there exists a basis $\{x_1, x_2, \ldots, x_s\}$ for $\mathbb{Z}^s$ and positive integers $n_1, n_2, \ldots, n_t$, where n_i divides n_{i+1} for all $1 \leq i \leq t-1$ such that $\{n_1 x_1, n_2 x_2, \ldots, n_t x_t\}$ is a basis of $\operatorname{Ker} h$. Then

$$G \cong (\mathbb{Z} \oplus \mathbb{Z} \oplus \cdots \oplus \mathbb{Z})/(n_1 \mathbb{Z} \oplus \cdots \oplus n_t \mathbb{Z} \oplus \{0\} \oplus \cdots \oplus \{0\})$$
$$\cong Z_{n_1} \oplus Z_{n_2} \oplus \cdots \oplus Z_{n_t} \oplus \mathbb{Z} \oplus \cdots \oplus \mathbb{Z}.$$

Now if $n_i = 1$, then $Z_{n_i} \cong \{0\}$, the trivial group. Let k be the number of indices such that $n_i > 1$. The theorem follows after setting $d_i = n_{t+1-i}$ for $1 \leq i \leq k$. $\qquad \square$

Definition 2.5.12

As with free groups, the integer r is called the *rank* or the *Betti number* of G. It is sometimes denoted by $\beta(G)$. The integers $d_1, d_2, \ldots, d_k$ are called the *invariant factors* of G and the expression (2.7) is called the *invariant factors decomposition* of G.

It is interesting to note that the proof of Theorem 2.5.11 is not constructive in the sense that it does not provide a method to find specific elements in G whose orders are the invariant factors of G. The invariant factors exist by virtue of the well-ordering principle of the integers.

Applied to finite groups (which obviously are finitely generated), Theorem 2.5.11 gives us an effective way to describe all abelian groups of a given order n. If G is finite, the rank of G is 0. Then we must find all finite sequences of integers $d_1, d_2, \ldots, d_k$ such that

- $d_i \geq 2$ for $1 \leq i \leq k$;
- $d_{i+1} \mid d_i$ for $1 \leq i \leq k-1$;
- $n = d_1 d_2 \cdots d_k$.

The first two conditions are explicit in the above theorem. The last condition follows from the fact that $d_i = |x_i|$ where the $\{x_1, x_2, \ldots, x_k\}$ is a list of corresponding generators of G. Then every element in G can be written uniquely as

$$g = \alpha_1 x_1 + \alpha_2 x_2 + \cdots + \alpha_k x_k$$

for $0 \leq \alpha_i < d_i$. Thus, the order of the group is $n = d_1 d_2 \cdots d_k$.

We list a few examples of abelian groups of a given order where we find the invariant factors decomposition. (Note that in the notation Z_d for a cyclic group, we assume the operation is multiplication but that is irrelevant to the theorem.)

Example 2.5.13. All the abelian groups of order 16 are Z_{16}, $Z_8 \oplus Z_2$, $Z_4 \oplus Z_4$, $Z_4 \oplus Z_2 \oplus Z_2$, and $Z_2 \oplus Z_2 \oplus Z_2 \oplus Z_2$. $\triangle$

Example 2.5.14. All the abelian groups of order 24 are Z_{24}, $Z_{12} \oplus Z_2$, and $Z_6 \oplus Z_2 \oplus Z_2$. $\triangle$

Example 2.5.15. All the abelian groups of order 360 are Z_{360}, $Z_{180} \oplus Z_2$, $Z_{120} \oplus Z_3$, $Z_{60} \oplus Z_6$, $Z_{90} \oplus Z_2 \oplus Z_2$, and $Z_{30} \oplus Z_6 \oplus Z_2$. $\triangle$

2.5.3 Elementary Divisors Decomposition

Though Theorem 2.5.11 gives a complete classification of finite abelian groups, finding all possible sequences of invariant factors that multiply to a given order is not always easy, especially if the order is high. There is an alternative characterization that is often easier to find, involving the so-called elementary divisors. The added benefit of this alternative decomposition is that it leads to a formula for the number of abelian groups of a given order.

To obtain the elementary divisors decomposition, we first recall Exercise 1.9.18 that states that if $\gcd(m, n) = 1$, then $Z_m \oplus Z_n \cong Z_{mn}$. We need one more lemma.

Lemma 2.5.16

Let q be a prime number and let G be an abelian group of order q^m. Then G is uniquely isomorphic to a group of the form

$$Z_{q^{\alpha_1}} \oplus Z_{q^{\alpha_2}} \oplus \cdots \oplus Z_{q^{\alpha_k}}$$

such that $\alpha_1 \geq \alpha_2 \geq \cdots \geq \alpha_k \geq 1$ and $\alpha_1 + \alpha_2 + \cdots + \alpha_k = m$.

Proof. (This follows as a corollary to Theorem 2.5.11 so we leave the proof as an exercise for the reader. See Exercise 2.5.13.) $\square$

Definition 2.5.17

Let m be a positive integer. Any non-increasing sequence of the form $\alpha_1 \geq \alpha_2 \geq \cdots \geq \alpha_k \geq 1$ such that $\alpha_1 + \alpha_2 + \cdots + \alpha_k = m$ is called a *partition* of m. The *partition function*, sometimes denoted by $p(m)$, is the number of partitions of m. The partition function is often extended to nonnegative integers by assigning $p(0) = 1$.

According to Lemma 2.5.16, if G is an abelian group of order q^m for some prime q, then there are $p(m)$ possibilities for G, each corresponding to a partition of m.

Theorem 2.5.18

Let G be a finite abelian group of order $n > 1$ and let $n = q_1^{\beta_1} q_2^{\beta_2} \cdots q_t^{\beta_t}$ be the prime factorization of n. Then G can be written in a unique way as

$$G \cong A_1 \oplus A_2 \oplus \cdots \oplus A_t \qquad \text{where } |A_i| = q_i^{\beta_i}$$

and for each $A \in \{A_1, A_2, \ldots, A_t\}$ with $|A| = q^m$,

$$A \cong Z_{q^{\alpha_1}} \oplus Z_{q^{\alpha_2}} \oplus \cdots \oplus Z_{q^{\alpha_l}}$$

with $\alpha_1 \geq \alpha_2 \geq \cdots \geq \alpha_l \geq 1$ and $\alpha_1 + \alpha_2 + \cdots + \alpha_l = m$.

Proof. According to Theorem 2.5.11,

$$G \cong Z_{d_1} \oplus Z_{d_2} \oplus \cdots \oplus Z_{d_k},$$

where $d_{i+1} \mid d_i$ for $i = 1, 2, \ldots, k-1$ and $n = d_1 d_2 \cdots d_k$. Consider the $k \times t$ matrix of nonzero integers (α_{ij}) defined such that

$$d_i = q_1^{\alpha_{i1}} q_2^{\alpha_{i2}} \cdots q_t^{\alpha_{it}} = \prod_{j=1}^{t} q_j^{\alpha_{ij}}$$

is the prime factorization of d_i, where by $\alpha_{ij} = 0$ we mean that q_j is not a prime factor of d_i.

The condition $d_{i+1} \mid d_i$ implies that for each j, the exponents on q_j satisfy $\alpha_{i+1,j} \leq \alpha_{ij}$. The condition that $n = d_1 d_2 \cdots d_k$ implies that for each j,

$$\beta_j = \alpha_{1j} + \alpha_{2j} + \cdots + \alpha_{sj}.$$

Note that $\gcd(q_j^a, q_{j'}^b) = 1$ for any nonnegative integers a and b if $j \neq j'$. Therefore,

$$Z_{d_i} \cong Z_{q_1^{\alpha_{i1}}} \oplus Z_{q_2^{\alpha_{i2}}} \oplus \cdots \oplus Z_{q_t^{\alpha_{it}}}.$$

The theorem follows from the identity $G_1 \oplus G_2 \cong G_2 \oplus G_1$ for groups G_1 and G_2. $\qquad \square$

Definition 2.5.19

The integers $q_i^{\alpha_i}$ that arise in the expression of G described in the above theorem are called the *elementary divisors* of G. The expression in Theorem 2.5.18 is the elementary divisors decomposition.

We use the terminology "elementary divisors" because every cyclic group Z_{p^α} where p is a prime number is not isomorphic to a direct sum of any smaller cyclic groups.

As a first example, notice that since 16 is a prime power, then because of Lemma 2.5.16, the list of groups given in Example 2.5.13 provides both the elementary divisor decompositions and the invariant factor decompositions of all 5 abelian groups of order 16.

Example 2.5.20. Let $n = 2160 = 2^4 3^3 5$. We find all abelian groups of order 2160. We remark that 3 has three partitions, namely

$$3, \quad 2+1, \quad 1+1+1,$$

while 4 has five partitions, namely

$$4, \quad 3+1, \quad 2+2, \quad 2+1+1, \quad 1+1+1+1.$$

Then A_1, which corresponds to the prime factor $q_1 = 2$, has the following possibilities

$$Z_{16}, \quad Z_8 \oplus Z_2, \quad Z_4 \oplus Z_4, \quad Z_4 \oplus Z_2 \oplus Z_2, \quad \text{or} \quad Z_2 \oplus Z_2 \oplus Z_2 \oplus Z_2.$$

The component A_2, which corresponds to the prime factor $q_2 = 3$, has the following possibilities

$$Z_{27}, \quad Z_9 \oplus Z_3, \quad \text{or} \quad Z_3 \oplus Z_3 \oplus Z_3.$$

The component A_3 has only one possibility, namely $A_3 \cong Z_5$.

We can now list all 15 abelian groups of order 2160. We have given both the elementary divisor decomposition on the left and the isomorphic invariant factor decomposition on the right.

$$Z_{16} \oplus Z_{27} \oplus Z_5 \cong Z_{2160}$$
$$Z_{16} \oplus Z_9 \oplus Z_3 \oplus Z_5 \cong Z_{720} \oplus Z_3$$
$$Z_{16} \oplus Z_3 \oplus Z_3 \oplus Z_3 \oplus Z_5 \cong Z_{240} \oplus Z_3 \oplus Z_3$$

$$Z_8 \oplus Z_2 \oplus Z_{27} \oplus Z_5 \cong Z_{1080} \oplus Z_2$$
$$Z_8 \oplus Z_2 \oplus Z_9 \oplus Z_3 \oplus Z_5 \cong Z_{360} \oplus Z_6$$
$$Z_8 \oplus Z_2 \oplus Z_3 \oplus Z_3 \oplus Z_3 \oplus Z_5 \cong Z_{120} \oplus Z_6 \oplus Z_3$$

$$Z_4 \oplus Z_4 \oplus Z_{27} \oplus Z_5 \cong Z_{540} \oplus Z_4$$
$$Z_4 \oplus Z_4 \oplus Z_9 \oplus Z_3 \oplus Z_5 \cong Z_{180} \oplus Z_{12}$$
$$Z_4 \oplus Z_4 \oplus Z_3 \oplus Z_3 \oplus Z_3 \oplus Z_5 \cong Z_{60} \oplus Z_{12} \oplus Z_3$$

$$Z_4 \oplus Z_2 \oplus Z_2 \oplus Z_{27} \oplus Z_5 \cong Z_{540} \oplus Z_2 \oplus Z_2$$
$$Z_4 \oplus Z_2 \oplus Z_2 \oplus Z_9 \oplus Z_3 \oplus Z_5 \cong Z_{180} \oplus Z_6 \oplus Z_2$$
$$Z_4 \oplus Z_2 \oplus Z_2 \oplus Z_3 \oplus Z_3 \oplus Z_3 \oplus Z_5 \cong Z_{60} \oplus Z_6 \oplus Z_6$$

$$Z_2 \oplus Z_2 \oplus Z_2 \oplus Z_2 \oplus Z_{27} \oplus Z_5 \cong Z_{270} \oplus Z_2 \oplus Z_2 \oplus Z_2$$
$$Z_2 \oplus Z_2 \oplus Z_2 \oplus Z_2 \oplus Z_9 \oplus Z_3 \oplus Z_5 \cong Z_{90} \oplus Z_6 \oplus Z_2 \oplus Z_2$$
$$Z_2 \oplus Z_2 \oplus Z_2 \oplus Z_2 \oplus Z_3 \oplus Z_3 \oplus Z_3 \oplus Z_5 \cong Z_{30} \oplus Z_6 \oplus Z_6 \oplus Z_2. \qquad \triangle$$

When listing out each possible isomorphism type for a group of given order, each group listed according to the invariant factors decomposition corresponds to a unique group in the list according to elementary divisors. The proof of Theorem 2.5.18 describes how to go from the invariant factors decomposition to the corresponding elementary divisors decomposition. To go in the opposite direction, collect the highest prime powers corresponding to each prime to get Z_{d_1}; collect the second highest prime powers corresponding to each prime to get Z_{d_2}; and so forth.

We often refer to Theorems 2.5.11 and 2.5.18 collectively as *the* Fundamental Theorem of Finitely Generated Abelian Groups. The two theorems provide alternative ways to uniquely describe the torsion part of the group.

The power of the Fundamental Theorem of Finitely Generated Abelian Groups (FTFGAG), and of classification theorems in general, is that in many applications of group theory, we encounter abelian groups for which we naturally know the order. Then the FTFGAG gives us a list of possible isomorphism types.

Example 2.5.21. Consider the group $U(20)$ of units in modular arithmetic modulo 20. We know that this multiplicative group is abelian and has order $\phi(20) = \phi(4)\phi(5) = 8$. According to FTFGAG, $U(20)$ may be isomorphic to Z_8, $Z_4 \oplus Z_2$, or $Z_2 \oplus Z_2 \oplus Z_2$. To determine which of these possibilities it is, consider powers of some elements. First, lets consider $\overline{3}$:

$$\overline{3}^1 = \overline{3}, \quad \overline{3}^2 = \overline{9}, \quad \overline{3}^3 = \overline{7}, \quad \overline{3}^4 = \overline{1}.$$

This actually gives us enough information to determine the isomorphism type. The element $\overline{3}$ has order 4 so $U(20) \not\cong Z_2 \oplus Z_2 \oplus Z_2$. Furthermore, both $\overline{9}$ and $\overline{19} = \overline{-1}$ have order 2. In Z_8, there is only one element of order 2. Hence, $U(20) \not\cong Z_8$. By FTFGAG and elimination of possibilities, $U(20) \cong Z_4 \oplus Z_2$. $\triangle$

Example 2.5.22. Consider the group $U(46)$. This is an abelian group of order $\phi(46) = \phi(2)\phi(23) = 22$. Using the elementary divisors decomposition of FTFGAG, we see that the only abelian group of order 22 is Z_{22}. Hence, $U(46)$ is cyclic. $\triangle$

> **Corollary 2.5.23**
>
> Let $n > 1$ be an integer and let $n = q_1^{\beta_1} q_2^{\beta_2} \cdots q_t^{\beta_t}$ be the prime factorization of n. There are
>
> $$p(\beta_1)p(\beta_2) \cdots p(\beta_t)$$
>
> abelian groups of order n, where p is the partition function.

Proof. (Left as an exercise for the reader. See Exercise 2.5.22.)   □

2.5.4   A Few Comments on Partitions of Integers (Optional)

Partitions of integers form a vast area of research and find applications in many areas of higher mathematics. See [19, Chapter XIX] or [1] for an introduction to partitions of integers. (The texts [13, 22] give the interested reader a glimpse of applications outside of besides number theory.)

Considering a partition α as any finite non-increasing sequence of positive integers, we use the notation

$$|\alpha| \stackrel{\text{def}}{=} \alpha_1 + \alpha_2 + \cdots + \alpha_k$$

and call this the *content* of α. The integer k is the *length* (or the *number of parts*) of α and is written $\ell(\alpha) = k$. In the terminology of partitions, each summand α_i of a partition is called a *part* of α. For example, if $\alpha = (5, 2, 2, 1)$, then the content of α is $|\alpha| = 10$, so α is a partition of 10 with 4 parts.

We often represent a partition by its so-called *Young diagram* in which each part α_i is represented by α_i boxes that are left aligned, descending on the page as i increases. The Young diagram for $(5, 2, 2, 1)$ is:

Like a matrix, we call the main diagonal of the Young diagram the set of ith boxes in the ith row.

The *conjugate* of a partition α is the partition α' obtained by reflecting the Young diagram of α through its main diagonal. Algebraically, the values of the conjugate partition are

$$\alpha'_j = \left|\{1 \le i \le k \,|\, \alpha_i \ge j\}\right|.$$

With $\alpha = (5, 2, 2, 1)$, the conjugate partition is $\alpha' = (4, 3, 1, 1, 1)$. The intuition of the Young diagram makes it clear that $|\alpha'| = |\alpha|$, so if α is a partition of n, then so is α'.

Example 2.5.24. Recall that the conjugacy classes in S_n correspond to all permutations with a given cycle type. Each cycle type corresponds uniquely to a partition of n. For example, in S_6, the cycle type $(a\,b\,c)(d\,e)$ corresponds to the partition $3 + 2 + 1$. More generally, each partition α with $|\alpha| = n$ corresponds to the conjugacy class of permutations that have disjoint cycles of length $\alpha_1, \alpha_2, \ldots, \alpha_k$. In particular, S_n has $p(n)$ conjugacy classes.   △

EXERCISES FOR SECTION 2.5

1. List all the possible invariant factors for: (a) 45; (b) 480; (c) 900.

2. List all the possible elementary divisors for: (a) 212; (b) 762; (c) 500.

3. List all the nonisomorphic abelian groups of order 945 both in invariant factors form and elementary divisors form. Show which decompositions correspond to which in the different forms.

4. List all the nonisomorphic abelian groups of order 864 both in invariant factors form and elementary divisors form. Show which decompositions correspond to which in the different forms.

5. What is the smallest value of n such that there exist 5 abelian groups of order n? List out the specific groups.

6. What is the smallest value of n such that there exist 4 abelian groups of order n? List out the specific groups.

7. What is the smallest value of n such that there exist at least 13 abelian groups of order n? List out the specific groups.

8. Suppose that an abelian group has order 100 and has at least 2 elements of order 2. What are the possible groups that satisfy these conditions?

9. List all the abelian groups of order 72 that contain an element of order 6.

10. Find the isomorphism type (by invariant factors decomposition) of $U(21)$.

11. Find the isomorphism type (by invariant factors decomposition) of $U(27)$.

12. Suppose that G is an abelian group of order 176 such that the subgroup $H = \{g^2 \mid g \in G\}$ has order 22. What are the possible groups G with this property?

13. Prove Lemma 2.5.16.

14. Let p be a prime number. For all abelian groups of order p^3, list how many elements there are of each order.

15. Let p and q be distinct prime numbers. For all abelian groups of order p^2q^2, list how many elements there are of each order.

16. Find all integers n such that there exists a unique abelian group of order n.

17. Let α be a partition of an integer and denote by G_α the group

$$G_\alpha = Z_{2^{\alpha_1}} \oplus Z_{2^{\alpha_2}} \oplus \cdots \oplus Z_{2^{\alpha_{\ell(\alpha)}}}.$$

   (a) Prove that G_α contains $2^{\ell(\alpha)} - 1 = 2^{\alpha'_1} - 1$ elements of order 2.
   (b) Prove that G_α contains $2^{|\alpha|} - 2^{|\alpha|-m}$ elements of order 2^{α_1}, where α_1 appears m times in the partition α. [Hint: Use Inclusion-Exclusion.]

18. Let $G = \langle x, y \mid x^{12} = y^{18} = 1, xy = yx \rangle$. Consider the subgroup $H = \langle x^4 y^{-6} \rangle$. Find the isomorphism type of G/H.

19. Let $G = (\mathbb{Z}/12\mathbb{Z})^2$ and let $H = \{(\bar{a}, \bar{b}) \mid 2\bar{a} + 3\bar{b} = \bar{0}\}$. Determine the isomorphism type of G/H.

20. Let $G = (\mathbb{Z}/12\mathbb{Z})^2$ and let $H = \{(\bar{a}, \bar{b}) \mid \bar{a} + 5\bar{b} = \bar{0}\}$. Determine the isomorphism type of G/H.

21. Prove Cauchy's Theorem for abelian groups. In other words, let G be an abelian group and prove that if p is a prime number dividing $|G|$ then G contains an element of order p.

22. Prove Corollary 2.5.23.

23. Let p be a prime number. Prove that $\mathrm{Aut}(Z_p \oplus Z_p) \cong \mathrm{GL}_2(\mathbb{F}_p)$.

24. Let G be a finite abelian group. Prove that $\mathrm{Aut}(G)$ is abelian if and only if G is cyclic.

25. What is the first integer n such that there exist two distinct partitions α and β of n such that $\alpha' = \alpha$ and $\beta' = \beta$? What is the first integer for which there exists three distinct partitions α with $\alpha' = \alpha$?

26. Consider the subgroup $H = \{(x_1, x_2, x_3) \in \mathbb{Z}^3 \mid 6 \text{ divides } 2x_1 + 3x_2 + 4x_3\}$ of the free abelian group $\mathbb{Z}^3$. Find a basis of H.

2.6   Projects

Investigative Projects

PROJECT I. **The Exponent of Matrix Groups.** Fermat's Little Theorem can be rephrased as: if p is a prime, then $\forall \bar{a} \in U(p)$, $\bar{a}^{p-1} = \bar{1}$. This is an immediate consequence of Lagrange's Theorem. The *exponent* of a group G, denoted $\exp(G)$, is the least positive integer m such that $g^m = 1$ for $g \in G$. Explore the exponent $k(n, p)$ of the group $G = \mathrm{GL}_n(\mathbb{F}_p)$.

PROJECT II. **Escher and Symmetry.** Consider some of M. C. Escher's artwork that depicts interesting tessellations of the plane. For each of these, consider their corresponding wallpaper group E.

   (1) Describe a set of natural (simple) generators for E.

   (2) Write down relations between the generators of E and thus give a presentation of E.

   (3) Find some normal subgroups N of E, and explain geometrically why N is normal and what information is contained in E/N.

   (4) Describe some subgroups of E that are not normal and explain geometrically why they are not normal.

(5) Find an Escher tessellation of the Poincaré disk and find generators and relations for that group.

PROJECT III. **The Special Projective Group** $PSL_2(\mathbb{F}_5)$. In Exercise 2.2.28, we proved that A_5 is simple and we know that it has order 60. In Example 2.3.9, we introduced the projective linear groups. Intuitively speaking, $PSL_n(\mathbb{F}_p)$, where p is prime, "removes" from $GL_n(\mathbb{F}_p)$ normal subgroups that are obvious. A quick calculation (that you should do) shows that $PSL_2(\mathbb{F}_5)$ has order 60. For this project, consider the following questions. Is $PSL_2(\mathbb{F}_5)$ simple? What are conjugacy classes in $PSL_2(\mathbb{F}_5)$? Is $PSL_2(\mathbb{F}_5)$ isomorphic to A_5? Can you generalize any of these investigations to other n and other p?

PROJECT IV. **Quotient Groups and Cayley Graphs.** Let G be a group and let N be a subgroup. Suppose that G has a presentation with a set of generators $\{g_1, g_2, \ldots, g_s\}$. Then we know that $\{\overline{g}_1, \overline{g}_2, \ldots, \overline{g}_s\}$, where $\overline{g}_i = g_i N$ is a generating set of G/N. Interpret geometrically how the Cayley graph of G/N with generators $\{\overline{g}_1, \overline{g}_2, \ldots, \overline{g}_s\}$ is related the Cayley graph of G with generators $\{g_1, g_2, \ldots, g_s\}$. Illustrate this relationship with examples that can be realized as polyhedra in $\mathbb{R}^3$.

PROJECT V. **Embeddings of S_n.** It is not hard to show that for every integer n, there is an embedding of S_n in $GL_n(\mathbb{F}_p)$ for all primes p. However, the fact that $GL_2(\mathbb{F}_2) \cong S_3$ gives an example where S_n is embedded in a group $GL_k(\mathbb{F}_p)$, where $k < n$. Investigate for what $k < n$ and what p it may be possible or is impossible to embed S_n into $GL_k(\mathbb{F}_p)$.

PROJECT VI. **Pascal's Triangle for Groups.** The entries of Pascal's (usual) Triangle are nonnegative integers. If we consider Pascal's Triangle in modular arithmetic $\mathbb{Z}/n\mathbb{Z}$, then all the operations in the triangle occur in the group $(\mathbb{Z}/n\mathbb{Z}, +)$. We generalize Pascal's Modular Triangle to a Pascal's Triangle for a group in the following way. Let a and b be elements in a group G. Start with a on the first row. On a diagonal going down from 1 to the left put a's. On a diagonal going down and to the right put b's. Then, in rows below, similar to the constructing of Pascal's Triangle, we fill in the rows by operating the element above to the left with the element above and to the right. The following diagram shows the first few rows.

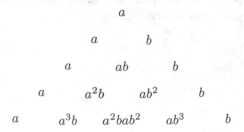

One way to visualize this Pascal's Triangle for a group is to color code boxes based on the group element. Write a program that draws the color-coded Pascal Triangle for a small group and certain elements a and b up to any number of rows you specify. Then, explore any patterns that emerge in Pascal's Triangle for the group. (Try, for example, cyclic groups, a dihedral group, or $Z_2 \oplus Z_2$.) Are there any patterns related to subgroup structure or quotient group structure? [This project was inspired by the article [4].]

Expository Projects

PROJECT VII. **Joseph-Louis Lagrange, 1736–1813.** The Lagrange Theorem is named in honor Italian-born but French-naturalized mathematician and astronomer Joseph-Louis Lagrange. Using reliable sources, explore Lagrange's personal background, his mathematical work, and the reception of his work in the mathematical community. Explain why, the Lagrange's Theorem bears his name, in light of the fact that modern group theory arose in its modern form after Lagrange's death.

PROJECT VIII. **Srinivasa Ramanujan, 1887–1920.** With little formal training in pure mathematics, Indian mathematician Srinivasa Ramanujan contributed many deep results in the areas of analysis and number theory, with emphases on infinite series and continued fractions. The depth of Ramanujan's leaps of intuition often stuns those who study his work, still a hundred years later. Using reliable sources, explore Ramanujan's personal background, his mathematical work, and the reception of his work in the mathematical community. Report on Ramanujan's work associated with the partition function $p(n)$.

PROJECT IX. **Birch and Swinnerton-Dyer Conjecture.** The Birch and Swinnerton-Dyer Conjecture is one of the seven Millennium Prize Problems. The conjecture involves the isomorphism type of the abelian group associated with the solutions of an elliptic curve over the rationals. Summarize the concept of an elliptic curve, describe the addition operation on an elliptic curve, describe a few examples of the abelian group $E(\mathbb{Q})$, and explain the statement of the Birch and Swinnerton-Dyer Conjecture.

PROJECT X. **Braid Groups.** Describe the braid group on n strands. State the definition, describe the original motivation, mention a few properties, and list a few applications of braid groups.

3

Rings

So far, the reader has encountered the algebraic structures of vector spaces, groups, semigroups, and monoids. We now turn to the study of another algebraic structure, that of rings. Generalizing the arithmetic of integers, rings involve two operations, often denoted $+$ and $\times$. However, ring axioms are loose enough to include many other algebraic contexts.

Sections 3.1 through 3.3 define rings, give initial key examples, and explore a classes of rings that are important in various applications. Section 3.4 introduces the concept of ring homomorphism as functions that preserve the ring structure.

Sections 3.5 and 3.6 define ideals of rings, offer methods to describe them, and describe natural operations on ideals. Section 3.7 discusses quotient rings and presents two important applications: the Chinese Remainder Theorem and the isomorphism theorems. Section 3.8 explores the concepts of maximal and prime ideals, critical concepts in applications to number theory and geometry.

3.1 Introduction to Rings

3.1.1 Ring Axioms and First Examples

> **Definition 3.1.1**
>
> A *ring* is a triple $(R, +, \times)$ where R is a set and $+$ and $\times$ are binary operations on R that satisfy the following axioms:
>
> (1) $(R, +)$ is an abelian group.
>
> (2) $\times$ is associative.
>
> (3) $\times$ is distributive over $+$, i.e., for all $a, b, c \in R$:
>
> > (a) $(a + b) \times c = a \times c + b \times c$ (right-distributivity);
> >
> > (b) $a \times (b + c) = a \times b + a \times c$ (left-distributivity).

As in group theory, we will often simply refer to the ring by R if there is no confusion about what the operations might be. In an abstract ring, we will denote the additive identity by 0 and refer to it as the "zero" of the ring. The

DOI: 10.1201/9781003299233-3

additive inverse of a is denoted by $-a$. As with typical algebra over the reals, we will often write the multiplication as ab instead of $a \times b$. Furthermore, if $n \in \mathbb{N}$ and $a \in R$, then $n \cdot a$ represents a added to itself n times,

$$n \cdot a \stackrel{\text{def}}{=} \overbrace{a + a + \cdots + a}^{n \text{ times}}.$$

We extend this notation to all integers by defining $0 \cdot a = 0$ and, if $n > 0$, then $(-n) \cdot a = -(n \cdot a)$.

Definition 3.1.2

- A ring $(R, +, \times)$ is said to have an *identity*, denoted by 1, if there is an element $1 \in R$ such that $1 \times a = a \times 1 = a$ for all $a \in R$.

- A ring is said to be *commutative* if $\times$ is commutative.

Since a ring always possesses an additive identity, when one simply says "a ring with identity," one refers to the multiplicative identity, the existence of which is not required by the axioms. A ring with an identity is also called a *unital ring*.

It is possible for the multiplicative identity 1 of a ring to be equal to the additive identity 0. However, according to part (1) in the following proposition, the ring would then consist of just one element, namely 0. This case serves as an exception to many theorems we would like to state about rings with identity. Consequently, since this is not a particularly interesting case, we will often refer to "a ring with identity $1 \neq 0$" to denote a ring with identity but excluding the case in which $1 \neq 0$.

We will soon introduce a number of elementary examples of rings. However, before we do so, we prove the following proposition that holds for all rings. Many of these properties, as applied to integers and real numbers, are rules that elementary school children learn early on.

Proposition 3.1.3

Let R be a ring.

(1) $\forall a \in R$, $0a = a0 = 0$.

(2) $\forall a, b \in R$, $(-a)b = a(-b) = -(ab)$.

(3) $\forall a, b \in R$, $(-a)(-b) = ab$.

(4) If R has an identity 1, then it is unique. Furthermore, $-a = (-1)a$.

Proof. For (1), note that $0 + 0 = 0$ since it is the additive identity. By distributivity,

$$a0 = a(0 + 0) = a0 + a0.$$

Adding $-(a0)$ to both sides of this equation, we deduce that $0 = a0$. A similar reasoning holds for $0a = 0$.

For (2), note that $0 = 0b = (a + (-a))b = ab + (-a)b$. Adding $-(ab)$ to both sides, we deduce that $(-a)b = -(ab)$. A similar reasoning holds for $a(-b) = -(ab)$.

An application of (2) twice gives $(-a)(-b) = -(a(-b)) = -(-(ab)) = ab$.

Finally, for part (4), suppose that e_1 and e_2 satisfy the axioms of an identity. Then $e_1 = e_1 e_2$ since e_2 is an identity but $e_2 = e_1 e_2$ since e_1 is an identity. Thus, $e_1 = e_1 e_2 = e_2$. Furthermore, $1a = a$ by definition so

$$0 = 0a = (1 + (-1))a = a + (-1)a.$$

Then adding $-a$ to both sides of this equation produces the result. □

In subsequent sections, we encounter many examples of rings and see a number of methods to define new rings from old ones. In this first section, however, we just introduce a few basic examples of rings.

Example 3.1.4. The triple $(\mathbb{Z}, +, \times)$ is a commutative ring. Note that $(\mathbb{Z} - \{0\}, \times)$ is not a group. In fact, the only elements in $\mathbb{Z}$ that have multiplicative inverses are 1 and -1. △

Example 3.1.5. The sets $\mathbb{Q}$, $\mathbb{R}$, and $\mathbb{C}$ with their usual operations of $+$ and $\times$ form commutative rings. In each of these, all nonzero elements have multiplicative inverses. △

Example 3.1.6 (Modular Arithmetic). For every integer $n \geq 2$, the context of modular arithmetic, namely the triple $(\mathbb{Z}/n\mathbb{Z}, +, \times)$ forms a ring. △

In a ring R with an identity $1 \neq 0$, by distributivity $n \cdot a = (n \cdot 1)a$ for all $n \in \mathbb{N}^*$ and all $a \in R$. The elements $n \cdot 1$ are important elements in R. However, as the example of modular arithmetic illustrates, $n \cdot 1$ could be 0. This leads to a fundamental property of rings with identity $1 \neq 0$.

Definition 3.1.7

Let R be a ring with identity $1 \neq 0$. The *characteristic* of R is the smallest positive integer n such that $n \cdot 1 = 0$. If such a positive integer does not exist, the characteristic of R is said to be 0. The characteristic of a ring is often denoted by char(R).

As elementary examples, the characteristic of $\mathbb{Z}/n\mathbb{Z}$ is n and the characteristic of $\mathbb{Z}$ is 0.

Example 3.1.8 (Quaternions). After the discovery complex of numbers and the development of complex analysis, mathematicians looked for other algebraic structures that extended the complex numbers. In this sense, they sought vector spaces over $\mathbb{R}$ that possessed a multiplication with nice properties, just as $\mathbb{C}$ has. In 1843, Hamilton defined the set of *quaternions* as

$$\mathbb{H} = \{a + bi + cj + dk \,|\, a, b, c, d \in \mathbb{R}\},$$

where elements add like vectors with $\{1, i, j, k\}$ acting as a basis. He defined multiplication $\times$ on $\mathbb{H}$ where arbitrary elements must satisfy distributivity and the elements $1, i, j, k$ multiply together as they do in the quaternion group Q_8. The triple $(\mathbb{H}, +, \times)$ is a ring. (We have not checked associativity but we leave this as an exercise for the reader. See Exercise 3.1.14.) It is obvious that $\mathbb{H}$ is not commutative since $ij = k$, whereas $ji = -k$.

To illustrate a few simple calculations, consider for example the elements $\alpha = 2 - 3i + 2k$ and $\beta = 5 + 4i - 4j$. The addition and multiplications are:

$$\alpha + \beta = 7 + i - 4j + 2k;$$
$$\alpha\beta = (2 - 3i + 2k)(5 + 4i - 4j)$$
$$= 10 + 8i - 8j - 15i - 12i^2 + 12ij + 10k + 8ki - 8kj$$
$$= 10 + 8i - 8j - 15i - 12(-1) + 12k + 10k + 8j - 8(-i)$$
$$= 22 + i + 22k;$$
$$\beta\alpha = (5 + 4i - 4j)(2 - 3i + 2k)$$
$$= 10 - 15i + 10k + 8i - 12i^2 + 8ik - 8j + 12ji - 8jk$$
$$= 10 - 15i + 10k + 8i - 12(-1) + 8(-j) - 8j + 12(-k) - 8i$$
$$= 22 - 15i - 16j - 2k.$$

It is crucial that we do not change the order of the quaternion basis elements in any product, in particular when applying distributivity. Hence, $(4i)(2k) = 8ik = -8j$, while $(2k)(4i) = 8ki = 8j$. $\triangle$

Example 3.1.9 (Ring of Functions). Let I be an interval in $\mathbb{R}$ and let $\mathrm{Fun}(I, \mathbb{R})$ be the set of functions from I to $\mathbb{R}$. Equipped with the usual addition and multiplication of functions, $\mathrm{Fun}(I, \mathbb{R})$ is a commutative ring. The properties of a ring are inherited from $\mathbb{R}$. In contrast, $\mathrm{Fun}(I, \mathbb{R})$ is not a ring when equipped with addition and composition because the axiom of distributivity fails. For example, consider the three functions $f(x) = x + 1$, $g(x) = x^2$, and $h(x) = x^3$. Then

$$(f \circ (g + h))(x) = f(x^2 + x^3) + 1 = x^3 + x^2 + 1$$
$$(f \circ g)(x) + (f \circ h)(x) = x^2 + 1 + x^3 + 1 = x^3 + x^2 + 2. \qquad \triangle$$

Definition 3.1.10 (Direct Sum)

Let $(R_1, +_1, \times_1)$ and $(R_2, +_2, \times_2)$ be two rings. The *direct sum* of the two rings is the triple $(R_1 \times R_2, +, \times)$, where $+$ and $\times$ are defined as

$$(a, b) + (c, d) = (a +_1 c, b +_2 d)$$
$$(a, b) \times (c, d) = (a \times_1 c, b \times_2 d).$$

The direct sum of the rings is denoted by $R_1 \oplus R_2$.

3.1.2 Units and Zero Divisors

Modular arithmetic presented examples of arithmetic with some properties that did not arise in the arithmetic in $\mathbb{Z}$ or in $\mathbb{Q}$. In particular, consider the examples of $\mathbb{Z}/5\mathbb{Z}$ and $\mathbb{Z}/6\mathbb{Z}$ introduced in Example A.6.6. There are qualitative differences between some of the arithmetic properties in $\mathbb{Z}/5\mathbb{Z}$ and in $\mathbb{Z}/6\mathbb{Z}$. These differences and more are common in ring theory. We introduce some key terminology.

Definition 3.1.11

Let R be a ring.

- A nonzero element $r \in R$ is called a *zero divisor* if there exists $s \in R \setminus \{0\}$ such that $rs = 0$ or $sr = 0$.
- Assume R has an identity $1 \neq 0$. An element of $u \in R$ is called a *unit* if it has a multiplicative inverse, i.e., $\exists v \in R$, such that $uv = vu = 1$. The element v is often denoted u^{-1}. The set of units in R is denoted by $U(R)$.

Note that the identity 1 is itself a unit, but that the 0 element is not a zero divisor. This lack of symmetry in the definitions may seem unappealing but this distinction turns out to be useful in all theorems that discuss units and zero divisors.

The notation $U(R)$ is reminiscent of the notation $U(n)$ as the set of units in $\mathbb{Z}/n\mathbb{Z}$. In the ring $(\mathbb{Z}/n\mathbb{Z}, +, \times)$, every nonzero element is either a unit or a zero divisor. Proposition A.6.8 established that the units in $\mathbb{Z}/n\mathbb{Z}$ are elements $\overline{a}$ such that $\gcd(a, n) = 1$. Now if $\gcd(a, n) = d \neq 1$, then there exists some $k \in \mathbb{Z}$ such that $ak = n\ell$. Then in $\mathbb{Z}/n\mathbb{Z}$ we have $\overline{a}\overline{k} = \overline{0}$.

In an arbitrary ring, it is not in general true that every nonzero element is either a unit or a zero divisor. We need look no further than the integers. The units in the integers are $U(\mathbb{Z}) = \{-1, 1\}$, and all the elements greater than 1 in absolute value are neither units nor zero divisors. On the other hand, as the following proposition shows, no element can be both.

Proposition 3.1.12

Let R be a ring with identity $1 \neq 0$. The set of units and the set of zero divisors in a ring are mutually exclusive.

Proof. Assume that a is both a unit and a zero divisor. Then there exists $b \in R - \{0\}$ such that $ba = 0$ or $ab = 0$. Assume without loss of generality that $ba = 0$. There also exists $c \in R$ such that $ac = 1$. Then

$$b = b(ac) = (ba)c = 0c = 0.$$

However, this is a contradiction since $b \neq 0$. $\square$

As mentioned above, the set of units contains the multiplicative identity. Furthermore, by definition, every element in $U(R)$ has a multiplicative inverse. This leads to the simple remark that we phrase as a proposition.

Proposition 3.1.13

Let R be a ring with identity $1 \neq 0$. Then $U(R)$ is a group.

Proposition 3.1.14

Let R_1 and R_2 be rings each with an identity $1 \neq 0$. Then, $R_1 \oplus R_2$ has an identity $(1, 1)$ and, as an isomorphism of groups,

$$U(R_1 \oplus R_2) \cong U(R_1) \oplus U(R_2).$$

Proof. (Left as an exercise for the reader. See Exercise 3.1.21.) □

The following definitions refer to elements with specific properties related to their powers. Properties of such ring elements are studied in the exercises.

Definition 3.1.15

Let R be a ring. An element $a \in R$ is called *nilpotent* if there exists a positive integer k such that $a^k = 0$. The subset of nilpotent elements in R is denoted by $\mathcal{N}(R)$.

Definition 3.1.16

Let R be a ring. An element $a \in R$ is called *idempotent* if $a^2 = a$.

3.1.3   Integral Domains, Division Rings, Fields

It is common in ring theory to define classes of rings in which the elements possess certain properties. Then it is convenient to state theorems for a particular class of rings. In fact, we have already defined the class of commutative rings. However, many classes possess particular terminology evocative of their properties. We illustrate this common habit by already introducing three important classes of rings defined in reference to the existence of units and zero divisors.

Definition 3.1.17

A ring R is called an *integral domain* if it is commutative, contains an identity $1 \neq 0$, and contains no zero divisors.

The terminology of integral domain evokes the fact that integral domains resemble the algebra of the integers. However, we will encounter many other integral domains besides the integers.

Example 3.1.18. The ring $\mathbb{Z} \oplus \mathbb{Z}$ is not an integral domain because, in particular, $(1,0) \cdot (0,1) = (0,0)$ so $(1,0)$ and $(0,1)$ are zero divisors. $\triangle$

Proposition 3.1.19 (Cancellation Law)

Let R be an integral domain. Then R satisfies the cancellation law, namely that for all $a, b, c \in R$,

$$ab = ac \implies b = c.$$

Proof. Adding $-(ac)$ to both sides, we have

$$ab = ac \Rightarrow ab - ac = 0 \Rightarrow a(b - c) = 0.$$

Since a is not a zero divisor, we must have $b - c = 0$. Thus, $b = c$. $\square$

One consequence of this proposition is that a does not have to be a unit. In fact, the cancellation law applies in any ring whenever a is not a zero divisor.

Definition 3.1.20

A ring R with identity $1 \neq 0$ is called a *division ring* if every nonzero element in R is a unit.

Example 3.1.21. The ring of quaternions $\mathbb{H}$ is a division ring. A simple calculation gives

$$
\begin{aligned}
&(a + bi + cj + dk)(a - bi - cj - dk) \\
&\quad = a^2 - abi - acj - adk + abi - b^2(-1) - bck - bd(-j) \\
&\qquad + acj - bc(-k) - c^2(-1) - cdi + adk - bdj - cd(-i) - d^2(-1) \\
&\quad = a^2 + b^2 + c^2 + d^2.
\end{aligned}
$$

For all quaternions $\alpha = a + bi + cj + dk \neq 0$, the sum of squares $a^2 + b^2 + c^2 + d^2 \neq 0$ and so the inverse of α is

$$\alpha^{-1} = \frac{1}{a^2 + b^2 + c^2 + d^2}(a - bi - cj - dk).$$

The quaternions $\mathbb{H}$ are an example of a noncommutative division ring. Because of the importance of the above calculation, if $\alpha = a + bi + cj + dk \in \mathbb{H}$ we define the notation $\overline{\alpha} = a - bi - cj - dk$ and we call

$$N(\alpha) \overset{\text{def}}{=} \alpha\overline{\alpha} = a^2 + b^2 + c^2 + d^2$$

the *norm* of α. $\triangle$

Definition 3.1.22

A commutative division ring is called a *field*.

A field is a set F that possesses an addition $+$ and a multiplication $\times$, in which $(F, +)$ is an abelian group, $(F - \{0\}, \times)$ is an abelian group, and in which $\times$ is distributive over $+$. In previous levels of algebra, students encounter the fields of $\mathbb{Q}$, $\mathbb{R}$, and $\mathbb{C}$. However, in the context of modular arithmetic, we have encountered other finite fields. When p is a prime number, $\mathbb{Z}/p\mathbb{Z}$ is a field containing p elements. We denote it by $\mathbb{F}_p$ to indicate the implied field structure.

3.1.4 Subrings

As in every algebraic structure, we define the concept of substructure.

Definition 3.1.23

Let $(R, +, \times)$ be a ring. A subset S is called a *subring* of R if $(S, +)$ is a subgroup of $(R, +)$ and if S is closed under $\times$. If R is a ring with an identity 1_R and if S is a subring with an identity $1_S = 1_R$, then S is called a *unital subring* of R.

Using the One-Step Subgroup Criterion from group theory, in order to prove that S is a subring of a ring R, we simply need to prove that S is closed under subtraction with the usual definition of subtraction

$$a - b \overset{\text{def}}{=} a + (-b)$$

and closed under multiplication.

As an example, if $R = \mathbb{Z}$, then for any integer n, consider the subset of multiples $n\mathbb{Z}$. We know that the subset $n\mathbb{Z}$ is a subgroup with $+$. Furthermore, for all $na, nb \in n\mathbb{Z}$ we have $(na)(nb) = n(nab) \in n\mathbb{Z}$, so $n\mathbb{Z}$ is closed under multiplication. Hence, $n\mathbb{Z}$ is a subring of $\mathbb{Z}$.

This first example illustrates that the definition of a subring makes no assumption that if R contains an identity $1 \neq 0$, then a subring S does also. In contrast, it is possible that a subring S contains an identity 1_S that is different from the identity 1_R. For example, consider the subring $S = \{\overline{0}, \overline{2}, \overline{4}\}$ of the ring $R = \mathbb{Z}/6\mathbb{Z}$. Obviously, $1_R = \overline{1}$ but it is easy to check that the identity of S is $1_S = \overline{4}$. With the above terminology, S is a subring of R but not a unital subring of R.

Example 3.1.24. Consider the set $C^0([a, b], \mathbb{R})$ of continuous real-valued functions from the interval $[a, b]$. This is a subset of the ring of functions $\text{Fun}([a, b], \mathbb{R})$ from the interval $[a, b]$ to $\mathbb{R}$. The reader should recall that some theorems, usually introduced in a first calculus course and proven in an analysis course, establish that subtraction and $\times$ are binary operations on $C^0([a, b], \mathbb{R})$. In particular, the proof that the product of two continuous functions is continuous is not trivial. Consequently, $C^0([a,b], \mathbb{R})$ is a subring of $\text{Fun}([a,b], \mathbb{R})$.   △

Some properties of a ring are preserved in subrings. For example, any subring of a commutative ring is again a commutative ring. Furthermore, if R is an integral domain, then any subring of R that contains 1 is also an integral domain. On the other hand, a subring of a field need not be a field even if it contains the identity

As an abstract example of subrings we discuss the center of a ring.

Definition 3.1.25

Let R be a ring. The *center* of R, denoted $C(R)$, consists of all elements that commute with every other element. In other words,

$$C(R) = \{z \in R \mid zr = rz \text{ for all } r \in R\}.$$

Proposition 3.1.26

Let R be a ring. The center $C(R)$ is a subring of R.

Proof. Let $z_1, z_2 \in C(R)$ and let $r \in R$. Then

$$r(z_1 + (-z_2)) = rz_1 + r(-z_2) = rz_1 + (-(rz_2)) = z_1 r + (-(z_2 r))$$
$$= z_1 r + (-(z_2)r) = (z_1 + (-z_2))r.$$

Hence, $z_1 + (-z_2) \in C(R)$ and thus $(C(R), +)$ is a subgroup of $(R, +)$. Furthermore,

$$r(z_1 z_2) = (rz_1)z_2 = (z_1 r)z_2 = z_1 (rz_2) = z_1(z_2 r) = (z_1 z_2)r,$$

so $z_1 z_2 \in C(R)$. This proves that $C(R)$ is a subring.   □

Exercises for Section 3.1

In Exercises 3.1.1 through 3.1.8, decide whether the given set R along with the stated addition and multiplication form a ring. If it is, prove it and decide whether it is commutative and whether it has an identity. If it is not, decide which axioms fail. You should always check that the symbol is in fact a binary operation on the given set.

1. Let $R = \mathbb{R}^{>0}$, with the addition of $x \boxplus y = xy$, and the multiplication $x \otimes y = x^y$.

2. Let $R = \mathbb{Q}$, with addition $x \boxplus y = x+y+1$, and multiplication $x * y = x+y+xy$. [See Exercise 1.2.2.]

3. Let $R = \mathbb{R}^3$ with addition as vector addition and multiplication as the cross product.

4. Let S be any set and consider $R = \mathcal{P}(S)$ with $\triangle$ as the addition operation and $\cap$ as the multiplication.

5. Let S be any set and consider $R = \mathcal{P}(S)$ with $\cup$ as the addition operation and $\cap$ as the multiplication.

6. Let S be any set and consider $R = \mathcal{P}(S)$ with $\overline{\triangle}$ as the addition operation (defined as $A\overline{\triangle}B = \overline{A\triangle B}$) and $\cup$ as the multiplication.

7. Let R be the set of finite unions of bounded intervals in $\mathbb{R}$ (possibly empty or singletons $\{a\}$). Let $A, B \in R$. Define the symmetric difference $A\triangle B$ on R as the addition and the convex hull of $A \cup B$ as the multiplication. (We define the convex hull of a subset S of $\mathbb{R}$ as the smallest bounded interval containing S.)

8. Let $R = \mathbb{Z} \times \mathbb{Z}$ and define $(a, b) + (c, d) = (a + c, b + d)$ and define also $(a, b) \times (c, d) = (ad - bc, bd)$.

9. Let R be a ring, let $r, s \in R$, and let $m, n \in \mathbb{Z}$. Prove the following formulas with the $\cdot$ notation.

    (a) $m \cdot (r + s) = (m \cdot r) + (m \cdot s)$

    (b) $(m + n) \cdot r = (m \cdot r) + (n \cdot r)$

10. Let R be a ring, let $r, s \in R$, and let $m, n \in \mathbb{Z}$. Prove the following formulas with the $\cdot$ notation.

    (a) $m \cdot (rs) = r(m \cdot s) = (m \cdot r)s$

    (b) $(mn) \cdot r = m \cdot (n \cdot r)$

11. Prove that in $C^0([a, b], \mathbb{R})$ the composition operation $\circ$ is right-distributive over $+$.

12. Let I be an interval of real numbers. Prove that the zero divisors in $\mathrm{Fun}(I, \mathbb{R})$ are nonzero functions $f(x)$ such that there exists $x_0 \in I$ such that $f(x_0) = 0$. Prove that all the elements in $\mathrm{Fun}(I, \mathbb{R})$ are either 0, a zero divisor, or a unit.

13. Prove (carefully) that the nonzero elements in $(C^0([a, b], \mathbb{R}), +, \times)$ that are neither zero divisors nor units are functions for which there exists an $x_0 \in [a, b]$ and an $\varepsilon > 0$ such that $f(x_0) = 0$ and for which $f(x) \neq 0$ for all x such that $0 < |x - x_0| < \varepsilon$.

14. Prove that multiplication in $\mathbb{H}$ is associative.

15. Let $\alpha = 1 + 2i + 3j + 4k$ and $\beta = 2 - 3i + k$ in $\mathbb{H}$. Calculate the following operations: (a) $\alpha + \beta$; (b) $\alpha\beta$; (c) $\beta\alpha$; (d) $\alpha\beta^{-1}$; (e) β^2.

16. Let $\alpha, \beta \in \mathbb{H}$ be arbitrary. Decide whether any of the operations $\alpha\beta^{-1}$, $\beta^{-1}\alpha$, $\beta\alpha^{-1}$, or $\alpha^{-1}\beta$ are equal.

17. Let $R = \{a + bi + cj + dk \in \mathbb{H} \mid a, b, c, d \in \mathbb{Z}\}$. Prove that R is a subring of $\mathbb{H}$ and prove that $U(R) = Q_8$, the quaternion group.

18. Prove that the equation over $x^2 + 3 = 0$ with $x \in \mathbb{H}$ has an infinite number of solutions, namely, $x = bi + cj + dk$ such that $b^2 + c^2 + d^2 = 3$.

19. Fix an integer $n \geq 2$. Let $R(n)$ be the set of symbols $\overline{a} + i\overline{b}$ where $\overline{a}, \overline{b} \in \mathbb{Z}/n\mathbb{Z}$. Define $+$ and $\times$ on R like addition and multiplication in $\mathbb{C}$.

    (a) Prove that $R(n)$ is a ring.

    (b) Set $n = 6$. Identify all the zero divisors in $R(6)$.

20. Define $\text{Hom}(V, W)$ as the set of linear transformations from a real vector space V to another real vector space W. Prove that $\text{Hom}(V, V)$, equipped with $+$ and $\circ$ (composition) is a ring.

21. Let R_1 and R_2 be rings with identity elements. Prove that $U(R_1 \oplus R_2) \cong U(R_1) \oplus U(R_2)$. Prove the equivalent result for a finite number of rings $R_1, R_2, \ldots, R_n$.

22. Prove that the characteristic $\text{char}(R)$ of an integral domain R is either 0 or a prime number. [Hint: By contradiction.]

23. Consider the ring $\mathbb{Z} \oplus \mathbb{Z}$ and consider the subset $R = \{(x, y) \mid x - y = 0\}$. Prove that R is a subring. Decide if R is an integral domain.

24. Prove that a finite integral domain is a field.

25. Let R_1 and R_2 be rings. Prove that $R_1 \oplus R_2$ is an integral domain if and only if R_1 is an integral domains and $R_2 = \{0\}$ or vice versa.

26. (*Binomial Formula*) Let R be a ring and suppose that x and y commute in R. Prove that for all positive integers n,

$$(x + y)^n = \sum_{i=0}^{n} \binom{n}{i} x^{n-i} y^i.$$

27. Let R be a ring and suppose that x and y commute in R. Prove that for all positive integers n,

$$x^n - y^n = (x - y)(x^{n-1} + x^{n-2}y + x^{n-3}y^2 + \cdots + y^{n-1}).$$

28. Prove that if $a \in R$ is idempotent, then $a^n = a$ for all positive integers n.

29. In the ring $M_2(\mathbb{Z})$,
    (a) find two nilpotent elements;
    (b) find two idempotent elements that are not the identity.

30. Prove that if R contains an identity, then all idempotent elements that are not the identity are zero divisors.

31. Prove that if $A \in M_n(\mathbb{R})$ is nilpotent, then all of its eigenvalues are 0.

32. Let R be a commutative ring.
    (a) Prove that the set of nilpotent elements, $\mathcal{N}(R)$, is closed under addition. [Hint: Binomial formula. See Exercise 3.1.26.]
    (b) Prove that $\mathcal{N}(R)$ is closed under multiplication.
    (c) Conclude that $\mathcal{N}(R)$ is a subring of R.

33. Let R be a commutative ring with an identity $1 \neq 0$. Let $x \in \mathcal{N}(R)$. Prove that $1 - x$ is a unit.

34. Let $R = \mathbb{Z}/81\mathbb{Z}$. Determine the elements in $\mathcal{N}(R)$. In particular, determine the cardinality of $\mathcal{N}(R)$.

35. Let $R = \mathbb{Z}/700\mathbb{Z}$. Determine the elements in $\mathcal{N}(R)$. In particular, determine the cardinality of $\mathcal{N}(R)$.

36. A *Boolean ring* is a ring R in which $r^2 = r$ for all $r \in R$.
    (a) Prove that the characteristic of a Boolean ring with an identity is 2.
    (b) Prove that every Boolean ring is commutative.

37. Let R be a ring and suppose that a and b are two elements such that $a^3 = b^3$ and $a^2 b = b^2 a$. Can $a^2 + b^2$ be a unit? [This exercise appeared in modified form as Problem A-2 on the 1991 Putnam Mathematics Competition.]

38. Let R be a ring such that $x^3 = x$ for all $x \in R$. Prove that R is commutative.

39. Prove that $\left\{ \frac{n}{k} \mid n, k \in \mathbb{Z} \text{ with } k \text{ odd} \right\}$ is a subring of $\mathbb{Q}$.

40. Prove that $\{a + bi \mid a, b \in \mathbb{Z}\}$ is a subring of $\mathbb{C}$.

41. Let R be any ring and let n be a positive integer. Prove that $\{n \cdot r \mid r \in R\}$ is a subring of R.

42. Determine with proof, which of the following subsets are subrings of $\mathbb{Z} \oplus \mathbb{Z}$.
    (a) $\{(a, b) \in \mathbb{Z} \oplus \mathbb{Z} \mid 2a + b = 0\}$
    (b) $\{(a, b) \in \mathbb{Z} \oplus \mathbb{Z} \mid a = b\}$
    (c) $\{(a, b) \in \mathbb{Z} \oplus \mathbb{Z} \mid a + b \text{ is even}\}$
    (d) $\{(a, b) \in \mathbb{Z} \oplus \mathbb{Z} \mid ab = 0\}$

43. Determine with proof, which of the following subsets are subrings of $\mathbb{Q}$.
    (a) Fractions, which when written in reduced form, have an odd denominator.
    (b) Fractions, which when written in reduced form, have an even denominator.
    (c) Fractions of the form $k2^n$, where k is odd and $n \in \mathbb{Z}$.
    (d) Fractions, which when written in reduced form, are $\frac{2^n}{k}$.

44. Prove that the set of periodic real-valued functions of period p is a subring of $\mathrm{Fun}(\mathbb{R}, \mathbb{R})$. [Note: Functions that are periodic with period p satisfy $f(x + p) = f(x)$ for all x. Such functions may be periodic with a lower period or even constant.]

45. Let $C^n([a, b], \mathbb{R})$ be the set of real-valued functions on $[a, b]$ whose first n derivatives exist and are continuous. Prove that $C^{n+1}([a, b], \mathbb{R})$ is a proper subring of $C^n([a, b], \mathbb{R})$.

46. Let R be any ring and let $\{S_i\}_{i \in \mathcal{I}}$ be a collection of subrings (not necessarily finite or countable). Prove that the intersection

$$\bigcap_{i \in \mathcal{I}} S_i$$

is a subring of R.

47. Let R be a ring and let R_1 and R_2 be subrings. Prove by a counterexample that $R_1 \cup R_2$ is in general not a subring.

48. Let R be a ring and let a be a fixed element of R. Define $C(a) = \{r \in R \mid ra = ar\}$. Prove that $C(a)$ is a subring of R.

49. Let R be a ring and let a be a fixed element of R.
    (a) Prove that the set $\{x \in R \mid ax = 0\}$ is a subring of R.
    (b) With $R = \mathbb{Z}/100\mathbb{Z}$, and $a = \bar{5}$, find the subring defined in part (a).

3.2 Rings Generated by Elements

Following the general outline presented in the preface, this section first introduces a particular method to efficiently describe certain types of subrings. Motivated by the notation, we introduce two important families of rings that build new rings from old ones.

3.2.1 Generated Subrings

Let A be a commutative ring. Let R be a subring of A and let S be a subset of A. The notation $R[S]$ denotes the smallest (by inclusion) subring of A that contains both R and S. Obviously, if $S \subset R$, the ring $R[S] = R$ so the notation is uninteresting. However, if elements of S are not in R, then R is a proper subring of $R[S]$.

Example 3.2.1. Consider the ring $\mathbb{Z}\left[\frac{1}{2}\right]$ as a subring of $\mathbb{Q}$. Since $\mathbb{Z}\left[\frac{1}{2}\right]$ is closed under multiplication, $\frac{1}{4} = \frac{1}{2} \times \frac{1}{2} \in \mathbb{Z}\left[\frac{1}{2}\right]$ and more generally $\frac{1}{2^n} \in \mathbb{Z}\left[\frac{1}{2}\right]$ for all nonnegative integers n. Also because the subring is closed under multiplication, for all integers k and n, the fraction $\frac{k}{2^n}$ is an element in $\mathbb{Z}\left[\frac{1}{2}\right]$. Hence, the set

$$R = \left\{ \frac{k}{2^n} \,\middle|\, k, n \in \mathbb{Z} \right\}$$

is a subset of $\mathbb{Z}[\frac{1}{2}]$ in $\mathbb{Q}$. However, if $k/2^m, \ell/2^n \in R$, then

$$\frac{k}{2^m} - \frac{\ell}{2^n} = \frac{2^n k - 2^n \ell}{2^{m+n}} \quad \text{and} \quad \frac{k}{2^m} \times \frac{\ell}{2^n} = \frac{k\ell}{2^{m+n}}.$$

Thus R is a subring of $\mathbb{Q}$. Since $\mathbb{Z}[\frac{1}{2}]$ is the smallest subring of $\mathbb{Q}$ containing both $\mathbb{Z}$ and $\frac{1}{2}$, then as a set, $\mathbb{Z}[\frac{1}{2}] = R$. $\triangle$

It is not uncommon for the ring A to be implied by the elements in the set S. The following two examples illustrate this habit of notation.

Example 3.2.2 (Gaussian Integers). Consider the ring $\mathbb{Z}[i]$. It is understood that i is the imaginary number that satisfies $i^2 = -1$. This notation assumes that the superset ring A is the ring $\mathbb{C}$. The ring $\mathbb{Z}[i]$ contains all the integers and, since it is closed under multiplication, it contains all integer multiples of i. Since $\mathbb{Z}[i]$ is closed under addition, it must contain the subset

$$\{a + bi \in \mathbb{C} \mid a, b \in \mathbb{Z}\}.$$

However, this subset is closed under subtraction and under multiplication with

$$(a + bi)(c + di) = (ac - bd) + (ad + bc)i.$$

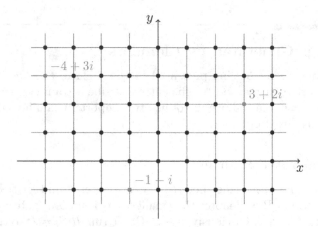

FIGURE 3.1: Gaussian integers $\mathbb{Z}[i]$.

Hence, this subset is the smallest subring in $\mathbb{C}$ containing both $\mathbb{Z}$ and the element i and so it is precisely $\mathbb{Z}[i]$. In the usual manner of depicting a complex number $a + bi$ as a point in the plane, the subring $\mathbb{Z}[i]$ consists of the points with integer coordinates. (See Figure 3.1.)

The ring $\mathbb{Z}[i]$ is called the ring of *Gaussian integers* and is important in elementary number theory.

In $\mathbb{C}$, the multiplicative inverse of an element is

$$(a + bi)^{-1} = \frac{a - bi}{a^2 + b^2}.$$

The group of units $U(\mathbb{Z}[i])$ consists of elements $a + bi \in \mathbb{Z}[i]$ such that

$$\frac{a}{a^2 + b^2}, \frac{b}{a^2 + b^2} \in \mathbb{Z}.$$

If $|a| \geq 2$, then $a^2 > |a|$, in which case $a^2 + b^2 > |a|$ and hence $a^2 + b^2$ could not divide a. A symmetric result holds for b. Consequently, if $a + bi \in U(\mathbb{Z}[i])$, then $|a| \leq 1$ and $|b| \leq 1$. However, if $|a| = 1$ and $|b| = 1$, then $a^2 + b^2 = 2$, while $a = \pm 1$ and so $a/(a^2 + b^2) \notin \mathbb{Z}$. Thus, we see that the only units in $\mathbb{Z}[i]$ have $|a| = 1$ and $b = 0$ or $a = 0$ and $|b| = 1$. Hence,

$$U(\mathbb{Z}[i]) = \{1, -1, i, -i\}.$$

This group of units is isomorphic to Z_4. $\triangle$

Example 3.2.3. Consider as another example the ring $\mathbb{Z}[\sqrt{2}]$. Obviously, $\mathbb{Z}$ is a subring of $\mathbb{R}$ and $\sqrt{2} \in \mathbb{R}$ so $\mathbb{Z}[\sqrt{2}]$ is the smallest subring of $\mathbb{R}$ containing the subring of integers and $\sqrt{2}$.

Following a similar process as in Example 3.2.2, it is easy to find that

$$\mathbb{Z}[\sqrt{2}] = \{a + b\sqrt{2} \in \mathbb{R} \mid a, b \in \mathbb{Z}\}.$$

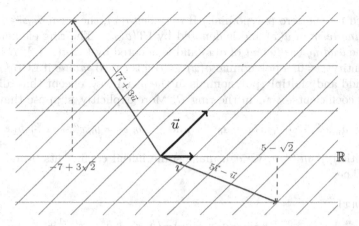

FIGURE 3.2: A representation of $\mathbb{Z}[\sqrt{2}]$.

We comment that all the powers of $\sqrt{2}$ must be in $\mathbb{Z}[\sqrt{2}]$ but all even powers of $\sqrt{2}$ are powers of 2 but for all odd integers $2k+1$, the powers $\sqrt{2}^{2k+1} = 2^k\sqrt{2}$.

The group of units $U(\mathbb{Z}[\sqrt{2}])$ is more complicated than $U(\mathbb{Z}[i])$ and we leave this investigation as a project. (See Project I.)

One way to depict the elements of $\mathbb{Z}[\sqrt{2}]$ is to represent the number $a+b\sqrt{2}$ in the plane as the point

$$a\vec{i} + b\vec{u}, \qquad \text{where } \vec{i} = (1,0) \text{ and } \vec{u} = (\sqrt{2}, \sqrt{2}).$$

The actual real number $a + b\sqrt{2}$ is obtained by projecting the vector onto the real line. (See Figure 3.2.) $\triangle$

3.2.2 Polynomial Rings

The notation we introduced from subrings generated by elements motivates another important construction for rings, that of polynomial rings. If R is any commutative ring, the notation $R[x]$ denotes the set of polynomials in the variable x and with coefficients in R. More precisely, polynomials are finite expressions of the form

$$a(x) = a_m x^m + \cdots + a_1 x + a_0$$

with $m \in \mathbb{N}$, the coefficients $a_i \in R$ for all $i = 0, 1, \ldots, m$, and $a_m \neq 0$.

A summand $a_i x^i$ of the polynomial is called a *term*. We call two terms *like terms* if in each, the variable x has the same degree. The element a_i is called the *coefficient* of the term $a_i x^i$. The integer m, i.e., the highest power of x appearing in a nonzero term of a nonzero polynomial, is called the *degree* of the polynomial, and is denoted $\deg a(x)$. (The concept of degree of a polynomial is undefined for the zero polynomial.) If $m = \deg a(x)$, the term $a_m x^m$ is called the *leading term* of $a(x)$ and is denoted by $\mathrm{LT}(a(x))$. The ring element a_m is called the *leading coefficient* of $a(x)$ and is denoted by $\mathrm{LC}(a(x))$. If R is a ring with identity $1 \neq 0$, a polynomial $a(x) \in R[x]$ is called *monic* if $\mathrm{LC}(a(x)) = 1$.

We add and multiply polynomials in the usual way, except that all operations on coefficients occur in the ring R. More explicitly, suppose that

$$a(x) = a_m x^m + \cdots + a_1 x + a_0 \qquad \text{and} \qquad b(x) = b_n x^n + \cdots + b_1 x + b_0.$$

For addition, if $m \neq n$, we can make some initial coefficients 0 and assume $n = m$. Then

$$
\begin{aligned}
a(x) &+ b(x) \\
&= (a_n x^n + a_{n-1} x^{n-1} + \cdots + a_1 x + a_0) + (b_n x^n + b_{n-1} x^{n-1} + \cdots + b_1 x + b_0) \\
&= (a_n + b_n) x^n + (a_{n-1} + b_{n-1}) x^{n-1} + \cdots + (a_1 + b_1) x + (a_0 + b_0).
\end{aligned}
$$

For multiplication, one distributes the terms of $a(x)$ with the terms of $b(x)$, multiplies terms and variable powers as appropriate, and then gathers like terms. We can write succinctly

$$a(x)b(x) = \sum_{k=0}^{m+n} \left(\sum_{i+j=k} a_i b_j \right) x^k. \tag{3.1}$$

The following proposition is the main point of this subsection. The result may feel obvious but we provide the proof to illustrate that the details are not so obvious.

Proposition 3.2.4

Let R be a commutative ring. Then with the operations of addition and multiplication defined as above, $R[x]$ is a commutative ring that contains R as a subring.

Proof. Let $a(x), b(x), c(x)$ be polynomials in $R[x]$. Let $n = \max\{\deg a, \deg b, \deg c\}$. If $k > \deg p(x)$ for any polynomial, we will assume $p_k = 0$. Then

$$
\begin{aligned}
a(x) &+ (b(x) + c(x)) \\
&= a(x) + ((b_n + c_n) x^n + \cdots + (b_1 + c_1) x + (b_0 + c_0)) \\
&= (a_n + (b_n + c_n)) x^n + \cdots + (a_1 + (b_1 + c_1)) x + (a_0 + (b_0 + c_0)) \\
&= ((a_n + b_n) + c_n) x^n + \cdots + ((a_1 + b_1) + c_1) x + ((a_0 + b_0) + c_0) \\
&= ((a_n + b_n) x^n + \cdots + (a_1 + b_1) x + (a_0 + b_0)) + c(x) \\
&= (a(x) + b(x)) + c(x).
\end{aligned}
$$

So $+$ is associative on $R[x]$.

The additive identity is the 0 polynomial. The additive inverse of a polynomial $a(x)$ is $-a_n x^n - \cdots - a_1 x - a_0$. The addition is commutative so $(R[x], +)$ is an abelian group.

To show that polynomial multiplication is associative, we use (3.1). If $\deg a(x) = m$, $\deg b(x) = n$, and $\deg c(x) = \ell$, then

$$(a(x)b(x))\, c(x)$$

$$= \left(\sum_{q=0}^{m+n} \left(\sum_{i+j=q} a_i b_j \right) x^k \right) c(x) = \sum_{h=0}^{m+n+\ell} \left(\sum_{q+k=h} \left(\sum_{i+j=q} a_i b_j \right) c_k \right) x^h$$

$$= \sum_{h=0}^{m+n+\ell} \left(\sum_{i+j+k=h} a_i b_j c_k \right) x^h = \sum_{h=0}^{m+n+\ell} \left(\sum_{i+r=h} a_i \left(\sum_{j+k=r} b_j c_k \right) \right) x^h$$

$$= a(x) \left(\sum_{r=0}^{n+\ell} \left(\sum_{j+k=r} b_j c_k \right) x^r \right) = a(x)\, (b(x)c(x))\,.$$

Since R is commutative, we have

$$a(x)b(x) = \sum_{k=0}^{m+n} \left(\sum_{i+j=k} a_i b_j \right) x^k = \sum_{k=0}^{m+n} \left(\sum_{i+j=k} b_j a_i \right) x^k = b(x)a(x).$$

Thus, the multiplication in $R[x]$ is commutative.

Finally, to prove distributivity, by virtue of commutativity in $R[x]$ we only need to prove left-distributivity. Assume without loss of generality that $\deg b(x) = \deg c(x) = n$. Then

$$a(x)\, (b(x) + c(x)) = a(x)\, ((b_n + c_n)x^n + \cdots + (b_1 + c_1)x + (b_0 + c_0))$$

$$= \sum_{k=0}^{m+n} \left(\sum_{i+j=k} a_i(b_j + c_j) \right) x^k = \sum_{k=0}^{m+n} \left(\sum_{i+j=k} (a_i b_j + a_i c_j) \right) x^k$$

$$= \sum_{k=0}^{m+n} \left(\sum_{i+j=k} a_i b_j \right) x^k + \sum_{k=0}^{m+n} \left(\sum_{i+j=k} a_i c_j \right) x^k$$

$$= a(x)b(x) + a(x)c(x).$$

This proves all the axioms of a commutative ring.

The subset of $R[x]$ of constant polynomials add and multiply just as in the ring R, so R is a subring. □

In elementary algebra, we regularly work in the context of $\mathbb{Z}[x]$, $\mathbb{Q}[x]$, or $\mathbb{R}[x]$, which are polynomial rings with integer, rational, or real coefficients respectively. However, consider the following example where the ring of coefficients is a finite ring.

Example 3.2.5. Consider the polynomial ring $S = (\mathbb{Z}/3\mathbb{Z})[x]$. As examples of operations in this ring, let $p(x) = x^2 + 2x + 1$ and $q(x) = x^2 + x + 1$ be two polynomials in S. (For brevity, we omit the bar and write 2 instead of $\overline{2}$.) We calculate the addition and multiplication:

$$p(x) + q(x) = (x^2 + 2x + 1) + (x^2 + x + 1) = 2x^2 + 0x + 2 = 2x^2 + 2,$$

$$\begin{aligned} p(x)q(x) &= (x^2 + 2x + 1)(x^2 + x + 1) \\ &= x^4 + x^3 + x^2 + 2x^3 + 2x^2 + 2x + x^2 + x + 1 \\ &= x^4 + (1 + 2)x^3 + (1 + 2 + 1)x^2 + (2 + 1)x + 1 \\ &= x^4 + x^2 + 1. \end{aligned}$$

$\triangle$

Polynomial rings are important families of rings and find applications in countless areas. We will use them for many examples.

Proposition 3.2.6

Let R be an integral domain.

(1) Let $a(x), b(x)$ be nonzero polynomials in $R[x]$. Then $\deg a(x)b(x) = \deg a(x) + \deg b(x)$.

(2) The units of $R[x]$ are the units of R, i.e., $U(R[x]) = U(R)$.

(3) $R[x]$ is an integral domain.

Proof. If $\deg a(x) = m$ and $\deg b(x) = n$, then $a(x)b(x)$ has no terms of degree higher than $m + n$. However, since $a_m \neq 0$ and $b_n \neq 0$, the product contains the term $a_m b_n x^{m+n}$ as long as $a_m b_n \neq 0$. Since R contains no zero divisors, $a_m b_n \neq 0$. Hence, $\deg a(x)b(x) = m + n = \deg a(x) + \deg b(x)$.

The multiplicative identity in $R[x]$ is the degree 0 polynomial 1. Suppose $a(x) \in U(x)$ and let $b(x) \in U(x)$ with $a(x)b(x) = 1$. Then $\deg(a(x)b(x)) = 0$. Since the degree of a polynomial is a nonnegative integer, part (1) implies that $\deg a(x) = 0$. Hence, $a(x) \in R$ and part (2) follows.

Since R is an integral domain it contains an identity $1 \neq 0$. This is also the multiplicative identity for $R[x]$. Let $a(x), b(x)$ be nonzero polynomials. Let $a_m x^m$ be the leading term of $a(x)$ and let $b_n x^n$ be the leading term of $b(x)$. Then $a_m b_n x^{m+n}$ is the leading term of their product. Since the coefficients a_m and b_n are nonzero and R contains no zero divisors, $a_m b_n \neq 0$ and hence $a(x)b(x) \neq 0$. Thus, $R[x]$ is an integral domain. $\square$

Proposition 3.2.7

Let R be a commutative ring. A polynomial $a(x) \in R[x]$ is a zero divisor if and only if $\exists r \in R - \{0\}$ such that $r\, a(x) = 0$.

Proof. ($\Longleftarrow$). This direction is obvious, since r can be viewed as a polynomial (of degree 0).

($\Longrightarrow$). Suppose that $a(x)$ is a zero divisor in $R[x]$. This means that there exists a polynomial $b(x) \in R[x]$ such that $a(x)b(x) = 0$. We write

$$a(x) = a_m x^m + \cdots + a_1 x + a_0,$$
$$b(x) = b_n x^n + \cdots + b_1 x + b_0.$$

We will show that $r = b_0^{m+1}$ satisfies $ra(x) = 0$. More precisely, we show by (strong) induction the claim that $a_i b_0^{i+1} = 0$ for all $0 \leq i \leq m$.

The term of degree 0 in the product $a(x)b(x)$ has the coefficient $a_0 b_0$. We must have $a_0 b_0 = 0$ since $a(x)b(x) = 0$. This gives the basis step of our proof by induction. Now suppose that $a_i b_0^{i+1} = 0$ for all $0 \leq i \leq k$. The term of degree $k + 1$ in $a(x)b(x) = 0$ is

$$0 = a_{k+1}b_0 + a_k b_1 + \cdots + a_1 b_k + a_0 b_{k+1} = \sum_{i=0}^{k+1} a_i b_{k+1-i}.$$

Multiplying this equation by b_0^{k+1}, we have

$$0 = a_{k+1}b_0^{k+2} + a_k b_0^{k+1}b_1 + \cdots + a_1 b_0^{k+1}b_k + a_0 b_0^{k+1}b_{k+1}.$$

Using the induction hypothesis, we get $0 = a_{k+1}b_0^{k+2}$ because all the other terms are 0. By strong induction, we establish the claim. Since $\deg a(x) = m$, we have $b_0^{m+1}a_i = 0$ for all $0 \leq i \leq m$, so $b_0^{m+1}a(x) = 0$. $\square$

Having described the construction of a polynomial ring in one variable, the construction extends naturally to polynomial rings in more than one variable. If R is a commutative ring, then $R[x]$ is another commutative ring and $R[x][y]$ is then a polynomial ring in the two variables x and y. We typically write $R[x, y]$ instead of $R[x][y]$ for the polynomial ring with coefficients in R and with the two variables x and y. More generally, we denote by $R[x_1, x_2, \ldots, x_n]$ the polynomial ring with variables $x_1, x_2, \ldots, x_n$ and coefficients in R. When necessary, we can view this as a polynomial ring in one variable but with coefficients in $R[x_1, x_2, \ldots, x_{n-1}]$.

3.2.3 Group Rings

As a third class of examples in this section we introduce group rings.

Let G be a finite group and list the elements out as $G = \{g_1, g_2, \ldots, g_n\}$. Let R be a commutative ring. We define the set $R[G]$ as the set of formal sums

$$a_1 g_1 + a_2 g_2 + \cdots + a_n g_n,$$

where $a_i \in R$. Note that if g_1 is the group identity, we usually write $a_1 g_1$ as just a term a_1. As with polynomials, we call any summand $a_i g_i$ a *term* of the formal sum.

Addition of formal sums is done component-wise:

$$\left(\sum_{i=1}^{n} a_i g_i\right) + \left(\sum_{i=1}^{n} b_i g_i\right) = \sum_{i=1}^{n}(a_i + b_i)g_i.$$

We define the multiplication $\cdot$ of formal sums by distributing $\cdot$ over $+$ and then rewriting terms as

$$(a_i g_i) \cdot (b_j g_j) = (a_i b_j)(g_i g_j) = (a_i b_j)g_k,$$

where the product $a_i b_j$ occurs in R and the operation $g_i g_j = g_k$ corresponds to the group operation in G. Then, just as with polynomials, one gathers like terms.

We illustrate the operations defined on formal sums with a few examples.

Example 3.2.8. Let $G = D_5$ and consider the set of formal sums $\mathbb{Z}[D_5]$. Let

$$\alpha = r^2 + 2r^3 - s \qquad \text{and} \qquad \beta = -r^2 + 7sr.$$

Then $\alpha + \beta = 2r^3 - s + 7sr$ and

$$\begin{aligned}
\alpha\beta &= (r^2 + 2r^3 - s)(-r^2 + 7sr) \\
&= -r^4 + 7r^2 sr - 2r^5 + 14r^3 sr + sr^2 - 7ssr \\
&= -r^4 + 7sr^4 - 2 + 14sr^3 + sr^2 - 7r \\
&= -2 - 7r - r^4 + sr^2 + 14sr^3 + 7sr^4. \qquad\qquad \triangle
\end{aligned}$$

It is common to use the group itself as the indexing set for the coefficients of the terms. Hence, we often denote a generic group ring element as

$$\alpha = \sum_{g \in G} a_g g.$$

In the proof of the following proposition, establishing associativity is the most challenging part. We use the above notation with iterated sums. The notation $(x, y) : xy = g$ stands for all pairs $x, y \in G$ such that $xy = g$.

Proposition 3.2.9

> Let R be a commutative ring and let G be a finite group. The set $R[G]$, equipped with addition and multiplication as defined above, is a ring and is called the *group ring* of R and G. Furthermore, R is a subring of $R[G]$.

Proof. If $|G| = n$, then the group $(R[G], +)$ is isomorphic as a group to the direct sum of $(R, +)$ with itself n times. Hence, $(R[G], +)$ is an abelian group. We need to prove that multiplication is associative and that multiplication is distributive over the addition.

Let $\alpha = \sum_{g \in G} a_g g$, $\beta = \sum_{g \in G} b_g g$, and $\gamma = \sum_{g \in G} c_g g$ be three elements in $R[G]$. Then

$$(\alpha\beta)\gamma = \left(\sum_{g \in G} \left(\sum_{\substack{(x,y): \\ xy=g}} a_x b_z \right) g \right) \gamma = \sum_{h \in G} \left(\sum_{\substack{(g,z): \\ gz=h}} \left(\sum_{\substack{(x,y): \\ xy=g}} a_x b_y \right) c_z \right) h$$

$$= \sum_{h \in G} \left(\sum_{\substack{(x,y,z): \\ (xy)z=h}} a_x b_y c_z \right) h = \sum_{h \in G} \left(\sum_{\substack{(x,y,z): \\ x(yz)=h}} a_x b_y c_z \right) h$$

$$= \sum_{h \in G} \left(\sum_{\substack{(x,g'): \\ xg'=h}} a_x \left(\sum_{\substack{(y,z): \\ yz=g'}} b_y c_z \right) \right) h = \alpha \left(\sum_{g' \in G} \left(\sum_{\substack{(y,z): \\ yz=g'}} b_y c_z \right) g' \right)$$

$$= \alpha(\beta\gamma).$$

This proves associativity of multiplication. Also

$$\alpha(\beta + \gamma)$$

$$= \sum_{g \in G} \left(\sum_{(x,y): \, xy=g} a_x (b_y + c_y) \right) g = \sum_{g \in G} \left(\sum_{(x,y): \, xy=g} (a_x b_y + a_x c_y) \right) g$$

$$= \sum_{g \in G} \left(\sum_{(x,y): \, xy=g} a_x b_y + \sum_{(x,y): \, xy=g} a_x c_y \right) g$$

$$= \sum_{g \in G} \left(\sum_{(x,y): \, xy=g} a_x b_y \right) g + \sum_{g \in G} \left(\sum_{(x,y): \, xy=g} a_x c_y \right) g$$

$$= \alpha\beta + \alpha\gamma.$$

This establishes left-distributivity. Right-distributivity is similar and establishes that $R[G]$ is a ring.

The subset $\{r \cdot 1 \in R[G] \mid r \in R\}$ is a subring that is equal to R. $\qquad \square$

Note that even if R is commutative, $R[G]$ is not necessarily commutative. Most of the examples of rings introduced so far in the text have been commutative rings. The construction of group rings gives a wealth of examples of noncommutative rings.

Example 3.2.10. Consider the group ring $(\mathbb{Z}/3\mathbb{Z})[S_3]$. The elements in $(\mathbb{Z}/3\mathbb{Z})[S_3]$ are formal sums

$$\alpha = a_1 + a_{(12)}(12) + a_{(13)}(13) + a_{(23)}(23) + a_{(123)}(123) + a_{(132)}(132),$$

where each $a_\sigma \in \mathbb{Z}/3\mathbb{Z}$. Since there are 3 options for each a_i, there are 3^6 elements in this group ring. As a simple illustration of some properties of elements, we point out that $(\mathbb{Z}/3\mathbb{Z})S_3$ contains zero divisors. For example,

$$(1 + 2(12))(1 + (12)) = 1 + (12) + 2(12) + 2(12)^2 = 1 + 0(12) + 2 = 0.$$

The ring has the identity element 1, which really is $1 \cdot 1$ and it contains units that are not the identity since $1(12) \cdot 1(12) = 1$ as an operation in the group ring.   △

Proposition 3.2.11

Let R be a commutative ring with an identity $1 \neq 0$ and let G be a group. Then the element $1 \cdot 1_G$ is the identity in $R[G]$ and G is a subgroup of $U(R[G])$.

Proof. (Left as an exercise. See Exercise 3.2.19.)   □

Proposition 3.2.11 along with Proposition 3.2.9 together show that the group ring $R[G]$ is a ring that includes the ring R as a subring and the group G as a subgroup of $U(R[G])$. This observation shows in what sense $R[G]$ is a ring generated by R and G.

Proposition 3.2.12

Let G be a finite group with $|G| > 1$ and R a commutative ring with more than one element. Then $R[G]$ always has a zero divisor.

Proof. Let $r \in R \setminus \{0\}$ and suppose that the element in $g \in G$ has order $m > 1$. Then

$$(r - rg)(r + rg + rg^2 + \cdots + rg^{m-1})$$
$$= r^2 + r^2g + r^2g^2 + \cdots + r^2g^{m-1} - (r^2g + r^2g^2 + r^2g^3 + \cdots + r^2g^m)$$
$$= r^2 - r^2g^m = r^2 - r^2 = 0.$$

Thus, $r - rg$ is a zero divisor.   □

Among the examples of rings we have encountered so far, group rings are likely the most abstract. They do not, in general, correspond to certain number sets, modular arithmetic, polynomials, functions, matrices, or any other mathematical object naturally encountered so far. However, as with any other mathematical object, we develop an intuition for it as we use it and finds applications.

3.2.4 Useful CAS Commands

Due to differences in how computer algebra systems deal with class objects, different CASs often deal with rings in quite unlike ways.

Maple does not require the user to clearly define the object class but assigns it a class based on how the user types it in. For example, typing p:=3*x^2+5*x+1/2, because the coefficients are rational, defines the polynomial $p = 3x^2 + 5x + \frac{1}{2}$ as an element of $\mathbb{Q}[x]$. Using the palettes, typing q:=x^2+3*$\sqrt{2}$x+1/2 defines a polynomial in $\mathbb{R}[x]$.

To perform operations in a ring of the form $(\mathbb{Z}/n\mathbb{Z})[x]$ where n is an integer, *Maple* offers the command modp1. We encourage the user/reader to study the help files for this command to see its various uses.

Maple's Application Center[1] offers a large variety of user created packages that the public can download. For example, there are various packages that implement the quaternions. Many of these packages offer many more methods associated with the geometry of quaternions that we have discussed so far.

When working with rings in SAGE, the user usually must define the ring carefully before being able to do computations in that ring. The constructor commands attach to the named variable how that variable should behave and what type of coefficients it can have with it.

_____ Sage _____

```
sage: R=PolynomialRing(QQ, 't')
sage: S=QQ['t']
sage: T.<t>=PolynomialRing(QQ)
sage: A=PolynomialRing(GF(11),'x')
sage: B.<y>=PowerSeriesRing(ZZ)
sage: H.<ii,jj,kk> = QuaternionAlgebra(RR,-1,-1)
sage: (1+2*ii+3*jj-kk)*(ii+2*kk)
7*ii-5*jj-kk
```

The first three commands offer three different ways to define R, S, and T as the polynomial ring $\mathbb{Q}[t]$. Subsequently, when SAGE sees the variable t, it understands it as the variable in $\mathbb{Q}[t]$. The fourth line defines A as $\mathbb{F}_{11}[x]$, which also defines how x should behave. The fifth line illustrates how SAGE defines a power series ring, defined in Exercise 3.2.21. SAGE offers many other constructors in ring theory; the last two lines in the above code define the ring of Hamilton's quaternions $\mathbb{H}$ and illustrate a multiplication in this ring.

The following bit of SAGE code defines R as the polynomial ring $\mathbb{F}_5[x, y]$ and performs a polynomial multiplication. Note that in the second line we are labeling the variables. In the third line, we are assigning the values of two variables in the same line.

_____

[1] https://www.maplesoft.com/applications/

—————————————————————— Sage ——————————————————————

```
sage: R=GF(5)['x','y']
sage: x,y=R.gens()
sage: a,b=1+3*x+2*x*y,3-2*y
sage: a*b
x*y^2 - x - 2*y - 2
```

EXERCISES FOR SECTION 3.2

1. Prove that $\mathbb{Z}[i]$ is an integral domain.

2. Prove that $\mathbb{Z}[\sqrt{2}, \sqrt{5}]$ consists of the following subring of $\mathbb{R}$,

$$\{a + b\sqrt{2} + c\sqrt{5} + d\sqrt{10} \in \mathbb{R} \mid a, b, c, d \in \mathbb{Z}\}.$$

Write out the multiplication between $a + b\sqrt{2} + c\sqrt{5} + d\sqrt{10}$ and $a' + b'\sqrt{2} + c'\sqrt{5} + d'\sqrt{10}$ and collect like terms.

3. Let $r_1, r_2, \ldots, r_n \in \mathbb{Q}$. Prove that $\mathbb{Z}[r_1, r_2, \ldots, r_n] = \mathbb{Z}\left[\frac{1}{m}\right]$ for some integer m.

4. Let p be a prime number. Consider the subset R_p defined by

$$R_p = \{r \in \mathbb{Q} \mid r = a/b \text{ with } p \nmid b\}.$$

   (a) Prove that R_p is a subring of $\mathbb{Q}$.
   (b) Prove that R_p cannot be written as $\mathbb{Z}[S]$ for any finite set $S \subseteq \mathbb{Q}$.

5. Prove that for all primes p, the ring $\mathbb{Q}[\sqrt{p}]$ is a field.

6. Consider the ring $\mathbb{Q}[\sqrt[3]{2}]$ as a subring of $\mathbb{R}$.

   (a) Prove that it consists of elements of the form $a + b\sqrt[3]{2} + c(\sqrt[3]{2})^2$ with $a, b, c \in \mathbb{Q}$.
   (b) Prove that every element of the form $a + b\sqrt[3]{2}$ with $(a, b) \neq (0, 0)$ is a unit. [Hint: Exercise 3.1.27.]

7. Consider the ring $(\mathbb{Z}/2\mathbb{Z})[x]$. Let $\alpha(x) = x^3 + x + 1$ and $\beta(x) = x^2 + 1$. Calculate: (a) $\alpha(x) + \beta(x)$; (b) $\alpha(x)\beta(x)$; (c) $\alpha(x)^2$.

8. Consider the ring $(\mathbb{Z}/6\mathbb{Z})[x]$. Let $\alpha(x) = 2x^3 + 3x + 1$ and $\beta(x) = 2x^2 + 5$. Calculate: (a) $\alpha(x) + \beta(x)$; (b) $\alpha(x)\beta(x)$; (c) $\alpha(x)^3$.

9. For all n, calculate $(2x + 3)^n$ in $\mathbb{Z}/6\mathbb{Z}[x]$. Repeat the same question but in $\mathbb{Z}/12\mathbb{Z}[x]$.

10. Suppose that R is a ring with identity $1 \neq 0$ of characteristic n. Prove that $R[x]$ is also of characteristic n.

11. Let p be a prime number. Prove that for all $a \in \mathbb{Z}/p\mathbb{Z}$, the following identity holds in the ring $(\mathbb{Z}/p\mathbb{Z})[x]$,

$$(x + a)^p = x^p + a.$$

Prove that n is prime if and only if $(x + a)^n = x^n + a$ in $\mathbb{Z}/n\mathbb{Z}[x]$.

12. Let $G = S_3$ and let $R = \mathbb{Z}/3\mathbb{Z}$. In the group ring $\mathbb{Z}/3\mathbb{Z}[S_3]$ consider $\alpha = 1 + (1\,2) + 2(1\,3) + (2\,3)$ and $\beta = 2 + 2(1\,3) + (1\,2\,3)$. Calculate: (a) $\alpha + 2\beta$; (b) $\alpha\beta$; and (c) $\beta\alpha$.

13. Let $G = Z^4$, generated by the element z. In the group ring $\mathbb{Z}[Z_4]$ consider $\alpha = 1 - z + 2z^2$ and $\beta = 3 + 2z + z^3$. Calculate (a) $3\alpha + 2\beta$; (b) $\alpha\beta$; (c) $\beta\alpha$.

14. Show that the element $(1\,2) + (1\,3) + (2\,3)$ is in the center of the group ring $\mathbb{Z}/3\mathbb{Z}[S_3]$.

15. Show that in the ring $(\mathbb{Z}/5\mathbb{Z})[Z_4]$, every element α satisfies $\alpha^5 = \alpha$. Decide with proof or counterexample whether this property is still true in $(\mathbb{Z}/5\mathbb{Z})[Z_5]$ or in $(\mathbb{Z}/5\mathbb{Z})[Z_7]$?

16. In the ring $\mathbb{Z}[Q_8]$, find $(i + j)^n$ for all positive integers n. [Hint: Note that in $\mathbb{Z}[Q_8]$, the element $(-1)k$ is not the same as $(-k)$. In Q_8, k and $(-k)$ are distinct group elements so they are not integer multiples of each other.]

17. In the subring $\{a + bi + cj + dk \in \mathbb{H} \mid a, b, c, d \in \mathbb{Z}\}$ of $\mathbb{H}$, find $(i + j)^n$ for all positive integers n. [Compare to the previous exercise.]

18. Let R be a commutative ring and G a group. Prove that $R[G]$ is commutative if and only if G is abelian.

19. Let R be a commutative ring with an identity $1 \neq 0$. Show that there is an embedding of G in $U(R[G])$. Find an example of a ring R and a group G in which G is a strict subgroup of $U(R[G])$.

20. Let R be a commutative ring and let G be a group. Prove that α is in the center of $R[G]$ if and only if $g\alpha = \alpha g$ for all $g \in G$.

21. Let R be a commutative ring. We denote by $R[[x]]$ the set of formal power series

$$\sum_{n=0}^{\infty} a_n x^n$$

with coefficients in R. In $R[[x]]$, we do not worry about issues of convergence. Addition of power series is performed term by term and for multiplication

$$\left(\sum_{n=0}^{\infty} a_n x^n\right)\left(\sum_{n=0}^{\infty} b_n x^n\right) = \sum_{n=0}^{\infty} c_n x^n \quad \text{where} \quad c_n = \sum_{k=0}^{n} a_k b_{n-k} = \sum_{i+j=n} a_i b_j.$$

(a) Prove that $R[[x]]$ with the addition and the multiplication defined above is a commutative ring.

(b) Suppose that R has an identity $1 \neq 0$. Prove that $1 - x$ is a unit.

(c) Prove that a power series of $\sum_{n=0}^{\infty} a_n x^n$ is a unit if and only if a_0 is a unit.

22. Consider the power series ring $\mathbb{Q}[[x]]$. (See Exercise 3.2.21.)

(a) Suppose that c_0 is a nonzero square element. Prove that that there exists a power series $\sum_{n=0}^{\infty} a_n x^n$ such that

$$\left(\sum_{n=0}^{\infty} a_n x^n\right)^2 = \sum_{n=0}^{\infty} c_n x^n.$$

(b) Find a recurrence relation for the terms a_n such that

$$\left(\sum_{n=0}^{\infty} a_n x^n\right)^2 = 1 + x.$$

23. Let R be a ring and let X be a set. Prove that $\mathrm{Fun}(X, R)$, the set of functions from X to R, is a ring with $+$ and $\times$ of functions defined by

$$(f_1 + f_2)(x) \stackrel{\mathrm{def}}{=} f_1(x) + f_2(x)$$
$$(f_1 \times f_2)(x) \stackrel{\mathrm{def}}{=} f_1(x) f_2(x).$$

24. Let R be a ring and let X be a set. The *support* of a function $f \in \mathrm{Fun}(X, R)$ is the subset

$$\mathrm{Supp}(f) = \{x \in X \mid f(x) \neq 0\}.$$

Consider the subset $\mathrm{Fun}_{fs}(X, R)$ of functions in $\mathrm{Fun}(X, R)$ that are of finite support, i.e., that are 0 except on a finite subset of X. Prove that $\mathrm{Fun}_{fs}(X, R)$ is a subring of $\mathrm{Fun}(X, R)$ as defined in Exercise 3.2.23.

3.3   Matrix Rings

This section introduces an important family of examples of noncommutative rings, that of matrix rings.

3.3.1   Matrix Rings

Let R be an arbitrary ring. We define $M_n(R)$ as the set of $n \times n$ matrices with entries from the ring R. As in linear algebra, we typically denote elements of $M_n(R)$ with a capital letter, say A, and we denote the entries of the matrix with the corresponding lowercase letters $a_{ij} \in R$ with the index i indicating the row and the index j indicating the column.

Let $A = (a_{ij})$ and $B = (b_{ij})$ be two elements in $M_n(R)$. The sum $A + B$ is defined as the $n \times n$ matrix whose (i, j)th entry is

$$a_{ij} + b_{ij}.$$

Inspired by the usual matrix product as defined in linear algebra, the product AB is defined as the $n \times n$ matrix whose (i, j)th entry is

$$\sum_{k=1}^{n} a_{ik} b_{kj}. \tag{3.2}$$

Since R need not be commutative, the order given in (3.2) is important.

The addition on $M_n(R)$ has the same properties as the direct sum group $(R^{n^2}, +)$. Hence, $(M_n(R), +)$ is an abelian group.

In linear algebra courses, students usually encounter matrices as representing linear transformations with respect to certain bases. The product of two matrices is defined as the matrix representing the composition of the linear transformations. Since the composition of functions is always associative (see Proposition A.2.12), it follows that the product of matrices over the real (or complex) numbers is associative. In order to prove that the multiplication in $M_n(R)$ is associative, we can only use (3.2) as the definition.

Let R be any ring. Let $A = (a_{ij})$, $B = (b_{ij})$, and $C = (c_{ij})$ be matrices in $M_n(R)$. Then the (i, j)th entry of $(AB)C$ is

$$\sum_{\ell=1}^{n} \left(\sum_{k=1}^{n} a_{ik} b_{k\ell} \right) c_{\ell j} = \sum_{\ell=1}^{n} \sum_{k=1}^{n} a_{ik} b_{k\ell} c_{\ell j} = \sum_{k=1}^{n} \sum_{\ell=1}^{n} a_{ik} b_{k\ell} c_{\ell j}$$
$$= \sum_{k=1}^{n} a_{ik} \left(\sum_{\ell=1}^{n} b_{k\ell} c_{\ell j} \right).$$

This is the (i, j) entry of $A(BC)$. Hence, $(AB)C = A(BC)$ and matrix multiplication is associative.

The (i, j)th entry of $A(B + C)$ is

$$\sum_{k=1}^{n} a_{ik}(b_{kj} + c_{kj}) = \sum_{k=1}^{n} (a_{ik} b_{kj} + a_{ik} c_{kj}) = \left(\sum_{k=1}^{n} a_{ik} b_{kj} \right) + \left(\sum_{k=1}^{n} a_{ik} c_{kj} \right).$$

This is the (i, j)th entry of $AB + AC$ so $A(B + C) = AB + AC$. This shows the matrix multiplication is left-distributive over addition. Right-distributivity is proved in a similar way. We have proven the key theorem of this section.

Proposition 3.3.1

The set $M_n(R)$ equipped with the operations of matrix addition and matrix multiplication is a ring.

In a first linear algebra course, students encounter matrices with real or complex coefficients. As one observes with $M_n(\mathbb{R})$, the multiplication in $M_n(R)$ is not commutative even if R is. With a little creativity, we can think of all manner of matrix rings. Consider for example,

$$M_2(\mathbb{Z}/2\mathbb{Z}); \qquad M_n(\mathbb{Z}[x]); \qquad M_n(\mathbb{Z}); \qquad M_n(C^0([0,1], \mathbb{R})); \quad \text{or} \quad M_n(\mathbb{H}).$$

Example 3.3.2. As an example of a matrix product in $M_n(R)$ where R is not commutative, consider the following product in $M_2(\mathbb{H})$.

$$\begin{pmatrix} i & 1+2j \\ i-k & 3k \end{pmatrix} \begin{pmatrix} i+j & k \\ 2+i & 2i-j \end{pmatrix}$$

$$= \begin{pmatrix} i(i+j) + (1+2j)(2+i) & ik + (1+2j)(2i-j) \\ (i-k)(i+j) + 3k(2+i) & (i-k)k + 3k(2i-j) \end{pmatrix}$$

$$= \begin{pmatrix} 1+i+4j-k & 2+2i-2j-4k \\ -1+i+2j+7k & 1+3i+5j \end{pmatrix}. \qquad \triangle$$

Rings of square matrices $M_n(R)$ naturally contain many subrings. We mention a few here but leave the proofs as exercises. Given any ring R, the following are subrings of $M_n(R)$:

- $M_n(S)$, where S is a subring of R;
- the set of upper triangular matrices;
- the set of lower triangular matrices;
- the set of diagonal matrices.

We point out that all the algorithms introduced in linear algebra—Gauss-Jordan elimination, various matrix factorizations, and so on—can be applied without any modification to $M_n(F)$, where F is a field. However, when R is a general ring, because nonzero elements might not be invertible and elements might not commute, some algorithms are no longer guaranteed to work and some definitions no longer make sense.

3.3.2 Matrix Inverses

Suppose R is a ring with an identity $1 \neq 0$. Denote by I_n the $n \times n$ matrix (a_{ij}) with $a_{ii} = 1$ for all $i = 1, 2, \ldots, n$ and $a_{ij} = 0$ for $i \neq j$. Just as with matrices with real entries, the matrix I_n is the multiplicative identity matrix in $M_n(R)$.

Invertible matrices form an important topic in linear algebra. In ring theory language, invertible matrices are the units in $M_n(R)$. In particular, if F is a field, then the group of units $U(M_n(F))$ is the general linear group $\mathrm{GL}_n(F)$ introduced in Example 1.2.11, properties of which are studied in group theory. This inspires the following more general definition.

Definition 3.3.3

Let R be a ring. We denote by $\mathrm{GL}_n(R)$ the group of units $U(M_n(R))$ and call it the *general linear group* of index n on the ring R.

This definition gives meaning to groups such as $\mathrm{GL}_n(\mathbb{Z}/k\mathbb{Z})$ or $\mathrm{GL}_n(\mathbb{Z})$ but also general linear groups over noncommutative rings, such as $\mathrm{GL}_n(\mathbb{H})$.

3.3.3 Determinants

If a ring R is commutative it is possible to define the determinant and recover some of the properties of determinants we encounter in linear algebra. The propositions are well-known but the proofs given in linear algebra sometimes rely on the ring of coefficients being in a field. For completeness, we give proofs for the context of arbitrary rings.

Note, throughout this discussion on determinants, we assume that the ring R is commutative.

Definition 3.3.4

If R is a commutative ring, the *determinant* is the function $\det : M_n(R) \to R$ defined on a matrix $A = (a_{ij}) \in M_n(R)$ by

$$\det A = \sum_{\sigma \in S_n} (\operatorname{sign} \sigma) a_{1\sigma(1)} a_{2\sigma(2)} \cdots a_{n\sigma(n)}. \qquad (3.3)$$

Example 3.3.5. Let $R = \mathbb{Z}/6\mathbb{Z}$ and consider the matrix

$$A = \begin{pmatrix} 2 & 3 & 5 \\ 1 & 0 & 3 \\ 4 & 2 & 1 \end{pmatrix}.$$

Then the determinant of A is

$$\det A = 2 \times 0 \times 1 + 3 \times 3 \times 4 + 5 \times 1 \times 2 - 3 \times 1 \times 1 - 5 \times 0 \times 4 - 2 \times 3 \times 2$$
$$= 0 + 0 + 4 - 3 - 0 - 0 = 1.$$

In the above calculation, the products correspond (in order) to the following permutations: 1, $(1\,2\,3)$, $(1\,3\,2)$, $(1\,2)$, $(1\,3)$, and $(2\,3)$. $\triangle$

Definition 3.3.4 is called the *Leibniz formula* for the determinant. Many courses on linear algebra first introduce the determinant via the Laplace expansion. As we will see shortly, the two definitions are equivalent. The Leibniz definition for the determinant leads immediately to the following important properties of determinants.

Proposition 3.3.6

Let $\tau \in S_n$ be a permutation. If A' is the matrix obtained from A by permuting the rows (respectively the columns) of A according to the permutation τ, then

$$\det(A') = \operatorname{sign}(\tau) \det(A).$$

Remark 3.3.7. Before we give a proof for this theorem, we need to preface it with a comment about permutations. Let $\sigma \in S_n$ be an permutation and let $x = (x_1, x_2, \ldots, x_n) \in A^n$, where A is any set. Suppose we say that σ acts on the n-tuples of A by permuting them. This means that it sends the 1st entry x_1 to the $\sigma(1)$ position in the n-tuple, the 2nd entry x_2 to the $\sigma(2)$ position, and so on. The result is the n-tuple

$$\sigma \cdot (x_1, x_2, \ldots, x_n) = (x_{\sigma^{-1}(1)}, x_{\sigma^{-1}(2)}, \ldots, x_{\sigma^{-1}(n)}).$$

This is because the entry that ends up in the ith position must be x_j such that $\sigma(j) = i$, so $j = \sigma^{-1}(i)$. $\triangle$

Proof (of Proposition 3.3.6). We prove the proposition first for permutations of the rows. By the Leibniz definition and the above remark,

$$\det A' = \sum_{\sigma \in S_n} (\operatorname{sign} \sigma) a_{\tau^{-1}(1)\sigma(1)} a_{\tau^{-1}(2)\sigma(2)} \cdots a_{\tau^{-1}(n)\sigma(n)}.$$

Since R is commutative, we can permute the terms in each product so that the row indices are in increasing order. This amounts to reordering the product according to the permutation τ. Hence

$$a_{\tau^{-1}(1)\sigma(1)} a_{\tau^{-1}(2)\sigma(2)} \cdots a_{\tau^{-1}(n)\sigma(n)} = a_{1\sigma(\tau(1))} a_{2\sigma(\tau(2))} \cdots a_{n\sigma(\tau(n))}.$$

For any fixed τ, as σ runs through all permutations, so does $\sigma\tau$. Hence,

$$\det A' = \sum_{\sigma \in S_n} (\operatorname{sign}(\sigma\tau))(\operatorname{sign}(\tau^{-1})) a_{1\sigma(\tau(1))} a_{2\sigma(\tau(2))} \cdots a_{n\sigma(\tau(n))}$$

$$= (\operatorname{sign} \tau^{-1}) \sum_{\sigma' \in S_n} (\operatorname{sign} \sigma') a_{1\sigma'(1)} a_{2\sigma'(2)} \cdots a_{n\sigma'(n)}$$

$$= (\operatorname{sign} \tau^{-1})(\det A) = (\operatorname{sign} \tau)(\det A),$$

because $\operatorname{sign} \tau^{-1} = \operatorname{sign} \tau$ for all permutations $\tau \in S_n$.

If A' is obtained from A by permuting the columns of A by τ, then

$$\det A' = \sum_{\sigma \in S_n} (\operatorname{sign} \sigma) a_{1\sigma(\tau^{-1}(1))} a_{2\sigma(\tau^{-1}(2))} \cdots a_{n\sigma(\tau^{-1}(n))}.$$

By a similar reasoning as for the rows, it again follows that $\det A' = (\operatorname{sign} \tau)(\det A)$. $\square$

Proposition 3.3.8

For all $A \in M_n(R)$, the transpose $A^\top$ satisfies $\det(A^\top) = \det(A)$.

Proof. By definition,

$$\det(A^\top) = \sum_{\sigma \in S_n} (\operatorname{sign} \sigma) a_{\sigma(1)1} a_{\sigma(2)2} \cdots a_{\sigma(n)n}.$$

Since R is commutative, by permuting the coefficients in each product so that the row index is listed in sequential order, we have

$$\det(A^\top) = \sum_{\sigma \in S_n} (\text{sign } \sigma) a_{1\sigma^{-1}(1)} a_{2\sigma^{-1}(2)} \cdots a_{n\sigma^{-1}(n)}$$

$$= \sum_{\sigma \in S_n} (\text{sign } \sigma^{-1}) a_{1\sigma^{-1}(1)} a_{2\sigma^{-1}(2)} \cdots a_{n\sigma^{-1}(n)},$$

because $\text{sign}(\sigma^{-1}) = \text{sign } \sigma$. However, the inverse function on group elements is a bijection $S_n \to S_n$ so as σ runs through all the permutations in S_n, the inverses σ^{-1} also run through all the permutations. Hence, $\det(A^\top) = \det(A)$. □

Note that neither Proposition 3.3.6 nor 3.3.8 would hold if R were not commutative. Commutativity is also required for the following theorem. Though the Leibniz formula could be used for a definition of the determinant for a matrix with coefficients in a noncommutative ring, many if not most of the usual properties we expect for determinants would not hold. This is why we typically only consider the determinant function on matrix rings over a commutative ring of coefficients.

Theorem 3.3.9

Let R be a commutative ring, n a positive integer, and let $A \in M_n(R)$. Denote by A_{ij} the submatrix of A obtained by deleting the ith row and the jth column of A. For each fixed i,

$$\det A = \sum_{j=1}^{n} (-1)^{i+j} a_{ij} \det(A_{ij}) \qquad (3.4)$$

and for each fixed j,

$$\det A = \sum_{i=1}^{n} (-1)^{i+j} a_{ij} \det(A_{ij}). \qquad (3.5)$$

Formula (3.4) is called the Laplace expansion about row i and (3.5) is called the Laplace expansion about column j.

Proof. Fix an integer i with $1 \leq i \leq n$. Break the sum in (3.3) by factoring out each matrix entry with a row index of i. Then (3.3) becomes

$$\det A = \sum_{j=1}^{n} a_{ij} \left(\sum_{\substack{\sigma \in S_n \\ \sigma(i)=j}} (\text{sign } \sigma) \overbrace{a_{1\sigma(1)} a_{2\sigma(2)} \cdots a_{n\sigma(n)}}^{a_{ij} \text{ removed}} \right).$$

In the product inside the nested summation, all terms with row index i and with column index j have been removed. Consequently, the inside summation resembles the Leibniz formula (3.3) of the submatrix A_{ij}, though we do not know if the sign of the permutation σ corresponds to that required by (3.3).

Let $\sigma \in S_n$ with $\sigma(i) = j$. Then the permutation

$$\sigma_{ij} = (j\, j+1 \ldots n)^{-1} \sigma(i\, i+1 \ldots n)$$

leaves n fixed but has the same number of inversions as σ does if we remove i from the domain $\{1, 2, \ldots, n\}$ of σ and remove j from the codomain. Since the sign of an m-cycle is $(-1)^{m-1}$, then

$$\operatorname{sign} \sigma_{ij} = (-1)^{n-i}(\operatorname{sign}\sigma)(-1)^{n-j} = (-1)^{2n-i-j}(\operatorname{sign}\sigma) = (-1)^{i+j}(\operatorname{sign}\sigma).$$

Thus, we also have $\operatorname{sign}\sigma = (-1)^{i+j} \operatorname{sign}\sigma_{ij}$ and so

$$\det A = \sum_{j=1}^{n} a_{ij}(-1)^{i+j}\left(\sum_{\substack{\sigma \in S_n \\ \sigma(i)=j}} (\operatorname{sign}\sigma_{ij})\overbrace{a_{1\sigma(1)}a_{2\sigma(2)}\cdots a_{n\sigma(n)}}^{a_{ij}\ \text{removed}}\right)$$

$$= \sum_{j=1}^{n} a_{ij}(-1)^{i+j}\det(A_{ij}).$$

This proves (3.4).

Laplace expansion about column j in (3.5) follows immediately from Proposition 3.3.8. □

Another property that follows readily from the Leibniz formula is that the determinant is linear by row and, by virtue of Proposition 3.3.8, linear by column as well. (See Exercise 3.3.11.) This property inspires us to consider other functions $F : M_n(R) \to R$ that are linear in every row and every column.

Proposition 3.3.10

A function $F : M_n(R) \to R$ is linear in every row and in every column if and only if there exists a function $f : S_n \to R$ such that

$$F(A) = \sum_{\sigma \in S_n} f(\sigma)a_{1\sigma(1)}a_{2\sigma(2)}\cdots a_{n\sigma(n)}. \tag{3.6}$$

Proof. Suppose first that F is linear in each row. By properties of linear transformations, if F is linear in row 1, then

$$F(A) = \sum_{j_1=1}^{n} c_{j_1}a_{1j_1}$$

for some elements c_{j_1} whose value may depend on the other rows. Furthermore, by picking appropriate values for the first row of A, we see that the functions c_{j_1} must be linear in all the other rows. Since each c_{j_1} is linear in row 2, then

$$F(A) = \sum_{j_1=1}^{n} \left(\sum_{j_2=1}^{n} d_{j_1,j_2} a_{2j_2} \right) a_{1j_1} = \sum_{1 \leq j_1 \leq j_2 \leq n} d_{j_1,j_2} a_{1j_1} a_{2j_2},$$

where d_{j_1,j_2} are ring elements that depend on the elements in rows 3 through n. Continuing until row n, we deduce that if F is linear in every row, then there exists a function $f : \{1, 2, \ldots, n\}^n \to R$ such that

$$F(A) = \sum_{1 \leq j_1 \leq j_2 \leq \cdots \leq j_n \leq n} f(j_1, j_2, \ldots, j_n) a_{1j_1} a_{2j_2} \cdots a_{nj_n}.$$

Now if F is also linear in each column, then $f(j_1, j_2, \ldots, j_n)$ must be 0 any time two of the indices $j_1, j_2, \ldots, j_n$ are equal, because otherwise, $F(A)$ would contain a quadratic term in one of the entries of the matrix. This proves that there exists a function $f : S_n \to R$ that satisfies (3.6).

Conversely, regardless of the function $f : S_n \to R$, the function F defined as in (3.6) is linear in every row and column. $\square$

Proposition 3.3.11

Suppose that $F : M_n(R) \to R$ is a function that is linear in every row, is linear in every column, and satisfies the alternating property that if A' is obtained from the matrix A by permuting the rows (or columns) according to the permutation τ, then $F(A') = (\text{sign}\,\tau)F(A)$. Then there exists a constant $c \in R$ such that $F(A) = c \det(A)$.

Proof. According to Proposition 3.3.10, there exists a function $f : S_n \to R$ such that (3.6) holds. Call $c = f(1)$. According to (3.6), $F(I) = f(1) = c$. Consider the permutation matrix E_σ, for $\sigma \in S_n$, whose entries are

$$e_{ij} = \begin{cases} 1 & \text{if } j = \sigma(i) \\ 0 & \text{otherwise.} \end{cases}$$

Then by the alternating property $f(\sigma) = F(E_\sigma) = (\text{sign}\,\sigma)f(I) = c(\text{sign}\,\sigma)$. Thus,

$$F(A) = \sum_{\sigma \in S_n} c(\text{sign}\,\sigma) a_{1\sigma(1)} a_{2\sigma(2)} \cdots a_{n\sigma(n)} = c \det A.$$

$\square$

The property described in Proposition 3.3.10 characterizes determinants. Indeed, the determinant is the unique function $M_n(R) \to R$ that is linear in the rows, linear in the columns, satisfies the alternating condition, and is 1 on the identity matrix. This characterization of the determinant leads to the following important theorem about determinants.

Proposition 3.3.12

Let R be a commutative ring. Then for any matrices $A, B \in M_n(R)$,

$$\det(AB) = (\det A)(\det B).$$

Proof. Given the matrix B, consider the function $F : M_n(R) \to R$ defined by $F(A) = \det(AB)$. For a fixed i, suppose that we can write the ith row of the matrix A as $a_{ij} = ra'_{ij} + sa''_{ij}$ with $1 \leq j \leq n$. We denote by A' the matrix of A but with the ith row replaced with the row $(a'_{ij})^n_{j=1}$ and denote by A'' the matrix of A but with the ith row replaced with the row $(a''_{ij})^n_{j=1}$. Denote by $C = AB$, $C' = A'B$, and $C'' = A''B$. Then the ith row of C can be written as

$$c_{ij} = \sum_{k=1}^{n} a_{ik}b_{kj} = \sum_{k=1}^{n}(ra'_{ik}b_{kj} + sa''_{ik}b_{kj})$$

$$= r\left(\sum_{k=1}^{n} a'_{ik}b_{kj}\right) + s\left(\sum_{k=1}^{n} a''_{ik}b_{kj}\right)$$

$$= rc'_{ij} + sc''_{ij}.$$

Since the determinant is linear in each row, then $\det C = r(\det C') + s(\det C'')$. Thus, $F(A) = rF(A') + sF(A'')$. Hence, F is linear in each row. By a similar reasoning, F is linear in each column.

We leave it as an exercise (Exercise 3.3.19) to prove that a function $F : M_n(R) \to R$ that is linear in each row and linear in each column satisfies the alternating property (described in Proposition 3.3.11) if and only if $F(A) = 0$ for every matrix A that has a repeated row or a repeated column. By the definition of matrix multiplication, if A has two repeated rows, then AB also has two repeated rows so and hence $F(A) = \det(AB) = 0$. Hence, F satisfies the alternating property.

Consequently, by Proposition 3.3.11, $F(A) = c\det(A)$ and $F(I) = c = \det(B)$. Thus, $\det(AB) = \det(A)\det(B)$. $\qquad \square$

Finally, if the ring R has an identity $1 \neq 0$, the determinant gives a characterization of invertible matrices.

Proposition 3.3.13

Let R be a commutative ring with an identity $1 \neq 0$. A matrix $A \in M_n(R)$ is a unit if and only if $\det A \in U(R)$. Furthermore, the (i, j)th entry of the inverse matrix A^{-1} is

$$(\det A)^{-1}(-1)^{i+j}\det(A_{ji}).$$

Proof. (Left as an exercise for the reader. See Exercise 3.3.17.) □

Proposition 3.3.12 generalizes the definition in Example 1.9.13 that discussed general linear groups over fields. Proposition 3.3.12 is equivalent to saying that the determinant function $\det : \mathrm{GL}_n(R) \to U(R)$ is a group homomorphism. We define the kernel of the homomorphism as the *special linear group*

$$\mathrm{SL}_n(R) = \{A \in M_n(R) \mid \det A = 1\}.$$

3.3.4 Useful CAS Commands

Maple uses the packages `LinearAlgebra` for commands associated with any matrix space where the ring of coefficients is R, a polynomial ring over R, or a ring of functions into R, where R is a subring of $\mathbb{C}$. The package `LinearAlgebra[Modular]` handles methods of linear algebra associated with matrices defined over $\mathbb{Z}/n\mathbb{Z}$, for some integer $n \geq 2$. We encourage the reader to consult the help files for these packages.

With SAGE, since the user creates different object classes by defining the ring, it is easy to define matrix "spaces" over any ring. For example, in the code below, we multiply two matrices in $M_2(\mathbb{H})$.

_____ Sage _____

```
sage: H.<ii,jj,kk> = QuaternionAlgebra(RR,-1,-1)
sage: M=MatrixSpace(H,2,2)
sage: A=M([-4+5*ii,-7*jj,-3+6*kk,7])
sage: B=M([-8+9*jj,3-4*kk,2*ii+jj,-kk])
[ 39 - 40*ii - 36*jj + 59*kk -12 + 22*ii + 20*jj + 16*kk]
[ 24 - 40*ii - 20*jj - 48*kk                15 + 23*kk]
```

In SAGE, it would be just as easy to define matrices in $M_3(\mathbb{F}_{127}[x,y])$ or sets of matrices with even more complicated rings of coefficients.

EXERCISES FOR SECTION 3.3

1. In $M_2(\mathbb{Z}/4\mathbb{Z})$, consider the matrices

$$A = \begin{pmatrix} 1 & 2 \\ 2 & 3 \end{pmatrix}, \quad B = \begin{pmatrix} 0 & 1 \\ 1 & 1 \end{pmatrix}, \quad C = \begin{pmatrix} 2 & 2 \\ 3 & 1 \end{pmatrix}.$$

Perform the following calculations, if they are defined: (a) $A + BC$; (b) ABC; (c) B^n for all $n \in \mathbb{N}$; (d) C^{-1}; (e) $A^{-1}B$.

2. Repeat Exercise 3.3.1 but with the ring of coefficient in $\mathbb{Z}/5\mathbb{Z}$.

3. Find the inverse of the matrix in Example 3.3.5.

4. Consider the following matrices in $M_2(\mathbb{H})$:

$$A = \begin{pmatrix} 1 & i \\ j & k \end{pmatrix}, \quad B = \begin{pmatrix} i+j & k \\ i & j \end{pmatrix}.$$

Calculate: (a) $A + B$; (b) AB; (c) B^3.

5. Let S be a subring of R. Prove that $M_n(S)$ is a subring of $M_n(R)$.

6. Prove that the subset of upper triangular matrices in $M_n(R)$ is a subring.

7. Prove that the subset of diagonal matrices in $M_n(R)$ is a subring.

8. Consider the ring $M_n(\mathbb{Z})$. Consider the subsets S of upper triangular matrices $A = (a_{ij})$ in which 2^{j-i} divides a_{ij} for all indices (i, j) with $j \geq i$. Prove that S is a subring of $M_n(\mathbb{Z})$.

9. (Multivariable calculus required) Recall the Hessian matrix of a real-valued function $f(x, y)$ defined on an open set $\mathcal{D} \subseteq \mathbb{R}^2$. In what algebraic structure does the Hessian matrix of f exist?

10. Suppose that R is a ring with identity $1 \neq 0$. Prove that the center $Z(M_n(R)) = \{aI_n \mid a \in Z(R)\}$ and I_n is the $n \times n$ identity matrix. Give an example where this result fails if R does not have an identity.

11. Let R be a commutative ring. Prove that the determinant is "linear by row." In other words, prove that

$$\det \begin{pmatrix} a_{11} & a_{12} & \cdots & a_{1n} \\ a_{21} & a_{22} & \cdots & a_{2n} \\ \vdots & \vdots & \ddots & \vdots \\ ra_{i1} + sa'_{i1} & ra_{i2} + sa'_{i2} & \cdots & ra_{in} + sa'_{in} \\ \vdots & \vdots & \ddots & \vdots \\ a_{n1} & a_{n2} & \cdots & a_{nn} \end{pmatrix}$$

$$= r \det \begin{pmatrix} a_{11} & a_{12} & \cdots & a_{1n} \\ a_{21} & a_{22} & \cdots & a_{2n} \\ \vdots & \vdots & \ddots & \vdots \\ a_{i1} & a_{i2} & \cdots & a_{in} \\ \vdots & \vdots & \ddots & \vdots \\ a_{n1} & a_{n2} & \cdots & a_{nn} \end{pmatrix} + s \det \begin{pmatrix} a_{11} & a_{12} & \cdots & a_{1n} \\ a_{21} & a_{22} & \cdots & a_{2n} \\ \vdots & \vdots & \ddots & \vdots \\ a'_{i1} & a'_{i2} & \cdots & a'_{in} \\ \vdots & \vdots & \ddots & \vdots \\ a_{n1} & a_{n2} & \cdots & a_{nn} \end{pmatrix}.$$

12. Let R be a commutative ring with an identity $1 \neq 0$. Let $\sigma \in S_n$ and defined the matrix E_σ as the $n \times n$ matrix with entries (e_{ij}) such that

$$e_{ij} = \begin{cases} 1 & \text{if } j = \sigma(i) \\ 0 & \text{otherwise.} \end{cases}$$

Prove that $\det E_\sigma = \text{sign } \sigma$.

13. In the ring $M_2(\mathbb{Q}[x])$, prove that the following matrix is invertible and find the inverse:

$$\begin{pmatrix} x+1 & x-2 \\ x+6 & x+3 \end{pmatrix}.$$

14. Let $A, B \in M_n(R)$, where R is a ring.

    (a) Prove that $(AB)^\top = B^\top A^\top$ if R is commutative.

    (b) Prove that this identity does not necessarily hold if R is not commutative.

15. Recall that the trace $\mathrm{Tr} : M_n(R) \to R$ of a matrix is the sum of its diagonal elements. Let R be a commutative ring. Prove that $\mathrm{Tr}(AB) = \mathrm{Tr}(BA)$ for all matrices $A, B \in M_n(R)$.

16. Let R be any ring. Suppose that $A \in M_n(R)$ is strictly upper triangular (upper triangular but with zeros on the diagonal). Prove that A is nilpotent. Prove that if $B \in \mathrm{GL}_n(R)$, then BAB^{-1} is also nilpotent.

17. Prove Proposition 3.3.13.

18. Find the number of units in $M_2(\mathbb{Z}/4\mathbb{Z})$.

19. Prove that a function $F : M_n(R) \to R$ that is linear in each row and linear in each column satisfies the alternating property (described in Proposition 3.3.11) if and only if $F(A) = 0$ for every matrix A that has a repeated row or a repeated column.

20. Let $F : M_n(R) \to R$ be a function that is linear in each row and linear in each column of an input matrix A. Prove that $F(A^\top) = F(A)$ for all $A \in M_n(R)$ if and only if the function f in Proposition 3.3.10 satisfies $f(\sigma^{-1}) = f(\sigma)$ for all $\sigma \in S_n$.

3.4 Ring Homomorphisms

In our study of groups, we emphasized that we do not typically study arbitrary functions between objects with a given algebraic structure but only functions that preserve the structure. In the context of rings, such functions are called ring homomorphisms.

3.4.1 Ring Homomorphisms

Definition 3.4.1

Let R and S be two rings. A *ring homomorphism* is a function $\varphi : R \to S$ satisfying

- $\forall a, b \in R$, $\varphi(a + b) = \varphi(a) + \varphi(b)$;
- $\forall a, b \in R$, $\varphi(ab) = \varphi(a)\varphi(b)$.

A bijective ring homomorphism is called a ring *isomorphism*. If there exists a ring isomorphism between rings R and S, we write $R \cong S$.

Example 3.4.2. The function $\varphi : \mathbb{Z} \to \mathbb{Z}/n\mathbb{Z}$ defined simply by $\varphi(a) = \bar{a}$ is a ring homomorphism. This statement is simply a rephrasing of the definition of the usual operations in modular arithmetic, namely that for all $a, b \in \mathbb{Z}$,

$$\overline{a + b} = \bar{a} + \bar{b} \qquad \text{and} \qquad \overline{ab} = \bar{a}\,\bar{b}. \qquad \triangle$$

Example 3.4.3. Let S be a subring of R. The inclusion function $i : S \to R$, simply defined by $i(a) = a$ is a ring homomorphism. △

Example 3.4.4. Let R be a commutative ring. Fix an element $r \in R$. We define the evaluation map $\mathrm{ev}_r : R[x] \to R$ by

$$\mathrm{ev}_r(a_n x^n + \cdots + a_1 x + a_0) = a_n r^n + \cdots + a_1 r + a_0.$$

Since, $\mathrm{ev}_r(a(x) + b(x)) = \mathrm{ev}_r(a(x)) + \mathrm{ev}_r(b(x))$ and $\mathrm{ev}_r(a(x)b(x)) = \mathrm{ev}_r(a(x)) \cdot \mathrm{ev}_r(b(x))$, the evaluation map is a ring homomorphism. △

Example 3.4.5. Let R be any ring. Let $U_2(R)$ be the subring of $M_2(R)$ of upper triangular matrices. Consider the function $\varphi : U_2(R) \to R \oplus R$ defined by

$$\varphi \begin{pmatrix} a & b \\ 0 & d \end{pmatrix} = (a, d).$$

Then

$$\varphi \left(\begin{pmatrix} a_1 & b_1 \\ 0 & d_1 \end{pmatrix} + \begin{pmatrix} a_2 & b_2 \\ 0 & d_2 \end{pmatrix} \right) = \varphi \begin{pmatrix} a_1 + a_2 & b_1 + b_2 \\ 0 & d_1 + d_2 \end{pmatrix} = (a_1 + a_2, d_1 + d_2)$$

$$= \varphi \begin{pmatrix} a_1 & b_1 \\ 0 & d_1 \end{pmatrix} + \varphi \begin{pmatrix} a_2 & b_2 \\ 0 & d_2 \end{pmatrix}.$$

Furthermore,

$$\varphi \left(\begin{pmatrix} a_1 & b_1 \\ 0 & d_1 \end{pmatrix} \begin{pmatrix} a_2 & b_2 \\ 0 & d_2 \end{pmatrix} \right) = \varphi \begin{pmatrix} a_1 a_2 & a_1 b_2 + b_1 d_2 \\ 0 & d_1 d_2 \end{pmatrix} = (a_1 a_2, d_1 d_2)$$

$$= \varphi \begin{pmatrix} a_1 & b_1 \\ 0 & d_1 \end{pmatrix} \varphi \begin{pmatrix} a_2 & b_2 \\ 0 & d_2 \end{pmatrix}.$$

The above two identities show that φ is a ring homomorphism. △

Example 3.4.6. Consider the operator $D : C^1([a, b], \mathbb{R}) \to C^0([a, b], \mathbb{R})$ defined by taking the derivative $D(f) = f'$. This is not a ring homomorphism. It is true that $D(f + g) = D(f) + D(g)$ but the product rule is $D(fg) = D(f)g + f D(g)$, which in general is not equal to $D(f)D(g)$. △

Example 3.4.7 (Reduction Homomorphism). Let n be an integer $n \geq 2$. Define the function $\pi : \mathbb{Z}[x] \to (\mathbb{Z}/n\mathbb{Z})[x]$ by

$$\pi(a_m x^m + \cdots + a_1 x + a_0) = \bar{a}_m x^m + \cdots + \bar{a}_1 x + \bar{a}_0.$$

It is not hard to show that π is a homomorphism. It is called the *reduction homomorphism*.

The reduction homomorphism leads to an important way to tell if an integer polynomial has no roots. The key result is that for all $a \in \mathbb{Z}$ the following diagram of ring homomorphisms is commutative:

$$\begin{array}{ccc} \mathbb{Z}[x] & \xrightarrow{\text{ev}_a} & \mathbb{Z} \\ \Big\downarrow{\pi} & & \Big\downarrow{\pi} \\ (\mathbb{Z}/n\mathbb{Z})[x] & \xrightarrow{\text{ev}_{\overline{a}}} & \mathbb{Z}/n\mathbb{Z} \end{array}$$

which means that $\pi \circ \text{ev}_a = \text{ev}_{\overline{a}} \circ \pi$. Consequently, if $q(x) \in \mathbb{Z}[x]$ is a polynomial such that $\pi(q(x))$ has no roots as a polynomial in $(\mathbb{Z}/n\mathbb{Z})[x]$, then $q(x)$ cannot have a root in $\mathbb{Z}[x]$. But to check that $\pi(q(x))$ has no roots in $(\mathbb{Z}/n\mathbb{Z})[x]$ we simply need to test all the congruence classes in $\mathbb{Z}/n\mathbb{Z}$. $\triangle$

Example 3.4.8. Consider the subset

$$R = \left\{ \begin{pmatrix} a & -b \\ b & a \end{pmatrix} \in M_2(\mathbb{R}) \,\middle|\, a, b \in \mathbb{R} \right\}$$

in $M_2(\mathbb{R})$. Consider the function $\varphi : \mathbb{C} \to M_2(\mathbb{R})$ defined by

$$\varphi(a + bi) = \begin{pmatrix} a & -b \\ b & a \end{pmatrix}.$$

We easily see that

$$\varphi((a + bi) + (c + di)) = \varphi((a + c) + (b + d)i) = \begin{pmatrix} a+c & -(b+d) \\ b+d & a+c \end{pmatrix}$$

$$= \begin{pmatrix} a & -b \\ b & a \end{pmatrix} + \begin{pmatrix} c & -d \\ d & c \end{pmatrix}$$

$$= \varphi(a + bi) + \varphi(c + di).$$

Now the multiplication in $\mathbb{C}$ is $(a + bi)(c + di) = (ac - bd) + (ad + bc)i$ while

$$\begin{pmatrix} a & -b \\ b & a \end{pmatrix} \begin{pmatrix} c & -d \\ d & c \end{pmatrix} = \begin{pmatrix} ac - bd & -ad - bc \\ ad + bc & ac - bd \end{pmatrix}.$$

These show that

$$\varphi((a + bi)(c + di)) = \begin{pmatrix} a & -b \\ b & a \end{pmatrix} \begin{pmatrix} c & -d \\ d & c \end{pmatrix} = \varphi(a + bi)\varphi(c + di).$$

We have proven that φ is a homomorphism. Clearly, φ is surjective onto $\text{Im}\,\varphi = R$, which is a subring of $M_2(\mathbb{R})$. However, $\text{Ker}\,\varphi = \{0\}$ so φ is also injective. Hence, φ is an isomorphism between $\mathbb{C}$ and R. $\triangle$

Since a ring homomorphism $\varphi : R \to S$ is a group homomorphism between $(R, +)$ and $(S, +)$, then by Proposition 1.9.9,

- $\varphi(0_R) = 0_S$

- $\varphi(-a) = -\varphi(a)$ for all $a \in R$;
- $\varphi(n \cdot a) = n \cdot \varphi(a)$ for all $a \in R$ and $n \in \mathbb{Z}$.

It is also true that for all $a \in R$ and all $n \in \mathbb{N}^*$, $\varphi(a^n) = \varphi(a)^n$. However, it is not necessarily true that $\varphi(1_R) = 1_S$, even if R and S both have identities. For example, if R and S are any rings, the function $\varphi : R \to S$ such that $\varphi(r) = 0$ is a homomorphism. As a nontrivial example, consider the function $f : \mathbb{Z} \to \mathbb{Z}/6\mathbb{Z}$ defined by $f(a) = \bar{3}\bar{a}$. It is not hard to check that f is a ring homomorphism but that the image is $\mathrm{Im}\, f = \{\bar{0}, \bar{3}\}$.

Following terminology introduced for groups (Definition 1.9.26), we call a homomorphism of a ring R into itself an *endomorphism* on R and an isomorphism of a ring onto itself an *automorphism* on R.

3.4.2   Kernels and Images

Definition 3.4.9

Let $\varphi : R \to S$ be a ring homomorphism. The *kernel* of φ, denoted $\mathrm{Ker}\, \varphi$, is the set of elements of R that get mapped to 0, namely

$$\mathrm{Ker}\, \varphi = \{r \in R \,|\, \varphi(r) = 0\}.$$

The *image* of φ, denoted $\mathrm{Im}\, \varphi$, is the range of the function, namely

$$\mathrm{Im}\, \varphi = \{s \in S \,|\, \exists r \in R, \varphi(r) = s\}.$$

Proposition 3.4.10

Let $\varphi : R \to S$ be a ring homomorphism.

(1) The image of φ is a subring of S.

(2) The kernel of φ is a subring of R. In fact, a stronger condition holds: if $a \in \mathrm{Ker}\, \varphi$, then for all $r \in R$, we also have $ra \in \mathrm{Ker}\, \varphi$ and $ar \in \mathrm{Ker}\, \varphi$.

Proof. For part (1), let $x, y \in \mathrm{Im}\, \varphi$. Then there exist $a, b \in R$ such that $x = \varphi(a)$ and $y = \varphi(b)$. Hence, $x - y = \varphi(a) - \varphi(b) = \varphi(a - b) \in \mathrm{Im}\, \varphi$. Furthermore, $xy = \varphi(a)\varphi(b) = \varphi(ab) \in \mathrm{Im}\, \varphi$. Since $\mathrm{Im}\, \varphi$ is closed under subtraction and multiplication, it is a subring of S.

For part (2), let $x, y \in \mathrm{Ker}\, \varphi$ and let $r \in R$. Then $\varphi(x) = \varphi(y) = 0$. Consequently, $\varphi(x - y) = \varphi(x) - \varphi(y) = 0 - 0 = 0$ so $x - y \in \mathrm{Ker}\, \varphi$ and $\mathrm{Ker}\, \varphi$ is closed under subtraction. Furthermore, $\varphi(rx) = \varphi(r)\varphi(x) = \varphi(r)0 = 0$ and also $\varphi(xr) = \varphi(x)\varphi(r) = 0\varphi(r) = 0$ so $\mathrm{Ker}\, \varphi$ is closed under multiplication within $\mathrm{Ker}\, \varphi$ but also closed under multiplication by any element in R. $\square$

Example 3.4.11. Consider Example 3.4.4. Let $s \in R$ be any element. Then $\mathrm{ev}_r(x + (s - r)) = s$ so ev_r is surjective and hence the image of ev_r is all of

R. The kernel $\mathrm{Ker}\,\mathrm{ev}_r$, however, is precisely the polynomials that evaluate to 0 at r, i.e., that have r as a root. $\triangle$

3.4.3 Convolution Rings (Optional)

We conclude this section with a discussion about a general class of rings that encompasses many examples we have discussed above.

The reader may have noticed that the multiplications in polynomial rings and in group rings look similar. However, the powers of x, namely $\{1, x, x^2, \ldots\}$, equipped with multiplication do not form a group. Therefore, a polynomial ring $R[x]$ is not in general a group ring. Nonetheless, it is possible to describe both constructions in a consistent way.

Let $(R, +, \times)$ be a ring and let $(S, \cdot)$ be a semigroup (a set equipped with an associative binary operation). Suppose that a subring $\mathcal{F}$ of $\mathrm{Fun}(S, R)$ is such that for any pair of functions $f_1, f_2 \in \mathrm{Fun}(S, R)$ the summation

$$(f_1 * f_2)(s) = \sum_{s_1 \cdot s_2 = s} f_1(s_1) f_2(s_2), \qquad (3.7)$$

involves only a finite number of terms for all $s \in S$. We call this condition the *convolution condition* and call the operation on functions the *convolution product* between f_1 and f_2.

Proposition 3.4.12

Let R be a ring, let $(S, \cdot)$ be a semigroup, and let $\mathcal{F}$ be a subring of $\mathrm{Fun}(S, R)$ that satisfies the convolution condition. Then $(\mathcal{F}, +, *)$ is a ring.

Proof. Since $(\mathcal{F}, +)$ is an abelian group by virtue of $(\mathcal{F}, +, \times)$ being a ring, we only need to check associativity of $*$ and the distributivity of $*$ over $+$.

Let $\alpha, \beta, \gamma \in \mathcal{F}$. Then for all $s \in S$, we have

$$(\alpha * (\beta * \gamma))(s)$$

$$= \sum_{s_1 \cdot s_2 = s} \alpha(s_1) \left(\sum_{t_1 \cdot t_2 = s_2} \beta(t_1)\gamma(t_2) \right) = \sum_{s_1 \cdot (t_1 \cdot t_2) = s} \alpha(s_1)\beta(t_1)\gamma(t_2)$$

$$= \sum_{(s_1 \cdot t_1) \cdot t_2 = s} \alpha(s_1)\beta(t_1)\gamma(t_2) = \sum_{q \cdot t_2 = s} \left(\sum_{s_1 \cdot t_1 = q} \alpha(s_1)\beta(t_1) \right) \gamma(t_2)$$

$$= ((\alpha * \beta) * \gamma)(s).$$

Hence, $\alpha * (\beta * \gamma) = (\alpha * \beta) * \gamma$.

Also for all $s \in S$,

$$
\begin{aligned}
(\alpha * (\beta + \gamma))(s) &= \sum_{s_1 \cdot s_2 = s} \alpha(s_1)(\beta(s_2) + \gamma(s_2)) \\
&= \sum_{s_1 \cdot s_2 = s} (\alpha(s_1)\beta(s_2) + \alpha(s_1)\gamma(s_2)) \\
&= \sum_{s_1 \cdot s_2 = s} \alpha(s_1)\beta(s_2) + \sum_{s_1 \cdot s_2 = s} \alpha(s_1)\gamma(s_2) \\
&= (\alpha * \beta)(s) + (\alpha * \gamma)(s).
\end{aligned}
$$

Hence, $\alpha * (\beta + \gamma) = \alpha * \beta + \alpha * \gamma$. This proves left-distributivity of $*$ over $+$. The proof for right-distributivity is similar and follows from right-distributivity of $\times$ over $+$ in R. $\qquad\square$

Definition 3.4.13

We call the ring $(\mathcal{F}, +, *)$ a *convolution ring* from $(S, \cdot)$ to R.

There are a few common situations in which a subring of $\mathrm{Fun}(S, R)$ satisfies the convolution condition. If the semigroup S is finite, then the condition is satisfied trivially. The semigroup $(\mathbb{N}, +)$ is such for all $n \in \mathbb{N}$ there is a finite number of pairs $(a, b) \in \mathbb{N}$ such that $a + b = n$. Hence, for any ring R, any subring of $\mathrm{Fun}(\mathbb{N}, R)$ satisfies the convolution condition.

As a third general example, we consider functions of finite support. For any function $f \in \mathrm{Fun}(S, R)$, we define the *support* as

$$
\mathrm{Supp}(f) = \{s \in S \mid f(s) \neq 0\}.
$$

In Exercise 3.2.24 we proved that the set $\mathrm{Fun}_{fs}(S, R)$ of functions from S to R of finite support, i.e., functions $f \in \mathrm{Fun}(S, R)$ such that $\mathrm{Supp}(f)$ is a finite set, is a subring of $\mathrm{Fun}(S, R)$. Furthermore, $\mathrm{Fun}_{fs}(S, R)$ satisfies the convolution condition because all the terms in the summation (3.7) are 0 except possibly for pairs $(s_1, s_2) \in \mathrm{Supp}(f_1) \times \mathrm{Supp}(f_2)$, which is a finite set.

We now give some specific examples of convolution rings that we have already encountered.

Example 3.4.14 (Polynomial Rings). Let R be a commutative ring and consider the semigroup $(\mathbb{N}, +)$. Note that $(\mathbb{N}, +)$ is isomorphic to $(\{1, x, x^2, \ldots\}, \times)$ as a monoid. Coefficients of polynomials are 0 except for a finite number of terms, so we consider the function $\psi : R[x] \to \mathrm{Fun}_{fs}(\mathbb{N}, R)$ where $\psi(a(x))$ is the function that to each integer n associates the coefficient a_n. The function ψ is obviously injective. Furthermore, given any $f \in \mathrm{Fun}_{fs}(S, R)$, if $n = \max\{i \mid f(i) \neq 0\}$, then

$$
\psi(f(n)x^n + f(n-1)x^{n-1} + \cdots + f(1)x + f(0)) = f.
$$

Hence, ψ is a bijection.

Suppose that $\psi(a(x)) = f$ so $f(i) = a_i$ and that $\psi(b(x)) = g$ so $g(i) = b_i$. Then

$$\psi(a(x) + b(x)) = (i \mapsto a_i + b_i) = f + g = \psi(a(x)) + \psi(b(x)).$$

According to (3.1),

$$\psi(a(x)b(x)) = \left(k \mapsto \sum_{i+j=k} a_i b_j \right) = \left(k \mapsto \sum_{i+j=k} f(i)g(j) \right)$$
$$= f * g = \psi(a(x)) * \psi(b(x)).$$

Hence, $R[x]$ is ring isomorphic to the convolution ring $(\mathrm{Fun}_{fs}(\mathbb{N}, R), +, *)$. $\triangle$

Example 3.4.15 (Group Rings). Let G be a finite group and let R be a commutative ring. If

$$\alpha = a_1 g_1 + a_2 g_2 + \cdots + a_n g_n$$
$$\beta = b_1 g_1 + b_2 g_2 + \cdots + b_n g_n$$

are two elements in $R[G]$ with $\alpha\beta = c_1 g_1 + c_2 g_2 + \cdots + c_n g_n$, then

$$c_k = \sum_{(i,j):\, g_i g_j = g_k} a_i b_j.$$

This is precisely the convolution product defined in (3.7). Consequently, as rings, $R[G]$ is isomorphic to the convolution ring $(\mathrm{Fun}(G, R), +, *)$. $\triangle$

The above example shows that $(\mathrm{Fun}(G, R), +, *)$ gives precisely the group ring that we defined in Section 3.2.3 for a commutative ring R and a finite group G. However, this construction can be generalized to infinite groups and noncommutative rings. For any group G and any ring R, we call the group ring $R[G]$ the convolution ring $(\mathrm{Fun}_{fs}(G, R), +, *)$.

EXERCISES FOR SECTION 3.4

1. Prove that there are only two ring homomorphisms $f : \mathbb{Z} \to \mathbb{Z}$.

2. Find all positive integers n and k such that the function $\mathbb{Z} \to \mathbb{Z}/n\mathbb{Z}$ defined by $f(a) = \overline{ka}$ is a homomorphism.

3. Prove that the "projection" function $\pi_1 : R \oplus R \to R$ given by $\pi_1(a, b) = a$ is a ring homomorphism and determine its kernel.

4. Prove that the function $f : \mathbb{Z} \oplus \mathbb{Z} \to \mathbb{Z}$ given by $f(m, n) = m - n$ is not a homomorphism.

5. Let D be an integer not divisible by a square integer (i.e., *square-free*). Prove that the function $f : \mathbb{Z}[\sqrt{D}] \to M_2(\mathbb{Z})$ defined by

$$f(a + b\sqrt{D}) = \begin{pmatrix} a & b \\ Db & a \end{pmatrix}$$

is an injective ring homomorphism. Deduce that $\mathbb{Z}[\sqrt{D}]$ is ring-isomorphic to $\mathrm{Im}\, f$.

6. Let R be a commutative ring and let $a \in R$ be a fixed element. Consider the function $f_a : R[x, y] \to R[x]$ defined by $f_a(p(x, y)) = p(x, a)$. Prove that f_a is a ring homomorphism.

7. Let R and S be commutative rings and let $\varphi : R \to S$ be a homomorphism. Prove that the function $\psi : R[x] \to S[x]$ defined by

$$\psi(a_n x^n + \cdots + a_1 x + a_0) = \varphi(a_n) x^n + \cdots + \varphi(a_1) x + \varphi(a_0)$$

is a homomorphism.

8. Let R be a commutative ring of prime characteristic p.

   (a) Prove that p divides $\binom{p}{k}$ for all integers $1 \le k \le p - 1$.

   (b) Prove that the function $f : R \to R$ given by $f(x) = x^p$ is a homomorphism. In other words, prove that

   $$(a + b)^p = a^p + b^p \qquad \text{and} \qquad (ab)^p = a^p b^p.$$

   [This function is called the *Frobenius homomorphism*.]

9. Given any set S, the triple $(\mathcal{P}(S), \triangle, \cap)$ has the structure of a ring. Let S' be any subset of S. Show that $\varphi(A) = A \cap S'$ is a ring homomorphism from $\mathcal{P}(S)$ to $\mathcal{P}(S')$.

10. Prove that $(5\mathbb{Z}, +)$ is group-isomorphic to $(7\mathbb{Z}, +)$ but that $(5\mathbb{Z}, +, \times)$ is not ring-isomorphic to $(7\mathbb{Z}, +, \times)$.

11. Show that $(\mathbb{Z} \oplus \mathbb{Z})[x]$ is not isomorphic to $\mathbb{Z}[x] \oplus \mathbb{Z}[x]$.

12. Consider the ring $(R, +, \times)$, where $R = \mathbb{Z}/2\mathbb{Z} \times \mathbb{Z}/2\mathbb{Z}$ as a set, where $+$ is the component-wise addition but the multiplication is done according to the following table:

$\times$	$(0,0)$	$(1,0)$	$(0,1)$	$(1,1)$
$(0,0)$	$(0,0)$	$(0,0)$	$(0,0)$	$(0,0)$
$(1,0)$	$(0,0)$	$(1,1)$	$(1,0)$	$(0,1)$
$(0,1)$	$(0,0)$	$(1,0)$	$(0,1)$	$(1,1)$
$(1,1)$	$(0,0)$	$(0,1)$	$(1,1)$	$(1,0)$

    Prove that $(R, +, \times)$ is a ring. Also prove that $(R, +, \times)$ is not isomorphic to $\mathbb{Z}/2\mathbb{Z} \oplus \mathbb{Z}/2\mathbb{Z}$.

13. Prove that $M_m(M_n(R))$ is isomorphic to $M_{mn}(R)$.

14. Show that the function $\varphi : \mathbb{H} \to M_2(\mathbb{C})$ defined by

    $$\varphi(a + bi + cj + dk) = \begin{pmatrix} a + bi & -c - di \\ c - di & a - bi \end{pmatrix}$$

    is an injective homomorphism. Conclude that $\mathbb{H}$ is ring-isomorphic to $\operatorname{Im} \varphi$.

15. Use the reduction homomorphism with $n = 5$ to show that the polynomial $x^4 + 3x^2 + 4x + 1$ has no roots in $\mathbb{Z}$.

16. Use the reduction homomorphism with $n = 3$ to show that the polynomial $x^5 - x + 2$ has no roots in $\mathbb{Z}$.

17. Let A be a matrix in $M_2(\mathbb{R})$. Define the function $\varphi : \mathbb{R}[x] \to M_2(\mathbb{R})$ defined by

$$\varphi(a_n x^n + a_{n-1} x^{n-1} + \cdots + a_1 x + a_0) = a_n A^n + a_{n-1} A^{n-1} + \cdots + a_1 A + a_0 I.$$

This looks like plugging A into the polynomial except that the constant term becomes the diagonal matrix $a_0 I$.

   (a) Only for this part, take $A = \begin{pmatrix} 1 & 2 \\ 3 & 4 \end{pmatrix}$. Calculate $\varphi(x^2 + x + 1)$ and $\varphi(3x^3 - 2x)$.

   (b) Show that φ is a ring homomorphism and that $\text{Im}\,\varphi$ is a commutative subring of $M_2(\mathbb{R})$.

   (c) Recall from linear algebra that the *characteristic polynomial* of a matrix A is $f_A(x) = \det(xI - A)$. For any $A \in M_2(\mathbb{R})$, show that the characteristic polynomial $f_A(x)$ of A is in $\text{Ker}\,\varphi$. [Note: The generalization of this result to $n \times n$ matrices is called the Cayley-Hamilton Theorem and requires a much more challenging proof than the 2×2 case.]

18. Suppose that R is a ring with identity $1 \neq 0$. Let $\varphi : R \to S$ be a nontrivial ring homomorphism (i.e., φ is not identically 0).

   (a) Suppose that $\varphi(1)$ is not the identity element in S (in particular, if S does not contain an identity element). Prove that $\varphi(1)$ is idempotent.

   (b) Prove that whether or not S has an identity, $\text{Im}\,\varphi$ has an identity, namely $1_{\text{Im}\,\varphi} = \varphi(1)$.

   (c) Suppose that S contains an identity 1_S and that $\varphi(1) \neq 1_S$. Prove that $\varphi(1)$ is a zero divisor.

   (d) Deduce that if S is an integral domain, then $\varphi(1) = 1_S$.

   (e) Suppose that R and S are integral domains. Prove that a nontrivial ring homomorphism $\varphi : R \to S$ induces a group homomorphism $\varphi^\times : (U(R), \times) \to (U(S), \times)$.

19. Show that the function $f : \mathbb{Z}/21\mathbb{Z} \to \mathbb{Z}/21\mathbb{Z}$ defined by $f(\bar{a}) = \overline{7}\bar{a}$ is an endomorphism on $\mathbb{Z}/21\mathbb{Z}$.

20. Let R be a commutative ring and let $R[x, y]$ be the polynomial ring on two variables x and y. Consider the function $f : R[x, y] \to R[x, y]$ such that $f(p(x, y)) = p(y, x)$. Prove that f is a nontrivial automorphism on $R[x, y]$.

21. Denote by $\text{End}(R)$ the set of endomorphisms on R (homomorphisms from R to itself).

   (a) Show that $\text{End}(R)$ is closed under the operation of function composition.

   (b) Show that $\text{End}(R)$ is closed neither under function addition nor under function multiplication.

22. *Augmentation map.* Let R be a commutative ring with identity $1 \neq 0$ and let G be a finite group. Define the function $\psi : R[G] \to R$ by

$$\psi(a_1 g_1 + a_2 g_2 + \cdots + a_n g_n) = a_1 + a_2 + \cdots + a_n.$$

Prove that ψ is a ring homomorphism. This function is called the *augmentation map* of the group ring $R[G]$.

23. The augmentation map for a group ring generalizes in the following way. Let R be a commutative ring with identity $1 \neq 0$ and let G be a finite group. Let $f : G \to U(R)$ be a group homomorphism. Prove that the function $\psi : R[G] \to R$ defined by

$$\psi(a_1 g_1 + a_2 g_2 + \cdots + a_n g_n) = a_1 f(g_1) + a_2 f(g_2) + \cdots + a_n f(g_n),$$

where $a_i f(g_i)$ involves a product in the ring R, is a ring homomorphism.

24. Let R be a commutative ring and let G be a group. Prove that if $\varphi : G \to G$ is a group automorphism, then the function $\Phi : R[G] \to R[G]$ defined by

$$\Phi \left(\sum_{g \in G} a_g g \right) = \sum_{g \in G} a_g \varphi(g)$$

is an automorphism of $R[G]$.

25. Let S be a set. For each subset $A \subseteq S$, define the *characteristic function* of A as the function $\chi_A : S \to \{0, 1\}$ such that

$$\chi_A(s) = \begin{cases} 1 & \text{if } s \in A \\ 0 & \text{if } s \notin A. \end{cases}$$

Prove the following.

  (a) The function $\psi : A \mapsto \chi_A$ is a bijection between $\mathcal{P}(S)$ and the set of functions from S to $\{0, 1\}$.

  (b) $\chi_{A \cap B}(s) = \chi_A(s) \cdot \chi_B(s)$ for all $s \in S$.

  (c) $\chi_{A \cup B}(s) = \chi_A(s) + \chi_B(s) - \chi_A(s) \cdot \chi_B(s)$ for all $s \in S$.

  (d) $\chi_{A-B}(s) = \chi_A(s)(1 - \chi_B(s))$ for all $s \in S$.

  (e) Prove that function ψ is a ring isomorphism from $(\mathcal{P}(S), \triangle, \cap)$ to the ring of functions $(\text{Fun}(S, \mathbb{F}_2), +, \times)$.

26. Let R be a ring with identity $1 \neq 0$ and let $(S, \cdot)$ be a monoid (semigroup with an identity e). Prove that the identity for a convolution ring $(\mathcal{F}, +, *)$ of functions from S to R is the function $i : S \to R$ defined by

$$i(s) = \begin{cases} 1 & \text{if } s = e \\ 0 & \text{if } s \neq e. \end{cases}$$

27. Let R be a ring and consider the monoid $S = (\{1, -1\}, \times)$. Prove that as a set $\text{Fun}(S, R)$ is in bijection with $R \times R$ but that the convolution ring $(\text{Fun}(S, R), +, *)$ is not isomorphic to $R \oplus R$.

28. Prove that ring of formal power series (see Exercise 3.2.21) over a commutative ring R is a convolution ring.

29. Let R be a commutative ring and consider the semigroup $\mathbb{Z}$. Prove that

$$\mathcal{F} = \{f \in \text{Fun}(\mathbb{Z}, R) \mid \exists N \in \mathbb{Z} \text{ such that } f(n) = 0 \text{ for all } n < N\}$$

is a subring of $\text{Fun}(\mathbb{Z}, R)$ that satisfies the convolution condition. [Compare to Exercise 3.4.28. If we view $\mathbb{Z}$ as isomorphic to the semigroup $(\{x^n \mid n \in \mathbb{Z}\}, \times)$, then the convolution ring $(\mathcal{F}, +, \times)$ is called the ring of *formal Laurent series* over R and is denoted by $R((x))$.]

3.5 Ideals

Ideals are an important class of subrings in a ring. In Section 3.7, we will see that they play the same role that normal subgroups do in group theory as it pertains to defining a quotient ring.

The concept of an ideal first arose in number theory in the context of studying properties of integer extensions, i.e., certain subrings of $\mathbb{C}$ that contain $\mathbb{Z}$. Gaussian integers and rings like $\mathbb{Z}[\sqrt[3]{2}]$ are some examples. In integer extensions, numbers do not always have certain desired divisibility properties but ideals do. (This is a result of Dedekind's Theorem from algebraic number theory, a topic beyond the scope of this book, but accessible with the preparation that this book provides.) This motivated the term "ideal." It also turns out that ideals play a pivotal role in algebraic geometry, a branch of mathematics where the tools of abstract algebra are brought to bear on the study of geometry.

3.5.1 Definition and Examples

Recall that given any subset S in a ring R, if $r \in R$, then the notation rS denotes the set $rS = \{rs \mid s \in S\}$ and the notation Sr denotes the set $Sr = \{sr \mid s \in S\}$.

Definition 3.5.1

Let R be a ring and let I be a subring of R.

(1) A subset $I \subseteq R$ is called a *left ideal* (resp. *right ideal*) of R if I is a subring of R and if $rI \subseteq I$ (resp. $Ir \subseteq I$). In other words, I is closed under left (resp. right) multiplication by elements of R.

(2) A subset I that is both a right ideal and a left ideal is called an *ideal* or a *two-sided ideal*.

A ring always contains at least two ideals, the subset $\{0\}$ and itself. The ideal $\{0\}$ is called the *trivial* ideal. Any ideal $I \subsetneq R$ is called a *proper* ideal.

Remark 3.5.2. In light of the One-Step Subgroup Criterion with addition, the definition of an ideal can be restated to say that an ideal of a ring R is a nonempty subset $I \subseteq R$ that is closed under subtraction and closed under multiplication by any element in R (i.e., ra and ar are in I for all $a \in I$ and $r \in R$). $\triangle$

If R is commutative, then left and right ideals are equivalent and hence are also two-sided ideals. Hence, the distinction between left and right ideals only occurs in noncommutative rings.

Though this does not illustrate the full scope of possible properties for ideals, it is important to keep as a baseline reference what ideals are in $\mathbb{Z}$.

Example 3.5.3 (Ideals in $\mathbb{Z}$). We claim that all ideals in $\mathbb{Z}$ are of the form $n\mathbb{Z}$, where n is a nonnegative integer. The subset $\{0\} = 0\mathbb{Z}$ is an ideal. Otherwise, let I be an ideal in $\mathbb{Z}$ and let n be the least positive integer in I. This exists by virtue of the Well-Ordering Principle of $\mathbb{Z}$.

Now let m be any integer in I. Integer division of m by n gives $m = nq + r$ for some integer q and some remainder $0 \le r < n$. However, since I is closed under multiplication by any element in the ring, then $nq \in I$ and since I is closed under subtraction, $r = m - nq \in I$. Since n is the least positive element in I and since r is nonnegative with $r < n$, then $r = 0$. We conclude that $m = qn$ and so every element in I is a multiple of n. Consequently, $I = n\mathbb{Z}.\triangle$

The style of proof in the above example (namely referring to an element that is minimal in some way) is not uncommon in ring theory. However, it will not apply in all rings. Indeed, the notion of minimality derives from the partial order $\le$ on $\mathbb{Z}$. In contrast, many rings do not possess such a partial order.

We now give an example where left and right ideals are not necessarily equal.

Example 3.5.4. Let $R = M_2(\mathbb{Z})$ and consider the subset

$$I = \left\{ \begin{pmatrix} a & a \\ c & c \end{pmatrix} \,\middle|\, a, c \in \mathbb{Z} \right\}.$$

Let $A, B \in I$ and let $C \in R$. Then

$$A - B = \begin{pmatrix} a_{11} & a_{11} \\ a_{21} & a_{21} \end{pmatrix} - \begin{pmatrix} b_{11} & b_{11} \\ b_{21} & b_{21} \end{pmatrix} = \begin{pmatrix} a_{11} - b_{11} & a_{11} - b_{11} \\ a_{21} - b_{21} & a_{21} - b_{21} \end{pmatrix}$$

so $A - B \in I$. Furthermore,

$$CA = \begin{pmatrix} c_{11} & c_{12} \\ c_{21} & c_{22} \end{pmatrix} \begin{pmatrix} a_{11} & a_{11} \\ a_{21} & a_{21} \end{pmatrix} = \begin{pmatrix} c_{11}a_{11} + c_{12}a_{21} & c_{11}a_{11} + c_{12}a_{21} \\ c_{21}a_{11} + c_{22}a_{21} & c_{21}a_{11} + c_{22}a_{21} \end{pmatrix}$$

so again $CA \in I$. Note that by assuming C is also in I, we can also deduce that I is closed under multiplication. So I is a subring and is also a left ideal. However,

$$AC = \begin{pmatrix} a_{11} & a_{11} \\ a_{21} & a_{21} \end{pmatrix} \begin{pmatrix} c_{11} & c_{12} \\ c_{21} & c_{22} \end{pmatrix} = \begin{pmatrix} a_{11}(c_{11} + c_{21}) & a_{11}(c_{12} + c_{22}) \\ a_{22}(c_{11} + c_{21}) & a_{22}(c_{12} + c_{22}) \end{pmatrix},$$

which in general is not in I. Hence, I is not a right ideal. $\triangle$

Example 3.5.5. We consider a few ideals and nonideals in the polynomial ring $R = \mathbb{R}[x]$.

Let I_1 be the set of polynomials whose nonzero term of lowest degree has degree 3 or greater. Let $a(x), b(x)$ be two polynomials. If $a_i = b_i = 0$ for $i = 0, 1, 2$, then $(a_i - b_i) = 0$ for $i = 0, 1, 2$ so $a(x), b(x) \in I_1$ implies that

$a(x) - b(x) \in I_1$. Now suppose that $a(x) \in I_1$ and $b(x)$ is arbitrary, then the first few coefficients of the terms in $a(x)b(x)$ are

$$\cdots + (a_2 b_0 + a_1 b_1 + a_0 b_2)x^2 + (a_0 b_1 + a_1 b_0)x + a_0 b_0.$$

Since $a_i = 0$ for $i = 0, 1, 2$, then the terms shown above are 0 and hence $a(x)b(x) \in I_1$. Hence, I_1 is a right ideal and hence is an ideal since $\mathbb{R}[x]$ is commutative.

Let I_2 be the set of polynomials $p(x)$ such that $p(2) = 0$ and $p'(2) = 0$. Let $a(x), b(x) \in I_2$. Then $a(2) - b(2) = 0$ and

$$\frac{d}{dx}(a(x) - b(x))\Big|_{x=2} = a'(2) - b'(2) = 0.$$

Thus, I_2 is closed under subtraction. Now let $a(x) \in I_2$ and let $p(x)$ be an arbitrary real polynomial. Then $a(2)p(2) = 0p(2) = 0$ and

$$\frac{d}{dx}(a(x)p(x))\Big|_{x=2} = a'(2)p(2) + a(2)p'(2) = 0p(2) + 0p'(2) = 0.$$

Hence, $a(x)p(x) \in I_2$ and we conclude that I_2 is an ideal. The ideal I_2 corresponds to polynomials that have a double root at 2. Figure 3.3 shows the graphs of just a few such polynomials.

In contrast, let S_1 be the subset of polynomials whose degree is 4 or less. It is true that S_1 is closed under subtraction but it is not closed under multiplication (e.g., $x \times x^4 = x^5 \notin S_1$) so it is not a subring and hence is not an ideal.

As another nonexample, let S_2 be the subset of polynomials whose terms of odd degree are 0. S_2 is closed under subtraction and under multiplication and hence S_2 is a subring. However, with $x \in \mathbb{R}[x]$ and $x^2 \in S_2$, the product $x \times x^2 = x^3 \notin S_2$ so S_2 is not an ideal. $\triangle$

Proposition 3.4.10 already provided an important class of ideals but we restate the result here.

Proposition 3.5.6

Let $\varphi : R \to S$ be a ring homomorphism. Then $\operatorname{Ker} \varphi$ is an ideal of R.

Example 3.5.7. Let R be a commutative ring with a $1 \neq 0$ and let G be a finite group with $G = \{g_1, g_2, \ldots, g_n\}$. Consider the group ring $R[G]$ and consider also the subset

$$I = \{a_1 g_1 + a_2 g_2 + \cdots + a_n g_n \mid a_1 + a_2 + \cdots + a_n = 0\}.$$

This subset I is an ideal by virtue of the fact that it is the kernel of the augmentation map defined in Exercise 3.4.22. $\triangle$

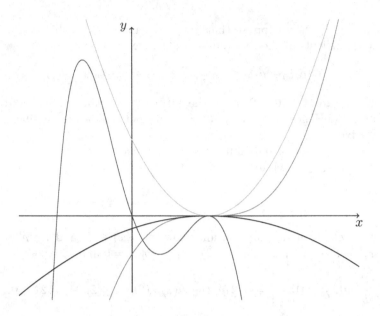

FIGURE 3.3: Some polynomials in the ideal $\{p(x) \in \mathbb{R}[x] \mid p(2) = p'(2) = 0\}$.

The following two propositions involve conditions when an ideal is the whole ring.

Proposition 3.5.8

Let R be a ring with an identity $1 \neq 0$. An ideal I is equal to R if and only if I contains a unit.

Proof. Suppose that $I = R$. Then I contains 1, which is a unit. Conversely, suppose that I contains a unit u. Call v the inverse of u. Now let $r \in R$ be any element. Then $r = r(vu) = (rv)u \in I$. Thus, $R \subseteq I$ and hence $R = I$. $\square$

Proposition 3.5.9

Let R be a commutative ring with an identity $1 \neq 0$. R is a field if and only if its only ideals are $\{0\}$ and R.

Proof. (Left as an exercise. See Exercise 3.5.14.) $\square$

EXERCISES FOR SECTION 3.5

1. Prove that $\mathbb{Z}$ is not an ideal of $\mathbb{Q}$. Find all the ideals of $\mathbb{Q}$.

2. List all the ideals of $\mathbb{Z}/12\mathbb{Z}$.

3. Let $R = M_2(\mathbb{Z})$. Let I be the set of matrices whose entries are multiples of 10. Show that I is an ideal of R.

4. Modify Example 3.5.4 to find an ideal in $M_2(\mathbb{Z})$ that is a right ideal but not a left ideal. Prove your claim.

5. Let $R = \mathbb{Z}[x]$. Which of the following subsets are ideals of R? Are any subrings of R but not ideals?

   (a) The set of polynomials whose coefficients are even.

   (b) The set of polynomials such that every other coefficient (starting at 0) is even.

   (c) The set of polynomials such that the 0th coefficient is even.

   (d) The set of polynomials with odd coefficients.

   (e) The set of polynomials whose terms all have even degrees.

6. Let $S = \{p(x) \in \mathbb{R}[x] \mid p'(3) = 0\}$. Prove that S is a subring of $\mathbb{R}[x]$ but not an ideal.

7. Consider the ring $C^0(\mathbb{R}, \mathbb{R})$ of continuous functions on $\mathbb{R}$. Let A be any subset of $\mathbb{R}$.

   (a) Let $S_1 = \{f \in C^0(\mathbb{R}, \mathbb{R}) \mid \forall a \in A, f(a) = 0\}$ is an ideal of $C^0(\mathbb{R}, \mathbb{R})$.

   (b) Let $S_1 = \{f \in C^0(\mathbb{R}, \mathbb{R}) \mid \forall a \in A, f(a) \in \mathbb{Z}\}$ is a subring of $C^0(\mathbb{R}, \mathbb{R})$, but not an ideal.

8. Let S be a set and consider the ring $R = (\mathcal{P}(S), \triangle, \cap)$. Prove that if $S' \subset S$, then the subring $R' = (\mathcal{P}(S'), \triangle, \cap)$ is an ideal.

9. Let R be a ring. Prove that the subring in $M_2(R)$ of upper triangular matrices is not an ideal of $M_2(R)$.

10. Let k be a positive integer. Prove that $M_n(k\mathbb{Z})$ is an ideal in $M_n(\mathbb{Z})$.

11. Prove that if I is an ideal in a ring R and S is a subring that $I \cap S$ is an ideal of S.

12. Let R be a commutative ring and let I be an ideal in R.

    (a) Prove that $I[x] = \{a_n x^n + \cdots + a_1 x + a_0 \mid a_i \in I \text{ for all } i\}$ is an ideal of $R[x]$.

    (b) Let $a \in R$ be fixed. Prove that $\{p(x) \in R[x] \mid p(a) \in I\}$ is an ideal of $R[x]$.

    (c) Let k be a nonnegative integer. Prove that $\{a_n x^n + \cdots + a_1 x + a_0 \mid a_i \in I \text{ for } 0 \leq i \leq k\}$ is an ideal of $R[x]$.

    (d) Prove that $\{a_n x^n + \cdots + a_1 x + a_0 \mid a_2 \in I\}$ is not an ideal of $R[x]$.

13. Let $I_1, I_2, \ldots, I_n$ be n ideals in a ring R. Consider the matrix ring $M_n(R)$.

    (a) Prove that the set of matrices in $M_n(R)$ where all the elements of the k'th column are elements of I_k is a left ideal of $M_n(R)$.

    (b) Prove that the set of matrices in $M_n(R)$ where all the elements of the k'th row are elements of I_k is a right ideal of $M_n(R)$.

14. Suppose that R is a commutative ring with $1 \neq 0$. Prove that R is a field if and only if its only ideals are $\{0\}$ and R.

15. Let $\varphi : R \to S$ be a ring homomorphism.

    (a) Show that if J is an ideal in S, then $\varphi^{-1}(J)$ is an ideal in R. [Note: This generalizes Proposition 3.5.6.]

    (b) Show that if I is an ideal in R, then $\varphi(I)$ is not necessarily an ideal in S.

    (c) Show that if φ is surjective and I is an ideal of R, then $\varphi(I)$ is an ideal of S.

16. Let ψ_1 and ψ_2 be two distinct ring homomorphisms $R \to S$. Let $J = \{r \in R \,|\, \psi_1(r) = \psi_2(r)\}$. Prove that J is a subring of R that is not an ideal.

3.6 Operations on Ideals

3.6.1 Ideals Generated by Subsets

There exists a convenient way to describe ideals in a ring using generating subsets.

Definition 3.6.1

Let A be a subset of a ring R.

(1) By the notation (A) we denote the smallest (by inclusion) ideal in R that contains the set A. We say that the ideal (A) is *generated by A*.

(2) An ideal that can be generated by a single element set is called a *principal* ideal.

(3) An ideal that is generated by a finite set A is called a *finitely generated* ideal.

The student should note that when we refer to the subset (A) of R it is *by definition* an ideal and hence there is no need to prove that it is.

The above notation, however, is not explicit since it does not directly offer a means of determining all elements in (A). There is an explicit way to create an ideal from a subset $A \subset R$. In order to describe it, we define the following sets.

Definition 3.6.2

Given a ring R and a subset A we can create some ideals from A.

(1) RA denotes the subset of finite left R-linear combinations of elements in A, i.e., $RA = \{r_1a_1 + r_2a_2 + \cdots + r_na_n \mid r_i \in R \text{ and } a_i \in A\}$.

(2) AR denotes the subset of finite right R-linear combinations of elements in A, i.e., $AR = \{a_1r_1 + a_2r_2 + \cdots + a_nr_n \mid r_i \in R \text{ and } a_i \in A\}$.

(3) RAR denotes the subset of finite dual R-linear combinations of elements in A, i.e., $RAR = \{r_1a_1s_1 + r_2a_2s_2 + \cdots + r_na_ns_n \mid r_i, s_i \in R \text{ and } a_i \in A\}$.

Proposition 3.6.3

Let R be a ring and let A be any subset.

(1) RA is a left ideal, AR is a right ideal, and RAR is a two-sided ideal.

(2) If R is a ring with an identity $1 \neq 0$, then $(A) = RAR$.

(3) If R is commutative with an identity $1 \neq 0$, then $RA = AR = RAR = (A)$.

Proof. For part (1), let $r_1a_1 + r_2a_2 + \cdots + r_ma_m$ and $r_1'a_1' + r_2'a_2' + \cdots + r_n'a_n'$ be two elements of RA. Then their difference

$$r_1a_1 + r_2a_2 + \cdots + r_ma_m - (r_1'a_1' + r_2'a_2' + \cdots + r_n'a_n')$$

is another finite linear combination of elements in RA. Also, for any $s \in R$,

$$s(r_1a_1 + r_2a_2 + \cdots + r_ma_m) = (sr_1)a_1 + (sr_2)a_2 + \cdots + (sr_m)a_m$$

is also a linear combination in RA. Hence, RA is a left ideal.

The proof that AR is a right ideal is similar.

Let $r_1a_1s_1 + r_2a_2s_2 + \cdots + r_ma_ms_m$ and $r_1'a_1's_1' + r_2'a_2's_2' + \cdots + r_n'a_n's_n'$ be two elements in RAR. Then their difference is

$$r_1a_1s_1 + r_2a_2s_2 + \cdots + r_ma_ms_m + (-r_1')a_1's_1' + (-r_2')a_2's_2' + \cdots + (-r_n')a_n's_n',$$

which is again an element in RAR. If t is any element in R, then

$$t(r_1a_1s_1 + r_2a_2s_2 + \cdots + r_ma_ms_m) = (tr_1)a_1s_1 + (tr_2)a_2s_2 + \cdots + (tr_m)a_ms_m$$
$$(r_1a_1s_1 + r_2a_2s_2 + \cdots + r_ma_ms_m)t = r_1a_1(s_1t) + r_2a_2(s_2t) + \cdots + r_ma_m(s_mt),$$

which are both elements of RAR. Hence, RAR is an ideal (two-sided).

For part (2), suppose that R has an identity $1 \neq 0$. By taking (r_i, s_i) as 1 or 0 as appropriate, we deduce that $A \subseteq RAR$. Hence, RAR is an ideal that contains A. By definition, $(A) \subseteq RAR$. However, by definition of ideals, $r_1 a_1 s_1 + r_2 a_2 s_2 + \cdots + r_m a_m s_m$ is an element of any ideal that also contains A. Thus, $RAR \subseteq (A)$. Hence, $(A) = RAR$.

Finally, for part (3) suppose that R is commutative with an identity $1 \neq 0$. By part (2) we already have $RAR = (A)$. Commutativity gives $RA = AR$. Also by commutativity,

$$r_1 a_1 s_1 + r_2 a_2 s_2 + \cdots + r_m a_m s_m = r_1 s_1 a_1 + r_2 s_2 a_2 + \cdots + r_m s_m a_m,$$

so $RAR \subseteq RA$. However, since R has an identity, by setting $s_i = 1$ for $i = 1, 2, \ldots, m$, we obtain all elements in RA as elements in RAR. Hence, $RA \subseteq RAR$ and thus we have $RA = AR = RAR$. $\qquad\square$

Example 3.6.4. Consider the ideal $I = (m, n)$ in $\mathbb{Z}$. Since $\mathbb{Z}$ is commutative,

$$I = \{sm + tn \mid s, t \in \mathbb{Z}\}.$$

By Proposition A.5.10, $I = (d)$ where $d = \gcd(m, n)$. Using this result, an induction argument shows that every finitely generated ideal in $\mathbb{Z}$ can be generated by a single element. However, in arbitrary rings there do exist ideals that are not finitely generated. The above induction argument would not be sufficient to conclude that all ideals in $\mathbb{Z}$ are generated by a single element. On the other hand, Example 3.5.3 gave a different argument that did prove that all ideals in $\mathbb{Z}$ are principle. $\qquad\triangle$

Example 3.6.5. Consider the ideal $I = (5, x^2 - x - 2)$ in $\mathbb{Z}[x]$. This ideal consists of polynomial linear combinations

$$5p(x) + (x^2 - x - 2)q(x) \qquad \text{with } p(x), q(x) \in \mathbb{Z}[x]. \qquad (3.8)$$

As in Example 3.6.4, just because the ideal is expressed using two generators does not mean that two generators are necessary. Assume that $I = (a(x))$ for some polynomial $a(x)$. Then since $5 \in I$, we have $a(x)r(x) = 5$ for some $r(x) \in \mathbb{Z}[x]$. Hence, by degree considerations, $\deg a(x) = 0$ and thus $a(x)$ must be either 1 or 5. Now $x^2 - x - 2$ has coefficients that are not multiplies of 5 so since $x^2 - x - 2 \in I$, we cannot have $I = (5)$. Hence, if I is a principal ideal, then $I = (1) = \mathbb{Z}[x]$. However, every polynomial $r(x)$ in the form of (3.8) has the property that $r(2)$ is a multiple of 5. This is not the case for all polynomials in $\mathbb{Z}[x]$. Hence, $I \neq \mathbb{Z}[x]$. The assumption that I is principle leads to a contradiction, so I requires two generators. $\qquad\triangle$

The following examples illustrate how the conditions in the various parts of Proposition 3.6.3 are required for the result to hold.

Example 3.6.6. Let R be the ring $2\mathbb{Z}$ and consider the subset $A = \{4\}$. The ideal $(4) = 4\mathbb{Z}$ whereas $RA = AR = 8\mathbb{Z}$ and $RAR = 16\mathbb{Z}$. $\qquad\triangle$

Example 3.6.7. Consider the ring $R = M_2(\mathbb{Z})$, which is a ring with an identity $1 \neq 0$ but a noncommutative ring. Let A be the set containing the single matrix

$$a = \begin{pmatrix} 0 & 1 \\ 0 & 0 \end{pmatrix}.$$

By properties of ranks of matrices, since $\operatorname{rank} a = 1$, the rank of any multiple of a (left or right) can be at most 1. Hence, Ra and aR are strict subsets of R. From Proposition 3.6.3(2), we know that $RaR = (a)$. However, it can be shown (see Exercise 3.6.3) that $(a) = M_2(\mathbb{Z})$ so that $Ra = aR \neq (a) = RaR.\triangle$

3.6.2 Principal Ideals

There are some common situations where we can immediately tell if a certain subset is a left or right ideal.

Proposition 3.6.8

Let R be a ring and let $a \in R$ be any element. Then the subset Ra is a left ideal and the subset aR is a right ideal of R. If R is commutative, then $Ra = aR$ and this is an ideal.

Proof. The set $Ra = \{ra \mid r \in R\}$. Suppose that $x, y \in Ra$. Then $x = r_1 a$ and $y = r_2 a$ for some $r_1, r_2 \in R$. Then $x - y = (r_1 - r_2)a$. So Ra is closed under subtraction. Furthermore, if $r' \in R$, then $r'x = (r'r_1)a \in Ra$. Hence, Ra is closed under multiplication on the left by any element in R. Hence, Ra is a left ideal. The reasoning is the same to show that aR is a right ideal. □

Definition 3.6.9

If R is a commutative ring, an ideal of the form aR for some $a \in R$ is called a *principal ideal*.

Example 3.5.3 established an important property of the integers, namely that every ideal in $\mathbb{Z}$ is principal. Rings with this property benefit from many associated nice properties. For this reason, we give a name to this particular class of rings.

Definition 3.6.10

A *principal ideal domain* is an integral domain R in which every ideal is principal. We often abbreviate the name and call such a ring a PID.

We will encounter properties of PIDs in the exercises and in subsequent sections.

3.6.3   Operations on Ideals

Definition 3.6.11

Let I, J be ideals in a ring R. We define the following operations on ideals:

(1) The sum, $I + J = \{a + b \,|\, a \in I \text{ and } b \in J\}$.

(2) The product, IJ consists of finite sums of elements $a_i b_i$ with $a_i \in I$ and $b_i \in J$. Thus,

$$IJ = \{a_1 b_1 + a_2 b_2 + \cdots + a_n b_n \,|\, a_i \in I,\ b_i \in J\}.$$

(3) The power I^k is the iterated product operation on I. We also define $I^0 = R$.

We point out that with operation of addition on ideals, if $A = \{a_1, a_2, \ldots, a_n\}$ is a subset of a ring R, then RA and AR can be restated as

$$RA = Ra_1 + Ra_2 + \cdots + Ra_n \quad \text{and} \quad AR = a_1 R + a_2 R + \cdots a_n R.$$

On the other hand, RAR is not equal to $Ra_1 R + Ra_2 R + \cdots + Ra_n R$, because in order to get the combination in Definition 3.6.2 of RAR, an element $a \in A$ may need to appear in more than one term $r_i a_i s_i$. However, by Exercise 3.6.15, if $I = (A)$ and $J = (B)$ for subsets $A, B \subseteq R$, then $I + J = (A \cup B)$.

Proposition 3.6.12

Let I, J be ideals in a ring R. Then $I + J$, IJ, and $I \cap J$ are ideals of R. Furthermore, the ideals satisfy the following containment relations.

$$
\begin{array}{ccccc}
 & & I & & \\
 & \subseteq & & \subseteq & \\
IJ & \subseteq & I \cap J & & I + J \\
 & & \subseteq & \subseteq & \\
 & & J & &
\end{array}
$$

Proof. Let $a_1 b_1 + a_2 b_2 + \cdots + a_m b_m$ and $a_1' b_1' + a_2' b_2' + \cdots + a_n' b_n'$ be two elements in the IJ. Then their difference

$$a_1 b_1 + a_2 b_2 + \cdots + a_m b_m + (-a_1') b_1' + (-a_2') b_2' \cdots + (-a_n') b_n'$$

is also a combination in IJ. Furthermore, for all $t \in R$,

$$t(a_1 b_1 + a_2 b_2 + \cdots + a_m b_m) = (ta_1) b_1 + (ta_2) b_2 + \cdots + (ta_m) b_m,$$
$$(a_1 b_1 + a_2 b_2 + \cdots + a_m b_m)t = a_1 (b_1 t) + a_2 (b_2 t) + \cdots + a_m (b_m t).$$

But for all $i = 1, 2, \ldots, m$ we have $ta_i \in I$ since I is an ideal and $b_i t \in J$ since J is an ideal. Hence, multiplying any element in IJ by any element in R produces an element in IJ. Thus, IJ is an ideal of R.

(We leave the proof that $I + J$ and that $I \cap J$ are ideals as an exercise. See Exercise 3.6.13.)

Some containments are obvious, namely, $I \cap J \subseteq I$ and $I \cap J \subseteq J$. Since 0 is an element of every ideal, $I \subseteq I + J$ and $J \subseteq I + J$. Finally, let $a \in I$ and $b \in J$. Then $ab \in I$ because I is an ideal and $a \in I$ but also $ab \in J$ because J is an ideal and $b \in J$. Thus, in a linear combination $a_1 b_1 + a_2 b_2 + \cdots + a_n b_n$ every product $a_i b_i \in I \cap J$ and hence the full linear combination is in $I \cap J$. We conclude that $IJ \subseteq I \cap J$. $\qquad\qquad\square$

Example 3.6.13. Let $R = \mathbb{Z}$ and let $I = 12\mathbb{Z}$ and $J = 45\mathbb{Z}$. The ideal operations listed in Definition 3.6.11 are

$$IJ = 540\mathbb{Z}, \qquad I \cap J = 180\mathbb{Z}, \quad \text{and} \quad I + J = 3\mathbb{Z}.$$

We observe that $IJ = (12 \times 45)$, $I \cap J = (\text{lcm}(12, 45))$, and $I + J = (\gcd(12, 45))$. Consequently, the operations on ideals directly generalize the notions of product, least common multiple, and greatest common divisor. $\triangle$

Since ideals generalize the notion of least common multiple and greatest common divisor, we should have an equivalent concept for relatively prime. Recall that $a, b \in \mathbb{Z}$ are called relatively prime if $\gcd(a, b) = 1$. In ring theory, the generalized notion is similar.

Definition 3.6.14

Let R be a ring with an identity $1 \neq 0$. Two ideals I and J of R are called *comaximal* if $I + J = R$.

Note that this definition only applies when a ring has an identity.

For example, (2) and (3) are comaximal in $\mathbb{Z}$. As another example, consider the ideals (x^2) and $(x + 1)$. We have $(x^2) + (x + 1) = (x^2, x + 1)$. But $(1 - x)(x + 1) + 1x^2 = 1$ is in $(x^2, x + 1)$ and hence, $(x^2, x + 1) = \mathbb{Z}[x]$. Thus, (x^2) and $(x + 1)$ are comaximal ideals.

It is easy to give generating sets of $I + J$ and IJ from generating sets of I and J. The proofs of these claims are left as exercises but we mention them here because of their importance. Suppose that I and J are generated by certain finite sets of elements, say $I = (a_1, a_2, \ldots, a_m)$ and $J = (b_1, b_2, \ldots, b_n)$. Then $I + J = (a_1, a_2, \ldots, a_m, b_1, b_2, \ldots, b_n)$ and the set IJ is generated by the set $\{a_i b_j \mid i = 1, 2, \ldots, m, \; j = 1, 2, \ldots, n\}$.

We conclude the section by mentioning two other operations on ideals, which, in the context of commutative rings, produce other ideals. The exercises explore some questions and examples related to these operations.

Definition 3.6.15

Let I be an ideal in a commutative ring R. The *radical ideal* of I, denoted by $\sqrt{I}$, is

$$\sqrt{I} = \{r \in R \mid r^n \in I \text{ for some } n \in \mathbb{N}^*\}.$$

Definition 3.6.16

Let I and J be ideals in a commutative ring R. The *fraction ideal* of I by J, denoted $(I : J)$ is the subset

$$(I : J) = \{r \in R \mid rJ \subseteq I\}.$$

In Section 3.1.2 we encountered the subset $\mathcal{N}(R)$ of nilpotent elements in a ring R. In Exercise 3.1.32, we showed that if R is a commutative ring, then $\mathcal{N}(R)$ is a subring of R. We could have also shown that $\mathcal{N}(R)$ is an ideal. We point out that by definition $\mathcal{N}(R)$ is the radical ideal of the trivial ideal (0). In other words,

$$\mathcal{N}(R) = \sqrt{(0)},$$

which shows directly that $\mathcal{N}(R)$ is an ideal of R, when R is commutative.

3.6.4 Useful CAS Commands

Most modern CAS implement operations on ideals in in multivariable polynomial ring $F[x_1, x_2, \ldots, x_n]$, where F is a field. The required algorithms use Gröbner bases, a technique that is beyond the scope of this book.

In *Maple*, the package `PolynomialIdeals` offers commands to compute operations on ideals in $F[x_1, x_2, \ldots, x_n]$. The default for F is a field of characteristic 0, so $\mathbb{Q}$, $\mathbb{R}$, or $\mathbb{C}$. However, it is possible to specify a field with a prime characteristic p. The example below illustrates operations on ideals generated by lists of polynomials.

—————————————— Maple ——————————————

$with(PolynomialIdeals)$:

$J := PolynomialIdeal([3 * x * y + y - z, 3 * x^2 * y - x * z + y])$:

$K := PolynomialIdeal([x - 1])$:

$Add(J, K);$
$$\langle x - 1, 3xy + y - z, 3x^2y - xz + y \rangle$$

$Multiply(J, K);$
$$\langle (3xy + y - z)(x - 1), (3x^2y - xz + y)(x - 1) \rangle$$

$Intersect(J, K)$
$$\langle xy - y, xz - z \rangle$$

$Quotient(J, K)$;

$$\langle y, z \rangle$$

$Radical(J)$

$$\langle -4y + z, xy - y \rangle$$

$Simplify(J)$;

$$\langle -4y + z, xy - y \rangle$$

In the last two lines, we calculated the radical ideal $\sqrt{J}$ and then decided to use the command Simplify, which returns the same ideal as J but generated by a simpler list of elements. With these last two lines, we observe that J is a radical ideal, i.e., an ideal such that $\sqrt{J} = J$.

Similarly, SAGE has object classes for ideals in various types of rings and methods on these classes that implement operations discussed in this section. However, here we need to define the field F first. The following code does exactly the same thing as the above *Maple* code.

─────────────────────── Sage ───────────────────────

```
sage: R=QQ['x,y,z']
sage: x,y,z=R.gens()
sage: J=ideal([3*x*y+y-z,3*x^2*y-x*z+y])
sage: K=(x-1)*R
sage: J+K
Ideal (3*x*y + y - z, 3*x^2*y - x*z + y, x-1) of
    Multivariate...
sage: I*J
Ideal (3*x^2*y - 2*x*y - x*z - y + z, 3*x^3*y - 3*x^2*y
    - x^2*z + x*y + x*z - y) of Multivariate...
sage: J.intersection(K)
Ideal (x*z - z, x*y - y) of Multivariate...
sage: J.quotient(K)
Ideal (z, y) of Multivariate...
sage: J.radical()
Ideal (4*y - z, x*z - z) of Multivariate...
```

Notice that in line 4, where we defined the ideal K, we explicitly used the notation aR, for a principal ideal in the commutative ring R. (In each response line above, Multivariate... is short for Multivariate Polynomial Ring in x, y, z over Rational Field.)

EXERCISES FOR SECTION 3.6

1. Let R be a commutative ring with $1 \neq 0$. Prove that $(a) \subseteq (b)$ if and only if there exists $r \in R$ such that $a = rb$.

2. Show that the ideal $(2, x)$ in $\mathbb{Z}[x]$ is not principal. Conclude that $\mathbb{Z}[x]$ is not a PID.

3. Let $R = M_2(\mathbb{Z})$ and let $A = \begin{pmatrix} 0 & 1 \\ 0 & 0 \end{pmatrix}$.

    (a) Determine all the matrices in RA and all the matrices in AR.

    (b) Obtain the matrices

    $$\begin{pmatrix} a & 0 \\ 0 & 0 \end{pmatrix}, \quad \begin{pmatrix} 0 & b \\ 0 & 0 \end{pmatrix}, \quad \begin{pmatrix} 0 & 0 \\ c & 0 \end{pmatrix}, \quad \begin{pmatrix} 0 & 0 \\ 0 & d \end{pmatrix}$$

    as elements in RAR.

    (c) Deduce that $RAR = (A) = M_2(\mathbb{Z})$.

4. Let $R = M_3(\mathbb{R})$ and let

    $$A = \left\{ \begin{pmatrix} 0 & 1 & 0 \\ 0 & 0 & 1 \\ 0 & 0 & 0 \end{pmatrix} \right\}.$$

    Determine explicitly RA, AR, RAR, and (A).

5. Let R be an integral domain. Prove that $(a) = (b)$ in R if and only if $a = bu$ for some unit $u \in U(R)$.

6. Consider the ideal $I = (13x + 16y, 11x + 13y)$ in the ring $\mathbb{Z}[x, y]$.

    (a) Prove that $I = (x - 2y, 3x + y)$. [Hint: By mutual inclusion.]

    (b) Prove that $(7x, 7y) \subseteq I$ but prove that this inclusion is strict.

7. Prove that $\mathbb{Q}[x, y]$ is not a PID.

8. In the ring $\mathbb{R}[x, y]$, let $I = (ax + by - c, dx + ey - f)$ where $a, b, c, d, e, f \in \mathbb{R}$.

    (a) Prove that if the lines $ax + by = c$ and $dx + ey = f$ intersect in a single point (r, s), prove that $I = (x - r, y - s)$.

    (b) Prove that if the lines $ax + by = c$ and $dx + ey = f$ are parallel, then $I = \mathbb{R}[x, y]$.

    (c) Prove that if the lines $ax + by = c$ and $dx + ey = f$ are the same, then $I = (ax + by - c)$.

9. Consider the ideal $I = (x, y)$ in the ring $\mathbb{R}[x, y]$. Find a generating set for I^k and show that I^k requires a minimum of $k + 1$ generators.

10. Let R be a commutative ring and let I be an ideal of R. Prove that the subset

    $$\{a_n x^n + \cdots + a_1 x + a_0 \in R[x] \mid a_k \in I^k\}$$

    is a subring of $R[x]$ but not necessarily an ideal.

11. Let $U_n(R)$ be the ring of upper triangular $n \times n$ matrices with coefficients in a ring R. Let I be an ideal in R. Prove that

    $$S = \{(a_{ij}) \in U_n(R) \mid a_{ij} \in I^{j-i}\}$$

    is a subring of $U_n(R)$ but not an ideal.

12. Suppose that R is a commutative ring with $1 \neq 0$. Prove that a principal ideal $(a) = R$ if and only if a is a unit. [Exercise 3.6.3 gives an example of a noncommutative ring where this result does not hold.]

13. Let I and J be ideals of a ring R. Prove that: (a) $I + J$ is an ideal of R and (b) $I \cap J$ is an ideal of R.

14. Let $\mathcal{C}$ be an arbitrary collection of ideals of a ring R. Prove that

$$\bigcap_{I \in \mathcal{C}} I$$

is an ideal of R.

15. Let R be a commutative ring. Suppose that I and J are ideals in R that are generated by certain finite sets of elements, say $I = (a_1, a_2, \ldots, a_m)$ and $J = (b_1, b_2, \ldots, b_n)$.
    (a) Prove that $I + J = (a_1, a_2, \ldots, a_m, b_1, b_2, \ldots, b_n)$.
    (b) Prove that IJ is generated by the set of mn elements, $\{a_i b_j \mid i = 1, 2, \ldots, m, \ j = 1, 2, \ldots, n\}$.

16. Let I and J be ideals of a ring R. Prove that $I \cup J$ is not necessarily an ideal of R.

17. Let I, J, K be ideals of a ring R. Prove that:
    (a) $I(J + K) = IJ + IK$;
    (b) $I(JK) = (IJ)K$.

18. Let $I_1 \subseteq I_2 \subseteq \cdots \subseteq I_k \subseteq \cdots$ be a chain (by in the partial order of inclusion) of ideals in a ring R. Prove that

$$\bigcup_{k=1}^{\infty} I_k$$

is an ideal in R.

19. Let R be a commutative ring with an identity $1 \neq 0$.
    (a) Prove that if I and J are comaximal ideals, then $IJ = I \cap J$.
    (b) Prove that if $I_1, I_2, \ldots, I_n$ is a finite collection of pairwise comaximal ideals, then

$$I_1 I_2 \cdots I_n = I_1 \cap I_2 \cap \cdots \cap I_n.$$

20. Let R be a commutative ring and let I be an ideal in R. Prove that the radical defined in Definition 3.6.15 is an ideal.

21. Let $R = \mathbb{Z}$. Calculate the radical ideals: (a) $\sqrt{(72)}$; (b) $\sqrt{(105)}$; (c) $\sqrt{(243)}$.

22. Let I be an ideal in a commutative ring R. Prove that $\sqrt{\sqrt{I}} = \sqrt{I}$.

23. Let R be a commutative ring. Show that the set $\mathcal{N}_R$ of nilpotent elements is equal to the ideal $\sqrt{(0)}$. [For this reason, the subring of nilpotent elements in a commutative ring is often called the *nilradical* of R.]

24. In the ring $\mathbb{Z}$, prove the following fraction ideal equalities. (a) $((2) : (0)) = \mathbb{Z}$; (b) $((24) : (4)) = (6)$; (c) $((17) : (15)) = (17)$.

3.7 Quotient Rings

The introduction to Chapter 2 motivated the construction of quotient groups with modular arithmetic. However, the triple $(\mathbb{Z}/n\mathbb{Z}, +, \times)$, which encompasses modular arithmetic, has a ring structure. It is an example of a quotient ring. This section parallels the discussion for quotient groups.

3.7.1 Quotient Rings

The concept of a quotient ring combines the process of taking a quotient set in such a way that the ring operations induce a ring structure on the quotient set.

Proposition 3.7.1

Let $(R, +, \times)$ be a ring. The cosets in the additive quotient group $(R/I, +)$ form a ring with $+$ and $\times$ defined by

$$(a+I)+(b+I) \overset{\text{def}}{=} (a+b)+I \quad \text{and} \quad (a+I)\times(b+I) \overset{\text{def}}{=} (ab)+I. \quad (3.9)$$

Proof. Since I is an ideal of R, it is a subring and hence a subgroup of $(R, +)$. Since $(R, +)$ for any ring, then $(I, +) \trianglelefteq (R, +)$. From group theory, we know that the set of cosets R/I is a group under the addition defined in (3.9). Since R is abelian, then R/I is as well.

We now need to show that the multiplication given in (3.9) is well-defined. Consider two cosets $a + I$ and $b + I$ in R/I. Suppose that $a + I = a' + I$ and $b + I = b' + I$. Then we know that $a' - a = r_1 \in I$ and $b' - b = r_2 \in I$. Then

$$a'b' - ab = (a + r_1)(b + r_2) - ab = ab + ar_2 + r_1 b + r_1 r_2 - ab$$
$$= ar_2 + r_2 b + r_1 r_2.$$

Now $ar_2 \in I$ since $r_2 \in I$ and I is a left ideal. Also $r_1 b \in I$ because $r_1 \in I$ and I is a right ideal. Finally $r_1 r_2 \in I$ by virtue of I being a subring of R. Thus $a'b' - ab \in I$, so $(ab) + I = (a'b') + I$. This shows that the product in (3.9) is well-defined.

The remaining ring axioms that we need to check follow immediately from (3.9). For example, to show associativity of multiplication, we have

$$((a + I) \times (b + I)) \times (c + I) = ((ab) + I) \times (c + I) = ((ab)c) + I$$
$$= (a(bc)) + I = (a + I) \times ((bc) + I)$$
$$= (a + I) \times ((b + I) \times (c + I)). \qquad \square$$

Definition 3.7.2

Let R be a ring and let I be a (two-sided) ideal in R. The ring R/I with addition and multiplication defined in Proposition 3.7.1 is the *quotient ring* of R with respect to I.

The careful reader might and should wonder why we needed to use ideals in rings to create quotient rings. Could we have created a quotient set from R that is still a ring, using say a subring S that is not an ideal or perhaps a more general subset of R? The following proposition answers that question.

Proposition 3.7.3

Let $(R, +, \times)$ be a ring and let $\sim$ be an equivalence relation on R. The operations $+$ and $\times$ on $R/\sim$ expressed by

$$[a] + [b] \overset{\text{def}}{=} [a + b] \quad \text{and} \quad [a] \times [b] \overset{\text{def}}{=} [a \times b] \quad (3.10)$$

are well-defined if and only if the equivalence classes of $\sim$ are the cosets $a + I$ of some (two-sided) ideal I.

Proof. (Left as an exercise for the reader. See Exercise 3.7.27.) □

This proposition illustrates the importance of ideals in ring theory: It is not possible to construct a quotient ring using just a subring. That subring must be an ideal.

We point out one minor technicality in the notation. As subsets of R, the set $(a + I) \times (b + I)$ is *not* necessarily equal to the subset $(ab) + I$ in R. As subsets of R, we can only conclude that $(a + I) \times (b + I) \subseteq (ab) + I$ and this inclusion is often proper. For example, let $R = \mathbb{Z}$, let $I = 11\mathbb{Z}$ and consider the cosets $2 + 11\mathbb{Z}$ and $6 + 11\mathbb{Z}$. In the quotient ring

$$(2 + 11\mathbb{Z}) \times (6 + 11\mathbb{Z}) = 12 + 11\mathbb{Z} = 1 + 11\mathbb{Z}.$$

As subsets in $\mathbb{Z}$,

$$(2 + 11\mathbb{Z})(6 + 11\mathbb{Z}) = \{(2 + 11k)(6 + 11\ell) \mid k, \ell \in \mathbb{Z}\}.$$

But $(2 + 11k)(6 + 11\ell) = 12 + 22\ell + 66k + 121k\ell$. Though $23 \in 1 + 11\mathbb{Z}$, we can show that there exist no integers $k, \ell \in \mathbb{Z}$ such that $23 = 12 + 22\ell + 66k + 121k\ell$. Assume that there did. Then $11 = 22\ell + 66k + 121k\ell$ so $1 = 2\ell + 6k + 11k\ell$. We deduce that $1 - 2\ell = k(6 + 11\ell)$ so $6 + 11\ell$ divides $2\ell - 1$. If $\ell > 0$, then $0 < 2\ell - 1 < 6 + 11\ell$ so $6 + 11\ell$ does not divide $2\ell - 1$. Obviously, if $\ell = 0$, then $6 + 11\ell$ does not divide $2\ell - 1$. If $\ell < 0$, then

$$7 < 9(-\ell) \implies 7 + 2(-\ell) < 11(-\ell) \implies 1 + 2(-\ell) < -6 + 11(-\ell)$$
$$\implies |1 - 2\ell| < |6 + 11\ell|$$

so again it is not possible for $6 + 11\ell$ to divide $2\ell - 1$. Hence, this contradicts our assumption that $23 \in (2 + 11\mathbb{Z})(6 + 11\mathbb{Z})$.

Partly because of this technicality, it is even more common in ring theory to borrow the notation from modular arithmetic and denote a coset $r + I$ in R/I by $\bar{r}$. As always, with this notation, the ideal I must be clear from context. Also inspired from modular arithmetic, we may say that r and s are *congruent* modulo I whenever $r + I = s + I$. Note that the operations in (3.9) are such that the correspondence $r \mapsto \bar{r}$ from R to R/I is a homomorphism.

Definition 3.7.4

Given a ring R and an ideal $I \subseteq R$, the homomorphism $\pi : R \to R/I$ defined by $\pi(r) = \bar{r} = r + I$ is called the *natural projection* of R onto R/I.

Example 3.7.5. The first and most fundamental example comes from the integers $R = \mathbb{Z}$. The subring $I = n\mathbb{Z}$ is an ideal. The quotient ring is $R/I = \mathbb{Z}/n\mathbb{Z}$, the usual ring of modular arithmetic. The reader can now see that our notation for modular arithmetic in fact comes from our notation for quotient rings.   △

Example 3.7.6. Example 3.5.5 presented two ideals I_1 and I_2 in $\mathbb{R}[x]$. We propose to describe the corresponding quotient rings.

Let I_1 be the ideal of polynomials whose nonzero term of lowest degree has degree 3 or greater. Using the generating subset notation, this ideal can be written as $I_1 = (x^3)$. In R/I_1, we have $\overline{a(x)} = \overline{b(x)}$ if and only if $b(x) - a(x) \in I_1$, so $b(x) - a(x) = x^3 p(x)$ for some polynomial $p(x)$. Thus, $\overline{a(x)} = \overline{b(x)}$ if and only if $a_0 = b_0$, $a_1 = b_1$ and $a_2 = b_2$. Hence, for every polynomial $a(x) \in \mathbb{R}[x]$, we have equality $\overline{a(x)} = \overline{b(x)}$ for some unique polynomial $b(x)$ of degree 2, namely $b(x) = a_2 x^2 + a_1 x + a_0$.

Because addition and multiplication behave well with respect to the quotient ring process, we can write any polynomial in R/I_1 as

$$\overline{a_2 x^2 + a_1 x + a_0} = \overline{a_2 x}^2 + \overline{a_1 x} + \overline{a_0}.$$

Now for any two real numbers, $\overline{a}\overline{b} = \overline{ab}$. However, the element $\bar{x}$ in the quotient ring has the particular property that $\bar{x}^3 = \overline{x^3} = \bar{0}$. In particular, any power of $\bar{x}$ greater than 2 gives $\bar{0}$. As a sample calculation,

$$\overline{(x^2 + 3x + 7)(2x^2 - x + 3)} = \overline{3}\bar{x}^2 - \overline{3}\bar{x}^2 + \overline{9}\bar{x} + \overline{14}\bar{x} - \overline{7}\bar{x} + \overline{21}$$

$$= \overline{14}\bar{x}^2 + \overline{2}\bar{x} + \overline{21}.$$

Now consider the ideal I_2 defined as the set of polynomials $p(x)$ such that $p(2) = 0$ and $p'(2) = 0$. In Section 4.3.3 we will offer a characterization of I_2 with generators. However, we can understand R/I_2 without it. Every polynomial $p(x) \in \mathbb{R}[x]$ satisfies

$$\overline{p(x)} = \overline{p(2) + p'(2)(x - 2)}.$$

Hence, every polynomial is congruent modulo I_2 to the 0 polynomial or a unique polynomial of degree 1 or less. Though we could write such polynomials as $\overline{a + bx}$, it is more convenient to write the polynomials in R/I_2 are $a + b(x - 2)$. Addition in R/I_2 is performed component-wise and the multiplication is

$$\overline{(a + b(x - 2))(c + d(x - 2))} = \overline{ac + (ad + bc)(x - 2) + bd(x - 2)^2}$$
$$= \overline{ac + (ad + bc)(x - 2)}, \tag{3.11}$$

where we replaced the product polynomial $p(x)$ of (possibly) degree 2 with the polynomial $p(2) + p'(2)(x - 2)$.

Since $\overline{p(x)} \times \overline{q(x)} = \overline{p(x)q(x)}$, the product in (3.11) shows that

$$\overline{p(x)q(x)} = \overline{p(2)q(2) + (p'(2)q(2) + p(2)q'(2))(x - 2)},$$

which recovers the product rule for derivatives.

Interestingly enough, the congruence class $\overline{a + bx}$ with respect to I_2 has a clear interpretation: It represents all polynomials that have $a + bx$ as the tangent line above $x = 2$. Figure 3.4 illustrates a few polynomials in $2 - \frac{1}{2}x$. Hence, all the polynomials shown are congruent to each other modulo I_2 and thus are equal in $\mathbb{R}[x]/I_2$. The addition (resp. the multiplication) between $\overline{a + bx}$ and $\overline{c + dx}$ in $\mathbb{R}[x]/I_2$ corresponds to the tangent line at $x = 2$ of the addition (resp. the product) any two polynomials with tangent lines $y = a + bx$ and $y = c + dx$ at $x = 2$. △

The ideal I_1 in Example 3.7.6 illustrates a common situation with quotient rings in polynomial rings. In the exercises, the reader is guided to prove the following results. (See Exercise 3.7.9.) Let R be a commutative ring with an identity $1 \neq 0$. Suppose that $a(x) = a_n x^n + \cdots + a_1 x + a_0 \in R[x]$ with $a_n \in U(R)$. In the quotient ring $R[x]/(a(x))$, for every polynomial $p(x)$ there exists a unique $q(x) \in R[x]$ with $\deg q(x) < n$ such $\overline{p(x)} = \overline{q(x)}$. Furthermore, in the quotient ring $R[x]/(a(x))$, the element $\overline{x}$ satisfies

$$\overline{x}^n = -a_n^{-1}(a_{n-1}x^{n-1} + \cdots + a_1 x + a_0).$$

Repeated application of this identity governs the multiplication operation in $R[x]/(a(x))$.

Example 3.7.7. Recall Example 3.6.5 which discussed the ideal $I = (5, x^2 - x - 2)$ in $\mathbb{Z}[x]$. We propose to describe $\mathbb{Z}[x]/I$. Since $5 \in I$, any element $a_0 \in \mathbb{Z}$ satisfies

$$\overline{a_0} = \overline{r}$$

in $\mathbb{Z}[x]/I$, where r is the remainder of a_0 when divided by 5. So

$$\mathbb{Z}[x]/I \cong (\mathbb{Z}/5\mathbb{Z})/(x^2 - x - 2) \cong (\mathbb{Z}/5\mathbb{Z})[x]/(x^2 + 4x + 3).$$

Then every element in $(\mathbb{Z}/5\mathbb{Z})[x]/(x^2 + 4x + 3)$ can be written as $\overline{a + bx}$, where $a, b \in \mathbb{Z}/5\mathbb{Z}$. It is also not uncommon to write the elements as $a + b\overline{x}$, where $\overline{x}$ satisfies $\overline{x}^2 = \overline{x} + 2$. △

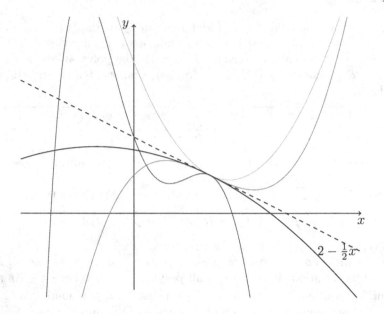

FIGURE 3.4: Some polynomials in the coset $2 - \frac{1}{2}x + I_2$.

The construction of quotient rings allows us to define many rings, the nature of whose elements seem less and less removed from something familiar. However, from an algebraic perspective, their ontology is no less strange than any other ring encountered "naturally."

When mathematicians first explored complex numbers, they dubbed the elements bi, where $b \in \mathbb{R}$, as *imaginary* because they considered such numbers so far removed from reality. With the formalism of quotient rings, complex numbers arise naturally as the quotient ring $\mathbb{R}[x]/(x^2 + 1)$. Indeed, every element in $\mathbb{R}[x]/(x^2 + 1)$ can be written uniquely as $a + b\overline{x}$ with $a, b \in \mathbb{R}$ and, since $\overline{x^2 + 1} = \overline{0}$, the variable satisfies $\overline{x}^2 = -1$. Hence, $\overline{x}$ has exactly the algebraic properties of the unit imaginary number i.

3.7.2   Isomorphism Theorems

In the study of groups, the isomorphism theorems established internal structure within groups and between groups related by homomorphisms. In a ring $(R, +, \times)$, the pair $(R, +)$ is an abelian group so the group-theoretic isomorphism theorems apply to any subgroup. As it turns out, if I is an ideal, the group isomorphisms are in fact isomorphisms of rings.

> **Theorem 3.7.8 (First Isomorphism Theorem for Rings)**
>
> If $\varphi : R \to S$ is a homomorphism of rings, then $\operatorname{Ker}\varphi$ is an ideal of R and $R/\operatorname{Ker}\varphi$ is isomorphic to $\varphi(R)$ as rings. Furthermore, if R is a ring and I is an ideal, then the natural projection (Definition 3.7.4) onto R/I is a surjective homomorphism and has $\operatorname{Ker}\varphi = I$.

Proof. We already saw that the kernel $\operatorname{Ker}\varphi$ is an ideal. The proof of the First Isomorphism Theorem of groups shows that the association $\Phi : R/(\operatorname{Ker}\varphi) \to S$ given by $\Phi(r + (\operatorname{Ker}\varphi)) = \varphi(r)$ is a well-defined function that is injective and that satisfies the homomorphism criteria for addition.

To establish this theorem, we only need to prove that Φ satisfies the homomorphism criteria for the multiplication. Let $\bar{r}_1, \bar{r}_2 \in R/(\operatorname{Ker}\varphi)$. Then

$$\Phi(\bar{r}_1\bar{r}_2) = \Phi(\overline{r_1 r_2}) = \varphi(r_1 r_2) = \varphi(r_1)\varphi(r_2) = \Phi(\bar{r}_1)\Phi(\bar{r}_2).$$

Thus, Φ is an injective ring homomorphism and hence establishes a ring isomorphism between $R/(\operatorname{Ker}\varphi)$ and the subring $\operatorname{Im}\varphi$ in S. $\qquad\square$

Example 3.7.9. For any ring R, the subset $\{0\}$ is an ideal of R and $R/\{0\}$ is isomorphic to R. To see this using the First Isomorphism Theorem, we use the identity homomorphism $i : R \to R$. It is obviously surjective and the kernel is $\operatorname{Ker}i = \{0\}$. Hence, $\{0\}$ is an ideal with $R/\{0\} \cong R$. $\qquad\triangle$

Example 3.7.10 (Augmentation Map Kernel). Let R be a commutative ring and let G be a finite group with $G = \{g_1, g_2, \ldots, g_n\}$. Let

$$I = \left\{ \sum_{i=1}^{n} a_i g_i \in R[G] \ \middle|\ a_1 + a_2 + \cdots + a_n = 0 \right\}.$$

We saw that this is an ideal by virtue of being the kernel of the augmentation map. Since the augmentation map is surjective, by the First Isomorphism Theorem, $R[G]/I \cong R$. $\qquad\triangle$

An important application of the First Isomorphism Theorem involves the so-called *reduction homomorphism* in polynomial rings. Let R be a commutative ring and let I be an ideal in R. The reduction homomorphism $\varphi : R[x] \to (R/I)[x]$ is defined by

$$\varphi(a_m x^m + \cdots + a_1 x + a_0) = \bar{a}_m x^m + \cdots + \bar{a}_1 x + \bar{a}_0,$$

where $\bar{a}$ is the coset of a in R/I. Because taking cosets into R/I behaves well with respect to $+$ and $\times$, it is easy to show that φ is in fact a homomorphism as claimed. The kernel of φ is $\operatorname{Ker}\varphi = I[x]$. The First Isomorphism Theorem leads to the following result.

Proposition 3.7.11

Let R be a ring and let I be an ideal. Then the subring $I[x]$ of $R[x]$ is an ideal and
$$R[x]/I[x] \cong (R/I)[x].$$

Listed below are the Second, Third, and Fourth isomorphism theorems for rings. We list them without proof and request the reader to prove them in the exercises. The proofs are very similar to the proofs for the corresponding group isomorphism theorems.

Theorem 3.7.12 (Second Isomorphism Theorem)

Let R be any ring, A a subring and let B be an ideal of R. Then $A + B$ is a subring of R, $A \cap B$ is an ideal of A and
$$(A + B)/B \cong A/(A \cap B).$$

Theorem 3.7.13 (Third Isomorphism Theorem)

Let R be any ring and let $I \subset J$ be ideals of R. Then J/I is an ideal of R/I and
$$(R/I)/(J/I) \cong R/J.$$

Theorem 3.7.14 (Fourth Isomorphism Theorem)

Let R be a ring and let I be an ideal. The correspondence $A \leftrightarrow A/I$ is an inclusion preserving bijection between the set of subrings of R/I and the set of subrings of R that contain I. Furthermore, a subring A of R is an ideal if and only if A/I is an ideal of R/I.

Example 3.7.15. A simple application of the Third Isomorphism Theorem for rings arises in modular arithmetic. Let $R = \mathbb{Z}$ and $I = (12) = 12\mathbb{Z}$. Now $J = (4) = 4\mathbb{Z}$ is also an ideal of $\mathbb{Z}$. The Third Isomorphism Theorem for this situation is
$$(\mathbb{Z}/12\mathbb{Z})/(4\mathbb{Z}/12\mathbb{Z}) \cong \mathbb{Z}/4\mathbb{Z}. \qquad \triangle$$

3.7.3 Chinese Remainder Theorem

The Chinese Remainder Theorem is a result in modular arithmetic that generalizes to quotient rings. In its form applied to modular arithmetic, it first appears circa the 4th century in the work of the Chinese mathematician Sun Tzu (not to be confused with the military general of the same name, known for *The Art of War*).

We first present the Chinese Remainder Theorem in its form applied to modular arithmetic.

Theorem 3.7.16

Let $n_1, n_2, \ldots, n_k$ be integers greater than 1 which are pairwise relatively prime, i.e., $\gcd(n_i, n_j) = 1$ for $i \neq j$. For any integers $a_1, a_2, \ldots, a_k \in \mathbb{Z}$, the system of congruences

$$\begin{cases} x \equiv a_1 \pmod{n_1} \\ x \equiv a_2 \pmod{n_2} \\ \vdots \\ x \equiv a_k \pmod{n_k} \end{cases}$$

has a unique solution x modulo $n = n_1 n_2 \cdots n_k$.

Proof. For each i, set $n_i' = n/n_i$. Now n_i and n_i' are relatively prime so there exist integers s_i and t_i such that

$$s_i n_i + t_i n_i' = 1.$$

Then for all i, we have $t_i n_i' \equiv 1 \pmod{n_i}$ so $a_i t_i n_i' \equiv a_i \pmod{n_i}$. Consider the integer given by

$$x = a_1 t_1 n_1' + a_2 t_2 n_2' + \cdots + a_k t_k n_k'.$$

Then, since $n_j' \equiv 0 \pmod{n_i}$ if $i \neq j$, we have $x \equiv a_i \pmod{n_i}$ for all i. Hence, x satisfies all of the congruence relations.

If another integer y satisfies all of the congruence conditions, then $x - y$ is congruent to 0 for all n_i, and hence $n \mid (x - y)$, establishing uniqueness of the solution. $\square$

Example 3.7.17. Find an x such that $x \equiv 3 \pmod 8$, $x \equiv 2 \pmod 5$, and $x \equiv 7 \pmod{13}$. The theorem confirms that there exists a unique solution for x modulo 520. The proof provides a method to find this solution. We have

$$n_1 = 8, \; n_2 = 5, \; n_3 = 13; \quad \text{and} \quad n_1' = 65, \; n_2' = 104, \; n_3' = 40.$$

We need to calculate t_i as the inverse of n_i' modulo n_i. Though we could use the Extended Euclidean Algorithm, in this example the integers n_i are small enough that trial and error suffices. We find

$$t_1 = 1, \quad t_2 = 4, \quad t_3 = 1.$$

Thus, according to the above proof, the solution to the system of congruence equations is

$$\begin{aligned} x &= a_1 t_1 n_1' + a_2 t_2 n_2' + a_3 t_3 n_3' \\ &= 3 \times 1 \times 65 + 2 \times 4 \times 104 + 7 \times 1 \times 40 = 1307. \end{aligned}$$

Hence, the solution to the system is $x \equiv 267 \pmod{520}$. $\triangle$

To generalize the above theorem to rings, a congruence is tantamount to equality in a quotient ring and the notion of relatively prime corresponds to comaximality of ideals.

Theorem 3.7.18 (Chinese Remainder Theorem)

Let R be a ring commutative ring with an identity $1 \neq 0$ and let $A_1, A_2, \ldots, A_k$ be ideals in R. The map

$$R \to R/A_1 \oplus R/A_2 \oplus \cdots \oplus R/A_k$$
$$r \mapsto (r + A_1, r + A_2, \ldots, r + A_k)$$

is a ring homomorphism with kernel $A_1 \cap A_2 \cap \cdots \cap A_k$. If the ideals are pairwise comaximal, then this map is surjective, $A_1 A_2 \cdots A_k = A_1 \cap A_2 \cap \cdots \cap A_k$, and

$$R/(A_1 A_2 \cdots A_k) \cong (R/A_1) \oplus (R/A_2) \oplus \cdots \oplus (R/A_k),$$

as a ring isomorphism.

Proof. We prove the result first for $k = 2$ and then extend by induction.

Let A_1 and A_2 be two comaximal ideals and consider $\varphi : R \to R/A_1 \oplus R/A_2$ defined as in the statement of the theorem. Since A_1 and A_2 are comaximal, there exist $a_1 \in A_1$ and $a_2 \in A_2$ such that $a_1 + a_2 = 1$. Let $r, s \in R$ be arbitrary and let $x = ra_2 + sa_1$. Then

$$\varphi(x) = (x + A_1, x + A_2) = (ra_2 + A_1, sa_1 + A_2)$$
$$= (r - ra_1 + A_1, s - sa_2 + A_2) = (r + A_1, s + A_2).$$

Hence, φ is surjective. The kernel of φ is $\{r \in R \mid r \in A_1 \text{ and } r \in A_2\} = A_1 \cap A_2$. Since A_1 and A_2 are comaximal, by the result of Exercise 3.6.19, $A_1 \cap A_2 = A_1 A_2$. By the First Isomorphism Theorem (for rings),

$$R/(\mathrm{Ker}\,\varphi) = R/(A_1 A_2) \cong (R/A_1) \oplus (R/A_2).$$

We have proved the theorem for any pair of comaximal ideals.

Now suppose that the theorem holds for some integer k. Suppose that we have $A_1, A_2, \ldots, A_{k+1}$ are pairwise comaximal. Since A_{k+1} is comaximal with A_i for $1 \leq i \leq k$, then for all i with $1 \leq i \leq k$, there exist $a_i \in A_i$ and $b_i \in A_{k+1}$ such that $a_i + b_i = 1$. Then

$$1 = 1^k = (a_1 + b_1)(a_2 + b_2) \cdots (a_k + b_k) = a_1 a_2 \cdots a_k + b, \text{ where } b \in A_{k+1}.$$

Thus, the ideals $A_1 A_2 \cdots A_k$ and A_{k+1} are comaximal. Thus, since the theorem holds for $k = 2$, then

$$R/(A_1 A_2 \cdots A_{k+1}) \cong (R/(A_1 A_2 \cdots A_k)) \oplus (R/A_{k+1}).$$

By the induction hypothesis, we deduce that

$$R/(A_1 A_2 \cdots A_{k+1}) \cong (R/A_1) \oplus (R/A_2) \oplus \cdots \oplus (R/A_{k+1}).$$

By induction, the theorem holds for all integers $k \geq 2$. $\qquad \square$

This theorem leads to the decomposition for the group of units in modular arithmetic.

Corollary 3.7.19

Let n be a positive integer and let $n = p_1^{\alpha_1} p_2^{\alpha_2} \cdots p_k^{\alpha_k}$. Then

$$\mathbb{Z}/n\mathbb{Z} \cong (\mathbb{Z}/p_1^{\alpha_1}\mathbb{Z}) \oplus (\mathbb{Z}/p_2^{\alpha_2}\mathbb{Z}) \oplus \cdots \oplus (\mathbb{Z}/p_k^{\alpha_k}\mathbb{Z}).$$

By Proposition 3.1.14, we have an isomorphism of groups

$$U(n) \cong U(p_1^{\alpha_1}) \oplus U(p_2^{\alpha_2}) \oplus \cdots \oplus U(p_k^{\alpha_k}).$$

It is still a number theory problem to determine $U(p^\alpha)$ for various primes p and powers α.

3.7.4 Useful CAS Commands

Not all computer algebra systems implement quotient rings or calculations within quotient rings for arbitrary rings. Consider the following code in SAGE.

--------------------------------- Sage ---------------------------------

```
sage: R.<x>=ZZ['x']
sage: J=(2*x^3-x+5)*R
sage: S=QuotientRing(R,J)
-----------
TypeError....
TypeError: polynomial must have unit leading coefficient
sage: I=(x^3-3*x+5)*R
sage: S=QuotientRing(R,I)
sage: S.gens()
(xbar,)
sage: a=S.gens()[0]
sage: a^5
-5*xbar^2 + 9*xbar - 15
```

This code first defines the ring $R = \mathbb{Z}[x]$. SAGE returns a TypeError when trying to construct R/J, not because the quotient ring does not exist but because SAGE only has implemented this situation when the polynomial generating the principal ideal J has a unit as a leading coefficient. Constructing

R/I causes no errors. In SAGE, the data type of S is a "Univariate Quotient Polynomial Ring in xbar over Integer Ring with modulus x^3 - 3*x + 5." Note that **xbar**, which refers to the element $\overline{x}$ in $\mathbb{Z}[x]/(x^3 - 3x + 5)$ is not a name; the line defining a gives it the role of $\overline{x}$. In the last line, we asked SAGE to calculate (in simplest terms) $\overline{x}^5$.

EXERCISES FOR SECTION 3.7

1. Refer to Exercises 3.5.5. For subsets that were ideals, describe the elements and operations in the corresponding quotient ring.

2. Prove that the quotient ring $\mathbb{Z}[x]/(5x - 1)$ is isomorphic to the subring $\mathbb{Z}[\frac{1}{5}] = \left\{ \frac{n}{5^k} \,\middle|\, n \in \mathbb{Z}, k \in \mathbb{N} \right\}$ of $\mathbb{Q}$.

3. Let $[a, b]$ be an interval in $\mathbb{R}$ and consider the ring $C^2([a, b], \mathbb{R})$ (real-valued functions over $[a, b]$ whose zeroth, first, and second derivatives are continuous). The operations are addition and multiplication of functions. Let

$$I = \{f(x) \in C^2([a, b], \mathbb{R}) \mid f(-1) = f'(-1) = f''(-1) = 0\}.$$

   Show that I is an ideal and describe the quotient ring $C^2([a, b], \mathbb{R})/I$.

4. Consider the ideal $I = (x+1, x^2+1)$ in the ring $\mathbb{R}[x]$. Prove that $2 \in (x+1, x^2+1)$. Deduce that $I = \mathbb{R}[x]$ so that $\mathbb{R}[x]/I \cong \{0\}$, the trivial ring.

5. Consider the ideal $I = (x+1, x^2+1)$ in the ring $\mathbb{Z}[x]$. Prove that $2 \in (x+1, x^2+1)$. Prove that $I = (x + 1, 2)$. Prove that $\mathbb{Z}[x]/I \cong \mathbb{Z}/2\mathbb{Z}$.

6. Consider the quotient ring $\mathbb{R}[x]/(x^3 + x - 2)$.

   (a) In this quotient ring, calculate and simplify as much as possible the sum and product of $\overline{x^2 + 7x - 1}$ and $\overline{2x^2 - x + 5}$.

   (b) Prove that this quotient ring is not an integral domain.

7. Consider the quotient ring $\mathbb{Z}[x]/(2x^3 + 5x - 1)$.

   (a) In this quotient ring, calculate and simplify as much as possible the sum and product of $\overline{4x^2 - 5x + 2}$ and $\overline{2x + 7}$.

   (b) Show that $\overline{x}$ is a unit.

8. Let $f(x) = x^2 + 2$ in $\mathbb{F}_5[x]$.

   (a) Prove that the quotient ring $\mathbb{F}_5[x]/(f(x))$ has 25 elements.

   (b) Prove that $\mathbb{F}_5[x]/(f(x))$ contains no zero divisor.

   (c) Deduce that $\mathbb{F}_5[x]/(f(x))$ is a field. [Hint: See Exercise 3.1.24.]

9. Let R be an integral domain and let $a(x) \in R[x]$ with $\deg a(x) = n > 0$ and $a_n \in U(R)$.

   (a) Prove that, in the quotient ring $R[x]/(a(x))$, the element $\overline{x}$ satisfies $\overline{x}^n = -\overline{a}_n^{-1}(\overline{a}_{n-1}\overline{x}^{n-1} + \cdots + \overline{a}_1\overline{x} + \overline{a}_0)$.

   (b) Prove that for every polynomial $p(x)$, there exists a polynomial $q(x)$ that is either 0 or has $\deg q(x) < n$ such that $\overline{p(x)} = \overline{q(x)}$ in $R[x]/(a(x))$. [Hint: Use induction on the degree of $p(x)$.]

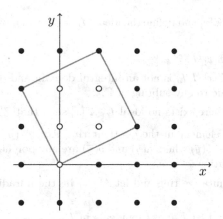

FIGURE 3.5: A visualization of $\mathbb{Z}[i]/(2+i)$.

(c) Prove that the polynomial $q(x)$ described above is unique.

[Remark: Because of this last result, in this case, we typically do not put ⁻ over the constants when considering an element of $R[x]/(a(x))$.]

10. Consider the ring $\mathbb{Z}[i]$ and the ideal $I = (2+i)$. We study the quotient ring $\mathbb{Z}[i]/(2+i)$.

    (a) Prove that $-1+2i$, $-2-i$, and $1-2i$ are in the ideal $(2+i)$.

    (b) Prove every element $a+bi \in \mathbb{Z}[i]$ is congruent modulo I to at least one element inside or on the boundary of the square with 0, $2+i$, $1+3i$, and $-1+2i$. Prove also that the vertices of the square are congruent to each other modulo I. [See Figure 3.5.]

    (c) Using division in $\mathbb{C}$, show that none of the five elements in 0, i, $2i$, $1+i$, $1+2i$ are congruent to each other and conclude that $\mathbb{Z}[i]/(2+i) = \{\overline{0}, \overline{i}, \overline{2i}, \overline{1+i}, \overline{1+2i}\}$.

    (d) Write down the multiplication table in $\mathbb{Z}[i]/(2+i)$ and deduce that this quotient ring is a field.

    (e) Find an explicit isomorphism between $\mathbb{Z}[i]/(2+i)$ and $\mathbb{F}_5$.

11. Some rings with eight elements.

    (a) Consider the (quotient) ring $R_1 = (\mathbb{Z}/2\mathbb{Z})[x]/(x^3+1)$. List all 8 elements of this ring and determine whether (and show how) they are units, zero divisors, or neither.

    (b) Repeat the same question with the ring $R_2 = (\mathbb{Z}/2\mathbb{Z})[x]/(x^3+x+1)$.

    (c) Consider also the ring $R_3 = \mathbb{Z}/8\mathbb{Z}$ of modular arithmetic modulo 8. The rings R_1, R_2, and R_3 all have 8 elements. Show that none of them are isomorphic to each other.

12. Show that $(\mathbb{Z}/7\mathbb{Z})[x]/(x^2-3)$ is a field.

13. Consider the ring $\mathbb{Z}[x]$ and define the ideals $I_n = (nx-1)$, where n is any positive integer.

   (a) Prove that $\mathbb{Z}[x]/I_n \cong \mathbb{Z}\left[\frac{1}{n}\right]$.

   (b) Prove that $\mathbb{Z}[x]/I_2I_3$ is not an integral domain and deduce that it cannot be isomorphic to any subring of $\mathbb{C}$.

   (c) Prove that there exists no ideal $I \in \mathbb{Z}[x]$ such that $\mathbb{Z}[x]/I \cong \mathbb{Q}$.

14. Prove that every element in the quotient ring $\mathbb{R}[x,y]/(y - x^2)$ can be written uniquely as $a(\bar{y})+\bar{x}b(\bar{y})$ where $a(\bar{y})$ and $b(\bar{y})$ are any polynomials in $\bar{y}$ and where $\bar{x}$ satisfies the relation $\bar{x}^2 = \bar{y}$.

15. Let R be a commutative ring and let $\mathcal{N}(R)$ be the nilradical of R. Prove that $\mathcal{N}(R/\mathcal{N}(R)) = 0$.

16. Let R be a ring and let I and J be ideals in R.

   (a) Prove that the function $f : R/I \to R/(I+J)$ defined by $f(r+I) = r+(I+J)$ is a well-defined function.

   (b) Show that f is a homomorphism.

   (c) Show that $\operatorname{Ker} f \cong J/(I \cap J)$.

17. Let R be a ring and let I be an ideal of R. Prove that $M_n(I)$ is an ideal of $M_n(R)$ and show that $M_n(R)/M_n(I) \cong M_n(R/I)$. [Hint: First Isomorphism Theorem.]

18. Let $U_n(R)$ be the set of upper triangular $n \times n$ matrices with coefficients in R. Prove that the subset

$$I = \{A \in U_n(R) \,|\, a_{ii} = 0 \text{ for all } i = 1, 2, \ldots, n\}$$

is an ideal in $U_n(R)$ and determine $U_n(R)/I$.

19. Prove the Second Isomorphism Theorem for rings. (See Theorem 3.7.12.)

20. Prove the Third Isomorphism Theorem for rings. (See Theorem 3.7.13.)

21. Prove the Fourth Isomorphism Theorem for rings. (See Theorem 3.7.14.)

22. Let R be a ring and let e be an idempotent element in the center $C(R)$. Observe that the ideals $(e) = Re$ and $(1-e) = R(1-e)$ and prove that $R \cong Re \oplus R(1-e)$.

23. Consider the group ring $\mathbb{Z}[S_3]$. Show that the set I of elements

$$\alpha = a_1 + a_2(1\,2) + a_3(1\,3) + a_4(2\,3) + a_5(1\,2\,3) + a_6(1\,3\,2)$$

satisfying

$$\begin{cases} a_1 + a_2 + a_3 + a_4 + a_5 + a_6 = 0 \\ a_1 - a_2 - a_3 - a_4 + a_5 + a_6 = 0 \end{cases}$$

is an ideal and show that $\mathbb{Z}[S_3]/I \cong \mathbb{Z} \oplus \mathbb{Z}$. Prove that every element in $\mathbb{Z}[S_3]/I$ can be written uniquely as $a_1\bar{1} + a_2\overline{(1\,2)}$, with $a_1, a_2 \in \mathbb{Z}$.

24. Let R be a PID and let I be an ideal in R. Prove that R/I is a PID.

25. Consider the subset $I = \{(2m, 3n) \,|\, m, n \in \mathbb{Z}\}$ in the ring $\mathbb{Z} \oplus \mathbb{Z}$. Prove that I is an ideal in $\mathbb{Z} \oplus \mathbb{Z}$ and that $(\mathbb{Z} \oplus \mathbb{Z})/I \cong \mathbb{Z}/2\mathbb{Z} \oplus \mathbb{Z}/3\mathbb{Z}$.

26. Let R and S be rings with identity $1 \neq 0$. Prove that every ideal of $R \oplus S$ is of the form $I \oplus J$ for some ideals $I \subseteq R$ and $J \subseteq S$.

27. In this exercise, we prove Proposition 3.7.3. Let $(R, +, \times)$ be a ring and let $\sim$ be an equivalence relation on R and call I the equivalence class $[0]$.

    (a) Using Exercise 2.3.30, show that if $\sim$ satisfies the addition part of (3.10), then I is a subgroup of $(R, +)$ and that the equivalence classes of $\sim$ are the cosets $a + I$, where $a \in R$.

    (b) Show that if in addition $\sim$ satisfies the multiplication criterion in (3.10), then I is a two-sided ideal of R.

    (c) Deduce Proposition 3.7.3.

28. Solve the following system of congruences in $\mathbb{Z}$:

$$\begin{cases} x \equiv 3 \pmod{5} \\ x \equiv 7 \pmod{11}. \end{cases}$$

29. Solve the following system of congruences in $\mathbb{Z}$:

$$\begin{cases} x \equiv 4 \pmod{10} \\ x \equiv 17 \pmod{21}. \end{cases}$$

30. Solve the following system of congruences in $\mathbb{Z}$:

$$\begin{cases} x \equiv 2 \pmod{3} \\ x \equiv 3 \pmod{5} \\ x \equiv 9 \pmod{13}. \end{cases}$$

3.8 Maximal Ideals and Prime Ideals

We now turn to two different classes of ideals in rings. Though it will not be obvious at first, they both attempt to generalize the notion and properties of prime numbers in $\mathbb{Z}$.

Recall that in elementary number theory there are two equivalent definitions of prime numbers.

(1) An integer $p > 1$ is a prime number if and only if p is only divisible by 1 and itself. (The definition of prime.)

(2) An integer $p > 1$ is a prime number if and only if whenever $p|ab$, then $p|a$ or $p|b$. (Euclid's Lemma, Proposition A.5.18.)

In ring theory in general, these two notions are no longer equivalent. The connection between properties of ideals and integer arithmetic comes from the fact that $a|b$ if and only if $b \in (a)$, if and only if $(b) \subseteq (a)$.

3.8.1   Maximal Ideals

Definition 3.8.1

An ideal I in a ring R is called *maximal* if $I \neq R$ and if the only ideals J such that $I \subseteq J \subseteq R$ are $J = I$ or $J = R$.

An arbitrary ring need not have maximal ideals. In many specific rings it is obvious or at least very simple to prove the existence of maximal ideals. However, the following general proposition proves the existence of maximal ideals in rings with a very few conditions. The proof relies on Zorn's Lemma, which is equivalent to the Axiom of Choice.

Theorem 3.8.2 (Krull's Theorem)

In a ring with an identity, every proper ideal is contained in a maximal ideal.

Proof. Let R be a ring with an identity and let I be any proper ideal. Let $\mathcal{S}$ be the set of all proper ideals that contain I. Then $\mathcal{S}$ is a nonempty set (since it contains I), which is partially ordered by inclusion. Let $\mathcal{C}$ be any chain of ideals in $\mathcal{S}$. We show that $\mathcal{C}$ has an upper bound.

Define the set

$$J = \bigcup_{A \in \mathcal{C}} A.$$

We prove that J is an ideal of R. Let $a, b \in J$. Then $a \in A_1$ and $b \in A_2$ where $A_1, A_2 \in \mathcal{C}$. But since $\mathcal{C}$ is a chain, either $A_1 \subseteq A_2$ or $A_2 \subseteq A_1$. Let's assume the former without loss of generality. Then both a and b are in the ideal A_2. Thus, $b - a \in A_2 \subset J$ so J is an additive subgroup of R by the one-step subgroup test. Now let $r \in R$ and let $a \in J$. Then a is an element of some $A \in \mathcal{C}$. Since A is an ideal we have $ra \in A \subset J$ and $ar \in A \subset J$. Thus, we have shown that J is an ideal.

Furthermore, since all ideals $A \in \mathcal{C}$ are proper, then none of them contains the identity. Thus, J does not contain the identity and hence J is a proper ideal. In particular, $J \in \mathcal{S}$. Thus, every chain of elements in $\mathcal{S}$ has an upper bound. By Zorn's Lemma, we conclude that $\mathcal{S}$ has a maximal element. A maximal element of $\mathcal{S}$ is a maximal ideal that contains I. $\square$

In the integers, suppose that (n) is a maximal ideal in $\mathbb{Z}$. Then any ideal $I = (m)$ satisfying $(n) \subseteq (m) \subseteq R$ is either $(n) = (m)$ or $(m) = R$. Expressing this in terms of divisibility, we deduce that $m | n$ implies $m = \pm 1$ or $m = \pm n$. Supposing that m is positive, then $m = 1$ or $m = n$. This corresponds to the definition of a prime number, listed above in (Criterion (1)). Hence, the maximal ideals in $\mathbb{Z}$ correspond to the ideals (p) where p is a prime number.

The next proposition offers a criterion for when an ideal is maximal.

Proposition 3.8.3

Suppose that M is an ideal such that R/M is a field. Then M is a maximal ideal.

Proof. By the Fourth Isomorphism Theorem for rings, there is a bijective correspondence between the ideals of R/M and the ideals of R that contain M. If R/M is a field, then any ideal $\bar{I}$ in R/M that contains more than 0 is all of R/M. Hence, R/M has exactly two ideals, namely (0) and R/M. By the Fourth Isomorphism, there are only two ideals in R that contain M. These must be R and M themselves, which implies that M is maximal. □

Proposition 3.8.4

Let R be a commutative ring with an identity $1 \neq 0$. Then an ideal M is maximal if and only if R/M is a field.

Proof. Proposition 3.8.3 established one direction. Under the assumption that R is commutative with an identity, the converse follows from Proposition 3.5.9. By the Fourth Isomorphism Theorem, when M is maximal, R/M contains only two ideals. Furthermore, R/M inherits commutativity from R, and $1 + M$ is an identity element in R/M. By Proposition 3.5.9, R/M is a field. □

If a ring is commutative with an identity, the above proposition offers a strategy to prove that an ideal I is maximal. Determine the quotient ring R/I, prove that R/I is a field, and then invoke the proposition. With the integers, we had previously seen that all ideals are of the form (n) with n nonnegative and that $\mathbb{Z}/n\mathbb{Z}$ is a field if and only if n is a prime number. Proposition 3.8.4 implies that (n) is maximal if and only if n is a prime number, which returns to prime numbers as the motivation for the notion of a maximal ideal.

Example 3.8.5. As a nonobvious example of Proposition 3.8.4, consider the ring of Gaussian integers $\mathbb{Z}[i]$ and the principal ideal $(2 + i)$. Exercise 3.7.10 shows that $\mathbb{Z}[i]/(2 + i)$ is isomorphic to $\mathbb{F}_5$, which is a field, so $(2 + i)$ is a maximal ideal. △

Example 3.8.6. Exercise 3.8.11 shows that the ring $M_2(\mathbb{R})$ has only two ideals, the whole ring itself and the zero ideal (0). So (0) is a maximal ideal in $M_2(\mathbb{R})$ but $M_2(\mathbb{R})/(0) \cong M_2(\mathbb{R})$. Consequently, the criterion that a ring be commutative is essential for the equivalence of Proposition 3.8.4. △

3.8.2 Prime Ideals

In the language of ring theory, the second criteria for primeness in the integers (Criterion (2)) can be restated to say that $p \geq 2$ is a prime number if and only if as ideals $(ab) \subseteq (p)$ implies that $(a) \subseteq (p)$ or $(b) \subseteq (p)$. This motivates the following definition.

Definition 3.8.7

Let R be any ring. An ideal $P \neq R$ is called a *prime* ideal if whenever two ideals A and B satisfy $AB \subseteq P$, then $A \subseteq P$ or $B \subseteq P$.

Many applications that involve prime ideals occur in commutative rings. In this context, Definition 3.8.7 has an equivalent statement.

Proposition 3.8.8

Let R be a commutative ring with an identity $1 \neq 0$. An ideal $P \neq R$ is prime if and only if $ab \in P$ implies $a \in P$ or $b \in P$.

Proof. Let P be a prime ideal. Let $a, b \in R$. Since R is commutative, then

$$(ab) = \{rab \mid r \in R\} \quad \text{and} \quad (a)(b) = \{r_1 r_2 ab \mid r_1 \in R, r_2 \in R\}.$$

Obviously $(a)(b) \subseteq (ab)$ but since R has an identity $1 \neq 0$, by setting $r_1 = 1$ or $r_2 = 1$ as needed, we see that $(ab) \subseteq (a)(b)$. Hence, $(ab) = (a)(b)$.

Now suppose that $ab \in P$. Then $(ab) = (a)(b) \subseteq P$ and since P is a prime ideal $(a) \subseteq P$ or $(b) \subseteq P$. Hence, $a \in P$ or $b \in P$.

Conversely, suppose that P satisfies the property that whenever $ab \in P$ then $a \in P$ or $b \in P$. Let A and B be two arbitrary ideals such that $AB \subseteq P$. For all pairs $(a, b) \in A \times B$, we have $ab \in P$ and, therefore by hypothesis, $a \in P$ or $b \in P$. Assume that $A \not\subseteq P$ and $B \not\subseteq P$. Then there exist $a_0 \in A - P$ and $b_0 \in B - P$ such that $a_0 b_0 \in P$. This leads to a contradiction, so we must conclude that $A \subseteq P$ or $B \subseteq P$. Hence, P is a prime ideal. $\square$

Prime ideals possess an equivalent characterization via quotient rings similar to Proposition 3.8.4 for maximal ideals.

Proposition 3.8.9

An ideal P in a commutative ring with an identity $1 \neq 0$ is prime if and only if R/P is an integral domain.

Proof. Suppose that P is a prime ideal. By definition, if $a, b \in R - P$, then $ab \in R - P$. Passing to the quotient ring, we see that $\overline{a}$ and $\overline{b}$ are not $\overline{0}$ in R/P implies that $\overline{ab} \neq 0$ in R/P. Thus, R/P has no zero divisors. The quotient ring R/P contains an identity $\overline{1}$ and inherits commutativity from R. Hence, , R/P is an integral domain.

Conversely, suppose that R/P is an integral domain. Let nonzero elements in R/P correspond to $\overline{a}, \overline{b}$ with $a, b \in R - P$. Since R/P is an integral domain, $\overline{ab} \neq \overline{0}$ so $ab \notin P$. We have shown that $a \notin P$ and $b \notin P$ implies that $ab \notin P$. The contrapositive to this is $ab \in P$ implies that $a \in P$ or $b \in P$. This proves that P is a prime ideal. $\square$

> **Corollary 3.8.10**
>
> In a commutative ring with an identity $1 \neq 0$, every maximal ideal is a prime ideal.

Proof. Let R be a commutative ring and let M be a maximal ideal. Then by Proposition 3.8.4, R/M is a field. Every field is an integral domain. So by Proposition 3.8.9, M is prime. $\square$

By Criteria (2) for primality in the integers, using either the definition or Proposition 3.8.9, we see that an ideal $(n) \subseteq \mathbb{Z}$ is prime if and only if n is a prime number or $n = 0$. The zero ideal (0) is the only ideal in $\mathbb{Z}$ that is prime but not maximal. However, as the following example illustrates, this similarity between prime and maximal ideals does not hold for arbitrary rings.

Example 3.8.11. Let $R = \mathbb{Z}[x]$ and consider the ideal $I_1 = (x^2 - 2)$. The quotient ring $R/I_1 \cong \mathbb{Z}[\sqrt{2}]$ is an integral domain but is not a field so I_1 is a prime ideal that is not maximal. $\triangle$

> **Proposition 3.8.12**
>
> Let R be a PID. Then every nonzero prime ideal P is a maximal ideal.

Proof. Let $P = (p)$ be a nonzero prime ideal in R and let $I = (m)$ be an ideal containing P.

Since $p \in (m)$ then there exists $r \in R$ such that $mr = p$. But then, $mr \in P$ so either $m \in P$ or $r \in P$. If $m \in P$, then $(m) \subseteq P$ which implies that $(m) = P$ since we already knew that $P \subseteq (m)$. Now suppose that $r \in P$ so that there exists some $s \in R$ such that $r = sp$. Then from $mr = p$ we deduce that $msp = p$. Since R is in an integral domain and $p \neq 0$, the cancellation law holds, so $ms = 1$. Thus, m is a unit and $(m) = R$. Hence, we have proved that either $I = P$ or $I = R$. Thus, we conclude that P is maximal. $\square$

Example 3.8.13. In this example, we revisit $\mathbb{Z}[x]$ and show a few different ideals that are prime, maximal, or neither. Consider the following chain of ideals

$$(x^3 + x^2 - 2x - 2) \subseteq (x^2 - 2) \subseteq (x^2 - 2, x^2 + 13) \subseteq (x^2 - 2, 5).$$

(It is not yet obvious that this is a chain but we shall see that it is.) The ideal $(x^3 + x^2 - 2x - 2)$ is not prime because $(x - 1)(x^2 - 2) = x^3 + x^2 - 2x - 2$ is in this ideal whereas neither $x - 1$ nor $x^2 - 2$ is in the ideal, since nonzero polynomials in $(x^3 + x^2 - 2x - 2)$ have degree 3 or higher.

The ideal $(x^2 - 2)$ is prime but not maximal because $\mathbb{Z}[x]/(x^2 - 2) \cong \mathbb{Z}[\sqrt{2}]$ is an integral domain but is not a field.

The ideal $I_3 = (x^2 - 2, x^2 + 13)$ is actually equal to $(x^2 - 2, 15)$. We see that $15 \in I_3$ because $x^2 + 13 - (x^2 - 2) = 15$. However, $x^2 + 13 \in (x^2 - 2, 15)$ so $I_3 = (x^2 - 2, 15)$. This is not a prime ideal because neither 3 nor 5 are in I_3 whereas $15 \in I_3$.

Finally, the ideal $(x^2 - 2, 5)$ is maximal. We can see this because

$$\mathbb{Z}[x]/(x^2 - 2, 5) \cong \mathbb{Z}/5\mathbb{Z}[x]/(x^2 - 2) \cong \mathbb{Z}/5\mathbb{Z}[x]/(x^2 + 3).$$

However, $\bar{a} \neq \bar{0}$ is not a zero divisor and if $\overline{a + bx}$ were a zero divisor with $\bar{b} \neq \bar{0}$ in $\mathbb{Z}/5\mathbb{Z}[x]/(x^2 + 3)$, then $\overline{b^{-1}a + x}$ would be a zero divisor. Then there would exist $\overline{x + c}$ and $\overline{x + d}$ with $c, d \in \mathbb{Z}/5\mathbb{Z}$ such that $(x + c)(x + d) = x^2 + 3$ in $\mathbb{Z}/5\mathbb{Z}[x]$. Then c and d would need to solve $c + d = 0$ and $cd = 3$. Hence, $d = -c$ and $-c^2 = 3$. Checking the five cases, we find that $-c^2 = 3$ has no solutions in $\mathbb{Z}/5\mathbb{Z}$. Hence, $\mathbb{Z}/5\mathbb{Z}[x]/(x^2 + 3)$ has no zero divisors. This quotient ring is commutative and has an identity so $\mathbb{Z}/5\mathbb{Z}[x]/(x^2 + 3)$ is an integral domain. It is finite so it is a field, so we conclude that $(x^2 - 2, 5)$ is a maximal ideal. $\triangle$

Example 3.8.14. Let $R = C^0([0, 1], \mathbb{R})$ be the ring of continuous real-valued functions on $[0, 1]$ and let a be a number in $[0, 1]$. Let M_a be the set of all functions such that $f(a) = 0$. Note that $M_a = \mathrm{Ker}(\mathrm{ev}_a)$ so M_a is an ideal. Also by the surjectivity of ev_a and the First Isomorphism Theorem, we have $R/M_a \cong \mathbb{R}$. Since $\mathbb{R}$ is a field, by Proposition 3.8.4, M_a is a maximal ideal.

Consider now the ideal

$$I = \{f \in C^0([0, 1], \mathbb{R}) \mid f(0) = f(1) = 0\}.$$

This ideal is not prime because, for example, the functions $g, h : [0, 1] \to \mathbb{R}$ given by $g(x) = x$ and $h(x) = 1 - x$ are such that $g, h \notin I$ but $gh \in I$. Consequently, it is not maximal either. $\triangle$

Prime ideals and maximal ideals possess many interesting properties. Properties of prime ideals in a commutative ring are a central theme in the study of commutative algebra, the branch of algebra that studies in depth the properties of commutative rings. The section exercises present some questions that ask the reader to determine whether a given ideal is prime or maximal but also investigate many of these properties. The reader is encouraged to at least skim the statements of exercises to acquire some intuition about the properties of prime ideals.

3.8.3   Useful CAS Commands

For ideals defined in the package `PolynomialIdeals` in *Maple* and for ideals in certain types rings in Sage there are commands to test whether an ideal is prime or maximal. The following commands test whether the ideal I has a certain property.

Test	*Maple*	Sage
If maximal	$IsMaximal(I);$	`I.is_maximal()`
If prime	$IsPrime(I);$	`I.is_prime()`

EXERCISES FOR SECTION 3.8

1. Show that the ideal $\{(2m, 3n) \mid m, n \in \mathbb{Z}\}$ is not a prime ideal.

2. Let R be a commutative ring. Consider the ideal (x) in the polynomial ring $R[x]$.
   (a) Prove that (x) is a prime ideal if and only if R is an integral domain.
   (b) Prove that (x) is a maximal ideal if and only if R is a field.

3. Let R be a ring with an identity $1 \neq 0$. Prove that if the set of nonunits is an ideal M, then M is the unique maximal ideal in R. Conversely, prove that if R is a commutative ring that contains a unique maximal ideal M, then $R - M$ is the set of all units. [Note: A commutative ring with a unique maximal ideal is called a *local* ring.]

4. Consider the subset $R \subset \mathbb{Q}$ of fractions $\frac{a}{b}$ for which, in reduced form, 19 does not divide b.
   (a) Show that R is a subring of $\mathbb{Q}$.
   (b) Determine the set of units in R.
   (c) Use Exercise 3.8.3 to conclude that the set of nonunits in R is the unique maximal ideal in R.

5. Consider the ring $U_n(\mathbb{R})$ of upper triangular $n \times n$ matrices with real coefficients. Fix an integer k with $1 \leq k \leq n$. Prove that the set

$$I_k = \{A \in U_n(\mathbb{R}) \mid a_{kk} = 0\}$$

is a maximal ideal in $U_n(\mathbb{R})$. [Hint: Use the First Isomorphism Theorem.]

6. Consider the ring $U_n(\mathbb{Z})$ of upper triangular $n \times n$ matrices with integer coefficients. Fix an integer k with $1 \leq k \leq n$. Prove that the set

$$I_k = \{A \in U_n(\mathbb{Z}) \mid a_{kk} = 0\}$$

is a prime ideal in $U_n(\mathbb{Z})$ that is not maximal. Explain how this differs from the previous exercise.

7. Let $R = C^0(\mathbb{R}, \mathbb{R})$ and consider the set of functions of bounded support,

$$I = \{f \in C^0(\mathbb{R}, \mathbb{R}) \mid \operatorname{Supp}(f) \text{ is bounded}\}.$$

[Recall that a subset S of $\mathbb{R}$ is bounded if there exists some c such that $S \subseteq [-c, c]$.]
   (a) Prove that I is an ideal.
   (b) Prove that any maximal ideal that contains I is not equal to any of the ideals M_a described in Example 3.8.14.

8. This exercise asks the reader to prove the following modification of Proposition 3.8.4. Let R be any ring. An ideal M is maximal if and only if the quotient R/M is a *simple ring*. [A simple ring is a ring that contains no ideals except the 0 ideal and the whole ring.]

9. In noncommutative rings, we call a left ideal a *maximal left ideal* (and similarly for right ideals) if it is maximal in the poset of proper left ideals ordered by inclusion. Prove that maximal left ideals (and maximal right ideals) exist under the same conditions as for Krull's Theorem.

10. Prove that the set of matrices $\{A \in M_2(\mathbb{R}) \mid a_{11} = a_{21} = 0\}$ is a maximal left ideal of $M_2(\mathbb{R})$. (See Exercise 3.8.9.) Find a maximal right ideal in $M_2(\mathbb{R})$.

11. Prove that the only two ideals in $M_2(\mathbb{R})$ are the 0 ideal and $M_2(\mathbb{R})$. (Hint: Refer to and generalize Exercise 3.6.3.)

12. Find all the prime ideals in $\mathbb{Z} \oplus \mathbb{Z}$.

13. Prove that $(y - x^2)$ is a prime ideal in $\mathbb{R}[x, y]$. Prove also that it is not maximal.

14. Prove that (y, x^2) is not prime in $\mathbb{R}[x, y]$.

15. Prove that the principal ideal $(x^2 + y^2)$ is prime in $\mathbb{R}[x, y]$ but not in $\mathbb{C}[x, y]$.

16. Show that in $C^1(\mathbb{R}, \mathbb{R})$ the ideal $I = \{f \mid f(2) = f'(2) = 0\}$ is not a prime ideal.

17. Show by example that the intersection of two prime ideals is in general not another prime ideal.

18. Let R be any ring.
    (a) Show that the intersection of two prime ideals P_1 and P_2 is prime if and only if $P_1 \subseteq P_2$ or $P_2 \subseteq P_1$.
    (b) Conclude that the intersection of two distinct maximal ideals is never a prime ideal.

19. Let R be a commutative ring. Prove that the nilradical $\mathcal{N}(R)$ is a subset of every prime ideal.

20. Show that a commutative ring with an identity $1 \neq 0$ is an integral domain if and only if $\{0\}$ is a prime ideal.

21. Let $\varphi : R \to S$ be a ring homomorphism and let Q be a prime ideal in S. Prove that $\varphi^{-1}(Q)$ is a prime ideal in R in two ways.
    (a) Suppose that R is commutative with an identity. Prove that $\varphi^{-1}(Q)$ is a prime ideal. [Hint: Proposition 3.8.9.]
    (b) Suppose that φ is surjective. Prove that $\varphi^{-1}(Q)$ is a prime ideal. [Hint: Exercise 3.5.15.]

22. Let $\varphi : R \to S$ be a ring homomorphism and let P be a prime ideal in R. Prove that the ideal generated by $\varphi(P)$ is not necessarily a prime ideal in S.

23. Let R be a ring and let S be a subset of R. Prove that there exists an ideal that is maximal (by inclusion) with respect to the property that it "avoids" (does not intersect) the set S.

24. Prove that the nilradical of a commutative ring is equal to the intersection of all the prime ideals of that ring. [Hint: Use Exercises 3.8.19 and 3.8.23.]

25. Let R be a commutative ring and let I be an ideal. Prove that the radical ideal $\sqrt{I}$ is equal to the intersection of all prime ideals that contain I. [Hint: Use the isomorphism theorems and Exercise 3.8.24.]

26. Consider the polynomial ring $\mathbb{R}[x_1, x_2, x_3, \ldots]$ with real coefficients but a countable number of variables. Define $I_1 = (x_1)$, $I_2 = (x_1, x_2)$, and $I_k = (x_1, x_2, \ldots, x_k)$ for all integers k. Prove that I_k is a prime ideal for all positive integers k and prove that

$$I_1 \subsetneq I_2 \subsetneq \cdots \subsetneq I_k \cdots$$

3.9 Projects

Investigative Projects

PROJECT I. **Roots of x^2+1 in $\mathbb{Z}/n\mathbb{Z}$.** Consider the ring of modular arithmetic $R = \mathbb{Z}/n\mathbb{Z}$. The goal of this project is to find the number of solutions to the equation $x^2 + 1 = 0$ in R. Determine the number of solutions for a large number of different values of n. Try to make (and if possible prove) a conjecture about the number of roots when n is prime, when n is a power of 2, when n is the product of two primes, and when n is general.

PROJECT II. **The Three-Sphere Group.** Recall (see Appendix A.1) that the set $\{z \in \mathbb{C} \mid |z| = 1\}$ is a subgroup of $(U(\mathbb{C}), \times)$. Furthermore, this set is isomorphic to the circle $\mathbb{S}^1$ where we locate point via and angle and we consider the addition operation by the addition of angles. Consider now the subset of the quaternions

$$S = \{\alpha \in \mathbb{H} \mid N(\alpha) = 1\},$$

where $N(\alpha)$ is the quaternion norm introduced in Example 3.1.21. This set S is a subgroup of $(U(\mathbb{H}), \times)$. Study this group. Show that geometrically we can view this set of a three-dimensional unit sphere $\mathbb{S}^3$ in $\mathbb{R}^4$. Study the group in comparison to $(\mathbb{S}^1, +)$. Are there subgroups, normal subgroups, etc., in $(S, \times)$? If there are normal subgroups, identify the corresponding quotient groups. Are there nontrivial homomorphisms with this group as its domain? Offer your own investigations or generalizations about this group.

PROJECT III. **Inverting Matrices of Quaternions.** Consider the ring $M_2(\mathbb{H})$. Attempt to find a criterion for when a matrix is invertible. Can you find a formula for the inverse of an invertible matrix in $M_2(\mathbb{H})$? Does your criteria extend to $M_n(\mathbb{H})$ for $n \geq 3$?

PROJECT IV. **Matrices of Quaternions.** Study the ring $M_2(\mathbb{H})$. Decide if you can find nilpotent elements, zero divisors, ideals, etc. Discuss solving systems of two equations in two variables but with variables and coefficients taken from $\mathbb{H}$.

PROJECT V. **Commutative Subrings in $M_{n \times n}(F)$.** Let F be a field and fix a matrix $A \in M_{n \times n}(F)$. Define the function $\varphi : F[x] \to M_{n \times n}(F)$ by

$$\varphi(a_n x^n + a_{n-1} x^{n-1} + \cdots + a_1 x + a_0) = a_n A^n + a_{n-1} A^{n-1} + \cdots + a_1 A + a_0 I.$$

First show that the function φ is a ring homomorphism and conclude that $\operatorname{Im} \varphi$ is a commutative subring of $M_{n \times n}(F)$. The matrices in the image of φ commute with A but this may not account for all matrices that commute with A. Setting $\mathrm{ad}_A : M_{n \times n}(F) \to M_{n \times n}(F)$ as the linear transformation such that $\mathrm{ad}_A(X) = AX - XA$, then $\mathrm{Ker}(\mathrm{ad}_A)$ is the subset of $M_{n \times n}(F)$ of matrices that commutes with A. What can be said about the relationship between $\mathrm{Ker}\,\mathrm{ad}_A$ and $\mathrm{Im}\,\varphi$? Are they always equal for any A? If not, what conditions on A make them equal? The set $\mathrm{Ker}\,\mathrm{ad}_A$ is a priori just a subspace of $M_{n \times n}(F)$; is it a subring or even an ideal of $M_{n \times n}(F)$? (Consider examples with 2×2 rings.)

PROJECT VI. **Subset Polynomial Ring.** If S is a set, then the power set $\mathcal{P}(S)$ has the structure of a ring when equipped with the operation $\triangle$ for addition and $\cap$ for the multiplication. Consider the polynomial ring $R = \mathcal{P}(S)[x]$. What are some properties (units, zero divisors, what ideals may be, what are the maximal ideals, is it an integral domain, is it a PID, etc.) of this ring? Discuss the same question with a quotient ring of R. For example, you could take $S = \{1, 2, 3, 4, 5\}$ and study properties of

$$\overline{R} = \mathcal{P}(S)[x]/(\{1, 3, 4\}x^3 + \{1, 2\}x + \{2, 3, 5\}),$$

but the choice of quotient ring is up to you.

PROJECT VII. **Application of FTFGAG.** Consider the group of units in quotient rings of the form $\mathbb{F}_p[x]/(n(x))$, where p is a prime and $n(x)$ is some polynomial in $\mathbb{F}_p[x]$. Since the quotient ring will be finite and abelian, the group of units will be finite and abelian, and hence is subject to the Fundamental Theorem of Finitely Generated Abelian Groups. Try a few examples and find out as much as you can about such groups. With examples, can you determine the isomorphism type of $U(\mathbb{F}_5[x]/(x^2 + x + 1))$ or $U(\mathbb{F}_2[x]/(x^3 + x + 1))$?

PROJECT VIII. **Quotient Rings and Calculus.** Revisit Example 3.7.6 and in particular how a product in a quotient ring recovers the product rule of differentiation. Generalize this observation to higher derivatives or to other function rings such as $C^n([a, b], \mathbb{R})$ or to even more general function rings. Are there other rules of differentiation that emerge from doing other operations in an appropriate quotient ring?

PROJECT IX. **Convolution Rings.** The construction of convolution rings is a general process with many natural examples subsumed them. Explore properties of convolution rings of your own construction. Consider using commutative or noncommutative rings, finite or infinite semigroups. Explore ring properties such as zero divisors, commutativity, units, ideals, and so on.

Expository Projects

PROJECT X. **Richard Dedekind, 1831–1916.** Richard Dedekind is generally credited with coining the term "ideal" in the context of ring theory. Using reliable sources, explore Dedekind's personal background, his mathematical work, and the reception of his work in the mathematical community. Discuss the mathematical context around why he used the term "ideal" as a label for the types of subrings of a ring now called ideals.

PROJECT XI. **Emmy Noether, 1882–1935.** At the time of her passing, many mathematicians and scientists considered Emmy Noether as the most influential woman in the history of mathematics. Using reliable sources, explore Noether's personal background, her mathematical work, and the reception of her work in the mathematical community. Discuss some of her contributions in abstract algebra and try to sketch the important of what are now called Noetherian rings.

PROJECT XII. **Logical Gates and Algebra.** Recall that a Boolean ring R is a ring with identity in which $r^2 = r$ for all $r \in R$. Discuss the connection between Boolean rings, Boolean algebra, the space $\mathcal{F}_n$ of functions $\mathbb{F}_2^n \to \mathbb{F}_2$, and logic gates in electrical circuits.

PROJECT XIII. **Zorn's Lemma.** Krull's Theorem (Theorem 3.8.2) was the first time we used Zorn's Lemma, a deep theorem from the theory of posets. Discuss the history of Zorn's Lemma, who first proved it, statements that are equivalent to it, and applications of Zorn's Lemma.

PROJECT XIV. **History of Ring Theory.** Answer the following questions as best you, using reliable sources. Who first wrote down the axioms of ring theory as we know them? What applications inspired mathematicians to work with such structures? When and why did mathematicians consider rings that were noncommutative or did not have an identity?

PROJECT XV. **The Chinese Remainder Theorem.** All modern texts in algebra and number theory name Theorem 3.7.16 the "Chinese Remainder Theorem." Explore why the theorem bears this name and who first dubbed it so.

4

Divisibility in Integral Domains

In elementary number theory, topics around divisibility play a central role. For example, divisibility, greatest common divisor, primes, the division algorithm and all the topics reviewed in Appendix A.5 figure prominently in nearly all areas of mathematics. Not only is it possible but very useful to generalize some of these notions from the ring of integers to arbitrary rings.

Often referring back to properties of the integers as a template, Sections 4.1 through 4.4 discuss the most general ring contexts in which we can talk about divisibility, the creation of rings of fractions, a context where we can perform something that resembles integer division, and unique factorization. These form the theoretical core to the chapter.

The second half of the chapter emphasizes applications of this ring theory. Section 4.5 introduces Gauss' Lemma and then explores consequences for factorization of polynomials. As a technical application, Section 4.6 introduces the RSA protocol for public key cryptography. Finally, Section 4.7 offers a brief introduction to algebraic number theory, a branch of mathematics that studies questions of interest in classical number theory but in extensions of the integers.

4.1 Divisibility in Commutative Rings

4.1.1 Divisors and Multiples

> **Definition 4.1.1**
>
> Let R be a commutative ring. We say that a nonzero element a *divides* an element b, and write $a|b$, if there exists $r \in R$ such that $b = ar$. The element a is called a *divisor* of b and b is called a *multiple* of a.

The reader may (and should) wonder why we restricted the concept of divisibility to commutative rings. If a ring R is not commutative, then given elements $a, b \in R$ it could be possible that there exists $r \in R$ such that $b = ar$ but that there does not exist $s \in R$ such that $b = sa$. Consequently, we would need to introduce notions of right-divisibility and left-divisibility. In the theory of noncommutative rings, we would take care to distinguish

DOI: 10.1201/9781003299233-4

these relations and develop appropriate theorems. This section restricts the attention to commutative rings.

When we say that an integer a is divisible by an integer b, there is an assumption that the $k \in \mathbb{Z}$ such that $b = ak$ is unique. The uniqueness follows immediately from Definition A.5.4 for divisibility over the integers. Consider now an arbitrary commutative ring R and suppose that $a = bk$ and $a = bk'$ in R. Then $0 = b(k - k')$. One way that $0 = b(k - k')$ could hold is if $b = 0$, which would imply that $a = 0$. This is why Definition 4.1.1 imposed the condition $a \neq 0$. However, if b is a zero divisor, then there exist distinct k and k' such that $b(k - k') = 0$. Hence, uniqueness of the factor k does not hold in rings with zero divisors.

In Section A.5.2, we pointed out that $(\mathbb{N}^*, |)$ is a partially ordered set. This remark inspires us to investigate whether and in what sense divisibility is close to being a partial order on a ring.

Proposition 4.1.2

Let R be a commutative ring. Divisibility is a transitive relation on R. Furthermore, if R has an identity, then divisibility is reflexive.

Proof. Suppose that $a, b, c \in R$ with a and b nonzero and $a|b$ and $b|c$. Then there exist $r, s \in R$ with $b = ar$ and $c = bs$. Then by associativity, $c = a(rs)$ so $a|c$.

Suppose also that R has an identity $1 \neq 0$. Then for all $a \in R$, we have $a = a1$ so $a \mid a$. Hence, divisibility is reflexive. $\qquad\square$

A discussion about antisymmetry is more involved. As a frame of reference, recall that divisibility is antisymmetric on $\mathbb{N}^*$ (Proposition A.5.5(4)). Now, $(\mathbb{N}^*, +, \times)$ is not a ring. On $\mathbb{Z}$, $\mid$ has the property that if $a|b$ and $b|a$, then $b = \pm a$.

Let R be a commutative ring. Suppose that $a, b \in R$ such that $a|b$ and $b|a$. Then there exist $r, s \in R$ with $b = ar$ and $a = bs$. Hence, $a = ars$. Without more information about the ring, there is not much else to be said. If the ring R has an identity $1 \neq 0$, then $a1 = ars$, which implies that $a(1 - rs) = 0$. If the ring R has zero divisors, then the cancellation law does not necessarily apply. However, if R contains no zero divisors, then we must have $rs = 1$, which means that r and s are units. We can see how this generalizes what happens in $\mathbb{Z}$ because the units of $\mathbb{Z}$ are 1 and -1.

The above discussion shows that for the usual notion of divisibility, it is preferable that the commutative ring have an identity $1 \neq 0$ and not include any zero divisors. These are precisely the properties of an integral domain. Consequently, from now on, unless we explicitly say so, we will discuss the notion of divisibility only in the context of integral domains.

Definition 4.1.3

Let R be an integral domain. Two elements a and b are called *associates* if there exists a unit u such that $a = bu$.

It is not hard to prove that the relation of associate is an equivalence relation on R. (See Exercise 4.1.6.) In this text, we will consistently use the following relation symbol on an integral domain

$$a \simeq b$$

to mean that a and b are associates. Notice that with this relation symbol, $r \simeq 1$ if and only if r is in the group of units $U(R)$.

We can now see in what sense divisibility is a partial order.

Proposition 4.1.4

Setting $[a] \mid [b]$ in the quotient set $(R \setminus \{0\})/ \simeq$ if and only if $a|b$ in R defines a relation on $(R \setminus \{0\})/ \simeq$. Furthermore, $\mid$ is a partial order on $(R \setminus \{0\})/ \simeq$.

Proof. We must first verify that setting $[a] \mid [b]$ is well-defined. Let a and a' be associates and let b and b' be associates so that $a = a'u$ and $b = b'v$ for $u, v \in U(R)$. But $a = br$ is equivalent to $a' = b'(vru^{-1})$ and $a' = b'r'$ is equivalent to $a = bv^{-1}r'u$. Hence, the choice of representatives from $[a]$ and $[b]$ is irrelevant, and our definition for divisibility on the set of associate classes is well-defined.

From Proposition 4.1.2, we already know that $\mid$ is transitive and reflexive on $(R \setminus \{0\})/ \simeq$. Furthermore, we saw that $a|b$ and $b|a$ in R if and only if a and b are associates. Hence, if $[a] \mid [b]$ and $[b] \mid [a]$ on $(R \setminus \{0\})/ \simeq$, then $[a] = [b]$. Hence, $\mid$ is antisymmetric on $(R \setminus \{0\})/ \simeq$. $\square$

Example 4.1.5. Consider the polynomial equation $x^3 - 1 = 0$. By the identity given in Exercise 3.1.27, this equation is equivalent to

$$(x - 1)(x^2 + x + 1) = 0.$$

The roots of $x^2 + x + 1 = 0$ are $\omega = \frac{-1+i\sqrt{3}}{2}$ and $\overline{\omega} = \frac{-1-i\sqrt{3}}{2}$.

Consider the ring $\mathbb{Z}[\omega]$. As a subring of $\mathbb{C}$ that includes the identity, it is an integral domain. It is not hard to show that in $\mathbb{C}$,

$$(a + b\omega)^{-1} = \frac{a + b\overline{\omega}}{a^2 - ab + b^2}.$$

In order for $a + b\omega$ to be a unit in $\mathbb{Z}[\omega]$, it is not hard to show that we need $a^2 - ab + b^2 = \pm 1$. This has six solutions, namely $(a, b) = (\pm 1, 0), (0, \pm 1), \pm(1, 1)$, or in other words

$$1, \frac{1 + i\sqrt{3}}{2}, \frac{-1 + i\sqrt{3}}{2}, -1, \frac{-1 - i\sqrt{3}}{2}, \frac{1 - i\sqrt{3}}{2}.$$

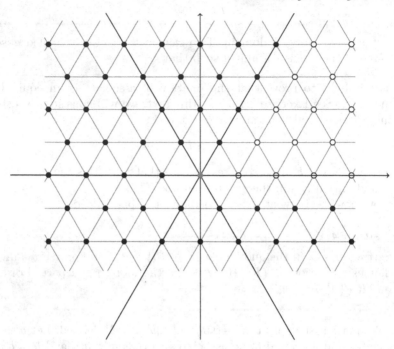

FIGURE 4.1: The elements and associates in $\mathbb{Z}[(-1 + i\sqrt{3})/2]$.

In polar coordinates, the units are

$$\cos\left(k\frac{\pi}{3}\right) + i\sin\left(k\frac{\pi}{3}\right) \qquad \text{for } k = 0, 1, \ldots, 5.$$

These correspond to the 6 distinct powers of ζ^k with $\zeta = \frac{1+i\sqrt{3}}{2}$.

In Figure 4.1, the dots • represent the elements in $\mathbb{Z}[\omega]$. Each element is an associate to a unique element in the sector defined in polar coordinates by $r > 0$ and $0 \le \theta < \frac{\pi}{3}$, represented by o. So according to Proposition 4.1.4, divisibility is a partial order on the elements in $\mathbb{Z}[\omega]$ in this sector. $\triangle$

4.1.2   Norms on Rings

In the study of rings, a comparison to the integers or other familiar rings is often fruitful. We commented above that most rings do not possess a natural total order that behaves well with respect to the operations of the ring. However, for some integral domains, the concept of a *norm* allows us to leverage properties of the integers to deduce results about the ring.

Definition 4.1.6

- Let R be an integral domain. Any function $N : R \to \mathbb{N}$ with $N(0) = 0$ is called a *norm* on R.
- A norm is called *positive* if $N(a) > 0$ for all $a \in R - \{0\}$.
- A norm is called *multiplicative* if $N(ab) = N(a)N(b)$.

An important class of rings that have norms are the rings of the form $\mathbb{Z}[\sqrt{D}]$, where D is a square-free integer (i.e., it is not divisible by a square integer greater than 1). Consider the function $N : \mathbb{Z}[\sqrt{D}] \to \mathbb{N}$ defined by

$$N(a + b\sqrt{D}) = |a^2 - Db^2|. \tag{4.1}$$

This is obviously a norm. Furthermore, since D is not a square, $\sqrt{D}$ is not a rational number so there exist no pairs $(a, b) \in \mathbb{Z}^2$ with $(a, b) \neq (0, 0)$ such that $a^2 - Db^2 = 0$. Consequently, the norm N is a positive norm. We now show that N is a multiplicative norm. Let $\alpha = a + b\sqrt{D}$ and $\beta = c + d\sqrt{D}$, then

$$N(\alpha)N(\beta) = |a^2 - Db^2| \, |c^2 - Dd^2| = |a^2c^2 - Da^2d^2 - Db^2c^2 + D^2b^2d^2|.$$

Furthermore, $\alpha\beta = (ac + Dbd) + (ad + bc)\sqrt{D}$ so

$$N(\alpha\beta) = |a^2c^2 + 2Dabcd + D^2b^2d^2 - D(a^2d^2 + 2abcd + b^2c^2)|$$
$$= |a^2c^2 - Da^2d^2 - Db^2c^2 + D^2b^2d^2|.$$

Consequently, for all $\alpha, \beta \in \mathbb{Z}[\sqrt{D}]$, $N(\alpha\beta) = N(\alpha)N(\beta)$.

A multiplicative norm on a ring is particularly useful when discussing divisibility because if a divides b in R, then $N(a)$ divides $N(b)$ as positive integers. If in addition the norm N is positive, then $N(1) = N(1^2) = N(1)N(1)$, which implies that $N(1) = 1$, since $N(1) \neq 0$. Then any unit $u \in U(R)$ satisfies $N(u) = 1$ because if $uv = 1$, then $N(u)N(v) = 1$ and the only multiplicative inverses in $\mathbb{N}$ is 1.

With the norm N in (4.1) defined on $\mathbb{Z}[\sqrt{D}]$, a stronger result holds.

Proposition 4.1.7

The element $\alpha \in \mathbb{Z}[\sqrt{D}]$ is a unit if and only if $N(\alpha) = 1$.

Proof. We already know that if α is a unit, then $N(\alpha) = 1$. Conversely, if $N(\alpha) = 1$, then $a^2 - Db^2 = \pm 1$ and so

$$\alpha^{-1} = \frac{a - b\sqrt{D}}{a^2 - Db^2} \in \mathbb{Z}[\sqrt{D}]. \qquad \square$$

4.1.3   Irreducible and Prime Elements

In Section 3.8 we remarked that the usual two equivalent criteria for primeness in $\mathbb{Z}$ are not necessarily equivalent in arbitrary rings. From those two criteria, Section 3.8 developed the notions of maximal ideals and prime ideals.

The reader may wonder why we defined a criteria of primeness in the context of ideals as opposed to ring elements. Primality on ideals is more general. This section began by showing that the relation of divisibility on a ring R has the properties that are similar to the integers only when R is an integral domain. In arbitrary rings, we can discuss prime ideals and consider ideal containment instead of divisibility on elements.

Though divisibility is a partial order on an integral domain in the sense of Proposition 4.1.4, the two criteria for primeness in $\mathbb{Z}$ are still not necessarily equivalent in integral domains. Hence, we need two separate definitions.

Definition 4.1.8

Let R be an integral domain.

(1) Suppose that r is nonzero and not a unit. Then r is called *irreducible* if whenever $r = ab$ either a or b is a unit. Otherwise, r is said to be *reducible*.

(2) A nonzero and nonunit element p is called *prime* if $p \mid ab$ implies that $p \mid a$ or $p \mid b$.

Note that p is a prime element if and only if the principal ideal (p) is a prime ideal. However, keep in mind that not all prime ideals are necessarily principal.

Proposition 4.1.9

In an integral domain, a prime element is always irreducible.

Proof. Suppose that (p) is a nonzero prime ideal and that $p = ab$. Then a or b is in (p). Without loss of generality, suppose that $a \in (p)$ so that $a = cp$ for some c. Then $p = pcb$ and hence $1 = cb$. Thus, b is a unit in R.   □

We show by way of example that the converse is not true.

Example 4.1.10. In the ring $\mathbb{Z}[\sqrt{10}]$ consider the four elements

$$\alpha = 6 + 2\sqrt{10}, \quad \beta = -15 + 5\sqrt{10}, \quad \gamma = 60 + 19\sqrt{10}, \quad \delta = -60 + 19\sqrt{10}.$$

It is easy to calculate that $\alpha\beta = 10 = \gamma\delta$. Using the norm N in (4.1), we have

$$N(\alpha) = 4, \qquad N(\beta) = 25, \qquad N(\gamma) = 10, \qquad N(\delta) = 10.$$

However, we claim that there are no elements in $\mathbb{Z}[\sqrt{10}]$ of norm either 2 or 5. Assume there exists an element $a + b\sqrt{10}$ of norm 2. Then $a^2 - 10b^2 =$

± 2. Modulo 10, this gives $a^2 \equiv \pm 2 \pmod{10}$. The squares modulo 10 are $0, 1, 4, 9, 6, 5$. Hence, we arrive at a contradiction and so we conclude there exists no element of norm 2. Assume now that there exists an element $a + b\sqrt{10}$ of norm 5. Then $a^2 - 10b^2 = \pm 5$. This implies that $5|a^2$ and thus that $5|a$. Writing $a = 5c$ leads to the equation $5c^2 - 2b^2 = \pm 1$. In modulo 5, this equation is $3b^2 \equiv \pm 1 \pmod 5$ which is equivalent to $b^2 \equiv \pm 2 \pmod 5$. The squares in modular arithmetic modulo 5 are $0, 1, 4$. Therefore, the assumption leads to a contradiction and thus there exists no element of norm 5.

Suppose that $\alpha = ab$. Then $N(ab) = 4$ so the pair $(N(a), N(b))$ is $(1, 4)$, $(2, 2)$, or $(4, 1)$. Since no element exists of norm 2, either $N(a) = 1$ or $N(b) = 1$. By Proposition 4.1.7, either a or b is a unit. Hence, α is irreducible. By a similar reasoning, we establish that β, γ, and δ are all irreducible elements. However, none of them are prime elements.

Notice that α divides $10 = \gamma\delta$. But $N(\alpha) = 4$ does not divide $N(\gamma) = 10$ or $N(\delta) = 10$, so α divides neither γ nor δ. Hence, α is an element that is irreducible but not prime. With a similar reasoning, we can show that β, γ, and δ are not prime elements either. $\triangle$

4.1.4 Greatest Common Divisors

The definition provided in Section A.5.3 for the greatest common divisor between two integers applies verbatim for greatest common divisors in an integral domain.

Definition 4.1.11

Let R be an integral domain. If $a, b \in R$ with $(a, b) \neq (0, 0)$, a *greatest common divisor* of a and b is an element $d \in R$ such that:

- $d|a$ and $d|b$ (d is a common divisor);

- if $d'|a$ and $d'|b$ (d' is another common divisor), then $d'|d$.

Proposition 4.1.12

If two elements a and b in an integral domain R have a greatest common divisor d, then any other greatest common divisor d' is an associate of d.

Proof. If both d and d' are greatest common divisors to a and b, then $d|d'$ and $d'|d$. Hence, there exist elements $u, v \in R$ such that $d' = ud$ and $d = vd'$. Then $d' = uvd'$ and since R is an integral domain, $1 = uv$, which implies u and v are units. Thus, $d \simeq d'$. $\square$

The hypothesis of Proposition 4.1.12 was careful to say, "if a and b have a greatest common divisor." In contrast to what happens in the ring of integers, two elements in arbitrary integral domains need not possess a greatest common divisor.

Example 4.1.13. The ring $\mathbb{Z}[\sqrt{10}]$ again offers an example of the nonexistence of greatest common divisors. Consider the elements $a = 12$ and $b = 24 + 6\sqrt{10}$. It is easy to see that 6 is a common divisor to a and b but $8 + 2\sqrt{10}$ is also because

$$12 = (4 - \sqrt{10})(8 + 2\sqrt{10}) \quad \text{and} \quad 24 + 6\sqrt{10} = 3(8 + 2\sqrt{10}).$$

Since neither $N(6) = 36$ nor $N(8+2\sqrt{10}) = 24$ divides the other, then neither 6 nor $8+2\sqrt{10}$ divides the other. Hence, neither of them is a greatest common divisor of 12 and $24 + 6\sqrt{10}$.

Assume there exists a greatest common divisor d to a and b. Then d is a multiple of 6 and of $8+2\sqrt{10}$ while it is a divisor of 12 and $24+6\sqrt{10}$. Hence, $N(6) = 36$ divides $N(d)$ which in turn divides $N(12) = 144$. In Example 4.1.10, we showed that $\mathbb{Z}[\sqrt{10}]$ has no elements with norm 2 so $N(d)/N(6) = 1$ or 4. If $N(d) = N(6)$, then d is an associate of 6 which leads to a contradiction since 6 is not a multiple of $8 + 2\sqrt{10}$. If $N(d) = 144$, then d is an associate of 12 which is a contradiction since 12 is not a divisor of $24 + 6\sqrt{10}$. Consequently, 12 and $24 + 6\sqrt{10}$ do not have a greatest common divisor.   △

We conclude the section with two definitions that adapt some more terminology of elementary number theory to integral domains. Results related to these concepts are left in the exercises.

Definition 4.1.14

Two elements a and b in an integral domain are said to be *relatively prime* if the only common divisors are units.

Definition 4.1.15

Let R be an integral domain. If $a, b \in R$ with $(a, b) \neq (0, 0)$, a *least common multiple* is an element $m \in R$ such that:

- $a|m$ and $b|m$ (m is a common multiple);
- if $a|m'$ and $b|m'$, then $m|m'$.

Proposition A.5.14 for least common multiples in $\mathbb{Z}$ carries over with only minor adjustments to any integral domain.

Proposition 4.1.16

Let R be an integral domain. Two nonzero elements a and b possess a greatest common divisor if and only if they possess a least common multiple.

Proof. (Left as an exercise for the reader. See Exercise 4.1.23.)   □

The reader may have noticed with some dissatisfaction that this section did not show how to determine if certain elements in an integral domain are irreducible or prime, or how to find a greatest common divisor of two elements if they exist. Such questions are much more difficult in arbitrary integral domains than they are in $\mathbb{Z}$. As subsequent sections attempt to generalize the results of elementary number theory to commutative rings, we will introduce subclasses of integral domains in which such question are tractable. In the meantime, we give the following definition.

Definition 4.1.17

An integral domain in which every two nonzero elements have a greatest common divisor is called a *gcd-domain*.

EXERCISES FOR SECTION 4.1

1. List all the divisors of $(6, 14)$ in $\mathbb{Z} \oplus \mathbb{Z}$.

2. Prove that if $a|b$ and $a|c$ in a commutative ring, then $a|(b + c)$.

3. Let R be a noncommutative ring. Prove that the relation on R of left-divisible is a transitive relation.

4. Prove that a subring of an integral domain that contains the identity is again an integral domain.

5. Let R be a commutative ring with a $1 \neq 0$. Prove or disprove that if R/I is an integral domain, then R is an integral domain (and I is prime).

6. Let R be an integral domain. Prove that the relation $a \simeq b$ defined by a is an associate to b is an equivalence relation.

7. Let D be a square-free integer and consider $\mathbb{Z}[\sqrt{D}]$ along with the norm defined in (4.1). Prove that if $N(\alpha)$ is a prime number, then α is irreducible in $\mathbb{Z}[\sqrt{D}]$.

8. Use Exercise 4.1.7 to find 5 irreducible elements in $\mathbb{Z}[i]$ that are not associates to each other.

9. Use Exercise 4.1.7 to find 5 irreducible elements in $\mathbb{Z}[\sqrt{3}]$ that are not associates to each other.

10. Find all the divisors of 21 in $\mathbb{Z}[\sqrt{-3}]$.

11. Consider the ring $\mathbb{Z}[i]$.
    (a) Let $a + bi \neq 0$. Prove that if $c + di$ is such that $(a + bi)(c + di) \in \mathbb{Z}$, then $c + di$ is $\pm(a - bi)$.
    (b) Prove that for all pairs $(a, b) \in \mathbb{Z}^2$, the expression $a^2 + b^2$ is never congruent to 3 modulo 4.
    (c) Prove that 2 is not irreducible.
    (d) Deduce that for a prime number p, the element $p = p + 0i$ is irreducible in $\mathbb{Z}[i]$ if $p \equiv 3 \pmod 4$.

12. Prove that $\mathbb{Z}[\sqrt[3]{2}]$ is an integral domain. Prove that if $a, b \in \mathbb{Z}$ such that $a^3 - 2b^3 = \pm 1$, then $a + b\sqrt[3]{2}$ is a unit.

13. Consider the ideal $I = (3, 2 + \sqrt{-5})$ in the ring $\mathbb{Z}[\sqrt{-5}]$.

(a) Prove that $1 \notin I$ so you can conclude that I is a proper ideal of the ring.

(b) Use the norm defined in (4.1) to show that the only elements $\alpha \in \mathbb{Z}[\sqrt{-5}]$ such $\alpha \mid 3$ and $\alpha \mid (2 + \sqrt{-5})$ are units.

(c) Deduce that I is not a principal ideal.

14. Let $\varphi : R \to S$ be an injective homomorphism between integral domains.

    (a) Prove that if $s \in \mathrm{Im}\,\varphi$ is an irreducible element in S, then $r = \varphi^{-1}(s)$ is an irreducible element in R.

    (b) Prove by a counterexample that if r is an irreducible element in R, then $\varphi(r)$ is not necessarily irreducible in S.

15. Prove that there are no elements $\alpha \in \mathbb{Z}[\sqrt{10}]$ with $N(\alpha) = 3$. Conclude that the elements $7 + 2\sqrt{10}$ and 3 are irreducible elements in $\mathbb{Z}[\sqrt{10}]$.

16. Consider with the ring $\mathbb{Z}[\sqrt[3]{2}]$.

    (a) Prove that the elements of $\mathbb{Z}[\sqrt[3]{2}]$ are of the form $\alpha = a + b\sqrt[3]{2} + c\sqrt[3]{4}$, where $a, b, c \in \mathbb{Z}$.

    (b) Prove that the function $\varphi : \mathbb{Z}[\sqrt[3]{2}] \to M_3(\mathbb{Z})$ defined by

$$\varphi(a + b\sqrt[3]{2} + c\sqrt[3]{4}) = \begin{pmatrix} a & 2c & 2b \\ b & a & 2c \\ c & b & a \end{pmatrix}$$

    is an injective ring homomorphism.

    (c) Deduce that $N : \mathbb{Z}[\sqrt[3]{2}] \to \mathbb{N}$ defined by $N(a + b\sqrt[3]{2} + c\sqrt[3]{4}) = |a^3 + 2b^3 + 4c^3 - 6abc|$ is a multiplicative norm and that $\alpha \in \mathbb{Z}[\sqrt[3]{2}]$ is a unit if and only if $N(\alpha) = 1$.

    (d) Deduce from (b) and (c) that $\alpha = 5 + 4\sqrt[3]{2} + 3\sqrt[3]{4}$ is a unit in $\mathbb{Z}[\sqrt[3]{2}]$ and find its inverse.

    (e) Prove that N is a positive norm. [Hint: Consider divisibility of a, b and c by 2 in the condition $N(a + b\sqrt[3]{2} + c\sqrt[3]{4}) = 0$.]

17. Let R be an integral domain. Prove that, if it exists, a least common multiple of $a, b \in R$ is a generator for the (unique) largest principle ideal containing $(a) \cap (b)$. Conclude that in a PID, if $(a) \cap (b) = (m)$, then m is a least common multiple of a and b.

18. Prove that in a PID, every irreducible element is prime.

19. Let R be an integral domain and let $a, b \in R$. Prove that a and b have a greatest common divisor d with $a = dk$ and $b = d\ell$, then k and ℓ are relatively prime.

20. Prove that Proposition A.5.12 does not hold when we replace $\mathbb{Z}$ with $\mathbb{Z}[\sqrt{10}]$.

21. Let p_1, p_2, q_1, q_2 be irreducible elements in an integral domain R such that none are associates to any of the others and $p_1 p_2 = q_1 q_2$. Prove that $p_1 q_1 q_2$ and $p_1 p_2 q_1$ do not have a greatest common divisor.

22. Prove that if a least common multiple m of a and b exists in an integral domain, then (m) is the largest principal ideal contained in $(a) \cap (b)$.

23. Prove Proposition 4.1.16. [Hint: For one direction of the if and only if statement, see Proposition A.5.14.]

In Exercises 4.1.24 through 4.1.29 the ring R is a gcd-domain. Furthermore, for two elements $a, b \in R$, we define $\gcd(a, b)$ as a greatest common divisor, well-defined up to multiplication of a unit and we also define $\operatorname{lcm}(a, b)$ as a least common multiple, well-defined up to multiplication of a unit.

24. Prove that $\gcd(a, b) \operatorname{lcm}(a, b) \simeq ab$ for all nonzero $a, b \in R$.

25. Prove that $\gcd(a, \gcd(b, c)) \simeq \gcd(\gcd(a, b), c)$ and also that $\operatorname{lcm}(a, \operatorname{lcm}(b, c)) \simeq \operatorname{lcm}(\operatorname{lcm}(a, b), c)$ for all nonzero $a, b, c \in R$.

26. Prove that $\gcd(ac, bc) \simeq \gcd(a, b)c$ for all nonzero $a, b, c \in R$. Prove also that $\operatorname{lcm}(ac, bc) \simeq \operatorname{lcm}(a, b)c$.

27. Prove that $\gcd(a, b) \simeq 1$ and $\gcd(a, c) \simeq 1$ if and only if $\gcd(a, bc) \simeq 1$.

28. Prove that if $\gcd(a, b) \simeq 1$ and $a | bc$, then $a | c$.

29. Prove that if $\gcd(a, b) \simeq 1$, $a | c$, and $b | c$, then $ab | c$.

30. A *Bézout domain* is an integral domain in which the sum of any two principal ideals is again a principal ideal. Prove that a Bézout domain is a gcd-domain.

4.2   Rings of Fractions

One way to deal with questions of divisibility in a ring is to force certain elements to be units. We already encountered this process in the definition of the rational numbers in reference to the integers. Most people first encounter fractions so early in their education that a precise construction was not appropriate at that time. We give one here.

A fraction $r = \frac{a}{b}$ consists of a pair of integers $(a, b) \in \mathbb{Z} \times \mathbb{Z}^*$. However, some fractions are considered equivalent. For example, since we use a fraction to represent a ratio, we must have

$$(an, bn) \sim (a, b) \qquad \text{for all } n \in \mathbb{Z}^*.$$

This is not yet a good definition for a relation since it does not give a criteria for when two arbitrary pairs are in relation. A complete expression of the equivalence relation $\sim$ is

$$(a, b) \sim (c, d) \qquad \Longleftrightarrow \qquad ad = bc. \qquad (4.2)$$

One can see this definition follows from the requirement from the simple calculation

$$(a, b) \sim (ac, bc) \sim (ac, ad) \sim (c, d).$$

Criterion (4.2) is the cross-multiplication method to identify equal fractions.

The set $\mathbb{Q}$ of rational numbers is defined as the equivalence classes of pairs $\mathbb{Z} \times \mathbb{Z}^*$ with the equivalence relation $\sim$ given in (4.2). We write $\frac{a}{b}$ as the equivalence class of the pair (a, b). Without belaboring the reasons for the arithmetic operations, the operations on fractions are

$$\frac{a}{b} + \frac{c}{d} = \frac{ad + bc}{bd} \qquad \text{and} \qquad \frac{a}{b} \frac{c}{d} = \frac{ac}{bd}. \qquad (4.3)$$

4.2.1   Issues with Dividing by Zero

Before we attempt to generalize the construction of $\mathbb{Q}$ from $\mathbb{Z}$ as much as possible to arbitrary commutative rings, we discuss a few issues with dividing by 0 or even defining a division by 0.

We begin with two important comments as to why we do not divide by 0 or even force a division by 0 in fractions of integers. Suppose that we attempted to make sense of fractions $\frac{a}{b}$ with $(a, b) \in \mathbb{Z} \times \mathbb{Z}$. We would have an equality of fractions

$$\frac{a}{0} = \frac{c}{d}$$

if and only if $ad = 0$. This leads to two cases.

If $a = 0$, then we would have a situation in which $\frac{0}{0} = \frac{c}{d}$ for all $c, d \in \mathbb{Z}$. This would be undesirable (or at least uninteresting) since then every fraction would be equivalent to the fraction $\frac{0}{0}$ and thus every fraction would be equivalent to every other fraction and the set of fractions would consist of the single element $\frac{0}{0}$. Consequently, we do not bother trying to construct fractions $\frac{0}{0}$.

If $d = 0$ in the product $ad = 0$, then we would have $\frac{a}{0} = \frac{b}{0}$ for all $a, b \in \mathbb{Z}^*$. Allowing or forcing this division is not completely uninteresting. In fact, this is reminiscent of real projective space discussed in Example A.3.9. We could define an equivalence relation $\sim$ on $\mathbb{Z}^2 - \{(0, 0)\}$ by

$$(a, b) \sim (c, d) \iff ad = bc.$$

Then the set of equivalence classes consists of all fractions along with one more element, the equivalence class of $(1, 0)$, which contains every pair $(a, 0)$. This is sometimes called the *integral projective line* and instead of writing the equivalence class of (a, b) with the fraction notation, they are written as $(a : b)$. We can also think of this as the set of all integer ratios. Even if this set may be interesting for certain applications, it does not carry a ring structure with the usual addition and multiplication because multiplication is not defined for $(1 : 0) \times (0 : 1)$.

Consequently, any construction of fractions that allows a 0 in the denominator either reduces the ring down to the trivial ring of one element or does not produce a ring structure. Despite this, it is possible to define rings of fractions in which the denominators are zero divisors.

4.2.2   Rings of Fractions

Let R be a commutative ring and let D be any nonempty subset of R that does not contain 0 and is closed under multiplication. D may contain zero divisors, but since $0 \notin D$, it cannot contain both a and b when $ab = 0$. The set D will be the set of denominators.

We define a relation $\sim$ on $R \times D$ such that $(a, d) \sim (au, du)$ for any $u \in D$. As mentioned in the introduction to this section, this does not directly define the relation between any two pairs in $R \times D$. Given any two pairs (a, d_1) and (b, d_2) we have

$$(a, d_1) \sim (ad_2, d_1 d_2) \qquad \text{and} \qquad (b, d_2) \sim (bd_1, d_1 d_2).$$

Because the cancellation law does not apply in rings with zero divisors, it is possible for $ad_2 u = bd_1 u$ for some u without $ad_2 = bd_1$. So the relation $\sim$ on $R \times D$ can be defined symmetrically by

$$(a, d_1) \sim (b, d_2) \iff (ad_2 - bd_1)u = 0 \text{ for some } u \in D. \qquad (4.4)$$

Proposition 4.2.1

The relation $\sim$ given in (4.4) is an equivalence relation on $R \times D$.

Proof. For all $(a, d) \in R \times D$, $ad - ad = 0$ so there does exist a $u \in D$ (any u in D) such that $(ad - ad)u = 0$. Hence, $\sim$ is reflexive.

Suppose that $(a, d_1) \sim (b, d_2)$. Then $(ad_2 - bd_1)u = 0$ for some $u \in D$. Then

$$-(ad_2 - bd_1)u = (bd_1 - ad_2)u = 0$$

so $(b, d_2) \sim (a, d_1)$, which shows that $\sim$ is symmetric.

Now suppose that $(a, d_1) \sim (b, d_2)$ and $(b, d_2) \sim (c, d_3)$. Then for some $u, v \in D$,

$$(ad_2 - bd_1)u = 0 \qquad \text{and} \qquad (bd_3 - cd_2)v = 0.$$

Hence, multiplying the first equation by vd_3 and the second by ud_1, we get $(ad_2 d_3 - bd_1 d_3)uv = 0$ and $(bd_3 d_1 - cd_2 d_1)uv = 0$. Adding these, and canceling $bd_1 d_2 uv$, we deduce that

$$0 = (ad_2 d_3 - cd_2 d_1)uv = (ad_3 - cd_1)d_2 uv.$$

Since $d_2 uv \in D$, we deduce that $(a, d_1) \sim (c, d_3)$. $\square$

Definition 4.2.2

Let R be a commutative ring. Let D be a nonempty multiplicatively closed subset with $0 \notin D$. The *set of fractions* of R with denominators in D, denoted $D^{-1}R$, is the set of $\sim$-equivalence classes on $R \times D$. We write the equivalence class for (a, d) by $\dfrac{a}{d}$.

Theorem 4.2.3

Let R be a commutative ring and D a nonempty multiplicatively closed set that does not contain 0. The operations defined in (4.3) are well-defined on $D^{-1}R$ (i.e., are independent of choice of representative for a given equivalence class). Furthermore, these operations give $D^{-1}R$ the structure of a commutative ring with a $1 \neq 0$.

Proof. Suppose that $\frac{a}{d_1} = \frac{b}{d_2}$ and $\frac{r}{d_3} = \frac{s}{d_4}$ are elements in $D^{-1}R$ with $(ad_2 - bd_1)u = 0$ and $(rd_4 - sd_3)v = 0$ for some $u, v \in D$. Adding the fractions gives

$$\frac{a}{d_1} + \frac{r}{d_3} = \frac{ad_3 + rd_1}{d_1 d_3} \quad \text{and} \quad \frac{b}{d_2} + \frac{s}{d_4} = \frac{bd_4 + sd_2}{d_2 d_4}.$$

To see that these two fractions are equal, note that

$$\begin{aligned} ((ad_3 + rd_1)d_2 d_4 &- (bd_4 + sd_2)d_1 d_3)uv \\ &= (ad_2 - bd_1)ud_3 d_4 v + (rd_4 - sd_3)vd_1 d_2 u \\ &= 0d_3 d_4 v + 0d_1 d_2 u = 0. \end{aligned}$$

Multiplying the two fractions gives

$$\frac{a}{d_1} \times \frac{r}{d_3} = \frac{ar}{d_1 d_3} \quad \text{and} \quad \frac{b}{d_2} \times \frac{s}{d_4} = \frac{bs}{d_2 d_4}.$$

To see that these two fractions are equal, note that

$$\begin{aligned} (ard_2 d_4 - bsd_1 d_3)uv &= (ad_2 u)(rd_4 v) - (bd_1 u)(sd_3 v) \\ &= (bd_1 u)(sd_3 v) - (bd_1 u)(sd_3 v) = 0. \end{aligned}$$

That addition and multiplication are commutative on $D^{-1}R$ follows from the commutativity of addition and multiplication on R. The addition of fractions is associative with

$$\frac{a}{d_1} + \frac{b}{d_2} + \frac{c}{d_3} = \frac{ad_2 d_3 + bd_1 d_3 + cd_1 d_2}{d_1 d_2 d_3}$$

regardless of which $+$ is performed first. Similarly, the multiplication of fractions is associative.

For any $d \in D$, an element of the form $\frac{0}{d}$ is the additive unit in $D^{-1}R$ because

$$\frac{a}{d'} + \frac{0}{d} = \frac{ad + 0}{dd'} = \frac{ad}{d'd} = \frac{a}{d'}.$$

The additive inverse to $\frac{a}{d}$ is just $\frac{-a}{d}$. Furthermore, any element of the form $\frac{d}{d}$ satisfies

$$\frac{a}{d'} \times \frac{d}{d} = \frac{ad}{dd'} = \frac{a}{d'}$$

for all $\frac{a}{d'} \in D^{-1}R$ so $\frac{d}{d}$ is a multiplicative unit.

The only remaining axiom, distributivity of $\times$ over $+$, is left as an exercise for the reader. (See Exercise 4.2.1.) $\square$

Definition 4.2.4

We call $D^{-1}R$, equipped with $+$ and $\times$ as given in (4.3), the *ring of fractions* of R with denominators in D.

We emphasize that the construction given in Theorem 4.2.3 directly generalizes the construction of $\mathbb{Q}$ from $\mathbb{Z}$ where $R = \mathbb{Z}$ and $D = \mathbb{Z}^*$. It is not hard to show that if we had taken $D = \mathbb{Z}^{>0}$, we would get a ring of fractions that is isomorphic to $\mathbb{Q}$.

There is always a natural homomorphism $\varphi : R \to D^{-1}R$ given by $\varphi(r) = \frac{rd}{d}$ for some $d \in D$. By the equivalence of fractions, $\frac{rd}{d} = \frac{rd'}{d'}$ for any $d, d' \in D$ so the choice of $d \in D$ is irrelevant for the definition of φ. In the case of $\mathbb{Z}$ and $\mathbb{Q}$, the homomorphism is $\varphi(n) = \frac{n}{1}$. However, in the general situation, we must define φ as above because D does not necessarily contain 1.

Though the natural homomorphism $\mathbb{Z} \to \mathbb{Q}$ is injective, the function φ is not necessarily injective in the case of arbitrary commutative rings. For $r, s \in R$, $\varphi(r) = \varphi(s)$ gives $\frac{rd}{d} = \frac{sd'}{d'}$ for some $d, d' \in D$, which implies that

$$(rdd' - sd'd)u = 0 \iff (r - s)dd'u = 0$$

for some $u \in D$. If D contains no zero divisors, then $\varphi(r) = \varphi(s)$ implies that $r - s = 0$ so $r = s$ and hence φ is injective. Conversely, if D does contain a zero divisor d with $bd = 0$ and $b \neq 0$, then

$$\varphi(b) = \frac{bd}{d} = \frac{0}{d} = \varphi(0)$$

and φ is not injective. This discussion establishes the following lemma.

Lemma 4.2.5

The function $\varphi : R \to D^{-1}R$ defined by $\varphi(r) = \frac{rd}{d}$ for some $d \in D$, is injective if and only if the multiplicatively closed subset D contains no zero divisors.

Note that when R is an integral domain, the condition in Lemma 4.2.5 is always satisfied.

Proposition 4.2.6

Let R be a commutative ring and D a multiplicatively closed subset that does not contain 0 or any zero divisors. Then $D^{-1}R$ contains a subring isomorphic to R. Furthermore, in this embedding of R in $D^{-1}R$, every element of D is a unit in $D^{-1}R$.

Proof. By Lemma 4.2.5, the function φ is injective so by the First Isomorphism Theorem, R is isomorphic to $\mathrm{Im}\,\varphi$. Let d be any element in D. We can view d in $D^{-1}R$ as the element $\frac{d^2}{d}$. But $D^{-1}R$ contains the element $\frac{d}{d^2}$ and it is easy to see that $\frac{d^2}{d} \times \frac{d}{d^2} = \frac{d^3}{d^3} = 1$ in $D^{-1}R$. $\square$

Example 4.2.7. Let $R = \mathbb{Z}$ and let $D = \{1, a, a^2, a^3, \ldots\}$ for some positive integer a. Then $D^{-1}R$ consists of all the fractions

$$\frac{n}{a^k} \qquad \text{with } n \in \mathbb{Z} \text{ and } k \in \mathbb{N}.$$

This is a ring that is also a subring of $\mathbb{Q}$. Recall that we denote this ring by $\mathbb{Z}\left[\frac{1}{a}\right]$.   △

Example 4.2.8. Let $R = \mathbb{Z}[x]$ and let $D = \{(1+x)^n\}_{n\geq0}$. The ring $D^{-1}R$ consists of all rational expressions of the form

$$\frac{p(x)}{(1+x)^n},$$

where $p(x) \in \mathbb{Z}[x]$ and $n \in \mathbb{N}$. The units in this ring are all polynomials of the form $\pm(1+x)^k$ for $k \in \mathbb{Z}$. The ring has no zero divisors.   △

Examples 4.2.7 and 4.2.8 are particular examples of a general construction. If R is a commutative ring and a is an element that is not nilpotent, then $R\left[\frac{1}{a}\right]$ is the ring of fractions $D^{-1}R$ where $D = \{a, a^2, a^3, \ldots\}$.

If R is an integral domain, then the set $D = R - \{0\}$ is multiplicatively closed and does not contain 0. In $D^{-1}R$, every nonzero element of R becomes a unit, so $D^{-1}R$ is a field.

Definition 4.2.9

> If R is an integral domain, and if $D = R - 0$, then $D^{-1}R$ is called *field of fractions* of R.

Since the natural homomorphism $\varphi : R \to D^{-1}R$ is an injection when R is an integral domain, we regularly view R as a subring of its field of fractions F.

The motivating example for this section is the construction of the rational numbers. Note that the set of rational numbers $\mathbb{Q}$ is the field of fractions of the integral domain $\mathbb{Z}$.

Example 4.2.10. Consider the integral domain $\mathbb{Z}[\sqrt{2}]$. The field of fractions F of $\mathbb{Z}[\sqrt{2}]$ consists of expressions of the form

$$\frac{a + b\sqrt{2}}{c + d\sqrt{2}} \qquad \text{with } a, b, c, d \in \mathbb{Z}.$$

It is clear that F is a subring of $\mathbb{R}$. Notice that with integers $a, b, c, d \in \mathbb{Z}$,

$$\frac{ad + bc\sqrt{2}}{bd} = \frac{ad}{bd} + \frac{bc\sqrt{2}}{bd} = \frac{a}{b} + \frac{c}{d}\sqrt{2}.$$

Hence, $\mathbb{Q}[\sqrt{2}] \subseteq F$. On the other hand, by equivalences of fractions,

$$\frac{a + b\sqrt{2}}{c + d\sqrt{2}} = \frac{(a + b\sqrt{2})(c - d\sqrt{2})}{c^2 - 2d^2} = \frac{(ac - 2bd) + (bc - ad)\sqrt{2}}{c^2 - 2d^2}.$$

Thus, $F \subseteq \mathbb{Q}[\sqrt{2}]$, so in fact $F = \mathbb{Q}[\sqrt{2}]$.   △

Example 4.2.11 (Rational Expressions). Let R be an integral domain. By Proposition 3.2.6, $R[x]$ is an integral domain as well. The field of fractions of $R[x]$ is the set of rational expressions,

$$\left\{ \frac{p(x)}{q(x)} \,\middle|\, p(x) \in R[x] \text{ and } q(x) \in R[x] - \{0\} \right\}.$$

This field of rational expressions over R is usually denoted by $R(x)$, in contrast to $R[x]$. △

Viewing an integral domain R as a subring in its field of fractions F, for $a, b \in R$ with $a \neq 0$, the element a divides b if and only if $\frac{b}{a} \in R$.

Though some texts only discuss rings of fractions in the context of integral domains, the definitions provided in this section only require R to be commutative. The following example presents a ring of fractions in which the denominator allows for zero divisors. Note, however, that since D must be multiplicatively closed and not contain 0, then D cannot contain nilpotent elements.

Example 4.2.12. Let R be the quotient ring $R = \mathbb{Z}[x]/(x^2 - 1)$. (For simplicity, we omit the over-line in $\overline{a + bx}$ notation.) We see that $x - 1$ and $x + 1$ are zero divisors. Consider the multiplicatively closed set $D = \{(1 + x)^k \mid 0 \leq k\}$. In R, we have $x^2 - 1 = 0$, so $x^2 = 1$. Then

$$(1 + x)^2 = x^2 + 2x + 1 = 1 + 2x + 1 = 2(x + 1).$$

Consequently, $(1 + x)^k = 2^{k-1}(x + 1)$ for $k \geq 1$.

Then the ring of fractions $D^{-1}R$ consists of expressions $\dfrac{a + bx}{1}$ or $\dfrac{a + bx}{2^n(1 + x)}$ for $n \geq 0$.

In $D^{-1}R$, we have the unexpected equality of fractions

$$\frac{x - 1}{(x + 1)^n} = \frac{x^2 - 1}{(x + 1)^{n+1}} = \frac{0}{(x + 1)^{n+1}} = \frac{0}{1}.$$

Then, for any fraction in $D^{-1}R$, we have

$$\frac{a + bx}{(x + 1)^k} = \frac{a + b + b(x - 1)}{(x + 1)^k} = \frac{a + b}{(x + 1)^k} + \frac{b(x - 1)}{(x + 1)^k} = \frac{a + b}{(x + 1)^k}.$$

Now consider the function $f : \mathbb{Z}\left[\frac{1}{2}\right] \to D^{-1}R$ given by $f\left(\frac{a}{2^n}\right) = \frac{a}{(x+1)^n}$ for all $n \in \mathbb{N}$. Then

$$f\left(\frac{a}{2^m} + \frac{b}{2^n}\right) = f\left(\frac{2^n a + 2^m b}{2^{m+n}}\right) = \frac{2^n a + 2^m b}{(x + 1)^{m+n}} = \frac{2^n a(x + 1) + 2^m b(x + 1)}{(x + 1)^{m+n+1}}$$

$$= \frac{2^n a(x + 1)}{(x + 1)^{m+n+1}} + \frac{2^m b(x + 1)}{(x + 1)^{m+n+1}} = \frac{a(x + 1)^{n+1}}{(x + 1)^{m+n+1}} + \frac{b(x + 1)^{m+1}}{(x + 1)^{m+n+1}}$$

$$= \frac{a}{(x + 1)^m} + \frac{b}{(x + 1)^n} = f\left(\frac{a}{2^m}\right) + f\left(\frac{b}{2^n}\right).$$

Furthermore,

$$f\left(\frac{a}{2^m}\frac{b}{2^n}\right) = \frac{ab}{(x+1)^{m+n}} = \frac{a}{(x+1)^m}\frac{b}{(x+1)^n} = f\left(\frac{a}{2^m}\right)f\left(\frac{b}{2^n}\right)$$

so f is a ring homomorphism. Since $\frac{a+bx}{(x+1)^k} = \frac{a+b}{(x+1)^k}$, then f is surjective. However, the kernel of f consists of all $a/2^k$ such that $a/(x+1)^k = 0/1$ in $D^{-1}R$, which means $a = 0$. Thus, $\operatorname{Ker} f = \{0\}$ so f is injective and thus an isomorphism. We have shown that $D^{-1}R$ is isomorphic to $\mathbb{Z}\left[\frac{1}{2}\right]$.

Even though R contains zero divisors, the ring of fractions $D^{-1}R = \mathbb{Z}\left[\frac{1}{2}\right]$ is an integral domain. By Proposition 3.1.12, zero divisors are not units in a ring. However, taking the ring of fractions construction forced the zero divisor $(x+1)$ to become a unit. In the process, the element $x - 1$ became 0 under the function $\varphi : R \to D^{-1}R$, which is not injective. $\triangle$

The above example illustrates that if D contains a zero divisor a with $ab = 0$ for some element b, in the usual homomorphism $\varphi : R \to D^{-1}R$, $\varphi(a)$ is a unit and $\varphi(b) = 0$.

The last example of the section is an important case of ring of fractions.

Example 4.2.13 (Localization). Let R be a commutative ring and let P be a prime ideal. The subset $D = R - P$ is multiplicatively closed. Indeed, since P is a prime ideal $ab \in P$ implies $a \in P$ or $b \in P$. Taking the contrapositive of this implication statement gives (and careful to use DeMorgan's Law), $a \notin P$ and $b \notin P$ implies $ab \notin P$. This means precisely $a \in D$ and $b \in D$ implies $ab \in D$, which means that D is multiplicatively closed.

The ring of fractions created from $D^{-1}R$, where $D = R - P$ and P is a prime ideal in R, is called the *localization* of R by P and is denoted R_P.

In the exercises, the reader is guided to prove that R_P contains a unique maximal ideal, which means that R_P is a local ring. (See Exercise 4.2.14.)

As an explicit example, the ring

$$\mathbb{Z}_{(19)} = \left\{\frac{a}{b} \,\Big|\, a, b \in \mathbb{Z} \text{ with } 19 \nmid b\right\}$$

is the localization of the integers $\mathbb{Z}$ by the prime ideal (19). $\triangle$

EXERCISES FOR SECTION 4.2

1. Prove that $\times$ is distributive over $+$ in any ring of fractions $D^{-1}R$.

2. Let $D = \{2^a 3^b \mid a, b \in \mathbb{N}\}$ as a subset of $\mathbb{Z}$. Prove that $D^{-1}\mathbb{Z}$ is isomorphic to $\mathbb{Z}\left[\frac{1}{6}\right]$ even though $D \neq \{1, 6, 6^2, \ldots\}$.

3. Let $D = \{2^a 3^b 5^c \mid a, b, c \in \mathbb{N}\}$ as a subset of $\mathbb{Z}$. Prove that $D^{-1}\mathbb{Z}$ is isomorphic to $\mathbb{Z}\left[\frac{1}{30}\right]$.

4. Let D be the subset in $\mathbb{Z}$ of all positive integers that are products of powers of primes of the form $4k + 1$. Prove that D is multiplicatively closed. Prove also that $D^{-1}\mathbb{Z}$ is neither isomorphic to $\mathbb{Z}\left[\frac{1}{n}\right]$ for any integer n nor isomorphic to $\mathbb{Z}_{(p)}$ for a prime number p.

5. Let $R = \mathbb{Z}/12\mathbb{Z}$ and let $D = \{\bar{3}, \bar{9}\}$. Determine the number of elements in the ring of fractions $D^{-1}R$ and exhibit a unique representative for all the fractions in $D^{-1}R$.

6. Let $R = \mathbb{Z}/100\mathbb{Z}$ and let $D = \{\bar{2}^a \mid a \geq 1\}$. Determine the number of elements in the ring of fractions $D^{-1}R$ and describe a unique representative for all the fractions in $D^{-1}R$.

7. Prove that if D contains only units, then $D^{-1}R$ is isomorphic to R.

8. Let R be an integral domain and let F be its field of fractions. Prove that as rings of rational expressions $F(x) = R(x)$.

9. Let R be a commutative ring and let $a \in R - \{0\}$ be an element that is not nilpotent, i.e., $a^m \neq 0$ for all $m \in \mathbb{N}$. Set

$$D_1 = \{a, a^2, a^3, \ldots\} \qquad \text{and} \qquad D_2 = \{a^k, a^{k+1}, a^{k+2}, \ldots\},$$

where k is any positive integer. Prove that $D_1^{-1}R$ is isomorphic to $D_2^{-1}R$. [Hint: Map $\frac{r}{a^m}$ to $\frac{ra^{k-1}}{a^{m+k-1}}$.]

10. Let $R = \mathbb{Z}[x]/(x^2 - (m+n)x + mn)$ for some integers $m \neq n$. Let $D = \{(x - n)^k \mid k \in \mathbb{N}\}$. Prove that $D^{-1}R$ is isomorphic to $\mathbb{Z}\left[\frac{1}{m-n}\right]$.

11. Let D be a multiplicatively closed subset of a commutative ring R, that does not contain the 0. Suppose that a is a zero divisor in D with $ab = 0$ for some element $b \in R$. If $\varphi : R \to D^{-1}R$ is the standard mapping of a ring into the ring of fractions $D^{-1}R$, show that $\varphi(a)$ is a unit and $\varphi(b) = 0$.

12. Prove that the ideals of $D^{-1}R$ are in bijection with the ideals of R that do not intersect D.

13. Let $\varphi : R \to D^{-1}R$ be the standard homomorphism of a commutative ring into a ring of fractions. Prove that if I is a principal ideal in $D^{-1}R$, then $\varphi^{-1}(I)$ is also a principal ideal.

14. Let R be a commutative ring and let P be a prime ideal. (a) Prove that the set of nonunits in R_P is the ideal P_P. (b) Deduce that R_P has a unique maximal ideal.

15. Consider the ring $R = \mathbb{R}[x, y]$ and the prime ideal $P = (x, y)$. Prove that the elements in the localization R_P are rational expressions of the form

$$\frac{r(x, y)}{1 + xp(x, y) + yq(x, y)} \qquad \text{for } p(x, y), q(x, y), r(x, y) \in \mathbb{R}[x, y].$$

16. Let $F[[x]]$ be the ring of formal power series with coefficients in a field F. We denote the field of fractions of $F[[x]]$ by $F((x))$. Prove that the elements of $F((x))$ can be written as

$$\sum_{k \geq -N} a_k x^k \qquad \text{for some integer } N.$$

[Such series are called *formal Laurent series* and $F((x))$ is called the *field of formal Laurent series* over F. See also Exercise 3.4.29.]

4.3   Euclidean Domains

As mentioned in its introduction, this chapter gathers together topics related to divisibility. In the previous section, we discussed rings of fractions, a construction that forces certain elements to be units. In particular, if R is an integral domain, a nonzero element a divides b if and only if, in the field of fractions, $\frac{b}{a}$ is in the subring R. However, much of the theory of divisibility of integers does not rely on the ability to take fractions of any nonzero elements.

This section introduces Euclidean domains, rings in which it is possible to perform something akin to the integer division algorithm.

4.3.1   Definition

Definition 4.3.1

> Let R be an integral domain. A *Euclidean function* on R is any function $d : R - \{0\} \to \mathbb{N}$ such that
> (1) For all $a, b \in R$ with $a \neq 0$, there exist $q, r \in R$ such that
>
> $$b = aq + r \qquad \text{with } r = 0 \text{ or } d(r) < d(a). \qquad (4.5)$$
>
> (2) For all nonzero $a, b \in R$, $d(b) \leq d(ab)$.
> R is called a *Euclidean domain* if it has a Euclidean function d.

We call any expression of the form (4.5) a *Euclidean division* of b by a. It is not uncommon to call q a *quotient* and r a *remainder* in the Euclidean division.

The Integer Division Theorem (Theorem A.5.6) states that for $a, b \in \mathbb{Z}$ with $a \neq 0$ there exist unique q, r such $b = aq + r$ and $0 \leq r < |a|$. The above definition does not require uniqueness of q and r, simply existence. The Integer Division Theorem establishes that the integers are a Euclidean domain with $d(x) = |x|$. However, according to Definition 4.3.1, for any given a and b either $b = aq$ or, the two possibilities of $b = aq + r$ with $0 < r < |a|$ or $b = aq' + r'$ with $-|a| < r' < 0$. In these latter two possibilities, $r' = r - |a|$ and $q' = q + \text{sign}(a)$.

Example 4.3.2 (Gaussian Integers). This example shows that the Gaussian integers $\mathbb{Z}[i]$ form a Euclidean domain with Euclidean function $d(z) = |z|^2$.

First, note that for all nonzero $\alpha, \beta \in \mathbb{Z}[i]$,

$$d(\alpha\beta) = |\alpha\beta|^2 = |\alpha|^2|\beta|^2 = d(\alpha)d(\beta).$$

Since $d(\alpha) \geq 1$ for all nonzero α, then $d(\beta) \leq d(\alpha\beta)$.

Let $\alpha, \beta \in \mathbb{Z}[i]$ with $\alpha \neq 0$. The ring $\mathbb{Z}[i]$ is a subring of the field $\mathbb{C}$. When we divide β by α as complex numbers, the result is of the form $r + si$ where $r, s \in \mathbb{Q}$. Let p and q be the closest integers to the r and s respectively so that $|p - r| \leq \frac{1}{2}$ and $|q - s| \leq \frac{1}{2}$. Then

$$\beta = (p + qi)\alpha + \rho,$$

where $\rho \in \mathbb{Z}[i]$. Let $\theta = \frac{\beta}{\alpha} - (p + qi)$ as an element in $\mathbb{Q}[i]$. Then $\alpha\theta = \rho$ and also $d(\theta) \leq \frac{1}{4} + \frac{1}{4} = \frac{1}{2}$. Hence, $d(\rho) = d(\alpha\theta) = d(\theta)d(\alpha) \leq \frac{1}{2}d(\alpha)$. In particular, $d(\rho) < d(\alpha)$.

This establishes that $d(z) = |z|^2$ is a Euclidean function and that $\mathbb{Z}[i]$ is a Euclidean domain.

We illustrate this Gaussian integer division with two explicit examples. Let $\beta = 19 - 23i$ and $\alpha = 5 + 3i$. In $\mathbb{Q}[i]$, we have

$$\frac{\beta}{\alpha} = \frac{13}{17} - \frac{86}{17}i.$$

The closest integers to the real and imaginary parts of this ratio are $p = 1$ and $q = -5$. Then

$$\rho = \beta - \alpha(p + qi) = 19 - 32i - (5 + 3i)(1 - 5i) = -2 - i$$

and we observe that $d(-2 - i) = 5 < d(5 + 3i) = 34$.

As another example, let $\gamma = 23 + 24i$ and keep $\alpha = 5 + 3i$. In $\mathbb{Q}[i]$, we have

$$\frac{\gamma}{\alpha} = \frac{11}{2} + \frac{3}{2}i.$$

In this case, we have two options for p and two options for q. All four options give us the following four possible Euclidean divisions:

$$23 + 24i = (5 + 3i)(5 + i) + (1 + 4i),$$
$$23 + 24i = (5 + 3i)(5 + 2i) + (4 - i),$$
$$23 + 24i = (5 + 3i)(6 + i) + (-4 + i),$$
$$23 + 24i = (5 + 3i)(6 + 2i) + (4 - i).$$

In the four possibilities, the different remainders are the four associates to $1 + 4i$ and have the Euclidean function $d(\rho) = 17 < d(\alpha) = 34$. $\triangle$

The Euclidean function in a Euclidean domain R offers some connection between R and the ring of integers. This connection, as loose as it is, is enough to establish some similar properties between R and $\mathbb{Z}$. The following proposition is one such example.

Proposition 4.3.3

Let I be an ideal in a Euclidean domain R. Then $I = (a)$ where a is an element of minimum Euclidean function value in the ideal. In particular, every Euclidean domain is a principal ideal domain.

Proof. Let d be the Euclidean function of R and consider $S = \{d(c) \,|\, c \in I\}$. By the well-ordering principle, since S is a subset of $\mathbb{N}$, it contains a least element n. Let $a \in I$ be an element such that $d(a) = n$.

Clearly $(a) \subseteq I$ since $a \in I$. Now let b be any element of I and consider the Euclidean division of b by a. We have $b = aq + r$ where $r = 0$ or $d(r) < d(a)$. Since $a, b \in I$, then $r = b - aq \in I$ so by the minimality of $d(a)$ in S it is not possible for $d(r) < d(a)$. Hence, $r = 0$, which implies that $b = aq$ and hence $b \in (a)$. This shows that $I \subseteq (a)$ and thus $I = (a)$. $\square$

4.3.2   Euclidean Algorithm

Euclidean domains derive their name from the ability to perform the Euclidean algorithm between two elements $a, b \in R$. The Euclidean algorithm involves performing successive Euclidean divisions in the following way. Set $b = r_0$ and $a = r_1$. Then

$$b = q_1 a + r_2, \qquad \text{where } r_2 = 0 \text{ or } d(r_2) < d(a)$$

$$\vdots$$

$$r_{i-1} = q_i r_i + r_{i+1}, \qquad \text{where } r_{i+1} = 0 \text{ or } d(r_{i+1}) < d(r_i) \qquad (4.6)$$

$$\vdots$$

$$r_{n-1} = q_n r_n.$$

This process must terminate because $d(r_1) = d(a)$ is finite and because the sequence $d(r_i)$ is a strictly decreasing sequence of nonnegative integers. It is possible that $r_{n-2} = q_{n-1} r_{n-1} + r_n$ with $d(r_n) = 0$ but $r_n \neq 0$; in this case, the axioms of Euclidean domains force $r_{n+1} = 0$.

Unlike the Euclidean Algorithm on the integers, there is no condition on the uniqueness of the elements that are involved in the Euclidean divisions in (4.6). As with integers, the Euclidean Algorithm leads to the following important theorem.

Theorem 4.3.4

Let R be a Euclidean domain and let a and b be nonzero elements of R. Let r be the last nonzero remainder in the Euclidean algorithm. Then r is a greatest common divisor of a and b. Furthermore, r can be written as $r = ax + by$ where $x, y \in R$.

Proof. Let $r = r_n$ be the final nonzero remainder in the Euclidean Algorithm. Then from the final step $r | r_{n-1}$. Suppose that $r | r_{n-i}$ and $r | r_{n-(i+1)}$. Then since

$$r_{n-(i+2)} = q_{n-(i+1)} r_{n-(i+1)} + r_{n-i},$$

by Exercise 4.1.2, r divides $r_{n-(i+2)}$. By induction, r divides r_{n-i} for all $i = 0, 1, \ldots, n$. Thus, r divides both a and b.

Now let s be any common divisors of a and b. Repeating an induction argument but starting at the beginning of the Euclidean Algorithm, it is easy to see that s divides r_k for all $k = 0, 1, \ldots, n$. In particular, s divides r. This proves that r is a greatest common divisor of a and b.

Again using an induction argument, starting from the beginning of the Euclidean Algorithm, it is easy to see that $r_k \in (a, b)$, the ideal generated by a and b. Hence, $r \in (a, b)$ and thus $r = ax + by$ for some $x, y \in R$. $\qquad \square$

Example A.5.11 illustrates the use of the Extended Euclidean Algorithm in $\mathbb{Z}$ to find $x, y \in \mathbb{Z}$ such that $r = ax + by$. The process described there, or any other implementation the algorithm, carries over identically to Euclidean domains.

4.3.3 Polynomial Rings over a Field

Another important example of Euclidean domains are polynomial rings over a field F. The following theorem not only establishes that $F[x]$ is a Euclidean domain with the degree $\deg : F[x] - \{0\} \to \mathbb{N}$ as the Euclidean function, but that the quotient and remainder of the Euclidean division are unique.

Theorem 4.3.5

Let F be a field. Then for all $a(x), b(x) \in F[x]$ with $a(x) \neq 0$, there exist unique polynomials $q(x)$ and $r(x)$ in $F[x]$ such that

$$b(x) = a(x)q(x) + r(x) \quad \text{with } r(x) = 0 \text{ or } \deg r(x) < \deg a(x). \quad (4.7)$$

Proof. We give the proof in two cases.

Case 1: Suppose that $a(x) \mid b(x)$. (This includes the situation $b(x) = 0$.) Then there exists $d(x)$ with $b(x) = a(x)d(x)$. This would mean that $(q(x), r(x)) = (d(x), 0)$ offers a solution to (4.7). Furthermore, for any other polynomial $q_1(x) \neq d(x)$, we have

$$b(x) - a(x)q_1(x) = a(x)d(x) - a(x)q_1(x) = a(x)(d(x) - q_2(x)).$$

So $\deg(b(x) - a(x)q_1(x)) \geq \deg a(x)$. Hence, the solution to (4.7) given by $(q(x), r(x)) = (d(x), 0)$ is the only solution.

Case 2: Suppose that $a(x)$ does not divide $b(x)$. Consider the set of non-negative integers $S = \{\deg(b(x) - q(x)a(x)) \mid q(x) \in F[x]\}$. By the Well-Ordering Principle, S has a least element. Let $q_0(x)$ be a polynomial that instantiates this least degree and call $r_0(x) = b(x) - a(x)q_0(x)$. Assume that $\deg r_0(x) \geq \deg a(x)$. Then $r_0(x) - \mathrm{LT}(r_0(x))/\mathrm{LT}(a(x))a(x)$ has degree lower than $r_0(x)$ because the subtraction cancels the leading term of $r_0(x)$. Thus,

$$r_2(x) = b(x) - \left(q(x) + \frac{\mathrm{LT}(r_0(x))}{\mathrm{LT}(a(x))} \right) a(x) \quad (4.8)$$

has degree strictly lower than $r_0(x)$, which contradicts the condition that $r_0(x)$ has minimal degree among polynomials of the form $b(x) - a(x)q(x)$. Hence, we can conclude that the polynomial $r_0(x)$ has degree strictly lower than $a(x)$. Thus $(q(x), r(x)) = (q_0(x), r_0(x))$ offers a solution to (4.7).

Let $q_1(x) \neq q_0(x)$ be any other polynomial in $F[x]$. Then

$$b(x) - a(x)q_1(x) = r_0(x) + a(x)(q_0(x) - q_1(x)).$$

Then $\deg(a(x)(q_0(x) - q_1(x))) \geq \deg a(x)$ and, in the above addition, $r_0(x)$ cannot cancel any leading term of $a(x)(q_0(x) - q_1(x))$, so $\deg(b(x) - a(x)q_1(x)) \geq \deg a(x)$. Thus, $r_0(x)$ is the only polynomial in S of degree less than $\deg a(x)$. Then $q_0(x) = (b(x) - r_0(x))/a(x)$, so the solution $(q(x), r(x)) = (q_0(x), r_0(x))$ to (4.7) is unique.   □

the hypothesis that the coefficients in the polynomial ring form a field came into play in the division of terms that appears in (4.8).

Corollary 4.3.6

For any field F, the polynomial ring $F[x]$ is a Euclidean domain with deg as the Euclidean function.

Proof. The only thing that Theorem 4.3.5 did not establish is that for any two nonzero polynomials $a(x), b(x) \in F[x]$, $\deg b(x) \leq \deg(a(x)b(x))$. However, this follows from the fact that the degree of a nonzero polynomial is nonnegative and that in $F[x]$,

$$\deg(a(x)b(x)) = \deg a(x) + \deg b(x).$$ □

Theorem 4.3.3 establishes that the polynomial ring $F[x]$ is also a principal ideal domain. In contrast, note that in the ring $\mathbb{Z}[x]$, the ideal $(2, x)$ is not principal so also by Theorem 4.3.3, the ring $\mathbb{Z}[x]$ is not a Euclidean domain.

An important consequence of this proposition is that it characterizes irreducible polynomials.

Proposition 4.3.7

Let F be a field and $p(x) \in F[x]$. The polynomial $p(x)$ is irreducible if and only if $F[x]/(p(x))$ is a field.

Proof. (Left as an exercise for the reader. See Exercise 4.3.18.)   □

The Euclidean division in $F[x]$ is called *polynomial division*. The proof of Theorem 4.3.5 is nonconstructive; the existence of $q(x)$ and $r(x)$ follow from the Well-Ordering Principle of the integers but the proof does not illustrate how to find $q(x)$ and $r(x)$. Polynomial division is sometimes taught in high school algebra courses but without justification of why it works. We review polynomial division here for completeness.

Polynomial division relies on the following fact. If $a(x), b(x) \in F[x]$ are such that $\deg b(x) \geq \deg a(x)$, then $\mathrm{LT}(b(x))/\mathrm{LT}(a(x))$ is a monomial and

$$p_1(x) = b(x) - \frac{\mathrm{LT}(b(x))}{\mathrm{LT}(a(x))} a(x)$$

has a degree lower than $b(x)$ because the leading term of $b(x)$ cancels out. By repeating this process on $p_1(x)$ to obtain $p_2(x)$ and so on, we ultimately obtain a polynomial that is 0 or is of degree less than $a(x)$. We illustrate the division with two examples.

Example 4.3.8. Consider the polynomials $a(x) = 2x^2 + 1$ and $b(x) = 3x^4 - 2x + 7$ in $\mathbb{Q}[x]$. At the first step, we find a monomial by which to multiply $a(x)$ in order to obtain the leading term of $b(x)$. This is $\frac{3}{2}x^2$. As with long division in base 10, we put the monomial on top of the quotient bar:

$$
\begin{array}{r}
\frac{3}{2}x^2 \qquad\qquad \\
2x^2 + 1 \,\overline{\big)\; 3x^4 \qquad -2x+7 }
\end{array}
$$

Then we subtract $\frac{3}{2}x^2(2x^2 + 1)$ from $b(x)$:

$$
\begin{array}{r}
\frac{3}{2}x^2 \qquad\qquad \\
2x^2 + 1 \,\overline{\big)\; 3x^4 \qquad\qquad -2x+7 } \\
-(3x^4 + \frac{3}{2}x^2) \qquad\qquad \\
\hline
-\frac{3}{2}x^2 \;\; -2x+7
\end{array}
$$

At this stage, we multiply $a(x)$ by the monomial $\mathrm{LT}(-\frac{3}{2}x^2 - 2x + 7)/\mathrm{LT}(a(x)) = -\frac{3}{4}$ to get a polynomial whose leading term is the same as the leading term of $-\frac{3}{2}x^2 - 2x + 7$. We add the monomial $-\frac{3}{4}$ to the quotient terms:

$$
\begin{array}{r}
\frac{3}{2}x^2 \qquad\qquad -\frac{3}{4} \\
2x^2 + 1 \,\overline{\big)\; 3x^4 \qquad\qquad -2x+7 } \\
-(3x^4 + \frac{3}{2}x^2) \qquad\qquad \\
\hline
-\frac{3}{2}x^2 \quad -2x+7 \\
-(-\frac{3}{2}x^2 \quad -\frac{3}{4}) \\
\hline
-2x + \frac{31}{4}
\end{array}
$$

This polynomial division algorithm ends at this stage since $\deg\left(-2x + \frac{31}{4}\right) < \deg(2x^2 + 1)$. This work shows that

$$3x^4 - 2x + 7 = (2x^2 + 1)\left(\frac{3}{2}x^2 - \frac{3}{4}\right) + \left(-2x + \frac{31}{4}\right)$$

is the polynomial division of $b(x)$ by $a(x)$. $\triangle$

Though we do not illustrate it here, we point out that since $F[x]$ is a Euclidean domain whenever F is a field, the Euclidean Algorithm applies and provides a method to calculate a greatest common divisor of two polynomials. With a minimum of experience performing polynomial division, the reader will notice how computationally intensive the Euclidean Algorithm on $F[x]$ if implemented by hand.

Example 4.3.9. As another example, we perform the polynomial division of $4x^4 + x^3 + 2x^2 + 3$ by $3x^2 + 4x + 1$ in $\mathbb{F}_5[x]$:

$$
\begin{array}{r}
3x^2 \quad +3x \quad +4 \\
3x^2 + 4x + 1 \;)\; \overline{\;4x^4 \quad +x^3 \quad +2x^2 \qquad +3\;} \\
-(4x^4 +2x^3 +3x^2) \\
\hline
4x^3 \quad +4x^2 \qquad +3 \\
-(4x^3 +2x^2 +3x) \\
\hline
2x^2 \quad +2x \quad +3 \\
-(2x^2 +x \quad +4) \\
\hline
x \quad +4
\end{array}
$$

We read this as $4x^4 + x^3 + 2x^2 + 3$ divided by $3x^2 + 4x + 1$ in $\mathbb{F}_5[x]$ has quotient $q(x) = 3x^2 + 3x + 4$ and remainder $r(x) = x + 4$. $\triangle$

4.3.4 Useful CAS Commands

The main CAS commands that we could hope for in relation to a Euclidean domain are quotient and remainder commands. The following blocks of code illustrate these commands in both *Maple* and then in SAGE.

───────────────── Maple ─────────────────

$iquo(17, 3)$

$$5$$

$irem(17, 3)$

$$2$$

$quo(3 \cdot x^2 + 7 \cdot x - 1, 2 \cdot x + 1, x);$

$$\frac{3x}{2} + \frac{11}{4}$$

$rem(3 \cdot x^2 + 7 \cdot x - 1, 2 \cdot x + 1, x);$

$$-\frac{15}{4}$$

$Quo(3 \cdot x^2 + 7 \cdot x - 1, 2 \cdot x + 1, x) \bmod 5;$

$$4x + 4$$

$Rem(3 \cdot x^2 + 7 \cdot x - 1, 2 \cdot x + 1, x) \bmod 5;$

$$0$$

The commands `iquo` and `rem` are for integers, whereas `quo` and `rem` are for polynomials. The last two lines illustrate the quotient and remainder over $\mathbb{F}_q$. We note that in *Maple* the commands for the quotient and the remainder are separate commands, whereas in SAGE they are provided by a single command.

──────────────── Sage ────────────────

```
sage: 17.quo_rem(3)
(5,2)
sage: A.<x>=11['x']
sage: a=A(3*x^2+7*x-1)
sage: b=A(2*x+1)
sage: a.quo_rem(b)
(3/2*x + 11/4, -15/4)
sage: a.change_ring(GF(5)).quo_rem(b.change_ring(GF(5)))
(4*x + 4, 0)
```

The integer division algorithm in Euclidean domains other than $\mathbb{Z}$ or $F[x]$, where F is a field, is specific to the ring. Hence, it is unlikely that a CAS has these code for any Euclidean domain. However, it is usually easy to define our own algorithm. We illustrate the use of procedures in *Maple* and SAGE for this purpose.

──────────────── Maple ────────────────

$ZIquorem := \mathbf{proc}(a, b)$
　　$\mathbf{local}\ q, r;$
　　$q := floor\left(Re\left(\frac{a}{b}\right) + \frac{1}{2}\right) + floor\left(Im\left(\frac{a}{b}\right) + \frac{1}{2}\right) * I;$
　　$r := a - b * q;$
　　$\mathbf{return}\ [q, r];$
$\mathbf{end}:$

ZIquorem(23+17*I,3-7*I);

$$[-1 + 4I, -2 - 2I]$$

——————————————— Sage (Jupyter) ———————————————

```
def ZIquo_rem(a,b):
    q=floor((a/b).real()+1/2)+floor((a/b).imag()+1/2)*I
    r=a-b*q;
    return [q,r]

ZIquo_rem(23+17*I,3-7*I)
```

[4*I - 1, -2*I - 2]

EXERCISES FOR SECTION 4.3

1. Perform the Euclidean division for $\beta = 32 + 8i$ divided by $\alpha = 3 + 8i$.

2. Perform the Euclidean division for $\beta = 719 - 423i$ divided by $\alpha = 24 - 38i$.

3. Perform the Euclidean Algorithm on $\beta = 24 + 17i$ and $\alpha = 13 - 16i$. Deduce the generator d of the ideal $I = (\alpha, \beta)$.

4. Perform the Euclidean Algorithm on $\beta = 14 + 23i$ and $\alpha = 42 + 3i$. Deduce the generator d of the ideal $I = (\alpha, \beta)$.

5. Perform the Extended Euclidean Algorithm associated with Exercise 4.3.3 to find a linear combination of α and β that give a greatest common divisor of α and β.

6. Perform the Extended Euclidean Algorithm associated with Exercise 4.3.4 to find a linear combination of α and β that give a greatest common divisor of α and β.

7. Perform polynomial division of $4x^4 + 3x^3 + 2x^2 + x + 1$ by $2x^2 + x + 1$ in $\mathbb{Q}[x]$.

8. Perform polynomial division of $4x^4 + 3x^3 + 2x^2 + x + 1$ by $2x^2 + x + 1$ in $\mathbb{F}_7[x]$.

9. Perform polynomial division of $4x^4 + 3x^3 + 2x^2 + x + 1$ by $2x^2 + x + 1$ in $\mathbb{F}_{13}[x]$.

10. Use the Euclidean Algorithm in $\mathbb{Q}[x]$ to find a generator of the principal ideal

$$I = (2x^4 + 7x^3 + 4x^2 + 13x - 10, 3x^4 + 5x^3 - 16x^2 + 14x - 4).$$

11. Prove that $\mathbb{Z}[\sqrt{2}]$ is a Euclidean domain with the norm $N(a + b\sqrt{2}) = |a^2 - 2b^2|$ as the Euclidean function.

12. Perform the Euclidean division of $\beta = 10 + 13\sqrt{2}$ by $\alpha = 2 + 3\sqrt{2}$. (See Exercise 4.3.11.)

13. Perform the Euclidean division of $\beta = 25 - 3\sqrt{2}$ by $\alpha = -1 + 13\sqrt{2}$. (See Exercise 4.3.11.)

14. Prove that $\mathbb{Z}[\sqrt{-2}]$ is a Euclidean domain with the norm $N(a + b\sqrt{-2}) = a^2 + 2b^2$ as the Euclidean function.

15. Let R be an integral domain. Prove that $R[x]$ is an Euclidean domain if and only if R is a field.

16. In $\mathbb{Q}[x]$, determine the monic greatest common divisor $d(x)$ of $a(x) = x^4 - 2x + 1$ and $b(x) = x^2 - 3x + 2$. Using the Extended Euclidean Algorithm, write $d(x)$ as a $\mathbb{Q}[x]$-linear combination of $a(x)$ and $b(x)$.

17. In $\mathbb{F}_2[x]$, determine the greatest common divisor $d(x)$ of $a(x) = x^4 + x^3 + x + 1$ and $b(x) = x^2 + 1$. Using the Extended Euclidean Algorithm, write $d(x)$ as a $\mathbb{F}_2[x]$-linear combination of $a(x)$ and $b(x)$.

18. Prove Proposition 4.3.7. [Hint: In a PID, an element is prime if and only if it is irreducible.] Conclude that for all nonzero polynomials $p(x) \in F[x]$, the quotient ring $F[x]/(p(x))$ is either a field or is not an integral domain.

19. Let R be a Euclidean domain with Euclidean function d. Let n be the least element in the set $S = \{d(r) \mid r \in R - \{0\}\}$. By the well-ordering of $\mathbb{Z}$, S has a least element n. Show that all elements $s \in R$ such that $d(s) = n$ are units.

20. *Least Common Multiples.* Let R be a Euclidean domain with Euclidean function d.

    (a) Prove that in R, any two nonzero elements a and b have a least common multiple.

    (b) Prove that least common multiples of a and b have the form $\dfrac{ab}{m}$ where m is a greatest common divisor.

4.4   Unique Factorization Domains

The Fundamental Theorem of Arithmetic states that any integer $n \geq 2$ can be written as a product of positive prime numbers and that any such product or primes is unique up to reordering. This property can also be stated by saying that integers have unique prime factorizations. The Fundamental Theorem of Arithmetic is taught early in a student's education, as soon as students know what prime numbers are. However, like many other algebraic properties of the integers, since students usually are not shown a proof of the Fundamental Theorem of Arithmetic it comes as a surprise that unique factorization does not hold in every integral domain.

As we did in Section 4.3, it is common in ring theory and in all of algebra more generally, to define a class of rings with specific properties and explore what further properties follow from the defining characteristics. This section introduces unique factorization domains: rings that possess a property like the Fundamental Theorem of Arithmetic.

4.4.1 Definition and Examples

Definition 4.4.1

A *unique factorization domain* (abbreviated as UFD) is an integral domain R in which every nonzero, nonunit element $r \in R$ has the following two properties:

- r can be written as a finite product of irreducible elements $r = p_1 p_2 \cdots p_n$ (not necessarily distinct);

- and for any other product into irreducibles, $r = q_1 q_2 \cdots q_m$, then $m = n$ and there is a reordering of the q_i such that each q_i is an associate to p_i.

A unique factorization domain is alternatively called a *factorial domain*.

In a UFD, we say that the factorization of an element r into irreducible elements is unique "up to reordering and associates." The condition of uniqueness can be restated more precisely as follows. If, as in the above definition,

$$r = p_1 p_2 \cdots p_n = q_1 q_2 \cdots q_m$$

then $m = n$ and there exists a permutation $\pi \in S_n$ such that $q_i \simeq p_{\pi(i)}$ for all $i = 1, 2, \ldots, n$ (i.e., q_i and $p_{\pi(i)}$ are associates).

Proposition 4.4.2

In any UFD, a nonzero element is prime if and only if it is irreducible.

Proof. By Proposition 4.1.9, every prime element is irreducible. We prove the converse in a UFD.

Let R be a UFD and let $r \in R$ be an irreducible element. Suppose that $a, b \in R$ and that $r|ab$. Then by definition of divisibility, there exists $c \in R$ such that $rc = ab$. Suppose that a, b, and c have the following factorizations into irreducible elements

$$a = p_1 p_2 \cdots p_m, \quad b = q_1 q_2 \cdots q_n, \quad c = r_1 r_2 \cdots r_k.$$

By definition of a UFD, since $ab = rc$, we have $k = m + n - 1$ and there is a reordering of the list

$$(p_1, p_2, \ldots, p_m, q_1, q_2, \ldots, q_n)$$

into

$$(r, r_1, r_2, \ldots, r_k)$$

possibly up to multiplication of units. Thus, there exists $r \mid p_i$ for some $i = 1, \ldots, m$ or $r \mid q_j$ for some $j = 1, \ldots, n$. If r divides some p_i, then $r \mid a$ and if r divides some q_j, then $r \mid b$. Hence, r is a prime element. $\square$

Because of this proposition, in a UFD we call the factorization of an element r into irreducible elements a *prime factorization* of r. Furthermore, the irreducible factors of an element are called *prime factors*.

Example 4.4.3. Every field is a UFD trivially since every nonzero element is a unit. $\triangle$

Example 4.4.4. The Fundamental Theorem of Arithmetic (Theorem A.5.19) is precisely the statement that $\mathbb{Z}$ is a UFD. $\triangle$

Example 4.4.5. We will soon see that every Euclidean domain is a unique factorization domain. Hence, $\mathbb{Z}[i]$ is a UFD. The norm function on rings of the form $\mathbb{Z}[\sqrt{D}]$, where D is square-free, helps in determining the prime factorization of elements. Recall that in such rings: (1) γ is a unit if and only if $N(\gamma) = 1$, and (2) γ is irreducible if $N(\gamma)$ is prime. (Note that (2) is not an if-and-only-if statement.) With this in mind, we propose to find a prime factorization of $\alpha = 6 + 5i$ and then of $\beta = 7 - 11i$.

First of all, note that $N(\alpha) = 6^2 + 5^2 = 61$ is a prime so α is irreducible and we are done.

For β, we have $N(\beta) = 7^2 + 11^2 = 170 = 2 \times 5 \times 17$. This is not sufficient to conclude that β is irreducible or not but it does tell us that an irreducible factor must have a norm that is a divisor of 170. We can try to find prime factors by trial and error. Note that $N(1 + i) = 2$, so $1 + i$ is irreducible and could possibly be a factor. Dividing β by $1 + i$ gives

$$\beta = (1 + i)(-2 - 9i)$$

so $1 + i$ is indeed a prime factor. The norm $N(-2 - 9i) = 85$ gives some possibilities on the prime factors of $-2 - 9i$. We observe that $N(2 + i) = 5$. However,

$$\frac{-2 - 9i}{2 + i} = -\frac{13}{5} - \frac{16}{5}i$$

so $2 + i$ is not a prime factor. On the other hand, $2 - i$, which is not an associate of $2 + i$, has norm 5 and we find that

$$\frac{-2 - 9i}{2 - i} = 1 - 4i.$$

Hence, $2 - i$ is a prime factor, as is $1 - 4i$ since $N(1 - 4i) = 17$. Hence, a prime factorization of β is

$$7 - 11i = (1 + i)(2 - i)(1 - 4i). \qquad \triangle$$

Example 4.4.6. Example 4.1.10 shows that $\mathbb{Z}[\sqrt{10}]$ is not a UFD. The example established that the elements $\alpha, \beta, \gamma, \delta$ are all irreducible, not associates of each other and that $10 = \alpha\beta = \gamma\delta$. $\triangle$

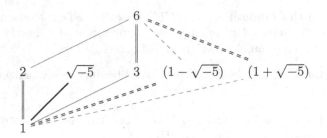

FIGURE 4.2: A visual interpretation of non-UFD.

Example 4.4.7. As a result of the main theorem (Theorem 4.5.5) in the next section, we will see that $F[x, y]$, where F is a field, is a UFD. However, it is rather easy to construct examples of subrings of $F[x, y]$ that are not UFDs. For example, consider the ring $R = F[x^2, xy, y^2]$. All the constants in $F \setminus \{0\}$ are units in R and R contains no polynomials of degree 1. Consequently, x^2, xy, and y^2 are irreducible elements in R. However, x^2y^2 can be factored into irreducible elements as

$$(x^2)(y^2) = x^2y^2 = (xy)(xy).$$

Furthermore, xy is not an associate of either x^2 or y^2. (If it were, x or y would need to be a unit, which is not the case.) Hence, this gives two nonequivalent factorizations of x^2y^2. $\triangle$

Example 4.4.8. For an easier example of a ring that is not a UFD, consider $R = \mathbb{Z}[\sqrt{-5}]$. Recall that an element $a + b\sqrt{-5} \in R$ is a unit if and only if $N(a + b\sqrt{-5}) = a^2 + 5b^2 = 1$. It is not hard to see that the only units in R are 1 and -1. It is also easy to see that there is no element $\gamma \in R$ such that $N(\gamma) = 2$ or $N(\gamma) = 3$. Now consider the following two factorizations of the element 6:

$$6 = 2 \times 3 = (1 + \sqrt{-5})(1 - \sqrt{-5}).$$

We have $N(2) = 4$, $N(3) = 9$, and $N(1 \pm \sqrt{-5}) = 6$. If $2, 3, 1 + \sqrt{-5}$ or $1 - \sqrt{-5}$ was reducible, then there would have to exist an element $\gamma \in R$ of norm 2 or 3. Since that is not the case, all four elements are irreducible. Furthermore, since the only units are 1 and -1, neither 2 nor 3 is an associate to either $1 + \sqrt{-5}$ or $1 - \sqrt{-5}$. Hence, we have displayed two distinct factorizations of 6 and hence $\mathbb{Z}[\sqrt{-5}]$ is not a UFD. $\triangle$

The ring $\mathbb{Z}[\sqrt{-5}]$ in the above example offers an opportunity to provide a visual interpretation of the criteria of a UFD.

Let R be an integral domain and consider the equivalence classes of associate elements. By Proposition 4.1.4, the set of equivalence classes $(R \setminus \{0\})/\simeq$ is a partial order under the relation of divisibility. Figure 4.2 illustrates a few of the elements and products involved in the example. Consider the paths from 1 to 6. If R is a unique factorization domain, then for each path from 1 to any element r will have a set of edges consisting of the same multiset of irreducible elements. That is not the case here because the path $1 \to 2 \to 6$ has edges corresponding to 2 and 3, whereas the path $1 \to 1 + \sqrt{-5} \to 6$ involves other factors.

4.4.2   UFDs and Other Classes of Integral Domains

For the following important proposition, we first need a lemma about PIDs. The lemma involves the so-called *ascending chain condition*.

Lemma 4.4.9

> If R is a PID, then every chain of ideals $I_1 \subseteq I_2 \subseteq \cdots \subseteq I_k \subseteq \cdots$ eventually terminates; i.e., there exists a k such that for all $n \geq k$, $I_n = I_k$.

Proof. Let R be a PID and let $I_1 \subseteq I_2 \subseteq \cdots \subseteq I_k \subseteq \cdots$ be an ascending chain of ideals. By Exercise 3.6.18, the set

$$I = \bigcup_{k=1}^{\infty} I_k$$

is an ideal of R. Since R is a PID, then $I = (a)$ for some a. However, $a \in I_k$ for some k. But then $I = (a) \subset I_n$ for all $n \geq k$ but since $I_n \subset I$ we must have $I_n = (a) = I$ for all $n \geq k$. Hence, the chain terminates. $\qquad\square$

Proposition 4.4.10

> Every PID is a UFD.

Proof. We first show by contradiction that every PID satisfies the first axiom for a UFD. Suppose that $r \in R$ cannot be written as a finite product of irreducibles. This tells us first that r is not irreducible so we can write $r = r_1 b_1$ where neither factor is a unit. The assumption also implies that at least one of these is not irreducible, say r_1. Then as ideals $(r) \subsetneq (r_1)$. Furthermore, we can write $r_1 = r_2 b_2$ and once again at least one of these is not an irreducible element, say r_2. Continuing with this reasoning, we define an ascending chain of ideals

$$(r) \subsetneq (r_1) \subsetneq (r_2) \subsetneq \cdots$$

which never terminates. This is a contradiction by Lemma 4.4.9. Hence, we conclude that every element in a PID can be written as a finite product of irreducible elements.

We now need to show the second axiom of UFD. One proceeds by induction on the minimum number of elements in a decomposition of an element r. Suppose that r has a factorization into irreducibles that has a single factor. Then r itself is an irreducible element. By the definition of an irreducible element, if $r = ab$, then either a or b is a unit and hence the other element is an associate of r. Hence, if there is a factorization with 1 element, all factorizations into irreducibles have 1 element.

For the induction hypothesis, suppose that if r has a factorization into n irreducibles, then all of its factorizations into irreducibles involve n irreducibles and that the irreducibles can be rearranged to be unique up to associates. Consider now an element that requires a minimum of $n+1$ irreducible factors for a factorization. Assume that we have two factorizations

$$r = p_1 p_2 \cdots p_{n+1} = q_1 q_2 \cdots q_m,$$

with $m \geq n+1$. Now p_1 divides $q_1 q_2 \cdots q_m$ or in other words, $q_1 q_2 \cdots q_m \in (p_1)$. However, in any PID, an irreducible is a prime element so since this is a prime ideal, we must have one of the $q_j \in (p_1)$, which means $q_j \mid p_1$. After renumbering we can assume that q_1 divides p_1. Thus, $q_1 a = p_1$ but since p_1 is irreducible and q_1 is not a unit, then a is a unit. We now have

$$p_1 p_2 \cdots p_{n+1} = a p_1 q_2 \cdots q_m \implies p_2 p_3 \cdots p_{n+1} = a q_2 q_3 \cdots q_m$$

since we are in an integral domain. We apply the induction hypothesis on the factorizations $p_2 \cdots p_{n+1} = a q_2 \cdots q_m$ and can conclude that the second criterion for UFD holds for $n+1$. By induction, the second axiom for UFDs hold for all $n \in \mathbb{N}^*$ and the proposition follows. $\qquad\square$

Proposition 4.3.3 along with Proposition 4.4.10 show that some of the classes of integral domains that we have introduced are subclasses of one another. We can summarize the containment of some of the ring classes introduced so far with the diagram

$$\text{fields} \subsetneq \text{Euclidean domains} \subsetneq \text{PIDs} \subsetneq \text{UFDs} \subsetneq \text{integral domains.} \qquad (4.9)$$

The propositions and examples provided so far have not shown strict containment between Euclidean domains and PIDs or between PIDs and UFDs. Example 4.5.7 gives examples of UFDs that are not PIDs.

As a consequence of Proposition 4.4.2, prime elements and irreducible elements are equivalent in Euclidean domains and in PIDs. Furthermore, as particular examples, since every Euclidean domain is a unique factorization domain, the ring of Gaussian integers $\mathbb{Z}[i]$ and the polynomial ring $F[x]$ over a field F are UFDs.

In the next section, we will establish a necessary and sufficient condition on R for $R[x]$ to be a UFD, but we can already give a characterization of when $R[x]$ is a PID.

Proposition 4.4.11

> If R is a commutative ring such that the polynomial ring $R[x]$ is a PID, then R is necessarily a field.

Proof. Suppose that $R[x]$ is a PID. Then the subring R is an integral domain. By Exercise 3.8.2, we see that (x) is a nonzero prime ideal. We can also see this by pointing out that two polynomials $a(x)$ and $b(x)$ with nonzero constant terms a_0 and b_0 are such that their product $a(x)b(x)$ has the nonzero constant term a_0b_0. Thus, $a(x) \notin (x)$ and $b(x) \notin (x)$ implies that $a(x)b(x) \notin (x)$. The contrapositive of this last statement establishes that (x) is a prime ideal.

By Proposition 3.8.12, since $R[x]$ is a PID then the nonzero prime ideal (x) is in fact a maximal ideal. Therefore, $R = R[x]/(x)$ is a field. $\square$

4.4.3 UFDs and Greatest Common Divisors

In Euclidean domains, we find the greatest common divisor between two elements a and b with the Euclidean algorithm. In a PID, if a and b are any two elements and if $(a) + (b) = (d)$, then d is a greatest common divisor of a and b. Greatest common divisors also exist in UFDs. Before providing a to find a greatest common divisor, we introduce the ord_r function.

Definition 4.4.12

> Let R be an integral domain and let $r \in R$ be an irreducible element. The function $\mathrm{ord}_r : R \setminus \{0\} \to \mathbb{N}$ is defined by $\mathrm{ord}_r(a) = n$ whenever n is the least positive integer such that r^{n+1} does not divide a.

Hence, $\mathrm{ord}_r(a) = n$ means that $r^k \mid a$ for $0 \leq k \leq n$ and $r^k \nmid a$ if $k > n$.

Lemma 4.4.13

> Let R be a UFD and let $a, b \in R \setminus \{0\}$. Then $a \mid b$ if and only if $\mathrm{ord}_r(a) \leq \mathrm{ord}_r(b)$ for all irreducible elements $r \in R$.

Proof. First suppose that $a \mid b$. Let r be any irreducible element of R and let $k = \mathrm{ord}_r(a)$. Then $r^k \mid a$ and by transitivity of divisibility, $r^k \mid b$. Hence, $\mathrm{ord}_r(a) \leq \mathrm{ord}_r(b)$.

Conversely, suppose that $\mathrm{ord}_r(a) \leq \mathrm{ord}_r(b)$ for all irreducible elements r. Let $a = p_1 p_2 \cdots p_m$ and $b = q_1 q_2 \cdots q_n$ be factorizations into irreducible elements. Let $a = p_1 p_2 \cdots p_m$ and $b = q_1 q_2 \cdots q_n$ be factorizations into irreducible elements. Let $S = \{r_1, r_2, \ldots, r_\ell\}$ be a (finite) set of irreducible elements, none of which are associates to each other and such that for each p_i and each q_j is associate to some element in S. Then we can write

$$a = u r_1^{\alpha_1} r_2^{\alpha_2} \cdots r_\ell^{\alpha_\ell} \qquad \text{and} \qquad b = v r_1^{\beta_1} r_2^{\beta_2} \cdots r_\ell^{\beta_\ell} \qquad (4.10)$$

for some units $u, v \in U(R)$ and where $\alpha_k = \mathrm{ord}_{r_k}(a)$ and $\beta_k = \mathrm{ord}_{r_k}(b)$. By hypothesis, $\alpha_k \leq \beta_k$ for all k so

$$b = a\big((u^{-1}v)r_1^{\beta_1-\alpha_1} r_2^{\beta_2-\alpha_2} \cdots r_\ell^{\beta_\ell-\beta_\ell}\big)$$

and thus $a \mid b$. $\qquad\qquad\square$

Some explanations behind the above proof are in order.

When working with factorizations into irreducibles in a UFD, it is convenient to write two elements a and b as in (4.10). In this expression, the integers α_k for $1 \leq k \leq \ell$ may be 0 if r_k is an associate to no p_i and just to a q_j. Similarly for β_k.

When considering prime factorization in $\mathbb{Z}$, we typically only use the positive prime numbers and the units can only be 1 or -1. However, in an arbitrary UFD, there may exist an infinite number of units. Consequently, there may not be a natural manner to select a preferred element out of each associate class.

Proposition 4.4.14

Let R be a unique factorization domain. For all $a, b \in R$, there exists a greatest common divisor of a and b.

Proof. Write a and b as in (4.10). By Lemma 4.4.13, the element

$$d = r_1^{\min(\alpha_1,\beta_1)} r_2^{\min(\alpha_2,\beta_2)} \cdots r_\ell^{\min(\alpha_\ell,\beta_\ell)}$$

is a divisor to both a and b. Let d' be any other common divisor of a and b. Each irreducible in a factorization of d' must be an associate to some p_i in the factorization of a. Hence, we can write

$$d' = w r_1^{\gamma_1} r_2^{\gamma_2} \cdots r_\ell^{\gamma_\ell},$$

where w is a unit. Furthermore, by Lemma 4.4.13, $\gamma_k \leq \alpha_k$ and $\gamma_k \leq \beta_k$. Consequently, $\gamma_k \leq \min(\alpha_k, \beta_k)$ and by Lemma 4.4.13 again, $d' | d$. Thus, d is a greatest common divisor. $\qquad\square$

By Proposition 4.1.16, any two elements a and b in a UFD have a least common multiple. In a parallel fashion, using the expressions in (4.10), it is easy to prove that

$$m = r_1^{\max(\alpha_1,\beta_1)} r_2^{\max(\alpha_2,\beta_2)} \cdots r_\ell^{\max(\alpha_\ell,\beta_\ell)}$$

is a least common multiple of a and b. (See Exercise 4.4.7.)

Proposition 4.4.14 states that UFDs, and consequently Euclidean domains and PIDs, are a subclass of gcd-domains. It turns out that there exist integral domains that are not UFDs but are gcd-domains. Hence, the class of UFDs is a strict subclass of gcd-domains.

4.4.4 Irreducible Gaussian Integers (Optional)

The ring of Gaussian integers, $\mathbb{Z}[i]$, is important for its applications to number theory. In this section, we study algebraic properties of $\mathbb{Z}[i]$ in further depth, in particular characterize the irreducible elements, and give a consequence for number theory.

First, note that since, when taking complex conjugation, $\overline{zw} = \overline{z}\,\overline{w}$, then an element in $\mathbb{Z}[i]$ is irreducible if and only if its complex conjugate is irreducible.

Lemma 4.4.15

> If an element $a + bi \in \mathbb{Z}[i]$ with $ab \neq 0$ is prime, then $a^2 + b^2$ is a prime number in $\mathbb{Z}$.

Proof. The element $a^2 + b^2 = (a + bi)(a - bi)$ is in the prime ideal $(a + bi)$. Assume that $a^2 + b^2 = mn$ as a composite integer with factors $m, n \geq 2$. Then since $(a + bi)$ is a prime ideal, $m \in (a + bi)$ or $n \in (a + bi)$. Without loss of generality, suppose that $m \in (a + bi)$. The integer m cannot be an associate of $a + bi$. Furthermore, there exists $z \in \mathbb{Z}[i]$ with $(a + bi)z = m$ so $zn = (a - bi)$. But n is not a unit either which makes $a - bi$ reducible. This is a contradiction and hence the assumption that $a^2 + b^2$ is a composite integer is false. The lemma follows. $\square$

Lemma 4.4.16

> Let $a + bi \in \mathbb{Z}[i]$ with $ab \neq 0$. Then $a + bi$ is prime if and only $a^2 + b^2 = p$ is a prime number with $p \equiv 1 \pmod 4$ or $p = 2$.

Proof. Suppose that $a + bi$ is prime in $\mathbb{Z}[i]$. Using modular arithmetic modulo 4, it is easy to that for $a, b \in \mathbb{Z}$, the sum $a^2 + b^2$ is never congruent to 3 modulo 4. (See Exercise A.6.8.) Furthermore, if $n \equiv 0 \pmod 4$, then n is divisible by 4, and hence is composite. By Lemma 4.4.15, we deduce that $a^2 + b^2$ is a prime integer congruent to 1 or 2 modulo 4. However, the only prime number that is congruent to 2 modulo 4 is 2.

Conversely, suppose that $a^2 + b^2 = p$ is a prime number. Then assume that $a + bi = \alpha\beta$ for some $\alpha, \beta \in \mathbb{Z}[i]$. Then $p = N(\alpha\beta) = N(\alpha)N(\beta)$ and hence either $N(\alpha) = 1$ or $N(\beta) = 1$. Hence, by Proposition 4.1.7, either α or β is a unit and we deduce that $a + bi$ is irreducible. $\square$

Note that if $a + bi$ is a prime element with $a^2 + b^2 = 2$, then $a + bi$ is one of the four elements $\pm 1 \pm i$.

Now we consider elements of the form a or bi. Obviously, bi is an associate to the integer b so without loss of generality we only consider elements $a + 0i \in \mathbb{Z}[i]$. Note that if an integer n is composite in $\mathbb{Z}$, it cannot be prime in $\mathbb{Z}[i]$ either. Hence, we restrict our attention to prime numbers.

Lemma 4.4.17

If p is a prime number in $\mathbb{Z}$ with $p \equiv 3 \pmod 4$, then p is also prime in $\mathbb{Z}[i]$.

Proof. Assume p not irreducible in $\mathbb{Z}[i]$. Then $p = \alpha\beta$ with neither α not β a unit. We have $p^2 = N(p) = N(\alpha)N(\beta)$ so $N(\alpha) = N(\beta) = p$ since neither $N(\alpha)$ nor $N(\beta)$ can be 1. However, $N(\alpha)$ is the sum of two squares but (by Exercise A.6.8) the sum of two squares cannot be congruent to 3 modulo 4. Hence, the assumption leads to a contradiction and the lemma follows. $\square$

Lemma 4.4.18

In $\mathbb{Z}[i]$, the integer 2 factors into two irreducibles $2 = (1+i)(1-i)$.

The last lemma turns out to be the most difficult and requires a reference to a result in number theory.

Lemma 4.4.19

If p is a prime number in $\mathbb{Z}$ with $p \equiv 1 \pmod 4$, then p is not prime in $\mathbb{Z}[i]$ but has the prime factorization of the form $p = (a+bi)(a-bi)$.

Proof. In 1770, Lagrange proved that if $p = 4n + 1$ is a prime number in $\mathbb{Z}$, then the congruence equation $x^2 \equiv -1 \pmod p$ has a solution. In other words, there exists an integer m such that $m^2 + 1$ is divisible by p. ([27, Theorem 11.5]) Now consider such a p as an element on $\mathbb{Z}[i]$. Note that $m^2 + 1 = (m+i)(m-i)$. The integer p cannot divide $m + i$ or $m - i$, because otherwise p would have to divide the imaginary part of $m + i$ or $m - i$, which is ± 1. Hence, since $\mathbb{Z}[i]$ is a unique factorization domain, p cannot be prime in $\mathbb{Z}[i]$.

The norm of p is $N(p) = p^2$ and since p is not prime in $\mathbb{Z}[i]$ then p must factor into $p = \alpha\beta$ with $N(\alpha) = N(\beta) = p$. Writing α and β in polar coordinates, $\alpha = \sqrt{p}e^\theta$ and $\beta = \sqrt{p}e^\phi$. Since $\alpha\beta = p$, we have $\theta + \phi = 2\pi k$ so $\phi = -\theta$ up to a multiple of 2π. Thus, β is the complex conjugate $\overline{\alpha}$. $\square$

The above five lemmas culminate in the following proposition.

Proposition 4.4.20

The prime (irreducible) elements in $\mathbb{Z}[i]$ are

- associates of integers of the form $p + 0i$, where p is prime with $p \equiv 3 \pmod 4$;

- associates of $1 + i$ or $1 - i$;

- elements $a + bi$ such that $a^2 + b^2$ is a prime number congruent to 1 modulo 4.

EXERCISES FOR SECTION 4.4

1. In the ring $\mathbb{Z}[i]$, write a factorization of $11 + 16i$.

2. In the ring $\mathbb{Z}[i]$, write a factorization of $9 - 19i$.

3. Write a factorization of $12 + 3\sqrt{2}$ in the UFD $\mathbb{Z}[\sqrt{2}]$.

4. Write a factorization of $-10 + 13\sqrt{2}$ in the UFD $\mathbb{Z}[\sqrt{2}]$.

5. Consider factorizations in $\mathbb{Z}[\sqrt{-5}]$. (See Example 4.4.8.)

   (a) Prove that there is no element $\alpha \in \mathbb{Z}[\sqrt{-5}]$ such that $N(\alpha) = 7$, $N(\alpha) = 11$ or $N(\alpha) = 13$.

   (b) Deduce that 7, 11, $3 + \sqrt{-5}$, $1 + 2\sqrt{-5}$ are irreducible elements.

   (c) Prove that $9 + 4\sqrt{-5}$ is an irreducible element.

6. Let R and S be two factorization domains and let $f : R \setminus \{0\} \to S \setminus \{0\}$ be a surjective multiplicative function ($f(ab) = f(a)f(b)$ for all $a, b \in R$) such that $f(1) = 1$.

   (a) Prove that for all $a \in R$, a is a unit in R if and only if $f(a)$ is a unit in S.

   (b) Prove that if $f(a)$ is irreducible in S, then a is irreducible in R.

   (c) Find an example in which a is irreducible but $f(a)$ is not.

   (d) Explain how a prime factorization of $f(a)$ may help in determining a prime factorization of a.

7. Use the factorizations in (4.10) to show that

$$m = r_1^{\max(\alpha_1, \beta_1)} r_2^{\max(\alpha_2, \beta_2)} \cdots r_\ell^{\max(\alpha_\ell, \beta_\ell)}$$

is a least common multiple of a and b.

8. Let p be a prime number in $\mathbb{Z}$. Recall that the localization of $\mathbb{Z}$ by the prime ideal (p), denoted by $\mathbb{Z}_{(p)}$, consists of all fractions whose denominator is not divisible by p. Discuss what variations can exist in the factorizations of elements in $\mathbb{Z}_{(p)}$.

9. Consider the quotient ring $R = \mathbb{R}[x, y]/(x^3 - y^5)$. Prove that $\bar{x}$ and $\bar{y}$ are irreducible elements in R. Deduce that R is not a UFD.

10. Consider the quotient ring $R = \mathbb{R}[x, y, z]/(x^2 - yz)$. Prove that $\bar{x}$, $\bar{y}$, and $\bar{z}$ are irreducible elements in R. Deduce that R is not a UFD.

11. Let F be a field. Show that the subring $F[x^4, x^2 y, y^2]$ of the polynomial ring $F[x, y]$ is not a UFD.

12. Let R be a UFD and let $p(x) \in R[x]$. Prove that if $p(c_i) = 0$ for n distinct constants $c_1, c_2, \ldots, c_n \in R$, then $p(x) = 0$ or $\deg p(x) \geq n$. Conclude that if $p(x)$ is a polynomial that is 0 or of degree less than n that has n distinct roots, then $p(x)$ is the 0 polynomial. [Hint: Work in the field of fractions of R and then use Gauss' Lemma.]

13. Let R be a UFD and let D be a multiplicatively closed set that does not contain 0. Prove that the ring of fractions $D^{-1}R$ is a UFD.

14. Let R be an integral domain and let $r \in R$ be an irreducible element. Prove that $\operatorname{ord}_r(ab) = \operatorname{ord}_r(a) + \operatorname{ord}_r(b)$ for all $a, b \in R \setminus \{0\}$.

15. Let R and r be as in Exercise 4.4.14. Let $(R \setminus \{0\})/\simeq$ be the set of equivalence classes of associate nonzero elements in R.

   (a) Prove that $\text{ord}_r : (R \setminus \{0\})/\simeq \, \to \mathbb{N}$ given by $\text{ord}_r([a]) \overset{\text{def}}{=} \text{ord}_r(a)$ for all $a \in R$ is a well-defined function.

   (b) Prove that $\text{ord}_r : (R \setminus \{0\})/\simeq \, \to \mathbb{N}$ is a monotonic function between the posets $((R \setminus \{0\})/\simeq, |)$ and $(\mathbb{N}, \leq)$.

16. Let R be a UFD and let $a \in R \setminus \{0\}$. Let S_a be a set of irreducible elements that divide a and such that no two elements in S_a are associates. Prove that S_a is a finite set and that
$$a \simeq \prod_{r \in S_a} r^{\text{ord}_r(a)}.$$

17. Let R be an integral domain. Suppose that for all $a \in R \setminus \{0\}$, any set S of irreducible elements that divide a in which no two elements are associates is finite. Prove that R is a unique factorization domain if and only if for all $a \in R \setminus \{0\}$ and all such sets S,
$$a \simeq \prod_{r \in S} r^{\text{ord}_r(a)}.$$

[Hint: Use Exercise 4.4.16.]

4.5   Factorization of Polynomials

Early on, students of mathematics learn strategies to factor polynomials or to find roots of a polynomial. Many of the theorems introduced in elementary algebra assume the coefficients are in $\mathbb{Z}$, $\mathbb{Q}$, or $\mathbb{R}$. In this section we review many theorems concerning factorization or irreducibility in $R[x]$ where R is an integral domain.

4.5.1   Polynomial Rings and Unique Factorization

It does not make sense to discuss factorization of polynomials unless $R[x]$ is a unique factorization domain. The first proposition begins with this condition.

Lemma 4.5.1

   If $R[x]$ is a UFD, then R is a UFD.

Proof. The ring R of coefficients is a subring of $R[x]$. By degree considerations and Proposition 3.2.6, if $p(x) = c$ is a constant polynomial and if $p(x) = a(x)b(x)$, then both $a(x)$ and $b(x)$ have degree 0. Thus, if $R[x]$ is a UFD, then every $c \in R$ has a unique factorization into irreducible elements of $R[x]$ but each of these elements has to have degree 0. Thus, R is a UFD.   □

Because of this lemma, we henceforth let R be a UFD.

Proposition 4.5.2 (Gauss' Lemma)

Let R be a UFD with field of fractions F. If $p(x) \in R[x]$ is reducible in $F[x]$, then $p(x)$ is reducible in $R[x]$. Furthermore, if $p(x) = A(x)B(x)$ in $F[x]$, then in $R[x]$, we have $p(x) = a(x)b(x)$, where $a(x) = uA(x)$ and $b(x) = vB(x)$, where $u, v \in F$.

Proof. Consider the equation $p(x) = A(x)B(x)$ where $A(x)$ and $B(x)$ are elements in $F[x]$. Let d_A be a least common multiple of all the denominators appearing in $A(x)$ and similarly for d_B. Set $a'(x) = d_A A(x)$ and $b'(x) = d_B B(x)$, which are polynomials in $R[x]$. Setting $d = d_A d_B$, we have $dp(x) = a'(x)b'(x)$ in $R[x]$.

If d is a unit in R, then we are done with $a(x) = d^{-1}a'(x) = d^{-1}d_A A(x)$ and $b(x) = b'(x) = d_B B(x)$.

If d is not a unit, consider the prime factorization of $d \in R$, namely $d = p_1 p_2 \cdots p_n$. For each $i \in \{1, 2, \ldots, n\}$, the ideal $p_i R[x]$ is a prime ideal in $R[x]$ since p_i is irreducible in R and therefore irreducible in the UFD $R[x]$. By Proposition 3.7.11, $R[x]/(p_i R[x]) \cong (R/p_i R)[x]$ and is an integral domain, since $p_i R[x]$ is a prime ideal. Considering the expression $dp(x) = a'(x)b'(x)$ reduced in the quotient ring, we have

$$\bar{0} = \overline{a'(x)}\ \overline{b'(x)}.$$

Thus, one of these polynomials in the quotient ring is 0. Therefore, it is possible to partition $\{1, 2, \ldots, n\} = \mathcal{I}$ into two subsets $\mathcal{I}_a$ and $\mathcal{I}_b$ such that if $i \in \mathcal{I}_a$, then $\overline{a'(x)} = 0$ in $(R/p_i R)[x]$ and if $i \in \mathcal{I}_b$, then $\overline{b'(x)} = 0$ in $(R/p_i R)[x]$. Then all the coefficients of $a'(x)$ are multiples of

$$\prod_{i \in \mathcal{I}_a} p_i$$

and similarly for $b'(x)$. Thus, the polynomials

$$a(x) = \left(\prod_{i \in \mathcal{I}_a} p_i\right)^{-1} a'(x) = d_A \left(\prod_{i \in \mathcal{I}_a} p_i\right)^{-1} A(x)$$

$$\text{and}\quad b(x) = \left(\prod_{i \in \mathcal{I}_b} p_i\right)^{-1} b'(x) = d_B \left(\prod_{i \in \mathcal{I}_b} p_i\right)^{-1} B(x),$$

satisfy $a(x), b(x) \in R[x]$ and $a(x)b(x) = \frac{d}{d}A(x)B(x) = A(x)B(x)$. $\qquad\square$

An immediate consequence of Gauss' Lemma is that if a polynomial with integer coefficients factors over $\mathbb{Q}$, then it factors over $\mathbb{Z}$ as well.

Corollary 4.5.3

Let R be a UFD and let F be its field of fractions. Let $p(x) \in R[x]$ be such that its coefficients have a greatest common divisor of 1. Then $p(x)$ is irreducible in $R[x]$ if and only if it is irreducible in $F[x]$. In particular, a monic polynomial is irreducible in $R[x]$ if and only if it is irreducible in $F[x]$.

Proof. (Left as an exercise for the reader. See Exercise 4.5.10.) □

Example 4.5.4. Consider the polynomial $p(x) = 6x^2 - x - 1$ in $\mathbb{Z}[x]$. If we consider $p(x)$ as an element of the bigger ring $\mathbb{Q}[x]$, it can be factored as

$$p(x) = \left(x - \frac{1}{2} \right) (6x + 2).$$

This factorization is not in $R[x]$ but it can be changed to $p(x) = (2x-1)(3x+1)$ in $\mathbb{Z}[x]$. Note that in $\mathbb{Q}[x]$, the unique factorization of $p(x)$ is

$$p(x) = 6 \left(x - \frac{1}{2} \right) \left(x + \frac{1}{3} \right),$$

where 6 is a unit in $\mathbb{Q}$. △

Theorem 4.5.5

R is a UFD if and only if $R[x]$ is a UFD.

Proof. Lemma 4.5.1 already gave one direction of this proof. Gauss' Lemma allows us to prove the converse.

Suppose now that R is a UFD and let F be its field of fractions. Recall that $F[x]$ is a Euclidean domain so it is a UFD. Let $p(x) \in R[x]$ and let d be the greatest common divisor of the coefficients of $p(x)$ so that $p(x) = dp_2(x)$ where the coefficients of $p_2(x)$ have a greatest common divisor of 1. Since R is a UFD, and d can be factored uniquely into irreducibles in R, it suffices to prove that $p_2(x)$ can be factored uniquely into irreducibles in $R[x]$.

Since $F[x]$ is a UFD, $p_2(x)$ can be factored uniquely into irreducibles in $F[x]$ and by Gauss' Lemma, there is a factorization of $p_2(x)$ in $R[x]$. Since the greatest common divisor of coefficients of $p_2(x)$ is 1, then the greatest common divisor of each of the factors of $p_2(x)$ in $R[x]$ is 1. By Corollary 4.5.3, the factors of $p_2(x)$ in $R[x]$ are irreducible. Thus, $p(x)$ can be written as a product of irreducible elements in $R[x]$.

Since $R[x]$ is a subring of $F[x]$, then the factorization of $p_2(x)$ in $R[x]$ is a factorization into irreducible elements in $F[x]$, which is unique up to rearrangement and multiplication by units. There exist fewer units in R than in F so the uniqueness of the factorization also holds in $R[x]$. □

Theorem 4.5.5 establishes that the algebraic context in which to discuss factorization of polynomials is when the ring of coefficients is itself a UFD. The theorem also has consequences for multivariable polynomial rings.

Corollary 4.5.6

If R is a UFD, then the polynomial ring $R[x_1, x_2, \ldots, x_n]$ with a finite number of variables is a UFD.

Proof. Theorem 4.5.5 establishes the induction step from $R[x_1, x_2, \ldots, x_{n-1}]$ to $R[x_1, x_2, \ldots, x_n]$ for all $n \geq 1$ and hence the corollary follows by induction on n. □

Example 4.5.7. Note that $\mathbb{Z}[x]$, $\mathbb{Z}[x, y]$, etc. are UFDs by the above theorems. However, they are not PIDs and thus give simple examples of rings that are UFDs but not PIDs. △

4.5.2 Irreducibility Tests

Unless otherwise stated, in the rest of the section, R is a UFD.

No discussion about factorization of polynomials is complete without some comment about how to determine if polynomials are irreducible. Determining if a polynomial $p(x) \in R[x]$ is irreducible is a challenging problem. Consequently, this brief section can only offer a few comments for polynomials of low degree and some strategies for polynomials of degree 4 or higher.

We first deal with irreducible factors of degree 0.

Definition 4.5.8

A polynomial $p(x) \in R[x]$ is called *primitive* if the coefficients of $p(x)$ are relatively prime.

In the language of unique factorization domains, a polynomial is primitive if and only if it does not have irreducible factors of degree 0. Note that if F is a field, then every polynomial in $F[x]$ is primitive.

With polynomials in $\mathbb{Z}[x]$, the *content* of a polynomial $p(x)$, denoted by $c(p)$ is defined as the greatest common divisor of the coefficients of $p(x)$ multiplied by the sign of the leading coefficient. Similarly, we may refer to the content of a polynomial $p(x) \in R[x]$, though this is only well-defined up to multiplication by a unit. Consequently, a polynomial $p(x) \in R[x]$ can be written $p(x) = c(p)q(x)$ where $q(x)$ is a primitive polynomial and is called the *primitive part* of $p(x)$.

Proposition 4.5.9

Let $p(x) \in R[x]$ and let F be the field of fractions of R. The polynomial $p(x)$ has a factor of degree 1 if and only if it has a root in F. Furthermore, if the root is $\frac{r}{s}$, then $p(x) = (sx - r)q(x)$ for some $q(x) \in R[x]$.

Proof. If $p(x)$ has a factor of degree 1, say $(sx - r)$, then $\frac{r}{s}$ is a root of $p(x)$ when viewed as an element in $F[x]$.

For the converse, suppose that $p(x)$ has a root $\alpha = \frac{r}{s}$ in F. Consider the polynomial division of $p(x)$ by $d(x) = x - \alpha$ in the Euclidean domain $F[x]$. Since the remainder must have degree less than $\deg d(x) = 1$, so there exist unique $q(x) \in F[x]$ and $r \in F$ such that $p(x) = (x - \alpha)q(x) + r$. However, $p(\alpha) = 0$ so $0 = 0q(\alpha) + r$ and thus $r = 0$. Therefore, $p(x) = (x - \alpha)q(x)$ in $F[x]$.

By Gauss' Lemma, $p(x) = (sx - r)q(x)$ for some polynomial $q(x) \in R[x]$. $\square$

Corollary 4.5.10

Let $p(x)$ be a nonzero polynomial in $R[x]$. The number of distinct roots in the field of fractions is less than or equal to the degree of $p(x)$.

Proof. Let F be the field of fractions of R. If $\alpha_1, \alpha_2, \ldots, \alpha_m$ are distinct roots of $p(x)$ in F with $\alpha_i = r_i/s_i$, then

$$p(x) = (s_1 x - r_1)(s_2 x - r_2) \cdots (s_m x - r_m)q(x)$$

for some $q(x) \in R[x]$. Hence, $\deg p(x) \geq m$. $\square$

The following proposition is sometimes presented in elementary algebra courses in the context of polynomials with integer coefficients. The proposition provides a short list of all the possible linear irreducible factors of a polynomial.

Proposition 4.5.11 (Rational Root Theorem)

Let F be the field of fractions of R. Suppose that $p(x) \in R[x]$ is written

$$p(x) = a_n x^n + \cdots + a_1 x + a_0.$$

Then the only roots of $p(x)$ in $F[x]$ are $u\frac{e}{d}$ where u is a unit, e divides a_0 and d divides a_n.

Proof. Let $\frac{r}{s}$ be a roots of $p(x)$ in F (with $p(x)$ viewed as an element of $F[x]$) and suppose that r and s have no common divisor. Then

$$a_n \frac{r^n}{s^n} + a_{n-1} \frac{r^{n-1}}{s^{n-1}} + \cdots + a_1 \frac{r}{s} + a_0 = 0$$

which, after multiplying by s^n, gives

$$a^n r^n + a_{n-1} s r^{n-1} + \cdots + a_1 s^{n-1} r + a_0 s^n = 0. \tag{4.11}$$

Then

$$s(a_{n-1} r^{n-1} + \cdots + a_1 s^{n-2} r + a_0 s^{n-1}) = -a_n r^n.$$

By unique factorization in R, all the prime factors of s divide $a_n r^n$ but since s and r are relatively prime, all the prime factors of s must be associates to the prime factors of a_n. Thus, $s|a_n$. With an identical argument applied to (4.11), we can show that $r|a_0$. The result follows. □

   Proposition 4.5.11 also makes it simple to determine if a given quadratic or cubic polynomial is irreducible.

Proposition 4.5.12

A primitive polynomial $p(x) \in R[x]$ of degree 2 or 3 is reducible in $R[x]$ if and only if it has a root in the field of fractions F.

Proof. Suppose a primitive polynomial $p(x)$ of degree 2 or 3 is reducible. Then $p(x) = a(x)b(x)$ with $a(x), b(x) \in R[x]$, not units. Since $\deg a(x), \deg b(x) \geq 1$ and since $\deg a(x) + \deg b(x) = \deg p(x) \leq 3$, then $\deg a(x) = 1$ or $\deg b(x) = 1$. By Proposition 4.5.9, $p(x)$ has a root in F.

   Conversely, suppose that $p(x)$ has a root $\frac{r}{s} \in F$. Then by Proposition 4.5.9, $p(x) = (sx - r)q(x)$. Since $p(x)$ is of degree 2 or 3, then $\deg q(x)$ is equal to 1 or 2, and hence $q(x)$ is not a unit. Thus, $p(x)$ is reducible. □

Example 4.5.13. We show that $p(x) = 2x^3 - 7x + 3$ is irreducible in $\mathbb{Z}[x]$. By Gauss' Lemma, $p(x)$ can factor over $\mathbb{Q}$ if and only if it can factor over $\mathbb{Z}$. We just need to check if it has roots in $\mathbb{Q}$ to determine whether it factors over $\mathbb{Q}$. The only possible roots according to Proposition 4.5.11 are

$$\pm 1, \ \pm 3, \ \pm\frac{1}{2}, \ \pm\frac{3}{2}.$$

It is easy to verify that none of these eight fractions are roots of $p(x)$. Then by Proposition 4.5.12, $p(x)$ is irreducible in $\mathbb{Z}[x]$. △

Example 4.5.14. Consider the polynomial $x^3 + 2x^2 + 2x + 3$ in $\mathbb{F}_5[x]$. As mentioned earlier, a field is trivially a UFD. Furthermore, the field of fractions of a field F is itself. Consequently, Proposition 4.5.12 still applies. In this situation, we could still refer to Proposition 4.5.11 to look for roots of the polynomial but, since every nonzero elements in a field is a unit, applying the proposition implies that we must simply check all elements of $\mathbb{F}_5$ to see if they are roots. Calculating directly,

$$0^3 + 2 \cdot 0^2 + 2 \cdot 0 + 3 = 3 \neq 0,$$
$$1^3 + 2 \cdot 1^2 + 2 \cdot 1 + 3 = 3 \neq 0,$$
$$2^3 + 2 \cdot 2^2 + 2 \cdot 2 + 3 = 3 \neq 0,$$
$$3^3 + 2 \cdot 3^2 + 2 \cdot 3 + 3 = 4 \neq 0,$$
$$4^3 + 2 \cdot 4^2 + 2 \cdot 4 + 3 = 2 \neq 0.$$

We observe that no element of the field is a root of the polynomial. So by Proposition 4.5.12, since the polynomial is a cubic and has no roots, then the polynomial is irreducible. △

Up to now, the propositions in this subsection have shown how to quickly determine if a polynomial of degree 3 or less is irreducible. For polynomials of degree 4 or more in $R[x]$, Proposition 4.5.11 helps quickly determine if a polynomial has a factor of degree 1. However, it requires more work to determine if a polynomial has an irreducible factor of degree 2 or more. The following examples illustrate what can be done for polynomials with coefficients in $\mathbb{Z}$ or in the finite field $\mathbb{F}_p$.

Example 4.5.15. We propose to show that $p(x) = x^4 + x + 2$ is irreducible as a polynomial in $\mathbb{F}_3[x]$. By checking the three field elements $0, 1, 2 \in \mathbb{F}_3$, it easy to see that none of them are roots. Hence, $p(x)$ has no linear factors. Assume that $p(x)$ is reducible. Then by degree considerations, $p(x)$ is the product of two quadratic polynomials. A priori, by considering the leading coefficient of $p(x)$, there appear to be two cases:

$$p(x) = (x^2 + ax + b)(x^2 + cx + d) \text{ and } p(x) = (2x^2 + ax + b)(2x^2 + cx + d).$$

However, by factoring out a 2 from each of the terms of the second case, we see that it is equivalent to the first case. Hence, we only need to consider the first situation. Expanding the product for $p(x)$ gives

$$p(x) = (x^2 + ax + b)(x^2 + cx + d) = x^4 + (a+c)x^3 + (d+ac+b)x^2 + (ad+bc)x + bd,$$

so the coefficients must satisfy

$$\begin{cases} a + c = 0 \\ b + d + ac = 0 \\ ad + bc = 1 \\ bd = 2. \end{cases} \tag{4.12}$$

The last of the four conditions gives $(b, d) = (1, 2)$ or $(2, 1)$. Applied to the second equation, we see that $ac = 0$, so $a = 0$ or $c = 0$. This last result applied to the first equation, shows that $a = c = 0$. But then in the third equation, we deduce that $0 + 0 = 1$, which is a contradiction. We conclude that $p(x)$ is irreducible. △

Example 4.5.16. We repeat the above example, but with $q(x) = x^4 + x + 2 \in \mathbb{Z}[x]$. By Proposition 4.5.11, if $q(x)$ has a linear factor, then it has one of the following roots: ± 1 or ± 2. A quick calculation shows that none of these four numbers is a root of $q(x)$, so $q(x)$ has no factors of degree 1. Assume that $q(x)$ is reducible. Then

$$q(x) = (ax^2 + bx + c)(dx^2 + ex + f).$$

From $ad = 1$, we deduce that $(a, d) = (1, 1)$ or $(-1, -1)$. As in the previous example, the case $(a, d) = (-1, -1)$ can be made equivalent to the case $(a, d) = (1, 1)$, by multiplying both quadratic factors by -1. Consequently, we again have equations similar to (4.12) but in $\mathbb{Z}$:

$$\begin{cases} b + e = 0 \\ c + f + be = 0 \\ bf + ce = 1 \\ cf = 2, \end{cases} \implies \begin{cases} e = -b \\ c + f - b^2 = 0 \\ b(f - c) = 1 \\ cf = 2. \end{cases}$$

From the third equation, $f - c = 1$ or -1. Furthermore, $b = \pm 1$ and hence the second equation gives $c + f = 1$. Since we are in $\mathbb{Z}$, the fourth equation implies that (c, f) can be one of the four pairs $(1, 2)$, $(2, 1)$, $(-1, -2)$, or $(-2, -1)$, but none of these four options satisfies $c + f = 1$. This implies a contradiction and hence we deduce that $q(x)$ is irreducible in $\mathbb{Z}[x]$. $\triangle$

The above two examples illustrate a similar strategy to checking if a quartic polynomial is irreducible but applied to different coefficient rings. However, we could have immediately deduced the result of Example 4.5.16 from Example 4.5.15 without as much work. The following proposition generalizes this observation.

Proposition 4.5.17

Let I be a proper ideal in the integral domain R and let $p(x)$ be a non-constant monic polynomial in $R[x]$. If the image of $p(x)$ in $(R/I)[x]$ under the reduction homomorphism is an irreducible element, then $p(x)$ is irreducible in $R[x]$.

Proof. Suppose that $p(x) = a(x)b(x)$, where $a(x)$ and $b(x)$ are not units in $R[x]$. Then the degrees of $a(x)$ and $b(x)$ must be positive since if either $a(x)$ or $b(x)$ were constant, then since $\mathrm{LC}(p(x)) = 1 = \mathrm{LC}(a(x))\mathrm{LC}(b(x))$, the constant polynomial would need to be a unit in R.

Now let $\varphi : R[x] \to (R/I)[x]$ be the reduction homomorphism. (Recall that φ maps the coefficients of $p(x)$ to its images in R/I.) Since φ is a homomorphism, then $\varphi(p(x)) = \varphi(a(x))\varphi(b(x))$. We already established that the leading coefficients of $a(x)$ and $b(x)$ must be units, since the leading coefficient of $p(x)$ is 1. However, since I is a proper ideal of R, it contains no units.

We deduce that $\deg \varphi(a(x)) = \deg a(x) \geq 1$ and $\deg \varphi(b(x)) = \deg b(x) \geq 1$. Hence, neither $\varphi(a(x))$ nor $\varphi(b(x))$ is a unit in $(R/I)[x]$ and thus $\varphi(p(x))$ is reducible.

We have proven that if $p(x)$ is reducible, then $\varphi(p(x))$ is reducible. The proposition is precisely the contrapositive of this statement.   □

The following example gives another application of Proposition 4.5.17 even as it illustrates some more reasoning with factorization of polynomials.

Example 4.5.18. We show that $q(x) = x^4 + 3x^3 + 22x^2 - 8x + 3$ is irreducible in $\mathbb{Z}$. We consider the polynomial modulo 2: $\overline{q(x)} = x^4 + x^3 + 1 \in \mathbb{F}_2[x]$. It is obvious that neither 0 nor 1 in $\mathbb{F}_2$ are roots of $\overline{q(x)}$. Hence, if $\overline{q(x)}$ is reducible in $\mathbb{F}_2[x]$, then it must be a product of two quadratics since it does not have a factor of degree 1. However, we point out that $\mathbb{F}_2[x]$ only has one irreducible polynomial of degree 2, namely $x^2 + x + 1$. (There are only 3 other quadratic polynomials in $\mathbb{F}_2[x]$, namely x^2, $x^2 + 1$, and $x^2 + x$, each of which is reducible.) The only quartic that is the product of two irreducible quadratics is

$$(x^2 + x + 1)(x^2 + x + 1) = x^4 + x^2 + 1.$$

Hence, $x^4 + x^3 + 1$ is irreducible. We point out that our reasoning also establishes that $x^4 + x + 1$ is irreducible in $\mathbb{F}_2[x]$.   △

For nearly all integral domains R, determining whether a polynomial in $R[x]$ is irreducible is a generally a difficult problem. Proposition 4.5.17 offers a sufficient condition for deciding if a polynomial is irreducible. Indeed, many theorems provide sufficient but not necessary conditions that are relatively easy to verify. We conclude the section with one more sufficient condition for irreducibility.

Theorem 4.5.19 (Eisenstein's Criterion)

Let P be a prime ideal in an integral domain R and let

$$f(x) = x^n + a_{n-1}x^{n-1} + \cdots + a_1 x + a_0$$

be a polynomial in $R[x]$. Suppose that $a_i \in P$ for all $i < n$ and $a_0 \notin P^2$. Then $f(x)$ is irreducible in $R[x]$.

Proof. Suppose that $f(x)$ can be written as $f(x) = a(x)b(x)$. Since $a_0 \notin P^2$, then $a_0 \neq 0$ and hence both of $a(x)$ and $b(x)$ have nonzero constant terms.

The reduction homomorphism, $R[x] \to R[x]/P[x] = (R/P)[x]$ gives

$$x^n = \overline{a(x)b(x)}$$

in $(R/P)[x]$. We prove by contradiction that $\overline{a(x)}$ and $\overline{b(x)}$ must each have a single term. Assume that $\overline{a(x)}$ or $\overline{b(x)}$ has more than one term. Since R/P

is an integral domain, the product of leading terms and the product of the terms of minimal degree in $\overline{a(x)}$ and $\overline{b(x)}$ give two (nonzero) terms of different degrees in the product $\overline{a(x)b(x)}$. This contradicts the fact that $\overline{a(x)b(x)}$ is a polynomial of a single term. Consequently, both $\overline{a(x)}$ and $\overline{b(x)}$ are polynomials of a single term, and thus all the nonleading terms of $a(x)$ and $b(x)$ are in P.

Since $f(x)$ is monic, the leading coefficients of $a(x)$ and $b(x)$ are units in R. Assume that $a(x)$ and $b(x)$ nonconstant polynomials. Then both have a unit as a leading coefficient and a constant term in P. But then, the constant term of $a(x)b(x)$ is in P^2 which contradicts the hypothesis of the theorem. Hence, $a(x)$ or $b(x)$ is a constant polynomial. Since the leading term of both of them is a unit, then $a(x)$ or $b(x)$ is a unit in $R[x]$. Thus, $f(x)$ is irreducible. □

Example 4.5.20. As an easy application of Eisenstein's Criterion, note that $f(x) = x^7 + 2x^6 + 12x^4 + 10 \in \mathbb{Z}[x]$ satisfies the condition of the theorem with the ideal $P = (2)$. Hence, $f(x)$ is irreducible. Obviously, it would have been difficult to eliminate all possible factorizations as we did in Examples 4.5.15 and 4.5.16. △

Example 4.5.21. Consider the monic polynomial $f(x) = x^5 + (-1 + 7i)x^3 + (4 + 7i)x + 5$ in $\mathbb{Z}[i][x]$. Since

$$-1 + 7i = (2 + i)(1 + 3i),$$
$$4 + 7i = (2 + i)(3 + 2i),$$
$$5 = (2 + i)(2 - i),$$

and $2 - i$ is not an associate of $2 + i$. All the nonleading coefficients of $f(x)$ are in the ideal $J = (2 + i)$ and $5 \notin J^2$. Hence, $f(x)$ is irreducible in $\mathbb{Z}[i][x]$. △

4.5.3 Useful CAS Commands

Because of the importance for applications, many CAS have many commands for factoring polynomials over different fields. Most systems offer a variety of commands for sorting, expanding products of polynomials, and collecting like terms with a stated variable as a reference. However, we list here just those commands for testing irreducibility and for factoring. The example below illustrates working over $\mathbb{Q}$, $\mathbb{Q}(\sqrt{2})$, and over $\mathbb{F}_7$.

———————————— Maple ————————————

$irreduc(x^4 + 1)$

$$true$$

$factor(x^4 + 1)$

$$x^4 + 1$$

$irreduc(x^4 + 1, \sqrt{2})$

$$false$$

$factor(x^4 + 1, \sqrt{2})$

$$-(x\sqrt{2} + x^2 + 1)(x\sqrt{2} - x^2 - 1)$$

$Irreduc(x^4 + 1) \bmod 7;$

$$false$$

$Factor(x^4 + 1) \bmod 7;$

$$(x^2 + 3x + 1)(x^2 + 4x + 1)$$

The following code below does the same thing in SAGE but implemented in a Sage/Jupyter notebook. (The top half is an input block and the bottom half illustrates the output.) In SAGE, to keeping using the same polynomial, we use a command to change the parent ring of a given polynomial.

———————————— Sage (Jupyter) ————————————

```
A.<x> = QQ[]
p = QQ['x'](x^4+1)
print(p.is_irreducible())
print(p.factor())
print(p.change_ring(QQ[sqrt(2)]).is_irreducible())
print(p.change_ring(QQ[sqrt(2)]).factor())
print(p.change_ring(GF(7)).is_irreducible())
print(p.change_ring(GF(7)).factor())
```

```
True
x^4+1
False
(x^2 - sqrt2*x + 1) * (x^2 + sqrt2*x+1)
False
(x^2 + 3*x +1) * (x^2 + 4*x +1)
```

4.5.4 Summary of Results

We conclude this section with a brief summary of results about polynomials rings over a ring. The following lists relevant connection between R and $R[x]$ for increasingly strict condition on $R[x]$.

- $R[x]$ is an integral domain if and only if R is an integral domain. (Proposition 3.2.6)

- $R[x]$ is a UFD if and only if R is a UFD. (Theorem 4.5.5)

- $R[x]$ is a PID only if R is a field. (Proposition 4.4.11)
- If F is a field, then $F[x]$ is not only a PID but a Euclidean domain. (Corollary 4.3.6)

EXERCISES FOR SECTION 4.5

1. Prove that the following polynomials in $\mathbb{Z}[x]$ are irreducible.
   (a) $9x^2 - 11x + 1$
   (b) $2x^2 + 5x + 7$
   (c) $x^3 + 4x + 3$

2. Prove that the following polynomials in $\mathbb{Z}[x]$ are irreducible.
   (a) $x^3 + 2x^2 + 3x + 4$
   (b) $x^4 + x^3 + 1$
   (c) $x^4 + 3x^2 - 6$

3. For each of the following polynomials in $\mathbb{F}_5[x]$, decide if it is irreducible and if it is reducible give a complete factorization.
   (a) $x^2 + 3x + 4$
   (b) $x^3 + x^2 + 2$
   (c) $x^4 + 3x^3 + x^2 + 3$

4. For each of the following polynomials in $\mathbb{F}_7[x]$, decide if it is irreducible and if it is reducible give a complete factorization.
   (a) $x^2 + 3x + 4$
   (b) $x^3 + x^2 + 2$
   (c) $x^4 + 3x^3 + x^2 + 3$

5. Let p be a prime number. Prove that in $\mathbb{Z}[x]$, the polynomial $x^3 + nx + p$ is irreducible for all but at most four values of n.

6. Prove that in $\mathbb{Z}[x]$, the polynomial $x^3 + px + q$, where p and q are odd primes, is irreducible.

7. Find all quadratic, cubic, and quartic irreducible polynomials in $\mathbb{F}_2[x]$.

8. Let F be a finite field with q elements. By determining all the monic reducible polynomials of degree 2, prove that there are $\frac{1}{2}(q^2 - q)$ monic irreducible quadratic polynomials in $F[x]$.

9. List all irreducible monic quadratic polynomials in $\mathbb{F}_5[x]$.

10. Prove Corollary 4.5.3.

11. Let F be a field and let a be a nonzero element in F. Prove that if $f(ax)$ is irreducible, then $f(x)$ is irreducible.

12. Prove a modification of Proposition 4.5.17 in which I is a prime ideal P in R and the polynomial $p(x)$ satisfies $\mathrm{LC}(p(x)) \notin P$.

13. Prove that $f(x) = x^3 + (2+i)x + (1+i)$ is irreducible in $\mathbb{Z}[i][x]$.

14. Let R be a UFD and let $a \in R$. Then $p(x) \in R[x]$ is irreducible if and only if $p(x + a)$ is irreducible.

15. Let p be a prime number in $\mathbb{Z}$. Use Exercise 4.5.14 to prove that the polynomial $x^{p-1} + x^{p-2} + \cdots + x + 1$ is irreducible in $\mathbb{Z}[x]$.

16. Consider the polynomial $p(x) = x^3 + 3x^2 + 5x + 5$ in $\mathbb{Z}[x]$. Find a shift of the variable x so that you can then use Eisenstein's Criterion to show that $p(x)$ is irreducible.

17. Let $c_1, c_2, \ldots, c_n \in \mathbb{Z}$ be distinct integers. Consider the polynomial

$$p(x) = (x - c_1)(x - c_2) \cdots (x - c_n) - 1 \in \mathbb{Z}[x].$$

   (a) Prove that if $p(x) = a(x)b(x)$, then $a(x) + b(x)$ evaluates to 0 at c_i for $i = 1, 2, \ldots, n$.

   (b) Deduce that if $a(x)$ and $b(x)$ are nonconstant, then $a(x) + b(x)$ is the 0 polynomial in $\mathbb{Z}[x]$.

   (c) Deduce that $p(x)$ is irreducible.

   [Hint: Exercise 4.4.12.]

18. Prove the following generalization of Eisenstein's Criterion. Let P be a prime ideal in an integral domain R and let

$$f(x) = a_n x^n + a_{n-1} x^{n-1} + \cdots + a_1 x + a_0$$

be a polynomial in $R[x]$. Suppose that: (1) $a_n \notin P$; (2) $a_i \in P$ for all $i < n$; and (3) $a_0 \notin P^2$. Then $f(x)$ is not the product of two nonconstant polynomials.

19. Let R be a UFD and let F be its field of fractions. Suppose that $p(x), q(x) \in F[x]$ and that $p(x)q(x)$ is in the subring $R[x]$. Prove that the product of any coefficient of $p(x)$ with any coefficient of $q(x)$ is an element of R.

20. Let F be a finite field of order $|F| = q$ and let $p(x) \in F[x]$ with $\deg p(x) = n$. Prove that $F[x]/(p(x))$ has q^n elements. [Hint: Exercise 3.7.9.]

21. Prove that for all primes p, there is a field with p^2 elements.

22. Let F be a field. Consider the derivative function $D : F[x] \to F[x]$ defined by $D(a_0) = 0$ and

$$D(a_n x^n + \cdots + a_1 x + a_0)$$
$$= (n \cdot a_n)x^{n-1} + ((n-1) \cdot a_{n-1})x^{n-2} + \cdots + (2 \cdot a_2)x + a_1.$$

   (a) From this definition, prove that D satisfies the differentiation rules

$$\begin{cases} D(p(x) + q(x)) = D(p(x)) + D(q(x)) & \text{and} \\ D(p(x)q(x)) = D(p(x))q(x) + p(x)D(q(x)) \end{cases}$$

   for all $p(x), q(x) \in F[x]$.

   (b) Prove that if $d(x)$ is an irreducible polynomial that is a common divisor to $p(x)$ and $D(p(x))$, then $d(x)^2$ divides $p(x)$.

23. Let R be a UFD and let $P(x) \in R[x]$. Let $n \in \mathbb{N}^*$ and define for this exercise $P^n(x)$ to be the polynomial $P(x)$ iterated n times, i.e.,

$$P^n(x) = \overbrace{P(P(\cdots P(x) \cdots))}^{n \text{ times}}.$$

Prove that if $d \mid n$ in $\mathbb{N}^*$, then $P^d(x) - x$ divides $P^n(x) - x$.

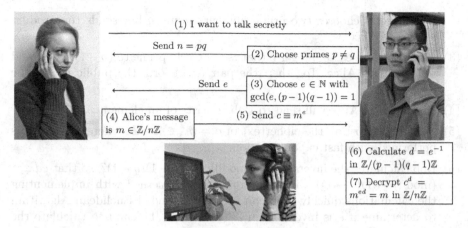

FIGURE 4.3: RSA diagram.

4.6 RSA Cryptography

In Section 1.12, we presented the idea of public key cryptography: a protocol for two parties who communicate entirely publicly to select a key that will nonetheless stay secret (not easily obtainable).

The Diffie-Hellman protocol relied on Fast Exponentiation to quickly calculate powers of group elements, whereas determining a from g and g^a is relatively slow. The RSA public key protocol, named after Ron Rivest, Adi Shamir, and Leonard Adleman, is a protocol between two parties A and B in which party B allows party A to generate a key that will allow party A to send a secret message to party B. Rivest, Shamir, and Adelman first introduced the protocol in the context of the integers but it can be generalized to other rings. We will also first describe the protocol over $\mathbb{Z}$ and then generalize.

4.6.1 RSA over $\mathbb{Z}$

As we did with Diffie-Hellman, we introduce three parties—Alice, Bob, and Eve—to describe the protocol. Alice wants to communicate secretly with Bob while Eve is attempting to intercept the communication. For the moment, Alice aims for the more humble goal of simply getting some information, possibly small, to Bob secretly. We indicate this message by m.

In Figure 4.3, that which is boxed stays secret to the individual and that which is not boxed is heard by everyone, including Eve.

(1) The protocol starts with Alice initiating and telling Bob that she wants to talk secretly.

(2) Bob secretly chooses two prime numbers p and q but sends the product $n = pq$ to Alice.

(3) Bob also chooses an integer e that is relatively prime to $(p-1)(q-1)$ and sends this to Alice. Together, the pair (n, e) form the public key of the protocol.

(4) The message Alice will send to Bob consists of an element $m \in \mathbb{Z}/n\mathbb{Z}$.

(5) Alice sends to Bob the ciphertext of $c = m^e \in \mathbb{Z}/n\mathbb{Z}$, where the power is performed with fast exponentiation.

(6) Bob calculates the inverse d of e modulo $\mathbb{Z}/(p-1)(q-1)\mathbb{Z}$ so that $ed \equiv 1$ (mod $(p-1)(q-1)$). Since we are always concerned with implementing the calculations quickly, Bob can use the Extended Euclidean Algorithm to determine if e is invertible in $\mathbb{Z}/(p-1)(q-1)\mathbb{Z}$ and to calculate the inverse d. (See Example A.6.10.)

(7) Then Bob calculates (using fast exponentiation) $c^d = m^{ed}$ in $\mathbb{Z}/n\mathbb{Z}$. By the Chinese Remainder Theorem, there is an isomorphism

$$\varphi : \mathbb{Z}/pq\mathbb{Z} \to (\mathbb{Z}/p\mathbb{Z}) \oplus (\mathbb{Z}/q\mathbb{Z}).$$

Hence, by Proposition 3.1.14, $U(pq) \cong U(p) \oplus U(q)$ and $|U(pq)| = (p-1)(q-1)$. If m and $(p-1)(q-1))$ are relatively prime, then $m \in U(\mathbb{Z}/n\mathbb{Z}) = U(n)$ so

$$m^{ed} = m^{1+k|U(pq)|} \qquad \text{for some } k \in \mathbb{Z}$$
$$= m,$$

by Lagrange's Theorem. If $\gcd(m, (p-1)(q-1)) \neq 1$, we still have the following cases. If $m = 0$ in $\mathbb{Z}/n\mathbb{Z}$, then $c = m^e = 0$ and $c^d = m^{ed} = 0$, so again $c^d = m$. If $p \mid m$ but $m \neq 0$ in $\mathbb{Z}/n\mathbb{Z}$, then under the isomorphism $\varphi(m) = (0, h)$ for some $h \in \mathbb{Z}/q\mathbb{Z}$. Then

$$(0, h)^{1+k(p-1)(q-1)} = \left(0, h^1 (h^{q-1})^{k(p-1)}\right) = (0, h\, 1^{k(p-1)}) = (0, h).$$

Hence, again $m^{ed} = m$ in $\mathbb{Z}/n\mathbb{Z}$. Similarly, if $q \mid m$, but $m \neq 0$ in $\mathbb{Z}/n\mathbb{Z}$, then $m^{ed} = m$ in $\mathbb{Z}/n\mathbb{Z}$. This allows Bob to recover m.

Because of their roles, the integer e is called the *encryption key* and d is called the *decryption key*.

From Eve's perspective, she knows everything that Bob does except what p and q are separately. If p and q are very large prime numbers, then it is very quick for Bob to calculate the product $n = pq$ but it is *very* slow for Eve to find the prime factorization, namely $p \times q$ of n. Furthermore, she would need to know p and q separately in order to calculate $(p-1)(q-1)$, which she needs to determine d.

As opposed to Diffie-Hellman, in which both Alice and Bob know the secret key g^{ab}, only Bob knows the secret keys p and q. Diffie-Hellman is symmetric in the sense that Alice and Bob could both use g^{ab} for communication. In RSA, Alice is in the same situation as Eve in that she cannot easily determine d. Hence, the RSA protocol sets up a one-way secret communication: Only Alice can send a secret message to Bob.

Again, as mentioned in the Diffie-Hellman protocol, it may seem unsatisfactory that it is theoretically possible for Eve to find d from the information passed in the clear. However, if the primes p and q are large enough, it may take over 100 years with current technology to find the prime factorization of n and hence determine p and q separately. Very few secrets need to remain secret for a century so the protocol is secure in this sense.

Even if n is large, it is not likely that a long communication can be encoded into a single element $m \in \mathbb{Z}/n\mathbb{Z}$. Alice and Bob can use two strategies to allow for Alice to send a long communication to Bob.

One strategy involves deciding upon an injective function H from the message space $\mathcal{M}$ into the set of finite sequences of elements in $\mathbb{Z}/n\mathbb{Z}$. Then the message, regardless of alphabet, can be encoded as a string of elements $(m_1, m_2, \ldots, m_\ell)$ in $\mathbb{Z}/n\mathbb{Z}$ and Alice sends the sequence $(m_1^e, m_2^e, \ldots, m_\ell^e)$ to Bob. Bob then decodes each element in the string as described above to recover $(m_1, m_2, \ldots, m_\ell)$. Then, since H is an injective function, Bob can find the unique preimage of $(m_1, m_2, \ldots, m_\ell)$ under H and thereby recover the message in the message space $\mathcal{M}$.

A second strategy uses the RSA protocol only to exchange a key for some subsequent encryption algorithm. In other words, Alice and Bob agree (publicly) on some other encryption algorithm that requires a secret key and they use RSA to decide what key to use for that algorithm. Essentially, Alice is saying to Bob: "Let's use m as the secret key."

As one last comment about speed of the protocol, prime factorization is always slow so whenever Bob needs to calculate the greatest common divisor of two integers, he uses the Euclidean Algorithm and when he calculates the inverse of e in $\mathbb{Z}/(p-1)(q-1)\mathbb{Z}$ he uses the Extended Euclidean Algorithm. The astute reader might wonder why we might not use Lagrange's Theorem to calculate d. Since $\gcd(e, (p-1)(q-1)) = 1$, then $e \in U((p-1)(q-1))$ so $d = e^{|U((p-1)(q-1))|-1}$. However, $|U((p-1)(q-1))| = \phi((p-1)(q-1))$, where ϕ is Euler's totient function. Calculating the totient function of an integer requires us to perform the prime factorization of that integer.

Example 4.6.1. For a first example, we use small prime numbers and illustrate the full process. In practice, of course, one uses computer programs to execute these calculations.

Suppose that Alice wants to communicate secretly with Bob and they agree to use RSA. Bob selects $p = 1759$ and $q = 2347$ and sends $n = 4128373$ to Alice. He also sends the encryption key $e = 72569$. Alice and Bob agree to encode strings of characters in the following way. Each character will correspond to a digit by

$$\{\_, A, B, \ldots, Z, \text{'.'}, \text{','}, \text{'} \text{'}\} \longrightarrow \{0, 1, 2, \ldots, 26, 27, 28\}.$$

Observe that the first character of the alphabet here is the space character. Note, they only use an alphabet of 29 characters. Since $29^4 = 707281 < n$, then each quadruple of four characters (c_1, c_2, c_3, c_4) is converted to the integer

$$c_4 \times 29^3 + c_3 \times 29^2 + c_2 \times 29 + c_1$$

and then viewed as an element of $\mathbb{Z}/4128373\mathbb{Z}$. If necessary, a message can be padded at the end with spaces so that the message has $4k$ characters in it, and then corresponds to a sequence of length k of elements in $\mathbb{Z}/4128373\mathbb{Z}$.

Alice wants to tell Bob "Hello there." Her string of character numbers is

$$(8, 5, 12, 12, 15, 0, 20, 8, 5, 18, 5, 27)$$

and her message as a string of elements in $\mathbb{Z}/4128373\mathbb{Z}$ is (m_1, m_2, m_3), where

$$m_1 = 12 \times 29^3 + 12 \times 29^2 + 5 \times 29 + 8 = 302913,$$
$$m_2 = 8 \times 29^3 + 20 \times 29^2 + 0 \times 29 + 15 = 211947,$$
$$m_3 = 27 \times 29^3 + 2 \times 29^2 + 18 \times 29 + 5 = 660712.$$

After using fast modular exponentiation, Alice sends to Bob

$$(c_1, c_2, c_3) = (m_1^e, m_2^e, m_3^e) = (1318767, 3245763, 2570792).$$

It is not difficult to do the Extended Euclidean Algorithm by hand but we can trust computer algebra system to give us the result efficiently. The Extended Euclidean Algorithm on the pair $(4124268, 72569)$ gives us the following answer $(1, -11549, 656357)$.[1] This means that

$$\gcd(4124268, 72569) = 1 \text{ and } -11549 \times 4124268 + 656357 \times 72569 = 1.$$

If Bob picked e at random, then it is possible that $\gcd(e, (p-1)(q-1)) > 1$. In this case, Bob would identify this error if the greatest common divisor at this stage did not come out to 1. To fix this, he would simply pick another e. The benefit of the Extended Euclidean Algorithm is that from the expression on the right in the above equation, we get that $1 \equiv 72549 \times 656357 \pmod{4124268}$ so $d = 656357$ is the inverse of 72549 modulo 4124268.

In order to decrypt Alice's message, using fast modular exponentiation, Bob calculates modulo $\mathbb{Z}/4128373\mathbb{Z}$ that

$$(c_1^d, c_2^d, c_3^d) = (302913, 211947, 660712).$$

Knowing the process by which Alice compressed her message into a string of elements in $\mathbb{Z}/4128373\mathbb{Z}$, Bob can recover Alice's message of "HELLO THERE." $\triangle$

---

[1] With *Maple* we can get this by [igcdex(4124268,72569,'s','t'),s,t]; and in SAGE, we get this by xgcd(4124268,72569).

4.6.2 RSA in Other Rings

We would like to generalize the RSA protocol to other rings. To do so, we must study the protocol to determine what procedures are involved and what is the proper algebraic context in which we can perform such procedures.

The concept of irreducible elements does not have an equivalent in group theory so we work first of all in the context of a commutative ring R. Also, the concept of irreducibility requires the existence of an identity. It might be possible to use a ring with zero divisors but the protocol assumes that for some m there is a power $k > 1$ high enough that $m^k = m$. Consequently, nilpotent elements would cause problems for this property and we are inclined to require that R be an integral domain.

In the algorithm, in order for $m^k = m$ for high enough k in R/I, we required that $R/(p)$ be a finite ring for the prime p and similarly for q. In attempting to generalize RSA, we could either base the definition on irreducible elements or maximal ideals. In this textbook, we use the following definition.

Definition 4.6.2

An integral domain that is not a field is called an *RSA domain* if for every maximal ideal $M \subseteq R$, the quotient ring R/M is a finite field.

The ring of integers $\mathbb{Z}$ is an RSA domain but $\mathbb{R}[x]$ is not. Indeed, for the irreducible element $x^2 + 1$, $\mathbb{R}/(x^2 + 1) \cong \mathbb{C}$, which is not finite. On the other hand, as Exercise 3.7.10 hints, for any prime element π in $\mathbb{Z}[i]$, the elements in $\mathbb{Z}[i]/(\pi)$ are the point in $\mathbb{C}$ with integral components that are contained in a square with edge along the segment from 0 to π. Hence, $\mathbb{Z}[i]/(\pi)$ is finite. Furthermore, for every finite field F, any maximal ideal M in $F[x]$ is principal so, by Exercise 4.6.5, $F[x]/M$ is finite.

With the notion of an RSA domain, we adapt the RSA algorithm as follows. (See Figure 4.4.)

(1) The protocol starts with Alice initiating and telling Bob that she wants to talk secretly.

(2) Bob secretly chooses two maximal ideals M_1 and M_2 but sends the product ideal $I = M_1 M_2$ to Alice. If R is a PID (as all of the above examples are), then Bob can work with the generating elements of any of the ideals.

(3) Bob also chooses an integer e that is relatively prime to $(|R/M_1| - 1)(|R/M_2| - 1)$ and sends this to Alice. Together, the pair (I, e) form the public key of the protocol.

(4) The message Alice will send to Bob consists of an element $m \in R/I$.

(5) Alice sends to Bob the ciphertext of $c = m^e \in R/I$, where the power is performed with fast exponentiation.

(6) Bob calculates the inverse d of e modulo $\mathbb{Z}/(|R/M_1| - 1)(|R/M_2| - 1)\mathbb{Z}$ so that $ed \equiv 1 \pmod{(|R/M_1| - 1)(|R/M_2| - 1)}$. Since we are always

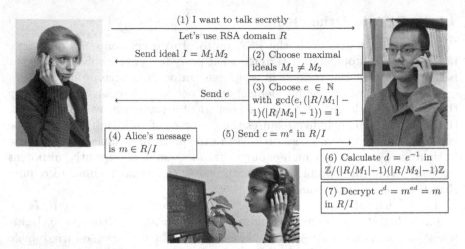

FIGURE 4.4: RSA diagram.

concerned with implementing the calculations quickly, Bob can use the Extended Euclidean Algorithm.

(7) Then Bob calculates (using fast exponentiation) $c^d = m^{ed}$ in R/I. Since M_1 and M_2 are distinct maximal ideals, then $M_1 + M_2 = R$ and hence they are comaximal. By the Chinese Remainder Theorem, there is an isomorphism
$$\varphi : R/I \to (R/M_1) \oplus (R/M_2).$$
Hence, by Proposition 3.1.14, $U(R/I) \cong U(R/M_1) \oplus U(R/M_2)$ and since R/M_1 and R/M_2 are fields, we have $|U(R/I)| = (|R/M_1|-1)(|R/M_2|-1)$. If $m = 0 \in R/I$, then $c = m^e = 0$ and $c^d = 0 = m$. If $m \in U(R/I)$, then
$$c^d = m^{ed} = m^{1+k|U(R/I)|} = m \qquad \text{for some } k \in \mathbb{Z}.$$

By Lagrange's Theorem, we deduce that $m^{ed} = m$. If $\varphi(m) = (0, h)$ with $h \in U(R/M_2)$, then
$$\varphi(m)^{ed} = (0, h)^{1+k(|R/M_1|-1)(|R/M_2|-1)} = \left(0, h(h^{|R/M_2|-1})^{k(|R/M_1|-1)}\right)$$
$$= (0, h\, 1^{k(|R/M_1|-1)}) = (0, h) = \varphi(m).$$

Hence, again $m^{ed} = m$ in R/I and similarly, if $\varphi(m) = (h, 0)$ with $h \in U(R/M_1)$. This allows Bob to recover m.

In practice, it is convenient to work in an algebraic context in which there is a simple and perhaps unique way to determine a representative of cosets in R/I. In particular, when calculating the powers m^e or c^d in R/I using fast exponentiation, it is convenient for memory storage to constantly reduce the power to a smallest equivalent expression in R/I. Euclidean domains offer such a context. Since Euclidean domains are PIDs, every ideal I is equal to (a) for some element a. Then, while performing the fast exponentiation algorithm, each time we take a power of m, we replace it with the remainder when we divide by a from the Euclidean division.

Some RSA domains that are also Euclidean domains include $\mathbb{Z}[i]$ and $F[x]$, where F is a finite field.

As an example, we adapt Example 4.6.1 to a scenario in $\mathbb{Z}[i]$ with different prime numbers. To simplify the example, we leave off the issue of encoding characters into the quotient ring R/I.

Example 4.6.3. Alice indicates that she wants to communicate secretly with Bob and they agree to use RSA over $\mathbb{Z}[i]$. Bob selects $p = 15 + 22i$ and $q = 7 + 20i$. According to Example 4.4.4, since $N(p) = 709$ and $N(q) = 449$ are prime, p and q are primes in the Gaussian integers. Bob sends $n = pq = -335+454i$ to Alice as well as the encryption key $e = 2221 = (100010101101)_2$, where the latter expression is in binary for use with the fast exponentiation algorithm (Section 1.12.2).

Alice wants Bob to receive the number $m = 67 + 232i$. Running the fast exponentiation algorithm, Alice sets the power variable π of $m \in \mathbb{Z}[i]/(n)$ initially as $\pi = 1$. Note that 11 is the highest nonzero power of 2 in the binary expansion of $e = 2221$. Also, in the following calculations, though we do not use the bar $\bar{a}$ notation, numbers are understood as elements in $\mathbb{Z}[i]/(n)$.

- $b_{11} = 1$ so $\pi := \pi^2 m = 67 + 232i$.
- $b_{10} = 0$ so $\pi := \pi^2 = (67 + 232i)^2 = -49335 + 31088i = 77 + 234i$, where the last equality holds after performing the Euclidean division of π^2 by n in $\mathbb{Z}[i]$.
- $b_9 = 0$ so $\pi := \pi^2 = -48827 + 36036i = 206 - 6i$.
- $b_8 = 0$ so $\pi := \pi^2 = 42400 - 2472i = -12 - 110i$.
- $b_7 = 1$ so $\pi := \pi^2 m = -1413532 - 2596912i = 154 + 67i$.
- $b_6 = 0$ so $\pi := \pi^2 = 19227 + 20636i = -4 + 135i$.
- $b_5 = 1$ so $\pi := \pi^2 m = -969443 - 4296848i = -207 + 24i$.
- $b_4 = 0$ so $\pi := \pi^2 = 42273 - 9936i = -192 - 100i$.
- $b_3 = 1$ so $\pi := \pi^2 m = 10708688 + 3659648i = 96 + 143i$.
- $b_2 = 1$ so $\pi := \pi^2 m = -7122403 - 766504i = -77 - 72i$.
- $b_1 = 0$ so $\pi := \pi^2 = 745 + 11088i = -132 - 77i$.
- $b_0 = 1$ so $\pi := \pi^2 m = -3945931 + 4028816i = 51 + 104i$.

This is the ciphertext $c = m^e = 51 + 104i$ in $\mathbb{Z}[i]/(-335 + 454i)$ that Alice sends in the clear to Bob.

On Bob's side, in order to calculate the decryption key d, he first needs to determine $|\mathbb{Z}[i]/(p)|$ and $|\mathbb{Z}[i]/(q)|$. By Exercise 4.6.6, the order of each of these quotient rings is $|p|^2$ and $|q|^2$, respectively. Hence, d is the inverse of $e = 2221$ modulo

$$(|p|^2 - 1)(|q|^2 - 1) = 317184.$$

Performing the extended Euclidean algorithm, Bob finds that $d = 208933$. In order to recover m, he will calculate c^d using the fast modular exponentiation in the finite ring $\mathbb{Z}[i]/(n)$, during which, like Alice, he takes Euclidean remainders when divided by n at each stage of the algorithm. Since in binary $d = (110011000000100101)_2$, the for loop in the algorithm with take only 18 steps. $\triangle$

EXERCISES FOR SECTION 4.6

1. Implementing RSA over $\mathbb{Z}$, take the role of Alice. Bob sends you $n = 28852217$ and $e = 33$. Suppose that you wish to send the plaintext number of $m = 45678$ to Bob. Calculate the corresponding ciphertext $c = m^e$ in $\mathbb{Z}/n\mathbb{Z}$.

2. Implementing RSA over $\mathbb{Z}$, take the role of Alice. Bob sends you $n = 5352499$ and $e = 451$. Suppose that you wish to send the plaintext number of $m = 87542$ to Bob. Calculate the corresponding ciphertext $c = m^e$ in $\mathbb{Z}/n\mathbb{Z}$.

3. Implementing RSA over $\mathbb{Z}$, take the role of Alice. Bob sends you $n = 5352499$ and $e = 451$. Suppose that you wish to send the message "FLY AT NIGHT" to Bob using the string-to-number sequence encoding as given in Example 4.6.1. Show that you can use 3 blocks of 4 letters. Then compute the three ciphertext numbers in $\mathbb{Z}/n\mathbb{Z}$.

4. Implementing RSA over $\mathbb{Z}$, take the role of Bob. You select $p = 131$ and $q = 211$ so that $n = 27641$.

   (a) Show that $e = 191$ is an acceptable encryption key for the RSA algorithm.

   (b) Find (expressed $0 \le d < (p-1)(q-1) = 27300$ the multiplicative inverse of e in $\mathbb{Z}/27300\mathbb{Z}$.

   (c) Suppose that Alice sends a message using 29 characters as in Example 4.6.1, also compressing strings of 3 letters in elements in $\mathbb{Z}/27641\mathbb{Z}$ as in the Example. From the following ciphertext

   $$(26799, 26841, 22169, 9764, 3426)$$

   recover the plaintext and interpret the result into English.

5. Let F be a finite field. Prove that for every $q(x) \in F[x]$, the order of the quotient ring $F[x]/(q(x))$ is $|F|^{\deg q(x)}$.

6. Consider the Euclidean domain $\mathbb{Z}[i]$ and let $I = (z)$ be a proper nontrivial ideal. (Since $\mathbb{Z}[i]$ is a Euclidean domain, it is a PID.)

   (a) Prove that

   $$\{m + ni \mid m, n \in \mathbb{Z} \text{ with } 0 \le m, n < |z|\}$$

   is a complete set of distinct representatives for the cosets in $\mathbb{Z}[i]/I$.

   (b) Conclude that the quotient ring contains $|\mathbb{Z}[i]/I| = z\bar{z} = |z|^2$ elements.

7. Implementing RSA over $\mathbb{Z}[i]$, you take the role of Alice. Bob sends you $n = 236 - 325i$ and $e = 14$.

(a) What is the ciphertext c for the message $m = 101 + 3i$?

(b) Suppose that you wish to send the message "FLY AT NIGHT" to Bob. You agree with Bob to convert strings of letters to strings of Gaussian integers by first mapping the characters in $\{\_, A,B,C,\ldots, Y,Z,',',','.'\}$ to the integers 0 through 28, and then mapping any pair of such integers (a,b) to $a + bi$. Convert the English message to strings of Gaussian integers and determine the string of ciphertext Gaussian integers.

8. Implementing RSA over $\mathbb{F}_{31}[x]$, you take the role of Bob. Suppose that you pick polynomials $f(x) = x^3 + x + 12$ and $g(x) = x^2 + 4x^2 + 8$.

(a) First prove that $f(x)$ and $g(x)$ are irreducible.

(b) Prove that any number e that is relatively prime to 2, 3, 5, and 331 is a suitable encryption key.

(c) Find the decryption key d given the encryption key $e = 17$.

9. Implementing RSA over $\mathbb{F}_{31}[x]$, you take the role of Alice. Suppose that Bob has sent you
$$n(x) = x^5 + 4x^4 + 9x^3 + 16x^2 + 25x + 3,$$
where all the coefficients are understood to be in $\mathbb{F}_{31}$. Let $m(x) = x^3 + 2x$. Calculate $m(x)^3$ in $\mathbb{F}_{31}[x]/(n(x))$.

4.7 Algebraic Integers

From a historical perspective, various other branches of mathematics—geometry, number theory, analysis, and so forth—shaped various areas in algebra. Number theory motivated considerable investigation in ring theory, and in particular topics related to divisibility. It is difficult to briefly summarize the goal of number theory since it includes many different directions of investigations: approximations of real numbers by rationals, distribution of the prime numbers in $\mathbb{Z}$, multivariable equations where we seek only integer solutions (Diophantine equations), and so on. In current terminology, algebraic number theory is a branch of algebra applied particularly to studying Diophantine equations and related generalizations.

This section offers a glimpse into algebraic number theory as an application of algebra internal to mathematics. A more complete introduction to algebraic number theory requires field theory (Chapter 5) including Galois theory, which is beyond this first course in algebra. For further study in algebraic number theory, the reader may consult [24].

Since algebraic number theory is a vast area, in this section, we content ourselves to introduce algebraic integers and illustrate how unique factorization in the Gaussian integers answers an otherwise challenging problem in number theory. In so doing, we will show the historical motivation behind certain topics in ring theory.

4.7.1   Algebraic Integers

Consider the integers $\mathbb{Z}$ as a subring in its field of fractions $\mathbb{Q}$. Now consider a larger field K containing $\mathbb{Q}$. We would like to imagine a ring R in K that includes $\mathbb{Z}$ and, in some intuitive way, serves the role in K that $\mathbb{Z}$ serves in $\mathbb{Q}$. Obviously, the integers $\mathbb{Z}$ form a subring of K. The field of fractions of $\mathbb{Z}$ is $\mathbb{Q}$ so, among other things, we would like R to be an integral domain whose field of fractions is K.

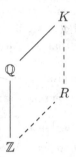

We typically do not study this construction where K is any field but we often restrict our attention to subfields of $\mathbb{C}$.

Definition 4.7.1

A field K that is a subfield of $\mathbb{C}$ and contains $\mathbb{Q}$ as a subfield is called a *number field*.

As we will see more in depth in Section 5.3, the adjective "algebraic" pertains to an element being the solution to a polynomial. So an algebraic relationship between $\mathbb{Z}$ would have to do with how integers arise as roots of polynomials in $\mathbb{Q}[x]$. Now the solution set to any polynomial $p(x) \in \mathbb{Q}[x]$ is equal to the set of solutions of $q(x) = cp(x) \in \mathbb{Z}[x]$ where c is the least common multiple of all the denominators of the coefficients in $p(x)$. Then Proposition 4.5.11 shows that if $q(x)$ has only integers for roots, then its leading coefficient is 1 or -1. This is tantamount to being monic. This motivates the following definition.

Definition 4.7.2

Let K be a number field. The *algebraic integers* in K, denoted $\mathcal{O}_K$, is the set of elements in K that are solutions to monic polynomials $p(x) \in \mathbb{Z}[x]$.

It is not at all obvious from this definition that $\mathcal{O}_K$ is a ring. We first establish an alternate characterization of algebraic integers in a number field K before establishing this key result.

Lemma 4.7.3

Suppose $\alpha \in K$. Then $\alpha \in \mathcal{O}_K$ if and only if $(\mathbb{Z}[\alpha], +)$ is a finitely generated free abelian group.

Proof. First suppose that $\alpha \in \mathcal{O}_K$. Then there exists a monic polynomial $p(x) \in \mathbb{Z}[x]$ such that $p(\alpha) = 0$. Then, there exists a positive integer n and $c_1, c_2, \ldots, c_n \in \mathbb{Z}$ such that

$$\alpha^n = -(c_{n-1}\alpha^{n-1} + \cdots + c_1\alpha + c_0).$$

Then by an induction argument, every power $k \geq n$ of α can be written as a $\mathbb{Z}$-linear combination of $1, \alpha, \alpha^2, \ldots, \alpha^{n-1}$. Thus,

$$\mathbb{Z}[\alpha] \subseteq \{c_{n-1}\alpha^{n-1} + \cdots + c_1\alpha + c_0 \mid c_1, c_2, \ldots, c_{n-1} \in \mathbb{Z}\}.$$

The right-hand side is a finitely generated free abelian group so by Theorem 2.5.9 is also a finitely generated free abelian group.

Conversely, suppose that $\mathbb{Z}[\alpha]$ is a finitely generated free abelian group. Then there exist $a_1, a_2, \ldots, a_n \in \mathbb{Z}[\alpha]$ such that every element $p(\alpha) \in \mathbb{Z}[\alpha]$ can be written as a linear combination

$$p(\alpha) = c_1 a_1 + c_2 a_2 + \cdots + c_n a_n \qquad \text{with } c_i \in \mathbb{Z}.$$

Note that for all a_i, the element $\alpha a_i \in \mathbb{Z}[\alpha]$. Hence, for each $i = 1, 2, \ldots, n$, there exist integers m_{ij} such that

$$\alpha a_i = \sum_{j=1}^{n} m_{ij} a_j. \tag{4.13}$$

Viewing $M = (m_{ij})$ as a matrix and $\vec{a} \in \mathbb{C}^n$ is the vector whose ith coordinate is a_i, then (4.13) can be written as

$$\alpha \vec{a} = M\vec{a}.$$

Since $\vec{a} \neq \vec{0}$, then $\vec{a}$ is an eigenvector of the matrix M and α is an eigenvalue. Thus, α is the root of the equation $\det(xI - M) = 0$. Since M is a matrix of integers, then $\det(xI - M)$ is a polynomial in $\mathbb{Z}[x]$. By considering the Lagrange expansion of the determinant, it is easy to see that $\det(xI - M)$ is monic. Hence, $\alpha \in \mathcal{O}_K$. $\qquad \square$

Theorem 4.7.4

The set $\mathcal{O}_K$ of algebraic integers in K is a subring of K.

Proof. Let α, β be arbitrary elements in $\mathcal{O}_K$. From Lemma 4.7.3, let $a_1, a_2, \ldots, a_n \in K$ and $b_1, b_2, \ldots, b_m \in K$ be such that every element in $\mathbb{Z}[\alpha]$

is an integer linear combination of the $a_1, a_2, \ldots, a_n$ and every element in $\mathbb{Z}[\beta]$ is an integer linear combination of $b_1, b_2, \ldots, b_m$. Then every element in the ring $\mathbb{Z}[\alpha, \beta]$ can be written as

$$q_n(\alpha)\beta^n + q_{n-1}(\alpha)\beta^{n-1} + \cdots + q_1(\alpha)\beta + q_0(\alpha)$$

for some $q_i(x) \in \mathbb{Z}[x]$. Also by Lemma 4.7.3, we deduce that every element in $\mathbb{Z}[\alpha, \beta]$ is an integer linear combination of the mn elements $a_i b_j$ with $1 \le i \le n$ and $1 \le j \le m$. In particular, every element in $\mathbb{Z}[\alpha - \beta]$ and in $\mathbb{Z}[\alpha\beta]$ can be written as a linear combination of $a_i b_j$. Hence, both $\alpha - \beta$ and $\alpha\beta$ are in $\mathcal{O}_K$ and so $\mathcal{O}_K$ is a subring of K. $\qquad\square$

By virtue of Theorem 4.7.4, we call $\mathcal{O}_K$ the *ring of algebraic integers* in K.

To illustrate Lemma 4.7.3, consider the element $\phi = (1 + \sqrt{5})/2$, often called the *golden ratio*, in the ring $\mathbb{Q}(\sqrt{5})$. One of the historical defining properties of the golden ratio is that $\frac{1}{\phi} = \phi - 1$, which can be rewritten as $\phi^2 = \phi + 1$. It is easy to show that

$$\phi^n = f_n \phi + f_{n-1},$$

where f_n is the nth term of the Fibonacci sequence. Consequently, for every polynomial $p(x) \in \mathbb{Z}[x]$, the element $p(\phi)$ can be written as

$$p(\phi) = c_1 + c_2 \phi, \qquad \text{for some } c_1, c_2 \in \mathbb{Z}.$$

Thus, $\mathbb{Z}[\phi]$ is a finitely generated free abelian group of rank 2. Lemma 4.7.3 allows us to conclude that ϕ is an algebraic integer in $\mathbb{Q}(\sqrt{5})$.

4.7.2 Quadratic Integer Rings

Definition 4.7.2 and Lemma 4.7.3 do not offer a tractable way of determining $\mathcal{O}_K$. We now determine all integer rings in quadratic extensions of $\mathbb{Q}$.

Let D be a square-free integer and consider the ring $\mathbb{Q}[\sqrt{D}]$. If $\alpha = a + b\sqrt{D} \in \mathbb{Q}[\sqrt{D}]$, we call the *conjugate* of α the element $\bar{\alpha} = a - b\sqrt{D}$. Note that $\alpha\bar{\alpha} = a^2 - Db^2 \in \mathbb{Q}$. Since D is square-free, we know that $\alpha\bar{\alpha} \ne 0$ if $\alpha \ne 0$. Therefore, every nonzero element in $\mathbb{Q}[\sqrt{D}]$ is invertible via

$$\alpha^{-1} = \frac{\bar{\alpha}}{\alpha\bar{\alpha}}.$$

This shows that $\mathbb{Q}[\sqrt{D}]$ is a field so we write $\mathbb{Q}(\sqrt{D})$.

The ring $\mathbb{Z}[\sqrt{D}]$ is a subring of $\mathbb{Q}(\sqrt{D})$. However, just because

$$\mathbb{Z}[\sqrt{D}] = \{a + b\sqrt{D} \mid a, b \in \mathbb{Z}\}$$

does not imply that it is the ring of integers in $\mathbb{Q}(\sqrt{D})$.

Theorem 4.7.5

Let D be a square-free integer. Then

$$\mathcal{O}_{\mathbb{Q}(\sqrt{D})} = \mathbb{Z}[\omega] = \{a + b\omega \mid a, b \in \mathbb{Z}\},$$

where

$$\omega = \begin{cases} \sqrt{D} & \text{if } D \equiv 2, 3 \pmod 4 \\ \frac{1+\sqrt{D}}{2} & \text{if } D \equiv 1 \pmod 4. \end{cases}$$

Proof. Let $\alpha = a + b\sqrt{D} \in \mathbb{Q}(\sqrt{D})$. Then α is a root of $m_\alpha(x) = x^2 - 2a + (a^2 - Db^2)$. Let $p(x) \in \mathbb{Q}[x]$ be any polynomial that has α as a root. Then we know from elementary algebra that the quadratic conjugate $\bar{\alpha}$ must also be a root. But

$$(x - \alpha)(x - \bar{\alpha}) = m_\alpha(x)$$

so from Proposition 4.5.9 we conclude that if $p(x)$ has α as a root, then $m_\alpha(x)$ divides $p(x)$. Write

$$p(x) = m_\alpha(x)q(x).$$

We now assume that $p(x)$ is monic and in $\mathbb{Z}[x]$. Gauss' Lemma implies that that $m_\alpha(x) \in \mathbb{Z}[x]$. Hence, instead of considering all polynomials $p(x)$, we only need to consider $m_\alpha(x)$. However, since we conclude that $m_\alpha(x) \in \mathbb{Z}[x]$, we must have $2a \in \mathbb{Z}$ and $a^2 - Db^2 \in \mathbb{Z}$.

Write $a = \frac{s}{2}$ for $s \in \mathbb{Z}$. Then we also have

$$\frac{s^2}{4} - Db^2 = t \in \mathbb{Z}. \tag{4.14}$$

If b is the fraction $b = \frac{p}{q}$, then after clearing the denominators we get $q^2 s^2 - 4Dp^2 = 4q^2 t$. Reducing this equation to mod 4, we find that $q^2 s^2 \equiv 0 \pmod 4$. This leads to two nonexclusive cases:

Case 1: $q \equiv 0 \pmod 2$. That t in (4.14) is an integer implies that $q = 2$. So

$$s^2 - Dp^2 = 4t,$$

where s and p are odd. Considering this equation mod 4 again, we see that this can only hold if $D \equiv 1 \pmod 4$. Hence, if $D \equiv 1 \pmod 4$, then $\mathcal{O}_{\mathbb{Q}(\sqrt{D})} = \frac{1}{2}\mathbb{Z}[\sqrt{D}]$, which we can write more succinctly as $\mathbb{Z}[\omega]$ where

$$\omega = \frac{1 + \sqrt{D}}{2}.$$

Case 2: $s \equiv 0 \pmod 2$. Then $a \in \mathbb{Q}$ is in fact an integer, which implies that b is also an integer. So if Case 1 does not hold, then $\mathbb{Z}[\sqrt{D}] = \mathcal{O}_{\mathbb{Q}(\sqrt{D})}$.

The theorem follows. $\qquad\square$

Rings of the form $\mathbb{Z}[\sqrt{D}]$ have entered into our study of rings and divisibility in integral domains. This theorem shows that only when $D \equiv 2, 3 \pmod{4}$ is $\mathbb{Z}[\sqrt{D}]$ the ring of algebraic integers in $\mathbb{Q}(\sqrt{D})$. For example, if $D = -1$, since $D \equiv 3 \pmod{4}$, we see that the Gaussian integers are the ring of algebraic integers in $\mathbb{Q}(i)$. In contrast, if $D = 5$, then the ring of algebraic integers in $\mathbb{Q}(\sqrt{5})$ is $\mathbb{Z}\left[\frac{1+\sqrt{5}}{2}\right]$.

We define the *field norm* $N : \mathbb{Q}(\sqrt{D}) \to \mathbb{Q}$ by

$$N(a + b\sqrt{D}) = N(\alpha) = \alpha\bar{\alpha} = (a + b\sqrt{D})(a - b\sqrt{D}) = a^2 - Db^2$$

it is easy to check that N is multiplicative in the sense that $N(\alpha\beta) = N(\alpha)N(\beta)$ for all $\alpha, \beta \in \mathbb{Q}$. It is also easy to check that for all square-free D, on the ring of integers $\mathcal{O}_{\mathbb{Q}(\sqrt{D})}$ the field norm is an integer.

The above two properties lead to the following fact: An element $\alpha \in \mathcal{O}_{\mathbb{Q}(\sqrt{D})}$ is a unit if and only if $N(\alpha) = \pm 1$ and the inverse of α is

$$\alpha^{-1} = \frac{\bar{\alpha}}{\alpha\bar{\alpha}}.$$

Example 4.7.6 (Eisenstein Integers). With $D = -3$, the algebraic integers in $\mathbb{Q}(\sqrt{-3})$ form the ring $\mathbb{Z}[\omega]$, where $\omega = \frac{1+i\sqrt{3}}{2}$. Note that the cube roots of unity are the roots of the equation $x^3 - 1 = 0$, which are given by

$$(x - 1)(x^2 + x + 1) = 0 \iff x = 1 \text{ or } x = \frac{-1 \pm i\sqrt{3}}{2}.$$

Writing $\zeta = (-1 + i\sqrt{3})/2$ we see that $\omega = \zeta + 1$ so $\mathcal{O}_{\mathbb{Q}[\sqrt{-3}]} = \mathbb{Z}[\zeta]$. As an element in $\mathbb{Z}[\zeta]$, the norm function is

$$N(a + b\zeta) = \left(a - \frac{b}{2}\right)^2 + 3\left(\frac{b}{2}\right)^2 = a^2 - ab + b^2.$$

The units in $\mathbb{Z}[\zeta]$ satisfy $a^2 - ab + b^2 = 1$. Solving as a quadratic for a in terms of b requires the discriminant $b^2 - 4(b^2 - 1) \geq 0$, which leads to $b^2 \leq \frac{4}{3}$. Similarly for a, we must have $a^2 \leq \frac{4}{3}$. This leads to only nine pairs (a, b) to check and we find that the six units are

$$\pm 1, \quad \pm\zeta = \pm\left(\frac{-1 + i\sqrt{3}}{2}\right), \quad \pm(1 + \zeta) = \pm\left(\frac{1 + i\sqrt{3}}{2}\right).$$

The ring $\mathbb{Z}\left[\frac{-1+i\sqrt{3}}{2}\right]$ is called the ring of *Eisenstein integers*. △

4.7.3 Integral Closure

From the perspective of abstract algebra, it is undesirable to study a construction on rings only on a particular case, the integers in this case. The following definitions generalize the construction of algebraic integers to integral closure of one ring in another. A more general treatment of integral closure over commutative rings requires module theory, another topic that may arise in a second course in abstract algebra.

Definition 4.7.7

Let $R \subseteq S$, where R and S are commutative rings with an identity. An element $s \in S$ is called *integral over* R if there exists a monic polynomial $f(x) \in R[x]$ such that $f(s) = 0$. The ring S is called integral over R if every element in S is integral over R.

Definition 4.7.8

If $R \subseteq S$ where R and S are as in the previous definition, then R is called *integrally closed in* S if every element in S that is integral over R belongs to R. If R is an integral domain, we simply say that R is integrally closed (without "in S") if R is integrally closed in its field of fractions.

Example 4.7.9. For example, $\mathbb{Z}$ is not integrally closed in $\mathbb{Q}(i)$ because $i \in \mathbb{Q}(i)$ is a root of the monic polynomial $x^2 + 1$ and $i \notin \mathbb{Z}$. In contrast, as we will see in the following proposition, $\mathbb{Z}[i]$ is integrally closed. $\triangle$

Proposition 4.7.10

If K is a number field, then the ring of algebraic integers $\mathcal{O}_K$ is integrally closed.

Proof. Since $\mathcal{O}_K$ is a subring of K, then the field of fractions of $\mathcal{O}_K$ is a subfield of K. Let c be an element in the field of fractions of $\mathcal{O}_K$ (and hence in K) that is the root of a monic polynomial

$$f(x) = x^n + a_{n-1}x^{n-1} + \cdots + a_1x + a_0 \qquad \text{such that } a_i \in \mathcal{O}_K.$$

By Lemma 4.7.3, each of the rings $\mathbb{Z}[a_i]$ is a finitely generated free abelian group. It is possible to prove by induction that $\mathbb{Z}[a_0, a_1, \ldots, a_{n-1}]$ is also a finitely generated free abelian group. (See Exercise 4.7.2.) Since $c^n = -(a_{n-1}c^{n-1} + \cdots + a_1c + a_0)$, then for every polynomial $p(x) \in \mathcal{O}_K[x]$, we can write $p(c)$ as

$$q_0 + q_1c + \cdots + q_{n-1}c^{n-1}$$

for some elements $q_0, q_1, \ldots, q_{n-1} \in \mathbb{Z}[a_0, a_1, \ldots, a_{n-1}]$. So if $\{\gamma_1, \gamma_2, \ldots, \gamma_\ell\}$ is a generating set of the finitely generated free abelian group $\mathbb{Z}[a_0, a_1, \ldots, a_{n-1}]$, then

$$\{\gamma_i c^j \mid 1 \leq i \leq \ell, \, 0 \leq j \leq n-1\}$$

is a generating set of $\mathbb{Z}[a_0, a_1, \ldots, a_{n-1}, c]$. Since $(\mathbb{Z}[c], +)$ is a subgroup of

$$(\mathbb{Z}[a_0, a_1, \ldots, a_{n-1}, c], +),$$

it is finitely generated by Theorem 2.5.9. By Lemma 4.7.3, we conclude that $c \in \mathcal{O}_K$. Hence, $\mathcal{O}_K$ is integrally closed. □

4.7.4 Sum of Squares

In 1640, Fermat proved which integers can be written as a sum of two squares. Though this represented a landmark result in number theory, mathematicians still actively study problems concerning the sum of squares or the sum of nth powers. Long before Fermat, mathematicians had found all Pythagorean triples, integer triples (a, b, c) that solve $a^2 + b^2 = c^2$. In 1770, Lagrange proved that every integer can be written as a sum of four squares. In the same year, Waring asked if for all integers $k \geq 2$, there was a number $g(k)$ such that every positive integer could be written as a sum of $g(k)$ kth powers. This became known as Waring's problem and was proved in the affirmative in 1909 by Hilbert. Many variations on sums of squares problems and Waring's problem still drive considerable investigation.

One variant to the sum of squares problem has to do with deciding in how many ways an integer can be written as a sum of two squares. Interestingly enough, the algebraic properties of $\mathbb{Z}[i]$ allow us to answer this question.

We have previously seen that $\mathbb{Z}[i]$ is a UFD by virtue of being a Euclidean domain. Furthermore, Section 4.4.4 gave a characterization of the prime elements in $\mathbb{Z}[i]$. The question of whether an integer n can be written as a sum of two squares is equivalent to asking whether there is a Gaussian integer $a + bi$ with norm $N(a + bi) = a^2 + b^2 = n$.

Since $\mathbb{Z}[i]$ is a UFD, then every element $a + bi$ can be written as a product of irreducible elements

$$a + bi = \rho_1 \rho_2 \cdots \rho_r$$

in a unique way (up to reordering and multiplication by units). Now consider also the unique factorization of n in $\mathbb{N}$

$$n = p_1 p_2 \cdots p_m = N(a + bi) = N(\rho_1) N(\rho_2) \cdots N(\rho_r).$$

If for some i, $p_i \equiv 3 \pmod 4$, then the prime p_i must divide some $N(\rho_j)$. By Proposition 4.4.20, we must have $\rho_j = p_i$ and $N(\rho_j) = p_i^2$. If $p_1 \equiv 1, 2 \pmod 4$, then also by Proposition 4.4.20, p_i factors into two irreducible elements that are complex conjugates to each other. This proves the following theorem by Fermat.

Theorem 4.7.11 (Fermat's Sum of Two Squares Theorem)

An integer n can be written as a sum of two squares if and only if $\mathrm{ord}_p(n)$ is even for all primes p with $p \equiv 3 \pmod 4$.

However, the prime factorization in $\mathbb{Z}[i]$ leads to a strong result.

Theorem 4.7.12

> Given a positive integer n satisfying Theorem 4.7.11, the equation $x^2 + y^2 = n$ with $(x, y) \in \mathbb{Z}$ has $4(a_1 + 1)(a_2 + 1) \cdots (a_\ell + 1)$ solutions, where $a_i = \mathrm{ord}_{p_i}(n)$ for all the primes p_i dividing n such that $p_i \equiv 1 \pmod 4$.

Proof. Counting the number of ways n can be written as a sum of two squares is the same problem of finding all distinct elements $A + Bi \in \mathbb{Z}[i]$ such that $N(A + Bi) = n$. Suppose that the prime factorization of n in $\mathbb{N}$ is

$$n = 2^k p_1^{a_1} \cdots p_\ell^{a_\ell} q_1^{2b_1} \cdots q_m^{2b_m},$$

where $p_i \equiv 1 \pmod 4$ and $q_j \equiv 3 \pmod 4$. Then the prime factorization of n in $\mathbb{Z}[i]$ is

$$n = (1+i)^k (1-i)^k \pi_1^{a_1} \overline{\pi}_1^{a_1} \cdots \pi_\ell^{a_\ell} \overline{\pi}_\ell^{a_\ell} q_1^{2b_1} \cdots q_m^{2b_m},$$

where each π_i is an irreducible element in $\mathbb{Z}[i]$ such that $N(\pi_i) = \pi_i \overline{\pi}_i = p_i$. We notice that $(1 - i) = -i(1 + i)$ so not only is $(1 - i)$ a conjugate to $1 + i$ it is also an associate. However, this does not occur for any of the irreducible π_i.

A Gaussian integer α such that $N(\alpha) = \alpha \overline{\alpha} = n$ can only be created by

$$\alpha = u(1+i)^k \pi_1^{c_1} \overline{\pi}_1^{d_1} \cdots \pi_\ell^{c_\ell} \overline{\pi}_1^{d_\ell} q_1^{b_1} \cdots q_m^{b_m},$$

where u is a unit and where for each $\leq i \leq \ell$, the nonnegative integers c_i and d_i satisfy $c_i + d_i = a_i$. For each i, there are $a_i + 1$ ways to choose the pairs (c_i, d_i). By the unique factorization in $\mathbb{Z}[i]$, for all such pairs (c_i, d_i) and for all i, the resulting Gaussian integers α are distinct. There are 4 units in $\mathbb{Z}[i]$ so there are exactly $4(a_1 + 1)(a_2 + 1) \cdots (a_\ell + 1)$ Gaussian integers α such that $N(\alpha) = n$. $\square$

To illustrate Theorem 4.7.12, consider $325 = 5^2 \times 13$. In $\mathbb{Z}[i]$, we have $5 = (2 + i)(2 - i)$ and $13 = (3 + 2i)(3 - 2i)$. The six nonassociate Gaussian integers with a norm of 325 are

$$(2+i)(2+i)(3+2i) = 1 + 18i,$$
$$(2+i)(2-i)(3+2i) = 15 + 10i,$$
$$(2-i)(2-i)(3+2i) = 17 - 6i,$$
$$(2+i)(2+i)(3-2i) = 17 + 6i,$$
$$(2+i)(2-i)(3-2i) = 15 - 10i,$$
$$(2-i)(2-i)(3-2i) = 1 - 18i.$$

Multiplying by the three nontrivial units -1, i, and $-i$, gives the $3 \times 6 = 18$ other solutions to $x^2 + y^2 = 325$.

This section illustrates the interaction between ring theory and number theory. Problems from classical number theory can sometimes find answers through algebra. In this case, working in the ring of Gaussian integers helps answer the question of how many ways an integer can be written as the sum of two squares.

EXERCISES FOR SECTION 4.7

1. Let R be a UFD and let F be its field of fractions. Prove that R is integrally closed in its field of fractions. [This result inserts in the chain in (4.9) the class of domains that are integrally closed between UFDs and integral domain.]

2. Suppose that $a_1, a_2, \ldots, a_n \in \mathbb{C}$ are such that the groups $(\mathbb{Z}[a_i], +)$ are finitely generated free groups of ranks $r_1, r_2, \ldots, r_n$. Prove that abelian group $(\mathbb{Z}[a_1, a_2, \ldots, a_n], +)$ is finitely generated of rank less than or equal to $r_1 r_2 \cdots r_n$.

3. This exercises guides a proof that for $D = -3, -7, -11$ the quadratic integer ring $\mathcal{O}_{\mathbb{Q}(\sqrt{D})}$ is a Euclidean domain with its norm N.

   (a) Let $D \equiv 1 \pmod 4$ be a negative integer and set $\omega = (1 + \sqrt{D})/2$. Show that as a subset of $\mathbb{C}$, the quadratic integer ring $\mathbb{Z}[\omega]$ consists of vertices of congruent parallelograms that cover the plane.

   (b) Consider a parallelogram with vertices 0, 1, ω, and $\omega + 1$. Show that the furthest an interior point P can be from any of the four vertices. [Hint: Show first that P must be the circumcenter for one of the triangles of the parallelogram.]

   (c) Use the previous part to prove that every element in $\mathbb{C}$ is at most $(1 + |D|)/(4\sqrt{|D|})$ away from some element in $\mathbb{Z}[\omega]$.

   (d) Show that $(1 + |D|)^2/16|D| < 1$ for $D = -3, -7, -11$.

   (e) Show that $d(z) = N(z) = |z|^2$ is a Euclidean function on $\mathbb{Z}[\omega]$.

   [With previous results, this exercise shows that $\mathcal{O}_{\mathbb{Q}(\sqrt{D})}$ is a Euclidean domain with the norm function as the Euclidean function for $D = -1, -2, -3, -7, -11$. It turns out, though it is more difficult to prove, that these are the only negative values of D for which $\mathcal{O}_{\mathbb{Q}(\sqrt{D})}$ is a Euclidean domain.]

4. Do the Euclidean division of $(21 + 13\sqrt{-7})/2$ by $(3 + 5\sqrt{-7})/2$ in $\mathbb{Z}\left[\frac{1+\sqrt{-7}}{2}\right]$. (See Exercise 4.7.3.)

5. Consider the element $\frac{1}{6}(4 + 4 \cdot 28^{1/3} + 28^{2/3})$ in the field $F = \mathbb{Q}(\sqrt[3]{28})$. Show that this is an algebraic integer in F by showing that it solves a monic cubic polynomial in $\mathbb{Z}[x]$.

6. Find all ways to write 91 as a sum of two squares.

7. Find all ways to write 338 as a sum of two squares.

8. Find all ways to write 61,000 as a sum of two squares.

9. Prove that if α is an algebraic integer in $\mathbb{C}$, then $\sqrt[n]{\alpha}$ is another algebraic integer for any positive n.

4.8 Projects

Investigative Projects

PROJECT I. **The Ring $\mathbb{Z}[\sqrt{2}]$.** In Example 3.2.3 we briefly encountered the ring $\mathbb{Z}[\sqrt{2}]$.

Recall that the *norm* on $\mathbb{Z}[\sqrt{2}]$ is $N(a + b\sqrt{2}) = |a^2 - 2b^2|$. Show first that $\alpha \in \mathbb{Z}[\sqrt{2}]$ is a unit if and only if $N(\alpha) = 1$. Primes in $\mathbb{Z}$ are not necessarily still irreducible in $\mathbb{Z}[\sqrt{2}]$. For example, $7 = (3 + \sqrt{2})(3 - \sqrt{2})$. However, $3 + \sqrt{2}$ is irreducible since $N(3 + \sqrt{2}) = 7$ so if $3 + \sqrt{2} = \alpha\beta$, then $N(\alpha)N(\beta) = 7$ so either $N(\alpha) = 1$ or $N(\beta) = 1$, and one of them would be a unit.

Here are a few questions to pursue: Find some units in this ring. Can you find patterns in the units, like some process that may give you many units? Investigate the irreducible elements in $\mathbb{Z}[\sqrt{2}]$. Try to find some, that not associate to each other. Try to find any patterns as to which elements in $\mathbb{Z}[\sqrt{2}]$ are irreducible.

PROJECT II. **Non-UFD Rings.** Find examples of non-UFD rings. Give a number of examples of nonequivalent factorizations into irreducibles. Prove that certain elements are irreducible. Can you find nonequivalent factorizations of the same number in which the number of irreducible factors is different? (Suggestion: Focus on rings of the form $\mathbb{Z}[\sqrt{D}]$ or $\mathcal{O}_{\mathbb{Q}(\sqrt{D})}$.)

PROJECT III. **Eisenstein Integers.** Revisit Example 4.7.6 about Eisenstein integers. Look for irreducible elements in the ring of Eisenstein integers. The Eisenstein integers form a hexagonal lattice in $\mathbb{C}$. Can you discern any patterns in the irreducible elements in this lattice?

PROJECT IV. **RSA in $\mathbb{F}_p[x]$.** (For students who have addressed Project VII in Chapter 2.) Modify the theory for RSA to work over the ring $\mathbb{F}_p[x]$. What takes the role of the primes and the product of two primes? Decide whether RSA over $\mathbb{Z}$ or RSA over $\mathbb{F}_p[x]$ is better and state your reasons.

PROJECT V. **Factorizations of Polynomials in a non-UFD.** Take R to be any polynomial ring that is not a factorization. Feel free to take a specific R. Study the factorization of polynomials in $R[x]$. (Discuss irreducibility in $R[x]$, illustrate nonunique factorization, and so forth.)

PROJECT VI. **Quadratic Integer Rings.** Consider some positive square-free integers D. Attempt to determine whether some $\mathcal{O}_{\mathbb{Q}(\sqrt{D})}$ might be Euclidean domains (for some Euclidean function) or PIDs.

PROJECT VII. **Power Series as Euclidean Domains.** If F is a field, then the ring of formal power series $F[[x]]$ is a Euclidean domain with the

Euclidean function d as the smallest power on x occurring in the power series. (Check this first.) Choose a field F. Give examples of interesting power series divisions and power series greatest common divisors using the division and the Euclidean Algorithm. Find some examples of pairs of power series p and q that are relatively prime and use the Extended Euclidean Algorithm to find power series s and t such that $sp + tq = 1$.

Expository Projects

PROJECT VIII. **Gabriel Lamé, 1795–1870.** French mathematician Gabriel Lamé is one of many mathematicians who thought they had found a proof for Fermat's Last Theorem, only later to learn of a key flaw in the "proof" which arose from an incorrect algebraic assumption. Using reliable sources, explore Lamé's mathematical work and describe the story of this flaw and its consequences in algebra.

PROJECT IX. **Sophie Germain, 1776–1831.** Sophie Germain famously corresponded with other mathematicians on a male pseudonym in order not to be ignored. Using reliable sources, explore Germain's personal background, her mathematical work, and the reception of her work in the mathematical community. Mention in particular various formulas, identities or concepts in algebra or number theory that bare her name.

PROJECT X. **Gotthold Eisenstein, 1823–1852.** In this section, we encountered the term Eisenstein Criterion and Eisenstein integers. Using reliable sources, explore Eisenstein's personal background, his mathematical work, and the reception of his work in the mathematical community. Find why the criterion and the particular ring $\mathbb{Z}[(1 + i\sqrt{3})/2]$ bear his name.

PROJECT XI. **Factorization of Multivariable Polynomials.** If F is a field and if $n \geq 2$, the polynomial ring $F[x_1, x_2, \ldots, x_n]$ is a unique factorization domain but neither a Euclidean domain, nor a principal ideal domain. Various computer algebra systems implement a factorization algorithm for elements in $F[x_1, x_2, \ldots, x_n]$. Find sources that explain the algorithm that some CAS uses and illustrate it by working out your own example by hand.

PROJECT XII. **Ideal Class Group.** Using level-appropriate sources, discuss the concept and application of the ideal class group of an algebraic number ring K. Give examples and mention a few currently unknown problems. Discuss the connection between the ideal class group of K and whether K is a unique factorization domain.

5

Field Extensions

Earlier in this book, we encountered fields as a particular class of rings: a commutative ring in which every nonzero element has a multiplicative inverse. Though fields are a particular class of rings, they possess unique properties that lead to many fruitful investigations. Since every nonzero element of a field is a unit, questions about divisibility among elements are not interesting. However, fields possess a rich structure, primarily because of the role that linear algebra plays.

The study of polynomial equations drove much of the development of field theory. However, fields lead to many application internal to mathematics and methods within the theory are valuable for digital communication and information security.

5.1 Introduction to Field Extensions

Though we have seen the definition of a field in Section 3.1.3, we have not to this point focused attention on homomorphisms between fields or how to concisely describe (or generate) a field. The particular properties of these aspects of field theory are what warrant studying fields in their own right.

5.1.1 Algebraic Structure of a Field Extension

The first distinctive aspect about fields is that homomorphisms between them have an elementary structure.

> **Proposition 5.1.1**
>
> A homomorphism of fields $\varphi : F \to F'$ either is identically 0 or is injective.

Proof. Let $\varphi : F \to F'$ be a homomorphism. The kernel $\mathrm{Ker}\,\varphi$ is an ideal in F. However, the only two ideals in F are (0) and F itself. If $\mathrm{Ker}\,\varphi = (0)$, then φ is injective. If $\mathrm{Ker}\,\varphi = F$, then φ is identically 0. $\square$

DOI: 10.1201/9781003299233-5

In previous sections, we termed an injective homomorphism (in group theory or ring theory) an *embedding*. So we call an injective homomorphism an embedding of F into F'. By the first isomorphism theorem for rings, when there is an embedding of F in F', there exists a subring of F' that is isomorphic to F. Consequently, we can view F as a subfield of F'. Thus, the existence of nontrivial homomorphisms between fields is tantamount to containment.

Recall that the characteristic char(R) of a ring R is either 0 or the least positive integer n such that $n \cdot 1 = 0$ if such an n exists. By Exercise 3.1.22, the characteristic of an integral domain is either 0 or a prime number p. Since fields are integral domains, this result applies. The concept of characteristic of a field leads to a slightly more nuanced concept.

Definition 5.1.2

The *prime subfield* of a field F is the subfield generated by the multiplicative identity.

In other words, the prime subfield is the smallest (by inclusion) field in F that contains the identity. Suppose that a field has positive characteristic p. Then the prime subfield must contain the elements

$$0, 1, 2 \cdot 1, \ldots, (p-1) \cdot 1$$

and $p \cdot 1 = 0$. However, the multiplication on these elements as defined by distributivity gives this set of elements the structure of $\mathbb{F}_p = \mathbb{Z}/p\mathbb{Z}$. On the other hand, if we suppose that the field F has characteristic 0, then F must contain

$$\ldots, -(3 \cdot 1), -(2 \cdot 1), -1, 0, 1, (2 \cdot 1), (3 \cdot 1), \ldots.$$

Therefore, $\mathbb{Z}$ is contained in F. But then the field F must also contain the field of fractions of $\mathbb{Z}$, namely $\mathbb{Q}$. Thus, $\mathbb{Q}$ is the prime field of F. We have proven the following proposition.

Proposition 5.1.3

Let F be a field. The prime subfield of F is $\mathbb{Q}$ if and only if char(F) $= 0$ and the prime subfield of F is $\mathbb{F}_p$ if and only if char(F) $= p$.

We have encountered a few fields before. For example, $\mathbb{Q}$, $\mathbb{R}$, and $\mathbb{C}$ are fields of characteristic 0 while the finite field $\mathbb{F}_p$ has characteristic p.

Definition 5.1.4

If K is a field containing F as a subfield, then K is called a *field extension* (or simply extension) of F. This relationship of extension is often denoted by K/F.

The notation for field extension, resembles the usual notation for a quotient ring but the two constructions are not related. There is never a confusion of notation because the construction of quotient fields never occurs in field theory. Indeed, the only ideals in a field are the trivial ideal or the whole field, so the only resulting quotient rings are the trivial field and the field itself.

By Proposition 5.1.3, every field is a field extension of either $\mathbb{Q}$ or $\mathbb{F}_p$ for some prime p.

Previous sections illustrated a few ways to construct some field extensions. These will become central to our study of field extensions.

The first method uses a commutative ring generated by elements and then passing to the associated field of fractions. In other words, suppose that F is a field contained in some integral domain R. If $\alpha \in R \setminus F$, then $F[\alpha]$ is the subring of R generated by R and α. See Section 3.2.1. Since $F[\alpha]$ is an integral domain, we can take the field of fractions $F(\alpha)$ of $F[\alpha]$. See Section 4.2. In this way, we construct the field extension $K = F(\alpha)$ of F. This construction extends to fields generated by F and subsets $S \subseteq R \setminus F$ such that $R(S)$ is the field of fractions of the integral domain $R[S]$. For example, the fields $\mathbb{Q}(\sqrt{2})$ or $\mathbb{Q}(\sqrt[3]{7}, i)$ are field extensions of $\mathbb{Q}$, inside $\mathbb{C}$.

This method presupposes that F is a subring of some integral domain R and hence is a subfield of the field of fractions of R.

The second method involves quotient rings of a polynomial ring. Let F be a field. Since $F[x]$ is a Euclidean domain, it is also a PID so every ideal I in $F[x]$ is of the form $I = (p(x))$ for some polynomial $p(x) \in F[x]$. By Proposition 3.8.12, the ideal $(p(x))$ is maximal if and only if it is prime if and only if $p(x)$ is irreducible. So if $p(x)$ is a irreducible polynomial of degree greater than 0, then the quotient ring $F[x]/(p(x))$ is a field. Furthermore, the inclusion F into $F[x]/(p(x))$ is an injection so F is a subfield of $F[x]/(p(x))$.

Example 5.1.5. Consider the polynomial $p(x) = x^2 - 5$. By Proposition 4.5.11, it has no rational roots so by Proposition 4.5.12 $p(x)$ is irreducible in $\mathbb{Q}[x]$. Then $\mathbb{Q}[x]/(p(x)) = \mathbb{Q}[\sqrt{5}]$ is a field. It is obvious by construction that $\mathbb{Q}[\sqrt{5}]$ is an integral domain. To show that every nonzero element is invertible, consider $\alpha = a + b\sqrt{5} \neq 0$. Then

$$\frac{1}{\alpha} = \frac{a - b\sqrt{5}}{(a + b\sqrt{5})(a - b\sqrt{5})} = \frac{a - b\sqrt{2}}{a^2 - 5b^2}. \tag{5.1}$$

Since there is no rational number $\frac{p}{q}$ such that $\frac{p^2}{q^2} = 5$, the denominator of this expression is a nonzero rational number so (5.1) gives a formula for the inverse. $\triangle$

One may attempt to generalize the idea in Example 5.1.5 and ask if a ring such as $\mathbb{Q}[\sqrt[3]{7}]$ is also a field. This turns out to be true but the proof becomes more difficult than the proof in the above Example 5.1.5. For example, finding the inverse to an element such as $1 + 3\sqrt[3]{7} - \frac{1}{2}(\sqrt[3]{7})^2$ is not as simple.

5.1.2 Field Extensions as Vector Spaces

In many linear algebra courses, because of the importance for applications to science, students usually encounter vector spaces with the assumption that the scalars are real numbers. However, the definition for a vector space over $\mathbb{R}$ can be generalized to a vector space V over a field F. We invite the reader to revisit the definition of a real vector space; then anywhere the definition says $\mathbb{R}$, replace the $\mathbb{R}$ with F. All of the definitions, algorithms, and constructions introduced for vector spaces over $\mathbb{R}$ can be adapted to vector spaces over a field. These include

- systems of linear equations (with coefficients and variables in F);
- the Gauss-Jordan elimination algorithm and the resulting reduced row echelon form for matrices in $M_{m \times n}(F)$;
- linear combinations;
- linear independence;
- subspaces and the span of sets of vectors;
- a basis of a subspace;
- the dimension of a vector space;
- coordinates with respect to a basis;
- linear transformations $T : V \to V$;
- the determinant of an $n \times n$ matrix;
- Cramer's rule.

In contrast, geometrical interpretations of linear transformations, of the determinant, and of bilinear forms do not directly generalize to vector spaces over arbitrary fields. Furthermore, though the definition of eigenvalues and eigenvectors generalizes immediately, the process of solving the characteristic equation requires more discussion over an arbitrary field than over $\mathbb{R}$.

Proposition 5.1.6

> Let K be a field extension of F. Then K is a vector space over the field F.

Proof. With the addition, $(K, +)$ is an abelian group. Furthermore, by distributivity and associativity of the multiplication in K, the following properties hold:

- $r(\alpha + \beta) = r\alpha + r\beta$ for all $r \in F$ and all $\alpha, \beta \in K$;
- $(r + s)\alpha = r\alpha + s\alpha$ for all $r, s \in F$ and all $\alpha \in K$.
- $r(s\alpha) = (rs)\alpha$ for all $r, s \in F$ and all $\alpha \in K$.
- $1\alpha = \alpha$ for all $\alpha \in K$.

These observations establish all the axioms of a vector space over F and the proposition follows. □

The great value of this simple proposition is that it makes it possible to use the theory of vector spaces to derive information and structure about extensions of F. The concept of degree is an important application of this connection between vector spaces over F and extensions of F.

Definition 5.1.7

If the extension K/F has a finite basis as an F-vector space, then the *degree* of the extension K/F, denoted $[K : F]$, is the dimension of K as a vector space over F. In other words $[K : F] = \dim_F K$. If the extension K/F does not have a finite basis, we say that the degree $[K : F]$ is infinite.

Example 5.1.8. Consider the extension $\mathbb{Q}(\sqrt{5})$ over $\mathbb{Q}$. According to Example 5.1.5, $\mathbb{Q}(\sqrt{5}) = \mathbb{Q}[\sqrt{5}]$ so every element in $\mathbb{Q}(\sqrt{5})$ can be written uniquely as $a + b\sqrt{5}$ for some $a, b \in \mathbb{Q}$. Hence, $\{1, \sqrt{5}\}$ is a basis for $\mathbb{Q}(\sqrt{5})$ over $\mathbb{Q}$ and therefore $[\mathbb{Q}(\sqrt{5}) : \mathbb{Q}] = 2$. △

Example 5.1.9 (Complex Numbers). We know that $\mathbb{R} \subseteq \mathbb{C}$ and $\mathbb{C}$ is a field so it is a field extension of $\mathbb{R}$. However, since every complex number can be written uniquely as $a + bi$ for $a, b \in \mathbb{R}$, then $\{1, i\}$ is an $\mathbb{R}$-basis of $\mathbb{C}$ over $\mathbb{R}$. Thus, $[\mathbb{C} : \mathbb{R}] = 2$. △

The two methods outlined above for constructing field extensions of a given field F appear quite different. However, a first key result that emerges from the identification of a field extension of F with a vector space over F is that these two methods are in fact the same. We develop this result in the theorems below.

Theorem 5.1.10

If $[F(\alpha) : F] = n$ is finite, then α is the root of some irreducible polynomial $m(x) \in F[x]$ of degree n. Furthermore, $F(\alpha) = F[\alpha]$.

Proof. If $[F(\alpha) : F] = 1$, then $F(\alpha) = F$ so $\alpha \in F$ and the corresponding polynomial $m(x)$ is $m(x) = x - \alpha$. We assume from now on that $[F(\alpha) : F] > 1$.

The field $F(\alpha)$ includes linear combinations of the form

$$a_0 + a_1\alpha + a_2\alpha^2 + \cdots + a_k\alpha^k,$$

where $a_i \in F$. All these linear combinations are polynomials in $F[x]$ evaluated at α. Since $F(\alpha)$ has dimension n as a vector space over F, the set $\{1, \alpha, \alpha^2, \ldots, \alpha^n\}$ is linearly dependent. Thus, there exists some nontrivial polynomial $q(x)$ of degree at most n such that $q(\alpha) = 0$.

Consider the set of polynomials $S = \{p(x) \in F[x] \setminus \{0\} \mid p(\alpha) = 0\}$. Since α the root of some polynomial over F, the set S is nonempty. By the well-ordering principle, S contains an element $m(x)$ of least degree. Suppose that $\deg m(x) = d$. We claim that $m(x)$ is irreducible. Assume that $m(x)$ is reducible with $m(x) = p_1(x)p_2(x)$, with $p_1(x), p_2(x) \in F[x]$ each of positive degrees. Then $0 = m(\alpha) = p_1(\alpha)p_2(\alpha)$ and since there are no zero-divisors in a field, $p_1(\alpha) = 0$ or $p_2(\alpha) = 0$. So $p_1(x) \in S$ or $p_2(x) \in S$ but this contradicts the minimality of $m(x)$ in S. This justifies the claim that $m(x)$ is irreducible.

Writing $m(x) = c_0 + c_1 x + \cdots + c_d x^d$, we note that $c_d \neq 0$, so

$$\alpha^d = -\frac{c_0}{c_d} - \frac{c_1}{c_d}\alpha - \cdots - \frac{c_{d-1}}{c_d}\alpha^{d-1}.$$

This expresses α^d as a linear combination of $\{1, \alpha, \ldots, \alpha^{d-1}\}$. By a recursion argument, we can see that for all $k \geq 0$, the element α^k can also be written as a linear combination of $\{1, \alpha, \ldots, \alpha^{d-1}\}$. Hence, the set of powers of α spans $F[\alpha]$ as a vector field over F. However, a priori $F[\alpha]$ is only a subset of $F(\alpha)$, and more precisely a subspace of dimension d over F.

By definition, every element in $F(\alpha)$ can be written as a rational expression of α, namely

$$\gamma = \frac{a(\alpha)}{b(\alpha)} \qquad \text{where } a(x), b(x) \in F[x] \text{ and } b(\alpha) \neq 0.$$

Suppose also that $a(x)$ and $b(x)$ are chosen such that $b(x)$ has minimal degree and $\gamma = a(\alpha)/b(\alpha)$. Performing the Euclidean division of $m(x)$ by $b(x)$ we get $m(x) = b(x)q(x) + r(x)$, where $\deg r(x) < \deg b(x)$ or $r(x) = 0$. Assume that $r(x) \neq 0$. Then

$$\gamma = \frac{a(\alpha)}{b(\alpha)} = \frac{a(\alpha)q(\alpha)}{b(\alpha)q(\alpha)} = \frac{a(\alpha)q(\alpha)}{p(\alpha) - r(\alpha)} = -\frac{a(\alpha)q(\alpha)}{r(\alpha)}.$$

Hence, $a(\alpha)/b(\alpha)$ can be written as $a_2(\alpha)/b_2(\alpha)$ where $\deg b_2(x) < \deg b(x)$. This contradicts the choice that $b(x)$ has minimal degree. Consequently, $r(x) = 0$ and hence $b(x)$ divides $m(x)$. However, $m(x)$ is irreducible so $b(x)$ is either a constant in F or a constant multiple of $m(x)$. On the other hand, $m(\alpha) = 0$, while $b(\alpha) \neq 0$. Consequently, $b(x)$ must be a constant. Consequently, $F(\alpha) \subset F[\alpha]$. Therefore, we conclude three things: $F(\alpha) = F[\alpha]$, $n = d$, and $m(x)$ is an irreducible polynomial such that $m(\alpha) = 0$. $\qquad\square$

Definition 5.1.11

Let F be a field. An extension field K over F is called *simple* if $K = F(\alpha)$ for some $\alpha \in K$.

It is important to note as a contrast that $F(\alpha)$ is not necessarily equal to $F[\alpha]$ if $[F(\alpha) : F]$ is infinite which may occur if α is not the root of a polynomial in $p(x)$. For example, keeping t as a free parameter, $F[t]$ is a subring of $F(t)$.

Furthermore, t is not a unit in $F[t]$ whereas in $F(t)$, the multiplicative inverse to the polynomial t is the rational expression $\frac{1}{t}$. Hence, $F[t]$ is a strict subring of $F(t)$.

Theorem 5.1.10 may feel unsatisfactory because the hypotheses assumed that α was some element in a field extension of F that remained unspecified. So one might naturally ask whether there exists a field extension of F that contains some α such that $p(\alpha) = 0$. We have already seen that the answer to this question is yes and we encapsulate the result in the following converse to Theorem 5.1.10.

Theorem 5.1.12

Let $p(x) \in F[x]$ be an irreducible polynomial of degree n. Then $K = F[x]/(p(x))$ is a field in which $\theta = \overline{x} = x + (p(x))$ that satisfies $p(\theta) = 0$. Furthermore, the elements

$$1, \theta, \theta^2, \ldots, \theta^{n-1}$$

form a basis of K as a vector space over F. So $[K : F] = n$ and $K = F[\theta]$.

Example 5.1.13. As an example of constructing a field extension, let $F = \mathbb{F}_2$ and consider the polynomial $p(x) = x^3 + x + 1$. Since $p(x)$ has no roots in $\mathbb{F}_2$ and since it is a cubic, it is irreducible by Proposition 4.5.12. Hence, $K = \mathbb{F}_2[x]/(x^3 + x + 1)$ is a field extension of $\mathbb{F}_2$ with $[K : \mathbb{F}_2] = 3$. Consequently, K is a finite field containing $2^3 = 8$ elements. $\triangle$

Let L be a field extension of F and let $\alpha \in L$. So if α is a root of the irreducible polynomial $p(x) \in F[x]$, then $F(\alpha) = F[\alpha]$. It is easy to understand the addition (and subtraction) in $F[\alpha]$ but the multiplication requires some simplifications and the process of finding inverses in $F[\alpha]$ is not obvious.

One method to find inverses to elements $q(\alpha) \in F[\alpha]$ come from the fact that $F[x]$ is a Euclidean domain. Every element in $K \cong F[x]/(p(x))$ is of the form $\overline{q(x)} = q(x) + (p(x))$ where $q(x)$ can be chosen so that $\deg q(x) < \deg p(x)$. Since $p(x)$ is irreducible in $F[x]$ then $p(x)$ and $q(x)$ do not have a common divisor of positive degree. Hence, performing the Extended Euclidean Algorithm leads us to find $a(x)$ and $b(x)$ such that

$$a(x)q(x) + b(x)p(x) = 1$$

in $F[x]$. In the quotient ring K, this implies that $\overline{a(x)q(x)} = 1$. Thus, in K, $a(\alpha)q(\alpha) = 1$, so that $a(\alpha)$ is the inverse to $q(\alpha)$. This method is not the simplest method to find inverses in K. The example below illustrates this method and a faster algorithm that uses linear algebra.

Example 5.1.14. Let $F = \mathbb{Q}$. Consider $p(x) = x^3 - 2$. This is irreducible in $\mathbb{Q}[x]$. Then $K = \mathbb{Q}[x]/(x^3 - 2)$ is a field and we denote by θ an element in K

such that $\theta^3 - 2 = 0$. (For simplicity, we could assume that $\theta = \sqrt[3]{2} \in \mathbb{C}$ but $\sqrt[3]{2}$ is not the only complex number that solves $x^3 - 2 = 0$.) Then from the above theorems,

$$K = \{a + b\theta + c\theta^2 \mid a, b, c \in \mathbb{Q}\}.$$

The fact that K is isomorphic to a subfield of $\mathbb{C}$ is irrelevant to this construction.

Let $\alpha = 3 - \theta + \theta^2$ and $\beta = 5 + 3\theta - \frac{1}{2}\theta^2$. The addition in K occurs component-wise so

$$\alpha + \beta = 8 + 2\theta + \frac{1}{2}\theta^2.$$

For the product, we remark that $\theta^3 = 2$ so also $\theta^4 = 2\theta$ and thus

$$
\begin{aligned}
\alpha\beta &= \left(3 - \theta + \theta^2\right)\left(5 + 3\theta - \frac{1}{2}\theta^2\right) \\
&= 15 + 9\theta - \frac{3}{2}\theta^2 - 5\theta - 3\theta^2 + \frac{1}{2}(2) + 5\theta^2 + 3(2) - \frac{1}{2}(2\theta) \\
&= 22 + 3\theta + \frac{1}{2}\theta^2.
\end{aligned}
$$

We now look for α^{-1}. There are at least two methods to find the inverse of an element. The first uses the Extended Euclidean Algorithm between $x^3 - 2$ and $x^2 - x + 3$. Since $x^3 - 2$ is irreducible in $\mathbb{Q}[x]$, the greatest common divisor between the polynomials is 1. The result of the algorithm is that

$$1 = \frac{1}{47}(11 + 5x - 2x^2)(x^2 - x + 3) - \frac{1}{47}(-2x + 7)(x^3 - 2).$$

Since $\theta^3 - 2 = 0$, plugging θ into this expression, we find that

$$\alpha^{-1} = (\theta^2 - \theta + 3)^{-1} = \frac{11}{47} + \frac{5}{47}\theta - \frac{2}{47}\theta^2.$$

Linear algebra offers an easier way to find inverses (and, more generally, compute division). Suppose that $\alpha^{-1} = a_0 + a_1\theta + a_2\theta^2$. Then this element satisfies

$$
\begin{aligned}
&(3 - \theta + \theta^2)(a_0 + a_1\theta + a_2\theta^2) = 1 \\
\Longleftrightarrow\ & 3a_0 + (3a_1 - a_0)\theta + (3a_2 - a_1 + a_0)\theta^2 + (-a_2 + a_1)\theta^3 + a_2\theta^4 \\
\Longleftrightarrow\ & 3a_0 + (3a_1 - a_0)\theta + (3a_2 - a_1 + a_0)\theta^2 + (-a_2 + a_1)(2) + a_2(2\theta) \\
\Longleftrightarrow\ & (-2a_2 + 2a_1 + 3a_0) + (2a_2 + 3a_1 - a_0)\theta + (3a_2 - a_1 + a_0)\theta^2.
\end{aligned}
$$

Consequently, the coefficients a_0, a_1, a_2 satisfy the system of linear equations

$$
\begin{cases}
-2a_2 + 2a_1 + 3a_0 = 1 \\
2a_2 + 3a_1 - a_0 = 0 \\
3a_2 - a_1 + a_0 = 0.
\end{cases}
\tag{5.2}
$$

Using a computer or calculator (or working by hand) gives

$$\text{rref} \begin{pmatrix} -2 & 2 & 3 & 1 \\ 2 & 3 & -1 & 0 \\ 3 & -1 & 1 & 0 \end{pmatrix} = \begin{pmatrix} 1 & 0 & 0 & -2/47 \\ 0 & 1 & 0 & 5/47 \\ 0 & 0 & 1 & 11/47 \end{pmatrix}.$$

Interpreting this calculation gives precisely the same result for $(3 - \theta + \theta^2)^{-1}$ as above. The fact that (5.2) has a unique solution follows from the fact that $p(x) = x^3 - 2$ us irreducible. $\triangle$

5.1.3 Isomorphisms of Fields

Theorems 5.1.12 and 5.1.10 indirectly lead to an interesting consequence. Let L be a field extension of a field F. Suppose that $\alpha, \beta \in L$ such that α and β both are roots of the same irreducible polynomial $p(x)$. Then we have isomorphisms of fields

$$F[\alpha] \cong F[x]/(p(x)) \cong F[\beta].$$

In fact, the (composition) isomorphism described above satisfies $f : F[\alpha] \to F[\beta]$ with $f(c) = c$ for all $c \in F$, $f(\alpha) = \beta$, and all other values of f resulting from the axioms of a homomorphism.

It is not at all uncommon that $F[\alpha] \neq F[\beta]$ so this isomorphism is not trivial or even an automorphism. Recall from group theory that an automorphism is an isomorphism from a field to itself.

We have shown that there is a close connection between properties of subfields and morphisms (homomorphisms between fields). The only nontrivial morphisms are either injections, which are embeddings, and isomorphisms. Despite or perhaps because of this restriction, the study of field extensions and automorphisms of fields is rich and has many applications.

5.1.4 Useful CAS Commands

SAGE offers a number of commands related to constructing field extensions. For example, the command `NumberField(p)` defines the extension $\mathbb{Q}(\alpha)$, where α is a root of the specified polynomial p. The following code computes α^{-1} from Example 5.1.14.

――――――――――――― Sage ―――――――――――――

```
sage: K.<t> = NumberField(x^3-2)
sage: a=t^2-t+3
sage: a^(-1)
-2/47*t^2 + 5/47*t + 11/47
```

EXERCISES FOR SECTION 5.1

1. Consider the ring $F = \mathbb{F}_5[x]/(x^2 + 2x + 3)$ and call θ the element $\bar{x}$ in F.
   (a) Prove that F is a field.
   (b) Prove that every element of F can be written uniquely as $a\theta + b$, with $a, b \in \mathbb{F}_5$.
   (c) Let $\alpha = 2\theta + 3$ and $\beta = 3\theta + 4$. Calculate (i) $\alpha + \beta$; (ii) $\alpha\beta$; (iii) α/β.

2. Let $\alpha = 1 + \sqrt[3]{2} - 3\sqrt[3]{2}^2$ and $\beta = 3 + \sqrt[3]{2}^2$ in $\mathbb{Q}(\sqrt[3]{2})$. Calculate (i) $\alpha\beta$; (ii) α/β; (iii) β/α.

3. Consider the ring $F = \mathbb{Q}[x]/(x^3 + 3x + 1)$ and called θ the element $\bar{x}$ in F.
   (a) Prove that F is a field (which we can write as $\mathbb{Q}(\theta)$).
   (b) Prove that every element of F can be written unique as $a\theta^2 + b\theta + c$, with $a, b \in \mathbb{Q}$.
   (c) Let $\alpha = 2\theta^2 - 1$ and $\beta = \theta^2 + 5\theta - 3$. Calculate (i) $\alpha + \beta$; (ii) $\alpha\beta$; (iii) α/β.

4. Let $F = \mathbb{F}_7$ and consider the irreducible polynomial $p(x) = x^3 - 2$. Write θ for the element $\bar{x}$ in $\mathbb{F}_7/(p(x))$. Find the inverse of $\theta^2 - \theta + 3$.

5. In $\mathbb{Q}(\sqrt[4]{10})$, find the inverse of $1 + \sqrt[4]{10}$.

6. Consider the field of order 8 constructed in Example 5.1.13. Call θ an element in F such that $\theta^3 + \theta + 1 = 0$, so that we can write $F = \mathbb{F}_2[\theta]$.
   (a) Let $\alpha = \theta^2 + 1$, $\beta = \theta^2 + \theta + 1$, and $\gamma = \theta$. Calculate: (i) $\alpha\gamma + \beta$; (ii) α/γ; (iii) $\alpha^2 + \beta^2 + \gamma^2$.
   (b) Solve for x in terms of y in the equation $y = \alpha x + \beta$.
   (c) Prove that the function $f : F \to F$ defined by $f(\alpha) = \alpha^3$ is a cyclic permutation on F.

7. Consider the field F of order 8 constructed in Example 5.1.13. Prove that $U(F)$ is a cyclic group.

8. Prove that $\{1, \sqrt{2}, \sqrt{3}\}$ is linearly independent in $\mathbb{C}$ as a vector space over $\mathbb{Q}$.

9. Prove that $\{1, \sqrt{3}, i, i\sqrt{3}\}$ is linearly independent in $\mathbb{C}$ as a vector space over $\mathbb{Q}$.

10. Consider the ring $K = \mathbb{Q}[\sqrt{2}, \sqrt{5}]$.
    (a) Prove that $K = \mathbb{Q}[\sqrt{2} + \sqrt{5}]$ and is a field. Determine $[K : \mathbb{Q}]$.
    (b) Set $\gamma = \sqrt{2} + \sqrt{5}$. Show that $\mathcal{B}_1 = \{1, \sqrt{2}, \sqrt{5}, \sqrt{10}\}$ and $\mathcal{B}_2 = \{1, \gamma, \gamma^2, \gamma^3\}$ are two bases of K as a vector space over $\mathbb{Q}$.
    (c) Determine the change of coordinate matrix from the basis $\mathcal{B}_2$ to $\mathcal{B}_1$ coordinates.
    (d) Use part (c) to write $2 + 3\gamma^2 - 7\gamma^3$ in the basis $\mathcal{B}_1$.
    (e) Use part (c) to write $-3 + \sqrt{2} - \sqrt{5} + 7\sqrt{10}$ as a linear combination of $\{1, \gamma, \gamma^2, \gamma^3\}$.

11. Construct a field of 9 elements and write down the addition and multiplication tables for this field.

12. Consider the field $\mathbb{F}_3(t)$ of rational expressions with coefficients in $\mathbb{F}_3$. Let

$$\alpha = \frac{2t + 1}{t + 2}, \qquad \beta = \frac{1}{2t^2 + 1}, \qquad \gamma = \frac{t + 1}{t^2 + 1}.$$

Calculate (a) $\alpha + \beta$; (b) $\beta\gamma$; (c) $\alpha\gamma/\beta$.

13. Prove that an automorphism of a field F leaves the prime subfield of F invariant.

14. Prove that the function $f : \mathbb{Q}[\sqrt{5}] \to \mathbb{Q}[\sqrt{5}]$ defined by $f(a + b\sqrt{5}) = a - b\sqrt{5}$ is an automorphism.

15. Prove that there exists an isomorphism of fields $f : \mathbb{R} \to \mathbb{R}$ that maps π to $-\pi$.

16. Let $K = F(\alpha)$ where α is the root of some irreducible polynomial $p(x) \in F[x]$. Suppose that $p(x) = a_n x^n + \cdots + a_1 x + a_0$. Show that the function $f_\alpha : K \to K$ defined by $f_\alpha(x) = \alpha x$ is a linear transformation and that the matrix of f_α with respect to the ordered basis $(1, \alpha, \alpha^2, \ldots, \alpha^{n-1})$ is

$$
\begin{pmatrix}
0 & 0 & 0 & \cdots & 0 & -a_0/a_n \\
1 & 0 & 0 & \cdots & 0 & -a_1/a_n \\
0 & 1 & 0 & \cdots & 0 & -a_2/a_n \\
\vdots & \vdots & \vdots & \ddots & \vdots & \vdots \\
0 & 0 & 0 & \cdots & 1 & -a_{n-1}/a_n
\end{pmatrix}.
$$

17. (*Analysis*) Prove that the only continuous automorphism on the field of real numbers is the identity function.

18. Let $\varphi : F \to F'$ be an isomorphism of fields. Let $p(x) \in F[x]$ be an irreducible polynomial and let $p'(x)$ be the polynomial obtained from $p(x)$ by applying φ to the coefficients of $p(x)$. Let α be a root of $p(x)$ in some extension of F and let β be a root of $p'(x)$ in some extension of F'. Prove that there exists an isomorphism

$$\Phi : F(\alpha) \to F'(\beta)$$

such that $\Phi(\alpha) = \beta$ and $\Phi(c) = \varphi(c)$ for all $c \in F$.

19. Let D be a square-free integer and let $K = \mathbb{Q}[\sqrt{D}]$. Prove that the function $f : \mathbb{Q}[\sqrt{D}] \to M_2(\mathbb{Q})$ defined by

$$f(a + b\sqrt{D}) = \begin{pmatrix} a & Db \\ b & a \end{pmatrix}$$

is an injective ring homomorphism. Conclude that $M_2(\mathbb{Q})$ contains a subring isomorphic to $\mathbb{Q}[\sqrt{D}]$.

20. Consider the field $\mathbb{Q}[\sqrt[3]{2}]$.

 (a) Prove that the function $\varphi : \mathbb{Q}[\sqrt[3]{2}] \to M_3(\mathbb{Q})$ defined by

$$\varphi(a + b\sqrt[3]{2} + c\sqrt[3]{2}^2) = \begin{pmatrix} a & 2c & 2b \\ b & a & 2c \\ c & b & a \end{pmatrix}$$

 is an injective homomorphism. Conclude that $M_3(\mathbb{Q})$ contains a subring that is a field isomorphic to $\mathbb{Q}[\sqrt[3]{2}]$.

 (b) Use part (a) to find the inverse of $3 - \sqrt[3]{2} + 5\sqrt[3]{2}^2$.

21. Consider the field of rational expressions $K_1 = \mathbb{Q}(x)$ with coefficients in $\mathbb{Q}$ and also the field $K_2 = \mathbb{Q}(\sqrt{p} \,|\, p \text{ is prime})$. Prove that K_1 and K_2 are extensions of $\mathbb{Q}$ of infinite degree. Prove also that K_1 and K_2 are not isomorphic.

5.2 Algebraic and Transcendental Elements

Section 5.1 introduced field extensions and emphasized the properties that follow from viewing an extension of a field F as a vector space over F. Theorem 5.1.10 brought together two disparate ways of constructing field extensions. It is also precisely this theorem that connects field theory so closely with the study of polynomial equations. This section further develops consequences of Theorem 5.1.10 by studying field extensions K/F as a field K containing roots of polynomials in $F[x]$.

5.2.1 Algebraic Elements

Let F be a field and let K be an extension of F.

Definition 5.2.1

An element $\alpha \in K$ is called *algebraic* over F if α is a root of some nonzero polynomial $f(x) \in F[x]$. If $\alpha \in K$ is not algebraic over F, then α is called *transcendental* over F.

Consider the fields $\mathbb{Q} \subseteq \mathbb{R}$. The element $\sqrt{2}$ in $\mathbb{R}$ is algebraic over $\mathbb{Q}$ because it is the root of $x^2 - 2$. As another example, note that it is easy to show that $\cos(3\theta) = 4\cos^3\theta - 3\cos\theta$. Hence, setting $\theta = \frac{\pi}{9}$, we see that

$$4\cos^3\left(\frac{\pi}{9}\right) - 3\cos\left(\frac{\pi}{9}\right) = \cos\left(\frac{\pi}{3}\right) = \frac{1}{2}.$$

Hence, though we do not know the value of $\cos(\pi/9)$, we see that it is a root of the cubic equation $4x^3 - 3x - \frac{1}{2} = 0$ and so it is algebraic over $\mathbb{Q}$. By an abuse of language, if we say that a number is algebraic (with no other qualifiers) we usually imply that $K = \mathbb{C}$ and $F = \mathbb{Q}$.

A first important property about algebraic elements is that for each algebraic element α there exists a naturally preferred polynomial with α as a root.

Proposition 5.2.2

Let K/F be a field extension and let $\alpha \in K$ be algebraic over F. There exists a unique monic irreducible polynomial $m_{\alpha,F}(x) \in F[x]$ such that α is a root of $m_{\alpha,F}(x)$.

Proof. Consider the set of polynomials $S = \{p(x) \in F[x] \setminus \{0\} \mid p(\alpha) = 0\}$. Borrowing from the proof of Theorem 5.1.10, any polynomial of minimum degree is irreducible. Let

$$a(x) = a_n x^n + \cdots + a_1 x + a_0 \qquad \text{and} \qquad b(x) = b_n x^n + \cdots + b_1 x + b_0$$

be two polynomials of least degree in S. Then α is also a root of $q(x) = b_n a(x) - a_n b(x)$ so $q(x)$ is either 0 or in S. However, the subtraction cancels the leading terms so $q(x)$ is either 0 or has $\deg q(x) < n$. Since n is the least degree of any polynomial in S, we conclude that $q(x) = 0$. Thus, $b_n a(x) = a_n b(x)$ and any two polynomials in S of least degree are multiples of each other. Consequently, there exists a unique monic irreducible polynomial in S. $\qquad\square$

The notation of $m_{\alpha,F}(x)$ indicates the dependence of the polynomial on the specific field of coefficients. The polynomial may change based on the context of the field of coefficients but, as the following corollary shows, the corresponding polynomials in different fields are related.

Corollary 5.2.3

Let $F \subseteq L \subseteq K$ be a chain of fields and suppose that $\alpha \in K$ is algebraic over F. Then α is algebraic over L and $m_{\alpha,L}(x)$ divides $m_{\alpha,F}(x)$ in $L[x]$.

Proof. Both polynomials $m_{\alpha,L}(x)$ and $m_{\alpha,F}(x)$ are in $L[x]$ and have α as a root. The polynomial division in $L[x]$ of the two polynomials gives

$$m_{\alpha,F}(x) = m_{\alpha,L}(x)q(x) + r(x),$$

where $r(x) = 0$ or $\deg r(x) < \deg m_{\alpha,L}(x)$. However, since α is a root of both polynomials, we deduce that $r(\alpha) = 0$. Since $\deg m_{\alpha,L}(x)$ is the least degree of a nonzero polynomial in $L[x]$ that has α as a root, then $r(x) = 0$ and hence $m_{\alpha,L}(x)$ divides $m_{\alpha,F}(x)$. $\qquad\square$

Definition 5.2.4

The polynomial $m_{\alpha,F}(x)$ is called the *minimal polynomial* for α over F. The *degree* of the algebraic element α over F is the degree $\deg m_{\alpha,F}(x)$.

The proof of Theorem 5.1.10 already established the following proposition but we restate it in this context to make the connection explicit.

Proposition 5.2.5

Let α be algebraic over F. Then $F(\alpha) \cong F[x]/(m_{\alpha,F}(x))$ and $[F(\alpha) : F] = \deg m_{\alpha,F}(x)$.

This proposition illustrates the reason for using the term "degree" as opposed to just "dimension" for the quantity $[F(\alpha) : F]$.

Example 5.2.6. Consider the element $\alpha = \sqrt[3]{7}$ over $\mathbb{Q}$. It is a root of $x^3 - 7$. By Eisenstein's Criterion, this polynomial is irreducible in $\mathbb{Z}[x]$ and hence by Gauss' Lemma it is irreducible in $\mathbb{Q}[x]$. Therefore the minimal polynomial of

$\alpha = \sqrt[3]{7}$ is $m_{\alpha,\mathbb{Q}}(x) = x^3 - 7$. We could have also used the Rational Root Theorem (Proposition 4.5.11) to tell that $x^3 - 7$ does not have a rational root. Since $x^3 - 7$ is a cubic without a root, we deduce that it is irreducible. △

Example 5.2.7. Consider the element $\alpha = \sqrt{2} + \sqrt{3} \in \mathbb{C}$. We determine the degree and the minimal polynomial over $\mathbb{Q}$. Note that $\alpha^2 = 2 + 2\sqrt{6} + 3$ so then $\alpha^2 - 5 = 2\sqrt{6}$. Hence,

$$(\alpha^2 - 5)^2 = 24 \implies \alpha^4 - 10\alpha^2 + 25 = 24 \implies \alpha^4 - 10\alpha^2 + 1 = 0.$$

So α is a root of $p(x) = x^4 - 10x^2 + 1$. We do not yet know if $p(x)$ is the minimal polynomial of α, since we have not checked if it is irreducible. By the Rational Root Theorem, and since neither 1 nor -1 is a root of $p(x)$, then $p(x)$ has not roots and hence no linear factors. If $p(x)$ is reducible, then it must be the product of two quadratic polynomials. Furthermore, without loss of generality, we can assume the polynomials are monic. Furthermore, by Gauss' Lemma, if $p(x)$ factors over $\mathbb{Q}$, then it must factor over $\mathbb{Z}$. Hence, we see that if $p(x)$ factors, then there are two cases:

$$p(x) = (x^2 + ax + 1)(x^2 + bx + 1) \qquad \text{or} \qquad p(x) = (x^2 + cx - 1)(x^2 + dx - 1).$$

These two cases require respectively

$$\begin{cases} a + b = 0 \\ ab + 2 = -10 \end{cases} \quad \text{or} \quad \begin{cases} c + d = 0 \\ cd - 2 = -10. \end{cases}$$

The first case leads to $(a, b) = \pm(2\sqrt{3}, -2\sqrt{3})$. The second possible factorization leads to coefficients of $(c, d) = \pm(2\sqrt{2}, -2\sqrt{2})$. None of these values are in $\mathbb{Q}$ so $p(x)$ is irreducible and $m_{\alpha,\mathbb{Q}}(x) = x^4 - 10x^2 + 1$. We deduce that the degree of $\sqrt{2} + \sqrt{3}$ over $\mathbb{Q}$ is 4.

Now consider the field $L = \mathbb{Q}(\sqrt{2})$. Assume that $\sqrt{3}$ is an element of $\mathbb{Q}(\sqrt{2})$. Then there exist rational numbers r and s such that $\sqrt{3} = r + s\sqrt{2}$. Squaring both sides and then rearranging gives $\sqrt{2} = (3 - a^2 - 2b^2)/(2ab)$. This is a contradiction because we know that $\sqrt{2}$ is not a rational number. Consequently, $\sqrt{3} \notin \mathbb{Q}(\sqrt{2})$, and thus $\alpha \notin \mathbb{Q}(\sqrt{2})$ so $\deg m_{\alpha,L}(x) > 1$. From our previous calculation, we see that in $L[x]$,

$$m_{\alpha,\mathbb{Q}}(x) = (x^2 - 2\sqrt{2}x - 1)(x^2 + 2\sqrt{2}x - 1).$$

Hence, α is a root of one of those two quadratics. By direct observation, we find that $m_{\alpha,L}(x) = x^2 - 2\sqrt{2}x - 1$.

To take a different approach in looking for $m_{\alpha,K}(x)$, where $K = \mathbb{Q}(\sqrt{3})$, notice that $\alpha - \sqrt{3} = \sqrt{2}$. After squaring, $\alpha^2 - 2\sqrt{3}\alpha + 3 = 2$ so α is a root of $x^2 - 2\sqrt{3}x + 1$. Since $\alpha \notin \mathbb{Q}(\sqrt{3})$, then α is not a root of a degree-1 polynomial so we must have $m_{\alpha,K}(x) = x^2 - 2\sqrt{3} + 1$. Again, we see that $m_{\alpha,K}(x)$ divides $m_{\alpha,\mathbb{Q}}(x)$ since

$$m_{\alpha,\mathbb{Q}}(x) = m_{\alpha,K}(x)(x^2 + 2\sqrt{3}x + 1). \qquad \triangle$$

To recapitulate some of our results, Theorem 5.1.10 established that if $[F(\alpha) : F]$ is finite, then α is algebraic. Conversely, if $[F(\alpha) : F] = n$ is finite, then for some n, the set $\{1, \alpha, \alpha^2, \ldots, \alpha^n\}$ is linearly dependent, so there exist nonzero $c_i \in F$ such that

$$c_n \alpha^n + \cdots + c_1 \alpha + c_0 = 0.$$

Thus, α is algebraic.

5.2.2 Transcendental Numbers (Optional)

The set of transcendental numbers in $\mathbb{C}$ are all numbers that are not solutions to algebraic equations with rational coefficients. At first pass, it is hard to imagine such numbers.

For a simple example of a transcendental element, consider the field of rational expressions $F(x)$ in the variable x over the field F. In this notation, it is understood that the variable is not the root of any polynomial equation. Hence, $x \in F(x)$ is transcendental over F. Though easy to understand, this example feels artificial since x is not a number and $\mathbb{Q}(x)$ is not a subfield of $\mathbb{C}$.

In general, it is hard to show that a given complex number is transcendental. In 1882, Lindemann proved that π is transcendental.[30] This is tantamount to saying that $[\mathbb{Q}(\pi) : \mathbb{Q}]$ is infinite. Since $\pi \in \mathbb{R}$, then $\mathbb{Q}(\pi)$ is a subfield of $\mathbb{R}$. Furthermore, $\mathbb{Q}[\pi]$ is a strict subring of $\mathbb{Q}(\pi)$. By Theorem 5.3.1, we deduce that $[\mathbb{R} : \mathbb{Q}]$ is infinite as well. The extension of $\mathbb{Q}(\pi)/\mathbb{Q}$ is an example of an extension that is finitely generated but of infinite degree.

Among the list of 23 Hilbert Problems, which David Hilbert posed to the mathematical community in 1900, the seventh problem asked: If a is algebraic with $a \neq 0, 1$ and if b is irrational, then is a^b transcendental? The Gelfond-Schneider Theorem, proved independently by the namesakes in 1934, answered Hilbert's seventh problem in the affirmative.[14] For example, $7^{\sqrt{3}}$ is transcendental.

The following theorem, Liouville's Theorem on Diophantine approximation, gives a strategy to find some transcendental numbers.

> **Theorem 5.2.8 (Liouville's Theorem)**
>
> Let α be an algebraic number (over $\mathbb{Q}$) of degree $n > 1$. There exists a real number $A > 0$ such that for all integers p and $q > 0$,
>
> $$\left| \alpha - \frac{p}{q} \right| > \frac{A}{q^n}.$$

Proof. Let $m_\alpha(x)$ be the minimal polynomial of α over $\mathbb{Q}$. Let c be the least common multiple of the denominators of the coefficients of $m_\alpha(x)$ and set $f(x) = cm_\alpha(x)$. Then $f(x) \in \mathbb{Z}[x]$ is a polynomial of degree n, with α as a root, with integer coefficients, and such that the greatest common divisor of the coefficients is 1.

Let δ be any positive real number less than the distance between α and any other root, namely

$$0 < \delta < \min(|\alpha - \alpha_1|, |\alpha - \alpha_2|, \ldots, |\alpha - \alpha_k|),$$

where $\alpha_1, \alpha_2, \ldots, \alpha_k$ are the roots of $f(x)$ that are different from α. Let M be the maximal value of $|f'(x)|$ over the interval $[\alpha - \delta, \alpha + \delta]$ and let A be a real number with $0 < A < \min(\delta, 1/M)$.

Let p/q be an arbitrary rational number. We consider two cases.

Case 1. Suppose $p/q \notin [\alpha - \delta, \alpha + \delta]$. Then $|\alpha - p/q| > \delta > A > \frac{A}{q^n}$.

Case 2. Now suppose that $p/q \in [\alpha - \delta, \alpha + \delta]$. By the Mean Value Theorem, there exists an $c \in [p/q, \alpha]$, such that

$$f'(c) = \frac{f(\alpha) - f(p/q)}{\alpha - p/q} = -\frac{f(p/q)}{\alpha - p/q},$$

which implies that

$$\left|\alpha - \frac{p}{q}\right| = \frac{\left|f\left(\frac{p}{q}\right)\right|}{|f'(c)|}.$$

Since $|f'(c)| \leq M$, we have $1/|f'(c)| \geq 1/M$. Hence,

$$\left|\alpha - \frac{p}{q}\right| \geq \frac{|f(p/q)|}{M} > A|f(p/q)|.$$

By the definition of δ, the only root of $f(x) \in [\alpha - \delta, \alpha + \delta]$ is α so in particular $f(p/q) \neq 0$. Since f is of degree n and since $f(x) \in \mathbb{Z}[x]$, the polynomial evaluated on the fraction satisfies $|f(p/q)| \geq 1/q^n$. Hence,

$$\left|\alpha - \frac{p}{q}\right| > A|f(p/q)| > \frac{A}{q^n},$$

and the theorem follows. $\square$

Liouville's Theorem offers a strategy to prove that some numbers are transcendental: by finding an irrational number α that violates the conclusion of the theorem for all positive n. The following proposition constructs a specific family of transcendental numbers using this strategy.

Corollary 5.2.9

Let b be a positive integer greater than 2 and let $\{a_k\}_{k\geq 1}$ be a sequence whose values are in $\{0, 1, 2, \ldots, b-1\}$. Then the series

$$\sum_{k=1}^{\infty} \frac{a_k}{b^{k!}}$$

converges to a transcendental number.

Proof. (Left as an exercise for the reader. See Exercise 5.2.15.) □

EXERCISES FOR SECTION 5.2

1. Find the minimal polynomial of $3 + 7\sqrt{5}$ over $\mathbb{Q}$.

2. Find the minimal polynomial of $5 + \sqrt[3]{2} + \sqrt[3]{2}^2$ over $\mathbb{Q}$.

3. Find the minimal polynomial of $\sqrt{10} + \sqrt{11}$ over $\mathbb{Q}$.

4. Find the minimal polynomial of $\frac{2+\sqrt{7}}{1+\sqrt{2}}$ over $\mathbb{Q}(\sqrt{2})$.

5. Determine the degree of $1 - 2\sqrt[3]{7} + 10\sqrt[3]{7}^2$ over $\mathbb{Q}$.

6. Consider the field $F = \mathbb{F}_2[x]/(x^3 + x + 1)$ with θ an element in F such that $\theta^3 + \theta + 1 = 0$. Find the minimal polynomial of $\theta^2 + 1$ over $\mathbb{F}_2$.

7. Consider the field $F = \mathbb{F}_2[x]/(x^3 + x + 1)$ with θ an element in F such that $\theta^3 + \theta + 1 = 0$. In $F[x]$, the polynomial $x + \theta$ divides $x^3 + x + 1$. Find the polynomial $q(x) \in F[x]$ such that $x^3 + x + 1 = (x + \theta)q(x)$.

8. Consider the field $F = \mathbb{F}_5[x]/(x^3 + 2x + 4)$ with θ an element in F such that $\theta^3 + 2\theta + 4 = 0$. In $F[x]$, the polynomial $x - \theta$ divides $x^3 + 2x + 4$. Find the polynomial $q(x) \in F[x]$ such that $x^3 + 2x + 4 = (x - \theta)q(x)$.

9. Let K/F be a field extension and let $\alpha \in K$. Prove that a polynomial $p(x) \in F[x]$ satisfies $p(\alpha) = 0$ if and only if $m_{\alpha,F}(x)$ divides $p(x)$ in $F[x]$.

10. (*Palindromic polynomials*) A palindromic polynomial in $\mathbb{Q}[x]$ is a polynomial

$$p(x) = a_n x^n + \cdots + a_1 x + a_0 \in \mathbb{Q}[x]$$

such that $a_i = a_{n-i}$ for all $i = 0, 1, \ldots, n$.

(a) Let $q(x) \in \mathbb{Q}[x]$ be a polynomial of degree n. Prove that $x^n q\left(x + \dfrac{1}{x}\right)$ is a palindromic polynomial of degree $2n$.

(b) Show that every palindromic polynomial $p(x) \in \mathbb{Q}[x]$ of even degree can be written $x^n q\left(x + \dfrac{1}{x}\right)$ for some $q(x) \in \mathbb{Q}[x]$.

(c) Use this to solve the equation $x^4 - 3x^3 - 2x^2 - 3x + 1 = 0$.

11. (*Even Quartic Polynomials*)

(a) Consider a complex number of the form $\sqrt{a + b\sqrt{c}}$, where $a, b, c \in \mathbb{Q}$. Prove that $\sqrt{a + b\sqrt{c}}$ is the root of an even (all powers are even) quartic polynomial.

(b) Prove that the roots of any even quartic polynomial are of the form $\pm\sqrt{a\pm b\sqrt{c}}$ with $a, b, c \in \mathbb{Q}$.

(c) Let $\alpha = \sqrt{a+b\sqrt{c}}$. Show that all the roots of $m_{\alpha,\mathbb{Q}}(x)$ are in $\mathbb{Q}(\alpha)$ if and only if $\sqrt{a^2 - cb^2} \in \mathbb{Q}(\alpha)$.

12. Let F be a field, let K/F be a field extension with $\alpha \in K$ an algebraic element over F. Consider the linear transformation $f_\alpha : K \to K$ defined by $f_\alpha(x) = \alpha x$ in Exercise 5.1.16.

(a) Prove that α is an eigenvalue of f_α.

(b) Deduce that α is a root of the characteristic polynomial for the linear transformation f_α.

(c) Deduce that $m_{\alpha,F}(x)$ divides the characteristic polynomials of f_α.

13. Use the result of Exercise 5.2.12 to find the minimal polynomial of $\beta = 1 + \sqrt[3]{7} - \sqrt[3]{7}^2$ over $\mathbb{Q}$. [Hint: View β as an element in $\mathbb{Q}(\sqrt[3]{7})$, with respect to the basis $\{1, \sqrt[3]{7}, \sqrt[3]{7}^2\}$.]

14. Use the result of Exercise 5.2.12 to find the minimal polynomial of $\beta = 2 - 3\sqrt{2} + \sqrt[4]{2}$ over $\mathbb{Q}$. [Hint: View β as an element in $\mathbb{Q}(\sqrt[4]{2})$, with respect to the basis $\{1, \sqrt[4]{2}, \sqrt{2}, \sqrt[4]{2}^3\}$.]

15. Prove Corollary 5.2.9.

5.3   Algebraic Extensions

Section 5.2 developed the concept of algebraic elements over a field. That section emphasized properties of divisibility of polynomials. In this section, we move into field theory deeper by taking advantage of the perspective of a field extension K over F as a vector space over F.

Even though it seems like a simple concept, properties of the degree of field extensions have far ranging consequences. As we will see, these consequences become the foundation for the solutions to mathematical questions that had been unsolved for hundreds of (and some over a thousand) years prior to the discovery of field theory. We begin to develop those properties now.

5.3.1   Properties of the Degree of an Extension

The key theorem of this section is the following.

Theorem 5.3.1

Let $F \subseteq K \subseteq L$ be fields. The degrees of the extensions satisfy

$$[L : F] = [L : K][K : F].$$

Proof. Suppose first that either $[L : K]$ or $[K : F]$ is infinite. If $[K : F]$ is infinite, then there does not exist a finite basis for K as a vector space over F. Since L contains K as a subspace then $\dim L \geq \dim K$ so L does not have a finite basis over F either so $[L : F]$ is infinite. If $[L : K]$ is infinite, then L does not have a finite basis as a vector space of K. If L possessed a finite basis as a vector space over F, then since $F \subset K$, this would serve as a finite basis of L over K. Hence, if $[L : K]$ is infinite, then so is $[L : F]$.

Now suppose that $[L : K] = n$ and $[K : F] = m$. Let $\{\alpha_1, \alpha_2, \ldots, \alpha_n\}$ be a basis of L over K and let $\{\beta_1, \beta_2, \ldots, \beta_m\}$ be a basis of K over F. Every element $x \in L$ can be written as

$$x = \sum_{i=1}^{n} c_i \alpha_i$$

for $c_i \in K$. Furthermore, for each i we can write the elements c_i as

$$c_i = \sum_{j=1}^{m} d_{ij} \beta_j$$

with $d_{ij} \in F$. Then

$$x = \sum_{i=1}^{n} \sum_{j=1}^{m} d_{ij} \alpha_i \beta_j.$$

Therefore, the field L is spanned as a vector space over F by $\{\alpha_i \beta_j \mid 1 \leq i \leq n, 1 \leq j \leq m\}$.

On the other hand, consider a linear combination of the form

$$0 = \sum_{i=1}^{n} \sum_{j=1}^{m} d_{ij} \alpha_i \beta_j = \sum_{j=1}^{m} \left(\sum_{i=1}^{n} d_{ij} \alpha_i \right) \beta_j.$$

Since $\{\beta_j\}_{j=1}^{m}$ forms a basis of K over F, then for each j,

$$\sum_{i=1}^{n} d_{ij} \alpha_i = 0.$$

Since the set $\{\alpha_i\}_{i=1}^{n}$ forms a basis of L over K, it is linearly independent so $d_{ij} = 0$ for all pairs (i, j). Thus,

$$\{\alpha_i \beta_j \mid 1 \leq i \leq n, \ 1 \leq j \leq m\}$$

is a linearly independent set. Hence, it forms a basis of L over F and so

$$[L : F] = \dim_F L = nm = [L : K][K : F]. \qquad \square$$

Though Theorem 5.3.1 describes how the degree evaluates on extensions of extensions, it can also be used to deduce information about field containment as the following example shows.

Example 5.3.2. We prove that $\sqrt{7} \notin \mathbb{Q}(\sqrt[3]{7})$. The minimal polynomial of $\sqrt[3]{7}$ over $\mathbb{Q}$ is $x^3 - 7$ so $[\mathbb{Q}(\sqrt[3]{7}) : \mathbb{Q}] = 3$. Assuming that $\sqrt{7} \in \mathbb{Q}(\sqrt[3]{7})$. Then

$$\mathbb{Q} \subseteq \mathbb{Q}(\sqrt{7}) \subseteq \mathbb{Q}(\sqrt[3]{7}).$$

So by Theorem 5.3.1,

$$[\mathbb{Q}(\sqrt[3]{7}) : \mathbb{Q}(\sqrt{7})][\mathbb{Q}(\sqrt{7}) : \mathbb{Q}] = [\mathbb{Q}(\sqrt[3]{7}) : \mathbb{Q}].$$

Hence, $3 = 2[\mathbb{Q}(\sqrt[3]{7}) : \mathbb{Q}(\sqrt{7})]$, which is a contradiction because degrees of extensions are integers. Thus, $\sqrt{7} \notin \mathbb{Q}(\sqrt[3]{7})$.

Note that this reasoning is much easier than proving directly that there do not exist $a, b, c \in \mathbb{Q}$ such that $\sqrt{7} = a + b\sqrt[3]{7} + c\sqrt[3]{7}^2$. $\triangle$

Example 5.3.3. In Example 5.2.7 we found that $m_{\alpha, \mathbb{Q}}(x) = x^4 - 10x^2 + 1$ for $\alpha = \sqrt{2} + \sqrt{3}$. Hence, $[\mathbb{Q}(\alpha) : \mathbb{Q}] = 4$. Since $\mathbb{Q} \subseteq \mathbb{Q}(\sqrt{2}) \subseteq \mathbb{Q}(\alpha)$, and $[\mathbb{Q}(\sqrt{2}) : \mathbb{Q}] = 2$, then by Theorem 5.3.1, $[\mathbb{Q}(\alpha) : \mathbb{Q}(\sqrt{2})] = 2$. Hence, α is the root of an irreducible quadratic polynomial in $\mathbb{Q}(\sqrt{2})[x]$. $\triangle$

5.3.2   Algebraic Extensions

In field theory, it is common to generalize concept about elements in field extensions to properties that apply to a field extension as a whole. The following definition illustrates this.

Definition 5.3.4

> A field extension K/F is called an *algebraic extension* if every element of K is algebraic over F.

The following theorems and propositions utilize the multiplicativity of degree extensions in Theorem 5.3.1 to provide many more results about algebraic elements and extensions.

Proposition 5.3.5

> If the extension K/F is of finite degree, then every element in K is algebraic over F and hence K/F is an algebraic extension.

Proof. Suppose that K/F is a finite extension and let $\alpha \in K$. Then $F \subseteq F(\alpha) \subseteq K$. Then by Theorem 5.3.1,

$$[K : F] = [K : F(\alpha)][F(\alpha) : F].$$

In particular, $[F(\alpha) : F]$ divides $[K : F]$ and so is finite. By Theorem 5.1.10, α is algebraic over F. $\square$

We should underscore that the implication in this proposition is not an "if and only if" statement. Indeed, it is easy to construct algebraic extensions that are not of finite degree. However, in order to make examples of this precise, we need a few more facts about degrees.

Definition 5.1.11 already gave the notion of a simple field extension. That notion generalizes to finitely generated.

Definition 5.3.6

An extension K of a field F is said to be *finitely generated* if $K = F(\alpha_1, \alpha_2, \ldots, \alpha_k)$ for some elements $\alpha_1, \alpha_2, \ldots, \alpha_k \in K$.

Note that if F is a field then $F(\alpha, \beta) = F(\alpha)(\beta)$, or more precisely the field extension over F generated by α and β is equal to the field extension over $F(\alpha)$ generated by β. Of course, this also implies that $F(\alpha, \beta) = F(\beta)(\alpha)$.

Suppose now that $\alpha_1, \alpha_2, \cdots, \alpha_k$ are all algebraic over a field F and have degree $\deg \alpha_i = n_i$. We define a chain of subfields F_i by $F_0 = F$ and

$$F_i = F(\alpha_1, \alpha_2, \cdots, \alpha_i) \qquad \text{for } 1 \leq i \leq k.$$

We then have

$$F = F_0 \subseteq F_1 \subseteq F_2 \subseteq \cdots \subseteq F_k.$$

For all i we have $n_i = \deg m_{\alpha_i, F}(x)$. Furthermore, $F_i = F_{i-1}(\alpha_i)$ so by Corollary 5.2.3, $m_{\alpha_i, F_{i-1}}(x)$ divides $m_{\alpha_i, F}(x)$. Thus, $[F_i : F_{i-1}] \leq n_i$. Therefore,

$$[F(\alpha_1, \alpha_2, \ldots, \alpha_k) : F] = [F_k : F_{k-1}] \cdots [F_2 : F_1][F_1 : F_0] \leq n_k \cdots n_2 n_1.$$

This give an upper bound for $[F_k : F]$. However, Theorem 5.3.1 gives a lower bound on $[F_k : F]$ because $F \subseteq F(\alpha_i) \subseteq F_k$. This establishes the important theorem.

Theorem 5.3.7

A field extension K/F is finite if and only if K is generated by a finite number of algebraic elements over F. If these algebraic elements have degrees $n_1, n_2, \ldots, n_k$ over F, then

$$\text{lcm}(n_1, n_2, \ldots, n_k) \,\big|\, [K : F] \quad \text{and} \quad [K : F] \leq n_1 n_2 \cdots n_k.$$

It is not always easy to determine when $[K : F]$ is strictly less than $n_1 n_2 \cdots n_k$ or when it is an equality. The following examples and some of the exercises explore a variety of situations related to this theorem.

Example 5.3.8. Consider the field $K = \mathbb{Q}(\sqrt{2}, \sqrt{5})$. The elements $\sqrt{2}$ and $\sqrt{5}$ both have degree 2 over $\mathbb{Q}$. By Theorem 5.3.7, $[K : \mathbb{Q}]$ is a multiple of 2 and less than or equal to 4. Hence, $[K : \mathbb{Q}]$ can be 2 or 4. Since $\sqrt{5} \notin \mathbb{Q}(\sqrt{2})$ the $[K : \mathbb{Q}(\sqrt{2})]$ is strictly greater than 1, so $[K : \mathbb{Q}] > 2$. Consequently, $[K : \mathbb{Q}(\sqrt{2})] = 2$ and $[K : \mathbb{Q}] = 4$. $\triangle$

Example 5.3.9. Following Example 5.2.7, it is straightforward to show that $\alpha = \sqrt{2} + \sqrt{3}$ and $\beta = \sqrt{2} + \sqrt{5}$ both have degree 4 over $\mathbb{Q}$. Theorem 5.3.7 tells us that $[\mathbb{Q}(\alpha, \beta) : \mathbb{Q}]$ is a multiple of 4 but is less or equal to 16. Hence, we deduce that $[\mathbb{Q}(\alpha, \beta) : \mathbb{Q}]$ is equal to 4, 8, 12, or 16.

Now β is the root of the polynomial $x^2 - 2\sqrt{2}x - 3 \in \mathbb{Q}(\alpha)[x]$. Without knowing whether this polynomial is irreducible in $\mathbb{Q}(\alpha)[x]$, we can deduce that $[\mathbb{Q}(\alpha, \beta) : \mathbb{Q}(\alpha)] \leq 2$. Hence, $[\mathbb{Q}(\alpha, \beta) : \mathbb{Q}]$ is equal to 4 or 8.

We can prove that $\sqrt{5} \notin \mathbb{Q}(\alpha)$. Assume that for some $a, b, c, d \in \mathbb{Q}$,

$$\sqrt{5} = a + b\sqrt{2} + c\sqrt{3} + d\sqrt{6}.$$

Then squaring this expression

$$5 = a^2 + 2b^2 + 3c^2 + 6d^2 + 2ab\sqrt{2} + 2ac\sqrt{3} + 2ad\sqrt{6} + 2bc\sqrt{6}$$
$$+ 4bd\sqrt{3} + 6cd\sqrt{2}$$
$$\implies (5 - a^2 - 2b^2 - 3c^2 - 6d^2) + (2ad + 2bc)\sqrt{6} = (2ab + 6cd)\sqrt{2}$$
$$+ (2ac + 4bd)\sqrt{3}.$$

After squaring once more, we obtain an equation for $\sqrt{6}$ in terms of rational numbers. This is a contradiction since $\sqrt{6}$ cannot be expressed as a rational number. Since $\sqrt{5} \notin \mathbb{Q}(\alpha)$, then $\beta \notin \mathbb{Q}(\alpha)$ and $[\mathbb{Q}(\alpha, \beta) : \mathbb{Q}(\alpha)] > 1$. This allows us finally to deduce that $[\mathbb{Q}(\alpha, \beta) : \mathbb{Q}] = 8$. $\triangle$

Example 5.3.10. As another example, it is not hard to show that $\mathbb{Q}(\sqrt[4]{2}, \sqrt[6]{2})$ is in fact equal to the simple extension $\mathbb{Q}(\sqrt[12]{2})$. Thus,

$$[\mathbb{Q}(\sqrt[4]{2}, \sqrt[6]{2}) : \mathbb{Q}] = 12,$$

which is less than $4 \times 6 = 24$, the product of the degrees of the generating algebraic elements. This gives another example where $[K : \mathbb{Q}]$ is strictly less than the product of the degrees of the generating algebraic elements. $\triangle$

Theorem 5.3.7, along with other theorems on degrees, lead to a powerful corollary that would be rather difficult to prove directly from the definition of algebraic elements.

Corollary 5.3.11

Let α and β be two algebraic elements over a field F. Then the following elements are also algebraic:

$$\alpha + \beta, \ \alpha - \beta, \ \alpha\beta, \ \frac{\alpha}{\beta} \quad (\text{for } \beta \neq 0).$$

Proof. Suppose that α and β are algebraic over F with degrees n_1 and n_2, respectively. By Theorem 5.3.7, $[F(\alpha, \beta) : F] \leq n_1 n_2$. Let γ be $\alpha + \beta, \ \alpha - \beta,$

$\alpha\beta$, or α/β. Then $F(\gamma)$ is a subfield of $F(\alpha, \beta)$. By Theorem 5.3.1, $[F(\gamma) : F]$ divides $[F(\alpha, \beta) : F]$ so $[F(\gamma) : F] = d$ is finite. By Theorem 5.1.10, γ is the root of an irreducible polynomial of degree d in $F[x]$. Hence, γ is algebraic. $\square$

Example 5.3.12. Consider the element $\gamma = \dfrac{1 + \sqrt{2}}{1 + \sqrt{3}}$, which is in $K = \mathbb{Q}(\sqrt{2}, \sqrt{3})$. The field K has degree 4 over $\mathbb{Q}$. Thus, γ is algebraic. For completeness, we can look for the minimal polynomial of γ. We first have

$$\gamma(1 + \sqrt{3}) = (1 + \sqrt{2}) \implies \gamma - 1 = \sqrt{2} - \gamma\sqrt{3}.$$

Squaring both sides gives

$$\gamma^2 - 2\gamma + 1 = 2 - 2\sqrt{6}\gamma + 3\gamma^2 \implies 2\sqrt{6}\gamma = 2\gamma^2 + 2\gamma + 1.$$

After squaring both sides again, we deduce that γ is the root of

$$x^4 + 2x^3 - \frac{9}{2}x + \frac{1}{2}x + \frac{1}{4}.$$

This is the unique monic irreducible polynomial $m_{\gamma,\mathbb{Q}}(x)$. $\triangle$

Example 5.3.13. We illustrate the use of the propositions in the first three sections of this chapter by determining the minimal polynomial of $\alpha = \sqrt{3} + \sqrt[3]{7}$ and also by proving that $\mathbb{Q}(\sqrt{3}, \sqrt[3]{7}) = \mathbb{Q}(\sqrt{3} + \sqrt[3]{7})$ in two different ways. the first method uses theorems from Section 5.1, while the second method uses Theorems from this section.

It is clear that $\alpha = \sqrt{3} + \sqrt[3]{7} \in \mathbb{Q}(\sqrt{3}, \sqrt[3]{7})$. Consequently, $\mathbb{Q}(\alpha)$, being defined as the smallest field containing $\mathbb{Q}$ and α must be contained in $\mathbb{Q}(\sqrt{3}, \sqrt[3]{7})$. Thus $\mathbb{Q}(\alpha) \subseteq \mathbb{Q}(\sqrt{3}, \sqrt[3]{7})$. Both methods involve proving the reverse inclusion.

Cubing $\alpha - \sqrt{3} = \sqrt[3]{7}$, we get $\alpha^3 - 3\sqrt{3}\alpha^2 + 9\alpha - 3\sqrt{3} = 7$. This leads to $\alpha^3 + 9\alpha - 7 = 3\sqrt{3}(\alpha^2 + 1)$ and after squaring we get

$$\alpha^6 - 9\alpha^4 - 14\alpha^3 + 27\alpha^2 - 126\alpha - 27 = 0.$$

Thus α is a root of $p(x) = x^6 - 9x^4 - 14x^3 + 27x^2 - 126x - 27$. However, Eisenstein's Criterion does not help use determine if this polynomial is irreducible and other methods require considerable amount of work. All that we conclude at this point is that the minimal polynomial $m_{\alpha,\mathbb{Q}}(x)$ divides $p(x)$.

Method 1: The method uses strategies from Section 5.1. By the proof methods for Theorem 5.1.10, $\{1, \sqrt[3]{7}, \sqrt[3]{7}^2\}$ is a basis of $\mathbb{Q}(\sqrt[3]{7})$ over $\mathbb{Q}$ and also $\{1, \sqrt{3}\}$ spans $\mathbb{Q}(\sqrt{3}, \sqrt[3]{7})$ over $\mathbb{Q}(\sqrt[3]{7})$. Consequently, the set

$$\mathcal{B} = \{1, \sqrt{3}, \sqrt[3]{7}, \sqrt{3}\sqrt[3]{7}, \sqrt[3]{7^2}, \sqrt{3}\sqrt[3]{7^2}\}$$

spans $\mathbb{Q}(\sqrt{3}, \sqrt[3]{7})$ over $\mathbb{Q}$. (Note: The set $\mathcal{B}$ is a basis of $\mathbb{Q}(\sqrt{3}, \sqrt[3]{7})$, but for what we do here we only need to assume or even hope that it spans $\mathbb{Q}(\sqrt{3}, \sqrt[3]{7})$.)

By taking successive powers, we can calculate that

$$\alpha = \sqrt{3} + \sqrt[3]{7}$$
$$\alpha^2 = 3 + 2\sqrt{3}\sqrt[3]{7} + \sqrt[3]{7^2}$$
$$\alpha^3 = 7 + 3\sqrt{3} + 9\sqrt[3]{7} + 3\sqrt{3}\sqrt[3]{7^2}$$
$$\alpha^4 = 9 + 28\sqrt{3} + 7\sqrt[3]{7} + 12\sqrt{3}\sqrt[3]{7} + 18\sqrt[3]{7^2}$$
$$\alpha^5 = 210 + 9\sqrt{3} + 45\sqrt[3]{7} + 35\sqrt{3}\sqrt[3]{7} + 7\sqrt[3]{7^2} + 30\sqrt{3}\sqrt[3]{7^2}.$$

We now try to decide if $\sqrt{3} \in \mathbb{Q}(\alpha)$. To do so, we consider the identity $\sqrt{3} = c_0 + c_1\alpha + c_2\alpha^2 + c_3\alpha^3 + c_4\alpha^4 + c_5\alpha^5$. This means that we want to solve the linear system (with respect to the basis $\mathcal{B}$)

$$\begin{pmatrix} 1 & 0 & 3 & 7 & 9 & 210 \\ 0 & 1 & 0 & 3 & 28 & 9 \\ 0 & 1 & 0 & 9 & 7 & 45 \\ 0 & 0 & 2 & 0 & 12 & 35 \\ 0 & 0 & 1 & 0 & 18 & 7 \\ 0 & 0 & 0 & 3 & 0 & 30 \end{pmatrix} \begin{pmatrix} c_0 \\ c_1 \\ c_2 \\ c_3 \\ c_4 \\ c_5 \end{pmatrix} = \begin{pmatrix} 0 \\ 1 \\ 0 \\ 0 \\ 0 \\ 0 \end{pmatrix}. \tag{5.3}$$

The reduce row echelon form of this system gives

$$c_0 = -\frac{637}{339}, \quad c_1 = \frac{311}{339}, \quad c_2 = -\frac{182}{339}, \quad c_3 = -\frac{80}{339}, \quad c_4 = \frac{7}{339}, \quad c_5 = \frac{8}{339}.$$

This shows that $\sqrt{3} \in \mathbb{Q}(\alpha)$. Then $\sqrt[3]{7} = \alpha - \sqrt{3}$, so $\sqrt[3]{7} \in \mathbb{Q}(\alpha)$ as well. Since $\mathbb{Q}(\sqrt{3}, \sqrt[3]{7})$ is the smallest field containing $\sqrt{3}$ and $\sqrt[3]{7}$, and since $\sqrt{3}, \sqrt[3]{7} \in \mathbb{Q}(\alpha)$, then $\mathbb{Q}(\sqrt{3}, \sqrt[3]{7}) \subseteq \mathbb{Q}(\alpha)$. By mutual containment, we conclude that $\mathbb{Q}(\sqrt{3}, \sqrt[3]{7}) = \mathbb{Q}(\alpha)$.

Though we he proven that $\mathbb{Q}(\sqrt{3}, \sqrt[3]{7}) = \mathbb{Q}(\alpha)$, we have not established that $\mathcal{B}$ is a basis of $\mathbb{Q}(\sqrt{3}, \sqrt[3]{7})$. So we do not yet know if $[\mathbb{Q}(\alpha) : \mathbb{Q}] = 6$. We can do this as follows. Assume that

$$a + b\sqrt{3} + c\sqrt[3]{7} + d\sqrt{3}\sqrt[3]{7} + e\sqrt[3]{7}^2 + f\sqrt{3}\sqrt[3]{7}^2 = 0$$

for some constants $a, b, c, d, e, f \in Q$ not all 0. Then

$$a + c\sqrt[3]{7} + e\sqrt[3]{7}^2 = -\sqrt{3}(b + d\sqrt[3]{7} + f\sqrt[3]{7}^2).$$

After squaring both sides, we deduce that $\sqrt[3]{7}$ solves a quadratic equation in rational coefficients. However, this is a contradiction because of Theorem 5.1.12 applied to the irreducible polynomial $x^3 - 7$. Hence $\mathcal{B}$ is a basis and we can now conclude that $[\mathbb{Q}(\sqrt{3}, \sqrt[3]{7}) : \mathbb{Q}] = 6$. Hence, $[\mathbb{Q}(\alpha) : \mathbb{Q}] = 6$ and by Theorem 5.1.10, we know that α is the root of an irreducible polynomial of degree 6.

Method 2: With this section, we can approach the problem from a different direction.

- $\sqrt{3}$ is a root of $x^2 - 3$. By Eisenstein's Criterion, this polynomial is irreducible in $\mathbb{Z}[x]$. By Gauss' Lemma, this is also irreducible in $\mathbb{Q}[x]$. Hence, $x^2 - 3$ is the minimal polynomial of $\sqrt{3}$ so $[\mathbb{Q}(\sqrt{3}) : \mathbb{Q}] = 2$.

- $\sqrt[3]{7}$ is a root of $x^3 - 7$. By Eisenstein's Criterion, this polynomial is irreducible in $\mathbb{Z}[x]$. By Gauss' Lemma, this is also irreducible in $\mathbb{Q}[x]$. Hence, $x^3 - 7$ is the minimal polynomial of $\sqrt[3]{7}$ so $[\mathbb{Q}(\sqrt[3]{7}) : \mathbb{Q}] = 3$.

- By Theorem 5.3.7, we deduce that $[\mathbb{Q}(\sqrt{3}, \sqrt[3]{7}) : \mathbb{Q}] = 6$.

- We point out that since $\mathbb{Q} \subseteq \mathbb{Q}(\sqrt{3}) \subseteq \mathbb{Q}(\sqrt{3}, \sqrt[3]{7})$, then by Theorem 5.3.1, we have

$$6 = [\mathbb{Q}(\sqrt{3}, \sqrt[3]{7}) : \mathbb{Q}] = [\mathbb{Q}(\sqrt{3}, \sqrt[3]{7}) : \mathbb{Q}(\sqrt{3})][\mathbb{Q}(\sqrt{3}) : \mathbb{Q}].$$

We deduce that $[\mathbb{Q}(\sqrt{3}, \sqrt[3]{7}) : \mathbb{Q}(\sqrt{3})] = 3$.

- By Theorem 5.3.1, considering the chain of fields $\mathbb{Q} \subseteq \mathbb{Q}(\alpha) \subseteq \mathbb{Q}(\sqrt{3}, \sqrt[3]{7})$, we deduce that

$$6 = [\mathbb{Q}(\sqrt{3}, \sqrt[3]{7}) : \mathbb{Q}] = [\mathbb{Q}(\sqrt{3}, \sqrt[3]{7}) : \mathbb{Q}(\alpha)][\mathbb{Q}(\alpha) : \mathbb{Q}].$$

From this we see that $[\mathbb{Q}(\alpha) : \mathbb{Q}]$ divides 6.

- We now eliminate cases 1, 2, and 3.

  - $[\mathbb{Q}(\alpha) : \mathbb{Q}] = 1$ if and only if α is a rational number. If we assume $\alpha \in \mathbb{Q}$, then $7 = (\alpha - \sqrt{3})^3$. Thus $7 = \alpha^3 - 3\sqrt{3}\alpha^2 + 9\alpha - 3\sqrt{3}$ and so $\sqrt{3} = (\alpha^3 + 9\alpha - 7)/(3\alpha^3 + 3)$. This means that $\sqrt{3} \in \mathbb{Q}$, which is a contradiction.

  - $[\mathbb{Q}(\alpha) : \mathbb{Q}] = 2$ if and only if α is the root of some quadratic polynomial in $\mathbb{Q}[x]$, say $x^2 + bx + c$. Hence,

  $$3 + 2\sqrt{3}\sqrt[3]{7} + \sqrt[3]{7^2} + b\sqrt{3} + b\sqrt[3]{7} + c = 0$$
  $$\Longleftrightarrow (\sqrt[3]{7})^2 + (2\sqrt{3} + b)\sqrt[3]{7} + (3 + b\sqrt{3} + c) = 0.$$

  This is a contradiction because it would mean that $\sqrt[3]{7}$ is the root of a quadratic polynomial over $\mathbb{Q}(\sqrt{3})[x]$. This would imply that $[\mathbb{Q}(\sqrt{3}, \sqrt[3]{7}) : \mathbb{Q}(\sqrt{3})] = 2$, even though we already saw that this must be 3.

  - $[\mathbb{Q}(\alpha) : \mathbb{Q}] = 3$ if and only if α is the root of some cubic polynomial in $\mathbb{Q}[x]$, say $x^3 + bx^2 + cx + d$. Once we expand out

  $$(\sqrt{3} + \sqrt[3]{7})^3 + b(\sqrt{3} + \sqrt[3]{7})^2 + c(\sqrt{3} + \sqrt[3]{7}) + d = 0,$$

  we see that it again implies that $\sqrt[3]{7}$ is the root of a quadratic polynomial over $\mathbb{Q}(\sqrt{3})$, which is a contradiction.

- The only case left is $[\mathbb{Q}(\alpha) : \mathbb{Q}] = 6$. Using concepts from linear algebra, since $\mathbb{Q}(\alpha)$ is a subspace of $\mathbb{Q}(\sqrt{3}, \sqrt[3]{7})$ but has the same dimension of $\mathbb{Q}(\sqrt{3}, \sqrt[3]{7})$ over $\mathbb{Q}$, then $\mathbb{Q}(\alpha) = \mathbb{Q}(\sqrt{3}, \sqrt[3]{7})$.

- From Theorem 5.1.10, we deduce that α is the root of some irreducible polynomial of degree 6.

At this point, whether we used Method 1 or Method 2, we can now deduce that since $m_{\alpha,\mathbb{Q}}(x)$ is monic, degree 6 and divides $p(x)$, then $m_{\alpha,\mathbb{Q}}(x) = p(x)$. In particular, $p(x) = x^6 - 9x^4 - 14x^3 + 27x^2 - 126x - 27$ is irreducible.

One of the key differences between the two methods is that Method 1 requires an explicit basis, whereas Method 2 uses Theorem 5.3.7 to quickly determine the degree $[\mathbb{Q}(\alpha) : \mathbb{Q}]$ without knowing a basis. $\triangle$

Let K be a field and consider the set of subfields of K. Consider the relation "is an extension of finite degree of" on the set of subfields of K. Theorem 5.3.1 proves that this relation is transitive. Since the relation of algebraic extension also satisfies antisymmetry and reflexivity, we deduce that it is a partial order on the subfields of K.

We can now mention a few field extensions that are algebraic and of infinite degree. For example, it is possible to show that $\mathbb{Q}(\sqrt{2}, \sqrt{3}, \sqrt{5}, \sqrt{7}, \ldots)$ is an algebraic extension of infinite degree over $\mathbb{Q}$. (See Exercise 5.3.10.) This also provides an example of a field extension that is not finitely generated but still algebraic.

Another important example of an algebraic extension of infinite degree is the subfield of $\mathbb{C}$ of all *algebraic numbers*, denoted $\overline{\mathbb{Q}}$, i.e., complex numbers that are roots of polynomials in $\mathbb{Q}[x]$. The set of algebraic numbers forms a field by virtue of Corollary 5.3.11. It is easy to see that $\overline{\mathbb{Q}}$ does not have finite degree over $\mathbb{Q}$ since, for all positive integers, it contains the field $\mathbb{Q}(\sqrt[n]{2})$ which has degree n over $\mathbb{Q}$. Thus, $[\overline{\mathbb{Q}} : \mathbb{Q}]$ is greater than every positive integer n so this degree is infinite.

An interesting property about algebraic numbers is that $\overline{\mathbb{Q}}$ is a countable set. (See Exercise 5.3.14.) To many who first encounter it, the result that $\mathbb{Q}$ is countable feels counterintuitive since the set of reals is uncountable and every real number can be approximated to arbitrary precision by rational numbers. As we think of all the possibilities covered by algebraic numbers and how few numbers we know for certain to be transcendental, it seems even more counterintuitive that $\overline{\mathbb{Q}}$ is a countable subset of the uncountable set $\mathbb{C}$.

5.3.3   The Lattice of Algebraic Extensions

The previous section established many properties about the degree of a finite field extension. We already pointed out that algebraic extensions need not have finite degree. The goal of this section is to study the relation of algebraic extensions regardless of whether the extensions are finite. The main theorem (Theorem 5.3.19) summarizes the section by establishing that the relation of algebraic extensions is a lattice.

Let L be an extension of a field F. Let $\mathrm{Alg}(L/F)$ be the set of all subfields of L that are algebraic extensions of F.

Proposition 5.3.14

Define the relation $\preccurlyeq$ on $\mathrm{Alg}(L/F)$ by $K_1 \preccurlyeq K_2$ if K_2 is an algebraic extension of K_1. Then $\preccurlyeq$ is a partial order on $\mathrm{Alg}(L/F)$.

Proof. For all $K \in \mathrm{Alg}(L/F)$, it is obvious that K is an algebraic extension of itself. Hence, the relation $\preccurlyeq$ is reflexive. If $K_1 \preccurlyeq K_2$ and $K_2 \preccurlyeq K_1$, then $K_1 \subseteq K_2$ and $K_2 \subseteq K_1$ so $K_1 = K_2$. Hence, $\preccurlyeq$ is antisymmetric.

Suppose that K_3 is an algebraic extension of K_2, which in turn is an algebraic extension of K_1. Let $\alpha \in K_3$. Since K_3 is algebraic over K_2, α is the root of a minimal polynomial

$$m_{\alpha, K_2}(x) = c_n x^n + \cdots + c_1 x + c_0 \in K_2[x].$$

Since K_2 is algebraic over K_1, then each coefficient c_i is algebraic over K_1. By Theorem 5.3.7, the degree $[K_1(c_0, c_1, \ldots, c_n) : K_1]$ divides

$$(\deg_{K_1} c_0)(\deg_{K_1} c_1) \cdots (\deg_{K_1} c_n).$$

In particular, this field is finite. Furthermore, by Theorem 5.3.1,

$$[K_1(\alpha) : K_1] = [K_1(\alpha) : K_1(c_0, c_1, \ldots, c_n)][K_1(c_0, c_1, \ldots, c_n) : K_1]$$
$$= n[K_1(c_0, c_1, \ldots, c_n) : K_1].$$

Hence, $K_1(\alpha)$ is a finite extension of K_1. Thus, K_3 is an algebraic extension of K_1 and $\preccurlyeq$ is transitive. $\qquad\square$

For any two subfields K_1 and K_2 of L that are algebraic extensions of F, the intersection $K_1 \cap K_2$ is again subfield of L that is an algebraic extension of F. It is the greatest lower bound between K_1 and K_2 with respect to the partial order of "algebraic extension of."

Let L be an extension of a field F and let K_1, K_2 be two subfields of L that are algebraic extensions of F. It is easy to show that the intersection $K_1 \cap K_2$ is a field extension of F. In general, the union $K_1 \cup K_2$ is not another field extension. In order to show that the partial order of algebraic extensions as a least upper bound for any two algebraic extensions of F, we need to introduce the composite of fields.

Definition 5.3.15

Let K_1 and K_2 be two subfields of any field E. Then the *composite field* $K_1 K_2$ is the smallest subfield of E (by inclusion) that includes both K_1 and K_2.

Proposition 5.3.16

Let K_1 and K_2 be two finite extensions of a field F, both contained in a field extension L. Then $[K_1 K_2 : F] \leq [K_1 : F][K_2 : F]$.

Proof. (Left as an exercise for the reader. See Exercise 5.3.3.)   □

Note that if K is a field extension of F and $\alpha, \beta \in K$, then the composite field is $F(\alpha)F(\beta) = F(\alpha, \beta)$. However, the construct of a composite field can be far more general than the composite of two simple field extensions.

We characterize elements in the composite of two subfields of a field L.

Proposition 5.3.17

Let K_1 and K_2 be two subfields of a field L. Let $\gamma \in K_1K_2$. Then

$$\gamma = \frac{\alpha_1\beta_1 + \alpha_2\beta_2 + \cdots + \alpha_m\beta_m}{a_1b_1 + a_2b_2 + \cdots + a_nb_n}. \tag{5.4}$$

for some integers m and n and for some elements $\alpha_1, \alpha_2, \ldots, \alpha_m \in K_1$, $a_1, a_2, \ldots, a_n \in K_1$, $\beta_1, \beta_2, \ldots, \beta_m \in K_2$ and $b_1, b_2, \ldots, b_n \in K_2$.

Proof. Let S be the set of all elements in L of the form (5.4), assuming the denominator is nonzero. For all $\alpha \in K_1$, we have $\alpha = (\alpha \cdot 1)/(1 \cdot 1) \in S$. Hence, $K_1 \subseteq S$ and similarly $K_2 \subseteq S$.

Since K_1K_2 contains K_1 and K_2 and is a field, $S \subseteq K_1K_2$. Performing distributivity on a product of linear combinations

$$(\alpha_1\beta_1 + \alpha_2\beta_2 + \cdots + \alpha_m\beta_m)(a_1b_1 + a_2b_2 + \cdots + a_nb_n)$$

produces a linear combination of products of elements from K_1 and K_2. Consider the difference of two elements in S. By performing cross-multiplication and distributivity on products of linear combinations, we recover another expression of the form (5.4). Hence, by the One-Step Subgroup Criterion, $(S, +)$ is a subgroup of $(L, +)$. Similarly, the division of two nonzero elements in S is again an element in S. Thus, S is a subring of L, containing the identity and closed under taking inverses and so S is a subfield of L. Since K_1K_2 is the smallest subfield of L containing both K_1 and K_2, then $K_1K_2 \subseteq S$. Consequently, $S = K_1K_2$.   □

Proposition 5.3.18

Let L be an extension of a field F and let K_1 and K_2 be two subfields of L that are algebraic over F. Then K_1K_2 is another algebraic extension of F.

Proof. Let $\gamma \in K_1K_2$. Then, by Proposition 5.3.17, γ is equal to an expression of the form (5.4). However, by a repeated application of Corollary 5.3.11, since $\alpha_i, \beta_i, a_j, b_j$ with $i = 1, 2, \ldots, m$ and $j = 1, 2, \ldots, n$ are algebraic, then γ is also algebraic. Thus, K_1K_2 is an algebraic extension of F.   □

We summarize the results of this section into a concise theorem.

Theorem 5.3.19

Let L be an extension of a field F. The relation "is an algebraic extension of" on the set of algebraic extensions of F in L is a partial order. Furthermore, this partial order is a lattice such that for any two algebraic extensions K_1 and K_2 of F in L, the least upper bound is $K_1 K_2$ and the greatest lower bound is $K_1 \cap K_2$.

For any two subfields K_1 and K_2 of L that are algebraic extensions of F, the Hasse diagram of the lattice $\mathrm{Alg}(L/F)$ includes the following subdiagram illustrating the least upper bound $K_1 K_2$ and the greatest lower bound $K_1 \cap K_2$.

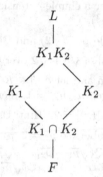

5.3.4 Useful CAS Commands

The Number Field type in Sage allows us to create extensions of $\mathbb{Q}$ that are not given as simple extensions. Furthermore, we can create towers of extensions and naturally performs computations in these field extensions.

──────────── Sage ────────────

```
sage: F.<a,b,c>=NumberField([x^2-2,x^3-3,x^4-5])
sage: (a+b-c)^3
(3*b^2 - 6*c*b + 3*c^2 + 2)*a - 3*c*b^2 + (3*c^2 + 6)*b
  - c^3 - 6*c + 3
sage:
sage: K.<cr3> = NumberField(x^3-3)
sage: L.<al> = K.extension(x^4-4*x^2-17)
sage: (al+cr3)^5
(10*cr3^2 + 4)*al^3 + (20*cr3 + 30)*al^2 + (15*cr3 + 17)*al
  + 3*cr3^2 + 85*cr3
```

──────────────────────────────

In these calculations, SAGE organizes the bases according to the proof of Theorem 5.3.1.

Exercises for Section 5.3

1. Let L be an algebraic extensions of a field F and let K_1, K_2 be two subfields of L containing F. Prove that $K_1 \cap K_2$ is a field extension of F and that $[K_1 \cap K_2 : F]$ divides $\gcd([K_1 : F], [K_2 : F])$.

2. Let $F = \mathbb{Q}(\sqrt{r_1}, \sqrt{r_2}, \ldots, \sqrt{r_n})$ where $r_i \in \mathbb{Q}$.
   (a) Prove that $[F : \mathbb{Q}] = 2^k$ for some nonnegative k.
   (b) Deduce that $\sqrt[3]{7} \notin F$.

3. Prove Proposition 5.3.16.

4. Suppose that L/F is a field extension of degree p, with p prime. Prove that any subfield K of L containing F is either L or F.

5. Let F be a field and consider a simple extension $F(\alpha)$ such that $[F(\alpha) : F]$ is odd. Prove that $F(\alpha) = F(\alpha^2)$.

6. Prove that the composite field of $\mathbb{Q}(\sqrt{2})$ and $\mathbb{Q}(\sqrt[3]{3})$ is $\mathbb{Q}(\sqrt[6]{72})$.

7. Find the minimal polynomial of $\sqrt{5} + \sqrt[3]{2}$ over $\mathbb{Q}$.

8. Let K/F be a field extension and let $\alpha, \beta \in K$ with degrees n_1 and n_2 respectively over F. Show that if $\gcd(n_1, n_2) = 1$, then $[F(\alpha, \beta) : F] = n_1 n_2$.

9. Prove that $[\mathbb{Q}(x, \sqrt{1-x^2}) : \mathbb{Q}(x)] = 2$. [Hint: Use the fact that $\mathbb{Q}[x]$ is a UFD.]

10. Let p_n be the nth prime number (so $p_1 = 2$, $p_2 = 3$, $p_3 = 5$, and so on).
    (a) Prove that $\sqrt{p_n} \notin \mathbb{Q}(\sqrt{p_1}, \sqrt{p_2}, \ldots, \sqrt{p_{n-1}})$. [Hint: Use a proof by induction on n. Use $F_{-1} = \mathbb{Q}(\sqrt{p_1}, \sqrt{p_2}, \ldots, \sqrt{p_{n-1}})$ and $F_0 = \mathbb{Q}(\sqrt{p_1}, \sqrt{p_2}, \ldots, \sqrt{p_n})$.]
    (b) Deduce that $[\mathbb{Q}(\sqrt{p_1}, \sqrt{p_2}, \ldots, \sqrt{p_n}) : \mathbb{Q}] = 2^n$ for all positive integers n.
    (c) Deduce that $\mathbb{Q}(\sqrt{p} \mid p \geq 2$ is a prime$)$ is an algebraic extension of $\mathbb{Q}$ of infinite degree.
    [This Exercise is motivated by [28].]

11. Show that $\sqrt[3]{2} \notin \mathbb{Q}(\sqrt[3]{3})$. Prove also that $[\mathbb{Q}(\sqrt[3]{2}, \sqrt[3]{3}) : \mathbb{Q}] = 9$.

12. Let $S = \{\sqrt[n]{2} \mid n \in \mathbb{Z} \text{ with } n \geq 2\}$. Prove that $\sqrt{3} \notin \mathbb{Q}[S]$.

13. Prove that $\cos(k\pi/n)$ is algebraic for all positive integers k and n.

14. In this exercise, we prove that the set of algebraic numbers $\overline{\mathbb{Q}}$ is countable. Recall that $\mathbb{Q}$ is countable and that if A and B are countable, then $A \times B$ is countable.
    (a) Prove that $A_1, A_2, \ldots, A_n$ are n countable sets, then the Cartesian product $A_1 \times A_2 \times \cdots \times A_n$ is a countable set.
    (b) Prove that the set of polynomials $\mathbb{Q}[x]$ is a countable set.
    (c) Since algebraic numbers consist of all the roots polynomials in $\mathbb{Q}[x]$, deduce with a proper proof that $\overline{\mathbb{Q}}$ is countable.

15. Prove that the field of fractions of the algebraic integers (introduced in Section 4.7) is the field of algebraic numbers $\overline{\mathbb{Q}}$.

5.4 Solving Cubic and Quartic Equations

As early as middle school and certainly in high school, students encounter the quadratic formula. The solutions to the generic quadratic equation $ax^2 + bx + c = 0$ with $a \neq 0$, are

$$x = \frac{-b \pm \sqrt{b^2 - 4ac}}{2a}.$$

The original interest in solving quadratic equations came from applications to geometry. There is historical evidence that as early as 1600 B.C.E. Babylonian scholars knew the strategy of completing the square to solve a quadratic equation [5]. Solutions to the quadratic equation appeared in a variety of forms throughout history.

Subsequent generations of scholars attempted to find formulas for the roots of equations of higher degree. One can approach the problem of finding solutions in a variety of ways: radical expressions, trigonometric sums, hypergeometric functions, continued fractions, etc. However, historically, by a formula for the roots of a polynomial equation, people understood an expression in terms of radicals of algebraic combinations of the coefficients of the generic polynomial. For centuries, mathematicians only made progress on particular cases. Then, in 1545, Cardano published formula solutions for both the cubic and the quartic equation in *Ars Magna*. Though Cardano often receives the credit, Tartaglia (a colleague) and Ferrari (a student) contributed significantly.

We propose to look at some formulas for the solutions to the cubic and the quartic equation and discuss the merits of the approach.

Throughout this section, we assume that the polynomials are in $\mathbb{R}[x]$ but the strategies can be generalized to $\mathbb{C}[x]$.

5.4.1 The Cubic Equation

Before we develop a method to solve the cubic equation, we should look at the simple case $x^3 = a$, where $a \in \mathbb{Q}$. Obviously, this equation has one real root, namely $x = \sqrt[3]{a}$. We then have

$$x^3 - a = 0 \iff (x - \sqrt[3]{a})(x^2 + \sqrt[3]{a}x + (\sqrt[3]{a})^2) = 0. \tag{5.5}$$

Solving the quadratic factor, we find that the three roots of this cubic are $x = \sqrt[3]{a}, \sqrt[3]{a}\omega, \sqrt[3]{a}\omega^2$, where ω is the complex number

$$\omega = \frac{-1 + i\sqrt{3}}{2}.$$

The complex number ω is called a (primitive) third *root of unity* since it satisfies $\omega^3 = 1$ and since ω does not solve the equation $x^n - 1$ for any n less than 3. Note that $\omega^2 = \omega^{-1} = \bar{\omega} = -1 - \omega$.

Without loss of generality, we can suppose that the cubic equation is already monic; i.e., one has already divided the polynomial equation by the leading coefficient. Therefore, we propose to solve

$$x^3 + ax^2 + bx + c = 0. \tag{5.6}$$

As a first step, we change the variables by setting $x = y - \frac{a}{3}$. This shift of variables has a similar effect as completing the square in solving the quadratic formula. We have

$$\left(y - \frac{a}{3}\right)^3 + a\left(y - \frac{a}{3}\right)^2 = y^3 - ay^2 + \frac{3a^2}{3^2} + \frac{a^3}{27} + ay^2 - \frac{2a^2}{3}y + \frac{a^3}{9}.$$

This change of variables leads to an equation in y equivalent to the original equation but that does not involve a quadratic term. We get

$$y^3 + py + q = 0, \tag{5.7}$$

where

$$p = b - \frac{a^2}{3} \quad \text{and} \quad q = c + \frac{2a^3 - 9ab}{27}.$$

Cardano's strategy introduces two variables u and v, satisfying

$$\begin{cases} u + v = y \\ 3uv + p = 0. \end{cases}$$

In other words, u and v are the two roots to the quadratic equation $t^2 - yt - \frac{p}{3} = 0$. Plugging $y = u + v$ into (5.7) gives

$$u^3 + 3uv(u + v) + v^3 + p(u + v) + q = 0$$
$$\iff u^3 + v^3 + (3uv + p)(u + v) + q = 0$$
$$\iff u^3 + v^3 + q = 0.$$

Multiplying through by u^3 gives $u^6 + u^3v^3 + qu^3 = 0$ but since $3uv + p = 0$, we get

$$u^6 + qu^3 - \frac{p^3}{27} = 0.$$

This becomes a quadratic equation in u^3 with the two solutions of

$$u^3 = -\frac{q}{2} \pm \sqrt{\frac{q^2}{4} + \frac{p^3}{27}}. \tag{5.8}$$

By (5.5), the possible values of u are

$$u = \omega^i \sqrt[3]{-\frac{q}{2} \pm \sqrt{\frac{q^2}{4} + \frac{p^3}{27}}} \quad \text{for } i = 0, 1, 2. \tag{5.9}$$

Note that u^3 and v^3 solve the system of equations

$$\begin{cases} u^3 + v^3 = q \\ 27u^3v^3 = -p^3, \end{cases}$$

from which we see that u^3 and v^3 are the two distinct roots of (5.8). We give u the $+$ and v the $-$ sign. However, the identity $3uv = -p$ leads to precisely three combinations of the possible powers on ω. The three roots of the cubic equation (5.7) are $y_1 = u_0 + v_0$, $y_2 = \omega u_0 + \omega^2 v_0$, and $y_3 = \omega^2 u_0 + \omega v_0$. Explicitly,

$$y_1 = \sqrt[3]{-\frac{q}{2} + \sqrt{\frac{q^2}{4} + \frac{p^3}{27}}} + \sqrt[3]{-\frac{q}{2} - \sqrt{\frac{q^2}{4} + \frac{p^3}{27}}},$$

$$y_2 = \left(\frac{-1 + i\sqrt{3}}{2}\right)\sqrt[3]{-\frac{q}{2} + \sqrt{\frac{q^2}{4} + \frac{p^3}{27}}} + \left(\frac{-1 - i\sqrt{3}}{2}\right)\sqrt[3]{-\frac{q}{2} - \sqrt{\frac{q^2}{4} + \frac{p^3}{27}}},$$

$$y_3 = \left(\frac{-1 - i\sqrt{3}}{2}\right)\sqrt[3]{-\frac{q}{2} + \sqrt{\frac{q^2}{4} + \frac{p^3}{27}}} + \left(\frac{-1 + i\sqrt{3}}{2}\right)\sqrt[3]{-\frac{q}{2} - \sqrt{\frac{q^2}{4} + \frac{p^3}{27}}}.$$

The three roots of the original cubic equation (5.6) are given by $x_i = y_i - \frac{a}{3}$.

The square root that appears in formula (5.9) indicates that there may be a bifurcation in behavior for the solutions for whether the expression under the square root is positive or negative. Indeed, the expression under the square root plays a similar role for the cubic equation as $b^2 - 4ac$ plays in the quadratic formula.

> **Definition 5.4.1**
>
> When a cubic equation is written in the form $y^3 + py + q = 0$, the expression
>
> $$\Delta = -27q^2 - 4p^3$$
>
> is called the *discriminant* of the cubic polynomial.

The reader might wonder why we define the discriminant as above rather than the quantity $q^2/4 + p^3/27$, which arose naturally from Cardano's method. The concept of discriminant has a more general definition (see Exercise 5.4.16), so we have stated the definition of the discriminant of a cubic to conform to the more general definition.

Theorem 5.4.2

Consider the cubic equation $y^3 + py + q = 0$ with $p, q \in \mathbb{R}$. Then

- if $\Delta > 0$, then the cubic equation has 3 real roots;
- if $\Delta = 0$, then the cubic equation has a double root;
- if $\Delta < 0$, then the cubic equation has 1 real root and 2 complex roots that are conjugate to each other.

Proof. Suppose that $\Delta > 0$. Then

$$\sqrt[3]{-\frac{q}{2} \pm \sqrt{\frac{q^2}{4} + \frac{p^3}{27}}} = \sqrt[3]{-\frac{q}{2} + \sqrt{-\frac{\Delta}{4 \cdot 27}}} = \sqrt[3]{-\frac{q}{2} + \sqrt{-\frac{\Delta}{108}}}$$

is a complex number with a nontrivial imaginary component. (In order to make sense of the cube root for complex numbers, we can assume that we choose the complex number with an angle θ such that $-\pi/3 \le \theta \le \pi/3$.) Furthermore, its complex conjugate is precisely

$$\sqrt[3]{-\frac{q}{2} - \sqrt{-\frac{\Delta}{108}}}.$$

Thus, $y_1 = u_0 + v_0 = 2\Re(u_0)$, twice the real part of u_0.

For the other roots, ωu_0 corresponds to the rotation of u_0 around the origin by an angle of $2\pi/3$ and we notice that $\omega^2 v_0$ is the complex conjugate of ωu_0. Hence, $y_2 = 2\Re(\omega u_0)$. Similarly, $y_3 = 2\Re(\omega^2 u_0)$. The values $u_0, \omega u_0,$ and $\omega^2 u_0$ are the vertices of a equilateral triangle with orthocenter at the origin. Since the point u_0 with polar angle satisfying $-\pi/3 \le \theta \le \pi/3$ is not on the x-axis, then the other two points are not reflected across the x-axis. Hence, $\Re(u_0)$, $\Re(\omega u_0)$, and $\Re(\omega^2 u_0)$ are all distinct real numbers. Hence, if $\Delta > 0$, the cubic equation has 3 real roots.

Suppose that $\Delta = 0$. Then $u_0 = v_0$, and they are both real. We obtain the following roots

$$y_1 = 2u_0, \qquad y_2 = \omega u_0 + \omega^2 v_0 = (\omega + \omega^2)u_0 = -u_0 = y_3.$$

If $u_0 = 0$, then we have a triple root at $y_i = 0$ but otherwise, we have two distinct real roots with one double root.

Suppose that $\Delta < 0$. Then $\sqrt{-\Delta/108}$ is a real number and both

$$u_0 = \sqrt[3]{-\frac{q}{2} + \sqrt{-\frac{\Delta}{108}}} \qquad \text{and} \qquad v_0 = \sqrt[3]{-\frac{q}{2} - \sqrt{-\frac{\Delta}{108}}}$$

are distinct positive real numbers. Then $y_1 \in \mathbb{R}$ whereas

$$y_2 = -\frac{1}{2}(u_0 + v_0) + i\frac{\sqrt{3}}{2}(u_0 - v_0) \quad \text{and} \quad y_3 = -\frac{1}{2}(u_0 + v_0) - i\frac{\sqrt{3}}{2}(u_0 - v_0)$$

are two complex roots that are conjugate to each other. $\qquad \square$

Theorem 5.4.2 assumes that p and q are real numbers. The solutions for the cubic equation also are correct when p and q are complex numbers. In this latter case, in the calculation for u_0, any value of the three possible values of the cube root of a complex will recover all three distinct roots.

Example 5.4.3. Consider the equation $x^3 - 3x - 1 = 0$. Cardano's solution for the cubic involves

$$u^3 = -\frac{q}{2} \pm \sqrt{\frac{q^2}{4} + \frac{p^3}{27}} = \frac{1}{2} \pm \frac{\sqrt{3}}{2}i.$$

Though Cardano did not have complex numbers at his disposal, we can do $u^3 = \cos(\frac{\pi}{3}) + i\sin(\frac{\pi}{3}) = e^{i\pi/3}$. Thus, $u_0 = e^{i\pi/9}$ and $v_0 = e^{-i\pi/9}$. So the roots are

$$x_1 = e^{i\pi/9} + e^{-i\pi/9} = 2\cos(\pi/9),$$
$$x_2 = e^{i2\pi/3}e^{i\pi/9} + e^{-i2\pi/3}e^{-i\pi/9} = 2\cos(7\pi/9),$$
$$x_3 = e^{-i2\pi/3}e^{i\pi/9} + e^{i2\pi/3}e^{-i\pi/9} = 2\cos(5\pi/9). \qquad \triangle$$

The proof of Theorem 5.4.2 indicates that Cardano's formula is not particularly easy to deal with. If the cubic has three real roots, then $\Delta > 0$, so $q^2/4 + p^3/27$ must be negative, which makes

$$\sqrt{\frac{q^2}{4} + \frac{p^3}{27}}$$

an imaginary number. It is precisely this case in which the solution to the cubic has three real roots. In particular, in order to find these real roots, we must pass into the complex numbers.

Example 5.4.4. Consider the cubic equation $x^3 - 15x - 20 = 0$. We have $\Delta = 2700$ so the equation should have three real roots. Also,

$$u_0 = \sqrt[3]{10 + 5i} \quad \text{and} \quad v_0 = \sqrt[3]{10 - 5i}.$$

Writing these complex numbers in polar form (see Appendix A.1) gives

$$10 + 5i = \sqrt{125}e^{i\arctan(1/2)} \implies \sqrt[3]{10 + 5i} = \sqrt{5}e^{i\arctan(1/2)/3},$$
$$10 - 5i = \sqrt{125}e^{-i\arctan(1/2)} \implies \sqrt[3]{10 - 5i} = \sqrt{5}e^{-i\arctan(1/2)/3}.$$

In particular, one of the solutions is

$$\sqrt[3]{10 + 5i} + \sqrt[3]{10 - 5i} = 2\sqrt{5}\cos\left(\frac{1}{3}\arctan\left(\frac{1}{2}\right)\right).$$

In a similar way, we can find trigonometric expressions for the other two roots. $\triangle$

Example 5.4.5. Consider the polynomial equation $x^3 + 6x^2 + 18x + 18 = 0$. Setting $x = y - 2$, we get the equation $y^3 + 6y - 2 = 0$. The discriminant is

$$\Delta = -4p^3 - 27q^2 = -4 \times 216 - 27 \times 4 = -972,$$

so there will be two complex roots and one real root. We calculate

$$u_0 = \sqrt[3]{-\frac{q}{2} + \sqrt{-\frac{\Delta}{108}}} = \sqrt[3]{1+3} = \sqrt[3]{4}, \quad v_0 = \sqrt[3]{1-3} = -\sqrt[3]{2},$$

so after some simplifying

$$x_1 = -2 + \sqrt[3]{4} - \sqrt[3]{2},$$

$$x_2 = -2 + \frac{1}{2}(-\sqrt[3]{4} + \sqrt[3]{2}) + i\frac{\sqrt{3}}{2}(\sqrt[3]{4} + \sqrt[3]{2}),$$

$$x_3 = -2 + \frac{1}{2}(-\sqrt[3]{4} + \sqrt[3]{2}) - i\frac{\sqrt{3}}{2}(\sqrt[3]{4} + \sqrt[3]{2}).$$

are the three roots of the original equation. $\qquad\qquad \triangle$

5.4.2 The Quartic Equation

Consider the generic quartic equation

$$x^4 + ax^3 + bx^2 + cx + d = 0, \tag{5.10}$$

where we can assume the polynomial is monic after dividing by the leading coefficient. As with the cubic equation, the change of variables $x = y - a/4$ eliminates the cubic term and changes (5.10) into

$$y^4 + py^2 + qy + r = 0, \tag{5.11}$$

for p, q, and r depending on a, b, c, and d. We propose to solve (5.11). We follow the strategy introduced by Ferrari in which we rewrite (5.11) as

$$y^4 = -py^2 - qy - r \tag{5.12}$$

and add an expression that simultaneously makes both sides into perfect squares. Because y^4 is alone on one side, we are limited to what we can add to create a perfect square. We choose to add the quantity

$$ty^2 + \frac{t^2}{4} \tag{5.13}$$

so that

$$y^4 + \left(ty^2 + \frac{t^2}{4}\right) = \left(y^2 + \frac{t}{2}\right)^2.$$

The trick to this method is to choose a value of t that makes the right-hand side into a perfect square as well. Adding (5.13) on the right-hand side of (5.12) gives

$$(t - p)y^2 - qy + \left(\frac{t^2}{4} - r\right). \qquad (5.14)$$

Now a quadratic expression $Ax^2 + Bx + C$ is the square of a linear expression if and only if $B^2 - 4AC = 0$. Hence, for (5.14) to be a perfect square, t must satisfy

$$q^2 - 4(t - p)\left(\frac{t^2}{4} - r\right) = 0 \iff t^3 - pt^2 - 4rt + (4rp - q^2) = 0.$$

This is called the *resolvent equation* for the quartic equation (5.11). We can solve for t using the solution method for the cubic, and in fact any of the three solutions work for the rest of the algorithm to finish solving the quartic. So when t solves the resolvent equation, (5.12) becomes

$$\left(y^2 + \frac{t}{2}\right)^2 = (my + n)^2$$

for some m and n that depend on p, q, and r. Then we have

$$\left(y^2 + \frac{t}{2}\right)^2 - (my+n)^2 = 0 \implies \left(y^2 - my + \frac{t}{2} - n\right)\left(y^2 + my + \frac{t}{2} + n\right) = 0.$$

So we now are reduced to solving two quadratic equations.

Example 5.4.6. Consider the quartic equation $y^4 + y^2 + 6y + 1 = 0$. The resolvent equation is $t^3 - t^2 - 4t - 32 = 0$. We could solve this via Cardano's method. However, by trying the rational roots that are possible by virtue of the Rational Root Theorem, we find that $t = 4$ is a solution to the resolvent equation. So following Ferrari's method, we add $4y^2 + 4$ to the equation

$$y^4 = -y^2 - 6y - 1$$

to get

$$y^4 + 4y^2 + 4 = 3y^2 - 6y + 3 \implies (y^2 + 2)^2 = (\sqrt{3}y - \sqrt{3})^2.$$

Hence,

$$(y^2 + 2)^2 - (\sqrt{3}y - \sqrt{3})^2 = 0 \implies (y^2 + \sqrt{3}y + 2 - \sqrt{3})(y^2 - \sqrt{3}y + 2 + \sqrt{3}) = 0.$$

Now applying the quadratic formula to two separate quadratic polynomials we get the four roots

$$y = \frac{1}{2}\left(-\sqrt{3} \pm \sqrt{3 - 4(2 - \sqrt{3})}\right) = \frac{1}{2}\left(-\sqrt{3} \pm \sqrt{4\sqrt{3} - 5}\right),$$

$$y = \frac{1}{2}\left(\sqrt{3} \pm \sqrt{3 - 4(2 + \sqrt{3})}\right) = \frac{1}{2}\left(\sqrt{3} \pm \sqrt{-4\sqrt{3} - 5}\right).$$

The first two roots are real and the last two roots are complex. $\triangle$

EXERCISES FOR SECTION 5.4

For Exercises 5.4.1 to 5.4.11 solve the equation using Cardano-Ferrari methods. For the solutions to a cubic equation, if all the roots are real, then write the solutions without reference to complex numbers.

1. $x^3 - 15x + 10 = 0$

2. $x^3 + 6x - 2 = 0$

3. $x^3 - 9x + 10 = 0$

4. $x^3 - 6x + 4 = 0$

5. $x^3 - 12x + 8 = 0$

6. $x^3 - 12x + 16 = 0$

7. $x^3 + 3x^2 + 12x + 4 = 0$

8. $x^3 - 9x^2 + 24x - 16 = 0$

9. $x^4 + 4x^2 + 12x + 7 = 0$

10. $x^4 + 4x^2 - 3x + 1 = 0$

11. $x^4 - 4x^3 + 4x^2 - 8x + 4 = 0$

12. Consider the polynomial $p(x) = x^3 - 6x^2 + 11x - 6$.

    (a) Solve the equation via Cardano's method.

    (b) Find the rational roots of this polynomial by the Rational Root Theorem.

    (c) Decide which rational root corresponds to which solution via Cardano's method.

13. Let n be a real number and consider the polynomial

    $$p(x) = x^3 - (3n+3)x^2 + (3n^2 + 6n + 2)x - (n^3 + 3n^2 + 2n).$$

    Apply Cardano's method to solve this equation. After finding the roots, explain why it was so easy to solve.

14. Prove that a palindromic polynomial of odd degree has -1 as a root. Use this and Exercise 5.2.10 to find all the roots to $x^5 + 2x^4 + 3x^3 + 3x^2 + 2x + 1 = 0$.

15. Consider the polynomial $p(x) = x^6 + 4x^4 + 4x^2 + 1$. Use either the strategy provided by Exercise 5.2.10 to find the roots or use Cardano's method to solve the equation in x^2 to find all the roots. Which do you think is easier?

16. The *discriminant* of a polynomial $a(x) = a_n x^n + \cdots + a_1 x + a_0$ is defined as

    $$\Delta = a_n^{2n-2} \prod_{i<j}(r_i - r_j)^2,$$

    where $r_1, r_2, \ldots, r_n$ are all the roots of $a(x)$, counted with multiplicity. Prove by this definition, the discriminant of the polynomial in (5.7) is indeed $-27q^2 - 4p^3$.

5.5 Constructible Numbers

5.5.1 Euclidean Geometry

Perhaps inspired by practical problems in surveying and architecture, many ancient civilizations possessed some form of geometry. Though the different cultures expressed their scholarship in different ways, most considered so-called construction problems.

Construction problems ask for a method to create a specific configuration (circle, triangle, line segment, point, etc.) with a specific property, and using specified tools (compass, straightedge, ruler, and so on). In some cultures, geometric knowledge would outline a recipe involving explicit numbers and then conclude, that they provided the desired construction. The numbers used needed to be generic enough that the validity of the construction did not rely on any particular properties of those numbers.

Greek mathematics, exemplified by Euclid's *Elements*, overlaid the practical geometric problems with a philosophical approach. Instead of providing a recipe for a geometric construction (and merely claim that it works), they defined their terms and common notions, and then, starting from a small list of five postulates, proved propositions about geometric objects using logic. Some geometry propositions in Euclid's *Elements* establish certain measure relationships while others state "it is possible to construct..." a specific configuration. For example, Proposition 12 in Book I states that it is possible to draw a straight line perpendicular to a given infinite straight line L through a given point not on L. The proof provides the construction using a compass and a straightedge and also establishes through logic that the construction produces the described configuration.

It is particularly interesting that propositions in the *Elements* never refer to specific distance values or angle values. (The *Elements* do refer to right angles and rational multiples thereof but never assigns a measure to the angle.) Perhaps because of this feature, the geometric constructions assumed the use of a straightedge, a ruler without distance markings.

Solutions to many such construction problems became jewels in the crown of Greek mathematics and served as examples for the purity of proofs for many generations of mathematics education. The ability to construct a circle inscribed in a triangle or the problem of constructing a regular pentagon are interesting, though still elementary, examples of these achievements.

A few problems stymied mathematicians for millennia. For example, Proposition 9 in Book I of the *Elements* gives a construction that bisects any angle α. However, the problem of trisecting the angle (constructing a line that cuts an angle by a third) remained an open problem for centuries. A few other problems that remained open for just as long included: to construct a regular heptagon; ("Squaring the circle") to construct a square with the same area as a given circle; and ("Doubling the cube") given a line segment a, to construct a line segment b such that the cube with side b has twice the volume as the cube with side a.

To the surprise of many, a large number of these open problems in geometry were either resolved or proved impossible using field theory.

5.5.2 Constructible Numbers

A first step to bringing algebra and geometry closer is to put a numerical value of the types of segments that can be created via a straightedge and compass construction. By labeling one preferred line segment as a reference length, we still do not need to refer to any particular units of measure.

Definition 5.5.1

The set of *constructible numbers* is the set of real numbers $a \in \mathbb{R}$ such that given a segment $\overline{OR}$, it is possible to construct with a straightedge and compass a segment $\overline{OA}$ such that as distances $OA = |a| \cdot OR$.

In this section, we will denote the set of constructible numbers by $\mathcal{C}$.

Recall that in a Euclidean construction, it is only possible to draw a line (with a straightedge) when we have two distinct points. Also, it is only possible to draw a circle (with a compass) with center O and with radius $\overline{OA}$ where O and A are points already obtained in the construction or specified in the hypotheses. In particular, it is not allowed in a construction to pick up the compass and retain the radius. Since a Euclidean construction problem only uses a straightedge and a compass, then the points that arise in a construction come from the intersection of two lines, a line and a circle, or two circles. In Definition 5.5.1, the only geometric objects specified in the hypothesis are two distinct points O and R.

By Proposition 2 in Book I of *Elements*, it suffices to construct any segment $\overline{CD}$ of length $|a| \cdot \overline{OR}$ to satisfy the requirement of Definition 5.5.1.

Proposition 5.5.2

Let a and b be any nonnegative constructible numbers. Then

$$a \pm b, \quad ab, \quad \frac{a}{b} \ (\text{if } b \neq 0), \quad \sqrt{a}$$

are also constructible.

Proof. In this proof, all references to propositions are from Euclid's *Elements*.

Fix two points O and R in the plane and let L be the line through O and R. Let A and B be two points on the line L with lengths $OA = aOR$ and $OB = bOR$ and such that A and B are on the same side of L from O as R is. Construct the circle Γ of center O and radius $\overline{OA}$. The circle Γ intersects L

in two points, one of which is A. Call A' the other point. Since O is between A' and B, then for distances

$$A'B = A'O + OB = (a+b)OR.$$

Hence, $a + b \in \mathcal{C}$. Suppose without loss of generality that A is between O and B. Then

$$b = OB = OA + AB \implies AB = (b-a)OR.$$

Hence, $b - a \in \mathcal{C}$.

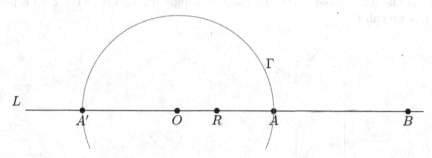

Next, we prove that if $a, b \in \mathcal{C}$, then $ab \in \mathcal{C}$. Let A and B be points on the line L with lengths $OA = aOR$ and $OB = bOR$ such that A and B are on the same side of L from O as R. Construct (Proposition I.11) the line L' that is perpendicular to L and that goes through O. Construct the circle Γ of center O and radius $\overline{OR}$. It intersects L' in two points. Pick one of these intersections and call it R'. Construct also the circle Γ' of center O and radius $\overline{OB}$. It intersects L' in two points. Call B' the point that is on the same side of L as R'.

Construct the line L_2 through R' and A. Construct (Proposition I.31) the line L_3 parallel to L_2 going through B'. Since L_2 intersects L (in A) and L_3 is parallel to L_2, then L_3 intersects L in a point we call C.

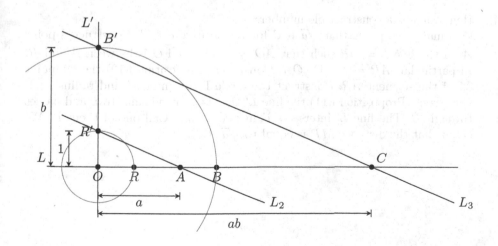

By Thales' Intercept Theorem,

$$\frac{OC}{OA} = \frac{OB'}{OR'} \implies \frac{OC}{aOR} = \frac{bOR}{OR} \implies OC = ab \cdot OR.$$

Hence, $ab \in \mathcal{C}$.

The proof that $\mathcal{C}$ is closed under division is similar. We suppose that we have already constructed the points O, R, R', A, B and B', and the lines L and L'. Now construct the line L_2 through A and B'. Construct also (Proposition I.31) the line L_3 parallel to L_2 going through R'. L_3 intersects L in a point that we call C.

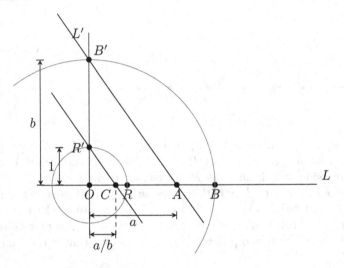

By Thales' Intercept Theorem,

$$\frac{OA}{OC} = \frac{OB'}{OR'} \implies OC = \frac{a \cdot OR \cdot OR}{b \cdot OR} = \frac{a}{b} \cdot OR.$$

Hence, a/b is a constructible number.

Finally, we prove that $\sqrt{a} \in \mathcal{C}$ for all positive $a \in \mathcal{C}$. Construct a point A on the line $L = \overleftrightarrow{OR}$ such that $AO = a \cdot OR$ and O is between A and R. In particular, $AR = (a + 1) \cdot OR$. Construct (Proposition I.10) the midpoint M of the segment $\overline{AR}$. Construct the circle Γ of center M and radius $\overline{MA}$. Construct (Proposition I.11) the line L' that is perpendicular to L and passes through R. The line L' intersects Γ in two points. Call one of them P. We claim that the distance MP is equal to $\sqrt{a} \cdot OR$.

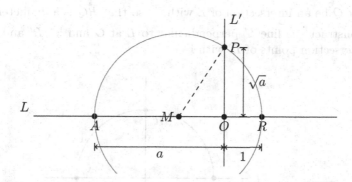

The radius MP is $MP = MA = \left(\frac{a+1}{2}\right) \cdot OR$ and we also have

$$\left(\frac{a+1}{2}\right) \cdot OR - OR = \left(\frac{a-1}{2}\right) \cdot OR.$$

Now MOP is a right triangle with $\angle MOP$ as the right angle. Hence, by Pythagoras,

$$OP = \sqrt{\left(\frac{a+1}{2}\right)^2 \cdot OR^2 - \left(\frac{a-1}{2}\right)^2 \cdot OR^2}$$

$$= \frac{1}{2}OR\sqrt{(a+1)^2 - (a-1)^2} = \frac{1}{2}OR\sqrt{4a} = \sqrt{a} \cdot OR.$$

Thus, $\sqrt{a}$ is a constructible number. $\qquad\square$

As an example of a construction problem, we provide a compass and straightedge construction for a regular pentagon.

Example 5.5.3. The angle $\alpha = \frac{2\pi}{5}$ satisfies $\cos 3\alpha = \cos(2\pi - 3\alpha) = \cos 2\alpha$. Using addition formulas, we find that $\cos 3\alpha = 4\cos^3 \alpha - 3\cos \alpha$ and $\cos 2\alpha = 2\cos^2 \alpha - 1$. Thus, $\cos \alpha$ solves the equation

$$4x^3 - 2x^2 - 3x + 1 = 0 \iff (x - 1)(4x^2 + 2x - 1) = 0.$$

Since $\cos(\alpha) \neq 1$, then $\cos(\alpha)$ must be a root of $4x^2 + 2x - 1$. Using the quadratic formula and reasoning that $\cos(2\pi/5) > 0$, we deduce that

$$\cos\left(\frac{2\pi}{5}\right) = \frac{-1 + \sqrt{5}}{4}.$$

This value suggests the following construction of the regular pentagon.

- Pick a center O and radius $\overline{OR}$.
- Construct a circle Γ_1 of center O and radius $\overline{OR}$.
- Let L be the line $\overleftrightarrow{OR}$.

- Let Q be an intersection of L with Γ_1 so that $\overline{RQ}$ is a diameter of Γ_1.
- Construct the line L' perpendicular to L at O and let R' and Q' be the intersection points of L' with Γ_1.

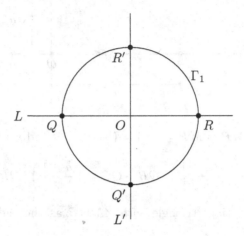

- Construct (via Proposition I.10 of the *Elements*) the midpoint M of $\overline{OQ}$.
- Construct the circle Γ_2 of center M and radius $\overline{MQ'}$.
- Γ_2 intersects L in two points. Call D the point between M and R.

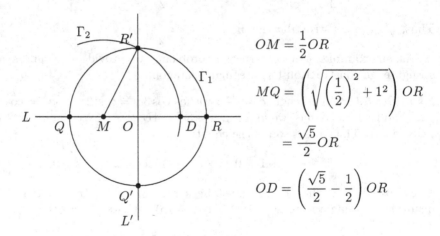

$$OM = \frac{1}{2}OR$$

$$MQ = \left(\sqrt{\left(\frac{1}{2}\right)^2 + 1^2} \right) OR$$

$$= \frac{\sqrt{5}}{2}OR$$

$$OD = \left(\frac{\sqrt{5}}{2} - \frac{1}{2} \right) OR$$

- Construct the midpoint E of the segment $\overline{OD}$.
- Construct the line L'' that is perpendicular to L through the point E.
- L'' intersects Γ_1 in two points. Call one of them A_1.

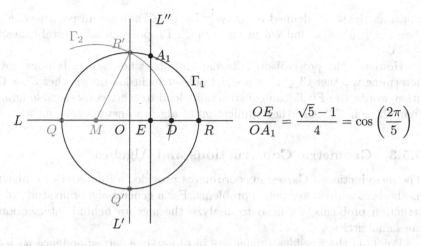

$$\frac{OE}{OA_1} = \frac{\sqrt{5}-1}{4} = \cos\left(\frac{2\pi}{5}\right)$$

- Since $\dfrac{OE}{OA_1} = \dfrac{\sqrt{5}-1}{4} = \cos\left(\dfrac{2\pi}{5}\right)$, then $\angle EOA_1 = \frac{2\pi}{5}$. Consequently, the segment RA_1 is one edge of a regular pentagon.

- Construct a circle of center A_1 and radius $\overline{A_1 R}$. This circle intersects Γ in two points: R and another point we call A_2. Note that $A_1 A_2 = A_1 R$.

- Repeat the previous step two more times to define A_3 and A_4, so that the polygon $RA_1 A_2 A_3 A_4$ is a regular pentagon.

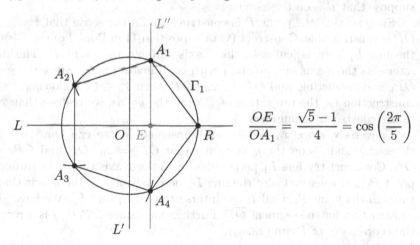

$$\frac{OE}{OA_1} = \frac{\sqrt{5}-1}{4} = \cos\left(\frac{2\pi}{5}\right)$$

$\triangle$

Proposition 5.5.2 is already quite interesting. Since $1 \in \mathcal{C}$, then using addition, subtraction and division repeatedly, we deduce that $\mathbb{Q} \subseteq \mathcal{C}$. Since $\mathcal{C}$ is closed under taking square roots of positive numbers, we also can construct numbers like

$$\sqrt{3}, \quad \sqrt{10 - 2\sqrt{3}}, \quad \text{or} \quad \frac{1}{7}\sqrt{2 + \sqrt{3 + \sqrt{5}}}.$$

Consider the set C' defined recursively by $1 \in C'$ and for all positive $a, b \in C'$, then $a \pm b$, ab, a/b, and $\sqrt{a}$ are also in C'. Proposition 5.5.2 establishes that $C' \subseteq C$.

However, the proposition falls short of the whole story. It does not yet determine whether $C' \subseteq C$ is a strict subset inclusion or whether $C' = C$. In other words, can Euclidean constructions lead to other constructible numbers than those in C'? A further application of algebra answers this question.

5.5.3   Geometric Constructions and Algebra

The introduction of Cartesian coordinates provided a framework for algebraic methods to address geometric problems. For a general algebraic study of construction problems, we need to analyze the algebra behind intersections of lines and circles.

Points in the Euclidean plane are in one-to-one correspondence with $\mathbb{R}^2$.

Proposition 5.5.4

Let P be a point with coordinates (x_0, y_0). The segment $\overline{OP}$ is constructible if and only if the numbers x_0 and y_0 are constructible.

Proof. We assume that O is the intersection of the x-axis and the y-axis and suppose that R is on the x-axis.

Suppose that the point P is constructible, in the sense that the segment $\overline{OP}$ is constructible. Construct (via Proposition 11 in Book I of the *Elements*) the line L_1 perpendicular to the x-axis through the point P. The line L_1 intersects the x-axis in a point P_1 with coordinates $(x_0, 0)$. Since the segment $\overline{OP_1}$ is constructible and $OP_1 = x_0 \cdot OR$, then $x_0 \in C$. Following a similar construction for the projection of P onto the y-axis, we deduce that x_0 and y_0 are constructible numbers.

Conversely, since x_0 and y_0 are constructible, we can construct P_1 on the x-axis and P_2 on the y-axis such that $OP_1 = x_0 \cdot OR$ and $OP_2 = y_0 \cdot OR$. Construct the line L_1 perpendicular to the x-axis that goes through the point P_1 and construct also the line L_2 perpendicular to the y-axis that goes through the point P_2. Call P the intersection of L_1 and L_2. We have given a construction for the segment $\overline{OP}$. Furthermore, since $OP_1 P P_2$ is a rectangle, the coordinates of P are (x_0, y_0). $\square$

When tracing a line with a straightedge, we always draw a line that passes through two already given points. The equation for a line through (x_1, y_1) and (x_2, y_2) is

$$y = y_1 + \frac{y_2 - y_1}{x_2 - x_2}(x - x_1) \iff (x_2 - x_1)(y - y_1) = (y_2 - y_1)(x - x_1).$$

When using a compass, we trace out a circle with center A with radius AB where A and B are points already obtained in the construction or specified in the hypotheses. The equation for a circle of center (x_0, y_0) and radius of length r is

$$(x - x_0)^2 + (y - y_0)^2 = r^2.$$

Beyond these tracing operations, we also will consider the intersection points between: (1) two lines; (2) two circles; (3) a line and a circle.

Let P_1, P_2, P_3, and P_4 be four points obtain by some compass and straightedge construction from the initial segment $\overline{OR}$. Suppose that the coordinates of P_i are (x_i, y_i). Now let L_1 be the line through P_1 and P_2 and let L_2 be the line through P_3 and P_4. Assuming that L_1 and L_2 are not parallel, the point Q of intersection between L_1 and L_2 satisfies the system of linear equations

$$\begin{cases} (y_2 - y_1)x - (x_2 - x_1)y = x_1(y_2 - y_1) - y_1(x_2 - x_1) \\ (y_4 - y_3)x - (x_4 - x_3)y = x_3(y_4 - y_3) - y_3(x_4 - x_3). \end{cases}$$

Via Cramer's rule, if $x_i, y_i \in F$ for $i = 1, 2, 3, 4$, where F is some field extension of $\mathbb{Q}$, then the point Q has coordinates in F.

Consider now the intersection of two circles. Let P_1 and P_2 be two distinct points obtained by some compass and straightedge construction from the initial segment $\overline{OR}$. Suppose that the coordinates of P_i are (x_i, y_i) where x_i, y_i are in some field extension F of $\mathbb{Q}$. Suppose also that r_1 and r_2 are two radii that are constructible numbers. The intersection points of the circle Γ_1 with center P_1 and radius r_1 and the circle Γ_2 with center P_2 and radius r_2 satisfy the system of equations

$$\begin{cases} (x - x_1)^2 + (y - y_1)^2 = r_1^2 \\ (x - x_2)^2 + (y - y_2)^2 = r_2^2. \end{cases}$$

The difference of the two equations is

$$(x^2 - 2xx_1 + x_1^2) + (y^2 - 2y_1y + y_1^2)$$
$$- (x^2 - 2xx_2 + x_2^2) - (y^2 - 2y_2y + y_2^2) = r_1^2 - r_2^2$$
$$\iff 2(x_2 - x_1)x + 2(y_2 - y_1)y = r_1^2 - r_2^2 + x_2^2 + y_2^2 - x_1^2 - y_1^2.$$

Depending on whether $x_2 - x_1 \neq 0$ or $y_2 - y_2 \neq 0$, it is possible to write y as an expression $ax + b$, where $a, b \in F$, or to write x as an expression $cy + d$, where $c, d \in F$. Without loss of generality, assume the former case. Replacing y with $ax + b$ in the equation for either of the circles leads to a quadratic equation in x. If the quadratic equation has no real solutions, it means that the two circles do not intersect. Otherwise, if the circles do intersect, both the points of intersection have coordinates in a field extension K of F with $[K : F] = 1$ or 2.

Now consider the intersection of a circle and a line. Let Γ be a circle with center (x_0, y_0) and radius r and let (x_1, y_1) and (x_2, y_2) be two distinct points. Suppose that there is a field extension F of $\mathbb{Q}$ such that $x_i, y_i \in F$ and $r \in F$.

The points of intersection of Γ and the line L through (x_1, y_1) and (x_2, y_2) satisfies the system of equations

$$\begin{cases} (x - x_1)^2 + (y - y_1)^2 = r^2 \\ (y_2 - y_1)x - (x_2 - x_1)y = x_1(y_2 - y_1) - y_1(x_2 - x_1). \end{cases}$$

As in the previous case, $x_1 \neq x_2$, in which case we can solve in the equation for the line, for y in terms of x, or $y_1 \neq y_2$, in which case we can solve in the equation for the line, for x in terms of y. The equation for the circle leads to a quadratic equation, either in y or in x. If there is no solution in real numbers to the equation, then we interpret this case to mean that Γ and L do not intersect. If the equation has solutions, then these solutions are in a field extension K of F with $[K : F] = 1$ or 2.

This discussion leads to the following strengthening of Proposition 5.5.2.

Theorem 5.5.5

If a real number $\alpha \in \mathbb{R}$ is constructible, then $[\mathbb{Q}(\alpha) : \mathbb{Q}] = 2^k$ for some nonnegative integer k. Furthermore, the set of constructible numbers is exactly the set $\mathcal{C}'$ of real numbers defined recursively as the set that contains 1 and for any positive elements $a, b \in \mathcal{C}'$, then $a \pm b$, ab, a/b and $\sqrt{a}$ are in $\mathcal{C}'$.

Proof. First, suppose that α is a constructible number. In the Euclidean construction of a segment $\overline{OP}$ with $OP = \alpha \cdot OR$, we construct a sequence of points $P_1, P_2, \ldots, P_n$ with $P_n = P$ and such that every controlling parameter of any geometric object (center and radius of a circle, two points of a line) is O, R, or one of these points. Let α_i be the constructible number such that $OP_i = \alpha_i \cdot OR$. Then

$$[\mathbb{Q}(\alpha) : \mathbb{Q}] = [\mathbb{Q}(\alpha) : \mathbb{Q}(\alpha_{n-1})] \cdots [\mathbb{Q}(\alpha_2) : \mathbb{Q}(\alpha_1)][\mathbb{Q}(\alpha_1) : \mathbb{Q}].$$

Furthermore, from the above discussion, we know that $[\mathbb{Q}(\alpha_i) : \mathbb{Q}(\alpha_{i-1})]$ is 1 or 2. Hence, $[\mathbb{Q}(\alpha) : \mathbb{Q}] = 2^k$ for some nonnegative integer k.

Proposition 5.5.2 established that $\mathcal{C} \subseteq \mathcal{C}'$. However, at each stage of a Euclidean construction since $[\mathbb{Q}(\alpha_i) : \mathbb{Q}(\alpha_{i-1})]$ is 1 or 2. If $[\mathbb{Q}(\alpha_i) : \mathbb{Q}(\alpha_{i-1})] = 1$, then a point obtained from previous points at the ith stage of the construction has coordinates that result from addition, subtraction, multiplication, or division of coordinates of previous points. If $[\mathbb{Q}(\alpha_i) : \mathbb{Q}(\alpha_{i-1})] = 2$, then α_i is the root of some quadratic polynomial with coefficients in $\mathbb{Q}(\alpha_{i-1})$. In particular, α_i is the sum of an element in $\mathbb{Q}(\alpha_{i-1})$ with the square root of an element in $\mathbb{Q}(\alpha_i)$. Hence, we conclude that $\mathcal{C} = \mathcal{C}'$. $\square$

One of the profound consequences of Theorem 5.5.5 is that it gives a way to show if certain geometric configurations cannot be obtained by a compass and straightedge construction. When first stated, the following three corollaries answered long-standing open problems in geometry.

Corollary 5.5.6

It is impossible to double the cube by a compass and straightedge construction.

Proof. Let OR be one edge of a cube C. A cube with double the volume would have an edge of length $\sqrt[3]{2} \cdot OR$. However, $[\mathbb{Q}(\sqrt[3]{2}) : \mathbb{Q}] = 3$. By Theorem 5.5.5, $\sqrt[3]{2}$ is not a constructible number so it is impossible to construct a segment of length $\sqrt[3]{2} \cdot OR$ with a compass and straightedge. $\qquad\square$

Corollary 5.5.7

It is impossible to square the circle by a compass and straightedge construction.

Proof. Recall that "squaring the circle" refers to the construction of starting from a circle Γ of center O and radius $\overline{OR}$, to construct a square whose area is equal to that of Γ. The area of Γ is πOR^2. The area of a square is a^2, where a is the length of the side. Constructing a square as desired would lead to constructing a line segment of length $\sqrt{\pi} OR$. However, by Lindemann's theorem that π is transcendental, $[\mathbb{Q}(\sqrt{\pi}) : \mathbb{Q}]$ is infinite. Hence, by Theorem 5.5.5, $\sqrt{\pi}$ is not a constructible number and thus it is impossible to construct a segment of the desired length. $\qquad\square$

For the last corollary we consider, we state a generalization of Theorem 5.5.5. The proof follows from the same procedure as that given for Theorem 5.5.5.

Theorem 5.5.8

Suppose that $O, R, C_1, C_2, \ldots, C_n$ are points given in the plane such that $OC_i = \gamma_i OR$. If a point A can be obtained from $O, R, C_1, C_2, \ldots, C_n$ with a compass and straightedge construction and $OA = \alpha OR$, then

$$[\mathbb{Q}(\alpha, \gamma_1, \gamma_2, \ldots, \gamma_n) : \mathbb{Q}(\gamma_1, \gamma_2, \ldots, \gamma_n)] = 2^k$$

for some nonnegative integer k.

Corollary 5.5.9

It is impossible to trisect every angle using a compass and straightedge construction.

Proof. Let $\theta = \angle AOR$ be an angle. There is no assumption that $\overline{OA}$ is constructible from $\overline{OR}$.

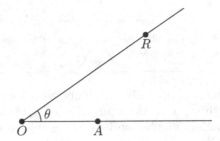

Using angle addition formulas, it is easy to show that for any angle α,

$$\cos(3\alpha) = 4\cos^3\alpha - 3\cos\alpha \implies \cos\theta = 4\cos^3\left(\frac{\theta}{3}\right) - 3\cos\left(\frac{\theta}{3}\right).$$

So $\cos(\theta/3)$ is a root of $4x^3 - 3x - \cos\theta = 0$ whose coefficients are in the field $\mathbb{Q}(\cos\theta)$. For an arbitrary angle θ, $4x^3 - 3x - \cos\theta$ is irreducible in $\mathbb{Q}(\cos\theta)[x]$. We conclude that

$$[\mathbb{Q}(\cos(\theta/3)) : \mathbb{Q}(\cos\theta)] = 3.$$

By Theorem 5.5.8, it is impossible to construct a point C from O, R, and A using a compass and straightedge such that $\angle ROC = \frac{1}{3}\angle ROA$. $\qquad\square$

EXERCISES FOR SECTION 5.5

1. Find a compass and straightedge construction for the number $\sqrt{3 + \sqrt{5}}$.

2. Discuss how to construct a regular dodecagon (12 sides) with a compass and straightedge.

3. This exercise guides a proof that it is impossible to construct a regular heptagon with a compass and straightedge. Set $\alpha = \frac{2\pi}{7}$.
   (a) Prove that $\cos(3\alpha) = \cos(4\alpha)$.
   (b) Deduce that $\cos\alpha$ solves $8x^3 + 4x^2 - 4x - 1 = 0$.
   (c) Deduce that $[\mathbb{Q}(\cos(\alpha)) : \mathbb{Q}] = 3$ and explain clearly why this means that the regular heptagon is not constructible with a compass and a straightedge.

4. Prove that it is impossible to construct a regular 9-gon with a compass and straightedge.

5. Given a circle Γ, is it possible to construct a circle with double the area? If so, provide a compass and straightedge construction.

6. For each nonnegative integer n, let $T_n(x)$ be the *Chebyshev polynomial* defined by $T_n(\cos\theta) = \cos(n\theta)$ for θ.
   (a) Prove $T_{mn}(x) = T_m(T_n(x))$.
   (b) Use this to find the four roots of $T_4(x)$.
   (c) Deduce that $T_4(x)$ is irreducible in $\mathbb{Q}[x]$.
   (d) Also use (a) to find all 6 roots of $T_6(x)$.

7. Is it possible to construct, using a compass and a straightedge, a triangle with sides in the ratio of 2 : 3 : 4 that has the same area of a given square? If so, suppose that one side of the square is a segment $\overline{OR}$; describe a compass and straightedge construction for the desired triangle. [Hint: Use Heron's Formula.]

8. Is it possible to construct, using a compass and a straightedge, a triangle with angles in the ratio of 2 : 3 : 4 that has the same area of a given square? If so, suppose that one side of the square is a segment $\overline{OR}$; describe a compass and straightedge construction for the desired triangle.

5.6 Cyclotomic Extensions

In the study of polynomial equations of higher order, there is arguably a simplest equation of a given degree, namely

$$z^n - 1 = 0.$$

The roots of this polynomial are called the nth *roots of unity*. Without an understanding of the properties of the roots of this polynomial, we should not expect to have a clear understanding of roots of polynomials of degree n. This section studies the roots of unity and extensions of $\mathbb{Q}$ that involve adjoining a root of unity.

5.6.1 Primitive Roots

Borrowing from the polar expression of complex numbers, we know that we can write 1 as $e^{2\pi ki}$ for any $k \in \mathbb{Z}$. If a complex number $z = re^{i\theta}$ satisfies $z^n = 1$, then

$$r^n e^{in\theta} = 1e^{2\pi ki}.$$

With the condition that r is a positive real number, $r = 1$ and $n\theta = 2\pi k$ for some k. Thus,

$$\theta = \frac{2\pi k}{n}, \qquad \text{for } k \in \mathbb{Z}.$$

The values $k = 0, 1, \ldots, n-1$ give n distinct complex numbers. Since a nonzero polynomial of degree n in $F[x]$ can have at most n roots in the field F, then this gives all the roots. The nth roots of unity are

$$e^{2\pi ik/n} = \cos\left(\frac{2\pi k}{n}\right) + i\sin\left(\frac{2\pi k}{n}\right) \qquad \text{for } k = 0, 1, 2, \ldots, n - 1.$$

We will often denote $\zeta_n = e^{2\pi i/n}$ so that the nth roots of unity are ζ_n^k. As in Figure 5.1, the elements ζ_n^k, with $k = 0, 1, \ldots, n - 1$, form the vertices of a regular n-gon on the unit circle.

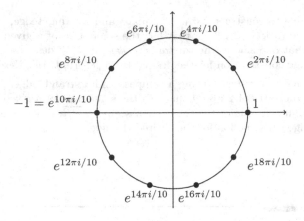

FIGURE 5.1: The 10th roots of unity.

The set of nth roots of unity, denoted by μ_n, forms a subgroup of $(\mathbb{C}^*, \times)$. Indeed, $1 \in \mu_n$, so μ_n is nonempty and $\zeta_n^a(\zeta_n^b)^{-1} = \zeta_n^{a-b} \in \mu_n$, so μ_n is a subgroup of $(\mathbb{C}^*, \times)$ by the One-Step Subgroup Criterion. Furthermore, μ_n is isomorphic to $\mathbb{Z}/n\mathbb{Z}$ via $\bar{a} \mapsto \zeta_n^a$.

Definition 5.6.1

A *primitive* nth root of unity is an nth root of unity that generates μ_n.

By Proposition 1.3.8, $\mathbb{Z}/n\mathbb{Z}$ can be generated by $\bar{a}$ if and only if $\gcd(a, n) = 1$. Therefore, the primitive roots of unity are of the form ζ_n^a where $1 \leq a \leq n$ with $\gcd(a, n) = 1$.

Any solution to the equation $x^n - 1 = 0$ over a field F is an element of finite order in the multiplicative group of units $U(F)$. Combining results from group theory with field theory gives the following result about the group of units in any field F.

Proposition 5.6.2

Let F be a field. Any finite group Γ in $U(F)$ is cyclic.

Proof. Let $|\Gamma| = n$. Since Γ is a finite abelian group, the Fundamental Theorem of Finitely Generated Abelian Groups applies. Suppose that as an invariant factors expression

$$\Gamma \cong Z_{n_1} \oplus Z_{n_2} \oplus \cdots \oplus Z_{n_\ell}.$$

Since $n_{i+1} \mid n_i$ for $1 \leq i \leq \ell - 1$, then $x^{n_1} - 1 = 0$ for all $x \in \Gamma$. Assume Γ is not cyclic. Then $n_1 < n$ and all n elements of Γ solve $x^{n_1} - 1 = 0$. This contradicts the fact that the number of distinct roots of a polynomial is less than or equal to its degree. (Corollary 4.5.10.) Hence, Γ is cyclic. $\qquad\square$

Definition 5.6.3

> Let p be a prime number. An integer a such that $\bar{a} \in \mathbb{Z}/p\mathbb{Z}$ generates $U(\mathbb{F}_p)$ is called a *primitive root modulo p*.

Primitive roots modulo p are not unique in $\mathbb{F}_p$. Since $U(\mathbb{F}_p)$ is cyclic, then $U(\mathbb{F}_p) \cong \mathbb{Z}_{p-1}$, so there are $\phi(p-1)$ generators. However, the existence of a primitive root modulo p is not at all directly obvious from modular arithmetic.

Example 5.6.4. Consider the prime $p = 17$. Then 2 is not a primitive root modulo 17 because $\bar{2}$ has order 8 and thus does not generate $U(17)$, which has order 16. In contrast, whether by hand calculation or assisted by computer, we can show that $\bar{3}$ has order 16 in $U(17)$. So, 3 is a primitive root modulo 17. $\qquad\triangle$

5.6.2 Cyclotomic Polynomials

Definition 5.6.5

> The nth *cyclotomic polynomial* $\Phi_n(x)$ is the monic polynomial whose roots are the primitive nth roots of unity. Namely,
>
> $$\Phi_n(x) = \prod_{\substack{1 \leq a \leq n \\ \gcd(a,n)=1}} (x - \zeta_n^a).$$

A priori, the cyclotomic polynomials are elements of $\mathbb{C}[x]$ but we will soon see that the $\Phi_n(x) \in \mathbb{Z}[x]$ and satisfy many properties.

We note right away that $\deg \Phi_n(x) = \phi(n)$ where ϕ is Euler's totient function. If we write

$$x^n - 1 = \prod_{1 \leq i \leq n} (x - \zeta_n^i) = \prod_{d|n} \prod_{\substack{1 \leq i \leq d \\ \gcd(i,d)=1}} (x - \zeta_n^{(n/d)i}) = \prod_{d|n} \prod_{\substack{1 \leq i \leq d \\ \gcd(i,d)=1}} (x - \zeta_d^i),$$

then we deduce the following implicit formula for the cyclotomic polynomials,

$$x^n - 1 = \prod_{d|n} \Phi_d(x). \tag{5.15}$$

The product identity (5.15) provides a recursive formula for $\Phi_n(x)$, starting with $\Phi_1(x) = x - 1$. For example, we have

$$x^2 - 1 = \Phi_1(x)\Phi_2(x) \implies \Phi_2(x) = \frac{x^2 - 1}{x - 1} = x + 1.$$

By the same token, we have

$$x^3 - 1 = \Phi_1(x)\Phi_3(x) \implies \Phi_3(x) = \frac{x^3 - 1}{x - 1} = x^2 + x + 1.$$

A few other examples of the nth cyclotomic polynomials are

$$\Phi_4(x) = \frac{x^4 - 1}{\Phi_2(x)\Phi_1(x)} = \frac{x^4 - 1}{x^2 - 1} = x^2 + 1,$$

$$\Phi_5(x) = \frac{x^5 - 1}{\Phi_1(x)} = x^4 + x^3 + x^2 + x + 1,$$

$$\Phi_6(x) = \frac{x^6 - 1}{\Phi_3(x)\Phi_2(x)\Phi_1(x)} = \frac{x^3 + 1}{x + 1} = x^2 - x + 1,$$

$$\Phi_7(x) = \frac{x^7 - 1}{\Phi_1(x)} = x^6 + x^5 + x^4 + x^3 + x^2 + x + 1,$$

$$\Phi_8(x) = \frac{x^8 - 1}{\Phi_4(x)\Phi_2(x)\Phi_1(x)} = \frac{x^8 - 1}{x^4 - 1} = x^4 + 1.$$

The first few calculations hint at a number of properties that cyclotomic polynomials might satisfy. We might hypothesize that the $\Phi_n(x)$ are in $\mathbb{Z}[x]$ and furthermore, that they only involve coefficients of 0, 1, and -1. We might also hypothesize that the polynomials are palindromic. Also, since each $\Phi_n(x)$ is obtained by dividing out of $x^n - 1$ all terms that we know must divide $x^n - 1$, we might also hope that the $\Phi_n(x)$ are irreducible. Some of these hypotheses are true and some are not.

Proposition 5.6.6

The cyclotomic polynomial $\Phi_n(x)$ is a monic polynomial in $\mathbb{Z}[x]$ of degree $\phi(n)$.

Proof. We have already shown that these polynomials have degree $\phi(n)$. We need to prove that $\Phi_n(x) \in \mathbb{Z}[x]$. We prove this by strong induction, noticing that $\Phi_1(x) = x - 1 \in \mathbb{Z}[x]$.

Suppose that $\Phi_k(x)$ is monic and in $\mathbb{Z}[x]$ for all $k < n$. Let $f(x)$ be the polynomial defined by

$$f(x) = \prod_{\substack{d \mid n \\ d < n}} \Phi_d(x).$$

By induction, the polynomial is monic and has coefficients in $\mathbb{Z}$. We know that $x^n - 1 = \Phi_n(x)f(x)$ and so by polynomial division in $\mathbb{Q}[x]$, $\Phi_n(x)$ is a polynomial with coefficients in $\mathbb{Q}$. By Gauss' Lemma we conclude that $\Phi_n(x) \in \mathbb{Z}[x]$. $\square$

Proposition 5.6.7

The polynomial $\Phi_n(x)$ is palindromic.

Proof. If $z = \zeta_n^a \in \mathbb{C}$ is a root of $\Phi_n(x)$, then $\frac{1}{z} = \zeta_n^{-a} = \zeta_n n - a$ is also a root of $\Phi_n(x)$, since $n - a$ is relatively prime to n whenever a is. Therefore, $\Phi_n(x) = x^{\varphi(n)} \Phi_n(1/x)$. However, given any polynomial $p(x)\mathbb{Q}[x]$, $x^{\deg p} p(1/x)$ is the polynomial obtained from $p(x)$ by reversing the order of the coefficients. $\qquad \square$

It is possible to prove that, as polynomials, $x^{\gcd(m,n)} - 1$ is the monic greatest common divisor of $x^m - 1$ and $x^n - 1$. We say that the sequence of polynomials $\{x^n - 1\}_{n=1}^{\infty}$ is a strong divisibility sequence in $\mathbb{Z}[x]$. In [6], the authors proved that the strong divisibility property is sufficient to define the cyclotomic polynomials $\Phi_n(x)$ in $\mathbb{Z}[x]$ that satisfy the recursive formula (5.15), without reference to roots of unity.

The hypothesis that all the coefficients of $\Phi_n(x)$ at -1, 0, or 1 is not true. The first cyclotomic polynomial that has a coefficient different from -1, 0, or 1 is $\Phi_{105}(x)$. As Exercise 5.6.14 shows, it is not a coincidence that the integer 105 happens to be the first positive integer that is the product of three odd primes. A direct calculation gives,

$$\Phi_{105}(x) = 1 + x + x^2 - x^5 - x^6 - 2x^7 - x^8 - x^9 + x^{12} + x^{13} + x^{14} + x^{15}$$
$$+ x^{16} + x^{17} - x^{20} - x^{22} - x^{24} - x^{26} - x^{28} + x^{31} + x^{32} + x^{33}$$
$$+ x^{34} + x^{35} + x^{36} - x^{39} - x^{40} - 2x^{41} - x^{42} - x^{43} + x^{46}$$
$$+ x^{47} + x^{48}.$$

The coefficients of x^7 and of x^{41} are -2.

Theorem 5.6.8

> For all $n \in \mathbb{N}^*$, the cyclotomic polynomial $\Phi_n(x)$ is an irreducible polynomial in $\mathbb{Z}[x]$ of degree $\phi(n)$.

Proof. What is left to show is that $\Phi_n(x)$ is irreducible. We will show that if p is any prime not dividing n, then $\Phi_n(x)$ is irreducible over $\mathbb{F}_p$.

Suppose that $\Phi_n(x)$ factors into $f(x)g(x)$ over $\mathbb{Q}$ and, without loss of generality, we suppose that $f(x)$ is irreducible with $\deg f(x) \geq 1$. Let ζ be a primitive nth root of unity. Then ζ^p is also a primitive root (since $p \nmid n$).

Assume now that $g(\zeta^p) = 0$. Then ζ is a root of $g(x^p)$ and since $f(x)$ is the minimal polynomial of ζ, we have $g(x^p) = f(x)h(x)$. Reduction modulo p into $\mathbb{F}_p$ we get

$$\bar{f}(x)\bar{h}(x) = \bar{g}(x^p) = (\bar{g}(x))^p,$$

where the last equality holds by the Frobenius homomorphism. Therefore, $\bar{f}(x)$ and $\bar{g}(x)$ have an irreducible factor in common in the UFD $\mathbb{F}_p[x]$. This implies that $\overline{\Phi_n}(x)$ has a multiple root in $\mathbb{F}_p$ and hence that $x^n - 1$ has a multiple root in the finite field. We prove that this leads to a contradiction. Recall the polynomial derivative D described in Exercise 4.5.22. If $x^n - 1$ has a multiple root in some field extension of $\mathbb{F}_p$, then this multiple root must be a root of $x^n - 1$ and of the derivative $D(x^n - 1) = nx^{n-1}$. However, since $p \nmid n$, then the only root of $D(x^n - 1)$ is 0, whereas 0 is not a root of $x^n - 1$. So by contradiction, we know that ζ^p is not a root of $g(x)$. Therefore, ζ^p must be a root of $f(x)$.

Now let a be any integer that is relatively prime to n. Then we can write $a = p_1 p_2 \cdots p_k$ as a product of not necessarily distinct primes that are relatively prime to n, and hence which do not divide n. From the above paragraph, if ζ is a root of $f(x)$, then ζ^{p_1} is also a root of $f(x)$. Then $\zeta^{p_1 p_2} = (\zeta^{p_1})^{p_2}$ is also a root of $f(x)$. By induction, we deduce that ζ^a is a root of $f(x)$. This now means that every primitive nth root is also a root of $f(x)$ so $g(x)$ is a unit and $f(x)$ is an associate of $\Phi_n(x)$. Hence, $\Phi_n(x)$ is irreducible. $\square$

Definition 5.6.9

The extension $\mathbb{Q}(\zeta_n)/\mathbb{Q}$ is called the nth *cyclotomic extension* of $\mathbb{Q}$.

Since the roots of $\Phi_n(x)$ involve powers of ζ_n, then all the roots of $\Phi_n(x)$ are in the cyclotomic extension $\mathbb{Q}(\zeta_n)$. Theorem 5.6.8 immediately gives the following result.

Corollary 5.6.10

For any integer $n \geq 2$, the degree of the cyclotomic field over $\mathbb{Q}$ is

$$[\mathbb{Q}(\zeta_n) : \mathbb{Q}] = \phi(n).$$

Example 5.6.11. In Exercise A.5.14, we proved that if $2^n - 1$ is prime, then n is prime. The converse implication is not necessarily true. With cyclotomic polynomials at our disposal, we see that

$$2^n - 1 = \prod_{d|n} \Phi_d(2) = (2-1) \prod_{\substack{d|n \\ d>1}} \Phi_d(2) = \prod_{\substack{d|n \\ d>1}} \Phi_d(2).$$

Therefore, if n is not prime, then $2^n - 1$ is certainly not prime because it is divisible by $\Phi_d(2)$ for $d > 1$ and $d|n$. $\triangle$

5.6.3   Möbius Inversion

Let $\{b_n\}_{n \geq 1}$ be a sequence in a ring R we define the associated product sequence $\{a_n\}_{n \geq 1}$ by

$$a_n = \prod_{d|n} b_d. \qquad (5.16)$$

Given a product sequence, it is possible to recover the original sequence b_n.

Theorem 5.6.12 (Möbius Inversion)

If a sequence $\{a_n\}_{n \geq 1}$ is given by (5.16), then

$$b_n = \prod_{d \mid n} (a_{n/d})^{\mu(d)},$$

where μ is the Möbius function defined on the positive integers by

$$\mu(n) = \begin{cases} 1 & \text{if } n = 1 \\ 0 & \text{if } n \text{ is not square-free} \\ (-1)^{\ell} & \text{if } n = p_1 p_2 \cdots p_\ell, \text{ where the } p_i \text{ are distinct primes.} \end{cases}$$

Proof. The proof involves manipulations of double products. We begin with the product on the right of the Möbius inversion formula,

$$\prod_{d \mid n} (a_{n/d})^{\mu(d)} = \prod_{d \mid n} \left(\prod_{e \mid (n/d)} b_e \right)^{\mu(d)} = \prod_{d \mid n} \left(\prod_{e \mid (n/d)} (b_e)^{\mu(d)} \right).$$

Note that the pairs (d, e) with $d \mid n$ and $e \mid (n/d)$ are the same as those with $e \mid n$ and $d \mid (n/e)$. Hence,

$$\prod_{d \mid n} a_{n/d}^{\mu(d)} = \prod_{e \mid n} \left(\prod_{d \mid (n/e)} (b_e)^{\mu(d)} \right) = \prod_{e \mid n} (b_e)^{\sum_{d \mid (n/e)} \mu(d)}. \qquad (5.17)$$

Now if $k = 1$, then $\sum_{d \mid k} \mu(d) = \mu(1) = 1$. On the other hand, suppose k is any integer with $k > 1$. Then the nonzero terms in $\sum_{d \mid k} \mu(d)$ correspond to products of distinct prime divisors of k. Suppose that $p_1, p_2, \ldots, p_r$ are the distinct prime divisors of k. In the sum $\sum_{d \mid k} \mu(d)$, we group together terms arising from products of i distinct prime divisors of k. Then

$$\sum_{d \mid k} \mu(d) = \sum_{i=0}^{r} \binom{r}{i} (-1)^i = (1 - 1)^r = 0.$$

Hence, we deduce that

$$\sum_{d \mid k} \mu(d) = \begin{cases} 1 & \text{if } k = 1 \\ 0 & \text{otherwise.} \end{cases}$$

We conclude from (5.17) that

$$\prod_{d \mid n} (a_{n/d})^{\mu(d)} = b_n$$

and the theorem follows. $\qquad \square$

Though the Möbius inversion formula has many applications in number theory, we presented it here to provide an alternative nonrecursive formula for the cyclotomic polynomials.

Proposition 5.6.13

For all positive integers n,

$$\Phi_n(x) = \prod_{d|n}(x^{n/d} - 1)^{\mu(d)}.$$

Proof. Follows from (5.15) and the Möbius inversion formula.   □

5.6.4   Useful CAS Commands

This section touches upon concepts closely connected to number theory. Most CAS offer plenty of commands that implement functions that are important for number theory. We list a few examples relevant to the content of this section, but point out that there exist many other useful commands. In *Maple*, most of what we need are in the `NumberTheory` package.

—————————————— Maple ——————————————

$with(NumberTheory):$

$CyclotomicPolynomial(10, x);$

$$x^4 - x^3 + x^2 - x + 1$$

$PrimitiveRoot(17);$

$$3$$

$Moebius(30);$

$$-1$$

———————————————————————————————

The command `PrimitiveRoot(n)` returns the smallest positive integer a such that $\bar{a}$ is a generator of $U(n)$. The following illustrates the same commands in SAGE.

—————————————— Sage ——————————————

```
sage: cyclotomic_polynomial(10,'x')
x^4 - x^3 + x^2 - x + 1
sage: primitive_root(17)
3
sage: moebius(30)
-1
```

———————————————————————————————

EXERCISES FOR SECTION 5.6

1. Show explicitly that 2 is a primitive root modulo 11. Use this to find all primitive roots modulo 11.

2. Use technology to find all primitive roots modulo 23.

3. Calculate $\Phi_{12}(x)$. Also find the roots of $\Phi_{12}(x)$ with radicals. [Hint: Exercise 5.2.10.]

4. Calculate $\Phi_{14}(x)$.

5. Prove that $\Phi_{2^n}(x) = x^{2^{n-1}} + 1$.

6. Prove that if $n > 1$ is odd, then $\Phi_{2n}(x) = \Phi_n(-x)$.

7. Show that if ζ_n is a primitive nth root of unity, then all the roots of $\Phi_n(x)$ are in $\mathbb{Q}(\zeta_n)$.

8. Suppose that m and n are positive relatively prime integers. Let ζ_n be a primitive nth root of unity and let ζ_m be a primitive mth root of unity. Prove that $\zeta_n\zeta_m$ is a primitive mnth root of unity.

9. Prove that if ζ_n is a primitive nth root of unity, then ζ_n^d is a primitive (n/d) root of unity.

10. Suppose that $p \mid n$. Prove that $\Phi_{pn}(x) = \Phi_n(x^p)$.

11. Suppose $n = p_1^{\alpha_1} p_2^{\alpha_2} \cdots p_\ell^{\alpha_\ell}$, where p_i are distinct primes. Prove that

$$\Phi_n(x) = \Phi_{p_1 p_2 \cdots p_\ell}\left(x^{p_1^{\alpha_1-1} p_2^{\alpha_2-1} \cdots p_\ell^{\alpha_\ell-1}}\right).$$

[Hint: Use Exercise 5.6.10.]

12. Use Exercise 5.6.11 to determine $\Phi_{225}(x)$.

13. Determine $\Phi_{60}(x)$. [Hint: Use previous exercises in this section.]

14. In 1883, Migotti proved that if p and q are distinct primes, then $\Phi_{pq}(x)$ only involves the coefficients of -1, 0, and 1. (In 1996, Lam and Leung offered a much shorter proof of this fact [21].) Use this theorem and previous exercises to prove that if $\Phi_n(x)$ has coefficients besides -1, 0, or 1, then it must be divisible by at least three distinct odd primes.

15. Suppose that p is a prime number such that $p \mid \Phi_n(2)$. Prove that $p \mid \Phi_{pn}(2)$. Deduce that $p \mid \Phi_{p^\alpha n}(2)$ for all positive integers α.

16. Let p be an odd prime dividing n. Suppose that $a \in \mathbb{Z}$ satisfying $\Phi_n(a) \equiv 0 \pmod{p}$. Prove that a is relatively prime to p and that the order of a in $U(p)$ is precisely n.

17. Let m and n be positive integers and let $l = \text{lcm}(m,n)$ and $d = \gcd(m,n)$. Prove that
    (a) $\mathbb{Q}(\zeta_m)\mathbb{Q}(\zeta_n) = \mathbb{Q}(\zeta_l)$ as a composite field;
    (b) $\mathbb{Q}(\zeta_m) \cap \mathbb{Q}(\zeta_n) = \mathbb{Q}(\zeta_d)$.

18. Prove that in the ring $\mathbb{Z}[x]$, the sequence of polynomial $\{x^n - 1\}_{n=1}^\infty$ satisfies the identity that
    $$\gcd(x^m - 1, x^n - 1) = x^{\gcd(n,m)} - 1,$$
    where in the first greatest common divisor we take the monic polynomial.

19. Prove that the Möbius inversion formula can also be written as

$$a_n = \prod_{d|n} b_d \Rightarrow b_n = \prod_{d|n} a_d^{\mu(n/d)}.$$

Exercises 5.6.20 through 5.6.25 deal with dynatomic polynomials *defined as follows. Let F be a field and let $P(x) \in F[x]$. Consider sequences in F that satisfy the recurrence relation $x_{n+1} = P(x_n)$. A fixed point is an element c in F or an extension of F that satisfies $P(x) - x = 0$. A 2-cycle is such a sequence that satisfies $x_2 = x_0$. An element on a 2-cycle satisfies the equation $P(P(x)) - x = 0$. However, fixed points also solve the equation $P(P(x)) - x = 0$. Consequently, elements that are on a 2-cycle but are not fixed points are solutions to the equation*

$$\frac{P(P(x)) - x}{P(x) - x}.$$

An n-cycle recurrence sequence as defined above such that $x_n = x_0$. Points on n-cycles satisfy $P^n(x) - x = 0$, where by $P^n(x)$ we mean $P(x)$ iterated n times. For example, $P^3(x) = P(P(P(x)))$. For any d that divides n, all the points on a d-cycle also satisfy $P^n(x) - x = 0$. Similar to cyclotomic polynomials, we define the nth dynatomic polynomial of $P(x)$ recursively by $\Phi_{P,1}(x) = P(x) - x$ and

$$P^n(x) - x = \prod_{d|n} \Phi_{P,d}(x).$$

An n-cycle that is not also a d-cycle for any d that divides n is called a primitive *n-cycle. Points on a primitive n-cycle must be roots of $\Phi_{P,n}(x)$. It is possible, though not easy to prove that $\Phi_{P,n}(x) \in F[x]$ for all $P(x) \in F[x]$ [6].*

20. Prove that polynomial iteration satisfies (a) $P^n(P^m(x)) = P^{m+n}(x)$; (b) $(P^n)^m(x) = P^{mn}(x)$; and (c) $\deg P^n(x) = k^n$ where $\deg P(x) = k$.

21. Suppose that $\deg P(x) = m > 1$. Prove that

$$\deg \Phi_{P,n}(x) = \sum_{d|n} \mu(d) m^{n/d}.$$

   Deduce that a sequence satisfying $x_{n+1} = P(x_n)$ can have at most $\frac{1}{n} \sum_{d|n} \mu(d) m^{n/d}$ primitive n-cycles.

22. Let $Q(x) = x^2 - 2$. Calculate $\Phi_{Q,2}(x)$, $\Phi_{Q,3}(x)$, and $\Phi_{Q,4}(x)$.

23. Let $P(x) = x^3 - 2$. Calculate $\Phi_{P,2}(x)$, $\Phi_{P,3}(x)$, and $\Phi_{P,4}(x)$.

24. Let $P(x) = x^2 - 2x + 2$. Calculate $\Phi_{P,2}(x)$, $\Phi_{P,3}(x)$, and $\Phi_{P,4}(x)$.

25. Let $Q(x) = x^2 - \frac{5}{4}$. Prove by direct calculation that for this particular polynomial, $\Phi_{Q,2}(x)$ divides $\Phi_{Q,4}(x)$.

5.7 Splitting Fields and Algebraic Closure

5.7.1 Splitting Fields

Let K be a field extension of F. Proposition 5.2.2, one of the key propositions of the section, established that for any element $\alpha \in K$ that is an algebraic element over F, there exists a unique monic polynomial $m_{\alpha,F}(x)$ of minimal degree in $F[x]$ such that α is a root of $m_{\alpha,F}(x)$. The motivating observation of this section is that though $F(\alpha)$ contains the root α, it does not necessarily contain all the roots of $m_{\alpha,F}(x)$.

Example 5.7.1. Let $F = \mathbb{Q}$ and consider $\alpha = \sqrt[3]{7} \in \mathbb{R}$. The minimal polynomial for α is $f(x) = x^3 - 7$. However, the three roots of this polynomial are

$$\sqrt[3]{7}, \quad \sqrt[3]{7}\left(\frac{-1 + i\sqrt{3}}{2}\right), \quad \sqrt[3]{7}\left(\frac{-1 - i\sqrt{3}}{2}\right).$$

Obviously, $\mathbb{Q}(\sqrt[3]{7})$ is a subfield of $\mathbb{R}$ but the other roots of the minimal polynomial are not in $\mathbb{R}$. △

Definition 5.7.2

A field extension K of F is called a *splitting field* for the polynomial $f(x) \in F[x]$ if $f(x)$ factors completely into linear factors in $K[x]$ but $f(x)$ does not factor completely into linear factors in $F'[x]$ where F' is any field with $F \subsetneq F' \subsetneq K$.

If K is an extension of F, we will also use the terminology that $f(x) \in F[x]$ splits completely in K to mean that $f(x)$ factors into linear factors in $K[x]$. However, this does not mean that K is a splitting field of $f(x)$ but simply that K contains a splitting field of $f(x)$.

Theorem 5.7.3

For any field F, if $f(x) \in F[x]$, there exists an extension K of F that is a splitting field for $f(x)$. Furthermore, $[K : F] \leq n!$ where $n = \deg f(x)$.

Proof. We proceed by induction on the degree of f. If $\deg f = 1$, then F contains the root of $f(x)$ so F itself is a splitting field for $f(x)$ and the degree of F over itself is 1.

Suppose that the theorem is true for all polynomials of degree less than or equal to n. Let $f(x)$ be a polynomial of degree $n + 1$. If $f(x)$ is reducible, then $f(x) = a(x)b(x)$ where $\deg a(x) = k$ with $1 \leq k \leq n$. By induction, both $a(x)$ and $b(x)$ have splitting fields, E_1 and E_2. Then the composite of these two fields, $E_1 E_2$, is a splitting field for $f(x)$. Furthermore, by the

induction hypothesis and Proposition 5.3.16, $[E_1 E_2 : F]$ is less than or equal
to $[E_1 : F][E_2 : F] = k!(n+1-k)! \le (n+1)!$.

Suppose now that $f(x)$ is irreducible. Then $F' = F[x]/(f(x))$ is a field
extension of F in which the element $\bar{x}$ is a root of $\overline{f(x)}$. Note that $[F' : F] =
(n+1)$. In $F'[t]$, $(t - \bar{x})$ is a linear factor of $f(t)$. One obtains the factorization
as follows. If

$$f(x) = \sum_{i=0}^{n+1} a_i x^i,$$

then $f(t) = 0$ is equivalent to $f(t) - f(\bar{x}) = 0$ so

$$0 = \sum_{i=1}^{n+1} a_i t^i - a_i \bar{x}^i = \sum_{i=1}^{n+1} (t - \bar{x})\left(a_i \sum_{j=0}^{i-1} t^j \bar{x}^{i-1-j}\right)$$

$$= (t - \bar{x}) \sum_{i=0}^{n} \sum_{j=0}^{i} a_{i+1} t^j \bar{x}^{i-j} = (t - \bar{x}) \sum_{j=0}^{n} \left(\sum_{i=j}^{n} a_{i+1} \bar{x}^{i-j}\right) t^j.$$

Therefore, in $F'[t]$, $f(t)$ factors into $f(t) = (t - \bar{x})q(t)$ where $q(t)$ has degree
n. By the induction hypothesis, $q(t)$ has a splitting field K over F' such that
$[K : F'] \le n!$. Therefore, $f(x)$ splits completely in K. Also,

$$[K : F] = [K : F'][F' : F] = (n+1)[K : F'] \le (n+1)\, n! \le (n+1)!.$$

By induction, the theorem holds for all fields and for all polynomials $f(x)$.$\square$

We would like to shift the notion of a splitting field of a polynomial $f(x)$
over F into a property of a field extension, without necessarily referring to a
specific polynomial $f(x)$.

Definition 5.7.4

A *normal extension* is an algebraic extension K of F that is the splitting field for some collection (not necessarily finite) of polynomials
$f_i(x) \in F[x]$.

Example 5.7.5. The splitting field of $f(x) = x^2 - x - 1$ over $\mathbb{Q}$ is $\mathbb{Q}\left(\frac{1+\sqrt{5}}{2}\right)$.
The degree of the extension is 2. $\triangle$

In fact, a splitting field of any quadratic polynomial $f(x) \in F[x]$ is $F(\alpha)$
where α is one of the roots.

Example 5.7.6. Consider the polynomial $f(x) = x^3 - 7$. Example 5.7.1 lists
the three roots of $f(x)$ in $\mathbb{C}$. A splitting field K for $f(x)$ must contain all three
of the roots. In particular, K must contain $\sqrt[3]{7}$ and $\zeta_3 = (-1 + \sqrt{-3})/2$. This
is sufficient so a splitting field is $K = \mathbb{Q}(\sqrt[3]{7}, \zeta_3)$. Recall that ζ_3 is algebraic
with degree 2 and with minimal polynomial $x^2 + x + 1$. Furthermore, ζ_3 is

not a real number so it is not an element of $\mathbb{Q}(\sqrt[3]{7})$ which is a subfield of $\mathbb{R}$. Hence, $[K : \mathbb{Q}(\sqrt[3]{7})] = 2$ and so the degree of the extension is

$$[K : \mathbb{Q}] = [\mathbb{Q}(\sqrt[3]{7}, \zeta) : \mathbb{Q}] = [K : \mathbb{Q}(\sqrt[3]{7})][\mathbb{Q}(\sqrt[3]{7}) : \mathbb{Q}] = 2 \cdot 3 = 6. \qquad \triangle$$

Example 5.7.7. We point out that the extension $\mathbb{Q}(\sqrt[3]{7})$ is not a normal extension of $\mathbb{Q}$ because it contains $\sqrt[3]{7}$ but not the other two roots of the minimal polynomial $x^3 - 7$. $\qquad \triangle$

Example 5.7.8. Consider the cubic polynomial $p(x) = x^3 + 6x^2 + 18x + 18 \in \mathbb{Q}[x]$ in Example 5.4.5. We prove that the splitting field K of $p(x)$ has degree $[K : \mathbb{Q}] = 6$. Let $x_1 = 2 + \sqrt[3]{4} - \sqrt[3]{2}$. Note that x_1 has degree 3 over $\mathbb{Q}$ since the polynomial is irreducible (by Eisenstein's Criterion modulo 2). However, $x_1 \in \mathbb{R}$, whereas x_2 and x_3 have imaginary components. Thus, $x_2, x_3 \notin \mathbb{Q}(x_1)$ and the splitting field of $p(x)$ is a nontrivial extension of $\mathbb{Q}(x_1)$. We can conclude from Theorem 5.7.3 that the splitting field of $p(x)$ has degree 6. However, we can also tell that

$$p(x) = (x - x_1)\left(x^2 + (6 + x_1)x - \frac{18}{x_1} \right).$$

So x_2 and x_3 are the roots of the quadratic polynomial. $\qquad \triangle$

Example 5.7.9. Consider the polynomial $g(x) = x^4 + 2x^2 - 2$. We can find the roots by first solving a quadratic polynomial for x^2. Thus,

$$x^2 = \frac{-2 \pm \sqrt{4 + 8}}{2} = -1 \pm \sqrt{3}.$$

Thus, the four roots of $g(x)$ are $\pm\sqrt{-1 + \sqrt{3}}$ and $\pm\sqrt{-1 - \sqrt{3}}$. Since $g(x)$ is irreducible, we have $[\mathbb{Q}(\sqrt{-1 + \sqrt{3}}) : \mathbb{Q}] = 4$. It is easy to tell that $\sqrt{-1 - \sqrt{3}} \notin \mathbb{Q}(\sqrt{-1 + \sqrt{3}})$ because $\mathbb{Q}(\sqrt{-1 + \sqrt{3}})$ is a subfield of $\mathbb{R}$ whereas $\sqrt{-1 - \sqrt{3}}$ is a complex number. Noticing first that $\sqrt{3} \in \mathbb{Q}(\sqrt{-1 + \sqrt{3}})$, as $\sqrt{3} = (\sqrt{-1 + \sqrt{3}})^2 + 1$, we see that $\sqrt{-1 - \sqrt{3}}$ is an algebraic element over $\mathbb{Q}(\sqrt{-1 + \sqrt{3}})$ satisfying the polynomial equation

$$x^2 + 1 + \sqrt{3} = 0.$$

Thus, a splitting field of $g(x)$ over $\mathbb{Q}$ is $K = \mathbb{Q}(\sqrt{-1 + \sqrt{3}}, \sqrt{-1 - \sqrt{3}})$. Furthermore,

$$[K : \mathbb{Q}] = \left[K : \mathbb{Q}\left(\sqrt{-1 + \sqrt{3}} \right) \right]\left[\mathbb{Q}\left(\sqrt{-1 + \sqrt{3}} \right) : \mathbb{Q} \right] = 2 \cdot 4 = 8.$$

This degree is a strict divisor of the upper bound $4! = 24$ as permitted by Theorem 5.7.3. $\qquad \triangle$

From the examples, the splitting field of some polynomials seems a natural construction so it may seem puzzling why we have been saying "a" splitting field. From the construction of a splitting field as described in Theorem 5.7.3 it is not obvious that splitting fields are unique. The following theorem establishes this important property.

Theorem 5.7.10

Let $\varphi : F \cong F'$ be an isomorphism of fields. Let $f(x) \in F[x]$ and let $f'(x) \in F'[x]$ be the polynomial obtained by applying φ to the coefficients of $f(x)$. Let E be a splitting field for $f(x)$ over F and let E' be a splitting field for $f'(x)$ over F'. Then the isomorphism φ extends to an isomorphism $\sigma : E \cong E'$ such that $\sigma|_F = \varphi$.

Proof. We proceed by induction on the degree n of $f(x)$.

If $n = 1$, then $E = F$ and $E' = F'$ and we can use $\sigma = \varphi$ as a trivial extension.

Suppose the induction hypothesis that the theorem holds for any field F, any isomorphism φ and any polynomial $f(x)$ with $\deg f(x) < n$. Let $p(x)$ be an irreducible factor of $f(x)$ in $F[x]$ with $\deg p(x) \geq 2$ and let $p'(x)$ be the corresponding irreducible factor of $f'(x)$ in $F'[x]$. If α is a root of $p(x)$ in E and β a root of $p'(x)$ in E', then by Exercise 5.1.18 there exists a field isomorphism $\sigma' : F(\alpha) \cong F'(\beta)$ that extends φ. Call $F_1 = F(\alpha)$ and $F_1' = F'[x]/(p'(x)) = F'(\beta)$. Then, over F_1, we have $f(x) = (x - \alpha)f_1(x)$ while over F_1' we have $f'(x) = (x - \beta)f_1'(x)$ for polynomials $f_1(x) \in F_1[x]$ and $f_1'(x) \in F_1'[x]$ both of degree $n - 1$.

We know that $f_1(x)$ splits completely in E but we can also conclude that the field E is a splitting field for $f_1(x)$ over F_1. Indeed, if L were any field $F_1 \subsetneq L \subsetneq E$ over which $f_1(x)$ split completely, then since $\alpha \in L$, $f(x)$ would also split completely in L. Since a splitting field is a minimal field extension over which a polynomial splits, E must be the splitting field of $f_1(x)$ over F_1. The same holds for E' as a splitting field of $f_1'(x)$ over F_1'.

By the induction hypothesis, there exists a field isomorphism $\sigma : E \cong E'$ that extends σ' and therefore σ also extends φ. $\square$

Corollary 5.7.11

Any two splitting fields for a polynomial $f(x) \in F[x]$ over a field F are isomorphic.

Because of Corollary 5.7.11, we talk about *the* splitting field.

Example 5.7.12 (Cyclotomic Fields). Recall that cyclotomic extensions are extensions of $\mathbb{Q}$ that contain the n-roots of unity, i.e., the roots of $x^n - 1 = 0$. As in Section 5.6, we call ζ_n the complex number

$$\zeta_n = e^{2\pi i/n} = \cos\left(\frac{2\pi}{n}\right) + i\sin\left(\frac{2\pi}{n}\right).$$

Then all the roots of unity are of the form ζ_n^k for $0 \le k \le n-1$. This shows, however, that all the roots of unity are in $\mathbb{Q}(\zeta_n)$. Consequently, $\mathbb{Q}(\zeta_n)$ is the splitting field of $x^n - 1$ and more precisely of the cyclotomic polynomial $\Phi_n(x)$. △

In some examples and exercises, we have seen that on occasion an extension $F(\alpha, \beta)$ of F is nonetheless a primitive extension with $F(\alpha, \beta) = F(\gamma)$. Splitting fields allow us to prove that this always happens under certain circumstances.

> ### Theorem 5.7.13 (Primitive Element Theorem)
>
> If char $F = 0$ and if α and β are algebraic over F, then there exists $\gamma \in F(\alpha, \beta)$ such that $F(\alpha, \beta) = F(\gamma)$.

Proof. Let K be the splitting field of $m_{\alpha,F}(x)m_{\beta,F}(x)$. Let $\alpha_1, \alpha_2, \ldots, \alpha_m$ be the roots of $m_{\alpha,F}(x)$ in K and let $\beta_1, \beta_2, \ldots, \beta_n$ be the roots of $m_{\beta,F}(x)$. We assume that $\alpha = \alpha_1$ and $\beta = \beta_1$. Every field of characteristic 0 has an infinite number of elements. Choose some element $d \in F$ such that

$$d \ne \frac{\alpha - \alpha_i}{\beta_j - \beta} \qquad \text{for } i \ge 1 \text{ and } j > 1.$$

Given this choice of d, set $\gamma = \alpha + d\beta$.

Obviously, $F(\gamma)$ is a subfield of $F(\alpha, \beta)$. We wish to show that the converse inclusion holds: $F(\alpha, \beta) \subseteq F(\gamma)$. In the polynomial ring $F(\gamma)[x]$, both $m_{\beta,F}(x)$ and $p(x) = m_{\alpha,F}(\gamma - dx)$ satisfy

$$m_{\beta,F}(\beta) = 0 \qquad \text{and} \qquad p(\beta) = m_{\alpha,F}(\alpha) = 0.$$

Hence, both polynomials are divisible by $m_{\beta,F(\gamma)}(x)$.

Now, since $m_{\beta,F(\gamma)}(x)$ divides $m_{\beta,F}(x)$ in $F(c)[x]$, then $m_{\beta,F}(x)$ splits completely in $K[x]$. Furthermore, the only zeros of $m_{\beta,F(\gamma)}(x)$ must also be zeros of $m_{\beta,F}(x)$, namely $\beta_1, \beta_2, \ldots, \beta_n$. However, a zero of $m_{\beta,F(\gamma)}(x)$ must also be a zero of $p(x)$, namely some x_0 such that $\gamma - dx_0 = \alpha_i$ or in other words, some x_0 such that

$$\alpha + d\beta = \alpha_i + dx_0 \iff \alpha - \alpha_i = d(x_0 - \beta).$$

By definition of d, the only x_0 that satisfies this and is a root of $m_{\beta,F}(x)$ is $x_0 = \beta$. Thus, in $K[x]$ and hence also in $F(\alpha, \beta)$, we deduce that β is the only root of $m_{\beta,F(\gamma)}(x)$. Since it is irreducible, $m_{\beta,F(\gamma)}(x) = x - \beta$. This shows that $\beta \in F(\gamma)$ From this, we also deduce that $\alpha = \gamma - d\beta \in F(\gamma)$. Hence, $F(\alpha, \beta) \subseteq F(\gamma)$ and the theorem follows. □

5.7.2   Algebraic Closure

As we consider algebraic extensions of a field, from an intuitive perspective, we often think of adjoining a collection of algebraic elements to some base field. Is there ever a situation in which there is nothing else we can adjoin? From another perspective, given a field F, is there some extension of $F[x]$ where every polynomial splits completely?

Definition 5.7.14

Let F be a field. A field L is called an *algebraic closure* of F if L is algebraic over F and if every polynomial $f(x) \in F[x]$ splits completely in L.

A related notion is the following property of a field within itself.

Definition 5.7.15

A field F is said to be *algebraically closed* if every polynomial $f(x) \in F[x]$ has a root in F.

Notice that if F is algebraically closed, then every polynomial $f(x) \in F[x]$ has a root α in F. Consequently, in $F[x]$, the polynomial factors $f(x) = (x - \alpha)p(x)$ for some $p(x) \in F[x]$. Since $p(x)$ and any subsequent factors must have a root, then by induction, every polynomial splits completely. Consequently, a field F is algebraically closed if it is an algebraic closure of itself. This remark motivates the following easy proposition.

Proposition 5.7.16

If L is an algebraic closure of F, then L is algebraically closed.

Proof. Let $f(x) \in L[x]$ and let α be a root of $f(x)$. Then α gives an algebraic extension $L(\alpha)$ of L. However, since L is algebraic over F, by Theorem 5.3.19, $L(\alpha)$ is an algebraic extension of F. Thus, α is algebraic over F and hence $\alpha \in L$. This shows that L is algebraically closed. $\qquad\square$

The concept of an algebraic closure of a field is a rather technical one. Though the previous portion of this section outlined how to construct a splitting field of a polynomial over a field, finding a field extension in which all polynomials split completely poses a problem of construction. Indeed, though Definition 5.7.14 defines the notion of algebraic closure, it is not at all obvious that an algebraic closure exists for a given field F. Also, if an algebraic closure of F exists, it is not readily apparent whether algebraic closures are unique. This section provides answers to these questions but the proofs of some of the results depend on Zorn's Lemma, which is equivalent to the Axiom of Choice.

It is also not at all clear that any algebraically closed fields exist. From the properties of complex numbers, the quadratic formula, and Cardano's cubic and quartic formula, one may hypothesize that the field of complex numbers is algebraically closed. Indeed, as early as the 17th century, mathematicians, including the likes of Euler, Laplace, Lagrange, and d'Alembert, attempted to prove this. The first rigorous proof was provided by Argand in 1806. Since then, mathematicians have discovered proofs involving techniques from disparate branches of mathematics. Because of its importance in algebra and the difficulty of proving it, the fact that $\mathbb{C}$ is an algebraically closed field became known as the Fundamental Theorem of Algebra.

Theorem 5.7.17 (Fundamental Theorem of Algebra)

The field $\mathbb{C}$ is algebraically closed.

The "simplest" proof of the Fundamental Theorem of Algebra uses theorems from complex analysis that are outside the scope of this text. This should not surprise us since the construction of $\mathbb{C}$ depends on the construction of $\mathbb{R}$ and properties of the reals and functions on them are precisely the purview of analysis. The simplest proof that is entirely algebraic uses Galois theory, a topic sometimes covered in a second course in abstract algebra, and hence just beyond this book. Consequently, for the moment, we accept this result without proof.

The proof of Theorem 5.7.10 showed how to construct a splitting field K of a single polynomial $f(x) \in F[x]$ over the field F. An algebraic closure of a field F must essentially be a splitting field for all polynomials in $F[x]$. It is hard to imagine what such a field would look like and how to describe such a field. If we had only a finite number of polynomials $f_1(x), f_2(x), \ldots, f_k(x)$, then the composite of the k splitting fields, which is also the splitting field of $f_1(x)f_2(x) \cdots f_k(x)$, contains all the roots of these polynomials. However, $F[x]$ contains an infinite number of polynomials, so we are faced with a problem of constructibility.

To keep track of *all* the polynomials in $F[x]$, Artin devised the strategy of introducing a separate variable for each polynomial. We give his proof here below.

Theorem 5.7.18

For any field F, there exists an algebraically closed field K containing F.

Proof. Let $\mathcal{P}$ be the set of associate classes of irreducible elements in $F[x]$. Every class in $\mathcal{P}$ can be represented by a unique monic nonconstant irreducible polynomial $p(x)$. Let S be a set of indeterminate symbols that is in bijection with $\mathcal{P}$ via $[p] \leftrightarrow x_p$, where $p(x)$ is monic. Consider the multivariable polynomial ring $F[S]$. In $F[S]$ consider the ideal

$$I = (p(x_p) \mid [p] \in \mathcal{P} \text{ and } p(x) \text{ is monic}).$$

We first prove that I is a proper ideal of $F[S]$. Assume that $I = F[S]$. Then there exist monic irreducible polynomials $p_1, p_2, \ldots, p_n(x)$ and polynomials

$g_1, g_2, \ldots, g_n \in F[S]$ such that

$$g_1 p_1(x_{p_1}) + g_2 p_2(x_{p_2}) + \cdots + g_n p_n(x_{p_n}) = 1. \tag{5.18}$$

The polynomials $g_1, g_2, \ldots, g_n$ can involve only a finite set of variables $x_1, x_2, \ldots, x_m$, some of which must be the variables $x_{p_1}, x_{p_2}, \ldots, x_{p_n}$. Let L be a field extension of F in which α_i is a root of $p_i(x)$ for each $i = 1, 2, \ldots, n$. Evaluating the expression (5.18) at some point $(c_1, c_2, \ldots, c_m \in L^m$ for which the variable corresponding to x_{p_i} is α_i, we get $0 = 1$ in L. This contradicts the assumption that $I = F[S]$. Hence, I is a proper ideal.

Since I is a proper ideal, by Krull's Theorem (Theorem 3.8.2), I is contained in a maximal ideal $M \in F[S]$. Then the quotient ring $K_1 = F[S]/M$ is a field containing F as a subfield. Furthermore, for each monic irreducible polynomial $p(x) \in F[x]$, the element $\overline{x_p}$ in K_1 is a root of $p(x)$. Hence, every polynomial $f(x) \in F[x]$ has a root in K_1. This does not yet show that F is algebraically closed since the roots are in a field extension.

We repeat the construction, now with K_1 serving the role of F. This gives a field extension K_2 of K_1 in which every polynomial $q(x) \in K_1[x]$ has a root in K_2. For all integers $i > 0$, construct the field extension K_{i+1} of K_i in the same way. This iterated construction creates a chain of nested field extensions of F,

$$F = K_0 \subseteq K_1 \subseteq K_2 \subseteq \cdots \subseteq K_n \subseteq \cdots$$

in which every polynomial $q(x) \in K_i[x]$ has a root in K_{i+1}. Let K be the union of all the fields

$$K = \bigcup_{i \geq 0} K_i.$$

The field K is an extension of F. Let $q(x) \in K[x]$ be a polynomial

$$q(x) = q_k x^k + \cdots + q_1 x + q_0.$$

For $0 \leq \ell \leq k$, the coefficient q_ℓ is in some K_{i_ℓ}. If N is the maximum of $\{i_1, i_2, \ldots, i_\ell\}$, then $q(x) \in K_N[x]$. Then $q(x)$ has a root in K_{N+1}, which is in K. Thus, K is an algebraically closed field that is an extension of F. $\quad\square$

It is interesting to observe that the existence of a maximal ideal M containing I follows from Zorn's Lemma.

The field K constructed in the above proof may seem woefully large. Indeed, the strategy of the proof simply provides a well-defined construction of a field extension that is large enough to be algebraically closed. However, this could be far larger than an algebraic closure. The following proposition pares down the algebraically closed field K to an algebraic closure of F.

Proposition 5.7.19

Let L be an algebraically closed field and let F be a subfield of L. The set K of elements in L that are algebraic over F is an algebraic closure of F.

Proof. By definition, K is algebraic over F. Furthermore, every polynomial $f(x) \in F[x] \subset L[x]$ splits completely over L. But each root α of $f(x)$ is algebraic of F so is an element of K. Therefore, all the linear factors $(x - \alpha)$ of the factorization of $f(x)$ are in $K[x]$. Hence, $f(x)$ splits completely in $K[x]$ and hence K is an algebraic closure of F. $\qquad\square$

Theorem 5.7.18 coupled with Proposition 5.7.19 establish the existence of algebraic closures for any field F. This has not yet answered the important question of whether algebraic closures of a field are unique (up to isomorphism). In order to establish this, we need an intermediate theorem.

Theorem 5.7.20

> Let F be a field, let E be an algebraic extension of F and let $f : F \to L$ be an embedding (injective homomorphism) of F into an algebraically closed field L. Then there exists an embedding $\lambda : E \to L$ that extends f, i.e., $\lambda|_F = f$.

Proof. Let S be the set of all pairs (K, σ), where K is a field with $F \subseteq K \subseteq E$ such that σ extends f to an embedding of K into L. We define a partial order $\preccurlyeq$ on S where $(K_1, \sigma_1) \preccurlyeq (K_2, \sigma_2)$ means if $K_1 \subseteq K_2$ and σ_2 extends σ_1, i.e., $\sigma_2|_{K_1} = \sigma_1$. The set S is nonempty since it contains the pair (F, f). For any chain

$$\{(K_i, \sigma_i)\}_{i \in \mathcal{I}}$$

in the poset $(S, \preccurlyeq)$ define $K' = \bigcup_{i \in \mathcal{I}}$. Every element $\alpha' \in K$ is in K_i for some $I \in \mathcal{I}$. Define the function $\sigma' : K' \to L$ by $\sigma'(\alpha) = \sigma_i$ if $\alpha \in K_i$. This function is well-defined because if $K_i \subseteq K_j$, then $\sigma_j|_{K_i} = \sigma_i$ so $\sigma_j(\alpha) = \sigma_i(\alpha)$. Therefore, the choice of index i to use for defining σ' is irrelevant. The pair (K', σ') is an upper bound for the described chain. Consequently, Zorn's Lemma applies and we conclude that S contains maximal elements.

For a maximal element (K, λ) in S, the field K is a subfield of E and the function $\lambda : K \to L$ is an embedding of K into L that extends f. Assume that there exists $\alpha \in E - K$. Since E is algebraic over F, by Corollary 5.2.3, it is algebraic over K. By Exercise 5.1.18, $\lambda : K \to L$ can be extended to an embedding $K(\alpha) \to L$, contradicting the maximality of the pair (K, λ). Hence, $E - K = \emptyset$, so $K = E$ and the function $\lambda : E \to L$ is an extension of $F \to L$. $\qquad\square$

Theorem 5.7.20 gives the following important Proposition.

Proposition 5.7.21

> Let F be a field and let E and E' be two algebraic closures of F. Then E and E' are isomorphic.

Proof. Let $f : F \to E'$ be an embedding of F into E'. By Proposition 5.7.16, E' is algebraically closed. Since E is algebraic over F, by Theorem 5.7.20, there exists an embedding $\lambda : E \to E'$ that extends f. Since E is algebraically closed and E' is algebraic over $f(F)$, then $\lambda(E)$ is algebraically closed. By Corollary 5.2.3, E' is algebraic over $\lambda(E)$. As an algebraic extension of an algebraically closed field, $E' = \lambda(E)$. Thus, λ is a surjective embedding, so λ is an isomorphism. $\square$

In light of this proposition, algebraic closures of a field are unique up to isomorphism. Consequently, we talk about *the* algebraic closure of a field F and we denote it by $\overline{F}$. The fact that the algebraic closure of the algebraic closure of F is just the algebraic closure of F can be succinctly stated by $\overline{\overline{F}} = \overline{F}$.

The field of complex numbers $\mathcal{C}$ is a field extension of $\mathbb{Q}$ that is algebraically closed. By Proposition 5.7.19, the set of algebraic elements in the extension $\mathbb{C}/\mathbb{Q}$ is the algebraic closure of $\mathbb{Q}$. The field $\overline{\mathbb{Q}}$ is called the field of *algebraic numbers*. Though this describes the algebraic closure of $\mathbb{Q}$, the extension $\mathbb{C}$ is not necessarily the field K constructed in the proof of Theorem 5.7.18. Hence, though we use Proposition 5.7.19 to show that the algebraic numbers are the algebraic closure $\overline{\mathbb{Q}}$, we did not need the construction in the proof of Theorem 5.7.18 but rather the Fundamental Theorem of Algebra.

In contrast, we point out that the Fundamental Theorem of Algebra does not help us to construct $\overline{\mathbb{F}_2}$. Instead, we must refer to Theorem 5.7.18. The field $\overline{\mathbb{F}_2}$ is an algebraically closed field of characteristic 2. Proposition 5.7.19 leads to the intuitive perspective that $\overline{\mathbb{F}_2}$ is a field that contains all the roots of all polynomials $f(x) \in \mathbb{F}_2[x]$. This is interesting because a priori, from the proof of Theorem 5.7.18, one needs to worry about the roots of polynomials with coefficients in *every* algebraic extension of $\mathbb{F}_2$ being back in $\overline{\mathbb{F}_2}$.

We observe that the algebraic closure of a field F does not necessarily have infinite degree. For example, the algebraic closure of $\mathbb{R}$ is $\mathbb{C}$ but $[\mathbb{C} : \mathbb{R}] = 2$.

5.7.3 Useful CAS Commands

Sage offers some commands associated with finding the splitting field of a polynomial. Consider the following command.

———————————— Sage ————————————

```
sage: (x^4-3).splitting_field('x')
Number Field in x with defining polynomial x^8 + 4*x^7
   + 10*x^6 + 16*x^5 + 13*x^4 + 4*x^3 + 28*x^2 + 28*x + 13
```

We interpret the answer to mean that the splitting field K of $x^4 - 3$ over $\mathbb{Q}$ is $\mathbb{Q}(\alpha)$, where α is the root of the given polynomial of degree 8. From the methods of this section, we know that the polynomial $x^4 - 3 \in \mathbb{Q}[x]$ has splitting field $\mathbb{Q}(\sqrt[4]{3}, i)$. Though it is not immediately obvious, these answers are the same. In fact, $\sqrt[4]{3} + i$ is a root of the given polynomial.

The answer provided by SAGE tells us that $[K : \mathbb{Q}] = 8$, which is considerably less than the possible degree 24, for the splitting field of a polynomial of degree 4.

EXERCISES FOR SECTION 5.7

1. Find the splitting field of $x^4 - 3x^2 + 1 \in \mathbb{Q}[x]$.

2. Find the splitting field of $x^6 - 2x^3 - 1 \in \mathbb{Q}[x]$.

3. Find the splitting field of $(x^2 - 2)(x^3 - 2) \in \mathbb{Q}[x]$.

4. Find the splitting field of $(x^3 - 2)(x^3 - 7) \in \mathbb{Q}[x]$.

5. Find the splitting field of $x^6 - 5 \in \mathbb{Q}[x]$ and determine the degree of the splitting field over $\mathbb{Q}$.

6. Describe the splitting field of $x^4 + 2x^2 + 1 \in \mathbb{F}_7[x]$ as a quotient ring of $\mathbb{F}_7[x]$.

7. Let F be a field and let $a \in F$. Show that the splitting field of a polynomial $p(x)$ is the same as the splitting field of the polynomial $q(x) = p(x - a)$.

8. Let p and q be prime numbers in $\mathbb{N}$. Prove that the splitting field of $x^p - q$ is an extension of degree $p(p - 1)$ for $\mathbb{Q}$.

9. Let F be a field and let $f(x)$ be an irreducible cubic polynomial in $F[x]$. Prove that the splitting field K of $f(x)$ has degree

$$[K : F] = \begin{cases} 3 & \text{if the discriminant } \Delta \text{ is the square of an element in } F, \\ 6 & \text{otherwise.} \end{cases}$$

10. Let $p(x) \in F[x]$ be a polynomial of degree n and let K be the splitting field of $p(x)$ over F. Prove that $[K : F]$ in fact divides $n!$.

11. Let $p(x), q(x) \in F[x]$ be two polynomials with $\deg p(x) = m$ and $\deg q(x) = n$. Notice that $p(q(x))$ is a polynomial of degree mn. Prove that the splitting field E of $p(q(x))$ has a degree that satisfies $[E : F] \leq m!(n!)^m$. Prove also that for $m, n \geq 2$, this quantity strictly divides $(mn)!$.

12. Let $P(x)$ be a polynomial in $F[x]$. Suppose that a dynatomic polynomial $\Phi_{P,n}(x)$ has degree k. (See Exercises 5.6.20 through 5.6.25.) Prove that if the roots of a dynatomic polynomial are only primitive n-cycles, then k is divisible by n and the degree of the splitting field E of $\Phi_{P,n}(x)$ has an index $[E : F]$ that is less than or equal to

$$k(k - n)(k - 2n) \cdots (2n) \cdot n \cdot 1.$$

13. Let $p(x) \in \mathbb{Q}[x]$ be a palindromic polynomial of even degree $2n$. Let K be the splitting field of $p(x)$. Prove that $[K : \mathbb{Q}] \leq 2^n n!$. [Hint: See Exercise 5.2.10.] [Note: This degree is less than the value of $(2n)!$ allowed by Theorem 5.7.3.]

14. Prove that a field F is algebraically closed if and only if the only the irreducible polynomials in $F[x]$ are precisely the polynomials of degree 1.

15. Prove that a field F is algebraically closed if and only if it has no proper algebraic extension.

16. Let K be an algebraic extension of a field F. Prove that $\overline{K} = \overline{F}$.

5.8   Finite Fields

Fields of characteristic 0 and fields of characteristic p have a number of qualitative differences. This section builds on theorems of Section 5.7 to analyze
finite fields. In particular, the main theorem of this section is that finite fields
of a given cardinality are unique up to isomorphism. However, in order to
establish this foundational result, we must take a detour into the concept of
separability.

5.8.1   Separable and Inseparable Extensions

Let F be a field. Let $f(x) \in F[x]$ be a polynomial of degree m and let K be
a splitting field of $f(x)$ over F. Suppose that $f(x)$ factors into linear terms in
$K[x]$ as

$$f(x) = a_m(x - \alpha_1)^{n_1}(x - \alpha_2)^{n_2} \cdots (x - \alpha_k)^{n_k},$$

where $\alpha_i \neq \alpha_j$ for $i \neq j$ and $n_1 + n_2 + \cdots + n_k = \deg f(x) = m$.

Definition 5.8.1

A root α_i is said to have *multiplicity* n_i if $(x - \alpha_i)^{n_i}$ divides $f(x)$ in
$K[x]$ but $(x - \alpha_i)^{n_i+1}$ does not divide $f(x)$. If $n_i > 1$, we will say that
α_i is a *multiple root*.

Since $K[x]$ is a UFD, we can use the $\operatorname{ord}_\pi : K[x] \to \mathbb{N}$ function and say
that α is a root of $f(x)$ if $\operatorname{ord}_{(x-\alpha)} f(x) > 0$ and that the multiplicity of α is
$n = \operatorname{ord}_{(x-\alpha)} f(x)$. According to the definition, α is a multiple root whenever
$\operatorname{ord}_{(x-\alpha)} f(x) > 1$.

Definition 5.8.2

A polynomial $f(x) \in F[x]$ is called *separable* if it has no multiple
roots in its splitting field over F.

Definition 5.8.3

An algebraic extension K/F is called *separable* if for all $\alpha \in K$, the
minimal polynomial $m_{\alpha,F}(x)$ is a separable polynomial. An algebraic
extension that is not separable is called *inseparable*.

It may at first seem difficult to imagine a field extension that is not separable. In this section, we will show that many field extensions that we have
studied so far are separable. The following example illustrates an inseparable
extension.

Example 5.8.4. Consider the infinite field $F = \mathbb{F}_3(x)$, which has characteristic 3. Consider also the field extension $K = F[\sqrt[3]{x}]$. The element $\sqrt[3]{x} \notin F$ and $\sqrt[3]{x}$ has minimal polynomial $m(t) = t^3 - x$. However,

$$m(t) = t^3 - 3t^2 \sqrt[3]{x} + 3tx^{2/3} - x = (t - \sqrt[3]{x})^3$$

so $\sqrt[3]{x}$ is a triple root of its own minimal polynomial. Hence, K is an inseparable extension. △

Exercise 4.5.22 presented the concept of a derivative of a polynomial. In essence, let $p(x) \in F[x]$. We define the derivative of $p(x)$ with respect to x as the polynomial $D_x(p(x)) \stackrel{\text{def}}{=} p'(x)$, where $p'(x)$ is the derivative encountered in calculus. We know that $\deg D_x(p(x)) < \deg p(x)$ regardless of the field. Furthermore, the derivative of a polynomial satisfies the addition rule and the Leibniz rule for multiplication. The polynomial derivative is particularly useful for the following proposition.

Proposition 5.8.5

A polynomial $f(x) \in F[x]$ is separable if and only if $f(x)$ and $D_x f(x)$ are relatively prime.

Proof. Suppose that $f(x)$ is not separable. Then there exists a root α of $f(x)$ such that $f(x) = (x - \alpha)^2 q(x)$ in the splitting field K of $f(x)$. Then by the properties of the derivative,

$$D_x(f(x)) = 2(x - \alpha)q(x) + (x - \alpha)^2 D_x(q(x)).$$

We see that α is a root of $D_x(f(x))$, so $m_{\alpha,F}(x)$ divides $D_x(f(x))$. Then $m_{\alpha,F}(x)$ divides both $f(x)$ and $D_x(f(x))$ so these two polynomials are not relatively prime.

Conversely, suppose that $f(x)$ and $D_x(f(x))$ are not relatively prime. Then there exists a monic irreducible polynomial $a(x)$ of degree greater than 1 that divides them both. Let α be a root of $a(x)$ in the splitting field K of $f(x)$. Then $f(x) = (x - \alpha)q(x)$ for some polynomial $q(x) \in K[x]$. Thus,

$$D_x(f(x)) = q(x) + (x - \alpha)D_x(q(x)).$$

Since $(x - \alpha)$ divides $D_x(f(x))$, then $(x - \alpha)$ divides

$$q(x) = D_x(f(x)) - a(x)D_x(q(x)).$$

Consequently, $q(x) = (x - \alpha)g(x)$ for some polynomial $g(x) \in K[x]$ and we deduce that

$$f(x) = (x - \alpha)^2 g(x).$$

Then $f(x)$ is not separable. □

This proposition leads immediately to the following proposition.

Proposition 5.8.6

> If char $F = 0$, then every irreducible polynomial is separable.

Proof. Let $a(x)$ be an irreducible polynomial in $F[x]$. If $\deg a(x) = 1$, then $a(x)$ is separable trivially.

Suppose that $\deg a(x) \geq 2$. If $\mathrm{LT}(a(x)) = a_n x^n$, then the leading term of $D_x(a(x))$ is $n a_n x^{n-1}$. Hence, $\deg D_x(a(x)) = n - 1 \geq 1$. Since $a(x)$ is irreducible, any polynomial $b(x)$ that divides $a(x)$ must either be a nonzero multiple of $a(x)$ or a nonzero constant. If $b(x)$ must also divide $D_x(a(x))$, then $\deg b(x) \leq n - 1$. Hence, $b(x)$ cannot be a nonzero constant multiple of $a(x)$. So it must be a nonzero constant. Thus, $D_x(a(x))$ and $a(x)$ are relatively prime and so by Proposition 5.8.5, $a(x)$ is separable.   □

Corollary 5.8.7

> Let F be a field of characteristic 0. Every algebraic extension of F is separable.

Proof. Let K be an algebraic extension of F and let $\alpha \in K$. Then $m_{\alpha,F}(x)$ is irreducible and by Proposition 5.8.6, $m_{\alpha,F}(x)$ is separable. Thus, K is separable.   □

The proof of Proposition 5.8.6 might not work on all polynomials in $F[x]$ when F has a positive characteristic. In characteristic 0, if $\deg a(x) = n \geq 1$, then we know that $\deg D_x(a(x)) = n - 1$. However, if char $F = p$ and $\deg a(x) = pk$, then the derivative of the leading term is

$$D_x(\mathrm{LT}(a(x))) = D_x(a_{pk} x^{pk}) = pk a_{pk} x^{pk-1} = 0.$$

Hence, $\deg D_x(a(x)) < n - 1$. Furthermore, the derivative of any monomial whose power is a multiple of p has a derivative that is identically 0. This leads to the following important point.

Proposition 5.8.8

> Let F be a field of characteristic p. Suppose that $a(x)$ is irreducible. Then $a(x)$ is separable if and only if one of the monomials of $a(x)$ has a degree that is not a multiple of p. Furthermore, for any irreducible polynomial, there exists an irreducible separable polynomial $b(x)$ and a nonnegative integer k
>
> $$a(x) = b(x^{p^k}).$$

Proof. By Proposition 5.8.5, $a(x)$ is not separable if and only if $a(x)$ and $D_x(a(x))$ are divisible by a factor of degree greater than 0 in $F[x]$. However, since $a(x)$ is irreducible, the only divisor of $a(x)$ of degree greater than 0 is any nonzero multiple of itself. Hence, $a(x)$ is not separable if and only if $a(x)$ divides $D_x(a(x))$. Since either $D_x(a(x)) = 0$ or $\deg D_x(a(x)) < \deg a(x)$, we conclude that $a(x)$ is not separable if and only if $D_x(a(x)) = 0$ if and only all monomials of $a(x)$ have a degree that is a multiple of p. This proves the first claim of the proposition.

Consequently, if $a(x)$ is not separable, then $a(x) = a_1(x^p)$ for some polynomial $a_1(x)$.

Let k be the greatest nonnegative integer such that p^k divides the degree of all monomial of $a(x)$. Then $a(x) = b\left(x^{p^\ell}\right)$ and at least one term of $b(x)$ has a degree not divisible by p.. By the first part of the theorem, then $b(x)$ is separable. Furthermore, if $b(x)$ is reducible with $b(x) = b_1(x)b_2(x)$, then $a(x)$ is reducible with $a(x) = b_1(x^{p^k})b_2(x^{p^k})$. By a contrapositive, since $a(x)$ is irreducible, then $b(x)$ is irreducible. $\qquad\square$

5.8.2 Classification of Finite Fields

As we will soon see, our strategy to classify all finite fields relies on the notion of separability. Consequently, we are now in a position to establish the main result of this section.

Every field F of a positive characteristic has char $F = p$ where p is a prime number. By Proposition 5.1.3, F is an extension of the finite field $\mathbb{F}_p$ of p elements and there is a unique field of order p up to isomorphism. Since a field extension K/F is a vector space, then a finite field K with index $[K : F] = n$ has $|K| = p^n$ elements. This proves the first important result on finite fields.

Proposition 5.8.9

Let F be a finite field with $|F| = q$. Then $q = p^n$ for some prime p and some positive integer n. In this case, F is an extension of $\mathbb{F}_p$ of degree n.

Exercise 3.4.8 introduced the Frobenius homomorphism $\sigma_p : R \to R$ on a ring R of characteristic p defined by $\sigma_p(\alpha) = \alpha^p$. If F is a field of characteristic p, then σ_p is an injective homomorphism. As we will see, this is an important function in the context of finite fields. If F is a finite field of characteristic p, then σ_p is also an automorphism.

Definition 5.8.10

If F is finite, the function $\sigma_p : F \to F$ is called the *Frobenius automorphism*. If F is not finite, σ_p is called the *Frobenius endomorphism*.

By Fermat's Little Theorem (Theorem A.6.15), the Frobenius automorphism σ_p is the identity function on $\mathbb{F}_p$. However, on field extensions of $\mathbb{F}_p$, the automorphism is nontrivial. For example, consider the field of order 9 defined by $F = \mathbb{F}_3[x]/(x^2 + x + 2)$. Let us call θ the element corresponding to $\bar{x}$ in F. Notice that $\theta^2 = 2\theta + 1$. Then $F = \{a + b\theta \mid a, b \in \mathbb{F}_3\}$. Obviously, $\sigma_3(a) = a$ for all $a \in \mathbb{F}_3$. However,

$$\sigma_3(\theta) = \theta^3 = (2\theta + 1)\theta = 2\theta^2 + \theta = 2(2\theta + 1) + \theta = 2\theta + 2.$$

The Frobenius automorphism helps us to establish the following proposition.

Proposition 5.8.11

Every irreducible polynomial over a finite field F is separable.

Proof. Let $a(x)$ be an inseparable polynomial over a finite field of characteristic p. By Proposition 5.8.8, $a(x) = b(x^p)$ for some polynomial $b(x)$. Then

$$a(x) = b(x^p) = b_n(x^p)^n + \cdots + b_1(x^p) + b_0.$$

However, since the Frobenius automorphism is a bijection on the finite field, for each $i = 0, 1, \ldots, n$, there exist $c_i \in F$ such that $c_i^p = b_i$. Hence,

$$\begin{aligned}
a(x) &= (c_n)^p(x^n)^p + \cdots c_1^p x^p + c_0^p \\
&= (c_n x^n)^p + \cdots + (c_1 x)^p + c_0^p \\
&= (c_n x^n + \cdots + c_1 x + c_0)^p = (c(x))^p.
\end{aligned}$$

In particular, $a(x)$ is reducible. By a contrapositive, if $a(x)$ is irreducible, then it is separable. $\qquad \square$

Proposition 5.8.9 pointed out that any finite field has order p^n. The converse is the main theorem of this section.

Theorem 5.8.12

For all primes p and for all positive integers n, there exists a unique (up to isomorphism) finite field of order p^n. Furthermore, every finite field is isomorphic to one of these.

Proof. Proposition 5.8.9 established the second part of the theorem. We need to prove the first part.

Consider the polynomial $x^{p^n} - x \in \mathbb{F}_p[x]$. Then

$$D_x(x^{p^n} - x) = -1,$$

which has no roots. Hence, $x^{p^n} - x$ and -1 are relatively prime and so $x^{p^n} - x$

is separable. Therefore, this polynomial has p^n distinct roots in its splitting field K.

Call S the set of distinct roots. Let $\alpha, \beta \in S$. Then

$$(\alpha - \beta)^{p^n} = \sigma_p^n(\alpha - \beta) = \alpha^{p^n} + (-1)^{p^n}\beta^{p^n} = \alpha - \beta,$$

where the last equality holds because α and β are in S and because $(-1)^{p^n} = -1$ for all primes p. Hence, $(S, +)$ is a subgroup of $(K, +)$ by the One-Step Subgroup Criterion. Similarly, if $\alpha, \beta \in S - \{0\}$, then

$$\left(\frac{\alpha}{\beta}\right)^p = \frac{\alpha^{p^n}}{\beta^{p^n}} = \frac{\alpha}{\beta},$$

where the last equality follows from the property that $\alpha, \beta \in S$. Hence, $(S - \{0\}, \times)$ is a subgroup of $(K - \{0\}, \times)$. Thus, S is a subring of K. However, since K is a smallest subfield by inclusion in which $x^{p^n} - x$ splits. Then $S = K$ and the roots of $x^{p^n} - x$ are precisely all the elements in the splitting field K. Thus, we have established the existence of a field of order p^n.

Conversely, let F be a field of cardinality p^n. The prime subfield of F is $\mathbb{F}_p$ and $[F : \mathbb{F}_p] = n$. By Proposition 5.6.2, $U(F)$ is a cyclic group of order $p^n - 1$. Thus, every element of $F - \{0\}$ satisfies the polynomial equation

$$x^{p^n - 1} - 1 = 0.$$

The polynomials $x^{p^n - 1} - 1$ and $x^{p^n} - x$ are in $\mathbb{F}_p[x]$. Therefore, F is the splitting field of $x^{p^n} - x$.

Finally, by Theorem 5.7.10, splitting fields are unique up to isomorphism so any two fields of cardinality p^n are isomorphic. $\qquad\square$

This theorem allows us to make the following definition.

Definition 5.8.13

If q is a prime power $q = p^n$, we denote by $\mathbb{F}_q$ or $\mathbb{F}_{p^n}$ the unique field of order q.

The uniqueness of finite fields of a given finite cardinality is not obvious from how we construct a finite field. As a simple example, let us consider the field of 8 elements. The polynomials $x^3 + x + 1$ and $x^3 + x^2 + 1$ are irreducible cubic polynomials in $\mathbb{F}_2[x]$. Consequently, we could construct a field of eight elements by

$$K_1 = \mathbb{F}_2[x]/(x^3 + x + 1) \quad \text{or} \quad K_2 = \mathbb{F}_2[x]/(x^3 + x^2 + 1).$$

Let us call $\alpha \in K_1$ as an element such that $\alpha^3 + \alpha + 1 = 0$ and $\beta \in K_2$ such that $\beta^3 + \beta^2 + 1 = 0$. It is easy to check that $\alpha + 1 \in K_1$ does not satisfy $x^3 + x + 1 = 0$ but rather $x^3 + x^2 + 1 = 0$. Similarly, $\beta + 1 \in K_2$ does not satisfy $x^3 + x + 1 = 0$ but rather $x^3 + x^2 + 1 = 0$. Consequently, $K_2 = \mathbb{F}_2[\alpha + 1]$ and

$K_1 = \mathbb{F}_2[\beta + 1]$. This shows that $K_1 \cong K_2$ via the isomorphism that extends the identity on $\mathbb{F}_2$ via $\alpha \mapsto \beta + 1$.

To see the strategy of Theorem 5.8.12 at work, we remark that $x^8 - 1$ factors into irreducibles in $\mathbb{F}_2[x]$ as

$$x^8 - 1 = x^8 + 1 = x(x+1)(x^3 + x + 1)(x^3 + x^2 + 1).$$

Theorem 5.8.12 established the unique field of 8 elements is the splitting field of $x^8 + 1$. Using K_1 as a reference,

- 0 is the root of $x = 0$;

- 1 is the root of $x + 1 = 0$;

- α, α^2, and $\alpha^2 + \alpha$ are roots of $x^3 + x + 1 = 0$;

- $\alpha + 1$, $\alpha^2 + 1$, and $\alpha^2 + \alpha + 1$ are roots of $x^3 + x^2 + 1 = 0$.

Theorem 5.8.12 affirms that a similar partitioning occurs in the construction of every finite field. The above remark about the factorization of $x^8 - 1$ generalizes to any prime p and any field extension of degree n. Denote by $\Psi_{p,n}(x)$ the product of all irreducible polynomials of degree n in $\mathbb{F}_p[x]$. Then

$$x^{p^n} - x = \prod_{d \mid n} \Psi_{p,n}(x).$$

By Theorem 5.6.12, applying Möbius inversion gives

$$\Psi_{p,n}(x) = \prod_{d \mid n} \left(x^{p^{n/d}} - x \right)^{\mu(d)},$$

where $\mu(n)$ is the Möbius function on positive integers. In particular, this implies that

$$\deg \Psi_{p,n}(x) = \sum_{d \mid n} \mu(d) p^{n/d}.$$

However, each irreducible factor of $\Psi_{p,n}(x)$ has degree n so we have proved the following result.

Proposition 5.8.14

There are

$$\frac{1}{n} \sum_{d \mid n} \mu(d) p^{n/d}$$

monic irreducible polynomials of degree n in $\mathbb{F}_p[x]$.

We conclude the section with a brief comment on the subfield structure of finite fields.

Exercise 5.8.8 asks the reader to prove that, for any prime p, the field $\mathbb{F}_{p^d}$ is a subfield of $\mathbb{F}_{p^n}$ if and only if $d \mid n$. Consequently, the Hasse diagram representing the subfield structure of $\mathbb{F}_{p^n}$ is the same as the Hasse diagram of the partial order of divisibility on the divisors of n. For example, if $n = 100$, for any prime p the subfield structure of $\mathbb{F}_{p^{100}}$ has the following Hasse diagram.

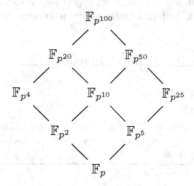

5.8.3 Useful CAS Commands

It is easy to define and work in any finite field in SAGE. The implementation uses a specific polynomial C_n of degree n to define the finite field $\mathbb{F}_{p^n} = \mathbb{F}_p[x]/(C_n(x))$. The code below illustrates enumerating the elements of the field $\mathbb{F}_8$, and then performing a calculation in $\mathbb{F}_{125}$.

––––––––––––––––––––––– Sage –––––––––––––––––––––––

```
sage: F8.<a>=GF(2^3)
sage: for i,x in enumerate(K): print("{} {}".format(i,x))
0 0
1 a
2 a^2
3 a + 1
4 a^2 + a
5 a^2 + a + 1
6 a^2 + 1
7 1
sage: F125.<b>=GF(5^3)
sage: (b^2+3*b+4)*(2*b^2+b+3)
3*b^2 + b + 1
sage: b^3
2*b+2
```

The last operation allows us to see the polynomial that SAGE uses to define $\mathbb{F}_{125}$. Since $[\mathbb{F}_{125} : \mathbb{F}_5] = 3$, we know that the algebraic generator must be the root of some irreducible polynomial of degree 3. In the above example, since

$b^3 = 2b + 2$, then we deduce that b is the root of the polynomial $x^3 + 3x + 3 \in \mathbb{F}_5[x]$.

We illustrate another interesting property of the choice SAGE makes for the algebraic generator $\overline{x}$ of $\mathbb{F}_{p^n} \cong \mathbb{F}_p[x]/(a(x))$, where $\deg a(x) = n$. The element $\overline{x}$ is sometimes, but not necessarily a generator of the group of units $U(\mathbb{F}_{p^n})$.

————————————— Sage (Jupyter) —————————————

```
for n in range(45,49):
    K.<c>=GF(5^n)
    print(n, ":", 5^n-1, "\n    :", c.multiplicative_order())
```

```
45 : 2842170943040400743484497070703124
   : 22920733411616135028100782825I
46 : 1421085471520200371742248535I5624
   : 47
47 : 71054273576010018587112426757812
   : 7105427357601001858711242675781
48 : 35527136788005000929355621337890624
   : 1110223024625156540423631668090
```

For each $n \in \mathbb{N}$, the group of units of $\mathbb{F}_{p^n}$ is cyclic with order $p^n - 1$. In the above code concerning $\mathbb{F}_{5^n}$ with $n \in \{45, 46, 47, 48\}$, only when $n = 47$ does the element $\overline{x}$ have a multiplicative order equal to the order of $|U(\mathbb{F}_{5^{47}})|$.

EXERCISES FOR SECTION 5.8

1. Give the addition and multiplication tables for $\mathbb{F}_9$.

2. Define $\mathbb{F}_8$ as $\mathbb{F}_2[x]/(x^3 + x + 1)$. Let $\theta \in \mathbb{F}_8$ be an element such that $\theta^3 + \theta + 1 = 0$.
   (a) Write down the addition and the multiplication tables for elements in this field.
   (b) In $\mathbb{F}_8$, solve the equation $(\theta^2 + 1)(\alpha - (\theta + 1)) = \theta\alpha + 1$.
   (c) Solve the following system of two linear equations in two variables x and y in the field $\mathbb{F}_8$:
   $$\begin{cases} \theta^2 x + (\theta + 1)y = \theta \\ (\theta^2 + 1)x + \theta^2 y = 1. \end{cases}$$

3. Write $x^9 - x$ as a product of irreducible polynomials in $\mathbb{F}_3[x]$.

4. Write $x^{16} - x$ as a product of irreducible polynomials in $\mathbb{F}_2[x]$.

5. Show that every element besides the 0 and 1 in $\mathbb{F}_{32}$ is a generator of $U(\mathbb{F}_{32})$.

6. Let p be an odd prime. Prove that

$$x^{p-1} - 1 = \prod_{\alpha \in U(\mathbb{F}_p)} (x - \alpha).$$

Deduce that $(p-1)! \equiv -1 \pmod{p}$. (This fact is called Wilson's Theorem.) Prove also that if n is a positive composite integer greater than 4, then $(n-1)! \equiv 0 \pmod{n}$.

7. Let $a \in \mathbb{F}_p \setminus \{0\}$. Prove that the polynomial $x^p - x + a \in \mathbb{F}_p[x]$ is irreducible.

8. Suppose that $d \mid n$. Prove that $\mathbb{F}_{p^d} \subseteq \mathbb{F}_{p^n}$ and that $[\mathbb{F}_{p^n} : \mathbb{F}_{p^d}] = \frac{n}{d}$.

9. Consider the polynomial $p(x) = x^4 + x + 1 \in \mathbb{F}_2[x]$.
   (a) Show that $p(x)$ is irreducible.
   (b) Show that $p(x)$ factors into two quadratics over $\mathbb{F}_4$ and exhibit these two quadratic polynomials.

10. Prove that a polynomial $f(x)$ over a field F of characteristic 0 is separable if and only if it is the product of irreducible polynomials that are not associates of each other. [Note: Consequently, separable polynomials over a field of characteristic 0 are polynomials that are square-free in $F[x]$.]

11. Prove that the finite field $\mathbb{F}_q$ has the property that every non-identity element in $U(\mathbb{F}_q)$ is a generator of $U(\mathbb{F}_q)$ if and only if $q = p + 1$, where p is a Mersenne prime.

12. Proposition 5.6.2 establishes the group of units in a finite field is a cyclic subgroup.
   (a) Find a generator of $U(\mathbb{F}_{16})$.
   (b) Find a generator of $U(\mathbb{F}_{27})$.

13. The polynomial $p_1(x) = x^2 + x + 1 \in \mathbb{F}_5[x]$ is irreducible. Call θ an element in $\mathbb{F}_{25} = \mathbb{F}_5[x]/(p_1(x))$ that satisfies $\theta^2 + \theta + 1 = 0$.
   (a) Find all other irreducible monic quadratic polynomials in $\mathbb{F}_5[x]$.
   (b) For each of the 10 polynomials found in the previous part, write the two roots in $\mathbb{F}_{25}$ as $a\theta + b$ for $a, b \in \mathbb{F}_5$.

14. Let $q = p^n$. Prove that the Frobenius automorphism $\varphi = \sigma_p : \mathbb{F}_q \to \mathbb{F}_q$ is a $\mathbb{F}_p$-linear transformation. Prove also that φ^n is the identity transformation.

15. Consider the Frobenius map φ from the previous exercise. Determine the eigenvalues and all corresponding eigenspaces for φ.

16. Consider the Frobenius automorphism $\sigma_3 : \mathbb{F}_9 \to \mathbb{F}_9$. Show how σ_3 maps the elements of $\mathbb{F}_9$. [Hint: Use the identification $\mathbb{F}_9 = \mathbb{F}_3[x]/(x^2 + x + 2)$.]

17. Prove that $(1 + x^p)^n = (1 + x)^{pn}$ in $\mathbb{F}_p[x]$. Deduce that $\binom{pn}{pk} \equiv \binom{n}{k} \pmod{p}$.

18. The polynomials $x^3 + 3x + 2$ and $2x^3 + 4x^2 + 1$ are irreducible in $\mathbb{F}_5[x]$. From the results of this section, we know that

$$\mathbb{F}_{125} \cong \mathbb{F}_5[x]/(x^3 + 3x + 2) \cong \mathbb{F}_5[x]/(2x^3 + 4x^2 + 1).$$

Let $\alpha \in \mathbb{F}_5[x]/(x^3 + 3x + 2) = K_1$ be an element satisfying $\alpha^3 + 3\alpha + 2 = 0$ and let $\beta \in \mathbb{F}_5[x]/(2x^3 + 4x^2 + 1) = K_2$ satisfy $2\beta^3 + 4\beta^2 + 1 = 0$. Explicitly find an isomorphism $\varphi : K_1 \to K_2$. In particular, find $\varphi(\alpha)$ as a linear combination of β.

19. Let p be a prime. By Exercise 5.8.7, the polynomial $x^p - x - 1$ is irreducible in $\mathbb{F}_p[x]$. Hence, we can represent $\mathbb{F}_{p^p}$ as the quotient ring $\mathbb{F}_p[x]/(x^p - x - 1)$ and so $\mathbb{F}_{p^p} = \mathbb{F}_p(\alpha)$, where α satisfies $\alpha^p = \alpha + 1$. We use the basis $\mathcal{B} = \{1, \alpha, \dots, \alpha^{p-1}\}$.

(a) Prove that for $0 \leq m \leq 2p - 2$, expressed in the basis $\mathcal{B}$, we have

$$\alpha^m = \begin{cases} \alpha^m & \text{if } m \leq p - 1 \\ \alpha^{m-p+1} + \alpha^{m-p} & \text{if } p \leq m \leq 2p - 2. \end{cases}$$

(b) Let

$$\beta = b_{p-1}\alpha^{p-1} + \cdots + b_1\alpha + b_0 \text{ and } \gamma = c_{p-1}\alpha^{p-1} + \cdots + c_1\alpha + c_0$$

be arbitrary elements in $\mathbb{F}_{p^p}$. Show that with respect to $\mathcal{B}$ we have

$$\beta\gamma = \left(b_0 c_0 + \sum_{i+j=p} b_i c_j \right)$$

$$+ \sum_{k=1}^{p-1} \left(\sum_{i+j=k} b_i c_j + \sum_{i+j=k+p-1} b_i c_j + \sum_{i+j=k+p} b_i c_j \right) \alpha^k.$$

20. Let $\mathbb{F}_q$ be a finite field and let $f(x)$ be an irreducible polynomial of degree n in $\mathbb{F}_q[x]$. Suppose that α is one of the roots of $f(x)$ in the field $\mathbb{F}_{q^n}$. Prove that

$$\alpha, \alpha^q, \alpha^{q^2}, \dots, \alpha^{q^{n-1}}$$

are the n distinct roots of $f(x)$ in $\mathbb{F}_{q^n}$.

21. Prove that a polynomial of degree $m = 2^k$ over $\mathbb{F}_2[x]$ is irreducible if and only if it divides

$$(x^{2^{2^k}} + x)/(x^{2^{2^{k-1}}} + x).$$

5.9   Projects

Investigative Projects

PROJECT I. **Field Extensions in $M_n(F)$.** Revisit Exercises 5.1.19 and 5.1.20 Try to generalize these results to other or any simple extension $F(\alpha)$ of a field F. Use your results to illustrate interesting multiplications and divisions in the field $F(\alpha)$.

PROJECT II. **Cardano's Triangle.** Recall Cardano's method to solve the cubic equation. When the discriminant is negative, so that the solution has three real roots, a geometric interpretation of the method shows the roots arising as the projections onto the x-axis of the vertices of some equilateral triangle rotated around some point on the x-axis. (For the equation, $x^3 + px + q = 0$, that point is the origin.) Explore the solution of the cubic from a geometric perspective. Can you see how to start from the geometry of projecting vertices of an equilateral triangle into a cubic equation? Explain the solution to a cubic equation from this geometric perspective.

PROJECT III. **Cardano's Method in $\mathbb{C}$.** Section 5.4 presented Cardano's method for solving the cubic and quartic equation with the assumption that the coefficients of the polynomial area real. Discuss the method and the content of the section assuming that the coefficients of the polynomial are complex numbers. How much changes and how much stays the same?

PROJECT IV. **Constructing a Regular 17-gon.** The prime number 17 has $\phi(17) = 16 = 2^4$. Hence, $[\mathbb{Q}(\zeta_{17}) : \mathbb{Q}] = 16$. Call $\zeta = \zeta_{17}$. Explain why Theorem 5.5.5 does not rule out the possibility of constructing $\cos(2\pi/17) = \frac{1}{2}(\zeta + \zeta^{-1})$. Show that:

- $\alpha_1 = \zeta + \zeta^2 + \zeta^4 + \zeta^8 + \zeta^9 + \zeta^{13} + \zeta^{15} + \zeta^{16}$ is real and is the root of a quadratic polynomial in $\mathbb{Q}$;

- $\alpha_2 = \zeta + \zeta^4 + \zeta^{13} + \zeta^{16}$ is real and is the root of a quadratic polynomial in $\mathbb{Q}(\alpha_1)$;

- $\alpha_3 = \zeta + \zeta^{16} = 2\cos(2\pi/17)$ is the root of a quadratic polynomial in $\mathbb{Q}(\alpha_2)$.

Use this sequence to write $\cos\left(\dfrac{2\pi}{17}\right)$ as a combination of nested square root expressions. Also use this sequence to find a straightedge and compass construction of the regular 17-gon. Justify your construction.

PROJECT V. **Irreducible Polynomials in $\mathbb{F}_2[x]$.** In certain applications of cryptography, it is particularly useful to have irreducible polynomials of degree n in $\mathbb{F}_2[x]$. Providing at least one for each n, attempt to find as many irreducible polynomials of degree $k = 2, 3, \ldots, n$. Do you see any patterns in which polynomials will be irreducible? Can you devise a fast algorithm to find an irreducible polynomial of degree n in $\mathbb{F}_2[x]$?

PROJECT VI. **Epicycloids in $\mathbb{Z}/n\mathbb{Z}$?** Let n be a somewhat large, say $n \geq 40$ integer and consider the group μ_n of nth roots of unity. This is a finite subgroup of $(U(\mathbb{C}), \times)$. Consider plotting the elements of μ_n on the unit circle in $\mathbb{C}$. For various n and for a given small positive integer m trace an edge between z and z^m for all $z \in \mu_n$. The edges create an envelope of a certain epicycloid. Explain why this is true? Study properties of the epicycloid depending on m and n. If this graph were created by nail and string artwork, is it ever possible to create the work with a single piece of string? (Why or why not?)

PROJECT VII. **Frobenius Automorphism.** For various values of p and n, find a matrix corresponding to the Frobenius automorphism σ_p on $\mathbb{F}_{p^n}$ as a linear transformation on $\mathbb{F}_{p^n}$ as a vector space over $\mathbb{F}_p$. Can you identify patterns in this associated matrix?

PROJECT VIII. **Dynatomic Polynomials.** Exercises 5.6.20 through 5.6.25 discussed the concept of dynatomic polynomials. In this project, explore the degrees and structure of the splitting fields of the family of polynomials of the form $P^n(x) - x$, where $P(x) \in F[x]$ and $P^n(x)$, means $P(x)$ iterated on itself n times.

Expository Projects

PROJECT IX. **Advanced Encryption Standard (AES).** Discuss the history behind the encryption algorithm referred to as the Advanced Encryption Standard (AES). Given an outline of the algorithm, including both encryption of plaintext and decryption of the ciphertext. Be precise concerning the use of finite field arithmetic in AES. Discuss its use in current systems and why many communication protocols favor this encryption system.

PROJECT X. **The Greatest Mathematician Who Never Was.** In the middle of the 20th century, various influential papers and mathematics books emerged penned by Nicolas Bourbaki. However, this person never existed but instead represented a group of people. Discuss the work of Bourbaki, the reason for writing under this pseudonym, and the influence of the group's work on mathematics as a whole.

PROJECT XI. **Cubic and Quartic Formulas.** Give an account of the discovery, dissemination, and reception in the mathematical community of the formulas using radicals of the cubic equation and of the quartic equation.

PROJECT XII. **Hilbert's 7th Problem.** In the year 1900, David Hilbert published 23 unsolved problems and posed them as the challenges for the 20th century. Indeed, efforts to solve these problems sparked considerable mathematical research for 100 years after their publication. Hilbert's 7th Problem pertained to transcendental numbers. Discuss the problem, why mathematicians considered it interesting, the approach taken by those who solved it, and any related theorems or related open conjectures.

6

Topics in Group Theory

Though this textbook waited until this point to introduce group actions, from a historical perspective, group actions came first and motivated group theory. Historically, mathematicians conceived of a group as a set S of bijective functions $f : X \to X$ (that perhaps preserved some interesting property), in which S is closed under composition and taking function inverses. In broad strokes, group actions involve viewing a group G as a subgroup of the group of permutations on a set.

Section 6.1 defines group actions in the modern sense and offers many examples. The perspective of group actions that simultaneously consider properties in the group G and in the set X on which it acts leads to more information that is available simply from the group itself. Section 6.2 presents orbits and stabilizers, which specifically considers this interplay between the set and the group. Section 6.3 presents some properties that are specific to transitive group actions, including block structures in group actions.

The general theory of group actions proves to be particularly fruitful when we consider a group acting on itself in some manner. Section 6.4 presents results pertaining to the action of a group on itself by left multiplication and by conjugation, resulting in Cayley's Theorem, Cauchy's Theorem, and the Class Equation. Section 6.5 introduces a specific action of a group on certain subsets of its subgroups, which leads to Sylow's Theorem, a profound result in group theory. We end the chapter with Section 6.6 on semidirect products followed by Section 6.7, which describes classification techniques to determine all non-isomorphic groups of a given order.

6.1 Introduction to Group Actions

To introduce group actions, we consider the dihedral group as first presented in Section 1.1. From the outset, we introduced D_n as a set of bijections on the vertices of the regular n-gon. Hence, if we label the vertices of the regular n-gon as $\{1, 2, \ldots, n\}$, then D_n can be viewed as a subgroup of S_n, the set of bijections on the vertices.

Group actions generalize as broadly as possible the perspective of viewing groups as sets of functions on a set. We warn the reader that since group

DOI: 10.1201/9781003299233-6

actions arise in so many different contexts within mathematics, there exist a variety of different notations and expressions.

6.1.1   Group Actions: Definition

Definition 6.1.1

A *group action* of a group G on a set X is a function from $G \times X \to X$, with outputs written as $g \cdot x$ or simply gx, satisfying

(1) (Compatibility) $g_1 \cdot (g_2 \cdot x) = (g_1 g_2) \cdot x$, for all $g_1, g_2 \in G$ and $x \in X$;

(2) (Identity) $1 \cdot x = x$, for all $x \in X$.

If there exists a group action of G on X, we say that G *acts* on X.

If a group G acts on a set X, then X is sometimes called a G-set. As another point of terminology, the function $G \times X \to X$ is sometimes also called a *pairing*. Some authors use the shorthand notation $G \circlearrowright X$ to mean "the group G acts on the set X."

The axioms capture the desired intuition for groups as sets of functions on X. In essence, every group element behaves like a function on X in such a way that function composition corresponds to the group operation and the identity of the group behaves as the identity function. More precisely, for each $g \in G$, the operation $g \cdot x$ is a function we can denote by $\sigma_g : X \to X$ with $\sigma_g(x) = g \cdot x$. Recall that we denote by S_X the set of bijective functions from a set X to itself.

Proposition 6.1.2

Let G be a group acting on a set X.

(1) For all $g \in G$, the function σ_g is a permutation of X.

(2) The map $\rho : G \to S_X$ defined by $\rho(g) = \sigma_g$ is a homomorphism.

Proof. For all $g \in G$, and for all $x \in X$,

$$\sigma_{g^{-1}}(\sigma_g(x)) = g^{-1} \cdot (g \cdot x) = (g^{-1}g) \cdot x = 1 \cdot x = x.$$

Similarly, $\sigma_g(\sigma_{g^{-1}}(x)) = x$. Hence, the function $\sigma_g : X \to X$ is bijective with inverse function $(\sigma_g)^{-1} = \sigma_{g^{-1}}$.

To show that ρ is a homomorphism, let $g_1, g_2 \in G$. Then $\rho(g_1) \circ \rho(g_2)$ is a bijection $X \to X$ such that for all x,

$$\rho(g_1) \circ \rho(g_2)(x) = \sigma_{g_1}(\sigma_{g_2}(x)) \qquad \text{by definition of } \rho$$
$$= \sigma_{g_1}(g_2 \cdot x)$$
$$= g_1 \cdot (g_2 \cdot x)$$
$$= (g_1 g_2) \cdot x \qquad \text{by compatibility axiom}$$
$$= \sigma_{g_1 g_2}(x)$$
$$= \rho(g_1 g_2)(x).$$

Thus, $\rho(g_1 g_2) = \rho(g_1) \circ \rho(g_2)$ and therefore ρ is a homomorphism. $\qquad \square$

In other words, actions of a group G on a set X are in one-to-one correspondence with homomorphisms from G to S_X. Any homomorphism $\rho : G \to S_X$ is called a *permutation representation* because it relabels the elements of G with permutations. We say that a group action *induces* a permutation representation of G. This inspires us to give an alternate definition for a group action that is briefer than Definition 6.1.1.

Definition 6.1.3 (Alternate)

A group action is a triple (G, X, ρ) where G is a group, X is a set, and $\rho : G \to S_X$ is a homomorphism. The image element $\rho(g)(x)$ is simply often denoted by gx.

In every group action, the group identity acts as the identity function on X. However, in an arbitrary group action, many other group elements could have no effect on X. It is an important special case when the group identity is the only group element that acts as the identity.

Definition 6.1.4

Suppose that a group action of G on X has a permutation representation of ρ. Then the action is called *faithful* if $\operatorname{Ker} \rho = \{1\}$. We also say that G acts faithfully on X.

Since $\operatorname{Ker} \rho = \{1\}$, then ρ is injective. This means that $\rho(g) \neq \rho(h)$ for all $g \neq h$ in G. Therefore, the action is faithful if and only if each distinct group element corresponds to a different function on X. Furthermore, by the First Isomorphism Theorem, if an action is faithful, then $G \cong G/(\operatorname{Ker} \rho) \cong \operatorname{Im} \rho$, which presents G as a subgroup of S_X.

Definition 6.1.1 is sometimes called a *left* group action of G on the set X to reflect the notational habit of applying functions on the left of the domain element. Beyond notation habit, in a left group action, the composed element $(g_1 g_2) \cdot x$ involves first acting on x by g_2 and then by g_1.

6.1.2   Examples of Group Actions

Arguably, every application that involves groups to study some other problem involves a group action. Consequently, there are countless examples of group actions. This section offers a few basic examples.

Example 6.1.5 (Dihedral Group). If we label the vertices as shown below, D_n acts on the vertices of a regular n-gon by performing the geometric transformation.

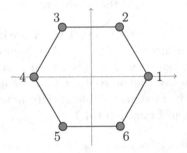

FIGURE 6.1: A hexagon.

If $n = 6$ and if we label the vertices of the hexagon $\{1, 2, 3, 4, 5, 6\}$ as in Figure 6.1, then the permutation representation ρ satisfies

$$\rho(r) = (1\,2\,3\,4\,5\,6),$$
$$\rho(s) = (2\,6)(3\,5).$$

Note that if we labeled the vertices of the hexagon differently, then we would induce a different homomorphism of D_6 into S_6.

When we defined D_n, we considered two dihedral symmetries distinct only if they acted differently on the vertices. In the language of group actions, this means that D_n acts faithfully on $\{1, 2, \ldots, n\}$. $\triangle$

Example 6.1.6 (Permutation Group). The permutation group S_n acts on the set $\{1, 2, \ldots, n\}$ by viewing each permutation $\sigma \in S_n$ as a bijection on $\{1, 2, \ldots, n\}$. This example is not surprising since S_n was essentially defined by bijections on $\{1, 2, \ldots, n\}$ compose with each other. The permutation action of S_n on $\{1, 2, \ldots, n\}$ is faithful. $\triangle$

Example 6.1.7 (Linear Algebra). Let F be a field and consider the vector space $V = F^n$ over the field F, for some positive integer n. Then the group $GL_n(F)$ acts on V by multiplying a vector by an invertible matrix. Explicitly, the pairing of the action $GL_n(F) \times V \to V$ is $A \cdot \vec{v} = A\vec{v}$, where the right-hand side is matrix-vector multiplication.

We can show that this action is faithful by considering how $GL_n(F)$ acts on the standard basis vectors e_i (the n-tuple that is all 0s except for a 1 in the ith entry). Recall that Ae_i is the ith column of A. Therefore, if $Ae_i = e_i$ for all $i = 1, 2, \ldots, n$, then the ith column of A is e_i (as a column). Hence, A acts trivially on V if and only if $A = I$. $\triangle$

Example 6.1.8 (Trivial Action). Let G be a group and X any set. Then the action $gx = x$ for all $g \in G$ and $x \in X$ is called the *trivial action* of G on X. Every group element g acts as the identity on X. In this sense, every group can act on every set. Intuitively, a trivial action is opposite from a faithful action in that $\operatorname{Ker}\rho = G$ for a trivial action, whereas $\operatorname{Ker}\rho = \{1\}$ for a faithful action. $\triangle$

Example 6.1.9. Consider the group D_6 and how it acts on the diagonals of the hexagon. The diagonals of the hexagon are $d_1 = \{1, 4\}$, $d_2 = \{2, 5\}$, and $d_3 = \{3, 6\}$. For any of these 2-element subsets of vertices, any dihedral symmetry of the hexagon maps a diagonal into another diagonal. Therefore, D_6 acts on $\{d_1, d_2, d_3\}$.

This action is not faithful because $r^3 \cdot d_i = d_i$ for $i = 1, 2, 3$. Note that $s \notin \operatorname{Ker}\rho$ because even though $s \cdot d_1 = d_1$, we also have $s \cdot d_2 = d_3$. The permutation representation ρ of this group action is completely defined by $\rho(r) = (1\,2\,3)$ and $\rho(s) = (2\,3)$. $\triangle$

Example 6.1.10. Consider the group D_5 and consider the set of 11 polygonal regions inside the pentagon bordered by the complete graph on the set of vertices as shown below. The dihedral group D_5 acts on the set of polygonal regions. If we label the regions with the integers $1, 2 \ldots, 11$, the action induces a homomorphism $\rho : D_5 \to S_{11}$.

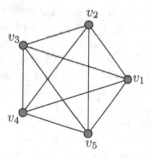

$\triangle$

Example 6.1.11 (Rigid Motions of the Cube). Consider the group G of rigid motions (solid rotations) of the cube. There are many actions that are natural to consider. G acts on:

- the set of vertices of the cube (8 elements);
- the set of edges of the cube (12 elements);
- the set of faces of the cube (6 elements);
- the set of diagonals on the faces of the cube (12 elements);

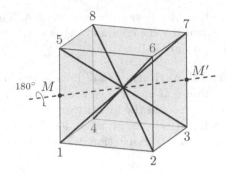

FIGURE 6.2: Rigid motion that transposes two diagonals.

- the set of diagonals through the centroid of the cube (4 elements);
- the set of segments that connect centroids of opposite faces on the cube (3 elements).

As one action of particular interest, consider the action of G on the diagonals through the center of the cube. This action correspond to a homomorphism $\rho : G \to S_4$. Let us label the vertices of the cube by $\{1, 2, 3, 4, 5, 6, 7, 8\}$ and label these long diagonals by $d_1 = \{1, 7\}$, $d_2 = \{2, 8\}$, $d_3 = \{3, 5\}$, and $d_4 = \{4, 6\}$. See Figure 6.2.

Pick an edge e. Let e' be the edge of the cube that centrally symmetric to e through the middle (centroid) O of the cube. Let M and M' be the midpoints of e and e' respectively. Let R_e be the rotation of the cube by $180°$ around the line (MM'). The rigid motion R_e interchanges the two diagonals that are on the square defined by the edge e and e' (the plane defined by e and O) but it leaves unchanged the diagonals that are in the plane perpendicular to the plane defined by e and e'. Thus, $\rho(R_e)$ is a transposition in S_4. There are six pairs of centrally symmetric edges, which lead to six distinct rigid motions of the form R_e, which induce the 6 transpositions in S_4. Since S_4 is generated by its transpositions, we deduce that $\rho(G) = S_4$.

From Exercise 1.3.26 we know that $|G| = 24$. Thus, the homomorphism ρ is a surjective function between two finite groups of the same size. We deduce that ρ is bijective, so ρ is an isomorphism. We conclude that the group of rigid motions of the cube is isomorphic to S_4.   △

Example 6.1.12 (Sets of Functions). Let X and Y be any sets and let $Y^X = \mathrm{Fun}(X, Y)$ be the set of functions from X to Y.

If G acts on Y, then there is a natural action of G on Y^X via

$$(g \cdot f)(x) = g \cdot (f(x)).$$

It is an easy exercise to see that this is a group action. If G acts on X, then there also exists a natural action of G on Y^X via

$$(g \cdot f)(x) = f(g^{-1} \cdot x). \tag{6.1}$$

It is crucial that the right-hand side involve g^{-1}. We check the compatibility axiom for this action. Let $g, h \in G$ and let $f \in Y^X$. Writing $h \cdot f = f'$, then the function $g \cdot (h \cdot f)$ satisfies

$$(g \cdot (h \cdot f))(x) = f'(g^{-1} \cdot x) = f(h^{-1} \cdot (g^{-1} \cdot x))$$
$$= f((h^{-1}g^{-1}) \cdot x) = f((gh)^{-1} \cdot x) = ((gh) \cdot f)(x).$$

It easy to check the identity axiom. Hence, (6.1) defines an action on Y^X. $\triangle$

Example 6.1.13 (Rearrangement of n-Tuples). Let A be a set, let $n \in \mathbb{N}^*$, and let $X = A^n$ be the set of n-tuples of A. Consider the pairing $S_n \times X \to X$ that permutes the entries of $(a_1, a_2, \ldots, a_n) \in A^n$ according to a permutation σ. In other words, in the action of σ on the n-tuple $(a_1, a_2, \ldots, a_n)$, the ith entry is sent to the $\sigma(i)$th position. Note that in $\sigma \cdot (a_1, a_2, \ldots, a_n)$, the ith entry is the $\sigma^{-1}(i)$th entry of $(a_1, a_2, \ldots, a_n)$. Thus,

$$\sigma \cdot (a_1, a_2, \ldots, a_n) = (a_{\sigma^{-1}(1)}, a_{\sigma^{-1}(2)}, \ldots, a_{\sigma^{-1}(n)}). \tag{6.2}$$

We show that this defines a group action of S_n on $X = A^n$. First, for all $\tau, \sigma \in S_n$, we have

$$\tau \cdot (\sigma \cdot (a_1, a_2, \ldots, a_n)) = \tau \cdot (a_{\sigma^{-1}(1)}, a_{\sigma^{-1}(2)}, \ldots, a_{\sigma^{-1}(n)})$$
$$= (a_{\sigma^{-1}(\tau^{-1}(1))}, a_{\sigma^{-1}(\tau^{-1}(2))}, \ldots, a_{\sigma^{-1}(\tau^{-1}(n))})$$
$$= (a_{(\tau\sigma)^{-1}(1)}, a_{(\tau\sigma)^{-1}(2)}, \ldots, a_{(\tau\sigma)^{-1}(n)})$$
$$= (\tau\sigma) \cdot (a_1, a_2, \ldots, a_n).$$

Also, $1 \cdot (a_1, a_2, \ldots, a_n) = (a_1, a_2, \ldots, a_n)$ since it does not permute the elements.

If (6.2) seems counterintuitive at first, observe that as sets A^n is equal to $\mathrm{Fun}(\{1, 2, \ldots, n\}, A)$ and the action described in (6.2) is precisely the action defined in Example 6.1.12.

In contrast, it is important to realize that the function $S_n \times X \to X$ defined

$$\sigma \cdot (a_1, a_2, \ldots, a_n) = (a_{\sigma(1)}, a_{\sigma(2)}, \ldots, a_{\sigma(n)})$$

is *not* a group action of S_n on X as it fails the compatibility axiom. $\triangle$

There are many other types of group actions of considerable interest. The examples provided so far just scratch the surface. The following subsection presents a few important actions of a group acting on itself. Following that, the reader is encouraged to peruse the exercises for many other examples.

6.1.3   Group Actions as an Algebraic Structure

With a fixed group G, we can view group actions of G as another algebraic structure. In this perspective, the focus lands on the set X so that when G acts on a set X, we refer to X as a G-set. In light of this, it is natural to discuss subobjects and morphisms related to group actions.

Definition 6.1.14

Let G be a group and let X be a G-set. A G-*subset* of X is a subset $A \subseteq X$ such that $g \cdot x \in A$ for all $g \in G$ and all $x \in A$. We also say that A is *closed* under the action of G or, equivalently, that A is *invariant* under G.

Whenever a subset A of X is closed under the action of G, the axioms of the action of G on X restrict to A, giving A the structure of a G-set.

As an example, consider the plane $\mathbb{R}^2$ equipped with an origin O and with a labeled x-axis. Consider the natural action of the dihedral group D_n on $\mathbb{R}^2$, where r corresponds to rotation by $2\pi/n$ around the origin O and where s corresponds to reflection about the x-axis. Any D_n-subset of $\mathbb{R}^2$ is a subset of $\mathbb{R}^2$ that has dihedral symmetry, i.e., is invariant under the action of D_n.

Example 6.1.15. Let G be the group of rigid motions of a cube (see Example 6.1.11) and let V be the set of vertices. There is a natural action of G on V by how the rotation maps the vertices. Namely for every vertex $v \in V$, $g \cdot v$ is the image of the vertex v under the rotation g. We equip $\mathcal{P}(V)$, the set of subsets of V with the G-action defined by

$$g \cdot \{x_1, x_2, \ldots, x_n\} = \{g \cdot x_1, g \cdot x_2, \ldots, g \cdot x_n\}.$$

The set of edges E of the cube is a G-subset of $\mathcal{P}(V)$ since every solid rotation of the cube maps an edge to another edge. The G-set $\mathcal{P}(V)$ has many other G-subsets, e.g., the set of faces, the set of long diagonals, the set $\mathcal{P}_k(V) = \{A \subseteq V \mid |A| = k\}$ of subsets of size k, etc. However, not all subsets of $\mathcal{P}(V)$ are G-subsets. For example, given a fixed vertex v_0, the singleton set $\{v_0\}$ is not a G-subset since not all $g \in G$ leave v_0 unchanged. $\triangle$

Definition 6.1.16

Let G be a group and let X and Y be two G-sets (i.e., there is an action of G on X and on Y). A G-set homomorphism between X and Y is a function $f : X \to Y$ such that

$$f(g \cdot x) = g \cdot f(x) \qquad \text{for all } g \in G \text{ and all } x \in X.$$

An isomorphism of G-sets if a G-set homomorphism that is also bijective.

Exercises 6.1.13 and 6.1.14 establish some results about G-set homomorphisms that we might expect from standard results about group homomorphisms and ring homomorphisms.

Note that in Definition 6.1.16 the group G acting on X and Y is the same. In this perspective, if G and G' are nonisomorphic groups, then we consider the collection of G-sets and G'-sets as two distinct algebraic structures. Because of this restriction, the above definition might feel unsatisfactory. For example, suppose that a group G acts on a set X and a group H acts on a set Y. We might consider the group actions as equivalent if that action is identical after a relabeling of the elements in G with elements in H and a parallel relabeling of elements in X with elements in Y. To name this desired phenomenon, we use the following definition.

Definition 6.1.17

A *group action homomorphism* between two group actions (G, X, ρ_1) and (H, Y, ρ_2) is a pair (φ, f), where $\varphi : G \to H$ is a homomorphism and $f : X \to Y$ is a function such that

$$f(g \cdot x) = \varphi(g) \cdot f(x) \qquad \text{for all } g \in G \text{ and all } x \in X.$$

A *group action isomorphism* (or *permutation isomorphism*) is a group action homomorphism (φ, f) such that φ is an isomorphism and f is a bijection.

If a group action isomorphism exists between two group actions, they are called *isomorphic* (or *permutation equivalent*).

Example 6.1.18. Consider the natural action of $\mathrm{GL}_2(\mathbb{F}_2)$ on the vector space $X = \mathbb{F}_2^2$ of four elements over $\mathbb{F}_2$. Also consider the action of S_3 on the set $Y = \{0, 1, 2, 3\}$ by fixing 0 and permuting $\{1, 2, 3\}$ as usual. Let $\varphi : \mathrm{GL}_2(\mathbb{F}_2) \to S_3$ be the isomorphism described in Example 1.9.20. Then the bijection $f : X \to Y$ that maps

$$\begin{pmatrix} 0 \\ 0 \end{pmatrix} \longmapsto 0 \qquad \begin{pmatrix} 1 \\ 0 \end{pmatrix} \longmapsto 3 \qquad \begin{pmatrix} 0 \\ 1 \end{pmatrix} \longmapsto 2 \qquad \begin{pmatrix} 1 \\ 1 \end{pmatrix} \longmapsto 1$$

makes the pair (φ, f) into a group action isomorphism. $\triangle$

EXERCISES FOR SECTION 6.1

1. Consider the natural action of D_7 on vertices of a regular heptagon. Consider the induced permutation representation $\rho : D_7 \to S_7$. Determine $\rho(r)$ and $\rho(s)$ with respect to your labeling of the vertices and choice of reflection for s.

2. Let n be a positive integer and consider the group $GL_n(\mathbb{R})$. Prove that the pairing $\mathrm{GL}_n(\mathbb{R}) \times \mathbb{R} \to \mathbb{R}$ defined by $A \cdot x = \det(A)x$ is a group action. Prove also that it is not faithful.

3. Let $G = D_6$ and let $H = \langle r^2 \rangle$. Since $H \trianglelefteq G$, conjugation of G on H is an action of G on H. Label the H elements 1, r^2, and r^4 as 1, 2, and 3 respectively and consider the induced permutation representation $\rho : D_6 \to S_3$.

   (a) Exhibit the images under ρ of all elements in D_6.

   (b) State the kernel of ρ and show that the action is not faithful.

4. Let G be a group acting on a set X. Show that defining

$$g \cdot S \stackrel{\text{def}}{=} \{g \cdot s \mid s \in S\}$$

   for all $g \in G$ and all $S \subseteq X$ induces an action of G on $\mathcal{P}(X)$.

5. Fix a positive integer n. Let $X = \{1, 2, \ldots, n\}$ and consider the mapping $S_n \times \mathcal{P}(X) \to \mathcal{P}(X)$ defined by

$$\sigma \cdot \{x_1, x_2, \ldots, x_k\} = \{\sigma(x_1), \sigma(x_2), \ldots, \sigma(x_k)\}.$$

   (a) Prove that this pairing defines an action of S_n on $\mathcal{P}(\{1, 2, \ldots, n\})$.

   (b) For a given k with $0 \le k \le n$, define $\mathcal{P}_k(X)$ as the set of subsets of X of cardinality k. Prove that $\mathcal{P}_k(X)$ are closed under the action of S_n on $\mathcal{P}(X)$.

   (c) Prove that a subset Y of $\mathcal{P}(X)$ is closed under the action of S_n if and only if Y is the union of some $\mathcal{P}_k(X)$.

6. Let F be a field, let $G = \mathrm{GL}_n(F)$, and let $X = M_n(F)$ be the set of $n \times n$ matrices with entries in F.

   (a) Show that the pairing $G \times X \to X$ given by $g \cdot A = gAg^{-1}$ is an action of $\mathrm{GL}_n(F)$ on $M_n(F)$.

   (b) Discuss how the relation of similarity on square matrices relates to this group action.

   (c) Prove that for all $g \in \mathrm{GL}_n(F)$ and for all $A, B \in M_n(F)$,

$$g \cdot (A + B) = g \cdot A + g \cdot B \quad \text{and} \quad g \cdot (AB) = (g \cdot A)(g \cdot B).$$

7. Consider the action defined in Exercise 6.1.5 where $n = 4$ and $k = 2$. The induced permutation representation is a homomorphism $\rho : S_4 \to S_6$. Label the elements in $\mathcal{P}_2(\{1, 2, 3, 4\})$ according to the following chart.

label	1	2	3	4	5	6
subset	$\{1,2\}$	$\{1,3\}$	$\{1,4\}$	$\{2,3\}$	$\{2,4\}$	$\{3,4\}$

   Give $\rho(\sigma)$ as a permutation in S_6 for $\sigma = 1$, $(1\,2)$, $(1\,2\,3)$, $(1\,2)(3\,4)$, and $(1\,2\,3\,4)$.

8. Let P_n be the of polynomials in $\mathbb{R}[x]$ that have degree n or less (including the 0) polynomial. Consider $\sigma \in S_{n+1}$ as a permutation on $\{0, 1, 2, \ldots, n\}$ and define

$$\sigma \cdot (a_n x^n + \cdots + a_1 x + a_0) = a_{\sigma^{-1}(n)} x^n + \cdots + a_{\sigma^{-1}(1)} x + a_{\sigma^{-1}(0)}.$$

   (a) Prove that this defines an action of S_{n+1} on P_n.

   (b) Decide with proof whether $\sigma \cdot (a(x) + b(x)) = \sigma \cdot a(x) + \sigma \cdot b(x)$.

   (c) Decide with proof whether $\sigma \cdot (a(x)b(x)) = (\sigma \cdot a(x))(\sigma \cdot b(x))$.

9. Suppose a group H acts on a set X. Let $\varphi : G \to H$ be a group homomorphism. Prove that the pairing $G \times X \to X$ defined by $g \cdot x = \varphi(g) \cdot x$, where the action symbol on the right is the action of H on X, defines an action of G on X.

10. Let G be a group acting on a set X and let $\rho : G \to S_X$ be the induced permutation representation. Prove that the pairing $(G/\operatorname{Ker}\rho) \times X \to X$ with

$$(g \operatorname{Ker}\rho) \cdot x = g \cdot x$$

is an action of $G/\operatorname{Ker}\rho$ on X. Prove also that this action is faithful.

11. In this exercise, we revisit Example 6.1.12. Let $X = \operatorname{Fun}(\{0,1,2\}, \{0,1,2\})$ be the set of functions from $\{0,1,2\}$ to itself. The set X has cardinality 27. We label the functions in X by f_k with $0 \le k \le 26$ as the function $f_k(i) = a_i$ where $k = (a_2 a_1 a_0)_3$ as an integer expressed in base 3. Hence, f_{14} satisfies

$$f_{14}(0) = 2, \quad f_{14}(1) = 1, \quad f_{14}(2) = 1$$

because in base 3, the integer 14 is $14 = (112)_3$. Consider elements in S_3 as permutations on $\{0,1,2\}$. Using the described labeling of functions in X,

(a) Give $\rho_1(\sigma)$ as a permutation on $\{0,1,2,\ldots,26\}$ for the permutation representation ρ_1 induced from the action of S_3 on X defined by $(\sigma \cdot f)(x) = \sigma \cdot f(x)$.

(b) Give $\rho_2(\sigma)$ as a permutation on $\{0,1,2,\ldots,26\}$ for the permutation representation ρ_2 induced from the action of S_3 on X defined by $(\sigma \cdot f)(x) = f(\sigma^{-1} \cdot x)$.

12. Suppose that a group G acts on a set X and let X^X be the set of all functions from X to X.

(a) Prove that the pairing $\star : G \times X^X \to X^X$ defined by

$$(g \star f)(x) = g \cdot f(g^{-1} \cdot x)$$

is an action of G on X^X.

(b) Prove that the subset of bijections on X is invariant.

(c) Let k be a nonnegative integer. Let $A_k \subseteq X^X$ be the set of functions $f : X \to X$ such that $f(x_0) = x_0$ for exactly k elements $x_0 \in X$. Prove that A_k is invariant.

13. Let G be a group. Let $\phi : X \to Y$ and $\psi : Y \to Z$ be G-set homomorphisms between G-sets. Prove that the composition $\psi \circ \phi$ is a G-set homomorphism.

14. Let $f : X \to Y$ be a G-set homomorphism.

(a) Prove that if S is a G-subset of X, then $f(S)$ is a G-subset of Y.

(b) Prove that if T is a G-subset of Y, then $f^{-1}(T)$ is a G-subset of X.

15. Let V be the set of vertices of a cube and let D be the long diagonals. Define the function $f : V \to D$ so that $f(v)$ is the unique long diagonal that contains the vertex v. Let G be the group of rigid motions of the cube and consider the natural actions of G on V and on D. Prove that f is a G-set homomorphism.

16. Let X be a set and let G and H be subgroups of S_X. Prove that the actions of G on X and H on X are permutation isomorphic if and only if G and H are conjugate subgroups in S_X.

6.2   Orbits and Stabilizers

Nontrivial actions of a group G on a set X connect information about the set X with information about the group G in interesting ways. This section begins to study this connection.

As a first example, suppose that p is prime and that Z_p acts on a set X with $|X| = n < p$. Let $\rho : Z_p \to S_X$ be the permutation representation induced from this action. By Lagrange's Theorem, $|\operatorname{Im}\rho|$ divides $|S_X| = n!$. By the First Isomorphism Theorem, $\operatorname{Im}\rho \cong G/\operatorname{Ker}\rho$ so $|\operatorname{Im}\rho| = |G|/|\operatorname{Ker}\rho|$ and thus $|\operatorname{Im}\rho|$ also divides $|G| = p$. Hence $|\operatorname{Im}\rho|$ divides $\gcd(p, n!) = 1$. Hence, the action of Z_p on X is trivial.

6.2.1   Orbits

One of the first nonobvious connections between groups and sets is that a group action of G on X defines an equivalence relation.

Proposition 6.2.1

Let G be a group acting on a nonempty set X. The relation defined by $x \sim y$ if and only if $y = g \cdot x$ for some $g \in G$ is an equivalence relation.

Proof. Let $x \in X$ and let 1 be the identity in G. Since $1 \cdot x = x$, then $x \sim x$ for all $x \in X$. Hence, $\sim$ is reflexive.

Suppose that $x, y \in X$ with $x \sim y$. Then there exists $g \in G$ such that $y = g \cdot x$. Consequently,

$$g^{-1} \cdot y = g^{-1} \cdot (g \cdot x) = (g^{-1}g) \cdot x = 1 \cdot x = x.$$

This shows that $y \sim x$ since $x = g^{-1} \cdot y$.

Suppose that $x, y, z \in X$ with $x \sim y$ and $y \sim z$. Then for some group elements $g, h \in G$, we have $y = g \cdot x$ and $z = h \cdot y$. Then

$$z = h \cdot (g \cdot x) = (hg) \cdot x.$$

This shows that $x \sim z$ and establishes transitivity. $\square$

Definition 6.2.2

Let G be a group acting on a nonempty set X. The $\sim$-equivalence class $\{g \cdot x \mid g \in G\}$, denoted by $G \cdot x$ (or more simply Gx), is called the *orbit* of G containing x.

Recall that the equivalence classes of an equivalence relation form a partition of X. Consequently, the orbits of G on X partition X.

Definition 6.2.3

Suppose that a group G acts on a set X. We say that G fixes an element $x - 0 \in X$ if $g \cdot x_0 = x_0$ for all $g \in G$. Furthermore, the action is called

(1) *free* if $g \cdot x = h \cdot x$ for some $x \in X$, then $g = h$;

(2) *transitive* if for any two $x, y \in X$, there exists $g \in G$ such that $y = g \cdot x$;

(3) *regular* if it is both free and transitive;

(4) *r-transitive* if for every two r-tuples $\{x_1, x_2, \ldots, x_r\}$ and $\{y_1, y_2, \ldots, y_r\}$ of distinct elements in X, there exists an element $g \in G$ such that

$$(y_1, y_2, \ldots, y_r) = (g \cdot x_1, g \cdot x_2, \ldots, g \cdot x_r).$$

Note that an element x is fixed by G if and only if $\{x\}$ is an orbit of G. On the opposite perspective, the action of G on X is transitive if and only if there is only one orbit, namely all of X. A group action is free if and only if the only element in G that fixes any element in X is the group identity.

Example 6.2.4. Consider the action of a group G on its set of subgroups $\mathrm{Sub}(G)$ by conjugation. (See Exercise 6.4.10.) There exists a natural bijection between H and gHg^{-1}. In particular, if G is a finite group, then the orbits of this action stay within subgroups of G of fixed cardinality. A subgroup H is fixed by this action if and only if $gHg^{-1} = H$ for all $g \in G$. Hence, the fixed subgroups are precisely the normal subgroups of G.

As a specific example, consider D_6, whose lattice of subgroups is given in Example 1.8.8. The orbits of the action of D_6 on $\mathrm{Sub}(D_6)$ are

$$\{D_6\}, \{\langle s, r^2 \rangle\}, \{\langle r^6 \rangle\}, \{\langle sr, r^2 \rangle\}, \{\langle s, r^3 \rangle\}, \{\langle sr, r^3 \rangle, \langle sr^2, r^3 \rangle\},$$

$$\{\langle r^2 \rangle\}, \{\langle s \rangle, \langle sr^2 \rangle, \langle sr^4 \rangle\}, \{\langle r^3 \rangle\}, \{\langle sr \rangle, \langle sr^3 \rangle, \langle sr^5 \rangle\}, \{\langle 1 \rangle\}.$$

Every singleton orbit corresponds to a normal subgroup in D_6. $\triangle$

Example 6.2.5. Consider the action of D_6 as linear transformations on $\mathbb{R}^2$, where r acts as a rotation around the origin O by $60°$ and s acts as a reflection through the x-axis. The orbit of O is just the singleton set $\{O\}$ and this is the only fixed point. The orbit of any point P whose polar coordinates (r, θ) has $\theta = k\frac{\pi}{3}$ for some integer k has exactly six points. The orbits of all other points in the plane have 12 points. See Figure 6.3.

Example 6.2.6. Continuing with examples associated with group actions of D_6 on various sets, consider the standard action of D_6 on the vertices of a regular hexagon, labeled as in Example 6.1.5. This action is transitive. In particular, if $a, b \in \{1, 2, 3, 4, 5, 6\}$, then the group element r^{b-a} maps a into

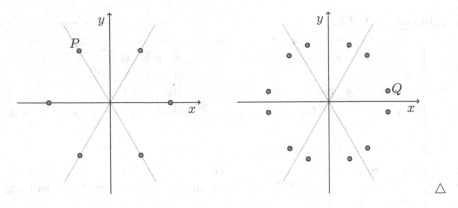

FIGURE 6.3: Some orbits of points in $\mathbb{R}^2$ under D_6 action.

b. However, the action is not 2-transitive (or r-transitive for $r \geq 2$). Indeed, if $g \cdot a = b$ with $a < 6$, then $g \cdot (a+1)$ can only be one of the two vertices on either side of b. $\triangle$

Example 6.2.7. Let F be a field and consider the action of $\mathrm{GL}_n(F)$ on the vector space $V = F^n$ by matrix-vector multiplication. See Example 6.1.7. This action has exactly two orbits. One orbit corresponds to the fixed point $\vec{0}$. On the other hand, consider any two nonzero vectors $\vec{a}$ and $\vec{b}$. Let $M_1, M_2 \in \mathrm{GL}_n(F)$ be matrices such that the first column of M_1 is $\vec{a}$ and the first column of M_2 is $\vec{b}$. Then $M = M_2 M_1^{-1} \in \mathrm{GL}_n(F)$ and $M\vec{a} = \vec{b}$. Thus, $V \setminus \{\vec{0}\}$ is an orbit of this action.

We point out that the action of $\mathrm{GL}_n(F)$ on $V \setminus \{\vec{0}\}$, though transitive is not 2-transitive. We can see this by taking $\vec{a}$ and $\vec{b}$ to be collinear vectors and $\vec{u}$ and $\vec{v}$ to be linearly independent vectors. There is no invertible matrix $g \in \mathrm{GL}_n(F)$ such that $g\vec{a} = \vec{u}$ and $g\vec{b} = \vec{v}$. $\triangle$

Example 6.2.8. Consider the set $X = M_{m \times n}(\mathbb{R})$ of real $m \times n$ matrices. The group $G = \mathrm{GL}_m(\mathbb{R}) \oplus \mathrm{GL}_n(\mathbb{R})$ acts on $M_{m \times n}(\mathbb{R})$ by

$$(A, B) \cdot M = AMB^{-1}. \tag{6.3}$$

Let $T : \mathbb{R}^n \to \mathbb{R}^m$ be a linear transformation and suppose that M is the matrix of T with respect to a basis of $\mathbb{R}^n$ and a basis of $\mathbb{R}^m$. Then the action described above corresponds to effecting a change of basis on $\mathbb{R}^m$ with basis change matrix A and a change of basis on $\mathbb{R}^n$ with basis change matrix B.

If $N = AMB^{-1}$, then $AM = NB$. Since A and B are invertible, $\mathrm{Im}\, B = \mathbb{R}^n$ and $\mathrm{Im}\, A = \mathbb{R}^m$. Hence, $\mathrm{rank}\, AM = \mathrm{rank}\, NB$ implies that $\mathrm{rank}\, M = \mathrm{rank}\, N$. We proceed to prove the converse.

Suppose that M has rank r. By definition, the linear transformation $T(\vec{x}) = M\vec{x}$, expressed with respect to the standard bases on $\mathbb{R}^n$ and $\mathbb{R}^n$

has $\dim \operatorname{Im} T = r$. By the Rank-Nullity Theorem $\dim \operatorname{Ker} T + \dim \operatorname{Im} T = n$ so $\dim \operatorname{Ker} T = n - r$. Let $\{\vec{u}_{r+1}, \vec{u}_{r+2}, \ldots, \vec{u}_n\}$ be a basis of $\operatorname{Ker} T$ as a subspace of $\mathbb{R}^n$. Complement this basis with a set of vectors $\{\vec{u}_1, \vec{u}_2, \ldots, \vec{u}_r\}$ to make an ordered basis $\mathcal{B} = (\vec{u}_1, \vec{u}_2, \ldots, \vec{u}_n)$ of $\mathbb{R}^n$.

Define $\vec{v}_i = T(\vec{u}_i)$ for $1 \le i \le r$ as vectors in $\mathbb{R}^m$. Consider the trivial linear combination

$$c_1 \vec{v}_1 + c_2 \vec{v}_2 + \cdots + c_r \vec{v}_r = \vec{0}.$$

Then, $T(c_1 \vec{u}_1 + c_2 \vec{u}_2 + \cdots + c_r \vec{u}_r) = \vec{0}$. However, $\operatorname{Span}(\vec{u}_1, \vec{u}_2, \ldots, \vec{u}_r) \cap \operatorname{Ker} T = \{\vec{0}\}$. So we deduce that $c_1 \vec{u}_1 + c_2 \vec{u}_2 + \cdots + c_r \vec{u}_r = \vec{0}$ in $\mathbb{R}^n$. Since the $\vec{u}_i$ are linearly independent, we deduce that $c_i = 0$ for $1 \le i \le r$. This shows that $\{\vec{v}_1, \vec{v}_2, \ldots, \vec{v}_r\}$ is a linearly independent set. We now complement this set with $m - r$ vectors to make an ordered basis $\mathcal{B}' = (\vec{v}_1, \vec{v}_2, \ldots, \vec{v}_m)$ on $\mathbb{R}^m$.

By construction, the matrix of T with respect to the basis $\mathcal{B}$ on $\mathbb{R}^n$ and $\mathcal{B}'$ on $\mathbb{R}^m$ is the matrix

$$\left(\begin{array}{c|c} I_r & 0 \\ \hline 0 & 0 \end{array} \right).$$

Consequently, every matrix M of rank r has the above matrix in its orbit. This example has shown that the orbits of $\operatorname{GL}_m(\mathbb{R}) \oplus \operatorname{GL}_n(\mathbb{R})$ acting on $M_{m \times n}(\mathbb{R})$ by (6.3) are the sets

$$\mathcal{O}_r = \{M \in M_{m \times n}(\mathbb{R}) \mid \operatorname{rank} M = r\},$$

where r is an integer satisfying $0 \le r \le \min(m, n)$. $\triangle$

6.2.2 Stabilizers

Let G act on a set X. An element $g \in G$ is said to *fix* an element $x \in X$ if $g \cdot x = x$. The axioms of group actions lead to strong interactions between groups and subsets of X, especially in relation to elements in G that fix an element $x \in X$ or conversely all the elements in X that are fixed by some element $g \in G$.

Proposition 6.2.9

Suppose that a group G acts on a set X. For any element $x \in X$, the subset $G_x = \{g \in G \mid g \cdot x = x\}$ is a subgroup of G.

Proof. Obviously, $1 \cdot x = x$, so $1 \in G_x$ and in particular G_x is nonempty. Suppose that $g, h \in G_x$. Then $g \cdot (h \cdot x) = g \cdot x = x$ but by the compatibility axiom $x = g \cdot (h \cdot x) = (gh) \cdot x$, so $gh \in G_x$ and G_x is closed under the group operation. Finally, if $g \in G_x$, then

$$g^{-1} \cdot (g \cdot x) = g^{-1} \cdot x \implies (g^{-1} \cdot g) \cdot x = g^{-1} \cdot x \implies x = g^{-1} \cdot x.$$

Therefore, G_x is closed under taking inverses. $\square$

Definition 6.2.10

Given $x \in X$, the subgroup $G_x = \{g \in G \mid g \cdot x = x\}$ is called the *stabilizer* of x in G.

Group properties lead to the following important theorem and the subsequent Orbit Equation.

Theorem 6.2.11 (Orbit-Stabilizer Theorem)

Let G be a group acting on a set X. The size of orbit $G \cdot x$ satisfies $|G \cdot x| = |G : G_x|$.

Proof. We need to show a bijection between the elements in the equivalence class $G \cdot x$ of x and left cosets of G_x.

Consider the function f from the orbit $G \cdot x$ to the set of cosets of G_x in G defined by

$$f : y \longmapsto gG_x \qquad \text{where } y = g \cdot x.$$

We first verify that this association is even a function. Suppose that $g_1 \cdot x = g_2 \cdot x$. Then $(g_2^{-1}g_1) \cdot x = x$ so $g_2^{-1}g_1 \in G_x$ and hence the cosets $g_1 G_x$ and $g_2 G_x$ are equal. This shows that any image of f is independent of the orbit representative so f is a function from G_x to the set of left cosets of G_x.

Now suppose that $f(y_1) = f(y_2)$ for two elements in $G \cdot x$, with $y_1 = g_1 \cdot x$ and $y_2 = g_2 \cdot x$. Then $g_1 G_x = g_2 G_x$ so $g_2^{-1}g_1 \in G_x$. Hence, $(g_2^{-1}g_1) \cdot x = x$ and, by acting on both sides by g_2, we get $g_1 \cdot x = g_2 \cdot x$. This proves that f is injective.

Finally, to prove that f is also surjective, let hG_x be a left coset of G_x in G. Then $hG_x = f(h \cdot x)$. The element $h \cdot x$ is in the orbit $G \cdot x$ so f is surjective. $\square$

Interestingly enough, the proof of the Orbit-Stabilizer Theorem does not assume that the group or the set X is finite. The equality of cardinality holds even if the cardinalities are infinite.

We notice as a special case that if G acts transitively on a set X, then $|G : G_x| = |X|$ for all elements $x \in X$. In particular, if X and G are both finite, then $|G| = |X||G_x|$, which implies that $|G|$ is a multiple of $|X|$. Furthermore, a group action can only be regular if $|G| = |X|$.

The Orbit-Stabilizer Theorem leads immediately to the following important corollary. The Orbit Equation is a generic equation, applicable to such an action, that flows from the fact that orbits of G partition X. However, the Orbit Equation often gives rise to interesting combinatorial formulas.

Corollary 6.2.12 (Orbit Equation)

Let G be a group acting on a finite set X. Let T be a complete set of distinct representatives of the orbits. Then

$$|X| = \sum_{x \in T} |G \cdot x| = \sum_{x \in T} |G : G_x|.$$

Example 6.2.13. As a simple illustration of the Orbit-Stabilizer Theorem, we prove that a group G of order 15 acting on a set X of size 7 has a fixed point. By the Orbit-Stabilizer Theorem, the orbit of an element x has order $|G : G_x|$, where G_x is the stabilizer of x. Hence, $|G : G_x|$ can be equal to 1, 3, 5, or 15. If $|G : G_x| = 1$, then $G = G_x$ so all of G fixes x and hence x is a fixed point of the action. Obviously, there can be no orbit of size 15 in a set of size 7. Assume there is no fixed point. There the orbits must have order 3 or 5. Hence, if there are r orbits of size 3 and s orbits of size 5, then $3r + 5s = 7$ for $r, s \in \mathbb{N}$. If $s = 0$, then $r = 7/3 \notin \mathbb{N}$; if $s = 1$, then $r = 2/3 \notin \mathbb{N}$; and if $s \geq 2$, then $r < 0$. We have shown that there exist no solutions in nonnegative integers to the equation $3r + 5s = 7$. We conclude by contradiction that there must be a fixed point. $\triangle$

Example 6.2.14. Consider the action of $G = S_n$ on $\mathcal{P}(\{1, 2, \ldots, n\})$ by

$$\sigma \cdot \{x_1, x_2, \ldots, x_k\} = \{\sigma(x_1), \sigma(x_2), \ldots, \sigma(x_k)\}.$$

We can interpret the result of Exercise 6.1.5 by saying that the orbits of this action consist of the set of subsets of a given cardinality k, for k ranging from 0 to n. For a fixed k, define $A = \{1, 2, \ldots, k\}$. The stabilizer G_A is the set of permutations $\sigma \in S_n$ that map $\{1, 2, \ldots, k\}$. Hence,

$$|G_A| = k!(n - k)!$$

because there are $k!$ ways σ can permute $\{1, 2, \ldots, k\}$ and $(n - k)!$ ways σ can permute the remaining elements of $\{1, 2, \ldots, n\}$. We recover the fact that there are

$$|G : G_A| = \frac{|G|}{|G_A|} = \frac{n!}{k!(n - k)!} = \binom{n}{k}$$

subsets of X of size k. Furthermore, since $|\mathcal{P}(\{1, 2, \ldots, n\})| = 2^n$, the orbit equation for this action is the well-known combinatorial formula

$$2^n = \sum_{k=0}^{n} \binom{n}{k}.$$

$\triangle$

6.2.3   The Lemma That Is Not Burnside's

The Orbit-Stabilizer Theorem establishes an connection between the size of
the set X and orders of stabilizers. Similarly, by considering the set of points in
X fixed by any given group element g, one arrives at another connection, called
the Cauchy-Frobenius Lemma. In the literature, this result is often called the
Burnside Lemma. Though Burnside presented it in his text [8, Chapter X], he
was not the first to prove it. In [25], Neumann chronicles how this incorrect
naming arose.

Definition 6.2.15

Let G be a group acting on a set X. For any $g \in G$, define the fixed
subset
$$X^g = \{x \in X \mid g \cdot x = x\}.$$

We note that the fixed subset X^g serves a parallel role as the stabilizer of
a set element x.

Theorem 6.2.16 (Cauchy-Frobenius Lemma)

Let G be a finite group acting on a finite set X. Suppose that G has
m orbits on X. Then
$$m|G| = \sum_{g \in G} |X^g|.$$

Proof. Consider the set $S = \{(x, g) \in X \times G \mid g \cdot x = x\}$. We determine $|S|$
in two different ways, by first summing through elements in G and then by
summing first through X. By summing first through G, we get

$$|S| = \sum_{g \in G} |X^g|.$$

By summing first through elements in x, we get

$$|S| = \sum_{x \in X} |G_x|.$$

Now let $\mathcal{O}_1, \mathcal{O}_2, \ldots, \mathcal{O}_m$ be the orbits of G on X and let $x_1, x_2, \ldots, x_m$ be a
complete set of distinct orbit representatives. By the Orbit-Stabilizer Theo-
rem, $|\mathcal{O}_i| = |G : G_{x_i}|$. Thus, $|G_{x_i}| = |G|/|\mathcal{O}_i|$. Hence,

$$|S| = \sum_{x \in X} |G_x| = \sum_{i=1}^{m} \sum_{x \in \mathcal{O}_i} |G_x| = \sum_{i=1}^{m} \sum_{x \in \mathcal{O}_i} \frac{|G|}{|\mathcal{O}_i|}$$

$$= \sum_{i=1}^{m} \frac{|G|}{|\mathcal{O}_i|} |\mathcal{O}_i| = \sum_{i=1}^{m} |G| = m|G|$$

and the result follows by identifying the two ways of counting $|S|$. $\square$

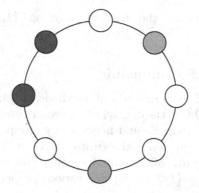

FIGURE 6.4: A colored-bracelet counting problem.

The Cauchy-Frobenius Lemma has many interesting applications in counting problems and combinatorics. In particular, if a counting problem can be phrased in a manner to count orbits of a group acting on a set, then the lemma provides a strategy to compute the number of orbits m.

Example 6.2.17. Suppose that we wish to design a bracelet with 8 beads using beads of 3 different colors, such as the one in Figure 6.4. We consider two bracelets equivalent if the arrangement of bead colors of one can be obtained from the other simply by rotating it. We propose to determine how many inequivalent bracelets of 8 beads can be made using 3 colors.

The rotation on a bracelet with 8 beads corresponds to the action of Z_8. There are 3^8 different bracelet without considering the equivalence. Any group elements in Z_8 of the same order d will fix the same number of bracelet colorings. Hence, instead of summing over all elements in Z_8, we can sum over the divisors of 8, corresponding to the order of various elements in Z_8. Note that for any $d|8$, there are $\phi(d)$ elements of order d. Finally, note that a bracelet coloring will be fixed by an element of order d if and only if the color pattern repeats every $\frac{8}{d}$ beads. But then a contiguous set of $\frac{8}{d}$ beads can be colored in any way. Hence, the number of inequivalent bracelets is m, where

$$8m = \sum_{d|8} \phi(d)3^{8/d} \implies m = \frac{1}{8}(3^8 + 3^4 + 2 \cdot 3^2 + 4 \cdot 3^1) = 834.$$

To connect this bead-coloring problem to more theoretical language of group actions, we can view the set of possible bracelets as the set of functions $X = \text{Fun}(\{1, 2, \ldots, 8\}, \{1, 2, 3\})$. The domain corresponds to the bead position and the codomain is the bead color. The action of rotating the bracelet corresponds to the action of Z_8 on X by

$$(\sigma \cdot f)(x) = f(\sigma^{-1} \cdot x),$$

where $\sigma \in Z_8$ acts in the natural way on $\{1, 2, \ldots, 8\}$, as $Z_8 = \langle (1\,2\,3\,4\,5\,6\,7\,8) \rangle$. △

6.2.4 Useful CAS Commands

Most CAS come with commands and methods to work with permutation groups. Furthermore, by virtue of Cayley's Theorem every finite group embeds into some symmetric group S_n and hence every group is isomorphic to some permutation group. In this sense, the group G acts on the set $X = \{1, 2 \ldots, n\}$

Both *Maple* and SAGE offer commands that test whether a permutation group G acting on $X = \{1, 2, \ldots, n\}$ has various properties.

Maple Function	SAGE
`IsRegular(G);`	`G.is_regular()`
Returns **true** or **false** whether the action is regular.	
`IsTransitive(G);`	`G.is_transitive()`
Returns **true** or **false** whether the action is transitive.	
`IsPrimitive(G);`	`G.is_primitive()`
Returns **true** or **false** whether the action is primitive.	

The topic of primitivity arises in the next section.

EXERCISES FOR SECTION 6.2

1. Let $X = \{1, 2, 3\}^3 = \{(i, j, k) \mid 1 \le i, j, k \le 3\}$.
   (a) Consider the action of S_3 on X by $\sigma \cdot (a_1, a_2, a_3) = (\sigma(a_1), \sigma(a_2), \sigma(a_3))$. Explicitly list all the orbits of this action.
   (b) Consider the action of S_3 on X of rearrangement from Example 6.1.13. Explicitly list all the orbits of this action.

2. Let X be the set of bit strings of length 6, namely $X = \{0, 1\}^6$.
   (a) Consider the action of S_6 on X of rearrangement from Examples 6.1.13. Give a unique representative for each orbit and determine how many elements are in each orbit.
   (b) View Z_6 as a subgroup of S_6 via $Z_6 \cong \langle (1\,2\,3\,4\,5\,6) \rangle$. Determine all the orbits of size 1, 2, 3, and 6.

3. Consider the group G of rigid motions of a cube and consider the action of G on the set of points C that make up the surface of the rotated cube. (See Example 6.1.11.) For each point of C, describe its orbits. In particular, for all divisors d of $|G| = 24$, determine which points of C have orbits of size d.

4. Let G be a group. Prove that the action of G on itself by left multiplication is a regular group action.

5. Consider the group $\mathcal{G}$ of invertible affine transformations of the plane acting on $\mathbb{R}^2$. Recall that in coordinates, an element of $\mathcal{G}$ has the form

$$\begin{pmatrix} x \\ y \end{pmatrix} \longmapsto A \begin{pmatrix} x \\ y \end{pmatrix} + \begin{pmatrix} e \\ f \end{pmatrix},$$

where $A \in GL_2(\mathbb{R})$ and $e, f \in \mathbb{R}$. Show that the natural action of $\mathcal{G}$ on $\mathbb{R}^2$ is 2-transitive.

6. Let G act on a nonempty set X. Suppose that $x, y \in X$ and that $y = g \cdot x$ for some $g \in G$. Prove that $G_y = gG_xg^{-1}$. Deduce that if the action is transitive, then the kernel of the action is

$$\bigcap_{g \in G} gG_xg^{-1}.$$

7. Let G be a group acting on a set X. Prove that a subset S of X is a G-invariant subset of X if and only if S is a union of orbits.

8. Suppose that a group G acts on a set X and also on a set Y. Prove that a G-set homomorphism $f : X \to Y$ maps orbits of G in X to orbits of G in Y.

9. Show that a group of order 55 acting on a set of size 34 must have a fixed point.

10. Suppose G is a group of order 21. Determine with proof the positive integers n such that an action of G on a set X of size n must have a fixed point.

11. Let $X = \mathbb{R}[x_1, x_2, x_3, x_4]$. Consider the action of $G = S_4$ on X by

$$\sigma \cdot p(x_1, x_2, x_3, x_4) = p(x_{\sigma(1)}, x_{\sigma(2)}, x_{\sigma(3)}, x_{\sigma(4)}).$$

   (a) Find the stabilizer of the polynomial $x_1 + x_2$ and give its isomorphism type.
   (b) Find a polynomial $q(x_1, x_2, x_3, x_4)$ whose stabilizer is isomorphic to D_4.
   (c) Explicitly list the elements in the orbit of $x_1x_2 + 5x_3$.
   (d) Explicitly list the elements in the orbit of $x_1x_2^2x_3^3$.

12. Let G be a group acting on a set X. Let H be a subgroup of G. It acts on X with the action of G restricted to H. Let $\mathcal{O}$ be an orbit of H in X.

   (a) Prove that for all $g \in G$, the set $g \cdot \mathcal{O} = \{g \cdot x \mid x \in \mathcal{O}\}$ is an orbit of the conjugate subgroup gHg^{-1}.
   (b) Deduce that if G is transitive on X and if $H \trianglelefteq G$, then all the orbits of H are of the form $g\mathcal{O}$.

13. Let $A = \{1, 2, \ldots, n\}$ and consider the set $X = \{(a, S) \in A \times \mathcal{P}(A) \mid a \in S\}$. Consider the action of S_n on X by $\sigma \cdot (a, S) = (\sigma(a), \sigma \cdot S)$, where $\sigma \cdot S$ is the power set action. (See Exercise 6.1.4.) Prove that the Orbit Equation for this action gives the combinatorial formula

$$\sum_{k=0}^{n} k \binom{n}{k} = n2^{n-1}.$$

14. Let A be a finite set of size k and consider the action of S_n on $X = A^n$ via

$$\sigma \cdot (a_1, a_2, \ldots, a_n) = (a_{\sigma^{-1}(1)}, a_{\sigma^{-1}(2)}, \ldots, a_{\sigma^{-1}(n)}).$$

(See Example 6.1.13.) Prove that the Orbit Equation for this action is

$$k^n = \sum_{s_1 + \cdots + s_k = n} \frac{n!}{s_1! s_2! \cdots s_k!},$$

where the summation is taken over all $(s_1, s_2, \ldots, s_k) \in \mathbb{N}^k$ that add up to n.

15. Let A and B be finite disjoint sets of cardinality a and b, respectively. Let $X = \mathcal{P}_k(A \cup B)$ be the set of subsets of $A \cup B$ of cardinality k. Let S_A and S_B be the groups of permutations on A and B, respectively. Consider the action of $G = S_A \oplus S_B$ on $A \cup B$ by

$$(\sigma, \tau) \cdot c = \begin{cases} \sigma(c) & \text{if } c \in A \\ \tau(c) & \text{if } c \in B. \end{cases}$$

Define the action of $G = S_A \oplus S_B$ on $\mathcal{P}_k(A \cup B)$ as the standard set of subsets action. Prove that the Orbit Equation of G on X gives the Vandermonde Identity

$$\binom{a+b}{k} = \sum_{i+j=k} \binom{a}{i} \binom{b}{j}.$$

16. Let X be the set of functions from $\{1, 2, \dots, n\}$ into itself and let $G = S_n \oplus S_n$. Define the pairing $G \times X \to X$ by

$$((\sigma, \tau) \cdot f)(a) = \sigma \cdot f(\tau^{-1} \cdot a) \qquad \text{for all } a \in \{1, 2, \dots, n\}.$$

   (a) Prove that this pairing is an action of G on X.

   (b) Prove that the set of orbits is in bijection with the partitions of n, where for any partition $\lambda = (\lambda_1, \lambda_2, \dots, \lambda_s)$, the corresponding orbit $\mathcal{O}_\lambda$ consists of functions in which the orders of the distinct fibers are $\lambda_1, \lambda_2, \dots, \lambda_s$.

   (c) Determine the number of functions in the orbit $\mathcal{O}_\lambda$.

   (d) Supposing the $n = 5$, find the stabilizer of the function f such that $f(1) = f(2) = 1$, $f(3) = f(4) = 2$ and $f(5) = 3$.

17. Consider the action of $G = S_3 \oplus S_3$ on $X = \{(i, j, k) \mid 1 \leq i, j, k \leq 3\}$ by

$$(\sigma, \tau) \cdot (a_1, a_2, a_3) = (\tau(a_{\sigma^{-1}(1)}), \tau(a_{\sigma^{-1}(3)}), \tau(a_{\sigma^{-1}(3)})).$$

This action has the effect of rearranging the elements in the triple according to σ and then permuting the outcome by τ. Explicitly calculate X^g for all $g \in G$ and verify the Cauchy-Frobenius Lemma.

18. Recall that the group of rigid motions of a tetrahedron is A_4. Suppose that we color the faces of a tetrahedron with colors red, green, or blue. We consider two colorings equivalent if one coloring can be obtained from another by rotating the tetrahedron. How many inequivalent such colorings are there?

19. We consider colorings of the vertices of a square as equivalent if one coloring can be obtained from the other by any D_4 action on the square. How many different colorings are there with (a) 3 colors; (b) 4 colors; (c) 5 colors?

20. We consider colorings of the vertices of an equilateral triangle as equivalent if one coloring can be obtained from the other by any D_3 action on the triangle. How many different colorings are there using p colors?

21. Repeat Example 6.2.17 but consider bracelet colorings equivalent if one is obtained from another under the action of some D_8 element. (Note that the bracelet in Figure 6.4 is only fixed by 1 under the Z_8 action but by $\{1, sr\}$ in D_8.)

6.3   Transitive Group Actions

The previous section discussed primarily the orbits of a group action on a set. From the emphasis of the section, the reader might get the impression that transitive group actions are not interesting since in that case there is only one orbit, namely the whole set. This could not be further from the truth.

There still exists considerable structure within a transitive group action. In fact, a group acts transitively on each orbit, so we may view any group action as a union of transitive group actions. In order to look deeper into the analysis of group actions, we must address properties of transitive group actions.

6.3.1   Blocks and Primitivity

Consider the Example 6.1.11, which discussed the group G of solid rotations of a cube. The example pointed out how G acts, among other things, on the faces of the cube, and on the long diagonals of the cube. There is a qualitative difference between the action of G on the set of faces and on the set of long diagonals. In particular, the intersection of any two distinct long diagonals is $\emptyset$ whereas some distinct faces intersect along an edge.

Definition 6.3.1

Let G be a group that acts transitively on a set X. A *block* is a nonempty subset $B \subseteq X$ such that for all $g \in G$, either $g \cdot B = B$ or $(g \cdot B) \cap B = \emptyset$.

For every transitive action of a group G on a set X, the singleton sets $\{x\}$ in X are blocks as is the whole set X. Since these subsets are always blocks, we call them *trivial*. However, as the group of rigid motions of the cube illustrates, some group actions may possess other blocks. Any of the long diagonals of a cube is a block whereas a face is not a block. Some group actions do not possess any nontrivial blocks. We give these a specific name.

Definition 6.3.2

A transitive action of a group G on a set X is called *primitive* if the only blocks of the action are the trivial ones.

Example 6.3.3. Consider the group Z_8 and its action on $X = \{1, 2, 3, 4, 5, 6, 7, 8\}$ where Z_8 is given as $\langle (1\,2\,3\,4\,5\,6\,7\,8) \rangle$ in S_8. This action is obviously transitive. Besides the trivial blocks and X, the subsets

$$\{1, 3, 5, 7\}, \ \{2, 4, 6, 8\}, \ \{1, 5\}, \ \{2, 6\}, \ \{3, 7\}, \ \{4, 8\}$$

are also blocks. The action of Z_8 on X has no other blocks.   △

Example 6.3.4. For $n \geq 3$, the action of S_n on $\{1, 2, \ldots, n\}$ is primitive. Let $\{a_1, a_2, \ldots, a_k\}$ be any subset of $\{1, 2, \ldots, n\}$ with $2 \leq k \leq n-1$. There exists $\ell \in X \setminus \{a_1, a_2, \ldots, a_k\}$ and $\sigma = (a_1 \, a_2 \ldots a_k \, \ell)$. Then

$$(\sigma \cdot \{a_1, a_2, \ldots, a_k\}) \cap \{a_1, a_2, \ldots, a_k\} = \{a_2, \ldots, a_k\},$$

which is neither empty, since $k \geq 2$, nor $\{a_1, a_2, \ldots, a_k\}$. $\triangle$

If B is a block of a transitive group action of G on X, then for every $y \in X$, there exists some $g \in G$ with $y \in g \cdot B$. Furthermore, since $B \cap (g \cdot B) = B$ or $\emptyset$, then the set of subsets $\Sigma = \{g \cdot B \mid g \in G\}$ is a partition of X. A set of subsets of X defined as $\{g \cdot B \mid g \in G\}$, where B is a block, is called a *system of blocks* on X.

For all $g \in G$, if B is a finite block, then $|g \cdot B| = |B|$. Consequently, in the partition on X induced from the system of blocks associated with B, all the other blocks (equivalence classes associated with the partition) have the same cardinality. Since the union of all the blocks in Σ is precisely X, we have proven the following result.

Proposition 6.3.5

If B is a block in the action of a group G on a finite set X, then $|B|$ divides $|X|$.

If Σ is a system of blocks, then the group G acts transitively in the obvious way on Σ, thereby inducing another group action but on a smaller set. Comparing the action of G on X with the action of G on a system of blocks Σ, we introduce the following two types of stabilizers.

Definition 6.3.6

Let G act on X and let $B \subseteq X$, not necessarily a block. We define the *pointwise stabilizer* of B as

$$G_{(B)} = \{g \in G \mid g \cdot x = x \text{ for all } x \in B\}$$

and the *setwise stabilizer* as

$$G_{\{B\}} = \{g \in G \mid g \cdot B = B\}.$$

Note that if B is a block in the action of G on X, then $G_{\{B\}}$ is the usual stabilizer of B in the action of G on Σ. In contrast, for the pointwise stabilizer

$$G_{(B)} = \bigcap_{x \in B} G_x.$$

It is not hard to see that $G_{(B)}$ and $G_{\{B\}}$ are both subgroups and that $G_{(B)} \trianglelefteq G_{\{B\}}$. (See Exercise 6.3.9.) Obviously, if B is a singleton $B = \{b\}$, then $G_{(B)} = G_{\{B\}} = G_b$.

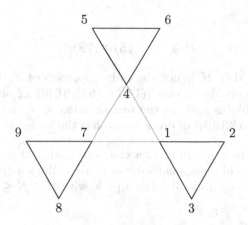

FIGURE 6.5: Visualizing a block structure.

Example 6.3.7. Consider the standard action of S_9 on $X = \{1, 2, \ldots, 9\}$. We propose to find a subgroup $H \leq S_9$ that acts transitively on X and has $\{1, 2, 3\}$ as a block.

Because of transitivity, the subgroup H must contain a permutation that maps 1 to 2 and a permutation that maps 1 to 3. The subgroup $\langle (1\,2\,3) \rangle$ is transitive on the block $B_1 = \{1, 2, 3\}$. Since H is transitive, it must contain a g such that $g \cdot 1 = 4$. But then the element g must map the block $\{1, 2, 3\}$ to another block of size 3. Without loss of generality, let us assume that this second block is $B_2 = \{4, 5, 6\}$. In that case, the third block is $B_3 = \{7, 8, 9\}$.

The permutation $\sigma = (1\,4\,7)(2\,5\,8)(3\,6\,9)$ cycles through blocks B_1, B_2, and B_3. Consequently, a subgroup of S_9 that is transitive on X and has $\{1, 2, 3\}$ as a block is

$$H = \langle (1\,2\,3),\ (1\,4\,7)(2\,5\,8)(3\,6\,9) \rangle.$$

The subgroup H has the effect of cycling through the blocks individually and independently cycling through the blocks as a whole.

Figure 6.5 gives an intuitive picture of the action of H on $\{1, 2, 3, 4, 5, 6, 7, 8, 9\}$. The permutation $(1\,2\,3)$ corresponds to a clockwise rotation of the triangle $\{1, 2, 3\}$; the permutation σ corresponds to a counterclockwise rotation of $120°$ of the whole figure; and the permutations

$$\sigma(1\,2\,3)\sigma^{-1} = (4\,5\,6) \qquad \text{and} \qquad \sigma^2(1\,2\,3)\sigma^{-2} = (7\,8\,9)$$

correspond to clockwise rotations by $120°$ individually in the triangles $\{4, 5, 6\}$ and in $\{7, 8, 9\}$.

The setwise stabilizer of B_1 is

$$H_{\{B_1\}} = \langle (1\,2\,3), (4\,5\,6), (7\,8\,9) \rangle$$

and this is the same setwise stabilizer for the other two blocks B_2 and B_3.

The pointwise stabilizer is

$$H_{(B_1)} = \langle (4\,5\,6), (7\,8\,9) \rangle.$$

We point out that H is not the only subgroup of S_9 that has $\{1, 2, 3\}$ as a block. The cyclic subgroup $\langle (1\,4\,7)(2\,5\,8)(3\,6\,9) \rangle$ of order 3 simply cycles through the blocks and has the same system of blocks as H. The cyclic subgroup $\langle (1\,4\,7\,2\,5\,8\,3\,6\,9) \rangle$ of order 9 also has the same system of blocks. $\triangle$

The following proposition gives a characterization of primitive groups. It relies on the notion of a maximal subgroup. We call a subgroup H of a group G a *maximal subgroup* if for all subgroups K with $H \leq K \leq G$, either $K = H$ or $K = G$.

Proposition 6.3.8

Let G act transitively on a set X with $|X| > 1$. Then the action is primitive if and only if G_x is maximal for all $x \in X$.

Proof. ($\Longleftarrow$) First, suppose that G has a nontrivial block B. Let $x \in B$. If $g \in G_x$, then $gB \cap B \neq \emptyset$ so $gB = B$ and thus $g \in G_{\{B\}}$. Consequently, $G_x \leq G_{\{B\}} \leq G$. Since $B \neq X$, we know that $G_{\{B\}} \neq G$. Furthermore, $|B| \neq 1$ so there exists another element $y \neq x$ in the block B. Again, since G acts transitively, there exists $h \in G$ such that $gx = y$. Since $y \in B$ and B is a block, $hB \cap B \neq \emptyset$ so $hB = B$. Therefore, $h \in G_{\{B\}}$ but $h \notin G_x$. Therefore, G_x is a strict subgroup of $G_{\{B\}}$. We have shown that if the action is not primitive, then there exists some x such that G_x is not a maximal subgroup and the contrapositive gives us the desired implication.

($\Longrightarrow$) Conversely, suppose that G_x is not maximal for some $x \in X$. Then let H be some subgroup such that $G_x \lneq H \lneq G$. Let B be the orbit $B = Hx$. Since $H - G_x \neq \emptyset$, then the orbit B contains more than one element, so $|B| > 1$. Assume that $B = X$. Then H would be transitive and hence $|X| = |G : G_x| = |H : G_x|$, which is a contradiction since $H \lneq G$. Thus, $B \subsetneq X$. Now suppose that $gB \cap B \neq \emptyset$ for some $g \in G$. Let $b \in gB \cap B$. Then there exist $h_1, h_2 \in H$ such that $b = h_1 x = gh_2 x$. Therefore, $h_1^{-1} gh_2 x = x$ so $h_1^{-1} gh_2 \in G_x \leq H$. Consequently, we deduce that if $gB \cap B \neq \emptyset$ then $g \in H$ so $gB = B$. Hence, B is a strict subset of X that is not a singleton set such that $gB = B$ or $gB \cap B = \emptyset$. Hence, B is a nontrivial block. Again, we have proven the contrapositive of the desired implication. $\square$

Recall that if $g \in G$ and $x \in X$, then the stabilizer of the element gx is

$$G_{gx} = gG_x g^{-1}. \tag{6.4}$$

Consequently, we can rephrase the above proposition with the stronger result.

> **Corollary 6.3.9**
>
> A transitive group action (G, X, ρ) is primitive if and only if G_x is maximal for some $x \in X$.

Proof. Suppose that for some x the stabilizer G_x is not maximal and that $G_x \lneq H \lneq G$ for some subgroup H. Since the action is transitive, for all $y \in X$, there exists $g \in G$ with $y = gx$. Then $G_y = G_{gx} = gG_xg^{-1}$ so $G_y \lneq gHg^{-1} \lneq G$ and G_y is not maximal. Hence, there exists x such that G_x is maximal if and only if G_x is maximal for all $x \in X$. □

6.3.2 Blocks and Normal Subgroups

If a group G acts transitively on a set X, then any subgroup $H \leq G$ also acts on X. The action of H on X might or might not be transitive. If H does not act transitively, the orbits of H partition X. However, more can be said if the subgroup is normal.

> **Proposition 6.3.10**
>
> Let (G, X, ρ) be a transitive group action and let $N \trianglelefteq G$. If N fixes a point x, then $N \leq \mathrm{Ker}\, \rho$.

Proof. If x is fixed by N, then $N_x = N$. Let $y \in X$ be arbitrary and let $g \in G$ with $y = gx$. Then by (6.4), $N_y = gN_xg^{-1} = gNg^{-1} = N$. The result follows. □

> **Proposition 6.3.11**
>
> Let (G, X, ρ) be a transitive group action and let $N \trianglelefteq G$. Then
> (1) the orbits of N form a system of blocks for G;
> (2) if G acts primitively, then N acts transitively on X or $N \leq \mathrm{Ker}\, \rho$;
> (3) if $\mathcal{O}$ and $\mathcal{O}'$ are two orbits of H, then they are permutation equivalent (but not necessarily isomorphic as N-set).

Proof. (1): Let $\mathcal{O}$ be an orbit of N action on X as a subset of G and define the set $\Sigma = \{g\mathcal{O} \mid g \in G\}$. By Exercise 6.2.12, all the orbits of H acting on X are of the form $g\mathcal{O}$. Hence, Σ is the set of orbits of N and, since G is transitive, Σ is a partition of X. Thus, Σ is a system of blocks for the action of G on X.

(2): If G acts primitively, then the action has two systems of blocks, namely $\{\{x\} \mid x \in X\}$ and $\{X\}$. By part (1), the set of orbits of N must be one of these two options. In the first case, N stabilizes all $x \in X$ so $N \leq \mathrm{Ker}\, \rho$. In the second case, N is transitive.

(3): By part (1), if $\mathcal{O}$ and $\mathcal{O}'$ are two orbits of N, then $\mathcal{O}' = g\mathcal{O}$ for some $g \in G$. Since N is a normal subgroup, conjugation on N given by $\psi_g(n) = gng^{-1}$ is an automorphism. Furthermore, the function between H-orbits $f : \mathcal{O} \to \mathcal{O}'$ defined by $f_g(x) = gx$ is a bijection. Then for all $n \in N$ and all $x \in \mathcal{O}$,

$$\psi_g(n) \cdot f_g(x) = (gng^{-1}) \cdot (gx) = (gn)x = g(nx) = f_g(nx).$$

Therefore, (ψ_g, f_g) is a group action isomorphism. $\square$

In Example 6.3.3, we saw that the natural action of $Z_8 = \langle z \mid z^8 = 1 \rangle$ on $\{1, 2, \ldots, 8\}$ has two nontrivial systems of blocks. They correspond directly to the proper (normal) subgroups of Z_8, namely $\langle z^2 \rangle$ and $\langle z^4 \rangle$.

Proposition 6.3.11 generalizes the defining property of normal subgroups. Consider the action of G on itself by left multiplication and let $H \leq G$ be any subgroup. Left multiplication is a transitive action. The orbits of the action of H on G by left multiplication are precisely the right cosets Hg. The right cosets form a system of blocks if and only if for all $g, x \in G$, we have $x(Hg) = Hg$ or $x(Hg) \cap Hg = \emptyset$. If $H \trianglelefteq G$, then $x(Hg) = x(gH) = (xg)H = H(xg)$ so $x(Hg)$ is another right coset, thereby satisfying the requirements for a block. Conversely, if $H \ntrianglelefteq G$, then there exists $g \in G$ such that $gH \neq Hg$. However, $gH \cap Hg$ is never the empty set since it contains g. Hence, H is not a block. Therefore, a subgroup H is normal if and only if its right cosets form a system of blocks under the action of left multiplication of G on itself.

Example 6.3.12 (Rigid Motions of the Cube). Consider again the group G of rigid motions of a cube and its action on the set of vertices. (See Example 6.1.11 and Figure 6.2.) We know that $G \cong S_4$. Furthermore, S_4 has only two normal subgroups, namely A_4 and $K = \langle (1\,2)(3\,4), (1\,3)(2\,4) \rangle$. As rigid motions, A_4 is generated by the rotations of degree $120°$ around the axes along long diagonals. Note that these rotations all have order 3 and correspond to the 3-cycles in S_4. Expressed as rigid motions, the subgroup K consists of the identity and the three rotations of $180°$ around the three different axes that join centers of opposite faces.

It is easy to check that (using the labeling as in Figure 6.2) the orbits of A_4 are $\{1, 3, 6, 8\}$ and $\{2, 4, 5, 7\}$. Interestingly enough, the orbits of K are also $\{1, 3, 6, 8\}$ and $\{2, 4, 5, 7\}$. Geometrically, these orbits correspond to two regular tetrahedra of vertices that are separated from each other by a diagonal edge in a face. We also observe that no normal subgroup of G has for orbits the system of blocks of long diagonals through the cube. $\cdot \,\triangle$

The above example shows that the converse to part (1) of Proposition 6.3.11 is not true. In other words, given a transitive group action with a system of blocks Σ, there does not necessarily exist a normal subgroup N whose orbits are the blocks in Σ.

When N is a normal subgroup of G, it is always natural to consider the quotient group G/N. Let Σ be the set of orbits of N in X. For all $n \in N$ and all $g \in G$,

$$(gn) \cdot (Nx) = g(Nx),$$

so the pairing on $G/N \times \Sigma$ defined by

$$(gN) \cdot (Nx) = (gN)x$$

is a transitive action of the quotient group G/N on the set orbits of N (which is the quotient set of the equivalence class of the action of N on X). This action might not be trivial but we are led to the following proposition.

Proposition 6.3.13

Let (G, X, ρ) be a transitive group action and let $N \trianglelefteq G$. Then N has at most $|G : N|$ orbits and if $|G : N|$ is finite, then the number of orbits of N divides $|G : N|$.

Proof. Since the action of G/N on Σ is transitive, the number of orbits of N, namely $|\Sigma|$, is less than $|G/N|$. Let Nx be one orbit. By the Orbit-Stabilizer Theorem

$$|\Sigma| = |(G/N) : (G/N)_{Nx}| = \frac{|G : N|}{|(G/N)_{Nx}|}.$$

Hence, $|G : N|$ divides $|\Sigma|$. $\qquad\square$

Proposition 6.3.11(2) gives a strategy for finding simple groups.

Corollary 6.3.14

If (G, X, ρ) is a faithful, transitive, primitive group action such that no normal subgroup $H \trianglelefteq G$ is transitive, then G is simple.

6.3.3 Multiple Transitivity

A transitive group action is called *multiply transitive* if it is r-transitive for some $r \geq 2$. We commonly say that an action is doubly transitive if it is 2-transitive and triply transitive if it is 3-transitive. Note that if a group action is r-transitive it is also k-transitive for all k with $1 \leq k \leq r$.

Example 6.3.15. In Exercise 6.2.5 the reader was invited to show that the group G of affine transformations of $\mathbb{R}^2$ acts doubly transitively on $\mathbb{R}^2$. However, invertible affine transformations cannot map a set of three collinear points into a nondegenerate triangle. Hence, the group does not act triply transitively on $\mathbb{R}^2$. Though this example is over the field $\mathbb{R}$, the same result holds over in the context of any field. $\qquad\triangle$

Example 6.3.16. We have already seen that the standard action of S_n on $X = \{1, 2, \ldots, n\}$ is n-transitive. However, consider the standard action of A_n on X. We can show that it is $(n - 2)$-transitive but not $(n - 1)$-transitive.

Let $(a_1, a_2, \ldots, a_{n-2})$ and $(b_1, b_2, \ldots, b_{n-2})$ be two $(n-2)$-tuples of distinct elements in X. Complete both to ordered n-tuples $(a_1, a_2, \ldots, a_n)$ and $(b_1, b_2, \ldots, b_n)$ of distinct elements. Consider the permutations

$$\sigma(a_i) = b_i \quad \text{and} \quad \tau(a_i) = \begin{cases} b_i & \text{if } 1 \leq i \leq n-2 \\ b_n & \text{if } i = n-1 \\ b_{n-1} & \text{if } i = n. \end{cases}$$

They both map the $(n-2)$-tuple $(a_1, a_2, \ldots, a_{n-2})$ to $(b_1, b_2, \ldots, b_{n-2})$. However, $\tau = (b_{n-1} b_n)\sigma$ so either one or the other is even. Hence, A_n is $(n-2)$-transitive on $\{1, 2, \ldots, n\}$. On the other hand, any subgroup of S_n that acts $(n-1)$-transitively on $\{1, 2, \ldots, n\}$ also acts n-transitively. However, A_n does not act n-transitively because the only permutation that maps the n-tuple $(1, 2, \ldots, n)$ into $(2, 1, 3, 4, \ldots, n)$ is odd. Hence, A_n does not act $(n-1)$-transitively. $\triangle$

> **Proposition 6.3.17**
>
> Let G act transitively on a set X. If $r \geq 2$, then G acts r-transitively if and only if for all $x \in X$, the stabilizer G_x acts $(r-1)$-transitively on $X \setminus \{x\}$.

Proof. First, suppose that G is r-transitive on X. Let $(a_1, a_2, \ldots, a_{r-1})$ and $(b_1, b_2, \ldots, b_{r-1})$ be two ordered $(r-1)$-tuples in $X \setminus \{x\}$. For all $x \in X$, there exists $g \in G$ such that

$$g \cdot (x, a_1, a_2, \ldots, a_{r-1}) = (x, b_1, b_2, \ldots, b_{r-1}).$$

Obviously, $g \in G_x$ and this g maps $(a_1, a_2, \ldots, a_{r-1})$ into $(b_1, b_2, \ldots, b_{r-1})$.

Conversely, suppose that G_x acts $(r-1)$-transitively on $X \setminus \{x\}$. Let $(a_1, a_2, \ldots, a_r)$ and $(b_1, b_2, \ldots, b_r)$ be two ordered r-tuples in $X \setminus \{x\}$. Since G acts transitively on X, then there exists g such that $b_1 = ga_1$. Since G_{b_1} acts $(r-1)$-transitively on $X - \{b_1\}$, there exists $h \in G_{b_1}$ such that $b_i = h(ga_i)$ for $2 \leq i \leq r$. Then the element (hg) satisfies

$$(hg) \cdot (a_1, a_2, \ldots, a_r) = (b_1, b_2, \ldots, b_r).$$

Hence, G acts r-transitively on all of X. $\square$

Exercise 6.3.13 asks the reader to prove that every 2-transitive group action is primitive. Consequently, every group action that is r-transitive with $r \geq 2$ is primitive. Again, coupled with Proposition 6.3.11, this result gives a strategy to prove that a group is simple.

EXERCISES FOR SECTION 6.3

1. Let G be a group acting transitively on a set X with $|X|$ prime. Prove that the action is primitive.

2. Show that if a finite group G acts transitively on a set X with $|X| \geq 2$, then there exists some $g \in G$ such that $X^g = \emptyset$.

3. We showed that the group G of rigid motions of a cube acting on its set of vertices has the long diagonals for blocks. We claim that this group action has more nontrivial blocks. Find all of them and prove you have found them all.

4. Find the largest subgroup of S_6 that acts transitively on $\{1, 2, 3, 4, 5, 6\}$ and has $\{1, 2, 3\}$ as a block.

5. Let $K = \langle (1\,2\,3)(4\,5\,6)(7\,8\,9), (1\,4\,7)(2\,5\,8)(3\,6\,9) \rangle$ be a subset of S_9 and consider its natural action on $X = \{1, 2, \ldots, 9\}$. Prove that K acts transitively and that it has the same system of blocks as the group H in Example 6.3.7.

6. The group described in Example 6.3.7 is not the largest subgroup of S_9 that acts on $X = \{1, 2, \ldots, 9\}$ with $\{1, 2, 3\}$ as a block. Find this largest subgroup.

7. Consider the group G of rigid motions on a cube. Show that G has a normal subgroup of order 4 and describe the associated system of blocks of vertices described in Proposition 6.3.11.

8. Show that no subgroup G of S_5 acts transitively on $X = \{1, 2, 3, 4, 5\}$ in such a way that G_x is an elementary abelian 2-group for any $x \in X$.

9. Let B be a nontrivial block of a transitive action of G on a set X. Prove that the pointwise stabilizer $G_{(B)}$ and the setwise stabilizer $G_{\{B\}}$ are subgroups of G and that $G_{(B)} \trianglelefteq G_{\{B\}}$.

10. Let G be a group acting transitively on a set X and let $\alpha \in X$. Let $\mathcal{B}$ be the set of all blocks B in the group action such that $\alpha \in B$, and let $\mathcal{S} = \{H \in \mathrm{Sub}(G) \mid G_\alpha \leq H\}$. The function $\Psi : \mathcal{B} \to \mathcal{S}$ defined by $\Psi(B) = G_{\{B\}}$ is a poset isomorphism between the posets $(\mathcal{B}, \subseteq)$ and $(\mathcal{S}, \leq)$.

11. Let G act transitively on a set X and let $x \in X$. Prove that the fixed set X^{G_x} is a block of X. Deduce that if G acts primitively, then either $X^{G_x} = \{x\}$ or else X is a finite set with $|X|$ prime.

12. Prove that if a finite group G acts r-transitively on a set X with $|X| = n$, then $|G|$ is divisible by $n!/(n-r)!$.

13. Prove that a 2-transitive group action is primitive.

14. Suppose that (G, X, ρ) is an r-transitive group action. Prove that if $N \trianglelefteq G$, then the action of N on X is $(r-1)$-transitive.

15. Let G act transitively on a finite set X. Suppose that for some element $x \in X$, the stabilizer G_x has s orbits on X. Prove that

$$\sum_{g \in G} |X^g|^2 = s|G|.$$

Deduce that G acts 2-transitively if and only if

$$\sum_{g \in G} |X^g|^2 = 2|G|.$$

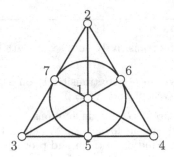

FIGURE 6.6: The Fano plane.

16. Let F be a finite field of order p^m for some prime p. Let G be the set of functions in $\mathrm{Fun}(F, F)$ of the form $f(x) = \alpha x + \beta$ such that $\alpha \in F - \{0\}$ and $\beta \in F$.

   (a) Prove that G is a nonabelian group of order $p^m(p^m - 1)$.

   (b) Prove that the action of G on F via $f \cdot \alpha = f(\alpha)$ is faithful and transitive.

   (c) Prove that G acts 2-transitively on F.

   (d) Prove that G contains a normal subgroup of order p^m that is abelian.

   (e) Determine all the maximal subgroups of G.

17. Consider the elements in $X = \{1, 2, \ldots, 7\}$ and the diagram shown in Figure 6.6. The diagram is called the *Fano plane* and arises in the study of finite geometries. Consider the subset L (called lines) of $\mathcal{P}(X)$ whose elements are the subsets of size 3 depicted in the Fano plane diagram either as a straight line or the circle. So for example $\{1, 2, 5\}$ and $\{5, 6, 7\}$ are in L. Let G be the largest subgroup of S_7 that maps lines to lines, i.e., acts on the set L.

   (a) Prove that the action of G on X is 2-transitive.

   (b) Prove that $|G| = 168$.

   (c) Prove that G is simple.

6.4   Groups Acting on Themselves

A fruitful area of investigation comes from considering ways in which groups can act on themselves or act on their own internal structure. Properties of group actions combined with considering actions of groups on themselves lead to new results about the internal structure of groups. Two natural actions of a group G on itself are the action of left multiplication and the action of conjugation.

6.4.1 Groups Acting on Themselves by Left Multiplication

A group acts on itself by left multiplication by the pairing $G \times G \to G$ given by $g \cdot h = gh$. In other words, every $g \in G$ corresponds to a function $\sigma_g : G \to G$ with $\sigma_g(x) = gx$. It is quite obvious that this is a group action since for all $g, h, x \in G$,

$$g \cdot (h \cdot x) = ghx = (gh) \cdot x$$

and $1x = x$ for all $x \in G$. Furthermore, the action of left multiplication if faithful because $\sigma_g(x) = gx = x$ for all $x \in G$ is the definition of $g = 1$.

If G is a finite group with $|G| = n$, we can label all the group elements as $G = \{g_1, g_2, \ldots, g_n\}$. Then the left multiplication action corresponds to an injective homomorphism $\rho : G \to S_n$, via $\rho(g) = \tau$ where $gg_i = g_{\tau(i)}$.

Example 6.4.1. Example 6.1.5 presented the standard action of D_n on the labeled set of vertices of the regular n-gon. This action is always different from the action of D_n on itself by left multiplication. To compare with that example, take $n = 6$ and label the elements in D_6 with integers $\{1, 2, \ldots, 12\}$ listed in the same order as

$$1, r, r^2, \ldots, r^5, s, sr, \ldots, sr^5.$$

Then the permutation representation of D_6 has

$$\rho(r) = (1\,2\,3\,4\,5\,6)(7\,12\,11\,10\,9\,8)$$
$$\rho(s) = (1\,7)(2\,8)(3\,9)(4\,10)(5\,11)(6\,12). \qquad \triangle$$

The group action by left multiplication on itself leads immediately to a powerful result about symmetric groups.

Theorem 6.4.2 (Cayley's Theorem)

> Every group is isomorphic to a subgroup of some symmetric group. If $|G| = n$, then G is isomorphic to a subgroup of S_n.

Proof. Since the permutation representation $\rho : G \to S_G$ is injective, then by the First Isomorphism Theorem, $G \cong G/\operatorname{Ker} \rho \cong \operatorname{Im} \rho$. $\qquad \square$

Because of this important theorem, the action of G on itself by left multiplication is also called the *Cayley representation*.

Cayley's Theorem is valuable for computational reasons. It is difficult to devise algorithms that perform group operations for an arbitrary group. However, it is easy to devise algorithms to perform the group operation in S_n. Cayley's Theorem guarantees that a group G can be embedded into some S_n, which reduces computations in G to computations in some corresponding S_n.

Since the action of G on itself by left conjugation is transitive, we are inspired to consider the possibility of blocks in this action. It is not hard to see that for any subset $H \leq G$, the set of left cosets forms a system of blocks for this action. This is because for any left coset, xH the product $g(xH) = (gx)H$ is another left coset. The set of left cosets of H forms a partition of G, so the set of left cosets of H forms a system of blocks in this action. However, the converse holds.

Proposition 6.4.3

Let a group G act on itself by left multiplication. A set Σ of subsets of G is a system of blocks for this action if and only if Σ is the set of left cosets of some subgroup $H \leq G$.

Proof. We have already shown one direction. We now assume that Σ is a system of blocks for this action. Let $H \in \Sigma$ be the block that contains the identity 1. Note that the set $g \cdot H$ contains the element g. Hence, if $g \cdot H = H$, then $g \in H$ since $1 \in H$. Conversely, suppose that $g \in H$. Then $g \cdot H$ is a block that contains the element $g \cdot 1 = g$. Since $g \cdot H \cap H \neq \emptyset$, we deduce that $g \cdot H = H$.

Let $x, y \in H$. Then $(xy) \cdot H = x \cdot (y \cdot H) = x \cdot H = H$. Hence, $xy \in H$. Finally, let $x \in H$. Then since $x \cdot H = H$, by multiplying on the left by x^{-1}, we get $H = x^{-1} \cdot H$. Thus, $x^{-1} \in H$. Therefore, H is a subgroup of G and the system of blocks is the set of left cosets of H. $\square$

6.4.2   Groups Acting on Themselves by Conjugation

Another type of action of a group on itself is by conjugation. In other words, let $X = G$ and consider the pairing which the pairing $G \times X \to G$ defined by $g \cdot x = gxg^{-1}$. It is easy to see that

$$g \cdot (h \cdot x) = g \cdot (hxh^{-1}) = ghxh^{-1}g^{-1} = (gh)x(gh)^{-1} = (gh) \cdot x.$$

Furthermore, $1 \cdot g = 1g1^{-1} = g$. This shows that conjugation satisfies the axioms of a group action. The permutation representation is a homomorphism $\rho : G \to S_G$. We have already seen that for each $g \in G$, the function $\psi_g(x) = gxg^{-1}$ is an automorphism on G so $\rho(G) \leq \mathrm{Aut}(G) \leq S_G$. In fact, we called the image subgroup $\rho(G)$, the group of inner automorphisms of G and denote it by $\mathrm{Inn}(G)$. (See Exercise 1.9.38.)

This action is not faithful in general. The kernel of the action is

$$\mathrm{Ker}\,\rho = \{g \in G \mid gxg^{-1} = x \text{ for all } x \in G\} = Z(G).$$

Hence, the action of G on itself by conjugation is faithful if and only if $Z(G) = \{1\}$.

Note that if A is a subset of a group G, then G might not necessarily act on A by conjugation. Indeed, gag^{-1} might not be in A for some $a \in A$. However, the normalizer $N_G(A)$ is the largest subgroup of G that acts on A by conjugation.

If G is not the trivial group, then the action of G on itself by conjugation is not a transitive action. The orbits are the conjugacy classes of G. The orbit equation for this action turns out to lead to another identity pertaining to the internal structure of a group that we could not get without the formalism of group actions.

The fixed elements (singleton orbits) in the action by conjugation are precisely the elements in the center $Z(G)$. Now suppose that $x \notin Z(G)$. The stabilizer of x is

$$G_x = \{g \in G \mid gxg^{-1} = x\} = C_G(x)$$

so by the Orbit-Stabilizer Theorem, the conjugacy class of x has order $|G : C_G(x)|$. This shows the surprising result that the cardinality of every conjugacy must divide $|G|$. Furthermore, by grouping the subset of fixed elements as a single term, the Orbit Equation immediately gives the following result.

Proposition 6.4.4 (Class Equation)

Let G be a finite group and let K be a complete set of distinct representatives of conjugacy classes that are of size 2 or greater. Then

$$|G| = |Z(G)| + \sum_{x \in K} |G : C_G(x)|.$$

As another example of a group action relevant to group theory, let G be a group and consider the associated group of automorphisms, $\text{Aut}(G)$. Of course, $\text{Aut}(G)$ acts on G by $\psi \cdot g = \psi(g)$ but $\text{Aut}(G)$ also acts on the set of subgroups $\text{Sub}(G)$ with the pairing $\text{Aut}(G) \times \text{Sub}(G) \to \text{Sub}(G)$ defined by $\psi \cdot H = \psi(H)$. Recall the concept of a characteristic subgroup of a group G: a subgroup H such that $\psi(H) = H$ for all automorphisms $\psi \in \text{Aut}(G)$. So a characteristic subgroup of G is a subgroup that remains unchanged by the action of $\text{Aut}(G)$ on $\text{Sub}(G)$ as we just described.

We can contrast the notion of characteristic subgroup to a normal subgroup by the fact that normal subgroups are the subgroups that remain unaffected by the action of $\text{Inn}(G)$, the subgroup of $\text{Aut}(G)$ that consists of automorphisms of the form ψ_g, where $\psi_g(x) = gxg^{-1}$.

6.4.3 Cauchy's Theorem

Another application of the Orbit-Stabilizer Theorem establishes Cauchy's Theorem, an important result in the classification of groups. We provide a clever proof given by James McKay [23]. The proof involves a group acting on a set closely related to itself.

Theorem 6.4.5 (Cauchy's Theorem)

If p is a prime number dividing the order of a finite group G, then G has an element of order p.

Proof. Consider the set $X = \{(g_1, g_2, \ldots, g_p) \in G^p \mid g_1 g_2 \cdots g_p = 1\}$. In the p-tuple $(g_1, g_2, \ldots, g_p) \in X$, the group elements $g_1, g_2, \ldots, g_{p-1}$ can be arbitrary and $g_p = (g_1, g_2, \ldots, g_{p-1})^{-1}$. Hence, $|X| = |G|^{p-1}$.

Consider the action of the cyclic group $Z_p = \langle z \mid z^p = 1 \rangle$ on X defined by

$$z \cdot (g_1, g_2, \ldots, g_p) = (g_2, g_3, \ldots, g_p, g_1).$$

An element in X that is fixed by the action of $H = Z_p$ on X has the form $(g, g, \ldots, g)$. Such elements have the property that $g^p = 1$. Let us suppose that there are r such fixed elements in X. If an element $x \in X$ is not fixed by the action of Z_p, then the stabilizer H_x is a proper subgroup of H, which implies that the stabilizer is trivial and that the orbit Hx has cardinality $|Hx| = |H : H_x| = p$. Let us suppose that there are s such nontrivial orbits in X. The Orbit-Stabilizer Theorem implies that

$$r + sp = |G|^{p-1}.$$

Since p divides $|G|$, then p divides r. We know that $(1, 1, \ldots, 1)$ is a fixed point of the Z_p action on X so $r \geq 1$. Since r is a nontrivial multiple of p, there are at least $p - 1$ more elements $g \in G$ satisfying $g^p = 1$. All such elements have order exactly p. $\square$

We have seen that Lagrange's Theorem does not have a full converse, in the sense that if d divides G, there does not necessarily exist a subgroup $H \leq G$ such that $|H| = d$. However, Cauchy's Theorem gives a partial converse in the sense that if d is a prime number dividing $|G|$, then there exists a subgroup $H \leq G$ such that $|H| = d$.

EXERCISES FOR SECTION 6.4

1. Consider the action of Q_8 on itself by left multiplication. Consider the induced permutation representation $\rho : Q_8 \to S_8$. Determine $\rho(-1)$, $\rho(i)$, $\rho(j)$, and $\rho(k)$.

2. Consider the action of D_4 on itself by conjugation. After labeling the elements in D_4, define the induced permutation representation $\rho : D_4 \to S_8$, and write down $\rho(g)$ for all $g \in D_4$.

3. Label the S_3 elements 1, (1 2), (1 3), (2 3), (1 2 3), and (1 3 2) with the integers 1, 2, 3, 4, 5, and 6 respectively. Consider the action of S_3 on itself by left multiplication and write the image of each element under the induced permutation representation as an element in S_6.

4. Let G be a group and let $X = G$. Show that the pairing $G \times X \to X$ given by $(g, x) \mapsto xg$ is not generally a (left) group action. Show, however, that the pairing $G \times X \to X$ given by $(g, x) \mapsto xg^{-1}$ does give a group action of G on itself.

5. Prove that the action of a group on itself by left multiplication is primitive if and only if $G \cong Z_p$, where p is prime.

6. Prove that the action of a group on itself by conjugation is trivial if and only if G is abelian.

7. Let $\rho : G \to S_G$ be the homomorphism induced from G acting on itself by left multiplication. Suppose $x \in G$ with $|x| = n$ and that $|G| = mn$. Prove that $\rho(x)$ is a product of m disjoint n-cycles.

8. Let G be a group and let H be a subgroup of G.

   (a) Prove that the mapping $g \cdot (xH) \overset{\text{def}}{=} (gx)H$ defines an action of G on the set of left cosets of H.

   (b) Denote by ρ_H the homomorphism of G onto the set of permutations on left cosets of H. Prove that the kernel of this action is

$$\text{Ker } \rho_H = \bigcap_{x \in G} xHx^{-1}.$$

   (c) Prove that $\text{Ker } \rho_H$ is the largest normal subgroup contained in H.

9. Use Exercise 6.4.8 to prove the following theorem. Suppose that p is the smallest prime dividing the order of G. Then if $H \leq G$ with $|G : H| = p$, then $H \trianglelefteq G$. [Hint: By contradiction. Assume $\text{Ker } \rho_H$ is a strict subgroup of H and then show that $|H : \text{Ker } \rho_H|$ must divide $(p-1)!$. Explain why this implies a contradiction.]

10. Let G be a group and let $\text{Sub}(G)$ be the set of subgroups of G. For all $H \in \text{Sub}(G)$ and all $g \in G$, define the pairing $g \cdot H = gHg^{-1}$. Prove that this pairing defines a group action of G on $\text{Sub}(G)$.

11. Consider the action of G on itself by conjugation. Prove that the invariant subsets of G are unions of conjugacy classes. Conclude that a subgroup of G is invariant under the action if and only if it is a normal subgroup.

12. Let G be a group. The automorphism group $\text{Aut}(G)$ acts on G by $\psi \cdot g = \psi(g)$ for all $\psi \in \text{Aut}(G)$ and all $g \in G$.

   (a) Find the orbits of this action if $G = Z_{12}$.

   (b) Find the orbits of this action if $G = Z_7$.

   (c) Find the orbits of this action if $G = D_4$. [Hint: First determine $\text{Aut}(D_4)$.]

13. Let $p < q$ be primes and let G be a group of order pq. Prove that there exists an injective homomorphism of G into S_q.

14. Let p be a prime. Use the Class Equation to prove that every p-group, i.e., a group of order p^k for some positive integer k, has a nontrivial center.

15. Let p be prime. Use Exercise 6.4.14 to prove that every group of order p^2 is abelian. In particular, if $|G| = p^2$, then G is isomorphic to Z_{p^2} or $Z_p \oplus Z_p$.

16. If G is a p-group and H is a proper subgroup, show that the normalizer $N_G(H)$ properly contains H. [Hint: Use Exercise 6.4.14.]

17. Let G be a group. Show that the pairing $(G \oplus G) \times G \to G$ defined by $(g, h) \cdot x = gxh^{-1}$ is an action of $G \oplus G$ on G. Show that the action is transitive. Also determine the stabilizer of the identity 1.

18. (*Cauchy's Theorem*) The original proof to Cauchy's Theorem did not use the group action described in the proof we gave, but it relied instead on the Class Equation. Let p be a prime that divides the order of a finite group G.

    (a) Prove Cauchy's Theorem for finite abelian groups. [Hint: Use induction on $|G|$.]

    (b) Prove Cauchy's Theorem for finite nonabelian groups by induction on $|G|$ and using the Class Equation.

19. Suppose that G is a finite group with m conjugacy classes. Show that the number of ordered pairs $(x, y) \in G \times G$ such that $yx = xy$ is equal to $m|G|$.

6.5   Sylow's Theorem

Sylow's Theorem is a partial converse to Lagrange's Theorem in that it states that a group has a subgroup of a given order. Sylow's Theorem leads to a variety of profound consequences for the internal structure of a group simply based on its order. Therefore, it helps with classification problems (Section 6.7)—theorems that find all groups of a given order.

    We present Sylow's Theorem in this section because it follows from a clever application of a group action on certain sets of subgroups within the group.

Example 6.5.1. Before presenting the necessary group action and proving the theorem, we illustrate Sylow's Theorem with an example. Consider the group $G = S_6$. Obviously, $|G| = 720 = 2^4 \cdot 3^2 \cdot 5$. Sylow Theorem will guarantee us that G has a subgroups of order 16, 9, and 5. That such subgroups exist is not immediately obvious. The theorem also gives us a condition on how many of such subgroups G has.

- Finding a subgroup of order 5 is easy. Indeed, $\langle (12345) \rangle$ works. There are $\binom{6}{5} 4!/4 = 36$ different groups of order 5.

- Finding a subgroup of order $9 = 3^2$ is not hard either. $H = \langle (123), (456) \rangle$ works. Since there are no 9-cycles in S_6, every subgroup of order 9 must be generated by two nonoverlapping 3-cycles. It is easy to show that there are $\frac{1}{2}\binom{6}{3} = 10$ such subgroups.

- Finding a subgroup of order $16 = 2^4$ is more challenging. Such a subgroup must be nonabelian without an element of order 8. We can obtain such a subgroup H by embedding D_4 into the subgroup of S_6 that fixes 5 and 6 and then adjoin the transposition $(5\,6)$. Hence,

$$H = \langle (1\,2\,3\,4), (2\,4), (5\,6) \rangle \cong D_4 \oplus Z_2.$$

There are 45 different subgroups of this form. $\triangle$

6.5.1 Sylow's Theorem

Definition 6.5.2

Let G be a group and p a prime.

- A group of order p^k for some $k \in \mathbb{N}^*$ is called a *p-group*. Subgroups of G that are p-groups are called *p-subgroup*.

- If G is a group of order $p^k m$, where $p \nmid m$, then a subgroup of order p^k is called a *Sylow p-subgroup* of G.

- The set of Sylow p-subgroups of G is denoted by $\mathrm{Syl}_p(G)$ and the number of Sylow p-subgroups of G is denoted by $n_p(G)$.

If p is a prime that does not divide $|G|$, then the notion of a p-subgroup is not interesting. However, to be consistent with notation, if p does not divide $|G|$ then trivially $\mathrm{Syl}_p(G) = \{\langle 1 \rangle\}$ and $n_p(G) = 1$.

Theorem 6.5.3 (Sylow's Theorem, Part 1)

For all groups G and all primes p that divide $|G|$, Sylow p-subgroups of G exist, i.e., $\mathrm{Syl}_p(G) \neq \emptyset$.

Proof. We use (strong) induction on the size of G. If $|G| = 1$, there is nothing to do, and the theorem is satisfied trivially. Assume that the theorem holds for all groups of size strictly less than n. We prove that the theorem holds for all groups G with $|G| = n$.

Let p be a prime and assume that $|G| = p^k m$ with $p \nmid m$. If p divides $Z(G)$, then by Cauchy's Theorem $Z(G)$ contains an element of order p and hence a subgroup N of order p. But then the group $|G/N| = p^{k-1} m$ is smaller than G; hence, by the induction hypothesis, G/N contains a Sylow p-subgroup $\bar{P}$ of order p^{k-1}. By the Fourth Isomorphism Theorem, there exists a subgroup P of G such that $\bar{P} = P/N$. Then $|P| = p^k$ and hence G contains a Sylow p-subgroup.

We are reduced now to the case where $p \nmid |Z(G)|$. Consider the Class Equation (Proposition 6.4.4),

$$|G| = |Z(G)| + \sum_{i=1}^{r} |G : C_G(g_i)|,$$

where $\{g_1, g_2, \ldots, g_r\}$ is a complete list of distinct representatives of the non-trivial conjugacy classes. Since p divides $|G|$ but not $|Z(G)|$, there exists some g_{i_0} such that p does not divide $|G : C_G(g_{i_0})|$. Then $C_G(g_{i_0})$ has order $p^k \ell$ where $p \nmid \ell$. Thus, again by strong induction, $C_G(g_{i_0})$ has a Sylow p-subgroup of order p^k, which is a subgroup of G. $\qquad\square$

Before establishing the rest of Sylow's Theorem, we require two lemmas. The first gives a property of Sylow-p subgroups concerning intersections with other p-subgroups.

Lemma 6.5.4

> Let $P \in \mathrm{Syl}_p(G)$. If Q is any p-subgroup of G, then $Q \cap N_G(P) = Q \cap P$.

Proof. Since $P \leq N_G(P)$, then $Q \cap P \leq Q \cap N_G(P)$ and so we need to show the opposite inclusion. Call $H = Q \cap N_G(P)$. Since $H \leq N_G(P)$, then the subset PH is a subgroup of G and

$$|PH| = \frac{|P| \cdot |H|}{|P \cap H|}.$$

Since $H \leq Q$, the prime p divides $|H|$ and hence all the orders on the right-hand side are powers of p. Hence, PH is a p-group. Furthermore, PH contains P and hence since P has a maximal p power dividing $|G|$, then $P = PH$. Thus, $P \cap H = H$ so $H \leq P$. Since $H \leq Q$, we conclude that $H = Q \cap N_G(P) \leq Q \cap P$. This proves the reverse inclusion so the result follows. $\square$

The second part of Sylow's Theorem follows from considering the action of G on $\mathrm{Syl}_p(G)$ by subgroup conjugation. Theorem 6.5.3 established the key step that $\mathrm{Syl}_p(G)$ is nonempty. Suppose that $P \in \mathrm{Syl}_p(G)$. This is the orbit of P under the conjugation action is

$$\mathcal{S}_P = \{gPg^{-1} \mid g \in G\} = \{P_1 = P, P_2, \ldots, P_r\}.$$

By definition of orbits, G acts transitively on $\mathcal{S}_P$. Let H be any subgroup of G. It also acts on $\mathcal{S}_P$ by conjugation but perhaps not transitively. Then under the action of H, the set $\mathcal{S}_P$ may get partitioned into $s(H)$ distinct orbits $\{\mathcal{O}_1, \mathcal{O}_2, \ldots, \mathcal{O}_{s(H)}\}$, where $s(H)$ is a positive integer that depends on H. Obviously, $r = |\mathcal{O}_1| + |\mathcal{O}_2| + \cdots + |\mathcal{O}_{s(H)}|$. The Orbit-Stabilizer Theorem applied to the action of H on $\mathcal{S}_P$ states that if P_i is any element in the orbit $\mathcal{O}_i$, then

$$|\mathcal{O}_i| = |H : N_H(P_i)|. \tag{6.5}$$

If H happens to be another p-subgroup, this formula simplifies.

Lemma 6.5.5

> Let $P \in \mathrm{Syl}_p(G)$ and let Q be any p-subgroup of G. Suppose that Q acts on the orbit $\mathcal{S}_P$ by subgroup conjugation. If the orbit of some $P_i \in \mathcal{S}_P$ is $\mathcal{O}_i$, then
> $$|\mathcal{O}_i| = |Q : Q \cap P_i|.$$

Proof. By (6.5) and Lemma 6.5.4,

$$|\mathcal{O}_i| = |Q : N_Q(P_i)| = |Q : N_G(P_i) \cap Q| = |Q : P_i \cap Q|$$

for $1 \leq i \leq s(Q)$. $\qquad\square$

We can now establish the second part of Sylow's Theorem.

Theorem 6.5.6 (Sylow's Theorem, Part 2)

Let G be a group of order $p^k m$, with $k \geq 1$ and where p is a prime not dividing m.

(1) If P is a Sylow p-subgroup, then any p-subgroup Q is a subgroup of some conjugate of P. (In other words, G acts transitively on $\mathrm{Syl}_p(G)$ and every p-subgroup is a subgroup of some Sylow p-subgroup.)

(2) The number of Sylow p-subgroups satisfies

$$n_p \equiv 1 \pmod{p}.$$

(3) Furthermore, $n_p = |G : N_G(P)|$, for any Sylow p-subgroup P, so n_p divides m.

Proof. By Theorem 6.5.3, we know that $\mathrm{Syl}_p(G)$ is nonempty. Let P be a Sylow p-subgroup of G and let $\mathcal{S}_P$ be the orbit of P in the action of G acting on the set of subgroups of G by conjugation. Let $r = |\mathcal{S}_P|$.

We first show that $r \equiv 1 \pmod{p}$ as follows. Apply Lemma 6.5.5 with $Q = P$ itself. Then $\mathcal{O}_1 = \{P\}$ so $|\mathcal{O}_1| = 1$. Then for all integers i with $1 < i \leq s(P)$, the orbit $\mathcal{O}_i$ satisfies

$$|\mathcal{O}_i| = |P : P_i \cap P|,$$

which is divisible by p. Thus,

$$r = |\mathcal{O}_1| + |\mathcal{O}_2| + \cdots + |\mathcal{O}_{s(P)}| \equiv 1 \pmod{p}.$$

We prove by contradiction that the action of G by conjugation on $\mathrm{Syl}_p(G)$ is transitive. As above, let P be an arbitrary Sylow p-subgroup. Suppose that there exists a Sylow p-subgroup P' that is not conjugate to P. Now consider the action of P' on $\mathcal{S}_P$ by conjugation and apply Lemma 6.5.5 with $Q = P'$.

For $1 \leq i \leq s(P')$, the p-group $P' \cap P_i$ is a strict subgroup of P', so

$$|\mathcal{O}_i| = |P' : P' \cap P_i| > 1.$$

Thus, p divides all $|P' : P' \cap P_i|$ which implies that p divides r. Since $r \equiv 1 \pmod{p}$, we have a contradiction. Thus, we conclude that there does not exist a Sylow p-subgroup that is not conjugate to P. This proves (1) and (2).

For part (3), notice that since the action of G on $\mathrm{Syl}_p(G)$ by conjugation is transitive, then $r = n_p$ so $n_p \equiv 1 \pmod{p}$. Also, the Orbit-Stabilizer Theorem tells us that $n_p = |G : N_G(P)|$. Then by the chain of subgroups

$$P \leq N_G(P) \leq G,$$

we deduce that

$$p^k m = |G| = |G : N_G(P)|\,|N_G(P) : P|\,|P| = n_p |N_G(P) : P| p^k$$

so n_p divides $|G : P| = m$. $\qquad\qquad\qquad\qquad\qquad\qquad\qquad\qquad\qquad\quad \square$

Part (1) implies that for a given prime p, all Sylow p-subgroups are conjugate to each other. This implies that they are all isomorphic to each other.

By part (3), $n_p = 1$ means that the one Sylow p-subgroup P has $N_G(P) = G$ so it is a normal subgroup. (Also, in Exercise 2.2.13, we saw that if there is only one subgroup of a given order, then that subgroup is normal.) In particular, if $n_p = 1$ for some prime p that divides $|G|$, then we immediately conclude that G is not simple. This result often gives a quick way to determine that no group of a certain order is simple. The following example illustrates this.

Example 6.5.7. We revisit Example 6.5.1, the motivating example for this section. Sylow's Theorem affirms that there exist subgroups of order 16, 9, and 5. Since for a given p, all Sylow p-subgroups are isomorphic, then every Sylow p-subgroup is isomorphic to the Sylow p-subgroups that we illustrated for $p = 2$, $p = 3$, and $p = 5$. We had determined that (a) there are 36 Sylow 5-subgroups, which conforms to $n_5 \equiv 1 \pmod{5}$; (b) there are 10 Sylow 9-subgroups, which conforms to $n_3 \equiv 1 \pmod{3}$; (c) there are 45 subgroups of order 16, which conforms to $n_2 \equiv 1 \pmod{2}$. $\qquad\qquad\qquad\qquad \triangle$

Example 6.5.8. Let G be a group with $|G| = 385 = 5 \cdot 7 \cdot 11$. By part (2), $n_{11} \equiv 1 \pmod{11}$ and while by part (3) we also have $n_{11} \mid 35$. The divisors of 35 are 1, 5, 7, and 35. The only divisor of 35 that satisfies both conditions is $n_{11} = 1$. Hence, every group of order 385 has a normal subgroup of order 11.

If we continue similar analysis with the other prime factors of 385, we notice that $n_7 \equiv 1 \pmod{7}$ and $n_7 \mid 55$. Again, the only possibility is $n_7 = 1$ so groups of order 385 must also possess a normal subgroup of order 7. However, for the prime $p = 5$, the conditions give $n_5 \equiv 1 \pmod{5}$ and $n_5 \mid 77$. Here, we have two possibilities, namely that $n_5 = 1$ or 11. $\qquad\qquad\qquad\qquad\qquad \triangle$

The situation in which $n_p(G) = 1$ is particularly important for determining the structure of the group G. We already commented that $n_p(G) = 1$ implies that G has a normal Sylow p-subgroup. However, the converse is also true.

> **Proposition 6.5.9**
>
> Let P be a Sylow p-subgroup of a group G. The following are equivalent:
>
> (1) $n_p(G) = 1$;
>
> (2) $P \trianglelefteq G$;
>
> (3) P is a characteristic subgroup of G.

Proof. (1) $\Longrightarrow$ (3): Since $n_p(G) = 1$, there is only one Sylow p-subgroup P of G. Every automorphism $\psi \in \text{Aut}(G)$ maps subgroups of G back into subgroups of the same cardinality. Hence, $\psi(P) = P$ so P is characteristic.

(3) $\Longrightarrow$ (2): Follows from the fact that conjugation by any $g \in G$ is an automorphism of G.

(2) $\Longrightarrow$ (1): By part (1) of Sylow's Theorem, the action of G on $\text{Syl}_p(G)$ is transitive. Since $gPg^{-1} = P$ for all $g \in G$, then $\text{Syl}_p(G) = \{P\}$ and also $n_p(G) = 1$. $\qquad\square$

6.5.2 Applications of Sylow's Theorem

Sylow's Theorem implies many profound consequences for the structure of a group simply from the order.

One of the simplest applications of Sylow's Theorem involves determining that a group is not simple by showing that $n_p(G) = 1$ for some prime p that divides $|G|$. Certain numerical situations allow us to conclude much more. For example, it may be possible to prove that a group of a certain order must be abelian; then the Fundamental Theorem of Finitely Generated Abelian Groups give us a classification of all groups of that order.

Example 6.5.10 (Groups of Order pq). Let G be a group of order pq where p and q are primes. As an application of the Class Equation, we saw in Exercise 6.4.15 that groups of order p^2 are isomorphic either to Z_{p^2} or to $Z_p \oplus Z_p$. We assume for now on that $p < q$.

Consider the center of the group $Z(G)$ and, in particular, its order $|Z(G)|$. If $|Z(G)| = pq$, then the group is abelian. By the FTFGAG, $G \cong Z_{pq}$.

In Exercise 2.3.21, we saw that if $G/Z(G)$ is cyclic, then G is abelian. If $|G| = pq$ with $p \neq q$, then we cannot have $|Z(G)| = p$ or q because otherwise $G/Z(G)$ would be isomorphic to Z_q or Z_p respectively, making G abelian and $|Z(G)| = pq$, contradicting $|Z(G)| = p$ or q.

Now assume that $|Z(G)| = 1$. By Sylow's Theorem, $n_q = 1 + kq$ (with $k \geq 0$) and n_q divides $|G|/q = p$. However, if $k > 0$, then $n_q > q > p$ which contradicts $n_q \mid p$, so we must have $n_q = 1$. Therefore, G contains one subgroup $Q \leq G$ of order q and it is normal. Similarly, $n_p \equiv 1 \pmod{p}$ and n_p must divide q. This leads to two cases.

Let us first suppose that $p \nmid (q - 1)$. Then we must have $n_p = 1$ and so G has a normal subgroup P of order p. Then by the Direct Sum Decomposition

Theorem (Theorem 2.3.10), $G \cong P \oplus Q$, so $G \cong Z_p \oplus Z_q \cong Z_{pq}$. Hence, G is abelian again, contradicting $Z(G) = \{1\}$.

Now suppose that $p \mid (q - 1)$. A priori, it is possible for $n_p > 1$. We now provide a constructive proof of a nonabelian group of order pq. Let x be a generator of Q, so of order q. Also, by Cauchy's Theorem, G has an element y of order p. Then $\langle y \rangle$ is a Sylow p-subgroup and all Sylow p-subgroups are conjugate to $P = \langle y \rangle$. By Proposition 2.1.18, we deduce that $PQ = G$. The subgroup P acts by conjugation on Q and this action defines a homomorphism of P into $\mathrm{Aut}(Q)$. Note that $\mathrm{Aut}(Q) \cong \mathrm{Aut}(Z_q) \cong U(q)$, the multiplicative group of units in $\mathbb{Z}/q\mathbb{Z}$. (See Exercise 1.9.40.)

Proposition 5.6.2 establishes that $U(q)$ is a cyclic group and hence as a generator of order $q - 1$. Since Q is cyclic, with generator x, automorphisms on Q are determined by where they map the generator, $\psi_k(x) = x^k$, where $\gcd(k, q - 1) = 1$. Let a be a positive integer such that $\psi_a(x) = x^a$ has order $q - 1$. If $d = (q - 1)/p$, then $\psi_a^d = \psi_{a^d}$ has order p. Then the action of conjugation of P on Q determined by the homomorphism $P \to \mathrm{Aut}(Q)$ given by $y \mapsto \psi_{a^d}$ is a nontrivial action. This gives a nonabelian group, which can be presented as

$$\langle x, y \mid x^q = y^p = 1, \ yxy^{-1} = x^\alpha \rangle,$$

where $\alpha \equiv a^d \pmod{q}$.

Finally, we wish to show that all nonabelian groups of order pq are isomorphic. We have seen that any two nonabelian groups of order pq have presentations of the form

$$G_1 = \langle x, y \mid x^q = y^p = 1, \ yxy^{-1} = x^\alpha \rangle \quad \text{and}$$
$$G_2 = \langle g, h \mid g^q = h^p = 1, \ hgh^{-1} = g^\beta \rangle,$$

where α and β both are elements of order p in the multiplicative group $U(q)$. Now in cyclic groups there exists a unique subgroup of any given order. Thus, $\langle \beta \rangle$ is the unique subgroup of order p in $U(q)$ and $\alpha \in \langle \beta \rangle$ so $\alpha = \beta^c$ for some integer $1 \leq c \leq p - 1$. Consider a function $\varphi : G_1 \to G_2$ that maps $\varphi(x) = g$ and $\varphi(y) = h^c$. Obviously, $g^q = 1$ and $(h^c)^p = 1$, but also

$$(h^c)g(h^c)^{-1} = h^c g h^{-c} = h^{c-1}g^\beta h^{-(c-1)} = h^{c-2}(g^\beta)^\beta h^{-(c-2)}$$
$$= h^{c-2}g^{\beta^2} h^{-(c-2)} = \cdots = g^{\beta^c} = g^\alpha.$$

By the Extension Theorem on Generators, we deduce that φ extends to a homomorphism from G_1 to G_2. However, it is easy to tell that φ is both surjective and injective so it is an isomorphism.

To recap, if G is a group of order pq, then

(1) if $p = q$, then G is isomorphic to Z_{p^2} or $Z_p \oplus Z_p$;

(2) if $p \neq q$ and $p \nmid (q - 1)$, then $G \cong Z_{pq}$;

(3) if $p \neq q$ and $p \mid (q - 1)$, then G is isomorphic to Z_{pq} or the unique nonabelian group of order pq. $\triangle$

Example 6.5.11 (Groups of Order 39). As a specific illustration of the previous classification result, consider $n = 39 = 3 \cdot 13$. The group Z_{39} is the only abelian group of order 39. Note that $3 \mid (13 - 1)$ so by the previous example there also exists a nonabelian group of order 39. The cyclic group $U(13)$ is generated by 2 because for example in modular arithmetic base 13, the powers of 2 are: $2, 4, 8, 3, 6, 12, 11, 9, 5, 10, 7, 1 \ldots$ We have $d = (13 - 1)/3 = 4$, so, using $a = 2$, we have $\alpha = a^d \bmod 13 = 2^4 \bmod 13 = 3$. The sequence of powers of 3 is $1, 3, 9, 1, 3, 9, \ldots$ As a presentation, the group

$$G = \langle x, y | x^{13} = y^3 = 1, yxy^{-1} = x^3 \rangle$$

is a nonabelian group of order 39. It is possible to construct this example even more explicitly as a subgroup of S_{13}. Let $\sigma = (1\,2\,3\ldots 13)$ and let τ be the permutation such that $\tau\sigma\tau^{-1} = \sigma^\alpha = \sigma^3$. Since $\sigma^3 = (1\,4\,7\,10\,13\,3\,6\,9\,12\,2\,5\,8\,11)$, by Example 2.2.13, we find that the appropriate permutation τ is $\tau = (2\,4\,10)(3\,7\,6)(5\,13\,11)(8\,9\,12)$. Then $\langle \sigma, \tau \rangle$ is isomorphic to this nonabelian group of order 39. △

Example 6.5.12 (Groups of Order 30). We now prove that by virtue of $|H| = 30$, H must have a normal (and hence unique) Sylow 5-subgroup and Sylow 3-subgroup. Let $Q_1 \in \mathrm{Syl}_3(H)$ and let $Q_2 \in \mathrm{Syl}_5(H)$. If either Q_1 or Q_2 are normal in H, then $Q_1 Q_2$ is a subgroup of H of order 15. Since 15 is half of 30, then $Q_1 Q_2 \trianglelefteq H$. Since Q_1 and Q_2 are characteristic subgroups of $Q_1 Q_2$ by the Corollary to Sylow's Theorem, then Q_1 and Q_2 are both normal subgroups of H. Therefore, we have proven that either both Q_1 and Q_2 are normal in H or neither are. If neither are, then $n_3(H) = 10$ and $n_5(H) = 6$. But this would lead to $10 \cdot 2 + 6 \cdot 4 = 44$ elements of order 3 or 5 whereas the group H has only 30 elements. This is a contradiction. Hence, both Q_1 and Q_2 are normal in H. △

The next example shows how counting elements of a given order may provide more information beyond that given immediately from Sylow's Theorem.

Example 6.5.13 (Groups of Order 105). Let G be a group of order $105 = 3 \cdot 5 \cdot 7$. Using the criteria of Sylow's Theorem, we find that $n_3(G) = 1$ or 7, that $n_5(G) = 1$ or 21 and that $n_7(G) = 1$ or 15. By these considerations alone, it would appear that G might not necessarily have a normal Sylow p-subgroups. However, that is not the case. Assume that $n_3(G) = 7$, that $n_5(G) = 21$, and that $n_7(G) = 15$. Each Sylow 5-subgroup would contain 4 elements of order 5 and these subgroups would intersect pairwise in the identity (since they are distinct cyclic subgroups of prime order). Hence, $n_5(G) = 21$ accounts for $4 \times 21 = 84$ elements of order 5. By a same reasoning, if $n_7(G) = 15$, then G contains 15 distinct cyclic subgroups of order 7, which accounts for $6 \times 15 = 60$ elements of order 7. However, this count gives us already $84 + 60 = 144$ elements of order 5 or 7 but this number is already greater than the order of the group, 105. Hence, every group of order 105 must contain a normal subgroup of order 5 or a normal subgroup of order 7. △

Example 6.5.14 (Groups of Order 2115). Let G be a group of order $2115 = 3^2 \cdot 5 \cdot 47$. It is easy to see that the conditions $n_{47} \equiv 1 \pmod{47}$ and $n_{47} \mid 45$ imply that $n_{47} = 1$. Hence, G must contain a normal subgroup N of order 47. However, because of the numerical relationships in this case, more can be said about N. Consider the action of G on N by conjugation. Since N is normal, this conjugation engenders a homomorphism $\psi : G \to \text{Aut}(N)$. However, $\text{Aut}(N) = \text{Aut}(Z_{47})$, which has order 46. But $\gcd(46, 2115) = 1$ so the only homomorphism $\psi : G \to \text{Aut}(N)$ is the trivial homomorphism. Thus, the action of conjugation of G on N is trivial and we conclude that N commutes with all of G so $N \leq Z(G)$. $\triangle$

EXERCISES FOR SECTION 6.5

1. Determine the isomorphism type of the Sylow 2-subgroups of A_6.

2. Let p be a prime and let suppose that $2 \leq k \leq p - 1$. Find a Sylow p-subgroup of S_{kp} by expressing it using generators.

3. Let p be a prime. Find a Sylow p-subgroup of S_{p^2} by expressing it using generators. Show also that it is a nonabelian group of order p^{p+1}.

4. Exhibit all Sylow 2-subgroups of S_4.

5. Exhibit a Sylow 2-subgroup of $\text{SL}_2(\mathbb{F}_3)$ by expressing it using generators.

6. Suppose that $G = \text{GL}_2(\mathbb{F}_p)$. Prove that $n_p(G) = p + 1$.

7. Exhibit a Sylow 3-subgroup of $\text{GL}_2(\mathbb{F}_{17})$ by expressing it using generators.

8. Let p be a prime number and consider the group S_{2p}.
   (a) Show that $n_p = \frac{1}{2}\binom{2p}{p}((p-2)!)^2$.
   (b) Use Sylow's Theorem to conclude that $\frac{1}{2}\binom{2p}{p}((p-2)!)^2 \equiv 1 \pmod{p}$.

9. Show that a group of order 418 has a normal subgroup of order 11 and a normal subgroup of order 19.

10. Prove that there is no simple group of order 225.

11. Prove that there is no simple group of order 825.

12. Prove that there is no simple group of order 2907.

13. Prove that there is no simple group of order 3124.

14. Prove that there is no simple group of order 4312.

15. Prove that there is no simple group of order 132.

16. Prove that there is no simple group of order 351.

17. Prove that a group of order 273 has a normal subgroup of order 91.

18. Prove that if $|G| = 2015$, then G contains a normal subgroup of order 31 and subgroup of order 13 in $Z(G)$.

19. Prove that if $|G| = 459$, then G contains a Sylow 17-subgroup in $Z(G)$.

20. Prove that every group of order 1001 is abelian.

21. Prove if $|G| = 9163$, then G has a Sylow 11-subgroup in $Z(G)$.

22. How many elements of order 7 must exist in a simple group of order 168?

23. Prove that $n_p(G) = 1$ is equivalent to the property that all subgroups of G generated by elements of order p^k are p-subgroups.

24. Let p be an odd prime. Show that every group of order $2p$ is isomorphic to Z_{2p} or to D_p.

25. Prove that if G is a group with $|G| = pqr$, where $p < q < r$ are primes, then G is not simple.

26. Suppose that for every prime p dividing $|G|$, the Sylow p-subgroups are non-abelian. Prove that $|G|$ is divisible by a cube.

27. Suppose that $|G| = p^2 q^2$ with p and q distinct primes. Prove that if $p \nmid (q^2 - 1)$ and $q \nmid (p^2 - 1)$, then G is abelian.

28. Suppose that H is a subgroup of G such that $\gcd(|\operatorname{Aut}(H)|, |G|) = 1$. Prove that $N_G(H) = C_G(H)$.

29. Prove that if $N \trianglelefteq G$, then $n_p(G/N) \leq n_p(G)$.

30. Let P be a normal Sylow p-subgroup of a group G and let $H \leq G$. Prove that $P \cap H$ is the unique Sylow p-subgroup of H.

31. Let $P \in \operatorname{Syl}_p(G)$ and let $N \trianglelefteq G$. Prove that $P \cap N \in \operatorname{Syl}_p(N)$. Prove also that PN/N is a Sylow p-subgroup of G/N.

32. Let G_1 and G_2 be two groups, both of which have orders divisible by a prime p. Prove that all Sylow p-subgroups of $G_1 \oplus G_2$ are of the form $P_1 \oplus P_2$, where $P_1 \in \operatorname{Syl}_p(G_1)$ and $P_2 \in \operatorname{Syl}_p(G_2)$.

33. Let G be a finite group and let M be a subgroup such that $N_G(P) \leq M \leq G$ for some Sylow p-subgroup P. Prove that $|G : M| \equiv 1 \pmod p$.

34. Let p be a prime dividing $|G|$. Prove that the intersection of all Sylow p-subgroups is the largest normal p-subgroup in G.

6.6 Semidirect Product

6.6.1 The Hölder Program

As early as Section 1.4, we introduced the idea of a classification theorem—a theorem that lists all the groups with a given property. A particular type of classification theorem involves finding all groups with a given cardinality.

With the concept of quotient groups, the effort to find all finite groups leads naturally to the Jordan-Hölder Program, which involves a two-pronged effort.

(1) Classify all finite simple groups.

(2) Find all methods such that given two groups H and K, we can construct G such that G contains a normal subgroup $N \cong H$ such that $G/N \cong K$.[1]

The group G arising in the second part of the Program satisfies $|G| = |H||K|$. Suppose we had solved the Jordan-Hölder Program. Given a positive integer n, we could use the first part of the Program to list all the simple groups of order n. Then, to find all non-simple groups, G would have a normal subgroup N with $|N| = d$ and $d \mid n$. If we knew all groups of order d and of order n/d, we could use part (2) of the Program to find all non-simple groups of order n with a normal subgroup of size d. So the construction of all groups of order n builds up recursively.

The Jordan-Hölder Program is very difficult but drove much of the 20th century research in finite group theory. The first part of the Program is completed and carries the name of "The Classification of Finite Simple Groups."

Theorem 6.6.1

Every finite simple group is isomorphic to one of the following:

(1) A cyclic group Z_p of prime order;

(2) An alternating group A_n with $n \geq 5$;

(3) A member of one of 16 infinite families of groups of Lie type over a finite field;

(4) One of 26 sporadic groups not isomorphic to any of the above groups; or

(5) The Tits group, sometimes called the 27th sporadic group.

The classification of finite simple groups "is generally regarded as a milestone of twentieth-century mathematics" [16]. How the theorem came about exemplifies the collaborative nature of mathematical investigation. Hundreds of mathematicians contributed to the effort.

Because of the extreme length and the number of disparate results necessary for a full classification, the realization that the classification of finite simple groups was within reach arose slowly. In 1972, when the classification felt close, Gorenstein laid out a 16-step program to break the project down into cases covering all possibilities [15]. In 1986, Gorenstein wrote a summary article declaring at long last that all finite simple groups had been found [16]. It was estimated at the time that the work spanned 15,000 pages of articles both published and unpublished.

However, as group theorists labored to synthesize the work, it became apparent that a gap remained in some unpublished material related to

---

[1]The usual way of describing this second part of the Jordan-Hölder Program, states "Classify all groups with a given set of composition factors." However, we have not discussed composition factors in this text. Nevertheless, our method of describing the Program is equivalent.

so-called quasithin groups. Aschbacher and Smith began working to rectify this gap and completed their work in 2004 in a pair of monographs [2, 3].

The form itself of the classification theorem is surprising. In retrospect, it is almost more surprising that so much work can be summarized in so brief a statement. Of course, to understand all parts of the theorem requires considerable advanced study. As of the publication of this textbook, group theorists believe that a complete proof might take around 5,000 pages. Efforts to find shorter proofs of the Classification of Finite Simple Groups continues to drive research.

As for the second part of the Jordan-Hölder Program, we already know one such method. Given two groups H and K, the direct sum $G = H \oplus K$ has a subgroup $\tilde{H} = \{(h, 1) \mid h \in H\} \cong H$ and $G/\tilde{H} \cong K$. Nevertheless, it is generally felt that the second part of the Program is more challenging than the first part.

In this section, following the vein of the second part of the Jordan-Hölder Program, we introduce a method that generalizes the direct sum construction. Before doing so, we mention one point of terminology. With algebraic structures, we talk about direct sums and direct products of families of objects. If the family of objects is a finite family, then the direct sum and direct product of that family are the same.

6.6.2 Semidirect Product

Suppose a group G contains a normal subgroup H. Suppose also that K is another subgroup of G with $H \cap K = 1$. Then K acts on H by conjugation. This action defines a homomorphism $\varphi : K \to \operatorname{Aut}(H)$. Furthermore, HK is a subgroup of G such that

$$khk^{-1} = \varphi(h) \iff kh = \varphi(h)k.$$

This remark sets up the construction of the semidirect product.

Proposition 6.6.2

Let H and K be groups and let $\varphi : K \to \operatorname{Aut}(H)$ be a homomorphism. The Cartesian product $H \times K$, equipped with the operation

$$(h_1, k_1) \cdot (h_2, k_2) = (h_1 \varphi(k_1)(h_2), k_1 k_2), \qquad (6.6)$$

is a group. Furthermore, if H and K are finite, then this group has order $|H||K|$.

Proof. Let (h_1, k_1), (h_2, k_2), and (h_3, k_3) be three elements in the Cartesian product $H \times K$. Then

$(h_1, k_1) \cdot ((h_2, k_2) \cdot (h_3, k_3))$

$\quad = (h_1, k_1) \cdot (h_2 \varphi(k_2)(h_3), k_2 k_3)$

$\quad = (h_1 \varphi(k_1)(h_2 \varphi(k_2)(h_3)), k_1(k_2 k_3))$

$\quad = (h_1 \varphi(k_1)(h_2) \varphi(k_1)(\varphi(k_2)(h_3)), (k_1 k_2) k_3) \qquad \varphi(k_1)$ is a homomorphism

$\quad = (h_1 \varphi(k_1)(h_2) \varphi(k_1 k_2)(h_3), (k_1 k_2) k_3) \qquad \varphi$ is a homomorphism

$\quad = (h_1 \varphi(k_1)(h_2), k_1 k_2) \cdot (h_3, k_3)$

$\quad = ((h_1, k_1) \cdot (h_2, k_2)) \cdot (h_3, k_3).$

This proves associativity.

The element $(1, 1)$ serves as the identity because

$$(1, 1) \cdot (h, k) = (1\varphi(1)(h), k) = (h, k) \qquad \text{and}$$
$$(h, k) \cdot (1, 1) = (h\varphi(k)(1), k) = (h, k),$$

where $\varphi(1)(h) = h$ because $\varphi(1)$ is the identity function and because $\varphi(k)(1) = 1$ since any homomorphism maps 1 to 1.

Let $(h, k) \in H \times K$. We prove that $(h, k)^{-1} = (\varphi(k^{-1})(h^{-1}), k^{-1})$. Indeed

$$(h, k) \cdot (\varphi(k^{-1})(h^{-1}), k^{-1}) = \left(h\varphi(k)\left(\varphi(k^{-1})(h^{-1})\right), kk^{-1}\right)$$
$$= (h\varphi(1)(h^{-1}), 1) = (1, 1)$$

and

$$(\varphi(k^{-1})(h^{-1}), k^{-1}) \cdot (h, k) = (\varphi(k^{-1})(h^{-1})\varphi(k^{-1})(h), k^{-1}k)$$
$$= (\varphi(k^{-1})(h^{-1}h), 1) = (1, 1).$$

Hence, the operation defined in (6.6) has inverses in the set.

Finally, note that the order of the group is the cardinality of $|H \times K|$, namely $|H||K|$. $\qquad\qquad\qquad\qquad\qquad\qquad\qquad\qquad\qquad\qquad\qquad\qquad\quad$ $\square$

Definition 6.6.3

Suppose that H and K are groups such that there exists a homomorphism $\varphi : K \to \mathrm{Aut}(H)$. The Cartesian product $H \times K$, equipped with the operation defined in (6.6), is called the *semidirect product* of H and K with respect to φ and is denoted by $H \rtimes_\varphi K$.

Proposition 6.6.4

Suppose that $G = H \rtimes_\varphi K$. Then $\tilde{H} = \{(h, 1) \,|\, h \in H\} \cong H$ and $\tilde{K} = \{(1, k) \,|\, k \in K\}$ are subgroups with $\tilde{H} \cong H$ and $\tilde{K} \cong K$. Furthermore, $\tilde{H} \trianglelefteq H \rtimes_\varphi K$ and $G/\tilde{H} \cong K$.

Proof. (Left as an exercise for the reader. See Exercise 6.6.6.) $\qquad\qquad\qquad$ $\square$

Because of this proposition, we will often abuse notation and write $H \trianglelefteq H \rtimes_\varphi K$ and $K \leq H \rtimes_\varphi K$ instead of $\tilde{H} \trianglelefteq H \rtimes_\varphi K$ and $\tilde{K} \leq H \rtimes_\varphi K$

Example 6.6.5. Let $H = Z_7$ and $K = Z_3$. Write $Z_3 = \langle x \,|\, x^3 = 1 \rangle$ and $Z_7 = \langle y \,|\, y^7 = 1 \rangle$. We know that $\mathrm{Aut}(Z_7) \cong U(7) \cong Z_6$. Explicitly, $\mathrm{Aut}(Z_7)$ contains the isomorphisms $\psi_a(g) = g^a$ for $a \in U(7)$. The homomorphisms $\varphi : Z_3 \to \mathrm{Aut}(Z_7)$ are determined by $\varphi(x)$, which must be an element of order 1 or 3. Hence, $\varphi(x)$ can be ψ_1 (the identity function), ψ_2, or ψ_4. We can see directly that the automorphism ψ_2 has order 3 because

$$\psi_2^3(g) = \psi_2(\psi_2(\psi_2(g))) = ((g^2)^2)^2 = g^8 = g$$

and so ψ_2^3 is the identity function.

If $\varphi(x) = \psi_1$, then $\varphi(x^a) = \psi_1$ for all a and from (6.6), we see that $H \rtimes_\varphi K \cong Z_7 \oplus Z_3$. $\triangle$

Before developing this example further and presenting more examples, it is useful to explore the relationship of H and K inside $H \rtimes_\varphi K$. However, implicit in Example 6.6.5 is that there always exist a homomorphism $K \to \mathrm{Aut}(H)$, namely the trivial homomorphism, which maps all elements in K to the identity automorphism. The following proposition describes this situation.

Proposition 6.6.6

The following are equivalent.

(1) φ is the trivial homomorphism into $\mathrm{Aut}(H)$;

(2) $H \rtimes_\varphi K \cong H \oplus K$ with the isomorphism being the set identity function;

(3) $K \trianglelefteq H \rtimes_\varphi K$.

Proof. (1) $\Longrightarrow$ (2) If φ is trivial, then $\varphi(k)$ is the identity function. Hence,

$$(h_1, k_1) \cdot (h_2, k_2) = (h_1 \varphi(k_1)(h_2), k_1 k_2) = (h_1 h_2, k_1 k_2).$$

Thus, $H \rtimes_\varphi K \cong H \oplus K$.

(2) $\Longrightarrow$ (3) We know that $K \trianglelefteq H \oplus K$.

(3) $\Longrightarrow$ (1) Suppose that $K \trianglelefteq H \rtimes_\varphi K$. Then for all $h_2 \in H$ and all $k_1, k_2 \in K$, the following element is in K:

$$(h_2, k_2) \cdot (1, k_1) \cdot (h_2, k_2)^{-1} = (h_2 \varphi(k_2)(1), k_2 k_1) \cdot (\varphi(k_2^{-1})(h_2^{-1}), k_2^{-1})$$
$$= (h_2 \varphi(k_2 k_1)(\varphi(k_2^{-1})(h_2^{-1})), k_2 k_1 k_2^{-1}).$$

Thus, $h_2 \varphi(k_2 k_1 k_2^{-1})(h_2^{-1}) = 1$ for all h_2 so $\varphi(k_2 k_1 k_2^{-1})$ is the identity function for all k_1, k_2. Setting $k_2 = 1$ shows that $\varphi(k_1)$ is the identity function for all $k_1 \in K$ so φ is trivial. $\square$

By virtue of this proposition, if we know that $\varphi : K \to \text{Aut}(H)$ is the trivial homomorphism, then we always write $H \oplus K$ instead of $H \rtimes_\varphi K$. If the homomorphism φ is understood by context or if $H \rtimes_\varphi K \cong H \rtimes_\psi K$ for any two nontrivial homomorphisms $\varphi, \psi : K \to \text{Aut}(H)$, then we simply write $H \rtimes K$ instead of $H \rtimes_\varphi K$.

From the identification of H with $H \times \{1\}$ and K with $\{1\} \times K$, we see that $H \rtimes_\varphi K = HK$. Furthermore,

$$(1,k) \cdot (h,1) \cdot (1,k)^{-1} = (\varphi(k)(h), k) \cdot (\varphi(k^{-1})(1), k^{-1}) = (\varphi(k)(h), k) \cdot (1, k^{-1})$$
$$= (\varphi(k)(h)\varphi(k)(1)kk^{-1}) = (\varphi(k)(h), 1).$$

This calculation shows that $\varphi(k)(h)$ corresponds to conjugation of the subgroup K on the normal subgroup H.

This inspires us to provide a characterization of groups that arise as semidirect products.

Proposition 6.6.7

Suppose that a group G contains a normal subgroup H and a subgroup K such that $G = HK$ and $H \cap K = \{1\}$. Then $G \cong H \rtimes_\varphi K$, where $\varphi : K \to \text{Aut}(H)$ is the homomorphism defined by conjugation $\varphi(k)(h) = khk^{-1}$.

Proof. Consider the function $f : H \rtimes_\varphi K \to G$ given by $f(h,k) = hk$. Since $G = HK$, this function is a surjection. If $f(h_1, k_1) = f(h_2, k_2)$, then $h_1 k_1 = h_2 k_2$ so $h_2^{-1} h_1 = k_2 k_1^{-1}$. Since $H \cap K = \{1\}$ then $h_2^{-1} h_1 = 1 = k_2 k_1^{-1}$ so $(h_1, k_1) = (h_2, k_2)$. Thus, this function is injective and therefore bijective.

Let $(h_1, k_1), (h_2, k_2) \in H \rtimes K$. Then

$$f((h_1, k_1) \cdot (h_2, k_2)) = f(h_1 \varphi(k_1)(h_2), k_1 k_2) = h_1 \varphi(k_1)(h_2) k_1 k_2$$

and

$$f(h_1, k_1) f(h_2, k_2) = h_1 k_1 h_2 k_2 = h_1(k_1 h_2 k_1^{-1}) k_1 k_2 = h_1 \varphi(k_1)(h_2) k_1 k_2.$$

We conclude that f is a homomorphism and thus an isomorphism. $\square$

Example 6.6.8. Let us revisit Example 6.6.5 in light of these propositions. We now give presentations for all three of the semidirect products.

Case 1. If φ_1 is such that $\varphi_1(x) = \psi_1$, then a presentation for $H \rtimes_{\varphi_1} K$ is

$$\langle x, y \mid x^3 = y^7 = 1, xyx^{-1} = y \rangle = Z_7 \oplus Z_3 \cong Z_{21}.$$

Case 2. If φ_2 is such that $\varphi_2(x) = \psi_2$, then a presentation for the semidirect product is

$$Z_7 \rtimes_{\varphi_2} Z_3 = \langle x, y \mid x^3 = y^7 = 1, xyx^{-1} = y^2 \rangle.$$

This is a nonabelian group so in particular it is not isomorphic to Z_{21}.

Case 3. If φ_3 is such that $\varphi_3(x) = \psi_4$, then a presentation for the semidirect product is

$$Z_7 \rtimes_{\varphi_3} Z_3 = \langle u, v \mid v^3 = v^7 = 1, uvu^{-1} = v^4 \rangle.$$

Again, this is a nonabelian group.

Now consider the mapping $f : Z_7 \rtimes_{\varphi_2} Z_3 \to Z_7 \rtimes_{\varphi_3} Z_3$ defined by $f(x) = u^2$ and $f(y) = v$. It is easy to check that $(u^2)^3 = v^7 = 1$ but we also have

$$u^2 v u^{-2} = u(uvu^{-1})u^{-1} = uv^4 u^{-1} = (uvu^{-1})^4 = (v^4)^4 = v^{16} = v^2.$$

Hence, u^2 and v satisfy the same relations as x and y so f defines a homomorphism by the Generator Extension Theorem. It is not hard to see that f is bijective and we deduce that the groups obtained from Case 2 and Case 3 are isomorphic. Consequently, we refer to the group with the simplified notation of $Z_7 \rtimes Z_3$ because it is the only nondirect semidirect product.

Proposition 6.6.7 allows us to take this example one step further to a classification result. Let G be a group of order 21. By Sylow's Theorem, we deduce that $n_7(G) = 1$ so G contains a normal subgroup H of order 7. G must also contain a subgroup K of order 3. The action of K on H by conjugation corresponds to a homomorphism $\varphi : K \to \mathrm{Aut}(H)$. We have seen that there are only two φ that lead to nonisomorphic groups, which are Z_{21} and $Z_7 \rtimes Z_3$.△

Example 6.6.9. A group H is abelian if and only if the inversion function $h \mapsto h^{-1}$ is an automorphism. As an automorphism, the inversion has order 2. Consider the cyclic group $Z_2 = \langle x \mid x^2 = 1 \rangle$ and the map $\varphi : Z_2 \to \mathrm{Aut}(H)$ such that $\varphi(x)(h) = h^{-1}$, and of course, $\varphi(1)(h) = h$. This allows us to construct the semidirect $H \rtimes_\varphi Z_2$. However, the function φ might not give the only nondirect semidirect product. △

Example 6.6.10 (Dihedral Groups). A particular example of the previous construction occurs with dihedral groups. Every dihedral group D_n is defined as $D_n = Z_n \rtimes_\varphi Z_2$ where $\varphi : Z_2 \to \mathrm{Aut}(Z_n)$ is defined by $\varphi(y)(x) = x^{-1}$ where x generates Z_n and y generates Z_2. This leads to the presentation

$$Z_n \rtimes_\varphi Z_2 = \langle x, y \mid x^n = y^2 = 1, \ yxy^{-1} = x^{-1} \rangle = D_n.$$

However, if n is not prime, there may be other nontrivial homomorphisms $\varphi : Z_2 \to \mathrm{Aut}(Z_n)$. For example, if $n = 15$, then $\mathrm{Aut}(Z_{15}) = U(15)$. We need to determine $U(15)$. By the Chinese Remainder Theorem, $\mathbb{Z}/15\mathbb{Z} = \mathbb{Z}/3\mathbb{Z} \oplus \mathbb{Z}/5\mathbb{Z}$ so

$$\mathrm{Aut}(Z_{15}) = U(\mathbb{Z}/3\mathbb{Z} \oplus \mathbb{Z}/5\mathbb{Z}) = U(3) \oplus U(5) \cong Z_2 \oplus Z_4.$$

Hence, $U(15)$ contains three elements of order 2, namely $\overline{4}$, $\overline{11}$, and $\overline{14} = -\overline{1}$. Writing Z_{15} as $Z_3 \oplus Z_5$, we see that these elements of order 2 in $\mathrm{Aut}(Z_{15})$ correspond to inversion on the Z_5 component alone, inversion on the Z_3 component alone, or inversion on both. The three resulting nondirect semidirect products of Z_{15} with Z_2 are $Z_3 \oplus D_5$, $Z_5 \oplus D_3$, and D_{15}. △

Remark 6.6.11. It is not always a simple task to determine if $H \rtimes_{\varphi_1} K$ and $H \rtimes_{\varphi_2} K$ are isomorphic, given two different homomorphisms $\varphi_1, \varphi_2 : K \to$ Aut(H). Exercises 6.6.3 and 6.6.4 show two general situations in which there does exist an isomorphism between semidirect products. $\triangle$

We reiterate that the semidirect product $H \rtimes K$ captures all situations in which G contains a normal subgroup N with $N \cong H$ and a subgroup $K' \leq G$ with $K' \cong K \cong G/N$. The semidirect product of H and K does not address the second part of the Jordan-Hölder Program when G does not contain a subgroup isomorphic to K with $G = HK$. The simplest example of this occurs with Q_8. Every subgroup of Q_8 is normal. Consider first the normal $N = \langle -1 \rangle$ with quotient group $Q_8/\langle -1 \rangle \cong Z_2 \oplus Z_2$. However, Q_8 has no subgroup isomorphic to $Z_2 \oplus Z_2$, so Q_8 does not arise as the semidirect product over N. Consider now the normal subgroup $N' = \langle i \rangle$ with quotient group $Q_8/\langle i \rangle \cong Z_2$. The only element of order 2 in Q_8 is in N' so there is no subgroup K such that $N'K = G$ and $N' \cap K = \{1\}$, because such a K would need to contain an element of order 2. A same reasoning holds for $N' = \langle j \rangle$ and $\langle k \rangle$. Hence, Q_8 does not arise as a semidirect product of its subgroups. In order to find such groups G, more advanced methods are necessary.

6.6.3   Some Automorphism Groups

In order to construct semidirect products $H \rtimes_\varphi K$ between two groups H and K, it is essential to know the automorphism group Aut(H). The only automorphism group we encountered so far is Aut$(Z_n) \cong U(n)$, where $U(n)$ is the group of units of $\mathbb{Z}/n\mathbb{Z}$. (See Exercise 1.9.40.) We need to explore the structure of $U(n)$ and other automorphism groups.

Proposition 6.6.12

If $n = p^k$ with p and odd prime and $k \in \mathbb{N}^*$, then Aut$(Z_{p^k}) \cong U(p^k)$ is a cyclic group of order $p^{k-1}(p - 1)$.

Proof. (Left as a guided exercise for the reader. See Exercise 6.6.10.) $\square$

Proposition 6.6.13

The group $U(2)$ is trivial. For $k \geq 2$, the automorphism group of the cyclic group is $U(2^k) \cong Z_2 \oplus Z_{2^{k-2}}$.

Proof. The proposition is obvious for $U(2^k)$ with $k = 1, 2$. We will suppose henceforth that $k \geq 3$.

We show that $U(2^k) = \langle \overline{5} \rangle \oplus \langle -\overline{1} \rangle$.

We first claim that $5^{2^{k-3}} \equiv 1 + 2^{k-1} \pmod{2^k}$. This is obvious for $k = 3$.

Suppose that it is true for some k. Note that

$$(1 + 2^{k-1} + c2^k)^2 = (1 + 2^{k-1})^2 + 2(1 + 2^{k-1})c2^k + c^2 2^{2k}$$
$$\equiv (1 + 2^{k-1})^2 \pmod{2^{k+1}}.$$

Therefore,

$$5^{2^{k-2}} \equiv (1 + 2^{k-1})^2 \equiv 1 + 2^k + 2^{2k-2} \equiv 1 + 2^k \pmod{2^{k+1}},$$

where the last congruence holds because $2k - 2 \geq k + 1$ for all $k \geq 3$. This establishes the claim by induction.

We see from the claim that 5 has order 2^{k-2}. Hence, the order $|\bar{5}|$ divides 2^{k-2} but if does not divide 2^{k-3}. So $|\bar{5}| = 2^{k-2}$.

Assume that $-1 \equiv 5^b \pmod{2^k}$ for some b. This would imply that $-1 \equiv 1 \pmod 4$, which is a contradiction. Hence, $-\bar{1} \notin \langle \bar{5} \rangle$. By the Direct Sum Decomposition Theorem for groups, we deduce that $U(2^k) \cong \langle -\bar{1} \rangle \oplus \langle \bar{5} \rangle$. Knowing the order of $\bar{5}$, the result follows. $\qquad\square$

The previous two propositions, coupled with the Chinese Remainder Theorem, give a complete description of the automorphism groups of cyclic groups. However, if a group is not cyclic, even if it is abelian, the automorphism groups can become rather complicated. The following proposition begins to show this.

Proposition 6.6.14

Let p be any primes and let Z_p^n be the elementary abelian group $Z_p \oplus Z_p \oplus \cdots \oplus Z_p$ with n copies of Z_p. Then $\mathrm{Aut}(Z_p^n) \cong \mathrm{GL}_n(\mathbb{F}_p)$.

Proof. Let V be the vector space of dimension n over the finite field $\mathbb{F}_p$. The group with addition $(V, +)$ is isomorphic to Z_p^n. An automorphism φ of $(V, +)$ is an invertible homomorphism with $\varphi(a + b) = \varphi(a) + \varphi(b)$ for all vectors a and b in $(V, +)$. Furthermore, for any positive integer k,

$$\varphi(k \cdot a) = \overbrace{\varphi(a) + \varphi(a) + \cdots + \varphi(a)}^{k \text{ times}} = k \cdot \varphi(a).$$

Since this holds for all $1 \leq k \leq p$, then we see that φ is an invertible linear transformation on $V = \mathbb{F}_p^n$. Hence, $\mathrm{Aut}(Z_p^n) = \mathrm{Aut}(\mathbb{F}_p^n, +) \cong \mathrm{GL}_n(\mathbb{F}_p)$. $\qquad\square$

Example 6.6.15. From the previous proposition, $\mathrm{Aut}(Z_5 \oplus Z_5) = \mathrm{GL}_2(\mathbb{F}_5)$. Note that $|\mathrm{GL}_2(\mathbb{F}_5)| = (25 - 5)(25 - 1) = 480$. By Cauchy's Theorem, $\mathrm{GL}_2(\mathbb{F}_5)$ contains an element of order 3. One such element is

$$g = \begin{pmatrix} 1 & 2 \\ 1 & 3 \end{pmatrix}.$$

If we write $Z_3 = \langle z \,|\, z^3 = 1\rangle$, then the homomorphism $\varphi : Z_3 \to GL_2(\mathbb{F}_5)$ defined by $\varphi(z) = g$ produces a nonabelian semidirect product $G = (Z_5 \oplus Z_5) \rtimes_\varphi Z_3$.

It is possible to give a presentation of this semidirect product. Note that for $Z_5 \oplus Z_5$ a presentation is

$$Z_5 \oplus Z_5 = \langle x, y \,|\, x^5 = y^5 = 1, \; xy = yx\rangle.$$

Now $Z_5 \oplus Z_5$ is isomorphic to the additive group $(\mathbb{F}_5^2, +)$ under the isomorphism $x^a y^b \leftrightarrow (a, b)$. Since

$$g\begin{pmatrix} a \\ b \end{pmatrix} = \begin{pmatrix} a + 2b \\ a + 3b \end{pmatrix},$$

a presentation for $(Z_5 \oplus Z_5) \rtimes_\varphi Z_3$ is

$$\langle x, y, z \,|\, x^5 = y^5 = z^3 = 1, \; xy = yx, \; zxz^{-1} = xy, \; zxz^{-1} = x^2 y^3 \rangle. \qquad \triangle$$

As a last example of an automorphism group, we prove in the exercises (Exercise 6.6.13) that if $n \neq 6$, then $\text{Aut}(S_n) = S_n$.

6.6.4 Wreath Product (Optional)

Consider the group described in Example 6.3.7. As a subgroup of S_9 it has generators

$$G = \langle (1\,2\,3), (1\,4\,7)(2\,5\,8)(3\,6\,9)\rangle.$$

It consists of permutations that cycle within the blocks $\{1, 2, 3\}$, $\{4, 5, 6\}$, and $\{7, 8, 9\}$ and permutations that cycle through the three blocks. The action of H that stays within the blocks is the subgroup

$$H = \langle (1\,2\,3), (4\,5\,6), (7\,8\,9)\rangle$$

and is isomorphic to $Z_3 \times Z_3 \times Z_3$. The generating permutation $\sigma = (1\,4\,7)(2\,5\,8)(3\,6\,9)$ satisfies

$$\sigma(1\,2\,3)\sigma^{-1} = (4\,5\,6), \quad \sigma(4\,5\,6)\sigma^{-1} = (7\,8\,9), \quad \text{and} \quad \sigma(7\,8\,9)\sigma^{-1} = (1\,2\,3).$$

Hence, $H \trianglelefteq G$. Setting $K = \langle \sigma\rangle$, the subgroups also satisfy $G = HK$. By Proposition 6.6.7, $G = H \rtimes_\varphi K$ where φ corresponds to K acting on H by conjugation. If x is a generator of Z_3, we can describe G as

$$(Z_3 \oplus Z_3 \oplus Z_3) \rtimes_\varphi Z_3,$$

where $\varphi : Z_3 \to \text{Aut}(Z_3 \oplus Z_3 \oplus Z_3)$ is defined by $\varphi(x)(g_1, g_2, g_3) = (g_3, g_1, g_2)$.

This is an example of a more general construction.

Let K and L be groups and let $\rho : K \to S_n$ be a homomorphism. Consider the action of S_n on

$$\overbrace{L \oplus L \oplus \cdots \oplus L}^{n \text{ times}}$$

by

$$\sigma \cdot (x_1, x_2, \ldots, x_n) = (x_{\sigma^{-1}(1)}, x_{\sigma^{-1}(2)}, \ldots, x_{\sigma^{-1}(n)}),$$

which corresponds to moving the ith entry to the $\sigma(i)$th location.

Definition 6.6.16

The *wreath product* of K on L by the homomorphism $\rho : K \to S_n$ is the semidirect product

$$L \wr_\rho K = (L \oplus L \oplus \cdots \oplus L) \rtimes_\varphi K,$$

where $\varphi : K \to \mathrm{Aut}(L^n)$ is the homomorphism $\varphi(k)(x_1, x_2, \ldots, x_n) = \rho(k) \cdot (x_1, x_2, \ldots, x_n)$.

We point out that the order of the wreath product is $|L \wr_\rho K| = |L|^n |K|$ and that elements of a wreath product are $(n + 1)$-tuples in the set $L^n \times K$.

It is possible to give an alternative approach to the wreath product. Set $\Gamma = \{1, 2, \ldots, n\}$. Consider the isomorphism between $\mathrm{Fun}(\Gamma, L)$ and L^n defined by $f \mapsto (f(1), f(2), \ldots, f(n))$ and where the group operation on functions $f, g \in \mathrm{Fun}(\Gamma, L)$ is

$$(f \cdot g)(i) = f(i)g(i),$$

where the latter operation is in the group L. Then elements of the wreath product $L \wr_\rho K$ are pairs $(f, k) \in \mathrm{Fun}(\Gamma, L) \times K$. The operation between elements in the wreath product is

$$(f_1, k_1) \cdot (f_2, k_2) = (i \mapsto f(i)g(\rho(k_1)^{-1}(i)), k_1 k_2).$$

This is called the *functional form* of the wreath product.

In the scenario when $n = |K|$ and $\rho : K \to S_K$ corresponds to the action of K on itself by left multiplication, the wreath product is called the *standard wreath product* of L by K and is denoted $L \wr K$.

Example 6.6.17. The motivating example for this subsection is a wreath product of Z_3 by Z_3. The homomorphism $\rho : Z_3 \to S_3$ sends the generator w of Z_3 to the 3-cycle $(1\,2\,3)$. So

$$\rho(w) \cdot (x_1, x_2, x_3) = (x_3, x_1, x_2).$$

We leave it as an exercise for the reader to prove that up to isomorphism there is only one nonabelian wreath product of Z_3 on Z_3 and that it has a presentation of

$$Z_3 \wr Z_3 = \langle x, y, z, w \mid x^3 = y^3 = z^3 = w^3 = 1,\ xy = yx,\ xz = zx,\ yz = zy,$$
$$wxw^{-1} = y,\ wyw^{-1} = z,\ wzw^{-1} = x \rangle. \qquad \triangle$$

EXERCISES FOR SECTION 6.6

1. Prove that $S_n = A_n \rtimes_\varphi Z_2$ for some appropriate φ.

2. Find a nontrivial homomorphism $\varphi : Z_3 \to \text{Aut}(Z_2 \oplus Z_2)$. Prove that the resulting semidirect product is $(Z_2 \oplus Z_2) \rtimes Z_3 \cong A_4$.

3. Let G be an arbitrary group and let n be a positive integer. Let φ_1 and φ_2 be homomorphisms $Z_n \to \text{Aut}(G)$ such that $\varphi_1(Z_n)$ and $\varphi_2(Z_n)$ are conjugate subgroups in $\text{Aut}(G)$. Suppose that Z_n is generated by z.

   (a) Prove that there exists an automorphism $\psi \in \text{Aut}(G)$ and $a \in U(n)$ such that $\varphi_2(z)^a = \psi \circ \varphi_1(z) \circ \psi^{-1}$.

   (b) Prove that the function $f : G \rtimes_{\varphi_1} Z_n \to G \rtimes_{\varphi_2} Z_n$ defined by $f(g, x) = (\psi(g), x^a)$ is an isomorphism.

4. Let P and Q be groups. Let φ_1 and φ_2 be homomorphisms $Q \to \text{Aut}(P)$ such that there exists an automorphism $\psi \in \text{Aut}\, Q$ such that $\varphi_1 \circ \psi = \varphi_2$. Show that the function $\Psi : P \rtimes_{\varphi_1} Q \to P \rtimes_{\varphi_2} Q$ defined by $\Psi(a, b) = (a, \psi^{-1}(b))$ is an isomorphism.

5. Suppose that $p < q$ are primes with $p \mid (q - 1)$. Prove that all nontrivial homomorphisms $\varphi : Z_p \to \text{Aut}(Z_q)$ lead to isomorphic semidirect product $Z_q \rtimes_\varphi Z_p$. [That is why the unique nonabelian group of order pq is written $Z_q \rtimes Z_p$.]

6. Prove Proposition 6.6.4.

7. Fix a positive integer n and let F be a field. Let T be the subgroup of $\text{GL}_n(F)$ of upper triangular matrices. Let $D \leq T$ be the subgroup of diagonal matrices and let $U = \{g \in T \mid g_{ii} = 1 \text{ for all } i\}$. Prove that T is a semidirect product $U \rtimes D$. Explicitly describe the relevant homomorphism $\varphi : D \to \text{Aut}(U)$ for $n = 2$ and $n = 3$.

8. Prove that the symmetries of the cube (including reflections) is a group of the form $S_4 \rtimes Z_2$.

9. Prove that there are 4 distinct homomorphisms from Z_2 into $\text{Aut}(Z_8)$. Show that the resulting semidirect products are $Z_8 \oplus Z_2$, D_8, the quasidihedral group QD_{16} (Exercise 1.10.9) and the modular group (Exercise 2.3.17).

10. In this exercise, we prove that is p is an odd prime then $U(p^k)$ is a cyclic group of order $p^{k-1}(p - 1)$.

   (a) Prove that if $k \geq 2$ and $a \in \mathbb{Z}$ with $p \nmid a$, then $(1 + ap)^{p^{k-2}} \equiv 1 + ap^{k-1}$ $(\text{mod } p^k)$.

   (b) Deduce that for any a with $p \nmid a$, the element $1 + ap$ has order p^{k-1} in $U(p^k)$.

   (c) By Proposition 5.6.2, $U(\mathbb{F}_p) = U(p)$ is a cyclic group. Show that there exists $g \in \mathbb{Z}$ that is a generator of $U(p)$ and such that $g^{p-1} \not\equiv 1$ $(\text{mod } p^2)$.

   (d) Prove that a g found in the previous part generates $U(p^k)$ to deduce that $U(p^k)$ is cyclic.

   [Hint: Recall that $p \mid \binom{p}{j}$ for all j with $1 \leq j \leq p - 1$.]

11. Determine the isomorphism type of $\text{Aut}(Z_{40})$ and express the result in invariant factors form.

12. Determine the isomorphism type of $\text{Aut}(Z_{210})$ and express the result in invariant factors form.

13. This exercise guides a proof that $\text{Aut}(S_n) = S_n$ for all $n \neq 6$.
    (a) Prove that for all $\psi \in \text{Aut}(S_n)$ and all conjugacy classes $\mathcal{K}$ of S_n, the subset $\psi(\mathcal{K})$ is another conjugacy class.
    (b) Let $\mathcal{K}$ be the conjugacy class of transpositions and let $\mathcal{K}'$ be another conjugacy class of elements of order 2 (e.g., cycle type like $(a\,b)(c\,d)$). Prove that $|\mathcal{K}| \neq |\mathcal{K}'|$, unless possibly if $n = 6$.
    (c) Prove that for each $\psi \in \text{Aut}(S_n)$ and for all k with $2 \leq k \leq n$, we have $\psi((1\,k)) = (a\,b_k)$ for some distinct integers $a, b_2, b_3, \ldots, b_n$ in $\{1, 2, \ldots, n\}$.
    (d) Show that the transpositions $(1\,2), (1\,3), \ldots, (1\,n)$ generate S_n.
    (e) Deduce that $\text{Aut}(S_n) = \text{Inn}(S_n) \cong S_n$.

14. Let G be a group. Consider the homomorphism $\varphi : G \to \text{Aut}(G)$ defined by $\varphi(g)(x) = gxg^{-1}$. Prove that the resulting semidirect product $G \rtimes_\varphi G$ is equal to $G \oplus G$ if and only if G is abelian. Find a presentation for $D_3 \rtimes_\varphi D_3$.

15. Give a presentation for a nonabelian semidirect product $(Z_7 \oplus Z_7) \rtimes_\varphi Z_3$.

16. Let $\varphi : Z_4 \to S_5$ be the homomorphism that sends the generator x of Z_4 to $\varphi(x) = (1\,2\,3\,4)$. Exercise 6.6.13 showed that $\text{Aut}(S_5) = \text{Inn}(S_5) = S_5$. Let $G = S_5 \rtimes_\varphi Z_4$. Perform the following calculations in G.
    (a) $((1\,4\,3)(2\,5), x^2) \cdot ((2\,4\,5\,3), x)$.
    (b) $((1\,4\,3)(2\,5), x^2)^{-1}$.
    (c) $((1\,3\,5\,2\,4), x^3) \cdot ((1\,3)(2\,4), 1) \cdot ((1\,3\,5\,2\,4), x^3)^{-1}$.

17. Let G be any group. We define the *holomorph* of G as the group $\text{Hol}(G) = G \rtimes \text{Aut}(G)$, where the semidirect product is the natural one where $\varphi : \text{Aut}(G) \to \text{Aut}(G)$ is the identity (not trivial) homomorphism.
    (a) Prove that the holomorph of Z_p is a nonabelian group and give a presentation of it.
    (b) Prove that $\text{Hol}(Z_2 \oplus Z_2) \cong S_4$.

18. Let ρ be the standard permutation representation of S_3 acting on $\{1, 2, 3\}$ and let $G = Z_5 \wr_\rho S_3$. Use the presentation of $Z_5 = \langle x \mid x^5 = 1 \rangle$.
    (a) Calculate the product in G of $(x, x^2, x, (1\,2)) \cdot (x^3, 1, x^2, (1\,2\,3))$.
    (b) Calculate the inverse $(x, x^2, x^4, (1\,3))^{-1}$.
    (c) Calculate the general conjugate $(x^a, x^b, x^c, \sigma) \cdot (x^p, x^q, x^r, 1) \cdot (x^a, x^b, x^c, \sigma)^{-1}$.

19. Give a presentation for the group $Z_5 \wr Z_3$.

20. Let n be a positive integer and let d be a nontrivial divisor. Show that the largest subgroup of S_n acting naturally on $X = \{1, 2 \ldots, n\}$ that has n/d blocks of size d is a wreath product $S_d \wr S_{n/d}$. Calculate the order of this group.

21. Prove that $Z_p \wr Z_p$ is a nonabelian group of order p^{p+1} that is isomorphic to the Sylow p-subgroup of S_{p^2}. (See Exercise 6.5.3.)

6.7 Classification Theorems

This section consists primarily of examples of classification of groups of a given order. For certain integers n, Sylow's Theorem may allow us to find a normal subgroup of a given order. Then, it may be possible to use a semidirect product construction and determine all possible groups of a given order.

Example 6.7.1 (Groups of Order pq). Example 6.5.10 and Exercise 6.6.5 already established this classification. We restate the results here. Let $|G| = pq$.

Case 1. If $p = q$, then G is isomorphic to Z_{p^2} or to Z_p^2.

Case 2. If $p < q$ and $p \nmid (q - 1)$, then G is isomorphic to Z_{pq}.

Case 3. If $p < q$ and $p \mid (q - 1)$, then G is isomorphic to Z_{pq} or to the only (nondirect) semidirect product $Z_q \rtimes Z_p$. $\triangle$

Example 6.7.2 (Groups of Order 12). Let G be a group of order 12.

If G is abelian, G is isomorphic to Z_{12} or $Z_6 \oplus Z_2$.

Suppose that G is not abelian. By Cauchy's Theorem, G contains elements of order 2 and 3. By Sylow's Theorem, n_3 is equal to 1 or 4 and n_2 is equal to 1 or 3. If $n_3 = 4$, then G must contain 8 elements of order 3 and the identity. A single Sylow 2-subgroup contributes 3 more elements that would account for all elements of G. Therefore, if $n_3 = 4$, we cannot have $n_2 > 1$. Hence, G cannot have both $n_3(G) = 4$ and $n_2(G) = 3$.

Note that if $n_3 = 1$ and $n_2 = 1$, then if P is a Sylow 2-subgroup and Q a Sylow 3-subgroup, then $P, Q \trianglelefteq G$ and $P \cap Q = 1$. By the Direct Sum Decomposition Theorem, $G = PQ \cong P \oplus Q$. However, P is isomorphic to Z_4 or $Z_2 \oplus Z_2$, while $Q \cong Z_3$. Therefore, if $n_3 = n_2 = 1$, we obtain the abelian cases. The nonabelian cases arise when: (1) $n_3 = 4$ and $n_2 = 1$; and (2) $n_3 = 1$ and $n_2 = 3$. However, in these nonabelian cases, the group can be expressed as $G = PQ$, with P a Sylow 2-subgroup and Q a Sylow 3-subgroup and where either P or Q is normal. Since $P \cap Q = \{1\}$, then by Proposition 6.6.7, G is a semidirect product $P \rtimes_\varphi Q$ or $Q \rtimes_\varphi P$.

Case 1. Let P be the normal Sylow 2-subgroup of G. P can be isomorphic to Z_4 or $Z_2 \oplus Z_2$. All the elements not in P must have order 3 since all elements of order 2 and 4 must be in the unique Sylow 2-subgroup of G. Note that $\mathrm{Aut}(Z_4) \cong U(4) \cong Z_2$, so there is no nontrivial homomorphism from Z_3 into $\mathrm{Aut}(Z_4)$. The only group of order 12 with a normal Sylow 2-subgroup that is isomorphic Z_4 is in fact abelian, Z_{12}. On the other hand, $\mathrm{Aut}(Z_2 \oplus Z_2) \cong GL_2(\mathbb{F}_2)$. This is nonabelian of order 6 so contains two elements of order 3, in particular

$$\begin{pmatrix} 0 & 1 \\ 1 & 1 \end{pmatrix} \quad \text{and} \quad \begin{pmatrix} 1 & 1 \\ 1 & 0 \end{pmatrix}.$$

So there exists a nontrivial homomorphism φ of $Z_3 = \langle x \rangle$ into $\text{Aut}(Z_2 \oplus Z_2)$ that sends the generator x to one of the above automorphisms, described by the matrices. With the first matrix, we have

$$\varphi(x)(1,0) = (0,1) \quad \varphi(x)(0,1) = (1,1) \quad \varphi(x)(1,1) = (1,0).$$

So we can create the group $(Z_2 \oplus Z_2) \rtimes_\varphi Z_3$. However, this is isomorphic to a group we already know, namely A_4. Furthermore, the two different homomorphisms into $\text{Aut}(P)$ given by the two different matrices both produce semidirect products that are isomorphic to A_4.

Case 2. Suppose now that Q is a normal subgroup of G of order 3. The quotient group G/Q is isomorphic to Z_4 or $Z_2 \oplus Z_2$ and we look for nontrivial homomorphisms of each of these groups into $\text{Aut}(Z_3) \cong U(3) \cong Z_2$. The only nontrivial element of $\text{Aut}(Z_3)$ is inversion which we will call λ.

If $G/Q \cong Z_4 = \langle x \rangle$, then the only nontrivial homomorphism into $\text{Aut}(Z_3)$ has $\varphi(x) = \lambda$, or in other words $\varphi(x)(h) = h^{-1}$ for all $h \in Q$. This is a new semidirect product with a presentation of

$$Z_3 \rtimes Z_4 = \langle x, y \mid x^4 = y^3 = 1, \ xyx^{-1} = y^{-1} \rangle.$$

On the other hand, if $G/Q \cong Z_2 \oplus Z_2$, then we have three choices for φ depending on which two (nontrivial) elements of $Z_2 \oplus Z_2$ get sent to λ. One can easily check that the resulting semidirect products are all isomorphic to $S_3 \oplus Z_2 \cong D_6$.

In conclusion, if $|G| = 12$, then G is isomorphic to one of the following (nonisomorphic) groups:

$$Z_{12}, \ Z_6 \oplus Z_2, \ D_6, \ A_4, \ Z_3 \rtimes Z_4. \qquad \triangle$$

Example 6.7.3 (Groups of Order 1225). Let G be a group of order $|G| = 1225 = 5^2 \cdot 7^2$. We prove that G is abelian. The index n_5 must satisfy $n_5 \mid 49$ and $n_5 \equiv 1 \pmod 5$. The divisors of 49 are 1, 7, and 49. Only 1 satisfies the second condition so $n_5 = 1$. Hence, G has a normal Sylow 5-subgroup P. Let Q be any Sylow 7-subgroup of G. Then $PQ \le G$ and

$$|PQ| = \frac{|P|\,|Q|}{|P \cap Q|} = \frac{5^2 \cdot 7^2}{1} = 1225,$$

so $PQ = G$. Since $P \cap Q = \{1\}$, then by Proposition 6.6.7, G is a semidirect product $P \rtimes_\varphi Q$.

Since $|P| = 5^2$, we know that $P \cong Z_{25}$ or $Z_5 \oplus Z_5$. Suppose first that $P \cong Z_{25}$. Consider the action of Q on P by conjugation. This induces a homomorphism from Q into $\text{Aut}(P) \cong U(25)$. However, $|U(25)| = 24$, which is relatively prime with $|Q| = 49$ so the only homomorphism from Q to $\text{Aut}(P)$

is trivial. Hence, in this cases, Q commutes with all of P. Similarly, suppose $P \cong Z_5 \oplus Z_5$ and consider the action of Q on P by conjugation. This induces a homomorphism $Q \to \mathrm{Aut}(Z_5 \oplus Z_5)$. However, $\mathrm{Aut}(Z_5 \oplus Z_5) \cong \mathrm{GL}_2(\mathbb{F}_5)$ which has order $(5^2 - 1)(5^2 - 5) = 480$, which is relatively prime to $|Q| = 49$. Hence, the only homomorphism $Q \to \mathrm{Aut}(P)$ is trivial, so again Q commutes with all of P. So we have shown that $G = P \oplus Q$ in both possibilities. We already know that if a group has order p^2, where p is prime, then the group is isomorphic to Z_{p^2} or Z_p^2.

We conclude that if G has order 1225, then G is abelian and the possible groups of order 1225 are given by FTFGAG.   △

Example 6.7.4 (Groups of Order 286). Let G be a group of order $|G| = 286 = 2 \cdot 11 \cdot 13$. By Sylow's Theorem, $n_{11}(G)$ divides 26 and $n_{11}(G) \equiv 1$ (mod 11). This implies that $n_{11}(G) = 1$ so G has a normal Sylow 11-subgroup P. Similarly, by Sylow's Theorem $n_{13}(G)$ divides 22 and $n_{13}(G) \equiv 1$ (mod 13). This implies that $n_{13}(G) = 1$ so G has a normal Sylow 13-subgroup Q. The subgroup PQ is a group of order 143. It is normal since $|G : PQ| = 2$. By Example 6.7.1, $PQ \cong Z_{143}$.

Write $PQ = \langle x \rangle$ with $|x| = 143$. By Cauchy's Theorem, G has an element of order 2, say the element y. Obviously, $PQ \langle y \rangle = G$ so G is a semidirect product $Z_{143} \rtimes_\varphi Z_2$.

Since as rings $\mathbb{Z}/143\mathbb{Z} = \mathbb{Z}/11\mathbb{Z} \oplus \mathbb{Z}/13\mathbb{Z}$, the automorphism group is

$$\mathrm{Aut}(Z_{143}) = U(\mathbb{Z}/11\mathbb{Z}) \oplus U(\mathbb{Z}/13\mathbb{Z}) = U(11) \oplus U(13) \cong Z_{10} \oplus Z_{12}.$$

The group $Z_{10} \oplus Z_{12}$ has three elements of order 2. In $U(143)$, these elements are $\overline{12}$, $-\overline{1} = \overline{142}$, and $-\overline{12} = \overline{131}$. Each of these elements leads to three homomorphisms $\varphi_i : Z_2 \to \mathrm{Aut}(Z_{143})$ with $\varphi_1(y)(x) = x^{12}$, $\varphi_2(y)(x) = x^{131}$, and $\varphi_3(y)(x) = x^{-1}$. These give 3 semidirect products

$$G_1 = Z_{143} \rtimes_1 Z_2 = \langle x, y \,|\, x^{143} = y^2 = 1,\ yxy^{-1} = x^{12} \rangle,$$
$$G_2 = Z_{143} \rtimes_2 Z_2 = \langle u, v \,|\, u^{143} = v^2 = 1,\ vuv^{-1} = u^{131} \rangle,$$
$$G_3 = Z_{143} \rtimes_3 Z_2 = \langle a, b \,|\, a^{143} = b^2 = 1,\ bab^{-1} = a^{-1} \rangle.$$

We could have approached the classification somewhat differently and we do so now to show the benefit. First note that $Z_{143} = Z_{11} \oplus Z_{13}$ and that $\mathrm{Aut}(Z_{143}) = U(11) \oplus U(13) = \mathrm{Aut}(Z_{11}) \oplus \mathrm{Aut}(Z_{13})$. Setting g_1 and g_2 as generators for Z_{11} and Z_{13} respectively, the homomorphisms φ_i correspond to

$$\varphi_1(y) : \begin{cases} g_1 \mapsto g_1 \\ g_2 \mapsto g_2^{-1}, \end{cases} \qquad \varphi_2(y) : \begin{cases} g_1 \mapsto g_1^{-1} \\ g_2 \mapsto g_2, \end{cases} \qquad \varphi_3(y) : \begin{cases} g_1 \mapsto g_1^{-1} \\ g_2 \mapsto g_2^{-1}. \end{cases}$$

Consequently,

$$G_1 \cong D_{13} \oplus Z_{11}, \qquad G_2 \cong D_{11} \oplus Z_{13}, \qquad \text{and} \qquad G_3 \cong D_{143}.$$

Furthermore, we easily determine that no two of these groups are isomorphic by counting the elements of order 2: 13 in G_1, 11 in G_2, and 143 in G_3.   △

In some of the examples above, we encountered a few common situations that we delineate in the following propositions.

> ### Proposition 6.7.5
>
> Suppose that G is a group with $|G| = p^a q^b$ where p and q are distinct primes and $a, b \geq 1$ are integers. Suppose also that $n_p(G) = 1$ or $n_q(G) = 1$. Then G is a semidirect product between a Sylow p-subgroup and a Sylow q-subgroup.

Proof. If $n_p(G) = 1$ or $n_q(G) = 1$, then there is a normal Sylow p-subgroup or a normal Sylow q-subgroup. Let P be a Sylow p-subgroup and let Q be a Sylow q-subgroup. Then $P \cap Q$ is a subgroup of P and of Q so its order must divide $\gcd(p^a, q^b) = 1$. Hence, $P \cap Q = \{1\}$. Also, since P or Q is normal, PQ is a subgroup of G and it has order $|P||Q|/|P \cap Q| = p^a q^b$. Thus, $PQ = G$. By Proposition 6.6.7, G is a semidirect product $P \rtimes_\varphi Q$ or $Q \rtimes_\varphi P$. $\square$

Example 6.7.6 (Groups of Order $p^2 q$, with $p \neq q$). Let p and q be primes and let G be a group of order $p^2 q$. Let $P \in \mathrm{Syl}_p(G)$ and let $Q \in \mathrm{Syl}_q(G)$. We break this example into cases.

Case 1: $p > q$. Since n_p divides q and $n_p \equiv 1 \pmod{p}$ then we must have $n_p = 1$. Thus, $P \trianglelefteq G$ and $G = P \rtimes_\varphi Q$ for some homomorphism $\varphi : Q \to \mathrm{Aut}(P)$. We now have two subcases:

- $P \cong Z_p \oplus Z_p$. Then $\mathrm{Aut}(P) \cong \mathrm{GL}_2(\mathbb{F}_p)$ and $|\mathrm{Aut}(P)| = p(p-1)^2(p+1)$. There exist nontrivial homomorphisms φ when $q|(p-1)$ or $q|(p+1)$. Note that if q^1 is the highest power of q dividing $p + 1$, then by Sylow's Theorem applied to $\mathrm{Aut}(P)$, all Sylow q-subgroups of $\mathrm{Aut}(P)$ are conjugate to each other. Hence, by Exercise 6.6.3, there exists a unique (up to isomorphism) semidirect product $P \rtimes_\varphi Q$.

- $P \cong Z_{p^2}$. Then $\mathrm{Aut}(P) = U(p^2) \cong Z_{p(p-1)}$. There exist nontrivial homomorphisms φ when $q|(p - 1)$.

Case 2: $p < q$. Then $n_q = 1 + kq$ and since n_q divides p^2 then n_q must be 1, p or p^2. If $n_q = 1$, then $Q \trianglelefteq G$. Since $q > p$ then if $n_q \neq 1$, we cannot have $n_q = 1 + kq = p$. Thus, $n_q = p^2$. We then have $kq = p^2 - 1 = (p-1)(p+1)$. Hence, q divides $p - 1$ or $p + 1$. Since $q > p$ then $q = p + 1$ which leads us to the case $p = 2$ and $q = 3$, so we are left with discussing the case $|G| = 12$. This case was settled in Example 6.7.2. We suppose now that $|G| \neq 12$.

We know from Proposition 6.7.5 that $G = Q \rtimes_\varphi P$ for some homomorphism $\varphi : P \to \mathrm{Aut}(Q)$. We have two subcases.

- $P \cong Z_p \oplus Z_p$. Since $\mathrm{Aut}(Q) = U(q) \cong Z_{q-1}$, then if $p|(q - 1)$, there exist nontrivial homomorphisms $\varphi : P \to \mathrm{Aut}(Q)$.

- $P \cong Z_{p^2}$. Again, $\text{Aut}(Q) = U(q) \cong Z_{q-1}$. A homomorphism $\varphi :$ $Z_{p^2} \to \text{Aut}(Z_q)$ is determined by where it maps the generator x of Z_{p^2}. By Lagrange's Theorem $|\varphi(x)|$ divides $\gcd(p^2, q-1)$, which, depending on p and q might be 1, p, or p^2. $\triangle$

Example 6.7.7 (Groups of Order p^3). Let p be an odd prime and let G be a group of order p^3. (If $p = 2$, it is easy to find all the groups of order 8 so we refer the reader to the table in Section A.7.)

For this example, we cannot play Sylow p-subgroups off themselves since the group itself is a p-group. By FTFGAG, we know that there are three nonisomorphic abelian groups of order p^3, namely Z_{p^3}, $Z_{p^2} \oplus Z_p$, and Z_p^3.

From now on assume G is nonabelian. By the Class Equation, it is possible to prove (Exercise 6.4.14) that every p-group has a nontrivial center. Also, we know that if $G/Z(G)$ is cyclic, then G is abelian. (See Exercise 2.3.21.) So we must have $Z(G) \cong Z_p$. Then $G/Z(G)$ is a p group of order p^2, which contains a normal $\overline{N}$ subgroup of order p. By the Fourth Isomorphism Theorem, G contains a normal subgroup N of order such that $N/Z(G) = \overline{N}$.

There are two cases for the isomorphism type of N.

Case 1. G has a normal subgroup N that is isomorphic to Z_{p^2}. Let $N = \langle x \rangle$. Assume that $G - N$ does not contain an element of order p so that all elements of order p are in N. In general, if g is an element of order p^2, then $\langle g \rangle$ contains $p(p-1)$ other generators (of order p^2). Also, if g_1 and g_2 are both of order p^2, then $g_1^k = g_2^\ell$ with k and ℓ bother relatively prime to p^2, then $\langle g_1 \rangle = \langle g_2 \rangle$. Hence, the assumption that N is the only subgroup that contains elements of order p implies that, if there are k distinct subgroups of order p^2, then $p^3 = 1 + (p-1) + kp(p-1)$. Thus, $p^2 = k(p-1)$. This is a contradiction. Hence, $G - N$ contains an element y of order p. Thus, by Proposition 6.6.7, G is a semidirect product of Z_{p^2} by Z_p.

We know that $\text{Aut}(N) \cong (\mathbb{Z}/p^2\mathbb{Z})^\times \cong Z_{p(p-1)}$. Hence, again by Cauchy's Theorem $\text{Aut}(N)$ contains an element of order p. By Exercise 6.6.10, the element $1 + p$ has order p modulo p^2. A nontrivial homomorphism $\varphi\langle y \rangle \to \text{Aut}(N)$ has $\varphi(y)(x) = x^{1+p}$. As a group presentation,

$$Z_{p^2} \rtimes_\varphi Z_p = \langle x, y \mid x^{p^2} = y^p = 1, \, yxy^{-1} = x^{1+p} \rangle.$$

Again, though there are choices for the homomorphism $\varphi : \langle y \rangle \to \text{Aut}(H)$, because of the options for generators of H, the different nontrivial semidirect products are all isomorphic.

Case 2. G does not contain a normal subgroup that is isomorphic to Z_{p^2}. Note that if $|x| = p^2$, then by Exercise 6.4.9, $\langle x \rangle \trianglelefteq G$. Hence, this case implies that all the nonidentity elements in G have order p. So the normal subgroup N of order p^2 has $N \cong Z_p \oplus Z_p$ and $G - N$ contains an element

z of order p. So $G = N\langle z \rangle$ and by Proposition 6.6.7, G is a semidirect product $(Z_p \oplus Z_p) \rtimes_\varphi Z_p$.

By Proposition 6.6.14, $\mathrm{Aut}(N) \cong \mathrm{GL}_2(\mathbb{F}_p)$, which has order $(p^2 - p)(p^2 - 1)$. By Cauchy's Theorem, since p divides $(p^2 - p)(p^2 - 1)$, then $\mathrm{Aut}(N)$ contains an automorphism ψ of order p. So we need a homomorphism $\varphi : \langle z \rangle \to \mathrm{Aut}(N)$ by $\varphi(z) = \psi$ for some such ψ.

To understand this group more explicitly, suppose that $N = \langle x \rangle \oplus \langle y \rangle$. An element in $\mathrm{GL}_2(\mathbb{F}_p)$ of order p is

$$A = \begin{pmatrix} 1 & 1 \\ 0 & 1 \end{pmatrix}.$$

If ψ corresponds to this matrix, then we have $\psi(x) = x$ and $\psi(y) = xy$. As a group presentation, we have with this specific ψ,

$$(Z_p \oplus Z_p) \rtimes_\varphi Z_p = \langle x, y, z \mid x^p = y^p = z^p = 1, \ xy = yx,$$
$$zxz^{-1} = x, \ zyz^{-1} = xy \rangle.$$

But p is the highest order of p dividing $|\mathrm{GL}_2(\mathbb{F}_p)|$ so, by Sylow's Theorem, all subgroups of order p in $\mathrm{GL}_2(\mathbb{F}_p)$ are conjugate. Consequently, Exercise 6.6.3, all nontrivial homomorphisms $\varphi : Z_p \to \mathrm{Aut}(A_p \oplus Z_p)$ produce isomorphic semidirect products. $\triangle$

We reiterate that the methods at our disposal up to this point to classify groups of a given order n often involve using Sylow p-subgroups, proving that all groups of order n must be a combination semidirect product of their Sylow p-subgroups, and then determining how many such products are distinct. Remark 6.6.11 pointed out that it is not always easy to tell when certain semidirect products are isomorphic. However, there are a few situations in which it is possible to tell, as seen in Exercises 6.6.3 and 6.6.4. In particular, in Exercise 6.6.3, the desired condition when comparing $H \rtimes_{\varphi_1} K$ and $H \rtimes_{\varphi_2} K$ is that $\mathrm{Im}\,\varphi_1$ and $\mathrm{Im}\,\varphi_2$ are conjugate subgroups in $\mathrm{Aut}(H)$. As an additional strategy using the result of this exercise, if $\mathrm{Im}\,\varphi_1$ and $\mathrm{Im}\,\varphi_2$ happen to be Sylow subgroups of $\mathrm{Aut}(H)$, then by Sylow's Theorem, they are conjugate subgroups.

EXERCISES FOR SECTION 6.7

1. Prove that there is only one nonabelian group of order $1183 = 7 \cdot 13^2$. Give an explicit presentation of it.

2. Prove that all groups of order 4225 are abelian.

3. Show that every group of order 30 has a normal subgroup of order 15. Then prove that there are exactly 4 nonisomorphic groups of order 30.

4. Prove that all groups of order $14161 = 7^2 \cdot 17^2$ are abelian.

5. Consider groups G of order $351 = 3^3 \cdot 13$. Prove that G has a normal Sylow 13-subgroup or a normal Sylow 3-subgroup. Show that there exists a unique nonabelian group of order 351 with a normal subgroup isomorphic to Z_3^3 and give its presentation. [Hint: The Sylow 13-subgroups of $\mathrm{Aut}(Z_3^3)$ are conjugate.]

6. Classify the groups of order 105. [There are 2 nonisomorphic groups.]

7. Classify the groups of order 20. [There are 5 nonisomorphic groups.]

8. Classify the groups of order 154.

9. Classify the groups of order 333.

10. Classify the groups of order 5819.

11. Classify the groups of order 1690.

12. Let p be an odd prime. Classify the groups of order $4p$. [Hint: The number of nonisomorphic groups is different depending on whether $p \equiv 1 \pmod 4$ or $p \equiv 3 \pmod 4$.]

13. Let $(p, p+2)$ be a prime pair, i.e., both p and $p+2$ are primes. Consider groups G of order $2p(p+2)$.
    (a) Prove that G contains a normal subgroup N of order $p(p+2)$.
    (b) Prove that G must be a semidirect product of N with another subgroup.
    (c) Prove that G must be isomorphic to one of the following four groups: $Z_{2p(p+2)}$, $D_p \oplus Z_{p+2}$, $D_{p+2} \oplus Z_p$, or $D_{p(p+2)}$.

14. Let p be an odd prime. Prove that every element in $\mathrm{GL}_2(\mathbb{F}_p)$ of order 2 is conjugate to a diagonal matrix with 1 or -1 on the diagonal. Use this result to classify the groups of order $2p^2$.

15. Recall the Heisenberg group $H(\mathbb{F}_p)$ introduced in Exercise 1.2.26. We observe that it is a group of order p^3. Referring to Example 6.7.7, determine the isomorphism type of $H(\mathbb{F}_p)$.

16. In this exercise, we classify groups G of order $56 = 2^3 \cdot 7$. Let P be a Sylow 2-subgroup and let Q be a Sylow 7-subgroup.
    (a) Prove that G has a normal Sylow 2-subgroup or a normal Sylow 7-subgroup. Deduce that G is $P \rtimes_\varphi Q$ or $Q \rtimes_\varphi P$ for an appropriate φ.
    (b) List all (three) abelian groups of order 56. (In the remainder of the exercise, assume G is nonabelian.)
    (c) Suppose that P is normal. Prove that there is: (i) one group when $P \cong Z_2^3$; (ii) two nonisomorphic groups when $P \cong Z_4 \oplus Z_2$; (iii) one group when $P \cong Z_8$; (iv) three non-isomorphic groups when $P \cong D_4$; and (v) two nonisomorphic groups when $P \cong Q_8$.
    (d) Suppose that Q is not a normal subgroup. Prove that the only nontrivial homomorphism $\varphi : Q \to \mathrm{Aut}(P)$ occurs when $P \cong Z_2^3$. Prove that there is only one nonisomorphic semidirect product $Z_2^3 \rtimes Z_7$. Give a presentation of this group.

6.8   Projects

Investigative Projects

PROJECT I.   **Sudoku and Group Actions.** Project IV in Chapter 1 discussed the group of symmetries within the set of permissible Sudoku fillings. Again, let S be the set of all possible solutions to a Sudoku puzzle, i.e., all possible ways of filling out the grid according to the rules, and G the group of transformations described in that project. By construction, the set G acts on S. With the formalism of group actions, we can attempt to address this problem more effectively. For example, we now call two Sudoku fillings equivalent if they are in the same G orbit.

Try to answer some of the following questions and any others you can imagine. Is the action of G on S faithful or transitive or free? If it is not transitive, are the orbits all the same size? If not, is there a range on what the sizes of the orbits can be? How many orbits are there?

PROJECT II.   **Invariants in Polynomial Rings.** Let $X = \mathbb{C}[x_1, x_2, \ldots, x_n]$ be the ring of multivariable polynomials and consider the action of S_n defined by

$$\sigma \cdot p(x_1, x_2, \ldots, x_n) = p(x_{\sigma(1)}, x_{\sigma(2)}, \ldots, x_{\sigma(n)}).$$

Find X^σ for various permutations σ. Can you deduce any conclusions about these fixed subsets? For various subgroups $H \leq S_n$, can you determine or say anything about the subset of X fixed by all of H?

PROJECT III.   **Quotient G-Sets.** Discuss the possibility of taking quotient objects (similar to quotient groups or quotient sets) in the context of G-sets. Given interesting examples. Try to prove as many interesting propositions that you can about quotient G-sets that illustrate the connection between this quotient process and the group action.

PROJECT IV.   **Young's Geometry.** The Fano plane described in Exercise 6.3.17 is an example of a finite geometry. In the study of finite geometries, one starts with a set of axioms and deduces as many theorems as possible. In a finite geometry, the axioms do not explicitly refer to a finite number of points or lines but lead to there only being a finite number. In Young's geometry, it is possible to prove that there are 9 points and 12 lines. Furthermore, a possible configuration representing the subset structure of points to lines is given in the picture below. Study the subgroup of the permutation group S_9 acting on the points that preserves the line structure in Young's geometry.

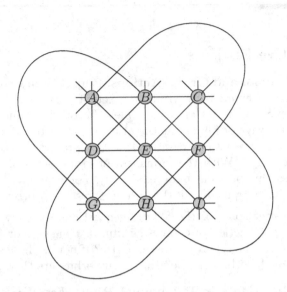

PROJECT V. **Coloring the Soccer Ball.** Use the Cauchy-Frobenius Lemma
to discuss the number of ways to color the soccer ball using 2, 3, 4, or more
colors, where we view two colorings as the same if one can be obtained
from the other by some rigid motion of the ball. The standard black-and-
white coloring of a soccer ball is invariant under the group of rigid motions
of the ball. Discuss the number of colorings that are invariant under some
subgroups of the group of rigid motions.

PROJECT VI. **Sylow p-Subgroup Conjecture.** Let G be a finite group. We
saw in Section 6.5 that if $n_p = 1$ for some prime p that divides $|G|$, then
that unique Sylow p-subgroup is normal. Therefore, if G is simple, then
$n_p > 1$ for all p dividing $|G|$. Discuss the conjecture of the converse: If
$n_p(G) > 1$ for all primes p dividing $|G|$, then G is simple.

PROJECT VII. **Simple Isn't So Simple.** Write a computer that uses Sylow's
Theorem to eliminate all orders between 1 and say 1000 (or more) for
which groups of that order cannot be simple. For any orders that could
have a simple group, list the n_p for all p dividing the order.

PROJECT VIII. **Sylow p-Subgroups Action.** Sylow's Theorem affirms that
a group G acts transitively by conjugation on $\mathrm{Syl}_p(G)$. Explore whether
this action is multiply transitive. If it is not, explore whether the action
of G on $\mathrm{Syl}_p(G)$ system of blocks. Discuss theoretically or with examples.

PROJECT IX. **The S_4 Tetrahedron.** In $(\mathbb{R}^+)^3$ (the first octant in Euclidean
3-space), there are 6 permutations of the inequalities $x_1 \leq x_2 \leq x_3$, each
one corresponding to the action of an element in S_3 on $0 \leq x_1 \leq x_2 \leq x_3$.

Each of the inequalities corresponds to a region of $(\mathbb{R}^+)^3$. Intersected with the plane $x_1 + x_2 + x_3 = 1$ produces 6 regions in an equilateral triangle.

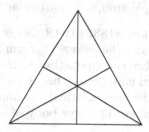

In this project, consider the action of S_4 on $(\mathbb{R}^+)^4$ that permutes the coordinates. There are 24 permutations of the inequalities $x_1 \leq x_2 \leq x_3 \leq x_4$, each one corresponding to a cone-shaped region in $(\mathbb{R}^+)^4$. The intersection of $(\mathbb{R}^+)^4$ with the plane $x_1 + x_2 + x_3 + x_4 = 1$ is a regular tetrahedron. The regions $0 \leq x_i \leq x_j \leq x_k \leq x_\ell$ for different indices cut this tetrahedron into 24 regions, each one corresponding to the action of an element in S_4.

Here are just a few ideas to explore: Physically construct a regular tetrahedron. Trace (at least on the surface) these 24 regions and show which ones correspond to which element in S_4. Discuss orbits of subgroups of S_4 or quotient groups of S_4 in reference to the tetrahedron. Use this tetrahedron to provide a faithful representation of S_4 in $\mathbb{R}^2$.

PROJECT X. **Combinatorial Identities.** Example 6.2.14 and Exercises 6.2.13, 6.2.14, and 6.2.15 established known combinatorial formulas from the Orbit Equation. Find a number of other combinatorial identities in a discrete mathematics or combinatorics textbook and try to recover these combinatorial identities as the orbit equations of some group actions.

PROJECT XI. **Automorphism Groups.** Section 6.6 gave the isomorphism type of $\text{Aut}(G)$ for a few groups G. Can you come up with others? For example, can you determine the automorphism group of $Z_3 \rtimes Z_4$, $Z_5 \rtimes Z_4$, $Z_7 \rtimes Z_3$, $(Z_3 \oplus Z_3) \rtimes Z_2$, or others?

PROJECT XII. **Groups of Order $p^3 q$.** Revisit Exercise 6.7.16. Explore how much of the exercise generalizes to groups of order $3^3 \cdot 37$. Explore how much of the exercise generalizes to groups of order $8p$, where p is an odd prime. Generalize the explorations as much as possible.

Expository Projects

PROJECT XIII. **Finite Simple Groups.** Many algebraists point to the Classification Theorem of Finite Simple Groups as the largest and most collaborative theorem in algebra. Write a summary and timeline of the steps behind the proof of this deep theorem. Identify leaders of the project and significant contributors.

PROJECT XIV. **Binary Golay Code.** Produce a report on the construction and applications of binary Golay codes. Emphasize connections between Golay codes and simple groups, in particular sporadic groups.

PROJECT XV. **Peter Sylow, 1832–1918.** No other famous theorems carry the name Sylow, besides the Sylow Theorem. Nevertheless, Sylow's Theorem is critical in advanced group theory. Using reliable sources, explore Peter Sylow's personal background, his mathematical work, and the reception of his work in the mathematical community. Decide if this theorem bearing his name is a part of larger body of work.

PROJECT XVI. **Agnes Meyer Driscoll, 1889–1971.** Few mathematicians are buried in the United States Arlington National Cemetery; even fewer still of those are women. Agnes Mere Driscoll is one such person. Discuss Driscoll's personal history, her mathematical work, and what earned her a place of honor in the national cemetery.

PROJECT XVII. **Kenjiro Shoda, 1902–1977.** Having studied under Emmy Noether, Japanese mathematician Kenjiro Shoda was one of the so-called "Noether boys." Using reliable sources, explore Kenjiro Shoda's personal background, his mathematical work, his influence on the mathematical community, and his influence in Japan more generally.

A

Appendix

This appendix provides six sections of background material: one section pertaining to complex numbers, a section of review for basic number theory, two further sections on equivalence relations and partial orders, and two sections of elementary number theory. The final section in this appendix provides a reference of all finite groups of orders up to 24.

A.1   The Algebra of Complex Numbers

A.1.1   Complex Numbers

When studying solutions to polynomials, one quickly encounters equations that do not have real roots. Consider the simple equation $x^2 + 1 = 0$. Since the square of every real number is nonnegative, there exists no real number such that $x^2 = -1$. The complex numbers begin with "imagining" that there exists a number i such that $i^2 = -1$. Mathematicians called i the imaginary unit. Mathematicians then assumed that for all other algebraic properties, i interacted with the real numbers just like any other real number. The powers of the imaginary unit i are

$$i^1 = i, \quad i^2 = -1, \quad i^3 = -i, \quad i^4 = 1, \quad i^5 = i, \ldots$$

and so forth.

An expression of the form bi, where b is a real number, is called an *imaginary number*. A complex number is an expression of the form $a + bi$, where $a, b \in \mathbb{R}$. We denote set of complex numbers by $\mathbb{C}$. We often denote a complex variable by the letter z. If $z = a + bi$, we call a the *real part* of z, denoted $a = \Re(z)$, and we call b the *imaginary part* of z, denoted by $b = \Im(z)$.

With this definition, every quadratic equation has a root. For example, applied to $2x^2 + 5x + 4 = 0$, the quadratic formula gives the following solutions:

$$x = \frac{-b \pm \sqrt{b^2 - 4ac}}{2a} = \frac{-5 \pm \sqrt{25 - 32}}{4} = -\frac{5}{4} \pm \frac{\sqrt{7}}{4}i.$$

The quadratic formula shows that whenever $a + bi$ is a root of an equation with real coefficients, then $a - bi$ is also a root. So in some sense, these two

DOI: 10.1201/9781003299233-A

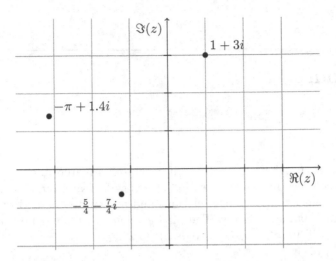

FIGURE A.1: Cartesian representation of $\mathbb{C}$.

complex numbers are closely related. If $z = a + bi$, we call the number $a - bi$, the *conjugate* of z, and denote it by $\bar{z}$.

Since a complex number involves two independent real numbers, we usually depict $\mathbb{C}$ by a Cartesian plane with $\Re(z)$ on the x-axis and $\Im(z)$ as the y-axis. By analogy with "the real line" we refer to this perspective as the *complex plane*. Figure A.1 shows a few complex numbers.

Polar coordinates are also useful in the study of complex numbers. The *absolute value* $|z|$, also called the *modulus*, of a complex number $z = a + bi$ is the distance r from the origin to z. Clearly, $|z| = r = \sqrt{a^2 + b^2}$. The *argument* of z is the angle θ of polar coordinates of the point (a, b). Converting between polar and Cartesian coordinates, we write

$$z = r(\cos\theta + i\sin\theta), \tag{A.1}$$

with $r \geq 0$. As with polar coordinates, though we typically consider $\theta \in [0, 2\pi)$, the argument θ can be any real number.

For example, the absolute value and the argument of $3 + 2i$ are

$$|3 + 2i| = \sqrt{3^2 + 2^2} = \sqrt{13} \quad \text{and} \quad \theta = \tan^{-1}\left(\frac{2}{3}\right).$$

The absolute value and the argument of $-12 + 5i$ are

$$|-12 + 5i| = \sqrt{12^2 + 5^2} = 13 \quad \text{and} \quad \theta = \tan^{-1}\left(-\frac{5}{12}\right) + \pi.$$

Without discussing the issue of convergence, using the power series of known functions, observe that

$$
e^{i\theta} = \sum_{k=0}^{\infty} \frac{(i\theta)^k}{k!} = \sum_{k=0}^{\infty} i^k \frac{\theta^k}{k!} = \left(\sum_{\substack{k \geq 0 \\ k \text{ even}}} (-1)^{k/2} \frac{\theta^k}{k!} \right) + i \left(\sum_{\substack{k \geq 0 \\ k \text{ odd}}} (-1)^{(k-1)/2} \frac{\theta^k}{k!} \right)
$$

$$
= \left(\sum_{n=0}^{\infty} (-1)^n \frac{\theta^{2n}}{(2n)!} \right) + i \left(\sum_{n=0}^{\infty} (-1)^n \frac{\theta^{2n+1}}{(2n+1)!} \right) = \cos\theta + i\sin\theta.
$$

Consequently, it is common to write the polar form (A.1) of a complex number more concisely as

$$
z = re^{i\theta},
$$

where $r = |z|$ and θ is the argument of z. A few examples of complex numbers in polar form are $-1 = e^{-i\pi}$ or $i = e^{i\pi/2}$.

A.1.2 Operations in $\mathbb{C}$

The addition of two complex numbers is defined as

$$
(a + bi) + (c + di) \stackrel{\text{def}}{=} (a + c) + (b + d)i.
$$

Since the real part and the imaginary part act as x-coordinates and y-coordinates for vectors in $\mathbb{R}^2$ and since the addition of complex numbers is done component-wise, the addition of complex numbers is identical to the addition of vectors in $\mathbb{R}^2$. The subtraction of two complex numbers is $(a + bi) - (c + di) = (a - c) + (b - d)i$.

The product of two complex numbers is

$$
(a + bi)(c + di) \stackrel{\text{def}}{=} ac + adi + bci + bd(-1) = (ac - bd) + (ad + bc)i.
$$

With this definition, if $z = a + bi$, then

$$
z\bar{z} = (a + bi)(a - bi) = a^2 + b^2 = |z|^2.
$$

Besides this identity, the formula for multiplication does not readily appear to have interesting geometric interpretation. However, the polar coordinate expression of complex numbers leads immediately to an interpretation of the product. We have

$$
(r_1 e^{i\theta_1})(r_2 e^{i\theta_2}) = r_1 r_2 e^{i(\theta_1 + \theta_2)},
$$

so the product of two complex numbers z_1 and z_2 has absolute value $|z_1 z_2| = |z_1||z_2|$ and has an argument that is the sum of the arguments of z_1 and z_2.

For division of two complex numbers, let $z_1 = a + bi = r_1 e^{\theta_1 i}$ and $z_2 = c + di = r_2 e^{\theta_2 i} \neq 0$ be two complex numbers. Then the two expressions of division are

$$
\frac{z_1}{z_2} = \frac{a + bi}{c + di} = \frac{(a + bi)(c - di)}{c^2 + d^2} \quad \text{and} \quad \frac{z_1}{z_2} = \frac{r_1}{r_2} e^{(\theta_1 - \theta_2)i}.
$$

Using the complex conjugate, we can express the inverse as

$$z^{-1} = \frac{1}{z} = \frac{\bar{z}}{z\bar{z}} = \frac{\bar{z}}{|z|^2}.$$

The complex conjugate interacts with the operations on $\mathbb{C}$ in the following way. For all $z_1, z_2 \in \mathbb{C}$,

$$\overline{z_1 + z_2} = \overline{z_1} + \overline{z_2} \quad \text{and} \quad \overline{z_1 z_2} = \overline{z_1}\,\overline{z_2}.$$

From this property, combined with the fact that the complex conjugate is a bijection from $\mathbb{C}$ to $\mathbb{C}$, we say that conjugation is an *automorphism* on $\mathbb{C}$. (See Section 3.4.1.)

From the multiplication operation, if $z = re^{i\theta}$, where $r \geq 0$ and θ is an angle, then for all integers $n \in \mathbb{Z}$, the powers of z are $z^n = r^n e^{in\theta}$. The argument θ is equivalent to any angle $\theta + 2\pi k$ for any $k \in \mathbb{Z}$. Consequently, all of the following complex numbers

$$\sqrt[n]{r}e^{i(\theta+2\pi k)/n} \qquad \text{for } k = 0, 1, 2, \ldots, n-1, \tag{A.2}$$

have an nth power equal to $z = re^{i\theta}$. Consequently, the n numbers in (A.2) are the nth roots of z.

As one example, consider the cube roots of $2i$. We write $2i = 2e^{i\pi/2}$ so the cube roots of $2i$ are $\sqrt[3]{2}e^{i\pi/6}$, $\sqrt[3]{2}e^{i(\pi/6+2\pi/3)}$, and $\sqrt[3]{2}e^{i(\pi/6+4\pi/3)}$. Expressed in their Cartesian form, these numbers are respectively

$$\sqrt[3]{2}\left(\frac{\sqrt{3}}{2} + \frac{1}{2}i\right), \quad \sqrt[3]{2}\left(-\frac{\sqrt{3}}{2} + \frac{1}{2}i\right), \quad \text{and} \quad -\sqrt[3]{2}i.$$

As another example, we calculate the square roots of $z = 1 + 3i$. Writing

$$z = \sqrt{10}e^{i\tan^{-1}(3)},$$

we find that the square roots of z are $\sqrt[4]{10}e^{i\tan^{-1}(3)/2}$ and $\sqrt[4]{10}e^{i\tan^{-1}(3)/2+i\pi} = -\sqrt[4]{10}e^{i\tan^{-1}(3)/2}$. Combining various trigonometric identities, we get

$$\cos\left(\frac{1}{2}\tan^{-1}r\right) = \sqrt{\frac{1}{2}\left(1 + \frac{1}{\sqrt{1+r^2}}\right)} \quad \text{and}$$

$$\sin\left(\frac{1}{2}\tan^{-1}r\right) = \sqrt{\frac{1}{2}\left(1 - \frac{1}{\sqrt{1+r^2}}\right)}.$$

Then the two square roots of $1 + 3i$ are

$$\pm\sqrt[4]{10}\left(\sqrt{\frac{1}{2}\left(1 + \frac{1}{\sqrt{1+3^2}}\right)} + i\sqrt{\frac{1}{2}\left(1 - \frac{1}{\sqrt{1+3^2}}\right)}\right)$$

$$= \pm\left(\sqrt{(\sqrt{10}+1)/2} + i\sqrt{(\sqrt{10}-1)/2}\right).$$

A.2 Set Theory Review

Built on the framework of Boolean logic, set theory sits at the foundation of all modern mathematics. This section and the next offer a brief review.

A.2.1 Sets and Operations

Definition A.2.1

(1) A *set* is a collection of objects for which there is a clear rule to determine whether an object is included or excluded.

(2) An object in a set is called an *element* of that set. We write $x \in A$ to mean "the element x is an element of the set A." We write $x \notin A$ if x is not an element of A.

The "clear rule" should have the same precision as truth values of logical statements in Boolean logic. Often, we describe this clear rule as a list, for example $A = \{1, 2, \pi, \sqrt{17}\}$, or with a constructor expression such as $B = \{x \text{ is a living person} \mid x \text{ is } 12 \text{ years or older}\}$.

Here are a few examples of common sets, along with their symbols.

- $\emptyset$ refers to the empty set.

- $\mathbb{N}$, $\mathbb{Z}$, $\mathbb{Q}$, $\mathbb{R}$, and $\mathbb{C}$ refer respectively to the set of nonnegative integers $\{0, 1, 2, \ldots\}$, the set of all integers, the set of rational numbers, the set of real numbers, and the set of complex numbers.

- We often use modifiers to the above sets. For examples, if A is one of the above number sets, then A^* means A without 0. As another example, $\mathbb{R}^{\geq 0} = \{x \in \mathbb{R} \mid x \geq 0\}$.

- Interval notation: $[a, b] = \{x \in \mathbb{R} \mid a \leq x \leq b\}$. The reader should be aware of other combinations.

- $\mathbb{R}^n$ is the set of n-tuples of real numbers.

- In linear algebra, $M_{m \times n}(\mathbb{C})$ is the set of $m \times n$ matrices with entries in $\mathbb{C}$.

- In analysis, if I is an interval of $\mathbb{R}$, then $C^0(I, \mathbb{R})$ means the set of all continuous functions with domain I and codomain $\mathbb{R}$.

Because of the close connection between sets and Boolean logic, there exist a few natural operations on sets. If A and B are any sets, we define

$$\text{union}: \quad A \cup B = \{x \mid x \in A \vee x \in B\}$$
$$\text{intersection}: \quad A \cap B = \{x \mid x \in A \wedge x \in B\}$$
$$\text{set difference}: \quad A \setminus B = \{x \mid x \in A \wedge x \notin B\}$$
$$\text{symmetric difference}: \quad A \triangle B = (A \setminus B) \cup (B \setminus A).$$

(Recall that $\vee$ means "or" and $\wedge$ means "and.") Note that the symmetric difference of two sets corresponds to the exclusive or operation applied to $x \in A$ or $x \in B$. Two sets A and B are called *disjoint* if $A \cap B = \emptyset$.

Definition A.2.2

A set A is called a *subset* of a set S if $x \in A \implies x \in S$. In other words, every element of A is an element of S. We write $A \subseteq S$.

For example, $\mathbb{N} \subseteq \mathbb{Z} \subseteq \mathbb{Q} \subseteq \mathbb{R} \subseteq \mathbb{C}$. We also use other symbols such as

$$A \subsetneq B \text{ to mean } A \subseteq B \text{ and } A \neq B; \text{ and}$$
$$A \not\subseteq B \text{ to mean } A \text{ is not a subset of } B.$$

When working with subsets of a given set S, we define one more operation on the subsets of S. The *complement* of A is $A^c = \{x \in S \mid x \notin A\}$.

It is often possible to express certain relationships of subset or equality between sets with identities on set operations. The following proposition lists a few of these.

Proposition A.2.3

Let A and B be sets.

(1) $A = B$ if and only if $A \subseteq B$ and $B \subseteq A$.

(2) $A \subseteq B$ if and only if $A \cap B = A$.

(3) $A \subseteq B$ if and only if $A \cup B = B$.

As a part of review, the reader should revisit the Boolean algebra identities on operations of sets. In particular, the associativity law for union (respectively intersection) states that for subsets A, B, C, the following holds

$$(A \cup B) \cup C = A \cup (B \cup C) \quad \text{and} \quad (A \cap B) \cap C = A \cap (B \cap C).$$

Therefore, we may commonly write $A \cup B \cup C$, instead of either $(A \cup B) \cup C$ or $A \cup (B \cup C)$, and similarly for intersection.

The associativity property leads to the ability to define generalized unions or intersection. Consider a set $\mathcal{I}$ and a collection of sets A_i for each $i \in \mathcal{I}$. We sometimes express this collection as $\mathcal{A} = \{A_i\}_{i \in \mathcal{I}}$. Then we define the generalized unions and intersections as

$$\bigcup_{i \in \mathcal{I}} A_i \stackrel{\text{def}}{=} \{x \mid \exists i \in \mathcal{I}, x \in A_i\}, \tag{A.3}$$

$$\bigcap_{i \in \mathcal{I}} A_i \stackrel{\text{def}}{=} \{x \mid \forall i \in \mathcal{I}, x \in A_i\}. \tag{A.4}$$

An interesting example of a generalized union is

$$\bigcup_{i \in \mathbb{N}^*} \left[\frac{1}{i}, 1\right] = (0, 1].$$

A.2.2   Cartesian Product; Power Sets

Definition A.2.4

Let A and B be sets. The *Cartesian product* of A and B, denoted $A \times B$ is the set that consists of ordered pairs (a, b), where $a \in A$ and $b \in B$. Hence,

$$A \times B \overset{\text{def}}{=} \{(a, b) \mid a \in A \text{ and } b \in B\}.$$

If $A_1, A_2, \ldots, A_n$ are n sets, the Cartesian product $A_1 \times A_2 \times \cdots \times A_n$ is the set of ordered n-tuples $(a_1, a_2, \ldots, a_n)$ with $a_i \in A_i$ for each i. Hence,

$$A_1 \times \cdots \times A_n \overset{\text{def}}{=} \{(a_1, \ldots, a_n) \mid a_i \in A_i, \text{ for } i = 1, \ldots, n\}.$$

If we take the Cartesian product of the same set A, we write

$$A^n \overset{\text{def}}{=} \overbrace{A \times A \times \cdots \times A}^{n \text{ times}}.$$

This notation is familiar to us from the notation $\mathbb{R}^2$ (respectively $\mathbb{R}^3$ or $\mathbb{R}^n$) for ordered pairs (resp. triples or n-tuples) of real numbers, which we regularly use to locate points in the Euclidean plane (resp. space or n-space).

The term "product" in the name Cartesian product comes from the fact that if A and B are finite sets with $|A|$ and $|B|$ elements respectively, then

$$|A \times B| = |A| \, |B|.$$

Taking the Cartesian product of two sets is one way of combining sets to create a new set. The concept of a power set is another way to create a new set from an old one.

Definition A.2.5

If S is a set, the *power set* of S, denoted by $\mathcal{P}(S)$ is the set of all subsets of S.

Note that $A \in \mathcal{P}(S)$ is equivalent to writing $A \subseteq S$.

Example A.2.6. Let $S = \{1, 2, 3\}$. Then the power set of S is

$$\mathcal{P}(S) = \{\emptyset, \{1\}, \{2\}, \{3\}, \{1, 2\}, \{1, 3\}, \{2, 3\}, \{1, 2, 3\}\}. \qquad \triangle$$

The terminology of "power" set comes from the following proposition.

Proposition A.2.7

Let n be a nonnegative integer. If S is a set with n elements, then $\mathcal{P}(S)$ has 2^n elements.

A.2.3   Relations and Functions

The everyday notion of a relationship between classes of objects is very general and somewhat amorphous. Mathematics requires a concept as broad as that of a relation but with rigor. Cartesian products offer a simple solution.

Definition A.2.8

A *relation from a set A to a set B* is a subset R of $A \times B$. A *relation on a set A* is a subset of A^2. If $(a, b) \in R$, we often write $a\,R\,b$ and say that "a is related to b via R." The set A is called the *domain* of R and B is the *codomain*.

At first sight, this definition may appear strange. We typically think of a relation as some statement about pairs of objects that is true or false. By gathering together all the true statements about a relation into a subset of the Cartesian product, this definition gives the notion of a relation (in mathematics) the same rigor as sets and as Boolean logic.

Example A.2.9. Let $\mathcal{C}$ be the set of circles in the Euclidean plane $\mathbb{E}^2$ and let $\mathcal{L}$ be the set of lines in $\mathbb{E}^2$. The concept of tangency of lines to circles is a relation from $\mathcal{L}$ to $\mathcal{C}$. According to Definition A.2.8, this relation is a subset R of $\mathcal{L} \times \mathcal{C}$ such that $(L, \Gamma) \in R$ if and only if L is tangent to Γ.

As other examples of relations, parallelism $\|$ is a relation on $\mathcal{L}$ (i.e., a relation from $\mathcal{L}$ to $\mathcal{L}$) as is the notion of perpendicularity, $\perp$. △

Note that the set of relations from a set A to a set B is simply $\mathcal{P}(A \times B)$.

Students of mathematics first encounter functions (defined over the reals) in algebra and explore their properties more deeply in pre-calculus, calculus and analysis. However, mathematical functions find their fullest definition in the context of set theory.

Definition A.2.10

Let A and B be two sets. A *function* f from A to B, written $f : A \to B$, is a relation from A to B such that for all $a \in A$, there exists a unique element of B related to a via f.

As with relations, the set A is called the domain of f and the set B is called the codomain. Unlike the notation for relation, we write $b = f(a)$ if b is the element in B related to $a \in A$ via f. It is not uncommon to say that f *maps* the element a to b. Also, the function is sometimes called a *mapping*, or more briefly, *map* from A to B.

We denote by B^A or alternatively by $\text{Fun}(A, B)$ the set of all functions from A to B. We can express the definition of a function as follows: if A and B are sets, then

$$B^A = \{f \in \mathcal{P}(A \times B) \mid \forall a \in A\, \exists! b \in B\, (a, b) \in f\}.$$

Though the notation $\text{Fun}(A, B)$ might evoke the term "function," the former notation B^A is more common because of the following combinatorial property. If A and B are finite sets with $|A|$ and $|B|$ elements respectively, by the multiplication counting rule, the number of functions from A to B is $|B|^{|A|}$.

Starting in precalculus, we study functions of the form $f : I \to \mathbb{R}$, where I is some interval of real numbers. Sequences of real numbers are also functions $f : \mathbb{N} \to \mathbb{R}$, though by a historical habit, we often write terms of a sequence as f_n instead of $f(n)$.

Definition A.2.11

Let A, B, and C be sets, and let $f : A \to B$ and $g : B \to C$ be two functions. The composition of g with f is the function $g \circ f : A \to C$ defined by for all $x \in A$,

$$(g \circ f)(x) = g(f(x)).$$

The following proposition establishes an identity about iterated composition of functions. Though simple, it undergirds desired algebraic properties for many later situations.

Proposition A.2.12

Let $f : A \to B$, $g : B \to C$, and $h : C \to D$ be functions. Then

$$h \circ (g \circ f) = (h \circ g) \circ f.$$

Proof. Let x be an arbitrary element in A. Then

$$
\begin{aligned}
(h \circ (g \circ f))(x) &= h((g \circ f)(x)) \\
&= h(g(f(x))) \\
&= (h \circ g)(f(x)) \\
&= ((h \circ g) \circ f)(x).
\end{aligned}
$$

Since the functions are equal on all elements of A, the functions are equal. $\square$

Definition A.2.13

We say that a function $f : A \to B$ is

(1) injective (one-to-one) if $f(a_1) = f(a_2) \implies a_1 = a_2$.

(2) surjective (onto) if for all $b \in B$, there exists $a \in A$ such that $f(a) = b$.

(3) bijective (one-to-one and onto), if it is both.

The contrapositive of the definition offers an alternative way to understand injectivity, namely that $a_1 \neq a_2 \implies f(a_1) \neq f(a_2)$.

A.2.4   Binary Operations

Among functions, much of algebra hinges on properties of particular classes of functions, namely binary operations.

Definition A.2.14

A *binary operation* on a set S is a function $\star : S \times S \to S$. We write $a \star b$ instead of $\star(a, b)$ for the image of the pair (a, b).

Binary operations are ubiquitous in mathematics. Some standard examples include $+$, $-$, and $\times$ on $\mathbb{Z}$ or $\mathbb{R}$. given any set $\cup$, $\cap$, $\triangle$ and $\setminus$ are operations on $\mathcal{P}(S)$. Note that division $\div$ is not a binary operator on $\mathbb{R}$ because for example $2 \div 0$ is not well-defined, while it is a binary operator on $\mathbb{R}^* = \mathbb{R} \setminus \{0\}$. Also $\div$ is not a binary operator on $\mathbb{Z}^*$ because for example, $5 \div 2$ is a rational number but not another element of $\mathbb{Z}^*$.

Taking a few examples from linear algebra the addition of vectors in $\mathbb{R}^n$ is a binary operation, as is the cross product of vectors in $\mathbb{R}^3$. In contrast, scalar multiplication is not a binary operation because it is a function $\mathbb{R} \times \mathbb{R}^n \to \mathbb{R}^n$. The dot product $\cdot$ in $\mathbb{R}^n$ is not a binary operation either because it is a function $\mathbb{R}^n \times \mathbb{R}^n \to \mathbb{R}$, where the codomain is not again $\mathbb{R}^n$.

Many algebraic structures are defined as sets equipped with some binary operations that satisfy certain properties. We list a few of the typical properties that figure prominently.

Definition A.2.15

Let S be a set equipped with a binary operation $\star$. We say that the binary operation is

(1) called *associative* if $\forall a, b, c \in S$, $(a \star b) \star c = a \star (b \star c)$;

(2) called *commutative* if $\forall a, b \in S$, $a \star b = b \star a$;

(3) said to have an *identity* if $\exists e \in S \, \forall a \in S, a \star e = e \star a = a$ (e is called an *identity element*);

(4) called *idempotent* if $\forall a \in S$, $a \star a = a$.

It is important to note the order of the quantifiers in (3). For example, let $\star$ be the operation of geometric average on $\mathbb{R}^{>0}$, i.e., $a \star b = \sqrt{ab}$. This operation is commutative and idempotent, but it does not have an identity: if we attempted to solve for b in $a \star b = a$, we would obtain $b = a$. However, because of the order of the quantifiers in the definition of identity, for b to be an identity, it cannot depend on the arbitrary element a.

Proposition A.2.16

Let S be a set equipped with a binary operation $\star$. If $\star$ has an identity element, then $\star$ has a unique identity element.

Proof. Suppose that there exist two identity elements e_1 and e_2 in S. Since e_1 is an identity element, then $e_1 \star e_2 = e_2$. Since e_2 is an identity element, then $e_1 \star e_2 = e_1$ Thus, $e_1 = e_2$. There do not exist two distinct identity elements so S has a unique identity element. □

Because of this proposition, we no longer say "an identity element" but "the identity element."

Definition A.2.17

A binary operation $\star$ on a set S with identity e is said to *have inverses* if $\forall a \in S\, \exists b \in S,\ a \star b = b \star a = e$.

For example, in $\mathbb{R}$ the operation $+$ has inverses because for all $a \in \mathbb{R}$, we have $a + (-a) = (-a) + a = 0$. So $(-a)$ is the (additive) inverse of a. In $\mathbb{R}^*$, the operation $\times$ also has inverses: For all $a \in \mathbb{R}^*$, we have $a \times \frac{1}{a} = \frac{1}{a} \times a = 1$. So $\frac{1}{a}$ is the (multiplicative) inverse of a.

Definition A.2.18

Let S be a set equipped with two binary operations $\star$ and $*$. We say that

(1) $\star$ is *left-distributive over* $*$ if $\forall a, b, c \in S$, $a \star (b * c) = (a \star b) * (a \star c)$;

(2) $\star$ is *right-distributive over* $*$ if $\forall a, b, c \in S$, $(b * c) \star a = (b \star a) * (c \star a)$;

(3) $\star$ is *distributive over* $*$ if $\star$ is both left-distributive and right-distributive over $*$.

The quintessential example for distributivity is that, as binary operations on $\mathbb{R}$, $\times$ is distributive over $+$. However, many other pairs of operations share this property.

The property that $(ab)^c = a^c b^c$, where $a, b, c \in \mathbb{R}^{>0}$ can be stated by saying that the power operation is right-distributive over multiplication (in the positive reals). However, the power operation is not also left-distributive because it is generally not true that $a^{(bc)} = a^b a^c$.

A.3   Equivalence Relations

A.3.1   Equivalence Relations

> **Definition A.3.1**
>
> An *equivalence relation* on a set S is a relation $\sim$ that is
>
> (1) reflexive, i.e. $\forall a \in S,\ a \sim a$;
>
> (2) symmetric, i.e. $\forall a, b \in S,\ a \sim b \longrightarrow b \sim a$; and
>
> (3) transitive, i.e., $\forall a, b, c \in S,\ ((a \sim b) \wedge (b \sim c)) \longrightarrow (a \sim c)$.

Equivalence relations generalize of the concept of equality. Intuitively speaking, an equivalence relation mentally models a notion of sameness or similarity.

Example A.3.2. Let S be any set. The equal relation is reflexive, symmetric, and transitive. In particular, $=$ is an equivalence relation. Two elements are in relation via $=$ if and only if they are the same object. $\triangle$

Example A.3.3. Let $\mathcal{L}$ be the set of lines in the Euclidean plane $\mathbb{E}^2$. Consider the relation of parallelism $\parallel$ on $\mathcal{L}$. Any line is parallel to itself so $\parallel$ is reflexive. If $L_1, L_2 \in \mathcal{L}$ with $L_1 \parallel L_2$, then $L_2 \parallel L_1$. Hence, $\parallel$ is symmetric. Finally, by Proposition I.30 in Euclid's *Elements*, for any lines $L_1, L_2, L_3 \in \mathcal{L}$, if $L_1 \parallel L_2$ and $L_2 \parallel L_3$, then $L_1 \parallel L_3$. This means that $\parallel$ is transitive, establishing that $\parallel$ is an equivalence relation on $\mathcal{L}$. $\triangle$

Example A.3.4. Let $X = \{1, 2, 3, \ldots, 10\}$. Define the relation $\sim_1$ on $\mathcal{P}(X)$ by $A \sim_1 B$ if $|A| = |B|$. It is easy to notice that this is an equivalence relation. We can define another equivalence relation $\sim_2$ on $\mathcal{P}(X)$ by $A \sim_2 B$ whenever their elements have the same sum. $\triangle$

Example A.3.5. Consider the relation $\sim$ on $\mathbb{N} \times \mathbb{N}$ defined by $(a_1, a_2) \sim (b_1, b_2)$ whenever

$$a_1 + b_2 = a_2 + b_1.$$

For all $(a_1, a_2) \in N^2$, we have $a_1 + a_2 = a_2 + a_1$ so $\sim$ is reflexive. If $a_1 + b_2 = a_2 + b_1$, then $b_1 + a_2 = b_2 + a_1$; in other words, $(a_1, a_2) \sim (b_1, b_2)$ implies that $(b_1, b_2) \sim (a_1, a_2)$. Thus $\sim$ is symmetric. Finally, to prove transitivity, suppose that $(a_1, a_2) \sim (b_1, b_2)$ and $(b_1, b_2) \sim (c_1, c_2)$. Then

$$a_1 + b_2 = a_2 + b_1 \quad \text{and} \quad b_1 + c_2 = b_2 + c_1. \tag{A.5}$$

Adding c_2, to the first equality in (A.5) gives $a_1 + b_2 + c_2 = a_2 + b_1 + c_2$. Using the second in (A.5) in the right side of this gives us $a_1 + b_2 + c_2 = a_2 + b_2 + c_1$. From the cancellation law for $+$ on $\mathbb{N}$, we deduce that $a_1 + c_2 = a_2 + c_1$. Hence, $(a_1, a_2) \sim (c_1, c_2)$.

Together, these show that $\sim$ is an equivalence relation on $\mathbb{N} \times \mathbb{N}$. $\triangle$

A.3.2 Equivalence Classes

Since an equivalence relation furnishes some notion of sameness, it is natural to gather similar elements into classes. Such classes formalize the sameness property.

> **Definition A.3.6**
>
> Let $\sim$ be an equivalence relation on a set S. For $a \in S$, the *equivalence class of a* is
> $$[a] \overset{\text{def}}{=} \{s \in S \mid s \sim a\}.$$
> If a subset $U \subseteq S$ satisfies $U = [a]$ for some $a \in S$, then U is called an *equivalence class* and a is called a *representative* of U.

We sometimes write $[a]_\sim$ to clarify if a certain context considers more than one equivalence relation at a time.

> **Definition A.3.7**
>
> Let S be a set and $\sim$ an equivalence relation on S. The set of $\sim$-equivalence classes on S is called the *quotient set of S by $\sim$* and is denoted by $S/\sim$.

Quotient sets occur often in mathematics. In this textbook, quotient groups and quotient rings play important roles. The connection between how the quotient set process interacts with some other structure on the set plays an important role in many other areas, including linear algebra, abstract algebra, topology, geometry, and analysis.

We call a subset T of S a *complete set of distinct representatives* of $\sim$ if any equivalence class U has $U = [a]$ for some $a \in T$ and for any two $a_1, a_2 \in T$, $[a_1] = [a_2]$ implies that $a_1 = a_2$. That every equivalence relation has a complete set of distinct representatives follows from the Axiom of Choice. We remark that there is a bijection between $S/\sim$ and a complete set of representatives T of $\sim$ via the function

$$\psi : T \to S/\sim$$
$$a \mapsto [a].$$

However, we do not consider these sets as equal since their objects are different.

Example A.3.8. The concept of negative integers appeared millennia after positive integers and even centuries after 0. Revisit the set $\mathbb{N} \times \mathbb{N}$ and the equivalence relation $\sim$ in Example A.3.5. Consider the quotient set $(\mathbb{N} \times \mathbb{N})/\sim$. Elements are equivalence classes $[(a, b)]$ with $a, b \in \mathbb{N}$. By considering various cases, we can find that $[(a, b)]$ accurately models a notion of "displacement" from a to b. In particular, it is possible to prove that every equivalence class has a representative of the form $(n, 0)$, $(0, 0)$, or $(0, n)$ for some positive integer

n, and also that none of these representatives are equivalent to each other (assuming different n). The quotient set $(\mathbb{N} \times \mathbb{N})/ \sim$ gives a set theoretic *definition* for $\mathbb{Z}$ from $\mathbb{N}$, where we identify the classes $[(n,0)]$, $[(0,0)]$, and $[(0,n)]$ as $-n$, 0, and n. $\triangle$

Example A.3.9 (Projective Space). Let $\mathcal{L}(\mathbb{R}^3)$ be the set of lines in $\mathbb{R}^3$ and consider the equivalence relation of parallelism $\parallel$ on $\mathcal{L}(\mathbb{R}^3)$. If L is a line in $\mathcal{L}(\mathbb{R}^3)$, then $[L]$ is the set of all lines parallel to L. The set $\mathcal{L}(\mathbb{R}^3)/ \parallel$ is called the *projective plane* and is denoted as $\mathbb{RP}^2$. It consists of all the directions lines can possess.

There are other ways to understand the projective plane. Every line in $\mathbb{R}^3$ is parallel to a unique line through the origin. Possible direction vectors for lines through the origin consist simply of nonzero vectors, so we consider the set $\mathbb{R}^3 \setminus \{(0,0,0)\}$. Lines through the origin are the same if and only if their given direction vector differs by a nonzero multiple. Hence, we define the equivalence relation $\sim$ on $\mathbb{R}^3 \setminus \{(0,0,0)\}$ by

$$\vec{a} \sim \vec{b} \Longleftrightarrow \vec{a} = \lambda \vec{b}, \text{ for some } \lambda \in \mathbb{R}^*.$$

Our comments on direction vectors show that $(\mathbb{R}^3 \setminus \{(0,0,0)\})/\sim = \mathbb{RP}^2$. $\triangle$

Example A.3.10 (Rational Numbers). It is surprising and interesting that the rational numbers are created from $\mathbb{Z}$ in a similar fashion that we created $\mathbb{Z}$ from $\mathbb{N}$ as in Example A.3.8. Consider the set of pairs $S = \mathbb{Z} \times \mathbb{Z}^*$ and the relation $\sim$ defined by

$$(a, b) \sim (c, d) \Longleftrightarrow ad = bc.$$

We leave it to the reader to show that this is an equivalence relation. (See Exercise A.3.18.) The quotient set $S/ \sim$ is a rigorous definition for $\mathbb{Q}$. The equivalence relation is precisely the condition that is given when two fractions are considered equal. Hence, the fraction notation $\frac{a}{b}$ for rational numbers represents the equivalence class $[(a,b)]_\sim$. Note that our habit of writing fractions in reduced form with a positive denominator is tantamount to selecting a complete set of distinct representatives. $\triangle$

Remark A.3.11. When working with functions whose domains are quotient sets, it is natural to wish to define a function of an equivalence class based on a representative of the class. More precisely, if S and U are sets and $\sim$ is an equivalence relation on S, we may wish to define a function

$$\begin{aligned} F : S/\sim &\longrightarrow U \\ [a] &\longmapsto f(a) \end{aligned} \tag{A.6}$$

where $f : S \to U$ is some function. This construction does not always produce a function. We say that a function defined according to (A.6) is *well-defined* if whenever $a \sim b$, then $f(a) = f(b)$. Thus, any representative of the equivalence of $[a]$ to define F will return the same value for $F([a])$. $\triangle$

As an example to illustrate the above remark, consider the equivalence relation $\sim$ on $\mathbb{Z} \times \mathbb{Z}^*$ described in Example A.3.10 to define $\mathbb{Q}$. Suppose that we attempted to define a function $f : \mathbb{Q} \to \mathbb{Z}$ by $f(\frac{a}{b}) = a + b$. This is not a well-defined function (hence, is not a function at all) because for example $\frac{1}{2} = \frac{3}{6}$ whereas $1 + 2 = 3$ and $3 + 6 = 9$. Depending on the representative chosen for the fraction $\frac{a}{b}$, the value of $a + b$ may differ.

A.3.3 Partitions

Let $\sim$ be an equivalence relation on a set S. For any two elements in $a, b \in S$, by definition $a \in [b]$ if and only if $a \sim b$. However, since $\sim$ is symmetric, this implies that $b \sim a$ and hence that $b \in [a]$. By transitivity, if $a \in [b]$, then $s \sim a$ implies that $s \sim b$, so $a \in [b]$ implies that $[a] \subseteq [b]$. Consequently, we have proven that the following statements are logically equivalent:

$$a \in [b] \iff [a] \subseteq [b] \iff b \in [a] \iff [b] \subseteq [a] \iff [a] = [b].$$

This observation leads to the following important proposition.

Proposition A.3.12

Let S be a set equipped with an equivalence relation $\sim$. Then
(1) distinct equivalence classes are disjoint;
(2) the union of distinct equivalence classes is all of S.

This property of equivalence classes described in Proposition A.3.12 has a particular name in set theory.

Definition A.3.13

Let S be a set. A collection $\mathcal{A} = \{A_i\}_{i \in \mathcal{I}}$ of nonempty subsets of S is called a *partition* of S if
(1) $A_i \cap A_j \neq \emptyset \implies i = j$ and
(2) $\displaystyle\bigcup_{i \in \mathcal{I}} A_i = S$.

Example A.3.14. Consider the set $\mathbb{Z}$. The notion of parity corresponds to a partition $\mathcal{P}_1 = \{\text{odds}, \text{evens}\}$. Another commonly used partition on $\mathbb{Z}$ is that of parity

$$\mathcal{P}_2 = \{\{1, 2, 3, \ldots\}, \{0\}, \{-1, -2, -3, \ldots\}\}.$$

(Note that parity consists of 3 subsets of $\mathbb{Z}$.) Though we can imagine many partitions on $\mathbb{Z}$, here is another $\mathcal{P}_3 = \{\{-a, a\} \mid a \in \mathbb{N}\}$. $\triangle$

The concept of a partition simply models the mental construction of subdividing a set into parts without losing any elements of the set and without any parts overlapping. Partitions and equivalence relations are closely connected. Proposition A.3.12 establishes that the quotient set of an equivalence relation is a partition. The following proposition establishes the converse.

Proposition A.3.15

Let $\mathcal{A} = \{A_i\}_{i \in \mathcal{I}}$ be a partition of a set S. Define the relation $\sim_{\mathcal{A}}$ on S by

$$a \sim_{\mathcal{A}} b \Longrightarrow \exists i \in \mathcal{I} \ \text{ with } a \in A_i \text{ and } b \in A_i.$$

Then $\sim_{\mathcal{A}}$ is an equivalence relation. Furthermore, the sets in $\mathcal{A}$ are the distinct equivalence classes of $\sim_{\mathcal{A}}$.

Proof. Let $a \in S$ be arbitrary. Since $\mathcal{A}$ is a partition, then $a \in A_i$ for some $i \in \mathcal{I}$. Hence, $a \sim_{\mathcal{A}} a$ and $\sim_{\mathcal{A}}$ is reflexive.

Suppose that $a \sim_{\mathcal{A}} b$. Then for some $i \in \mathcal{I}$, we have $a \in A_i$ and $b \in A_i$. Obviously, this implies that $b \sim_{\mathcal{A}} a$, showing that $\sim_{\mathcal{A}}$ is symmetric.

Suppose that $a \sim_{\mathcal{A}} b$ and $b \sim_{\mathcal{A}} c$. Then for some $i \in \mathcal{I}$, we have $a \in A_i$ and $b \in A_i$ and for some $j \in \mathcal{I}$, we have $b \in A_j$ and $c \in A_j$. However, since $b \in A_i \cap A_j$, then $A_i \cap A_j \neq \emptyset$. By definition of a partition, $i = j$ and so $a \in A_i$ and $c \in A_i$ and thus $a \sim_{\mathcal{A}} c$. This shows transitivity and establishes that $\sim_{\mathcal{A}}$ is an equivalence relation.

Let A_i be a set in $\mathcal{A}$ and let s be any element in A_i. By construction, $[s] = A_i$ and so the elements of $\mathcal{A}$ are the equivalence classes of $\sim_{\mathcal{A}}$. $\qquad\square$

EXERCISES FOR SECTION A.3

For Exercises A.3.1 through A.3.13, prove or disprove whether the described relation on the given set is an equivalence relation. If the relation is not an equivalence relation, determine which properties it lacks.

1. Let P be the set of living people. For all $a, b \in P$, define the relation $a \, R \, b$ if a and b have met.

2. Let P be the set of living people. For all $a, b \in P$, define the relation $a \, R \, b$ if a and b live in a common town.

3. Let $\mathcal{C}$ be the set of circles in $\mathbb{R}^2$ and let R be the relation of concentric on $\mathcal{C}$.

4. Let $S = \mathbb{Z} \times \mathbb{Z}$ and define the relation R by $(m_1, m_2) \, R \, (n_1, n_2)$ if $m_1 m_2 = n_1 n_2$.

5. Let $S = \mathbb{Z} \times \mathbb{Z}$ and define the relation R by $(m_1, m_2) \, R \, (n_1, n_2)$ if $m_1 n_1 = m_2 n_2$.

6. Let $S = \mathbb{Z} \times \mathbb{Z}$ and define the relation R by $(m_1, m_2) \, R \, (n_1, n_2)$ if $m_1 n_2 = m_2 n_1$.

7. Let P_3 be the set of polynomials with real coefficients and of degree 3 or less. Define the relation R by $p(x) \, R \, q(x)$ to mean that $q(x) - p(x)$ has 5 as a root.

8. Consider the set $C^0(\mathbb{R})$ of continuous functions over $\mathbb{R}$. Define the relation R by $f \, R \, g$ if there exist some $a, b \in \mathbb{R}$ such that

$$g(x) = f(x + a) + b \qquad \text{for all } x \in \mathbb{R}.$$

9. Let $\mathcal{P}_{\text{fin}}(\mathbb{R})$ be the set of finite subsets of $\mathbb{R}$ and define the relation $\sim$ on $\mathcal{P}_{\text{fin}}(\mathbb{R})$ by $A \sim B$ if the sum of elements in A is equal to the sum of elements in B. Prove that $\sim$ is an equivalence relation.

10. Let $\ell^{\infty}(\mathbb{R})$ be the set of sequences of real numbers. Define the relation R on $\ell^{\infty}(\mathbb{R})$ by $(a_n) R (b_n)$ if

$$\lim_{n \to \infty} (b_n - a_n) = 0.$$

11. Let $\ell^{\infty}(\mathbb{R})$ be the set of sequences of real numbers. Define the relation R on $\ell^{\infty}(\mathbb{R})$ by $(a_n) R (b_n)$ if the sequence $(a_n + b_n)_{n=1}^{\infty}$ converges.

12. Let S be the set of lines in $\mathbb{R}^2$ and let R be the relation of perpendicular.

13. Let $\mathcal{W}$ be the words in the English language (i.e., have an entry in the *Oxford English Dictionary*). Define the relation R on $\mathcal{W}$ by $w_1 R w_2$ is w_1 comes before w_2 in alphabetical order.

14. Let $C^0([0,1])$ be the set of continuous real-valued functions on $[0,1]$. Define the relation $\sim$ on $C^0([0,1])$ by

$$f \sim g \iff \int_0^1 f(x)\, dx = \int_0^1 g(x)\, dx.$$

Show that $\sim$ is an equivalence relation and describe (with a precise rule) a complete set of distinct representatives of $\sim$.

15. Let $C^{\infty}(\mathbb{R})$ be the set of all real-value functions on $\mathbb{R}$ such that all its derivatives exist and are continuous. Define the relation R on $C^{\infty}(\mathbb{R})$ by $f R g$ if $f^{(n)}(0) = g^{(n)}(0)$ for all positive, even integers n.

   (a) Prove that R is an equivalence relation.

   (b) Describe concisely all the elements in the equivalence class $[\sin x]$.

16. Let $S = \{1, 2, 3, 4\}$ and the relation $\sim$ on $\mathcal{P}(S)$, defined by $A \sim B$ if and only if the sum of elements in A is equal to the sum of elements in B, is an equivalence relation. List the equivalence classes of $\sim$.

17. Let $\mathcal{T}$ be the set of (nondegenerate) triangles in the plane.

   (a) Prove that the relation $\sim$ of similarity on triangles in $\mathcal{T}$ is an equivalence relation.

   (b) Concisely describe a complete set of distinct representatives of $\sim$.

18. Prove that the relation defined in Example A.3.10 is an equivalence relation.

19. Let $S = \{1, 2, 3, 4, 5, 6\}$. For the partitions of S given below, write out the equivalence relation as a subset of $S \times S$.

   (a) $\{\{1, 2\}, \{3, 4\}, \{5, 6\}\}$

   (b) $\{\{1\}, \{2\}, \{3, 4, 5, 6\}\}$

   (c) $\{\{1, 2\}, \{3\}, \{4, 5\}, \{6\}\}$

20. Let $C^1([a,b])$ be the set of continuously differentiable functions on the interval $[a, b]$. Define the relation $\sim$ on $C^1([a,b])$ as $f \sim g$ if and only if $f'(x) = g'(x)$ for all $x \in (a, b)$. Prove that $\sim$ is an equivalence relation on $C^1([a,b])$. Describe the elements in the equivalence class for a given $f \in C^1([a,b])$.

21. Let $M_{n \times n}(\mathbb{R})$ be the set of $n \times n$ matrices with real coefficients. For two matrices $A, B \in M_{n \times n}(\mathbb{R})$, we say that B is similar to A if there exists an invertible $n \times n$ matrix S such that $B = SAS^{-1}$.

(a) Prove that similarity $\sim$ is an equivalence relation on $M_{n \times n}(\mathbb{R})$.

(b) Prove that $\det(SAS^{-1}) = \det(A)$ for all invertible matrices S. Deduce that the function $f : (M_{n \times n}(\mathbb{R})/\sim) \to \mathbb{R}$ defined by $f([A]) = \det A$ is a well-defined function from the set of similarity classes of matrices to the reals.

(c) Determine with a proof or counterexample whether the function $g : (M_{n \times n}(\mathbb{R})/\sim) \to \mathbb{R}$ defined by $g([A]) = \operatorname{Tr} A$, the trace of A, is a well-defined function.

22. Define the relation $\sim$ on $\mathbb{R}$ by $a \sim b$ if and only if $b - a \in \mathbb{Q}$. Prove that for all real $x \in \mathbb{R}$, there exists $y \in [x]_\sim$ that is arbitrarily close to x. (In other words, for all $\varepsilon > 0$, there exists y with $y \sim x$ and $|x - y| < \varepsilon$.)

23. Let R_1 and R_2 be equivalence relations on a set S. Determine (with a proof or counterexample) which of the following relations are also equivalence relations on S. (a) $R_1 \cap R_2$; (b) $R_1 \cup R_2$; (c) $R_1 \triangle R_2$. [Note that $R_1 \cup R_2$, and similarly for the others, is a relation as a subset of $S \times S$.]

24. Which of the following collections of subsets of the integers form partitions? If it is not a partition, explain which properties fail.

(a) $\{p\mathbb{Z} \,|\, p \text{ is prime}\}$, where $k\mathbb{Z}$ means all the multiples of k.

(b) $\{\{3n, 3n+1, 3n+2\} \,|\, n \in \mathbb{Z}\}$.

(c) $\{\{k \,|\, n^2 \leq k \leq (n+1)^2\} \,|\, n \in \mathbb{N}\}$.

(d) $\{\{n, -n\} \,|\, n \in \mathbb{N}\}$.

25. Consider the relation $\sim$ on $\mathbb{R}$ defined by $x \sim y$ if $y - x \in \mathbb{Z}$.

(a) Prove that $\sim$ is an equivalence relation.

(b) Prove that if $a \sim b$ and $c \sim d$, then $(a + c) \sim (b + d)$.

(c) Decide with a proof or counterexample whether $ac \sim bd$, whenever $a \sim b$ and $c \sim d$.

26. Let S be a set and let $\mathcal{A} = \{A_i\}_{i \in \mathcal{I}}$ be a partition of S. Another partition $\mathcal{B} = \{B_j\}_{j \in \mathcal{J}}$ is called a *refinement* of $\mathcal{A}$ if

$$\forall j \in \mathcal{J}, \ \exists i \in \mathcal{I}, \quad B_j \subseteq A_i.$$

Let $\mathcal{A}$ and $\mathcal{B}$ be two partitions of a set S and let $\sim_\mathcal{A}$ (resp. $\sim_\mathcal{B}$) as the equivalence relation corresponding to $\mathcal{A}$ (resp. $\mathcal{B}$). Prove that $\mathcal{B}$ is a refinement of $\mathcal{A}$ if and only if $s_1 \sim_\mathcal{B} s_2 \implies s_1 \sim_\mathcal{A} s_2$.

27. Let S be a set and let $\mathcal{A} = \{A_i\}_{i \in \mathcal{I}}$ and $\mathcal{B} = \{B_j\}_{j \in \mathcal{J}}$ be two partitions of S. Prove that the collection of sets

$$\{A_i \cap B_j \,|\, i \in \mathcal{I} \text{ and } j \in \mathcal{J}\} \setminus \{\emptyset\}$$

is a partition of S that is a refinement of both $\mathcal{A}$ and $\mathcal{B}$.

A.4 Partial Orders

A.4.1 Definition and First Examples

Section A.3.1 introduced equivalence relations as a generalization of the notion of equality. Equivalence relations provide a mental model for calling certain objects in a set equivalent. Similarly, the concept of a partial order generalizes the inequality $\leq$ on $\mathbb{R}$ to a mental model of ordering objects in a set.

> **Definition A.4.1**
>
> A *partial order* on a set S is a relation $\preccurlyeq$ that is
>
> (1) reflexive, i.e., $\forall a \in S$, $a \preccurlyeq a$;
>
> (2) antisymmetric, i.e., $\forall a, b \in S$, $((a \preccurlyeq b) \wedge (b \preccurlyeq a) \rightarrow a = b)$; and
>
> (3) transitive, i.e., $\forall a, b, c \in S, ((a \preccurlyeq b) \wedge (b \preccurlyeq c)) \rightarrow (a \preccurlyeq c)$.
>
> A pair $(S, \preccurlyeq)$, where S is a set and $\preccurlyeq$ is a partial order on S is often succinctly called a *poset*.

The name *poset* abbreviates "**partially ordered set**." Motivated by the notations for inequalities over $\mathbb{R}$, we use the symbol $\prec$ to mean

$$x \prec y \Longleftrightarrow x \preccurlyeq y \text{ and } x \neq y$$

and the symbol $\npreceq$ to mean that it is not true that $x \preccurlyeq y$.

Example A.4.2. Consider the relation $\leq$ on $\mathbb{R}$. For all $x \in \mathbb{R}$, $x \leq x$ so $\leq$ is reflexive. For all $x, y \in \mathbb{R}$, if $x \leq y$ and $y \leq x$, then $x = y$ and hence $\leq$ is antisymmetric. It is also true that $x \leq y$ and $y \leq z$ implies that $x \leq z$ and hence $\leq$ is transitive. Thus, the inequality $\leq$ on $\mathbb{R}$ is a partial order. $\triangle$

Note that $\geq$ is also a partial order on $\mathbb{R}$ but that $<$ and $>$ are not. The strict inequality $<$ is not reflexive though it is both antisymmetric and transitive. ($<$ is antisymmetric vacuously: because there do not exist any $x, y \in \mathbb{R}$ such that $x < y$ and $y < x$, the conditional statement "$x < y$ and $y < x$ implies $x = y$" is trivially satisfied.)

Though modeled after the relation of $\leq$, a partial order on a set S has a number of additional possibilities. For example, in a general poset $(S, \preccurlyeq)$, given two arbitrary elements $a, b \in S$, it is possible that neither $a \preccurlyeq b$ nor $b \preccurlyeq a$.

> **Definition A.4.3**
>
> Let $(S, \preccurlyeq)$ be a poset. If for some pair $\{a, b\}$ of distinct elements, either $a \preccurlyeq b$ or $b \preccurlyeq a$, then we say that a and b are *comparable*; otherwise a and b are called *incomparable*. A partial order in which every pair of elements is comparable is called a *total order*.

The posets $(\mathbb{N}, \leq)$ and $(\mathbb{R}, \leq)$ are total orders. Many posets are not total orders as the following examples illustrate.

Example A.4.4. Consider the donor relation $\rightarrow$ defined on the set of blood types $B = \{\mathsf{o}, \mathsf{a}, \mathsf{b}, \mathsf{ab}\}$ where $t_1 \rightarrow t_2$ if and only if t_1 can (healthily) donate to t_2. As a subset of $B \times B$, this relation is

$$\{(\mathsf{o},\mathsf{o}), (\mathsf{o},\mathsf{a}), (\mathsf{o},\mathsf{b}), (\mathsf{o},\mathsf{ab}), (\mathsf{a},\mathsf{a}), (\mathsf{a},\mathsf{ab}), (\mathsf{b},\mathsf{b}), (\mathsf{b},\mathsf{ab}), (\mathsf{ab},\mathsf{ab})\}.$$

It is not hard to check that $\rightarrow$ is reflexive, antisymmetric, and transitive, showing that $(B, \rightarrow)$ is a poset. Note that a and b are not comparable, meaning that neither can donate to the other. $\triangle$

Example A.4.5. Let S be any set. The subset relation $\subseteq$ on $\mathcal{P}(S)$ is a partial order. In the partial order, many pairs of subsets in S are incomparable. In fact, two subsets A and B are incomparable if and only if $A \setminus B$ and $B \setminus A$ are both nonempty. $\triangle$

Example A.4.6. Consider the relation $\preccurlyeq$ on $\mathbb{R}^2$ defined by

$$(x_1, y_1) \preccurlyeq (x_2, y_2) \iff 2x_1 - y_1 < 2x_2 - y_2 \text{ or } (x_1, y_1) = (x_2, y_2).$$

That $(x_1, y_1) \preccurlyeq (x_1, y_1)$ is built into the definition so $\preccurlyeq$ is reflexive. It is impossible for $2x_1 - y_1 < 2x_2 - y_2$ and $2x_2 - y_2 \leq 2x_1 - y_1$ so the only way $(x_1, y_1) \preccurlyeq (x_2, y_2)$ and $(x_2, y_2) \preccurlyeq (x_1, y_1)$ can occur is if $(x_1, y_1) = (x_2, y_2)$. Finally, the relation is also transitive so that $\preccurlyeq$ is a partial order on $\mathbb{R}^2$.

In this poset on $\mathbb{R}^2$, two elements (x_1, y_1) and (x_2, y_2) are incomparable if and only if $2x_2 - y_2 = 2x_1 - y_1$ and $(x_1, y_1) \neq (x_2, y_2)$, namely they are distinct points on the same line of slope 2. $\triangle$

Besides the distinction between totally ordered posets and partial orders that are not total, another dichotomy arises when comparing properties of the posets $(\mathbb{N}, \leq)$ and $(\mathbb{R}, \leq)$. In $(\mathbb{R}, \leq)$, given any $x \leq y$ with $x \neq y$, there always exists an element z such that $x < z < y$; in contrast, in $(\mathbb{N}, \leq)$, for example $2 \leq 3$ but for all $z \in \mathbb{N}$, if $2 \leq z \leq 3$, then $z = 2$ or $z = 3$.

Definition A.4.7

Let $(S, \preccurlyeq)$ be a poset and let $x \in S$. We call $y \in S$ an *immediate successor* (resp. *immediate predecessor*) to x if $y \neq x$ with $x \preccurlyeq y$ (resp. $y \preccurlyeq x$) and for all $z \in S$ such that $x \preccurlyeq z \preccurlyeq y$ (resp. $y \preccurlyeq z \preccurlyeq x$), either $z = x$ or $z = y$.

In $(\mathbb{N}, \leq)$ all elements have both immediate successors and immediate predecessors, except for 0 that does not have an immediate predecessor. In $(\mathbb{Z}, \leq)$ all elements have both immediate successors and immediate predecessors. In contrast, as commented above, in $(\mathbb{R}, \leq)$ no element has either an immediate successor or an immediate predecessor. As a more interesting example, in the poset

$$\left\{ \frac{1}{n} \;\middle|\; n \in \mathbb{N}^* \right\} \cup \{0\}$$

equipped with $\leq$ as a subset of $\mathbb{Q}$, the element 0 has no immediate successor but every other element does.

A partial order does not have to be a total order to have immediate successors or predecessors. In the blood donor relation $(B, \rightarrow)$ in Example A.4.4, o has two immediate successors, namely a and b.

A.4.2 Subposets

If $(S, \preccurlyeq)$ is a poset and T any subset of S, then when we restrict $\preccurlyeq$ only to elements of T, the quantifiers in the definition of a partial order still hold when restricted to T. Hence $(T, \preccurlyeq)$ is also a poset. We sometimes call $(T, \preccurlyeq)$ a *subposet* of $(S, \preccurlyeq)$.

Though a generic poset $(S, \preccurlyeq)$ need not be a total order, many of the familiar terms associated with inequalities $\mathbb{R}$ have corresponding definitions in any poset.

Definition A.4.8

Let $(S, \preccurlyeq)$ be a poset, and let A be a subset of S.

(1) A *maximal* element of A is an $M \in A$ such that if $t \in A$ with $M \preccurlyeq t$, then $t = M$.

(2) A *minimal* element of A is an $m \in A$ such that if $t \in A$ with $t \preccurlyeq m$, then $t = m$.

As an example, consider the blood donor relation $(B, \rightarrow)$ described in Example A.4.4 and consider the subset $A = \{o, a, b\}$. Then A has one minimal element o and two maximal elements a and b.

Definition A.4.9

Let $(S, \preccurlyeq)$ be a poset, and let A be a subset of S.

(1) An *upper bound* of A is an element $u \in S$ such that $\forall t \in A, t \preccurlyeq u$.

(2) A *lower bound* of A is an element $\ell \in S$ such that $\forall t \in A, \ell \preccurlyeq t$.

(3) A *least upper bound* of A is an upper bound u of A such that for all upper bounds u' of A, we have $u \preccurlyeq u'$.

(4) A *greatest lower bound* of A is a lower bound ℓ of A such that for all lower bounds ℓ' of A, we have $\ell' \preccurlyeq \ell$.

We say that a subset $A \subseteq S$ is *bounded above* if A has an upper bound, is *bounded below* if A has a lower bound, and is *bounded* if A is bounded above and bounded below.

If u_1 and u_2 are two least upper bounds to A, then by definition $u_1 \preccurlyeq u_2$ and $u_2 \preccurlyeq u_1$. Thus, $u_1 = u_2$ and we conclude that least upper bounds are unique. Similarly, greatest lower bounds for a set are unique. Therefore, if a subset A has a least upper bound, we talk about *the* least upper bound of A and denote this element by $\mathrm{lub}(A)$. Similarly, if a subset A has a greatest lower bound, we talk about *the* greatest lower bound of A and denote this element by $\mathrm{glb}(A)$.

From the perspective of analysis, one of the most important differences between the posets $(\mathbb{R}, \leq)$ and its subposet $(\mathbb{Q}, \leq)$ is that any bounded subset of $\mathbb{R}$ has a least upper bound whereas this does not hold in $\mathbb{Q}$. Consider for example, the subset

$$A = \left\{ \frac{p}{q} \in \mathbb{Q} \,|\, p^2 < 2q^2 \right\}.$$

In $\mathbb{R}$, $\mathrm{lub}(A) = \sqrt{2}$ whereas in $\mathbb{Q}$ for any upper bound $u = \frac{r}{s}$ of A we have

$$\sqrt{2} < \frac{1}{2} \left(\frac{r}{s} + \frac{2s}{r} \right) < \frac{r}{s}.$$

(We leave the proof to the reader.) Hence, A has no least upper bound.

Example A.4.10. Let S be any set and consider the power set $\mathcal{P}(S)$ equipped with the $\subseteq$ partial order. Let X be a subset of $\mathcal{P}(S)$. Then a maximal element of X is a set M in X such that no other set in X contains M. Note that there may be more than one of these. An upper bound of X is any subset of S that contains every element in every set in X. The least upper bound of X is the union

$$\mathrm{lub}(X) = \bigcup_{A \in X} A.$$

$\triangle$

Definition A.4.11

In a poset $(S, \preccurlyeq)$ any subposet $(T, \preccurlyeq)$ that is a total order is called a *chain*.

The concept of a chain allows us to introduce an important theorem that is essential in a variety of contexts in algebra.

Theorem A.4.12 (Zorn's Lemma)

Let $(S, \preccurlyeq)$ be a poset. Suppose that every chain in S has an upper bound. Then S contains a maximal element.

In the context of **ZF**-set theory, Zorn's Lemma is equivalent to the Axiom of Choice. (See [31, Theorem 5.13.1] for a proof.)

Definition A.4.13

A poset $(S, \preccurlyeq)$ is called a *lattice* if for all pairs $(a, b) \in S \times S$, both $\text{lub}(a, b)$ and $\text{glb}(a, b)$ exist.

Lattices are a particularly nice class of partially order sets. They occur frequently in various areas of mathematics. Given any set S, the power set $(\mathcal{P}(S), \subseteq)$ is a lattice with $\text{lub}(A, B) = A \cup B$ and $\text{glb}(A, B) = A \cap B$. In Section 1.8 we show how to utilize the lattice structure on the set of subgroups of a group effectively to quickly answer questions about the internal structure of a group.

A.4.3 Hasse Diagrams

For partially ordered sets with a relatively small number of elements it is possible to easily visualize the relation via a *Hasse diagram*.

Let $(S, \preccurlyeq)$ be a poset in which S is finite. In a Hasse diagram, each element of S corresponds to a point in the plane, with the points placed on the page so that if $a \preccurlyeq b$, then b appears higher on the page. The points of the Hasse diagram are also called *nodes* or *vertices*. Finally, we draw an edge between two points (corresponding to) a and b with b above a if b is an immediate successor of a.

As a first example, Figure A.2 gives the Hasse diagram for the donor relation described in Example A.4.4.

FIGURE A.2: The Hasse diagram for the donor relation on blood types.

By reflexivity, we know that $a \preccurlyeq a$ for all $a \in S$, nevertheless we do not draw a loop at each vertex. Because of transitivity, $p \preccurlyeq q$ if and only if there is a (rising) path through the Hasse diagram from p to q. For example, in Figure A.2, the diagram does not show an explicit edge between o and ab but we can see that o $\rightarrow$ ab because there is a rising path in the diagram from o to ab.

Example A.4.14. Consider the partial order on $S = \{a, b, c, d, e, f, g, h, i\}$ described by the Hasse diagram shown in Figure A.3. The diagram makes it clear what relations hold between elements. For example, notice that all elements in $\{a, b, c, d, e, f, g\}$ are incomparable with the elements in $\{h, i\}$. The maximal elements in S are d and i. The minimal elements are a, e, g, and h. As a least upper bound calculation, $\text{lub}(a, f) = c$ because c is the first element in a chain above a that is also in a chain above f. We also see that $\text{lub}(e, h)$ does not exist because there is no chain above e that intersects with a chain above h. $\triangle$

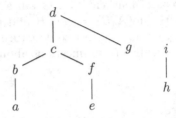

FIGURE A.3: An example of a poset defined by a Hasse diagram.

Hasse diagrams allow for easy visualization of properties of the poset. For example, a poset will be a lattice if and only if taking any two points p_1 and p_2 in the diagram, there exists a chain rising from p_1 that intersects with a chain rising from p_2 (existence of the least upper bound) and a chain descending from p_1 that intersects with a chain descending from p_2 (existence of the greatest upper bound).

Figure A.4 illustrates three different lattices. The reader is encouraged to notice that, the third Hasse diagram corresponds to the partial order $\subseteq$ on $\mathcal{P}(\{1, 2, 3\})$. (See Figure A.5.)

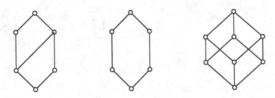

FIGURE A.4: Lattices.

A.4.4 Lexicographic Order

The term lexicographic order constructs an partial order on the Cartesian product of a finite collection of posets. The terminology comes from how we order words in the dictionary: When comparing two words w_1 and w_2, we put w_1 earlier in the dictionary if the first time a letter of w_1 differ from those of w_2, that letter in w_1 is earlier in the alphabet than the corresponding letter in w_2.

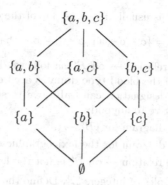

FIGURE A.5: Lattice of $(\mathcal{P}(\{a, b, c\}), \subseteq)$.

Definition A.4.15

Let $(A_1, \preccurlyeq_1), \ldots, (A_n, \preccurlyeq_n)$ be posets. The *lexicographic order* $\preccurlyeq_{\text{lex}}$ on $A_1 \times \cdots \times A_n$ is the reflexive relation that satisfies

$$(a_1, \ldots, a_n) \preccurlyeq_{\text{lex}} (b_1, \ldots, b_n)$$
$$\iff a_k \preccurlyeq b_k \text{ where } k = \min(i \in \{1, \ldots, n\} \mid a_i \neq b_i).$$

Exercises for Section A.4

1. Let $S = \{a, b, c, d, e\}$. In the following relations on S determine with explanation whether or not the relation is a partial order. If it fails antisymmetry, then remove a least number of pairs; if it fails transitivity, then add some pairs to make the relation a partial order.

   (a) $R = \{(a, a), (b, b), (c, c), (d, d), (e, e), (a, c)\}$

   (b) $R = \{(a, a), (b, b), (c, c), (d, d), (e, e), (a, c), (a, d)\}$

   (c) $R = \{(a, a), (b, b), (c, c), (d, d), (e, e), (a, c), (d, a)\}$

   (d) $R = \{(a, a), (b, b), (c, c), (d, d), (e, e), (b, c), (c, d), (d, e), (a, e)\}$

2. Let R_1 and R_2 be partial orders on a set S.

   (a) Prove that $R_1 \cap R_2$ is a partial order.

   (b) Show by a counterexample that $R_1 \cup R_2$ is not necessarily a partial order.

3. Let $(S, \preccurlyeq)$ be a partial order in which every element has an immediate successor. Prove that it is not necessarily true that for any two elements $a \preccurlyeq b$ that any chain between a and b has finite length.

4. Draw the Hasse diagram for the poset $(\{1, 2, 3, 4, 5, 6\}, \leq)$.

5. Draw the Hasse diagram of the partial order $\subseteq$ on $\mathcal{P}(\{1, 2, 3, 4\})$.

6. Let $A = \{a, b, c, d, e, f, g\}$. Draw the Hasse diagram for the partial order $\preccurlyeq$ given as a subset of $A \times A$ as

$$\preccurlyeq = \{(a, a), (b, b), (c, c), (d, d), (e, e), (f, f), (g, g), (a, c),$$
$$(b, c), (d, g), (a, e), (b, e), (c, e), (d, h), (g, h)\}.$$

7. A person's blood type is usually listed as one of the eight elements in the set

$$B' = \{o+, o-, a+, a-, b+, b-, ab+, ab-\}.$$

We define the donor relation $\rightarrow$ on B' as follows. The relation $t_1 \rightarrow t_2$ holds if the letter portion of the blood type donates according to Example A.4.4 and if someone with a $+$ designation can only give to someone else with $+$, while someone with $-$ can give to anybody.

(a) Draw the Hasse diagram for $(B', \rightarrow)$.

(b) Draw the Hasse diagram for the lexicographic order on $B' = B \times \{+, -\}$.

(c) Deduce that the relation $\rightarrow$ on B' is not the lexicographic order.

8. Consider the set of triples of integers $\mathbb{Z}^3$. Define the relation $\preccurlyeq$ on $\mathbb{Z}^3$ by

$$(a_1, a_2, a_3) \preccurlyeq (b_1, b_2, b_3)$$

$$\iff \begin{cases} a_1 + a_2 + a_3 < b_1 + b_2 + b_3 & \text{if } a_1 + a_2 + a_3 \neq b_1 + b_2 + b_3; \\ a_1 + a_2 + a_3 \preccurlyeq_{\text{lex}} b_1 + b_2 + b_3 & \text{if } a_1 + a_2 + a_3 = b_1 + b_2 + b_3, \end{cases}$$

where $\preccurlyeq_{\text{lex}}$ is the lexicographic order on $\mathbb{Z}^3$ (with each copy of $\mathbb{Z}$ equipped with the partial order $\leq$). Prove that $\preccurlyeq$ is a partial order on $\mathbb{Z}^3$. Prove also that $\preccurlyeq$ is a total order.

9. Let $(A_i, \preccurlyeq_i)$ be posets for $i = 1, 2, \ldots, n$ and define $\preccurlyeq_{\text{lex}}$ as the lexicographic order on $A_1 \times A_2 \times \cdots A_n$. Prove that $\preccurlyeq_{\text{lex}}$ is a total order if and only if $\preccurlyeq_i$ is a total order on A_i for all i.

10. Let $\preccurlyeq$ be the lexicographic order on $\mathbb{R}^3$, where each $\mathbb{R}$ is equipped with the usual $\leq$. Prove or disprove the following statement: For all vectors $\vec{a}, \vec{b}, \vec{c}, \vec{d}$, if $\vec{a} \preccurlyeq \vec{b}$ and $\vec{c} \preccurlyeq \vec{d}$, then $\vec{a} + \vec{c} \preccurlyeq \vec{b} + \vec{d}$.

11. Answer the following questions pertaining to the poset described by the Hasse diagram below.

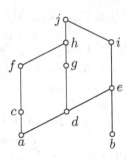

(a) List all the minimal elements.

(b) List all the maximal elements.

(c) List all the maximal elements in the subposet with $\{a, b, c, d, e, f, g\}$.

(d) Determine the length of the longest chain and find all chains of that length.

(e) Find the least upper bound of $\{a, b\}$, if it exists.

(f) Find the greatest lower bound of $\{b, c\}$, if it exists.

(g) List all the upper bounds of $\{f, d\}$.

12. Consider the partial order on $\mathbb{R}^2$ given in Example A.4.6. Let A be the unit disk

$$A = \{(x, y) \in \mathbb{R}^2 \mid x^2 + y^2 \leq 1\}.$$

(a) Show that A has both a maximal and minimal element. Find all of them.

(b) Find all the upper bounds and all the lower bounds of A.

13. Consider the lexicographic order on $\mathbb{R}^2$ coming from the standard $(\mathbb{R}, \leq)$. Let A be the closed disk of center $(1, 2)$ and radius 5.

(a) Show that A has both a maximal and minimal element. Find all of them.

(b) Find all the upper bounds and all the lower bounds of A.

(c) Show that A has both a least upper bound and a greatest lower bound.

14. Prove that in a finite lattice, there exists exactly one maximal element and one minimal element.

15. Determine whether the posets corresponding to the following Hasse diagrams are lattices. If they are not, explain why.

(a) (b) (c) (d)

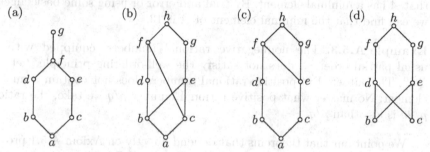

16. Let S be a set. Show that the relation of refinement is a partial order on the set of partitions of S. (See Exercise A.3.26.)

A.5   Basic Properties of Integers

The following results about integers arise in a discrete mathematics course, a basic course on number theory, or perhaps a transition to advanced mathematics course. Some of these topics are taught as early as late elementary school and developed to different degrees in high school. However, in those contexts, teachers generally teach many basic theorems in number theory without giving proofs. In this section, we list elementary divisibility properties of integers, first because we need these results, and second as a reference for how to develop similar topics in ring theory.

Because this appendix stands primarily as a review or reference, we state some of the theorems without proof. We provide proofs when the strategy reappears in a useful way in another part of this textbook.

A.5.1   Well-Ordering of the Integers

There exist a few different, though equivalent formulations of the set of axioms for the integers. We leave better formed discussion about these axioms for a course in number theory; [27, Appendix A] and [29, Section 2.1] give two slightly different formulations. See [11] for a brief synthesis of Peano's axioms for the integers.

Axiom A.5.1 (Well-Ordering Property on $\mathbb{N}$)

Let A be any nonempty subset of nonnegative integers. There exists $m \in A$ such that $m \le a$ for all $a \in A$. In other words, A contains a *minimal* element.

Example A.5.2. Let $A = \{n \in \mathbb{N} \mid n^2 > 167\}$. A is nonempty because for example $20^2 = 400 > 167$ so $20 \in A$. Theorem A.5.1 allows us to conclude that A has a minimal element. By trial and error or using some basic algebra, we can find that the minimal element of A is 13. $\triangle$

Example A.5.3. The nonnegative rational numbers, equipped with the usual partial order $\le$, does not satisfy the well-ordering principle. Let $S = \mathbb{Q}^{\ge 0}$. The subset A of positive rational numbers does not contain a minimal element. No matter what positive rational number p/q we take, the rational $p/2q$ is less than p/q. $\triangle$

We point out that theorems that depend directly on Axiom A.5.1 produce nonconstructive existence theorems because they affirm the existence of an element with a certain property without offering a method to "construct" or to find this element.

A.5.2   Divisibility

The notion of divisibility of integers is often introduced in elementary school. However, in order to prove theorems about divisibility, we need a rigorous definition.

Definition A.5.4

If $a, b \in \mathbb{Z}$ with $a \ne 0$, we say that a divides b if $\exists k \in \mathbb{Z}$ such that $b = ak$. We write $a \mid b$ if a divides b and $a \nmid b$ otherwise. We also say that a is a *divisor* (or a *factor*) of b and that b is a *multiple* of a.

We write $a\mathbb{Z}$ for the set of all multiples of the integer a.

Proposition A.5.5

Let a, b, and c be integers.

(1) Any nonzero integer divides 0.

(2) Suppose $a \neq 0$. If $a|b$ and $a|c$, then $a|(b+c)$.

(3) Suppose $a \neq 0$ and $b \neq 0$. If $a|b$ and $b|c$, then $a|c$.

(4) Suppose $a \neq 0$ and $b \neq 0$. If $a|b$ and $b|a$, then $a = b$ or $a = -b$.

Note that if we restrict ourselves to $\mathbb{Z}^*$, the set of nonzero integers, then divisibility $|$ is a relation on $\mathbb{Z}^*$. Divisibility is not a relation on $\mathbb{Z}$ because we do not ascribe any truth value to $0|a$ for any $a \in \mathbb{Z}$.

Suppose that we restrict the relation of divisibility to positive integers $\mathbb{N}^*$. Since any positive integer divides itself, divisibility on $\mathbb{N}^*$ is reflexive, antisymmetric (by Proposition A.5.5(4)), and transitive (by Proposition A.5.5(3)). Thus, $(\mathbb{N}^*, |)$ is a partially ordered set. This poset has a minimal element of 1, because $1|n$ for all positive integers n, but has no maximal element.

Theorem A.5.6 (Integer Division)

For all $a, b \in \mathbb{Z}$ with $a \neq 0$, there exist unique integers q and r such that

$$b = qa + r \qquad \text{where } 0 \leq r < |a|.$$

The integer q is called the *quotient* and r is called the *remainder*.

Proof. Given $a, b \in \mathbb{Z}$ with $a \neq 0$. Define the set

$$S = \{b - ka \in \mathbb{N} \,|\, k \in \mathbb{Z}\}.$$

Let $b - k_1 a$ and $b - k_2 a$ be two elements of S. We have

$$(b - k_1 a) - (b - k_2 a) = (k_2 - k_1)a,$$

so the difference of any two elements in S is a multiple of a and thus of $|a|$.

By the Well-Ordering Principle, S has a least element r. Suppose that we have $r = b - qa$ for some q. Assume that $r \geq |a|$, then if $\text{sign}(a)$ is the sign of a, we have

$$b - (q + \text{sign}(a))a = b - aq - |a| = r - |a| \geq 0.$$

Then $r - |a|$ contradicts the minimality of r as a least element of S. Hence, we know that $0 \leq r < |a|$. Since any two elements of S differ by a multiple of a, then r is the unique element $n \in S$ with $0 \leq n < a$. Thus, with the desired conditions of the theorem, r is unique and then so is q. $\qquad \square$

It is obvious that $d|n$ if and only if the remainder of the integer division of n by d is 0. As simple as this observation may seem, it offers a common strategy to show that one integer divides another.

Borrowing notation from some programming languages, it is common to write $b \bmod a$ to stand for the remainder of the division of b by a.

A.5.3   The Greatest Common Divisor

The concepts of greatest common divisor (or factor) and the least common multiple arise early in math education. However, we introduce these concepts in a manner that may appear novel but is equivalent with the elementary school formulation but has the benefit of allowing us to generalize to other algebraic contexts.

Definition A.5.7

If $a, b \in \mathbb{Z}$ with $(a, b) \neq (0,0)$, a *greatest common divisor* of a and b is an element $d \in \mathbb{Z}$ such that:

- $d|a$ and $d|b$ (d is a common divisor);
- if $d'|a$ and $d'|b$ (d' is another common divisor), then $d'|d$.

Because of condition (2) in this definition, it is not obvious that two integers not both 0 have a greatest common divisor. (If we had said $d' \leq d$ in the definition, the proof that two integers not both 0 have a greatest common divisor is a simple application of the Well-Ordering Principle on $\mathbb{Z}$.) The key to showing that integers possess a greatest common divisor relies on the *Euclidean Algorithm*, which we described here below.

Let a and b be two positive integers with $a \geq b$. The Euclidean Algorithm starts by setting $r_0 = a$ and $r_1 = b$ and then repeatedly perform the following integer divisions

$$r_0 = r_1 q_1 + r_2 \qquad \text{(where } 0 \leq r_2 < b)$$
$$r_1 = r_2 q_2 + r_3 \qquad \text{(where } 0 \leq r_3 < r_2)$$

$$\vdots$$

$$r_{n-2} = r_{n-1} q_{n-1} + r_n \qquad \text{(where } 0 \leq r_3 < r_2)$$
$$r_{n-1} = r_n q_n + 0 \qquad \text{(and } r_n > 0).$$

This process terminates because the sequence $r_1, r_2, r_3, \ldots$ is a strictly decreasing sequence of positive integers and hence has at most $r_1 = b$ terms in it. Note that if $b \mid a$, then $n = 1$ and the Euclidean Algorithm has one line.

To see what the Euclidean Algorithm tells us, consider the positive integer r_n. By the last line of the Euclidean Algorithm we see that $r_n | r_{n-1}$. From the second to last row, of the Euclidean Algorithm, $r_n \mid r_{n-1} q_{n-1}$ and by Proposition A.5.5(2), $r_n \mid r_{n-2}$. Repeatedly applying this process ($n-1$ times, and hence a finite number), we see that $r_n \mid r_1$ and $r_n \mid r_0$, so r_n is a common divisor of a and b.

Also, suppose that d' is a common divisor of a and b. Then $d'k_0 = a = r_0$ and $d'k_1 = b = r_1$. We have

$$r_2 = r_0 - r_1 q_1 = d'k_0 - d'k_1 q_1 = d'(k_0 - k_1 q_1).$$

Hence, d' divides r_2 with $d'k_2 = r_2$. Repeating this process ($n - 1$ times), we deduce that $d'|r_n$. Thus, r_n is a positive greatest common divisor of a and b. Though this does not yet complete a proof, the Euclidean Algorithm leads to the following theorem.

Proposition A.5.8

There exists a unique positive greatest common divisor for all pairs of integers $(a, b) \in \mathbb{Z} \times \mathbb{Z} - \{(0, 0)\}$.

Because of Proposition A.5.8, we regularly refer to "the" greatest common divisor of two integers as this unique positive one and we use the notation $\gcd(a, b)$. The proof of Proposition A.5.8 tells us how to calculate the greatest common divisor: (1) if $a \neq 0$, then $\gcd(a, 0) = |a|$; (2) if $a, b \neq 0$, then $\gcd(a, b)$ is the result of the Euclidean Algorithm applied to $|a|$ and $|b|$.

Example A.5.9. We perform the Euclidean Algorithm to find $\gcd(522, 408)$.

$$522 = 408 \times 1 + 114$$
$$408 = 114 \times 3 + 66$$
$$114 = 66 \times 1 + 48$$
$$66 = 48 \times 1 + 18$$
$$48 = 18 \times 2 + 12$$
$$18 = 12 \times 1 + 6$$
$$12 = 6 \times 2 + 0$$

By Proposition A.5.8 and the Euclidean Algorithm, $\gcd(522, 408) = 6$. $\triangle$

Two integers always have 1 and -1 as common divisors. However, if $\gcd(a, b) = 1$, then we say that a and b are *relatively prime*.

There is an alternative characterization of the greatest common divisor. Let $a, b \in \mathbb{Z}^*$ and define $S_{a,b}$ as the set of integer linear combinations in integers of a and b, i.e.,

$$S_{a,b} = \{sa + tb | s, t \in \mathbb{Z}\}.$$

Proposition A.5.10

The set $S_{a,b}$ is the set of all integer multiples of $\gcd(a, b)$. Consequently, $\gcd(a, b)$ is the least positive linear combination in integers of a and b.

Proof. By Proposition A.5.5, any common divisors to a and b divides sa, tb, and $sa + tb$. This shows that $S_{a,b} \subseteq \gcd(a, b)\mathbb{Z}$. We need to show the reverse inclusion.

By the Well-Ordering Principle, the set $S_{a,b}$ has a least positive element. Call this element d_0 and write $d_0 = s_0 a + t_0 b$ for some $s_0, t_0 \in \mathbb{Z}$. We show by contradiction that d_0 is a common divisor of a and b. Suppose that d_0 does not divide a. Then by integer division,

$$a = qd_0 + r \qquad \text{where } 0 < r < d_0.$$

Then $a - r = qd_0 = qs_0 a + qt_0 b$. Then after rearranging, we get

$$r = (1 - qs_0)a + (-qt_0)b.$$

This writes r, which is positive and less than d_0, as a linear combination of a and b. This contradicts the assumption that d_0 is the minimal positive element in $S(a,b)$. Hence, the assumption that d_0 does not divide a is false so d_0 divides a. By a symmetric argument, d_0 divides b as well. Since $kd_0 = (ks_0)a + (kt_0)b$, then every multiple of d_0 is in $S_{a,b}$ in particular every multiple of $\gcd(a,b)$ is too. Hence, $\gcd(a,b)\mathbb{Z} \subseteq S_{a,b}$. We conclude that $S_{a,b} = \gcd(a,b)\mathbb{Z}$ and the proposition follows. $\qquad\qquad\square$

Proposition A.5.10 does not offer a way to find the integers s and t such that $\gcd(a,b) = sa + tb$. If a and b are small, then one can find s and t by inspection. For example, by inspecting the divisors of 22, it is easy to see that $\gcd(22, 14) = 2$. A linear combination that illustrates Proposition A.5.10 for 22 and 14 is

$$2 \times 22 - 3 \times 14 = 44 - 42 = 2.$$

However, it is possible to backtrack the steps of the Euclidean Algorithm and find s and t such that $sa + tb = \gcd(a,b)$. This is called the *Extended Euclidean Algorithm*. The following example illustrates this.

Example A.5.11. In Example A.5.9, we saw that the Euclidean Algorithm gave $\gcd(522, 408) = 6$. We start from the penultimate line in the algorithm and work backward, in such a way that each line gives 6 as a linear combination of the intermediate remainders r_i and r_{i+1}.

$$
\begin{aligned}
6 &= 18 - 12 \times 1 \\
&= 18 - (48 - 18 \times 2) \times 1 = 18 \times 3 - 48 \times 1 \\
&= (66 - 48 \times 1) \times 3 - 48 \times 1 = 66 \times 3 - 48 \times 4 \\
&= 66 \times 3 - (114 - 66 \times 1) \times 4 = 66 \times 7 - 114 \times 4 \\
&= 114 \times 4 - (408 - 114 \times 3) \times 7 = 408 \times 7 - 114 \times 25 \\
&= (522 - 408 \times 1) \times 25 - 408 \times 7 = 408 \times 32 - 522 \times 25
\end{aligned}
$$

Hence, this last line gives $6 = (-25) \times 522 + 32 \times 408$. $\qquad\qquad \triangle$

We encourage the reader to consult [11] or [27] for a more organized form of the Extended Euclidean Algorithm.

The characterization of the greatest common divisor as in Proposition A.5.10 leads to many consequences about the greatest common divisor. The following proposition gives one such example.

Proposition A.5.12

Let a and b be nonzero integers that are relatively prime. For any integer c, if $a|bc$, then $a|c$.

Proof. Since a and b are relatively prime, then $\gcd(a, b) = 1$. By Proposition A.5.10, there exist integers $s, t \in \mathbb{Z}$ such that $sa + tb = 1$. Since $a \mid bc$, there exists $k \in \mathbb{Z}$ such that $ak = bc$. Then

$$atk = tbc = c(1 - as) = c - cas,$$

which implies that

$$c = atk + acs = a(tk + cs).$$

From this we conclude that $a \mid c$. □

A.5.4 The Least Common Multiple

Definition A.5.13

If $a, b \in \mathbb{Z}^*$, a *least common multiple* is an element $m \in \mathbb{Z}$ such that:

- $a|m$ and $b|m$ (m is a common multiple);
- if $a|m'$ and $b|m'$, then $m|m'$.

Similar to our presentation of the greatest common divisor, we should note that from this definition, it is not obvious that a least common multiple always exists. The following proposition states the key result about least common multiples and in fact gives a formula for it based on the greatest common divisor.

Proposition A.5.14

There exists a unique positive least common multiple m for all pairs of integers $(a, b) \in \mathbb{Z} \times \mathbb{Z} - \{(0, 0)\}$. Furthermore,

$$m = |ab| / \gcd(a, b).$$

We regularly call this unique positive least common multiple of a and b "the" least common multiple. We denote this positive integer as $\operatorname{lcm}(a, b)$.

When restricted to the positive integers $\mathbb{N}^*$, the existence theorems given in Propositions A.5.8 and A.5.14 show that the poset $(\mathbb{N}^*, |)$ is a lattice. The greatest common divisor $\gcd(a, b)$ of two positive integers a and b is the greatest lower bound of $\{a, b\}$ in the terminology of posets and the least common multiple is the least upper bound of $\{a, b\}$. Remark that $(\mathbb{N}^*, |)$ is an infinite lattice while for $n \geq 3$, the subposet $(\{1, 2, \ldots, n\}, |)$ is not a lattice.

A.5.5 Prime Numbers

Definition A.5.15

An element $p \in \mathbb{Z}$ is called a *prime number* if $p > 1$ and the only divisors of p are 1 and itself. If an integer $n > 1$ is not prime, then n is called *composite*.

For short, we often say "p is prime" instead of "p is a prime number." As simple as the concept of primality is, properties about prime numbers have intrigued mathematicians since Euclid and before. The distribution of prime numbers in $\mathbb{N}$ or in sequences of integers, additive properties of prime numbers, fast algorithms for checking if a number is prime, and many other questions still offer active areas of research in number theory.

One of the earliest results about prime numbers dates back to Euclid, who used a clever argument by contradiction.

Theorem A.5.16 (Euclid's Prime Number Theorem)

The set of prime numbers is infinite.

By definition, every positive integer is either 1, a prime number, or a composite number. Let S be the set of composite integers that are not divisible by a prime number. Suppose that S is nonempty. By the well-ordering of the integers, S has a least element m. Since m is composite, we can write $m = ab$, where neither a nor b is 1. Then either a is prime or a is composite. Now a cannot be prime because m is not divisible by a prime number. Hence, a is composite. Since $a < m$, then $a \notin S$ so a is divisible by a prime number. By Proposition A.5.5(3), m must also be divisible by a prime number. This contradicts the assumption that $S \neq \emptyset$. This reasoning establishes the fundamental result that every positive integer greater than 1 is divisible by a prime number. In fact, the following simple proposition tells us a slightly tighter result about the prime factors of numbers.

Proposition A.5.17

If n is composite, then it has a divisor d such that $1 < d \leq \sqrt{n}$.

Proof. Suppose that all the divisors of n are greater than $\sqrt{n}$. Since n is composite, there exist positive integers a and b greater than 1 with $n = ab$. The supposition that $a > \sqrt{n}$ and $b > \sqrt{n}$ implies that $ab = n > n$, a contradiction. The proposition follows. $\square$

We mention two other key properties about prime numbers. We omit here a proof for the second theorem since we will discuss these topics in greater generality in Section 4.4.

The following proposition gives an alternative definition for primality.

Proposition A.5.18 (Euclid's Lemma)

If $p > 1$, then p is prime if and only if for all $a, b \in \mathbb{Z}$, $p|ab$ implies $p|a$ or $p|b$.

Proof. Suppose that $p|ab$. If $p|a$, then the proof is done. Suppose instead that $p \nmid a$. Then, since the only divisors of p are 1 and itself, $\gcd(p, a) = 1$. By Proposition A.5.12, $p|b$ and the proposition is still true. $\qquad\square$

Theorem A.5.19 (Fundamental Theorem of Arithmetic)

If $n \in \mathbb{Z}$ and $n \geq 2$, then there is a unique factorization (up to rearrangement) of n into a product of prime numbers. More precisely, if n can be written as the product of primes in two different ways as

$$n = p_1 p_2 \cdots p_r = q_1 q_2 \cdots q_s \qquad (A.7)$$

with p_i and q_j primes, then $r = s$ and there is a bijective function $f : \{1, 2, \ldots, r\} \to \{1, 2, \ldots, r\}$ such that $p_i = q_{f(i)}$.

In the factorizations in (A.7), we do not assume that the p_i are all unique. It is common to write the generic factorization of integers as

$$n = p_1^{\alpha_1} p_2^{\alpha_2} \cdots p_k^{\alpha_k} \qquad (A.8)$$

with the primes p_i all distinct and α_i are nonzero integers. It is also common to list the primes in increasing order. Using these latter two habits, we call the expression in (A.8) *the prime factorization* of n.

A.5.6   The Euler ϕ-Function

As we will soon see, given a positive integer n, counting the number of integers less than n that are relatively prime to n appears in numerous contexts.

Definition A.5.20

Euler's *totient function* (or Euler's ϕ-function) is the function $\phi : \mathbb{N}^* \to \mathbb{N}^*$ such that $\phi(n)$ is the number of positive integers less than n that are relatively prime to n. In other words,

$$\phi(n) \stackrel{\text{def}}{=} |\{a \in \mathbb{N}^* \,|\, 1 \leq a \leq n \text{ and } \gcd(a, n) = 1\}|. \qquad (A.9)$$

Example A.5.21. A few sample calculations of Euler's totient function:

(1) $\phi(8) = 4$, because in $\{1, 2, 3, 4, 5, 6, 7, 8\}$ only $1, 3, 5$, and 7 are relatively prime to 8.

(2) $\phi(20) = 8$, because for $1 \leq a \leq n$, the integers relatively prime to 20 are those that are not divisible by 2 or by 5. Thus,

$$\{a \in \mathbb{Z} \mid 1 \leq a \leq n \text{ and } \gcd(a, n) = 1\} = \{1, 3, 7, 9, 11, 13, 17, 19\}.$$

(3) $\phi(243) = \varphi(3^5) = 3^5 - 3^4 = 243 - 81 = 162$, because the positive integers not relatively prime to 243 consist of the 81 integers less than 243 that are divisible by 3. $\triangle$

The following proposition gives a formula for Euler's totient function. The proof of this formula is subtle. See [27, Theorem 7.5].

Proposition A.5.22

If a positive integer n has the prime decomposition of $n = p_1^{\alpha_1} p_2^{\alpha_2} \cdots p_\ell^{\alpha_\ell}$, then

$$\phi(n) = \left(p_1^{\alpha_1} - p_1^{\alpha_1 - 1}\right) \left(p_2^{\alpha_2} - p_2^{\alpha_2 - 1}\right) \cdots \left(p_\ell^{\alpha_\ell} - p_\ell^{\alpha_\ell - 1}\right). \quad \text{(A.10)}$$

EXERCISES FOR SECTION A.5

1. Find the prime factorization of the following integers: (a) 56; (b) 97; (c) 126; (d) 399; (e) 255; (f) 1728.

2. Find the prime factorization of the following integers: (a) 111; (b) 470; (c) 289; (d) 743; (e) 2345; (f) 101010.

3. Draw the Hasse diagram of $(\{1, 2, 3, \ldots, 12\}, |)$.

4. Let n be a positive integer. Show that the number of edges in the Hasse diagram of $(\{1, 2, 3, \ldots, n\}, |)$ is

$$\sum_{p: \text{ primes} \leq n} \left\lfloor \frac{n}{p} \right\rfloor.$$

5. Use the Euclidean Algorithm to find the greatest common divisor of the following pairs of integers.
   (a) $a = 234$, and $b = 84$
   (b) $a = 5241$, and $b = 872$
   (c) $a = 1010101$, and $b = 1221$

6. Use the Euclidean Algorithm to find the greatest common divisor of the following pairs of integers.
   (a) $a = 55$, and $b = 34$
   (b) $a = 4321$, and $b = 1234$
   (c) $a = 54321$, and $b = 1728$

7. Define the Fibonacci sequence $\{f_n\}_{n \geq 0}$ by $f_0 = 0$, $f_1 = 1$ and $f_n = f_{n-1} + f_{n-2}$ for all $n \geq 2$. Let f_n and f_{n+1} be two consecutive terms in the Fibonacci sequence. Prove that $\gcd(f_{n+1}, f_n) = 1$ and show that for all $n \geq 2$, the Euclidean algorithm requires exactly $n - 1$ integer divisions (including the last one that has a remainder of 0).

8. Let $a, b, c \in \mathbb{Z}$. Prove that $a|b$ implies that $a|bc$.

9. Perform the Extended Euclidean Algorithm on the three pairs of integers in Exercise A.5.5.

10. Perform the Extended Euclidean Algorithm on the three pairs of integers in Exercise A.5.6.

11. Suppose that $a, b \in \mathbb{Z}^*$ and that $s, t \in \mathbb{Z}^*$ such that $sa + tb = \gcd(a, b)$. Show that s and t are relatively prime.

12. Let a, b, c be positive integers. Prove that $\gcd(ab, ac) = a \gcd(b, c)$.

13. Let a and b be positive integers. Show that the set of common multiples of a and b is $\mathrm{lcm}(a, b)\mathbb{Z}$, i.e., the set of multiples of $\mathrm{lcm}(a, b)$.

14. Prove that if $2^n - 1$ is prime, then n is prime. [Hint: Recall that for all real numbers,

$$a^n - b^n = (a - b)(a^{n-1} + a^{n-2}b + a^{n-3}b^2 + \cdots + b^{n-1}).$$

Prime numbers of the form $2^p - 1$, where p is prime, are called Mersenne primes and have been historically of great research interest. The converse implication is not true however. For example, $2^{11} - 1 = 2047 = 23 \times 89$.]

15. Prove that the product of two consecutive positive integers is even.

16. Prove that the product of four consecutive positive integers is divisible by 24.

17. Suppose that a and b are positive integers with prime factorizations written as

$$a = p_1^{\alpha_1} p_2^{\alpha_2} \cdots p_n^{\alpha_n} \quad \text{and} \quad b = p_1^{\beta_1} p_2^{\beta_2} \cdots p_n^{\beta_n},$$

where p_i distinct primes and $\alpha_i, \beta_i \geq 0$.
    (a) Prove that $\gcd(a, b) = p_1^{\min(\alpha_1, \beta_1)} p_2^{\min(\alpha_2, \beta_2)} \cdots p_n^{\min(\alpha_n, \beta_n)}$.
    (b) Prove that $\mathrm{lcm}(a, b) = p_1^{\max(\alpha_1, \beta_1)} p_2^{\max(\alpha_2, \beta_2)} \cdots p_n^{\max(\alpha_n, \beta_n)}$.

18. Use the previous exercise to prove the formula implied by Proposition A.5.14, namely that $\gcd(a, b) \, \mathrm{lcm}(a, b) = ab$ for all positive integers a and b.

19. Prove that $\sqrt{2}$ is not a rational number. [Hint: Assume $\sqrt{2} = a/b$ as a reduced fraction and argue by contradiction.]

20. Prove that for all primes p and all integers $k \geq 2$, the number $\sqrt[k]{p}$ is irrational. [Hint: See the previous exercise.]

21. Find the greatest number of 5^k that divides $200!$. Use this to deduce the number of 0s to the right in the decimal expansion of $200!$.

22. For the following integers, calculate $\phi(n)$ by directly listing the set in (A.9):
    (a) $\phi(30)$; (b) $\phi(33)$; (c) $\phi(12)$.

23. Prove that for all integers n,

$$n = \sum_{d \mid n} \phi(d),$$

where this summation notation means we sum over all positive divisors d of n. [Hint: Consider the set of fractions $\left\{ \frac{1}{n}, \frac{2}{n}, \frac{3}{n}, \ldots, \frac{n}{n} \right\}$ written in reduced form.]

24. Without using Proposition A.5.22, prove the following identities for any prime p.

    (a) $\phi(p) = p - 1$
    (b) $\phi(p^k) = p^k - p^{k-1}$

A.6 Modular Arithmetic

In this section, we assume that n is an integer with $n \geq 2.0$

We review modular arithmetic here because modular arithmetic will provide relatively easy examples for groups, rings, and fields. As in other sections of this appendix, we do not always provide proofs for the propositions.

A.6.1 Congruence

Definition A.6.1

Let a and b be integers. We say that a *is congruent to b modulo n* if $n \mid (b - a)$ and we write

$$a \equiv b \pmod{n}.$$

If n is understood from context, we simply write $a \equiv b$. The integer n is called the *modulus*.

Proposition A.6.2

The congruence modulo n relation on $\mathbb{Z}$ is an equivalence relation.

Proof. We assume n is fixed. For any integer $a \in \mathbb{Z}$, we have $n \mid (a - a) = 0$ so the congruence relation is reflexive.

Suppose that $a \equiv b$. Then $n \mid (b - a)$, so there exists $k \in \mathbb{Z}$ with $nk = b - a$. Then $n(-k) = a - b$ and so $n \mid (a - b)$ and hence $b \equiv a$. This shows that the congruence relation is symmetric.

Suppose that $a \equiv b$ and $b \equiv c$. Then $n \mid (b - a)$ and $n \mid (c - b)$. By Proposition A.5.5(2),

$$n \mid ((b - a) + (c - b)) \quad \text{so} \quad n \mid (c - a).$$

Hence, $a \equiv c$ and we deduce that congruence is transitive. The result follows.□

Section A.3.1 introduced notation that is standard for equivalence classes and quotient sets in the context of generic equivalence relations. However, the congruence relation has such a long history, that it carries its own notations.

When the modulus n is clear from context, we denote by $\bar{a}$ the equivalence class of $a \bmod n$ and call it the *congruence class* of a. If we consider the integer division of a by n with $a = nq + r$, we see that $n \mid a - r$. Hence, $r \in \bar{a}$. In fact, we can characterize the equivalence class of a in a few different ways:

$$\bar{a} = \{b \in \mathbb{Z} \mid b \equiv a \pmod{n}\} = \{b \in \mathbb{Z} \mid a \bmod n = b \bmod n\}$$
$$= \{a + kn \mid k \in \mathbb{Z}\} = a + n\mathbb{Z}.$$

It is important for applications of congruences to note that

$$a \equiv 0 \pmod{n} \quad \Longleftrightarrow \quad a \mid n.$$

Instead of writing $\mathbb{Z}/\equiv$ for the quotient set for the congruence relation, we always write

$$\mathbb{Z}/n\mathbb{Z} \overset{\text{def}}{=} \{\bar{0}, \bar{1}, \ldots, \overline{n-1}\}$$

for the set of equivalence classes modulo n. We pronounce this quotient set as "Z mod n Z."

Example A.6.3. Suppose $n = 15$, then we have the equalities $\bar{2} = \overline{17} = \overline{-13}$ and many others because these numbers are all congruent to each other. We will also say that $2, 17, -13$, are representatives of the congruence class $\bar{2}$. △

A.6.2   Modular Arithmetic

Proposition A.6.4

Fix a modulus n. Let $a, b, c, d \in \mathbb{Z}$ such that $a \equiv c$ and $b \equiv d$. Then

$$a + b \equiv c + d \qquad \text{and} \qquad ab \equiv cd.$$

Proof. By definition, $n \mid (c - a)$ and $n \mid (d - b)$, so there exist k, ℓ such that

$$c - a = nk \tag{A.11}$$
$$d - b = n\ell. \tag{A.12}$$

Adding these two expressions, we get

$$(d + c) - (b + a) = nk + n\ell = n(k + \ell).$$

This shows that $n \mid (d + c) - (b + a)$ so $a + b \equiv c + d$.

To show the multiplication, multiply Equation (A.12) by c and subtract from it Equation (A.11) multiplied by b. We obtain

$$c(d - b) - b(c - a) = cn\ell - bnk \Longleftrightarrow cd - ab = n(c\ell - bk).$$

This illustrates that $n \mid (cd - ab)$, which means that $ab \equiv cd \pmod{n}$.   □

Recall by Remark A.3.11 that we must verify that a function defined on a quotient set is well-defined. This is important for modular arithmetic but Proposition A.6.4 leads to the following vital corollary.

Corollary A.6.5

Let n be a modulus and let $a, b \in \mathbb{Z}$. The operations $+$ and $\cdot$ on $\mathbb{Z}/n\mathbb{Z}$, given as

$$\overline{a} + \overline{b} \overset{\text{def}}{=} \overline{a+b} \quad \text{and} \quad \overline{a} \cdot \overline{b} \overset{\text{def}}{=} \overline{a \cdot b}$$

are well-defined.

Modular arithmetic modulo n is the arithmetic arising from the addition and multiplication operations as defined in Corollary A.6.5 on the set $\mathbb{Z}/n\mathbb{Z}$.

Example A.6.6. To illustrate a few examples of modular arithmetic, we show the addition and multiplication tables corresponding to $\mathbb{Z}/5\mathbb{Z}$ and $\mathbb{Z}/6\mathbb{Z}$. In $\mathbb{Z}/5\mathbb{Z} = \{\overline{0}, \overline{1}, \overline{2}, \overline{3}, \overline{4}\}$, the tables of operations are:

$+$	$\overline{0}$	$\overline{1}$	$\overline{2}$	$\overline{3}$	$\overline{4}$
$\overline{0}$	$\overline{0}$	$\overline{1}$	$\overline{2}$	$\overline{3}$	$\overline{4}$
$\overline{1}$	$\overline{1}$	$\overline{2}$	$\overline{3}$	$\overline{4}$	$\overline{0}$
$\overline{2}$	$\overline{2}$	$\overline{3}$	$\overline{4}$	$\overline{0}$	$\overline{1}$
$\overline{3}$	$\overline{3}$	$\overline{4}$	$\overline{0}$	$\overline{1}$	$\overline{2}$
$\overline{4}$	$\overline{4}$	$\overline{0}$	$\overline{1}$	$\overline{2}$	$\overline{3}$

$\cdot$	$\overline{0}$	$\overline{1}$	$\overline{2}$	$\overline{3}$	$\overline{4}$
$\overline{0}$	$\overline{0}$	$\overline{0}$	$\overline{0}$	$\overline{0}$	$\overline{0}$
$\overline{1}$	$\overline{0}$	$\overline{1}$	$\overline{2}$	$\overline{3}$	$\overline{4}$
$\overline{2}$	$\overline{0}$	$\overline{2}$	$\overline{4}$	$\overline{1}$	$\overline{3}$
$\overline{3}$	$\overline{0}$	$\overline{3}$	$\overline{1}$	$\overline{4}$	$\overline{2}$
$\overline{4}$	$\overline{0}$	$\overline{4}$	$\overline{3}$	$\overline{2}$	$\overline{1}$

$$(A.13)$$

In $\mathbb{Z}/6\mathbb{Z} = \{\overline{0}, \overline{1}, \overline{2}, \overline{3}, \overline{4}, \overline{5}\}$, the tables of operations are:

$+$	$\overline{0}$	$\overline{1}$	$\overline{2}$	$\overline{3}$	$\overline{4}$	$\overline{5}$
$\overline{0}$	$\overline{0}$	$\overline{1}$	$\overline{2}$	$\overline{3}$	$\overline{4}$	$\overline{5}$
$\overline{1}$	$\overline{1}$	$\overline{2}$	$\overline{3}$	$\overline{4}$	$\overline{5}$	$\overline{0}$
$\overline{2}$	$\overline{2}$	$\overline{3}$	$\overline{4}$	$\overline{5}$	$\overline{0}$	$\overline{1}$
$\overline{3}$	$\overline{3}$	$\overline{4}$	$\overline{5}$	$\overline{0}$	$\overline{1}$	$\overline{2}$
$\overline{4}$	$\overline{4}$	$\overline{5}$	$\overline{0}$	$\overline{1}$	$\overline{2}$	$\overline{3}$
$\overline{5}$	$\overline{5}$	$\overline{0}$	$\overline{1}$	$\overline{2}$	$\overline{3}$	$\overline{4}$

$\cdot$	$\overline{0}$	$\overline{1}$	$\overline{2}$	$\overline{3}$	$\overline{4}$	$\overline{5}$
$\overline{0}$	$\overline{0}$	$\overline{0}$	$\overline{0}$	$\overline{0}$	$\overline{0}$	$\overline{0}$
$\overline{1}$	$\overline{0}$	$\overline{1}$	$\overline{2}$	$\overline{3}$	$\overline{4}$	$\overline{5}$
$\overline{2}$	$\overline{0}$	$\overline{2}$	$\overline{4}$	$\overline{0}$	$\overline{2}$	$\overline{4}$
$\overline{3}$	$\overline{0}$	$\overline{3}$	$\overline{0}$	$\overline{3}$	$\overline{0}$	$\overline{3}$
$\overline{4}$	$\overline{0}$	$\overline{4}$	$\overline{2}$	$\overline{0}$	$\overline{4}$	$\overline{2}$
$\overline{5}$	$\overline{0}$	$\overline{5}$	$\overline{4}$	$\overline{3}$	$\overline{2}$	$\overline{1}$

$$(A.14)$$

The addition and multiplication tables for $\mathbb{Z}/5\mathbb{Z}$ and $\mathbb{Z}/6\mathbb{Z}$ display some similarities but also some differences. The patterns in the addition tables are similar. In $\mathbb{Z}/5\mathbb{Z}$ every nonzero element $\overline{a}$ has a multiplicative inverse, i.e., some $\overline{b}$ such that $\overline{a}\overline{b} = \overline{1}$. However, in $\mathbb{Z}/6\mathbb{Z}$ the nonzero elements $\overline{2}$, $\overline{3}$, and $\overline{4}$ do not have inverses. Furthermore, in $\mathbb{Z}/6\mathbb{Z}$, there exist nonzero elements $\overline{a}$ and $\overline{b}$ such that $\overline{a}\overline{b} = \overline{0}$. $\triangle$

To use the term "arithmetic" connotes the ability to do addition, multiplication, subtract, and division; to solve equations; and to study various properties among these operations. Subtraction of two elements is defined

$$\overline{a} - \overline{b} \overset{\text{def}}{=} \overline{a} + (-\overline{b}),$$

where $-\overline{b}$ is the additive inverse of $\overline{b}$. The additive inverse of $\overline{b}$ is an element $\overline{c}$ such that $\overline{b} + \overline{c} = \overline{b+c} = \overline{0}$. We can take $-\overline{b} = \overline{-b}$. If we use $\{0, 1, 2, \ldots, n-1\}$ as the complete set of distinct representatives, then we would write $-\overline{b} = n - b$.

However, as the multiplication table for $\mathbb{Z}/6\mathbb{Z}$ in Example A.6.6 illustrates, there exist nonzero elements that do not have multiplicative inverses. This is just one of the differences between arithmetic in $\mathbb{Z}$ and $\mathbb{Q}$ and modular arithmetic.

A.6.3 Units and Powers

In this section, we work in the modular arithmetic of $\mathbb{Z}/n\mathbb{Z}$.

Definition A.6.7

If $\overline{a}$ has a multiplicative inverse, it is called a *unit*. We denote the set of units in $\mathbb{Z}/n\mathbb{Z}$ as

$$U(n) = \{\overline{a} \in \mathbb{Z}/n\mathbb{Z} \mid \exists \overline{c} \in \mathbb{Z}/n\mathbb{Z}, \overline{ac} = \overline{1}\}.$$

We denote the inverse of $\overline{a}$ by $\overline{a}^{-1}$.

Proposition A.6.8

As sets, $U(n) = \{\overline{a} \in \mathbb{Z}/n\mathbb{Z} \mid \gcd(a, n) = 1\}$.

Proof. Suppose that $\overline{ac} = \overline{1}$. Then $ac \equiv 1 \pmod{n}$ and so there exists $k \in \mathbb{Z}$ such that $ac = 1 + kn$. Thus, $ac - kn = 1$. Hence, there is a linear combination of a and n that is 1. The number 1 is the least positive integer so by Proposition A.5.10, $\gcd(a, n) = 1$. So far, this shows that if $\overline{a}$ is a unit modulo n, then a is relatively prime to n.

To show the converse, suppose now that $\gcd(a, n) = 1$. Then again by Proposition A.5.10, there exists $s, t \in \mathbb{Z}$ such that $sa + tn = 1$. Then $sa = 1 - tn$ and so $sa \equiv 1 \pmod{n}$. The proposition follows. $\square$

Corollary A.6.9

The number of units in $\mathbb{Z}/n\mathbb{Z}$ is $|U(n)| = \varphi(n)$ (Euler's totient function).

Note that it does not make sense to say, for example, that the inverse of 2 modulo 5 is $\frac{1}{2}$. The fraction $\frac{1}{2}$ is a specific element in $\mathbb{Q}$. The following sentences are proper. In $\mathbb{Q}$, $2^{-1} = \frac{1}{2}$. In $\mathbb{Z}/5\mathbb{Z}$, $\overline{2}^{-1} = \overline{3}$. In $\mathbb{Z}/6\mathbb{Z}$, $\overline{2}^{-1}$ does not exist.

Finding the inverse of $\bar{a}$ in $\mathbb{Z}/n\mathbb{Z}$ is not easy, especially for large values of n. If n is small, then we can find an inverse by inspection. The proof of Proposition A.6.8 shows that $\bar{s} = \bar{a}^{-1}$ in the linear combination

$$sa + tn = 1,$$

which must hold for some integers s and t if a has an inverse modulo n. The Extended Euclidean Algorithm described in Example A.5.11 provides a method to find such s and t.

Example A.6.10. We look for the inverse of $\overline{79}$ in $\mathbb{Z}/123\mathbb{Z}$. We write the Euclidean Algorithm and, to the right, the Extended Euclidean Algorithm applied to 123 and 79. (The following should be read top to bottom down the left half and then bottom to top on the right half.)

$$
\begin{array}{ll|l}
123 & = 79 \times 1 + 44 & 1 = (123 - 79) \times 9 - 79 \times 5 = 123 \times 9 - 79 \times 14 \\
79 & = 44 \times 1 + 35 & 1 = 44 \times 4 - (79 - 44) \times 5 = 44 \times 9 - 79 \times 5 \\
44 & = 35 \times 1 + 9 & 1 = (44 - 35) \times 4 - 35 \times 1 = 44 \times 4 - 35 \times 5 \\
35 & = 9 \times 3 + 8 & 1 = 9 - (35 - 9 \times 3) \times 1 = 9 \times 4 - 35 \times 1 \\
9 & = 8 \times 1 + 1 & 1 = 9 - 8 \times 1 \\
8 & = 1 \times 8 + 0 &
\end{array}
$$

By Proposition A.6.8, since the Euclidean Algorithm gives $\gcd(123, 79) = 1$, we know that $\overline{79}$ is a unit in $\mathbb{Z}/123\mathbb{Z}$. The identity $1 = 123 \times 9 - 79 \times 14$ gives that $1 \equiv -14 \times 79 \equiv 109 \times 79 \pmod{123}$. Thus, in $\mathbb{Z}/123\mathbb{Z}$, we have $\overline{79}^{-1} = \overline{109}$. $\triangle$

Example A.6.11. Let $n = 13$. We calculate $\overline{3}^{-1}(\overline{6} - \overline{11})$. First note that $\overline{3}^{-1} = \overline{9}$ because $3 \times 9 = 27 \equiv 1 \pmod{13}$. Thus,

$$\overline{3}^{-1}(\overline{6} - \overline{11}) = \overline{9} \times (\overline{6} - \overline{11}) = \overline{9}(-\overline{5}) = -\overline{45} = -\overline{6} = \overline{7}. \qquad \triangle$$

Example A.6.12. Suppose we are in $\mathbb{Z}/15\mathbb{Z}$. We show how to solve the equation $\overline{7}x + \overline{10} = y$.

Note first that $\overline{2} \cdot \overline{7} = \overline{14} = -\overline{1}$. So $-\overline{2} = \overline{13}$ is the multiplicative inverse of $\overline{7}$ modulo 15. Now we have

$$\overline{7}x + \overline{10} = y \implies \overline{7}x = y - \overline{10} \implies -\overline{2} \cdot \overline{7}x = -\overline{2}(y - \overline{10})$$
$$\implies x = -\overline{2}y + \overline{20} = \overline{13}y + \overline{5}. \qquad \triangle$$

We notice that if p is a prime number, then $U(p) = \{\overline{1}, \overline{2}, \ldots, \overline{p-1}\}$. Because of the arithmetic properties and because every nonzero element in $\mathbb{Z}/p\mathbb{Z}$ has a multiplicative inverse, $\mathbb{Z}/p\mathbb{Z}$ with addition and multiplication is called a *field*. We define fields in Section 3.1.3 and then study them in detail in Chapter 5. Because of the specific importance of fields, we often use a different notation.

If p is prime, we denote $\mathbb{Z}/p\mathbb{Z}$ by $\mathbb{F}_p$ and call it the *field* of p elements.

We end the section by discussion what happens when we take powers of elements in modular arithmetic. If $a \geq 2$ is an integer, then in $\mathbb{Z}$ the powers a^k increase without bound. However, since $\mathbb{Z}/n\mathbb{Z}$ is a finite set, powers of elements in $\mathbb{Z}/n\mathbb{Z}$ must stay in $\mathbb{Z}/n\mathbb{Z}$. Furthermore, they demonstrate interesting patterns.

Example A.6.14. We calculate the powers of $\bar{2}$ and $\bar{3}$ in $\mathbb{Z}/7\mathbb{Z}$.

k	0	1	2	3	4	5	6	7	8
$\overline{2}^k$	$\bar{1}$	$\bar{2}$	$\bar{4}$	$\bar{1}$	$\bar{2}$	$\bar{4}$	$\bar{1}$	$\bar{2}$	$\bar{4}$
$\overline{3}^k$	$\bar{1}$	$\bar{3}$	$\bar{2}$	$\bar{6}$	$\bar{4}$	$\bar{5}$	$\bar{1}$	$\bar{3}$	$\bar{2}$

We notice that the powers follow a repeating pattern. This is because if $a \in \mathbb{Z}/n\mathbb{Z}$ and $a^k = a^{k+l}$, then

$$a^{k+2l} = a^{k+l}a^l = a^k a^l = a^{k+l} = a^k$$

and, by induction, we can prove that $a^{k+ml} = a^k$ for all $m \in \mathbb{N}$. Observing the pattern for $\bar{3}$, we see for example that, in congruences,

$$3^{3201} \equiv 3^{6 \times 533 + 3} \equiv (3^6)^{533} \cdot 3^3 \equiv 3^3 \equiv 6 \pmod{7}.$$

In modular arithmetic notation, we have $\overline{3}^{3201} = \bar{6}$ in $\mathbb{Z}/7\mathbb{Z}$. It is interesting that we have easily calculated the remainder of 3^{3201} when divided by 7, without ever calculating 3^{3201}, an integer with $\lfloor \log_{10}(3^{3201}) \rfloor + 1 = \lfloor 3201 \log_{10} 3 \rfloor = 1,528$ digits. $\triangle$

Some of the patterns in powers of a number in a given modulus are not always easy to detect. The following theorem gives a general result.

Theorem A.6.15 (Fermat's Little Theorem)

Let p be a prime number and a an integer with $p \nmid a$. Then

$$a^{p-1} \equiv 1 \pmod{p}.$$

A course in elementary number theory will provide a direct proof of Fermat's Little Theorem. However, it is interesting that this theorem follows as a quick corollary to Lagrange's Theorem, Theorem 2.1.10.

EXERCISES FOR SECTION A.6

1. List ten elements in the congruence class $\bar{3}$ in modulo 7.

2. List all the elements in $\mathbb{Z}/13\mathbb{Z}$.

3. List all the elements in $\mathbb{Z}/24\mathbb{Z}$ and in $U(24)$.

4. Perform the following calculations in the modular arithmetic of the given modulus n.
   (a) $\bar{3} + \bar{5} \cdot \bar{7}$ with $n = 9$
   (b) $(\bar{5} \cdot \bar{4} - \overline{72} \cdot \bar{3})^2$ with $n = 11$
   (c) $\overline{13} \cdot \overline{42} \cdot \overline{103}$ with $n = 15$

5. Write out the elements in the set $U(30)$.

6. In $\mathbb{Z}/17\mathbb{Z}$, solve for x in terms of y in $y = \bar{2}x + \bar{3}$.

7. In $\mathbb{Z}/29\mathbb{Z}$, solve for x in terms of y in $y = \overline{17}x + \overline{20}$.

8. Show that for all integers a, we have $a^2 \equiv 0$ or $1 \pmod 4$. Show how this implies that for all integers $a, b \in \mathbb{Z}$, the sum of squares $a^2 + b^2$ never has a remainder of 3 when divides by 4.

9. Prove that if $d|m$ and $a \equiv b \pmod m$, then $a \equiv b \pmod d$.

10. Prove that if a, b, c, and m are integers with $m \geq 2$ and $c > 0$, then $a \equiv b \pmod m$ implies that $ac \equiv bc \pmod{mc}$.

11. Perform the Extended Euclidean Algorithm to calculate $\overline{52}^{-1}$ in $\mathbb{Z}/101\mathbb{Z}$.

12. Perform the Extended Euclidean Algorithm to calculate $\overline{72}^{-1}$ in $\mathbb{Z}/125\mathbb{Z}$.

13. Find the smallest positive integer n such that $2^n \equiv 1 \pmod{17}$.

14. Find the smallest positive integer n such that $3^n \equiv 1 \pmod{19}$.

15. Show that the powers of $\bar{7}$ in $\mathbb{Z}/31\mathbb{Z}$ account for exactly half of the elements in $U(31)$.

16. Show that a number is divisible by 11 if and only if the alternating sum of its digits is divisible by 11. (An alternating sum means that we alternate the signs in the sum $+ - + - \cdots$.) [Hint: $10 \equiv -1 \pmod{11}$.]

17. Prove that if n is odd, then $n^2 \equiv 1 \pmod 8$.

18. Show that the difference of two consecutive cubes (an integer of the form n^3) is never divisible by 3.

19. Use Fermat's Little Theorem to determine the remainder of 73^{4171} modulo 13.

20. Find the units digit of 78^{357}.

21. Let $\{b_n\}_{n \geq 1}$ be the sequence of integers defined by $b_1 = 1$, $b_2 = 11$, $b_3 = 111$, and in general
$$b_n = \overbrace{111 \cdots 1}^{n \text{ digits}}.$$

   Prove that for all prime numbers p different from 2 or 5, there exists a positive n such that $p \mid b_n$.

22. Show that $3 \mid n(n+1)(n+2)$ for all integers n.

23. Let p be a prime. Prove that p divides the binomial coefficient $\binom{p}{k}$ for all k with $1 \leq k \leq p-1$. Use the binomial theorem to conclude that

$$(a+b)^p \equiv a^p + b^p \pmod{p} \tag{A.15}$$

for all integers $a, b \in \mathbb{Z}$.

24. Prove that if $ac \equiv bc \pmod{m}$, then $a \equiv b \pmod{\frac{m}{d}}$ where $d = \gcd(m, c)$.

25. Consider the sequence of integers $\{c_n\}_{n \geq 0}$ defined by

$$c_0 = 1, \quad c_1 = 101, \quad c_2 = 10101, \quad c_3 = 1010101, \quad \ldots$$

Prove that for all integers $n \geq 2$, the number c_n is composite.

A.7 Lists of Groups

This following table provides a list of all groups, organized by their order, and up to order 24. The table also indicates where the group (or family of groups) first appears in the text.

Order	Abelian Y/N	Groups and Notes
1	Abelian	$\{1\}$
2	Abelian	Z_2 (see Examples 1.3.15 and 1.10.2 for Z_n)
3	Abelian	Z_3
4	Abelian	Z_4, $Z_2 \oplus Z_2$ ($\cong V_4$, called the *Klein-4 group*)
5	Abelian	Z_5
6	Abelian	Z_6
	Nonabelian	D_3 (see Section 1.1 for D_n)
7	Abelian	Z_7
8	Abelian	Z_8, $Z_4 \oplus Z_2$, $Z_2 \oplus Z_2 \oplus Z_2$
	Nonabelian	D_4, Q_8 (see Example 1.4.6)
9	Abelian	Z_9, $Z_3 \oplus Z_3$
10	Abelian	Z_{10}
	Nonabelian	D_5
11	Abelian	Z_{11}
12	Abelian	Z_{12}, $Z_6 \oplus Z_2$
	Nonabelian	D_6, A_4 (see Example 1.6.12 for A_n), $Z_3 \rtimes Z_4$ (Exercise 2.3.15)
13	Abelian	Z_{13}
Continued on next page		

Order	Abelian Y/N	Groups and Notes
14	Abelian	Z_{14}
	Nonabelian	D_7
15	Abelian	Z_{15}
16	Abelian	Z_{16}, $Z_8 \oplus Z_2$, $Z_4 \oplus Z_4$, $Z_4 \oplus Z_2 \oplus Z_2$, $Z_2 \oplus Z_2 \oplus Z_2 \oplus Z_2$
	Nonabelian	$D_4 \oplus Z_2$, $Q_8 \oplus Z_2$, D_8, QD_{16} (Exercise 1.10.9), modular group M_{16} (Exercise 2.3.17), $(Z_4 \oplus Z_2) \rtimes Z_2$, $Z_4 \rtimes Z_4$, $(D_4 \oplus Z_4)/\langle(r^2, z^2)\rangle$, generalized quaternion Q_{16}
17	Abelian	Z_{17}
18	Abelian	Z_{18}, $Z_6 \oplus Z_3$
	Nonabelian	D_9, $D_3 \oplus Z_3$, $(Z_3 \oplus Z_3) \rtimes Z_2$
19	Abelian	Z_{19}
20	Abelian	Z_{20}, $Z_{10} \oplus Z_2$
	Nonabelian	D_{10}, $Z_5 \rtimes Z_4$, F_{20} (Exercise 2.3.16)
21	Abelian	Z_{21}
	Nonabelian	$Z_7 \rtimes Z_3$ (G_2 in Example 1.10.6)
22	Abelian	Z_{22}
	Nonabelian	D_{11}
23	Abelian	Z_{23}
24	Abelian	Z_{24}, $Z_{12} \oplus Z_2$, $Z_6 \oplus Z_2 \oplus Z_2$
	Nonabelian	S_4, D_{12}, $D_3 \oplus Z_4$, $D_4 \oplus Z_3$, $Q_8 \oplus Z_3$, $D_6 \oplus Z_2$, $A_4 \oplus Z_2$, $(Z_3 \rtimes Z_4) \oplus Z_2$, $Z_3 \rtimes D_4$, $Z_3 \rtimes Z_8$, $SL_2(\mathbb{F}_3)$, $\langle a, b, c \mid a^6 = b^2 = c^2 = abc = 1\rangle$

Useful classification results that support the above table are

- groups of order 4 (Example 1.4.4);

- groups of order 8 (Example 1.4.5);

- groups of order p, where p is prime (Proposition 2.1.14): Z_p;

- groups of order $2p$, where p is prime (Exercise 2.1.35): Z_{2p} and D_p;

- Fundamental Theorem of Finitely Generated Abelian Groups: All abelian groups of a given order are determined by Theorems 2.5.11 and 2.5.18.

The notation $N \rtimes H$ stands for the semidirect product between two groups. For the definition of this construction, see Section 6.6.

Bibliography

[1] George E. Andrews. *The Theory of Partitions*. Encyclopedia of Mathematics and its Applications. Cambridge University Press, Cambridge, U.K., 1998.

[2] Michael Aschbacher and Stephen D. Smith. *The Classification of Quasithin Groups. I. Structure of Strongly Quasithin K-groups*, volume 111 of *Mathematical Surveys and Monographs*. American Mathematical Society, Providence, RI, 2004.

[3] Michael Aschbacher and Stephen D. Smith. *The Classification of Quasithin Groups. II. Main Theorems: The Classification of Simple QTKE-groups*, volume 112 of *Mathematical Surveys and Monographs*. American Mathematical Society, Providence, RI, 2004.

[4] Michale J. Bardzell and Kathleen M. Shannon. The PascGalois triangle: A tool for visualizing abstract algebra. In Allen C. Hibbard and Ellen J. Maycock, editors, *Innovations in Teaching Abstract Algebra*, number 60 in MAA Notes, pages 115–123. Mathematical Association of America, Providence, RI, 2002.

[5] A. E. Berriman. The babylonian quadratic equation. *The Mathematical Gazette*, 40:185–192, 1956.

[6] Nathan Bliss, Ben Fulan, Stephen Lovett, and Jeff Sommars. Strong divisibility, cyclotomic polynomials, and iterated polynomials. *The American Mathematical Monthly*, 120(6):519–536, 2013.

[7] William W. Boone. The word problem. *Proceedings of the National Academy of Sciences*, 44:1061–1065, 1958.

[8] William Burnside. *Theory of Groups of Finite Order*. Cambridge University Press, Cambridge, 2nd edition, 1911.

[9] John Clough. A rudimentary geometric model for contextual transposition and inversion. *Journal of Music Theory*, 42(2):297–306, 1998.

[10] Alissa S. Crans, Thomas M. Fiore, and Ramon Satyendra. Musical actions of a dihedral group. *The American Mathematical Monthly*, 116(6):479–495, 2009.

[11] Danilo R. Diedrichs and Stephen Lovett. *Transition to Advanced Mathematics*. Taylor & Francis, Boca Raton, FL, 2021.

[12] Allen Forte. *The Structure of Atonal Music*. Yale University Press, New Haven, CT, 1973.

[13] William Fulton. *Young Tableaux: With Applications to Representation Theory and Geometry*, volume 35 of *London Mathematical Society Student Texts*. Cambridge University Press, Cambridge, U.K., 1996.

[14] Aleksandr Gelfond. Sur le septième problème de Hilbert. *Bulletin de l'Académie des Sciences de l'URSS*, 4:623–634, 1934.

[15] Daniel Gorenstein. The classification the finite simple groups. I. simple groups and local analysis. *Bulletin of the AMS. New Series*, 1(1):43–199, 1979.

[16] Daniel Gorenstein. Classifying the finite simple groups. *Bulletin of the AMS*, 14(1):1–98, 1986.

[17] Branko Grunbaum and Geoffrey Shephard. *Tilings and Patterns*. W.H. Freeman, New York, 1990.

[18] Ján Haluska. *The Mathematical Theory of Tone Systems*. CRC Press, New York, 2003.

[19] G. H. Hardy and E. M. Wright. *An Introduction to the Theory of Numbers*. Oxford University Press, New York, 6th edition, 2008.

[20] James E. Humphreys. *Reflection Groups and Coxeter Groups*. Cambridge University Press, Cambridge, U.K., 1992.

[21] T. Y. Lam and K. H. Leung. On the cyclotomic polynomial $\phi_{pq}(x)$. *The American Mathematical Monthly*, 103(7):562–564, 1996.

[22] I. G. Macdonald. *Symmetric Functions and Hall Polynomials*. Oxford Mathematical Monographs. Oxford University Press, New York, 1999.

[23] James McKay. Another proof of Cauchy's group theorem. *The American Mathematical Monthly*, 66(2):119, 1959.

[24] Richard A. Mollins. *Algebraic Number Theory*. Chapman & Hall, Boca Raton, FL, 1999.

[25] Peter M. Neumann. A lemma that is not Burnside's. *Mathematical Scientist*, 4(2):133–141, 1979.

[26] P. S. Novikov. On the algorithmic unsolvability of the word problem in group theory. *Proceedings of the Steklov Institute of Mathematics*, 44:1–143, 1955. In Russian.

[27] Kenneth H. Rosen. *Elementary Number Theory and Its Applications.* Addison Wesley, New York, 5th edition, 2005.

[28] R. L. Roth. On extensions of $\mathbb{Q}$ by square roots. *The American Mathematical Monthly*, 78(4):392–393, 1971.

[29] Robert R. Stoll. *Introduction to Set Theory and Logic.* W.H. Freeman, San Francisco, 1963.

[30] C. L. F. von Lindemann. Über die Zahl π. *Mathematische Annalen*, 20:213–225, 1882.

[31] Martin M. Zuckerman. *Sets and Transfinite Numbers.* Macmillan Publishing, New York, 1974.

Index

Printed in the United States
by Baker & Taylor Publisher Services